FRONTIERS IN TISSUE ENGINEERING

FRONTIERS IN TISSUE ENGINEERING

Edited by

CHARLES W. PATRICK JR.

Laboratory of Reparative Biology and Bioengineering
Department of Plastic Surgery, M.D. Anderson Cancer Center
Houston, Texas, USA

ANTONIOS G. MIKOS

and

LARRY V. MCINTIRE

Institute of Biosciences and Bioengineering
Rice University, Houston, Texas, USA

PERGAMON

U.K.   Elsevier Science Ltd, The Boulevard, Langford Lane, Kidlington, Oxford OX5 1GB, U.K.

U.S.A.   Elsevier Science Inc., 655 Avenue of the Americas, New York NY 10010, U.S.A.

JAPAN   Elsevier Science Japan, Higashi Azabu 1-chome Building 4F, 1-9-15, Higashi Azabu, Minato-ku, Tokyo 106, Japan

First edition 1998

Library of Congress Cataloging in Publication Data
A catalog record for this book is available from the Library of Congress

British Library Cataloguing in Publication Data
A catalogue record for this book is available from the British Library

ISBN 0 08 042689 1

Printed in Great Britain by Redwood Books Ltd.

Cover Illustration: Tissue engineering involves a synergistic coupling of the disparate fields of engineering, life science, and clinical science. The illustration depicts a pancreas viewed from each field's vantage point in the following order, left to right: clinical science (anatomy and transplantation of pancreas), engineering (encapsulated islets releasing insulin), and life science (upregulation of insulin secretion via genetic engineering). Illustration by C.W. Patrick Jr.

Contents

Contributors

Tahsin Oguz Acarturk
Division of Plastic and
  Maxillofacial Reconstructive
  Surgery,
University of Pittsburgh Medical
  Center,
676 Scaife Hall,
3550 Terrace Street,
Pittsburgh, Pennsylvania 15261, USA

James M. Anderson
Institute of Pathology,
Case Western Reserve University,
Cleveland, Ohio 44106, USA

Anthony Atala
Department of Urology,
Children's Hospital and
Harvard Medical School,
300 Longwood Ave.,
Boston, Maryland 02115, USA

Julia E. Babensee
Department of Chemical Engineering
  and Applied Chemistry
  and Centre for Biomaterials,
University of Toronto,
Toronto, Ontario, M5S, 3E5, Canada

You Han Bae
Department of Materials Science and
  Engineering,
Kwangju Institute of Science and
  Technology,
Kwangju, 506-303, Korea

Jennifer A. Baumgartner
Department of Chemical Engineering,
University of Delaware,
Newark, Delaware, 19716, USA

Derek C. Blakeney
Department of Internal Medicine
  and Department of Biomedical
  Engineering,
The University of Michigan Medical
  Center,

Veterans Administration Medical
  Center and
The University of Michigan at Ann
  Arbor,
Ann Arbor, Michigan 48105, USA

Torsten Blunk
Department of Chemical Engineering,
Massachusetts Institute of
  Technology,
77 Massachusetts Ave.,
Cambridge, Massachusetts 02139,
USA

Linda G. Braddon
Parker H. Petit Institute for
  Bioengineering and Bioscience,
Georgia Institute of Technology,
Atlanta, Georgia 30332-0363, USA

Scott P. Bruder
Osiris Therapeutics, Inc.,
2001 Aliceanna Street,
Baltimore, Maryland 21231, USA

Elisa A. Burgess
Oregon Health Sciences University,
Department of Surgery,
3181 SW Sam Jackson Park Rd L352A,
Portland, Oregon 97201-3098, USA

Arnold I. Caplan
Skeletal Research Center,
Biology Department,
Case Western Reserve University,
2080 Adelbert Road,
Cleveland, Ohio 44106, USA

Rosa S. Choi
Department of Surgery,
Harvard Medical School,
Children's Hospital
300 Longwood Ave.,
Boston, Massachusetts 02115, USA

Priscilla B. Chauvin
Laboratory of Reparative Biology and
  Bioengineering,

Department of Plastic Surgery,
The University of Texas M.D. Anderson Cancer Center
1515 Holcombe Blvd., Box 62,
Houston, Texas 77030, USA

Stuart L. Cooper
Department of Chemical Engineering,
University of Delaware,
Newark, Delaware 19716, USA

James J. Cunningham
Department of Chemical Engineering,
The University of Michigan,
Ann Arbor, Michigan 48109, USA

Paul Dimilla
Pittsburgh Tissue Engineering
Initiative, Inc.,
Center for Biotechnology and
Bioengineering,
300 Telephone Drive,
Pittsburgh, Pennsylvania 15219, USA

Paul D. Drumheller
Gore Hybrid Technologies, Inc.,
Flagstaff, Arizona 86003, USA

Charles Durfor
Center for Devices and Radiological
Health,
United States Food and Drug
Administration,
5600 Fishers Lane,
Rockville, Maryland 20857, USA

Dwaine F. Emerich
Cyto Therapeutics, Inc.,
2 Richmond Square,
Providence, Rhode Island 02906, USA

Gregory R. D. Evans
Laboratory of Reparative Biology and
Bioengineering,
Department of Plastic Surgery,
The University of Texas M.D.
Anderson Cancer Center,
1515 Holcombe Blvd, Box 62
Houston, Texas 77030, USA

David J. Fink
Osiris Therapeutics, Inc.,

2001 Aliceanna Street,
Baltimore, Maryland 21231, USA

Lisa E. Freed
Department of Chemical Engineering,
Massachusetts Institute of
Technology,
77 Massachusetts Ave.,
Cambridge, Massachusetts 02139,
USA

Julie R. Friend
Department of Chemical Engineering
and Materials Science,
University of Minnesota,
421 Washington Ave. SE,
Minneapolis, Minnesota 55455-0132,
USA

Elizabeth J. Furnish
Department of Chemical Engineering,
The University of Texas-Austin,
Austin, Texas 78712, USA

Frank T. Gentile
Cyto Therapeutics, Inc.,
2 Richmond Square,
Providence, Rhode Island 02906, USA

Keith J. Gouch
Department of Chemical Engineering
and Division of Health Sciences and
Technology,
Massachusetts Institue of Technology,
77 Massachusetts Ave.,
Cambridge, Massachusetts 02139,
USA

J. Hardin-Young
Organogenesis Inc.,
150 Dan Road,
Canton, Massachusetts 02021, USA

Kiki B. Hellman
Center for Devices and Radiological
Health,
United States Food and Drug
Administration,
5600 Fishers Lane,
Rockville, Maryland 20857, USA

Jeffrey O. Hollinger
Oregon Health Sciences University,
Department of Surgery,
3181 SW Sam Jackson Park Rd L352A,
Portland, Oregon 97201-3098, USA

Wei-Shou Hu
Department of Chemical Engineering
and Materials Science,
University of Minnesota,
421 Washington Ave. SE,
Minneapolis, Minnesota 55455-0132,
USA

H. David Humes
Department of Internal Medicine and
Department of Biomedical
Engineering,
The University of Michigan Medical
Center,
Veterans Administration Medical
Center and
The University of Michigan at Ann
Arbor,
Ann Arbor, Michigan 48105, USA

Yoshito Ikada
Research Center for Biomedical
Engineering,
Kyoto University,
53 Kawahara-cho, Shogoin, Sakyo-ku,
Kyoto 606, Japan

Peter C. Johnson
Division of Plastic and Maxillofacial
Reconstructive Surgery,
University of Pittsburgh Medical
Center, 676 Scaife Hall,
3550 Terrace Street,
Pittsburgh, Pennsylvania 15261, USA
and Pittsburgh Tissue Engineering
Initiative, Inc.,
Center for Biotechnology and
Bioengineering,
300 Technology Drive,
Pittsburgh, Pennsylvania 15219, USA

Sung Wan Kim
Center for Controlled Chemical
Delivery,

University of Utah,
Salt Lake City, Utah 84112, USA

Timothy W. King
Laboratory of Reparative Biology and
Bioengineering,
Department of Plastic Surgery,
The University of Texas M.D. Ander-
son Cancer Center,
1515 Holcombe Blvd., Box 62,
Houston, Texas 77030, USA

Emma Knight
Center for Biologics Evaluation and
Research,
United States Food and Drug Admin-
istration,
1401 Rockville Pike,
Rockville, Maryland 20852, USA

Konstantinos Konstantopoulos
Cox Laboratory for Biomedical
Engineering,
Institute of Biosciences and
Bioengineering,
Rice University
6100 S. Main
Houston, Texas 77005, USA

Sharad Kukreti
Cox Laboratory for Biomedical
Engineering,
Institute of Biosciences and
Bioengineering,
Rice University
6100 S. Main
Houston, Texas 77005, USA

Nina M. K. Lamba
Department of Chemical Engineering,
University of Delaware,
Newark, Delaware, 19716, USA

Robert Langer
Department of Chemical Engineering,
Massachusetts Institute of
Technology,
77 Massachusetts Ave.,
Cambridge, Massachusetts 02139,
USA

Won-Kyoung Lee
Department of Internal Medicine and
Department of Biomedical Engineer-
ing,
The University of Michigan Medical
Center,
Veterans Administrations Medical
Center and
The University of Michigan at Ann
Arbor,
Ann Arbor, Michigan 48105, USA

L. Louie
Department of Materials Science and
Engineering,
Massachusetts Institute of
Technology,
Cambridge, Massachusetts 02139,
USA

Michael H. May
Department of Chemical Engineering
and Applied Chemsistry
and Center for Biomedicals,
University of Toronto,
Toronto, Ontario, M5S, 3E5, Canada

Larry V. McIntire
Institute of Biosciences and
Bioengineering,
Rice University,
6100 S. Main,
Houston, Texas 77005, USA

Antonios G. Mikos
Institute of Biosciences and
Bioengineering,
Rice University,
6100 S. Main,
Houston, Texas 77005, USA

Michael J. Miller
Laboratory of Reparative Biology and
Bioengineering,
Department of Plastic Surgery,
The University of Texas M.D.
Anderson Cancer Center,
1515 Holcombe Blvd, Box 62
Houston, Texas 77030, USA

David J. Mooney
Department of Biologic and Materials
Science,
The Univesity of Michigan,
Ann Arbor, Michigan 48109, USA
and
Department of Chemical Engineering,
The University of Michigan,
Ann Arbor, Michigan 48109, USA

Jeffrey R. Morgan
Center for Engineering in Medicine,
Massachusetts General Hospital,
Shriners Burns Institute and
Harvard Medical School,
Boston, Massachusetts 02115, USA

Robert M. Nerem
Parker H. Petit Institute for
Bioengineering and Bioscience,
Georgia Institute of Technology,
Atlanta, Georgia 30332-0363, USA

Janeta Nikolovski
Department of Internal Medicine
and Department of Biomedical
Engineering,
The University of Michigan Medical
Center, Veterans Administration
Medical Center and
The University of Michigan at Ann
Arbor,
Ann Arbor, Michigan 48105, USA

Gregory M. Organ
Columbia Michael Reese Hospital,
Medical Center,
Chicago, Illinois 60680, USA

Bernhard Palsson
Department of Bioengineering,
University of California–San Diego,
La Jolla, California 92093-0412, USA

N.L. Parenteau
Organogenesis Inc.,
150 Dan Road,
Canton, Massachusetts 02021, USA

Charles W. Patrick Jr.
Laboratory of Reparative Biology and

Bioengineering,
Department of Plastic Surgery,
The University of Texas M.D. Anderson Cancer Center,
1515 Holcombe Blvd., Box 62
Houston, Texas 77030, USA

Patricia Petrosko
Division of Plastic and Maxillofacial Reconstructive Surgery,
University of Pittsburgh Medical Center,
676 Scaife Hall,
3550 Terrace Street,
Pittsburgh, Pennsylvania 15261, USA

Gregory P. Reece
Laboratory of Reparative Biology and Bioengineering,
Department of Plastic Surgery,
The University of Texas M.D. Anderson Cancer Center,
1515 Holcombe Blvd., Box 62
Houston, Texas 77030, USA

Geoffrey L. Robb
Laboratory of Reparative Biology and Bioengineering,
Department of Plastic Surgery,
The University of Texas M.D. Anderson Cancer Center,
1515 Holcombe Blvd., Box 62,
Houston, Texas 77030, USA

Angela M. Rodriguez
University of Massachusetts Medical Center,
S-2, Room 751, 55 Lake Avenue North,
Worcester, Massachusetts 01655, USA

Julia M. Ross
Department of Chemical and Biochemical Engineering,
University of Maryland-Baltimore Maryland,
Baltimore, Maryland 21250, USA

Alan S. Rudolph
Center for Biomolecular Science and Engineering,

Naval Research Laboratory,
Washington DC 20375, USA

Amarpreet S. Sawhney
Focal, Inc.,
Lexington, Massachusetts 02173, USA

Christine E. Schmidt
Department of Chemical Engineering,
The University of Texas-Austin,
Austin, Texas 78712, USA

Timothy Scott-Burden
Vascular Cell Biology Laboratory,
Texas Heart Institute,
Houston, Texas 77030, USA

Michael V. Sefton
Department of Chemical Engineering and Applied Chemistry
and Center for Biomaterials,
University of Toronto,
Toronto, Ontario, M5S 3E5, Canada

Dror Seliktar
Parker H. Petit Institute for Bioengineering and Bioscience,
Georgia Institue of Technology,
Atlanta, Georgia 30332-0363, USA

Lonnie D. Shea
Department of Chemical Engineering,
The University of Michigan,
Ann Arbor, Michigan 48109, USA

M. Spector
Department of Orthopedic Surgery,
Brigham and Women's Hospital,
Harvard Medical School,
Boston, Massachusetts 02115, USA
and
Rehabilitation Engineering Research and Development,
Brockton/West Roxbury VA Medical Center,
West Roxbury, Massachusetts 02115, USA

Christopher J. Tennant
University of Maryland-College Park
Baltimore, Maryland, USA

J. Teumer
Organogenesis Inc.,
150 Dan Road,
Canton, Massachusetts 02021, USA

Christine L. Tock
Vascular Cell Biology Laboratory,
Texas Heart Institue,
Houston, Texas 77030, USA

Charles A. Vacanti
University of Massachusetts Medical
    Center,
S-2, Room 751, 55 Lake Avenue North,
Worcester, Massachusetts 01655, USA

Joseph P. Vacanti
Department of Surgery,
Harvard Medical School,
Children's Hospital
300 Longwood Ave.,
Boston, Massachusetts 02115, USA

Gordana Vunjak-Novakovic
Department of Chemical Engineering,
Massachusetts Institute of
    Technology,
77 Massachusetts Ave.,
Cambridge, Massachusetts 02139,
USA

Jennifer L. West
Department of Bioengineering,
Rice University,
Houston, Texas 77005, USA

Markus S. Widmer
Institute of Biosciences and
    Bioengineering,
Rice University,
6100 Main Street,
Houston, Texas 77005, USA

I.V. Yannas
Department of Mechanical
    Engineering,
Massachusetts Institute of
    Technology,
Cambridge, Massachusetts 02139,
USA

Martin L. Yarmush
Center for Engineering in Medicine,
Massachusetts General Hospital,
Shriners Burns Institute and
    Harvard Medical School,
Boston, Massachusetts 02115, USA

Alan W. Yasko
Orthopaedic Surgical Oncology,
The University of Texas M.D. Ander-
    son Cancer Center,
1515 Holcombe Blvd.,
Houston, Texas 77030, USA

Michael J. Yaszemski
Orthopaedic Surgery,
Adult Reconstruction and Spine
    Surgery,
Mayo Clinic,
200 First Street SW,
Rochester, Minnesota 55905, USA

Isaac C. Yue
Department of Chemical Engineering,
The University of Michigan,
Ann Arbor, Michigan 48109, USA

Randell G. Young
Osiris Therapeutics, Inc.,
2001 Aliceanna Street,
Baltimore, Maryland 21231, USA

Preface

The beauty of the human form has been the subject of countless artistic masterpieces through the ages, the focus of exquisite poetry and central to every art form. However, it can also be considered the most elegant and sophisticated of machines. As with all machines, wear and tear occurs with aging and accidents can happen. With living machines, disease processes can also occur which can lead to destruction of tissues. It is the role of reconstructive surgeons to help patients who are afflicted with disease, trauma, or malfunction due to aging. Reconstructive surgical procedures have been developed over the last century to replace human tissues and organs so that people can return to a good quality of life or, in the case of many operations, to save life. Current reconstructive strategies include replacement of defective body parts with totally prosthetic devices, transfer of one tissue to another site or transplantation of organs and tissue from one individual to another when that becomes necessary. These techniques have improved and saved countless lives. However, all have limitations.

The fundamental roadblock to future improvements is the lack of sufficient tissue for ideal structural replacement to produce functional normalcy. It has been estimated that in the United States alone, the care of patients with tissue malfunction and loss can exceed one-half of a trillion dollars annually. It is this need that has driven the emergence of the field of Tissue Engineering. The concept of the rational design and fabrication of living tissues and organs for repair and replacement is relatively new, emerging within the last quarter of this century. The field has brought together diverse scientific and clinical areas. Clinical physicians and surgeons are working hand-in-hand with basic scientists in the fields of molecular and cell biology, biochemistry and the engineering fields of chemical engineering and materials science. This unique grouping has produced dramatic experimental examples that are shedding knowledge in the areas of new tissue fabrication, implantation and function. From its inception, the field has also brought the business community in contact with the academic community. As well, the regulatory agencies have been involved with the rigorous design and safe implementation of clinical trials. This broad representation of the medical, scientific, industrial and regulatory community is detailed in this book. The book is divided into three parts including the fundamentals and the methods of tissue engineering, tissue engineering applied to specialized tissues and tissue engineering applied to organs. The book has taken on an enormous task of providing a comprehensive overview of this emerging field and communicating the excitement felt by its many participants. No one can predict the future of any new scientific endeavor at its early stages, but if enthusiasm were an important component, then this field promises to bring hope to many people who are afflicted with serious disease or injury.

J.P. VACANTI,
Department of Surgery, Children's Hospital,
300 Longwood Ave., Boston, MA 02115, USA

R.S. LANGER,
Department of Chemical Engineering, Massachusetts Institute of
Technology, 77 Massachusetts Ave., Boston, MA 02139, USA

SECTION I
PROSPECTUS OF TISSUE ENGINEERING

CHAPTER I

Prospectus of Tissue Engineering

CHARLES W. PATRICK JR.
Laboratory of Reparative Biology and Bioengineering,
Department of Plastic Surgery,
M.D. Anderson Cancer Center,
1515 Holcombe Blvd., Box 62,
Houston, Texas 77030, USA

ANTONIOS G. MIKOS
Institute of Biosciences and Bioengineering,
Rice University,
6100 S. Main,
Houston, Texas 77005-1892, USA

and

LARRY V. MCINTIRE
Institute of Biosciences and Bioengineering,
Rice University,
6100 S. Main,
Houston, Texas 77005-1892, USA

1 Tissue engineering defined

Tissue engineering is a relatively new and emerging interdisciplinary field that applies the knowledge of bioengineering, the life sciences, and the clinical sciences towards solving the critical medical problems of tissue loss and organ failure. It involves applying engineering principles of transport and reaction phenomena as well as methods of analysis towards understanding the complex biological processes that occur in tissue development and repair. Frequently, knowledge of molecular phenomena and cellular interactions with surface, biochemical, and mechanical environments are employed. Tissue engineering has been formally defined as 'the application of the principles and methods of engineering and the life sciences toward the fundamental understanding of structure-function relationships in normal and pathological mammalian tissues and the development of biological substitutes that restore, maintain, or improve tissue function' [1]. Implied in the above is the essence of tissue engineering: the use of living cells, together with either natural or synthetic extracellular components, in the development of implantable parts or devices for the restoration or replacement of function [2]. Ideally, tissue engineering leads to new medical therapies customized to match the biology of specific patients. To meet this goal, tissue engineering builds on recent advances in biochemistry as well as cell, molecular, and developmental biology to make rapid progress in research and development. Tissue engineering integrates this knowledge using a systems approach to engineer complex, interacting, three dimensional biological equivalents. As will become evident in the following chapters, tissue engineering combines interdisciplinary skills from such fields as disparate as biological science, modern material science, computer science, and systems modeling. In the end, researchers in tissue engineering must be able to bridge the two worlds of engineering and biology and, just as importantly, design tissue engineered products within the framework of clinical utilization.

2 Historical perspective

The application of tissue engineering predates the development of its conventional definition. The idea of using biological substitutes to repair or replace damaged tissues goes back to earliest records. According to the book of *Genesis*, a rib was harvested from Adam, the first donor, and was used to fashion Eve [3]. The Etruscans and Romans prepared artificial teeth and dentures to replace damaged teeth. Written evidence cites medical treatment for facial injuries more than 4,000 years ago. Physicians in ancient India were utilizing skin grafts for reconstructive work as early as 800 B.C. Organ transplants, however, had to await the development of modern surgical procedures. Figure 1 illustrates the transplant milestones in the United States and Canada. Organ transplants increased dramatically in the late 1980's after the Food and Drug Administration (FDA) approved the use of cyclosporine, an immunosuppressive drug used to reduce organ rejection, in 1983 [4]. It is possible to transplant approximately 25 different organs and tissues, including bone and cartilage, bone marrow, cornea, hearts, heart-lung, kidney, liver, lung, and pancreas. However, due to supply, technical, and economic factors (discussed below), there has been a surge of university, industrial, and clinical interest to develop new medical therapies based on biological equivalents to replace or repair damaged tissue and organs. This interest in tissue engineering formally began in 1987.

In the Spring of 1987, the Engineering Directorate of the National Science Foundation (NSF) held a Panel discussion focusing on future directions in bioengineering. Researchers from the Bioengineering and Research to Aid the Handicapped (BRAH) and Biotechnology (BIO-

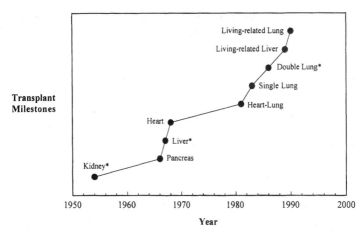

Fig. 1. Transplant milestones in the US and Canada. *Denotes this transplant was the first of its kind in the world. Kidney was conducted by Dr. J. E. Murray, Brigham and Women's Hospital, Boston, MA. Pancreas was conducted by Drs. W. Kelly and R. Lillehei, University of Minnesota, Minneapolis, MN. Liver was conducted by Dr. T. Starzl, University of Colorado Health Sciences Center, Denver, CO. Heart was conducted by Dr. N. Shumway, Stanford University Hospital, Stanford. CA. Heart–lung was conducted by Dr. B. Reitz, Stanford University Hospital, Stanford. CA. Single lung and double lung were conducted by Dr. J. Cooper, Toronto Lung Transplant Group, Toronto General Hospital, Canada. Living-related liver was conducted by Dr. C. Broelsch, University of Chicago Medical Center, Chicago, IL. Living-related lung was conducted by Dr. V. A. Starnes, Stanford University Hospital, Stanford, CA.

TECH) programs were asked to contrive new target research areas. BRAH wanted to develop whole organ replacement systems whereas BIOTECH wanted to develop new methods of homologous cell culture [5]. The target research areas appeared to overlap and the Panel coined the term 'tissue engineering' to consolidate their efforts. On the basis of the initial Panel discussion, a Panel meeting on Tissue Engineering was held at NSF in October 1987. The panel consisted of representatives from NSF, National Institutes of Health (NIH), Office of Naval Research (ONR), Department of Energy (DOE), National Aeronautics and Space Administration (NASA), Red Cross, and University representatives from areas of cell and molecular biology, medicine, and bioengineering [6]. On October 29–30, 1987, a Forum on Issues, Expectations and Prospects for Emerging Technology Initiation was held at NSF under the auspices of the Division of Emerging Engineering Technologies. The Forum recommended Micromechanical Technology and Tissue Engineering as areas of emerging technology with potential for rapid industrial expansion. Furthermore, the Forum recommended that a Workshop be held to identify appropriate areas for engineering research in Tissue Engineering. This Workshop was held at Lake Tahoe, CA on February 26–29, 1988. It was at this Workshop that the term 'tissue engineering' was formally defined. In addition, participants of the Workshop provided NSF with recommendations regarding technological goals. Pursuant to these recommendations, NSF created an initiative for tissue engineering research and the first research proposals were received and funded in 1988.

Since then, there has been a multitude of research projects carried out whose focus is in developing and understanding the mechanisms underlying tissue engineered tissues and organs. To be sure, tissue engineering has impacted global biotechnology and increased our knowledge base. This is attested to by the numerous tissue engineering sessions presently held at national and international conferences. In addition, there are several biotechnology companies whose business scope is focused on tissue engineered products. The overall market for engineered tissues has been estimated to be in the $400B range, of which $80B is estimated to be the value of the engineered tissues themselves [7]. It is no surprise the senior editor of BIO/TECHNOLOGY designated tissue engineering as 'a new force in biotech' [8].

3 Necessity of tissue engineering

To date, the clinical solution to restoring structure and function of impaired tissues and organs relies on 'robbing Peter to pay Paul'. That is, tissue defects resulting from congenital abnormalities (birthmarks, cleft-lip and palate, hand deformities, abnormal breast development, skull and facial bone deformities, prominent and deformed ears) and disease and trauma (scars, wounds and soft tissue deformity from trauma or disease, burn scars, growths and tissue defects including cancer treatment and mastectomy, lacerations, severed limbs, fingers, or toes, skull and jaw injuries, hand injuries and acquired problems) are repaired by transfering healthy donor tissue as an autograft, isograft, allograft, or xenograft to a tissue defect. Likewise, diseased, injured, or abnormal organs are conventionally repaired by performing an organ transplant using living or post-mortem donor organs, or by mechanically assisting an impaired organ. Although these solutions have resulted in tremendous strides in increasing patient survival and quality of life, there are inherent limitations.

It is estimated that more that one million reconstructive procedures are performed by plastic and reconstructive surgeons every year to repair tissue defects. The transfer of tissue grafts requires the sacrifice of tissue from a donor site, resulting in certain associated morbidity (scar formation, potential injury and impairment to donor site, etc.). For autografts, the volume of donor tissue that can be harvested is dependent on blood supply and the need to avoid visceral injuries and contour deformities. Allografts and xenografts possess the risks of immune rejection and infectious disease transmission. In addition to these limitations, there is presently a growing donor availability crisis (Fig. 2). As of February 5, 1997 there were 50,613 U.S. patients waiting for organ transplants [9]. Every 18 minutes a new name is added to the United Network for Organ Sharing (UNOS) national transplant waiting list. The total number of organ transplants has increased 57% between 1988 and 1995. Concomitantly, the waiting list has almost tripled between 1988 and 1995 (from 16,026 to 43,983). Similarly, the time spent waiting for a transplant has increased over time. In 1988, 56% of registrants received a transplant within 1 year; by 1995, the rate had decreased to 44%. There were 3,549 reported deaths on the waiting list in 1995, which is more than double the 1,507 deaths reported in 1988. The number of patients on the waiting list has also more than doubled, from 27,883 in 1988 to 65,677 in 1995. Approximately 50,000 patients with heart failure are denied potentially life-saving transplants due to donor organ shortage [10, 11]. Moreover, there are approximately 17 million diabetics in the United States, but donor scarcity precludes the use of pancreatic transplantation [12]. The worldwide need among insulin-dependent diabetes mellitus patients alone for pancreatic transplants is at a level at least 1,000 times the current donor organ supply [13]. Approximately eight million surgical procedures are performed annually in the United States to treat tissue loss and end-stage organ failure, requiring 40–90 million hospital days [14]. Also, 500,000 arthoplastic procedures and total joint replacements are performed each year in the United States to repair damaged bone and cartilage [15]. In addition, 7.5 million patients per year rely on nonbiological implants, ranging from artificial hearts to artificial hips, to replace the structure and function of any impaired organ or tissue [16]. Five million units of blood are used each year for transfusions during surgical and trauma situations [17]. Besides the fact that blood is normally in short supply, there is the risk of viral infection during a transfusion. For instance, the likelihood of contracting hepatitis and AIDS from a three unit transfusion is 1/70 and 1/50,000, respectively.

In addition to the availability crisis, tissue and organ transplantation is expensive. Each year in the United States, organ failure and tissue loss cost an estimated $400 B [18]. In 1990, the median cost of a heart-lung transplant was $240,000 for initial care and $47,000 per year for follow-up medication and care [19]. In addition, in 1991 it costs Medicare $56,000 for the 1st year of a kidney transplant and $6,000 per year subsequently [20]. In 1987 the average cost of a heart, kidney, liver, or a pancreas transplant was estimated to be $57,000–$110,000, $25,000–$30,000, $135,000–$238,000, and $30,000–$40,000, respectively [21]. Restenosis costs to the healthcare system for angioplasty and bypass surgery is $5 and $4.3 B,

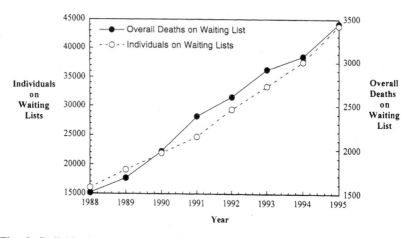

Fig. 2. Individuals on organ waiting lists and overall deaths occurring while patients are waiting continue to increase.

respectively [2]. At present, free muscle flaps used for breast reconstruction are $17,000. These costs could be reduced by using tissue engineered constructs based on autologous cells and tissues. Costs can be reduced by decreasing the time required in the operating room (currently charges are approximately $40/minute), the number of hospital days required for patient recovery, patient back-to-work time, and follow-up complications that require additional surgery.

4 Closing remarks

The donor scarcity, technical difficulties, expense, and complex labor intensive care associated with conventional tissue and organ transplantation provides a significant impetus for development of tissue engineered therapies. Tissue engineering is a field with enormous potential to make truly significant contributions to mankind over the next decades. As life expectancy continues to rise, the potential pool of recipients of tissue engineered constructs is growing as well. The number of people in the U.S. projected in 2040 to be 85 or older ranges from 15–43 million (compared to 3 million for 1990) [22]. A tissue engineered product will be more available and cost-effective than donor organs. Availability of a tissue engineered product can normally be resolved by using cell expansion culturing techniques. For example, one company is attempting to expand reassembled pancreatic islets using primary human cell expansion to produce 10,000–100,000 gland equivalents from every endocrine pancreas

Table 1. Organ transplant waiting list and transplants performed.

Organ transplant	No. of registrations*	No. performed*	% unfulfilled
Kidney	34,780	10,892	69
Liver	7,622	3,925	49
Pancreas	329	110	67
Pancreas islets	72		
Kidney–pancreas	1,467	918	37
Intestine	86		
Heart	3,711	2,361	36
Heart–lung	232	68	71
Lung	2,314	871	62

*UNOS; DOT/HRSA/DHHS [9]. The data and analyses reported in the 1996 Annual Report of the U.S. Scientific Registry of Transplantation Recipients and the Organ Procurement and Transplantation Network have been supplied by UNOS. The authors alone are responsible for the reporting and interpretation of these data.

obtained [13]. In addition, tissue engineering will allow one to intervene before patients are critically ill. For instance, there are about 95,000 total knee replacements and 41,000 other procedures to repair defects of the knee [23]. Treatment of knee injuries at an early stage could prevent the development of cartilage inflammation, eliminating or postponing the need for a total joint replacement. To be sure, the field of tissue engineering will have social, legal, ethical, and regulatory challenges to address. Moreover, as funding from governmental agencies become increasingly scarce and competitive, the need for industrial funding will increase. Hence, increased cooperation and sensitivity to the needs and goals of industry and academics needs to take place to ensure that innovative tissue engineering modalities become clinically translatable.

Table 2. Future challenges in tissue engineering.

Novel applications for reconstructive surgery*
Union of genetic engineering and tissue engineering*
Novel delivery approaches*
Functional artificial organs*
Biomimetic approaches*
Embracing nanotechnology
Commercial mentoring
Translate from lab to clinic

*From K. Hellman, Inaugural Meeting of the Tissue Engineering Society [24].

References

1. Skalak, R. and Fox, C. F., *Tissue Engineering*, Alan R. Liss, New York, 1988, preface.
2. Nerem, R. M., Tissue engineering in the USA. *Medical and Biological Engineering Computing*, 1992, **30**, CE8–CE12.
3. Genesis 2:21.
4. Fox, R. C. and Swazey, J. P., *Spare Parts: Organ Replacement in American Society*, Oxford University Press, Oxford, 1992, pp. 1–244.
5. Edgington, S. M., 3-D biotech: Tissue engineering. *Bio / Technology*, 1992, **10**, 855–859.
6. Heineken, F. G. and Skalak, R., Tissue engineering: A brief overview. *Journal of Biomechanical Engineering*, 1991, **113**, 111–112.
7. Pittsburgh Tissue Engineering Initiative, Pittsburgh, PA, 1996, http://www.pittsburgh-tissue.net.
8. Edington, S. M., A new force in biotech: Tissue engineering. *Bio / Technology*, 1994, **12**, 361–364.
9. 1996 Annual Report of the U. S. Scientific Registry for Transplant Recipients and the Organ Procurement and Transplantation Network–Transplant Data: 1988–1995. UNOS Richmond, VA, and the Division of Transportation, Bureau of Health Resources Development, Health Resources and Services Administration, U.S. Department of Health and Human Services, Rockville, MD.
10. Nowak, R., Xenotransplants set to resume. *Science*, 1994, **266**, 1148–1151.
11. Lelkes, P. I. and Samet, M. M., Endothelialization of the luminal sac in artificial cardiac prostheses: A challenge for both biologists and engineers. *Journal of Biomechanical Engineering*, 1991, **113**, 132–142.
12. Colton, C. K. and Avgoustiniatos, E. S., Bioengineering in development of the hybrid artificial pancreas. *Journal of Biomechanical Engineering*, 1991, **113**, 152–170.
13. Dimond, P. F., Human primary cells *in vitro* offer the promise of 'cell therapies'. *Genetic Engineering News*, 1994, June 15, 16–17.
14. Langer, R. and Vacanti, J. R., Tissue engineering. *Science*, 1993, **260**, 920–926.
15. Praemer, A., Furner, S. and Rice, D. P., Medical implants and major join procedures. In *Musculoskeletal Conditons in the United States*, ed. Praemer, A., Furner, S. and Rice, D. P., American Academy of Orthopaedic Surgeons, Park Ridge, IL, 1992, pp. 125–141.
16. Service, R. F., Liability concerns threaten medical implant research. *Science*, 1994, **266**, 726–727.
17. Carter, T. H., Biotechnology, economics, and the business of blood. In *Biotechnology of Blood*, ed. Goldstein, J., Butterworth–Heinemann, 1991.
18. Lipkin R., Tissue engineering: Replacing damaged organs with new tissues. *Science News*, 1995, **148**, 24–26.
19. Theodore, J. and Lewiston, N., Lung transplantation comes of age. *New England Journal of Medicine*, 1990, **322**, 772–774.
20. Levinsky, N. G. and Rettig, R. A., The Medicare end-stage renal disease program: a report from the Institute of Medicine. *New England Journal of Medicine*, 1991, **324**, 1143–1148.

21. Cooper, T., Survey of development, current status, and future prospects for organ transplantation. In *Human Organ Transplantation: Societal, Medical-Legal, Regulatory, and Reimbursement Issues*, ed. Cowan, D. H., Kantorowitz, J. A., Moskowitz, J. and Rheinstein, P. H., Health Administration Press, Ann Arbor, MI, 1987, pp. 19–26.

22. Kirkland R. I., Why we will live longer... and what it will mean. *Fortune*, 1994, **129**, 66–78.

23. Brittberg, M., Lindahl, A., Nilsson, A., Ohlsson, C., Isaksson, O. and Peterson, L., Treatment of deep cartilage defects in the knee with autologous chondrocyte transplantation. *New England Journal of Medicine*, 1994, **331**, 889–895.

24. Hellman, K., Directions in tissue engineering: Technologies, applications, and regulatory issues. Paper presented at the Inaugural Meeting of the Tissue Engineering Society, Orlando, FL, December 13–15, 1996.

SECTION II
FUNDAMENTALS AND METHODS OF TISSUE ENGINEERING

CHAPTER II.1

Cell–Extracellular Matrix Interactions

JULIA M. ROSS
Department of Chemical and Biochemical Engineering,
University of Maryland-Baltimore County,
Baltimore, MD 21250, USA

1 Introduction

The goal of tissue engineering is to regenerate living tissue substitutes in order to replace or enhance lost tissue function or structure. To accomplish this, tissue cells must be organized and behave as if they are part of the original tissue *in vivo*. In other words, the cells must 'believe' that they are in their native environment, and receive all the signals from that environment that allow for differentiation and proliferation, cell migration and cell adhesion. How do cells organize themselves and sense their environment? What guides tissue regeneration and repair? In most cases, a critical component of these processes is the way in which cells interact with the surrounding extracellular matrix (ECM).

The extracellular matrix is a structural material that lies beneath the epithelia and surrounds the connective tissue cells. The ECM acts as a support material created by the cells as a scaffolding on which to reside. However, a cell can interact with its own ECM products as well as ECM produced by other cells via specific cell surface receptors, thereby linking the ECM to the cell interior via intracellular signaling processes. These cell surface receptors can respond to cell signaling events as well as transduce specific cell signals. In this way, various cell functions are continuously related to, and dependent on the composition and organization of the matrix. *In vivo*, the interaction between a cell and the ECM can affect cell and tissue differentiation, cell migration, cell proliferation, cell adhesion and tissue regeneration and repair. Local variations in the composition or organization of the ECM can therefore give rise to spatial and temporal variations in tissue structure and function. In developing organisms, the ECM is constantly being degraded, remodeled and resynthesized locally, changing the environment in which tissue cells reside.

In attempting to regenerate tissues, the ECM can be used to organize transplanted cells. Natural ECM molecules can be isolated and used in tissue engineered constructs or synthetic materials can be manufactured to simulate the natural ECM environment experienced by cells. Furthermore, by understanding and manipulating cellular processes controlled by the ECM, the *clever* tissue engineer may be able to 'trick' the developing tissue into displaying desirable properties, while diminishing undesirable traits.

The roles of cellular interactions with the ECM in tissue engineering are both vast and complex. The aim of this review is to give a very basic, broad overview of the cell biology of the ECM as well as an introduction to cell–ECM adhesive receptors/ligands. Due to the abundance of research literature in this area, the references in this work are limited to several excellent review articles and books. The reader interested in more depth is encouraged to consult the original literature referenced in the reviews.

2 Composition of the ECM

The extracellular matrix is composed of a variety of macromolecules which can be grouped into four major classes, each of which is responsible for specific ECM characteristics:

1. Collagens: source of strength, many developmental and physiological roles including involvement in cell attachment and differentiation;
2. Proteoglycans: source of matrix resiliency, participate in cell adhesion, migration, and proliferation;
3. Cell interactive glycoproteins: help create tissue cohesiveness through cell adhesion capabilities;
4. Elastic fibers: tissue flexibility.

Of these four classes, collagens, proteoglycans, and cell interactive glycoproteins are the most critical in cell–ECM interactions and will be reviewed here (see [1] for review).

2.1 The collagen family

Collagen is the most abundant protein in the human body and collagenous proteins are a major constituent of all ECM (see [2] for review). Traditionally, a structural role has been attributed to collagen. However, collagen comprises a heterogeneous class of molecules. Some types of collagen exhibit the classical structural properties attributed to collagen, while others demonstrate different mechanical and chemical properties. Vertebrates contain at least 15 different types of collagen that are found in unique temporal and spatial tissue specific patterns and exhibit different functional properties and molecular characteristics (see Table 1). For example, type I collagen is the most abundant type of collagen isolated from many adult connective tissues (skin, bone, tendon), while type II

Table 1. The collagen family. Subgroups are determined by supramolecular structure or by molecular size as compared to type I collagen. Modified from [2].

Class	Type	Chain composition	Common distribution
Fibrillar	I	$[\alpha 1(I)]_2[\alpha 2(I)]$	Fibrous stromal matrices, primary collagen in skin, bone, tendon, cornea, not found in cartilage
	II	$[\alpha 1(II)]_3$	Cartilage, corneal stroma, notochord
	III	$[\alpha 1(III)]_3$	With type I in hetertypic fibrils, walls of arteries and other hollow organs, not found in bone
	V	$[\alpha 1(V)]_2[\alpha 2(V)]$	Most interstitial tissues, with type I in hetertypic fibrils
	XI	$[\alpha 1(XI)][\alpha 2(XI)][\alpha 3(X)]$	With type II in heterotypic fibrils
Fibril associated	IX	$[\alpha 1(IX)][\alpha 2(IX)][\alpha 3(IX)]$	Associated with type II fibrils in cartilage, corneal stroma, notochord
	XII	$[\alpha 1(XII)]_3$	Certain type I containing matrices
Network forming	IV	$[\alpha 1(IV)]_2[\alpha 2(IV)]$	All basal laminas
Filamentous	VI	$[\alpha 1(VI)][\alpha 2(VI)][\alpha 3(VI)]$	Most interstitial tissues, associated with type I
Short chain	VIII	$[\alpha 1(VIII)]_3$	Descemet's membrane, subendothelial matrices
	X	$[\alpha 1(X)]_3$	Hypertrophic and mineralizing cartilage
Long chain	VII	$[\alpha 1(VII)]_3$	Anchoring filaments, epithelia

collagen is the primary component of cartilage and type IV collagen is the only collagen present in the basement membrane.

The shape and most of the structural properties attributed to collagen are determined by triple helical domains. The triple helical domain consists of 3 separate chains (α chains) twisted into a left-handed helix. These domains may then wrap around one another in a higher order rope-like fashion. The folding of the chains into the correct helical conformation requires that glycine be present at every third amino acid. The α chains are therefore composed of Gly–Xaa–Yaa repeats, in which Xaa and Yaa can be any amino acids. While there is no recognizable pattern to the Xaa and Yaa positions in collagen, Xaa is frequently proline and Yaa is frequently hydroxyproline. The α – chains of each type of collagen are unique and are different gene products. In fibril forming collagens, a single triple helical domain comprises > 95% of the molecule. Other collagens may have multiple triple helical domains and these may comprise only a fraction of the molecule's mass.

In addition to having a structural function, collagen is also known to be important in cell attachment and differentiation. In some cases, cells are able to adhere to collagen specifically via a cell surface receptor molecule. In other cases, attachment of cells to collagenous structures is indirect and occurs via cell interactive glycoproteins as described below.

2.2 Proteoglycans

Proteoglycans are complex macromolecules that each contain a core protein with one or more covalently bound linear polysaccharide chains known as glycosaminoglycans (GAG) (see [3] for review). The number of GAG chains on a core protein may vary from 1–100 and their length may vary from a few disaccharides to hundreds, thereby allowing for a great deal of diversity in proteoglycan structure. There are four major classes of GAG disaccharide backbones based on their chemical structure. They include:

1. Hyaluronic acid (HA) (can exist as a simple GAG with no covalently bound protein);
2. Chondroitin sulfate/dermatan sulfate (CS/DS);
3. Keratan sulfate (KS);
4. Heparan sulfate/heparin (HS).

Proteoglycans are found inside cells, on cell surfaces and are ubiquitous in connective tissues and extracellular matrices. For example, HA molecules self-associate to produce entangled networks that trap large amounts of water, and are highly elastic, yet easily deformed. Due to these properties, HA based matrices can provide pathways through with cells move and can define the space in which cells differentiate and form new matrices. It is, therefore a major component of the ECM that surrounds migrating and proliferating cells. In embryonic tissues, HA (with little or no associated proteoglycans) can constitute the major structural macromolecule in the ECM. A second example of an ECM proteoglycan is Aggrecan, the large CS/KS proteoglycan in cartilage. This complex proteoglycan has several HA binding sites in its core protein causing the formation of proteoglycan-HA aggregates in the ECM. The proteoglycans cause a swelling pressure on the collagen network, allowing cartilage to withstand compressive forces and cushion and protect underlying bone, thereby imparting both stiffness and resilience to the tissue. Similarly, the ECM of most tissues contain a variety of specific complex and low-molecular weight proteoglycans. In addition, both HS and CS proteoglycans are integral components of the ECM of the basal lamina.

In general, proteoglycans participate in molecular events that regulate cell adhesion, migration and proliferation and also have mechanical roles in matrix assembly, tissue architecture and growth factor binding.

2.3 Cell interactive glycoproteins

Cell interactive glycoproteins (protein molecules with carbohydrate chains bound to them) are ECM molecules that bind not only collagen and proteoglycans, but also cell surface receptors (see [4] for review). Due to multiple binding domains, these proteins provide considerable complexity to the potential interactions between cells and the ECM, as well as between various matrix molecules. They are important for organizing the other components of the ECM and for regulating cell-surface interactions. All the cell interactive glycoproteins share several characteristics, including a distinct region specialized for binding to cell surfaces as well as to other extracellular molecules. Most of these domains contain a specific amino acid sequence that is recognized specifically by a cell surface receptor. These polypeptide sequences often serve as adhesion recognition signals and may provide cell-type specificity. Many cell interactive glycoproteins exist in several variant forms (families of closely related proteins) and can form oligomers or polymers. Finally, these molecules are generally regulated in location, quantity and type during tissue development and remodeling.

2.3.1 Fibronectins

Fibronectins are multifunctional cell interaction glycoproteins in the ECM that are known to play important roles in cell adhesion, migration and wound healing. It is found in blood as a soluble plasma glycoprotein and is also secreted by various cells into the ECM where it interacts with collagens and specific proteoglycans. The primary role of fibronectin is to attach cells to various extracellular matrices (matrices other than type IV collagen). Due to their cell adhesive capability, fibronectins are responsible for organizing cells and are essential in differentiation during embryogenesis.

Fibronectin exists as a dimer of two similar peptides as shown in Fig. 1. It is encoded by a single gene, but exists in a number of variant forms due to alternative splicing in its precursor mRNA. All fibronectin molecules have the same basic functional domains. In some cases the alternative splicing is in a cell adhesion sequence, potentially regulating fibronectin cell-type specificity.

Fibronectin contains specific high-affinity binding sites for cell surface receptors and also possesses regions that bind collagen, heparan sulfate, fibrin, complement protein, bacteria and DNA (see Fig. 1). The cell-binding domains of fibronectin contain multiple Arg–Gly–Asp (RGD) and Arg–Gly–Asp–Ser (RGDS) as well as Leu–Asp–Val (LDV) and Arg–Glu–Asp–Val (REDV) cell recognition motifs. The presence of several different cell binding regions suggests broad cell adhesion properties for fibronectin. Finally, fibronectin may act to link the cell cytoskeleton to the extracellular matrix via membrane receptors. For example, in fibroblast culture, fibronectin has been shown to co-localize with actin stress fibers where the cells attach to the substrate.

2.3.2 Laminin

Laminin is a major component of basement membranes that has been implicated in a variety of biological processes including cell adhesion, migration and differentiation. Laminin is primarily responsible for anchoring cells to the basement membrane (adhesion to type IV collagen). In other words, laminin acts as a bridging molecule between the basal lamina and cells.

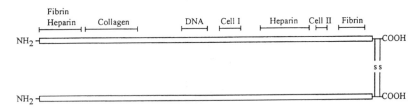

Fig. 1. The general structure and functional domains of fibronectin. The glycoprotein is depicted by a dimer of similar subunits linked by disulfide bonds. The major ligands bound in each region of the molecule are listed. Cells interact with the fibronectin molecule in 2 specific regions. Cell I contains the RGD(S) recognition motif, while Cell II contains LDV and REDV motifs and two active cell binding sequences in the heparin binding domain. Modified from [4].

The laminin molecule consists of three chains (called A, B1 and B2) and is cross-shaped as shown in Fig. 2. Laminin can interact directly with cells in several regions of the molecule due to specific adhesive recognition sequences that bind cell-surface receptors. These sequences include Arg–Gly–Asp–Asn (RGDN) and Ile–Lys–Val–Ala–Val (IKVAV) in the A chain, Tye–Ile–Gly–Ser–Arg (YIGSR) and Pro–Asp–Ser–Gly–Arg (PDSGR) in the B1 chain, and Arg–Asn–Ile–Ala–Glu–Ile–Ile–Lys–Asp–Ile (peptide 20) in the B2 chain. In addition, laminin has specific regions that bind heparin (RYVVLPR in the B1 chain and region G of the A chain) and type IV collagen (B1 and B2 globular regions).

Laminin also has several epidermal growth factor (EGF)-like domains that are exposed upon degradation. These domains have been shown to stimulate cell proliferation in a manner similar to EGF. Degradation of laminin *in vivo* could lead to liberation of growth factor activity that stimulates growth in regions of tissue damage. Such activity could result in tissue remodeling involving only those cells sensitive to these signals.

Unlike fibronectin, variant forms of laminin are assembled from closely related genes encoding isoforms of certain chains. The form of laminin present in the ECM is a function of both tissue type and developmental stage suggesting that the function of variant forms of laminin are distinct.

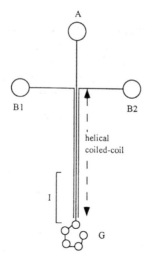

Fig. 2. The structure of the laminin molecule. (G) binds heparin and type IV collagen; (I) major cell binding region with 2 recognition sequences; (Center of cross) minor cell binding region with several peptide sequences including 1 RGD recognition motif; (B1 and B2) globular regions bind type IV collagen. Modified from [4].

2.3.3 Vitronectin

Vitronectin (also known as S-protein) is a cell interactive glycoprotein that is found in soluble form in blood plasma and in fibrillar form in the ECM of a variety of tissues. In the ECM, it sometimes co-localizes with fibronectin. The cell attachment activity of vitronectin is based on the Arg–Gly–Asp–Val (RGDV) recognition motif and cell adhesion to vitronectin occurs via any of several receptors of the integrin family. In addition to cell binding capability, vitronectin also binds collagen, heparin, and plasminogen activator inhibitor (PAI-1), thus suggesting roles for the molecule in tissue remodeling and cell migration.

2.3.4 Thrombospondin

Thrombospondin is a large, trimeric multifunctional glycoprotein that is released from activated platelets and also secreted into the ECM by a variety of cell types during growth. The level of thrombospondin in the ECM is highest prior to differentiation. It contains at least three cell-binding domains, including a heparin (proteoglycan) binding site and an RGD recognition sequence. Thrombospondin also binds other ECM molecules including collagen, laminin and fibrinogen. Depending on culture conditions, thrombospondin may co-localize with fibronectin or with heparan sulfate proteoglycan in the ECM. Depending on the type of cell, adhesion to thrombospondin can lead to cell spreading or aggregation, or to negative effects on cell adhesion (e.g. inhibition of cell attachment to fibronectin). Thrombospondin has also been implicated in regulating the growth of smooth muscle cells.

2.3.5 Tenascin

Tenascin is a 6 subunit oligomer characterized by structural units homologous to those of other proteins. The molecule contains 13 EGF repeats, 8–11 fibronectin repeats and a globular terminus homologous to fibrinogen. It is commonly found at sites of tissue remodeling and can demonstrate either adhesive (weak interactions in the fibronectin-like region which includes the RGD motif) or anti-adhesive activity (e.g. inhibits adhesion to fibronectin and laminin). The mechanism of inhibition is not yet clear. Finally, tenascin is capable of binding fibronectin weakly and chondroitin sulfate proteoglycan.

2.3.6 Other cell interactive glycoproteins

The cell interactive glycoproteins mentioned above are some of the most widely distributed and well-studied molecules. However, other interesting and important cell interactive glycoproteins are known to exist, many of which are specific to certain tissues. For example, bone sialoprotein and osteopontin are believed to mediate osteoblast and osteoclast adhesion in bone. Von Willebrand factor (vWf) is a large multifunctional glycoprotein found in the extracellular matrix of blood vessels that is critical to platelet adhesion and aggregation under high shear conditions. The functions of other potentially interesting molecules in the ECM, such as epinectin, chondronectin and fibrillin, are not as well understood as those mentioned above. In addition, many other ECM molecules contain cell recognition regions (such as the RGD motif or heparin binding sites), some of which may have cell interactive capability. Because of the number and diversity of molecules that comprise the ECM, it is likely that new cell interactive glycoproteins will be identified in the future. The complex interactions between these numerous molecules with one another and with various types of cells remain to be elucidated.

3. Cell–ECM adhesion and adhesion receptors

Cell adhesion is critical for the assembly of cells into tissues and the maintenance of tissue integrity [5]. In order to form a three-dimensional structure, cells must not only adhere to one another (see Chapter II.2), but also to the underlying support structure. The way in which the cells organize themselves into specific three-dimensional patterns is dependent on the specific cell adhesion mechanisms. Cell adhesion is accomplished via transmembrane receptors (typically glycoproteins) that mediate binding at the extracellular surface and determine the specificity of the adhesive interaction. Typically, a domain (binding region) within the receptor will specifically recognize and bind to a short peptide sequence in the extracellular matrix. As mentioned in the preceding section, collagens, proteoglycans, and the cell adhesive glycoproteins have all been identified as cell binding molecules in the ECM. For details on how this binding is translated into an intracellular signal, the reader is referred to Chapter II.2.

3.1 Integrins

Cells use a variety of different cell-surface adhesion receptors to interact with the ECM, the most prominent of which belong to the integrin superfamily [6–8]. Integrins are heterodimeric transmembrane glycoproteins which consist of noncovalently associated α and β subunits (see Fig. 3 and Table 2). Each subunit consists of an extracellular domain, a membrane spanning domain, and in most cases a short cytoplasmic region. Currently, at least 15 distinct α subunits and 8 β subunits have been identified. These subunits can combine to form at least 21 distinct integrin receptors [9]. The β subunits of all integrins show a great deal of sequence homology, while the α subunits exhibit greater heterogeneity. The interaction of integrins with the ECM requires both subunits and the specific α/β pairing determines the integrin's ligand binding ability.

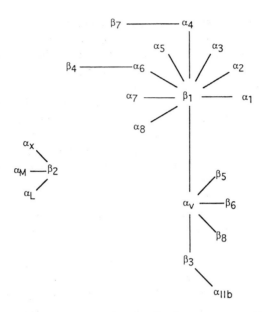

Fig. 3. The integrin superfamily of cell adhesion receptors. Each integrin is composed of an α and β subunit that are non-covalently coupled. Each α and β subunit is a transmembrane protein. Modeled after [8].

Table 2. Selected integrin receptors and their known extracellular matrix ligands.

Subunit composition	Known ECM ligands
β_1 family	
$\alpha_1\beta_1$	Laminin (collagen)
$\alpha_2\beta_1$	Collagen (laminin)
$\alpha_3\beta_1$	Fibronectin, laminin, collagen
$\alpha_4\beta_1$	Fibronectin
$\alpha_5\beta_1$ ('fibronectin receptor')	Fibronectin
$\alpha_6\beta_1$ ('laminin receptor')	Laminin
$\alpha_7\beta_1$	Laminin
$\alpha_v\beta_1$	Fibronectin
β_3 family	
$\alpha_{IIb}\beta_3$	Fibrinogen, fibronectin, vitronectin, von Willebrand factor
$\alpha_v\beta_3$ ('vitronectin receptor')	Vitronectin, fibrinogen, fibronectin, von Willebrand factor, thrombospondin, laminin
Other	
$\alpha_v\beta_5$	Vitronectin
$\alpha_v\beta_6$	Fibronectin
$\alpha_4\beta_7$	Fibronectin

Integrins recognize specific peptide sequences in their ligands, the best studied of which is the RGD sequence found in a number of ECM proteins including type I collagen, fibronectin, fibrinogen, vitronectin, laminin and thrombospondin. However, not all integrins bind ligands via the RGD sequence. In addition, while many ECM proteins have multiple RGD sequences, it is common for a specific integrin to interact with only 1 specific RGD sequence, suggesting that the protein sequence and conformation in the immediate vicinity of the RGD site is also important.

Most cells in multicellular organisms express at least one integrin, and in adult mammals, most cells constitutively express multiple integrins. However, the spatial and temporal pattern of the integrin expression is distinct. To add to the complexity, most cells simultaneously display multiple integrins that can interact with the same ECM molecule and most integrins can recognize several ECM ligands. The same integrin receptor can be expressed on cells with markedly different functions and it is likely that specific integrins perform different functions in different cell types. Integrin receptors have been shown to interact with a wide variety of ECM molecules including collagen, fibronectin, fibrinogen, laminin, vitronectin, won Willebrand factor and thrombospondin. The reason for the apparent redundancy in integrin ligand specificity is not clear.

In addition to functioning as cell adhesion molecules, integrins may also function as signaling molecules allowing cells to detect and respond to local changes in the composition of ECM ligands. In addition to constitutively expressed integrins, the local concentration and expression of other integrin molecules on the surface of epithelial cells can be modulated by growth factors [10]. Therefore, the signals generated by integrins on epithelial cells may play important roles in modulating epithelial cell differentiation, proliferation, cell spreading and migration, and gene expression, including genes that encode cytokines and locally active proteases.

3.2 Cell surface proteoglycans

In addition to the integrin receptors, cells also use cell surface proteoglycans for attachment to the ECM [3]. Heparan sulfate proteoglycans are the most common proteoglycans associated with cell surfaces. In general, these proteoglycans are unique from those found in the extracellular matrix. The best studied transmembrane proteoglycans belong to the syndecan family and are located on the surface of epithelial cells. These molecules are comprised of a cytoplasmic domain, a hydrophobic membrane spanning region (both of which are highly conserved) and an extracellular domain (which shows variation between members of this family) with multiple attachment regions for glycosaminoglycan chains. Heparan sulfate chains attached to the syndecan receptor surface are capable of binding to collagens I, III, and V, fibronectin and thrombospondin. The intracellular domain of syndecan is known to interact with actin fibers, suggesting that this receptor may act as a bridge between the extracellular matrix and the cell cytoskeleton and have a role in maintaining cell shape. The expression of syndecan is transient and developmentally regulated during organogenesis. Several other transmembrane proteoglycans have been identified, many of which are expressed only on specific cell types.

4. Summary

Cellular interactions with the extracellular matrix are complex. Not only does the ECM provide structural support to cells and tissues, but it is also responsible for generating cell signals (via transmembrane receptors on cell surfaces) that are capable of affecting cell proliferation and differentiation, cell migration, and cell adhesion. In tissue engineering, where one is attempting to regenerate a specific tissue or replace a specific tissue function, it is critical to understand the interactions between the tissue cells of interest and their natural ECM environment. A thorough understanding of these interactions may allow the tissue engineer to purposefully and systematically regulate cellular function. Furthermore, specific cell adhesion mechanisms could be exploited by incorporating cell recognition peptide motifs (such as RGD) into synthetic polymeric materials for implantation. Such a system would allow the engineer to develop biomaterials with specific mechanical properties that also promote a desirable cellular behavior. In summary, nature has developed an intricate environment in which normal tissue cells live and grow. By elucidating the ways in which the cells and their ECM environment interact, the tissue engineer may begin to systematically and purposefully design and grow new tissues.

References

1. Hay, E. D., *Cell Biology of the Extracellular Matrix*. 2nd edn, Plenum Press, New York, NY, 1991, pp. 468.
2. Linsenmayer, T. F., Collagen. In *Cell Biology of the Extracellular Matrix*, 2nd edn, ed. E. D. Hay. Plenum Press, New York, 1991, pp. 7–44.
3. Wight, T. N., Heinegard, D. K. and Hascall, V. C., Proteoglycans: Structure and function. In *Cell Biology of the Extracellular Matrix*, 2nd edn, ed. E. D. Hay. Plenum Press, New York, 1991, pp. 45–78.
4. Yamada, K. M., Fibronectin and other cell interactive glycoproteins. In *Cell Biology of the Extracellular Matrix*, 2nd edn, ed. E. D. Hay. Plenum Press, New York, 1991, pp. 111–146.
5. Gumbiner, B. M., Cell adhesion: The molecular basis of tissue architecture and morphogenesis. *Cell*, 1996, **84**, 345–357.

6. Hynes, R. O., Integrins: a family of cell surface receptors. *Cell*, 1987, **48**, 549–554.

7. Hynes, R. O., Integrins: Versatility, modulation, and signaling in cell adhesion. *Cell*, 1992, **69**, 11–25.

8. Albelda, S. M. and Buck, C. A., Integrins and other cell adhesion molecules. *FASAB Journal*, 1990, **4**, 2868–2880.

9. Rosales, C., O'Brien, V., Kornberg, L. and Juliano, R., Signal transduction by cell adhesion receptors. *Biochem. Biophys. Acta*, 1995, **1242**, 77–98.

10. Sheppard, D., Epithelial integrins. *Bioessays*, 1996, **18**(8), 655–660.

CHAPTER II.2

Cell–Cell Interactions

SHARAD KUKRETI,
KONSTANTINOS KONSTANTOPOULOS,
LARRY V. MCINTIRE
Cox Laboratory for Biomedical Engineering,
Institute of Biosciences and Bioengineering,
Rice University,
Houston, TX 77005-1892, USA

1 Introduction

Cell adhesion is instrumental in diverse biological phenomena, including tissue morphogenesis, inflammation, and thrombosis. Cell–cell interactions play a vital role in assembling cells together and in determining the overall tissue architecture during tissue generation. A relevant example is neovascularization or formation of new blood vessels. The endothelium which forms the inner layer of the vessel wall is a biologically active monolayer presenting a dynamic interface between flowing blood and tissues of the body. Adhesive junctions between endothelial cells not only help in maintaing the integrity of the vessel structure experiencing hemodynamic forces due to blood flow, but also actively regulate the permeability of bioactive molecules and transmigration of various leukocyte populations across the endothelial monolayer during inflammation. In the vascular system, cell attachment during inflammatory or thrombotic processes occurs under flow conditions and depends on the balance between the dispersive hydrodynamic forces and the adhesive forces generated by the interaction of membrane-bound receptors and their ligands. Cellular interactions are required to be highly specific in nature for proper regulation of events to result in homeostatic inflammation and coagulation. Inflammatory processes normally protect the body from infections by foreign pathogens. However, when dysregulated, leukocyte migration to healthy tissues can lead to unwanted disorders and pathologic conditions. Similarly, defective modulation of the hemostatic mechanism can lead to the formation of platelet rich thrombus which can compromise the patency of the blood vessel. Understanding the complex interplay among blood flow, cell adhesion, and vascular biology at the molecular level is crucial for developing tissue engineering based approaches for therapeutic interventions.

This chapter reviews families of adhesion receptors involved in inflammatory or thrombotic responses and endothelial—endothelial cell adhesive junctions. We also discuss the signaling mechanisms associated with the adhesion processes, particularly those mediated by integrins and cadherins. Finally we discuss cell—cell adhesion in graft rejection and platelet—biomaterial interactions as relevant examples in current tissue engineering research.

2 Families of adhesion receptors

2.1 Selectins

All three known members of this family share a similar cassette structure of three ordered domains.

(i)   N-terminal Ca^{2+} dependent lectin or carbohydrate binding motif.
(ii) Epidermal Growth Factor (EGF)-like motif.
(iii) Complement regulatory (CR) motif.

Selectin binding is completely dependent on the conformational changes associated with filling of two Ca^{2+} binding sites and is also sensitive to pH, with decreased binding below physiological pH.

2.1.1 P-selectin
Monoclonal antibodies recognizing activated but not resting platelets led to the discovery of P-selectin molecule. This 140 kDa integral membrane protein resides in the α-granules of unactivated platelets. Subsequently, it was also identified in the Wiebel—Palade bodies of endothelial cells [1]. Sorting of P-selectin into these granules is determined by sequences in its cytoplasmic tail which interact with the sorting machinery in cells. Upon stimulation with agonists like histamine or thrombin, P-selectin gets rapidly distributed to the cell plasma membrane [2]. On endothelial cells, P-selectin expression declines to basal levels after 30 minutes of stimulation. Internalized P-selectin molecules are degraded in the lysosomes, a process that is also regulated by specific amino acid sequences in its cytoplasmic domain. Some P-selectin may be recycled back to storage granules or secreted consistent with an alternatively spliced form lacking the transmembrane domain. This rapid expression kinetics on endothelial cells indicates a role of P-selectin in recruiting of neutrophils in the early stages of acute inflammatory reactions. However, there are reports indicating

chronic expression of P-selectin by IL-3 [3] and IL-4 [4] stimulation of human umbilical vein endothelial cells. Murine brain microvascular endothelial cells lack constitutive P-selectin and hence cannot support leukocyte adhesion via the rapid mobilization pathway. However, stimulation with IL-1β, LPS or TNF-α upregulates P-selectin expression within 3–4 hours thus providing a potent mechanism for mediating leukocyte adhesion [5, 6]. Immunohistochemical analysis of human atherosclerotic plaques has also revealed a strong expression of P-selectin in arterial endothelial cells overlying active atherosclerotic plaques, thus providing a mechanism for monocyte recruitment into sites of atherosclerosis [7].

Recently, a glycoprotein P-selectin ligand (PSGL-1) has been reported [8–10]. PSGL-1 is a highly extended molecule with a ~50 nm long extracellular domain and binds to P-selectin near the N-terminus [11]. It has been shown that both amino terminal tyrosine sulfation and presence of O-linked sialyl Lewis structures (Sle^x) [12] on PSGL-1 are essential for high affinity binding to P-selectin. The electrostatic binding interaction mediated by the sulfated region of PSGL-1 may contribute to the high tensile strength of the unimolecular interaction between purified P-selectin and PSGL-1 on neutrophils [13].

2.1.2 L-selectin

Experiments demonstrating inhibition of binding of murine and human lymphocytes to postcapillary high endothelial venules (HEVs) of peripheral lymph nodes with simple monomeric sugars like mannose-6-phosphate led to cloning of L-selectin. It is present on the surface of all leukocytes but gets rapidly shed upon cell activation. The mechanism of shedding is not completely understood. It has been hypothesized that the EGF-like domain might be involved in the binding of proteases, which could lead to activation-induced shedding of L-selectin [14]. L-selectin is also detected at high concentrations in the serum of healthy individuals [15], and could increase or decrease in certain diseased states. Histochemical studies have revealed that L-selectin is preferentially concentrated on tips of leukocyte surface microvilli, hence making it an ideal candidate for establishing leukocyte attachment under flow [16].

There is evidence that terminal sialic acid residues on O-linked oligosaccharides are important for ligand binding activity to L-selectin [17]. CD34 and GlyCAM-1, expressed on human peripheral lymph node HEV and recognized by monoclonal antibody MECA-79, present ligands to L-selectin and have been shown to mediate lymphocyte recirculation and leukocyte rolling. Several studies [18] suggest PSGL-1 is also a ligand for L-selectin on neutrophils, monocytes and CD34$^+$ hematopoeitic progenitor cells. Interestingly, it has been found that L-selectin on neutrophils and monocytes but not lymphocytes is also decorated with terminal sialyl Lewis structures and hence capable of acting as a ligand for other selectins [19].

2.1.3 E-selectin

Antibodies against leukocyte integrins or their endothelial ligands were unable to completely block adhesion to IL-1β or TNF-α stimulated endothelial cells. Hybridoma screening methods applied to activated umbilical vein endothelial cells produced mAbs that identified a 115 kDa protein now called E-selectin. The expression of E-selectin on endothelial cells peaks within 4–6 hours of stimulation with IL-1β or TNF-α and then slowly declines to basal levels [20]. Cytokine stimulation of endothelial cells causes increased gene transcription and protein synthesis leading to E-selectin expression. It has been observed that E-selectin gene transcription is downregulated within 6–9 hours after induction [20].

Moreover, E-selectin mRNA has a short half life [21] and in addition E-selectin also gets internalized and degraded in lysosomes [22]. Such a combination of processes ensures that expression of E-selectin on the endothelial cell surface is transient. However, *in vivo*, particularly in blood vessels of skin, a much longer half life of E-selectin mRNA has been observed [23].

Sialylation and fucosylation appear to be essential for the binding activity of E-selectin ligands [24, 25]. Recently [26], an E-selectin ligand (ESL-1) has been cloned from mouse neutrophilic progenitor cells. Fucosylation of ESL-1 is essential to support adhesion to E-selectin. ESL-1 protein has 5 putative sites for N-linked glycosylation but is weakly, if at all, modified by O-linked carbohydrate side chains. Using affinity isolation experiments, it has been shown that human but not mouse L-selectin binds to E-selectin [27]. Prior work has also suggested involvement of PSGL-1 in binding to E-selectin [8, 28]. However, it was recently shown that FucTVII but not PSGL-1 is essential for ligand binding activity to E-selectin [29].

A unique feature of the interaction supported by selectin family of receptors is the requirement of a minimum threshold hydrodynamic shear stress [30]. It has been observed that at shear stress levels below 0.5 dynes/cm^2, there is inhibition of selectin mediated adhesion. This could provide an additional mechanism for regulating intravascular plugging of microvessels by leukocytes during periods of transient ischemia.

The physiological relevance of selectin mediated adhesion is undescored in a recently described syndrome called leukocyte adhesion deficiency type II (LAD-II) [31]. This syndrome is characterized by impaired immunity and developmental anomalies which can be linked to defective fucose metabolism. In particular, the fucose containing oligosaccharide moiety sialyl Lewis^x, which serves as ligand for vascular selectins is absent in these patients.

2.2 Integrins

The term integrin was coined to describe membrane receptors which integrate the extracellular environment with the intracellular cytoskeleton. The integrins comprise a large family of cell membrane heterodimeric glycoproteins consisting of noncovalently attached α and β subunits. There are 8 known β subunits and 14 known α subunits. It is now evident that a single α subunit can pair with more than one β subunit. α-subunits average approximately 1100 amino acids and β subunits average approximately 750 amino acids. The α subunits contain several divalent cation binding sites. As a result of this, the function of these receptors is cation dependent. Certain α subunits also contain an I domain (for 'inserted', or 'interacting' domain), which apparently provides binding specificity to those integrins which have it. In general, integrins without the I domain bind principally to Arg–Gly–Asp (RGD)-containing peptide sequences in extracellular matrix proteins, whereas presence of I domain seems to allow more extensive non-RGD binding. The I domain is also associated with conformational changes involved in affinity modulation [32].

2.2.1 Leukocyte integrins

On the leukocyte surface, β_2 family of integrins consists of three members while the α_4 family has two members.

LFA-1

Screening mAbs for the ability to inhibit cytotoxic T- lymphocyte mediated killing of tumor targets led to the identification of LFA-1 (CD11a/CD18, $\alpha_L \beta_2$) molecule. It is present on virtually all immunue cells and participates in a broad spectrum of both antigen-dependent and antigen-independent interactions of immune cells. Stimulation of endothelial cells with IL-1β or TNF-α or LPS leads to upregulation of ICAM-1 (intercellular adhesion molecule-1), which has been shown to bind LFA-1. Presence of an I domain is presumably important in binding of LFA-1 to ICAM-1 since no RGD sequences are present in ICAM-1. Observations indicating alternative ligands for LFA-1 led to cloning of a novel integral membrane protein, intracellular adhesion molecule 2 (ICAM-2). ICAM-2 is constitutively expressed on endothelial cells, but unlike ICAM-1 it is not elicited by inflammatory mediators. The relative role of ICAM-2 in inflammatory reactions *in vivo* remains to be determined.

Mac-1

Mac-1 (CD11b/CD18, $\alpha_M \beta_2$) was first identified by monoclonal antibodies as a marker for myeloid cells. It can mediate phagocytosis and lysis of C3bi-coated erythrocytes and contributes to elevated natural killer (NK) activity against C3bi-coated target cells. Recent evidence suggests that Mac-1, like LFA-1, may play a more general role in mediating adhesive interactions of myeloid cells.

Several reports have documented that Mac-1 presents a ligand to ICAM-1. It has been suggested that chemotactic stimuli modify, either functionally or spatially, the Mac-1 heterodimer to allow its recognition by ICAM-1. Mac-1 also binds the C3bi fragment of complement C3.

p150,95

The functional importance of the third $\beta 2$ integrin p150,95 (CD11c/CD18, $\alpha_X \beta_2$) in cellular interactions remains less well defined. It has been shown to be a complement receptor.

2.2.2 VLA-4

Present on mononuclear cells but not on resting neutrophils, VLA-4 (CD49d/CD29, $\alpha_4 \beta_1$) mediates cellular trafficking to inflammatory sites. Its major ligand on the endothelial cells is the vascular cell adhesion molecule-1 (VCAM-1). VLA-4 also binds to the alternatively spliced connecting segment-1 (CS1) domain of fibronectin which is expressed on cytokine stimulated endothelial cells. In general, cell adhesion to fibronectin CS1 requires higher levels of $\alpha_4 \beta_1$ integrin activity than adhesion to VCAM-1 [33]. VLA-4 appears to be an exception to the general rules stated above in that, although it lacks an I domain, it can bind both VCAM-1 and fibronectin at non-RGD sites.

2.2.3 $\alpha_4 \beta_7$

Expressed on lymphocytes, this integrin mediates lymphocyte homing by binding to mucosal addressin cell adhesion molecule 1 (MAdCAM-1), a vascular ligand selectively expressed in gut-associated lymphoid tissues and in the intestinal lamina propria. Like L-selectin, $\alpha_4 \beta_7$ integrin is also concentrated on the tips of microvilli indicating that this might be a common characteristic of adhesion molecules specializing in initiating leukocyte–endothelial contact under physiologic flow conditions [16].

The regulation of integrin-mediated binding occurs at three levels: modulation of receptor synthesis, mobilization of an intracellular pool of receptors and modulation of binding affinity of individual receptors. Modulation of receptor synthesis is usually associated with cell differentiation rather than with the rapid response of terminally differentiated cells to particular challenges. For example, LFA-1 and VLA-4 increase two-to four fold upon conversion of naive to memory T-lymphocytes [34, 35], which could influence localization and recirculation of these subpopulations. Also differentiation of myeloid lineages is associated with early expression of LFA-1 and later expression of Mac-1, and the subsequent maturation of monocytes to tissue macrophages is associated with a decrease in Mac-1 and VLA-4 expression [36, 37].

Cell adhesion can be modulated within minutes by mobilizing intracellular pools of integrin receptors [38]. For example, in neutrophils and monocytes, chemotactic factors result in up to a 10-fold increase in surface expression of Mac-1 [39]. The mechanism of this form of upregulation involves fusion of leukocyte granules containing the receptors with the cell surface [40]. An even more rapid mechanism of cell adhesion modulation is the ability of some integrins to increase their binding affinity through changes in the molecular conformation upon stimulation of the cell. In addition to serving as effector molecules of leukocyte activation, they can also serve as sensors, triggering a variety of intracellular leukocyte activation responses upon ligand binding.

The importance of integrins is dramatically highlighted by type I leukocyte adhesion deficiency (LAD-I), a disease involving recurrent life-threatening bacterial and fungal infections [41]. This syndrome is attributed to defects in the β_2-subunit of the leukocyte integrins [42]. In these patients neutrophils are unable to localize to sites of infection.

2.2.4 Platelet integrins

Platelets have at least five different integrins [43]. Three of them share the β_1-subunit, also found in cell surface receptors on mononuclear leukocytes. $\alpha_2\beta_1$ integrin (GPIa-IIa; CD49b/CD29) is present at the level of ~ 1,000 copies per platelet, and is required for platelet adhesion to collagen types I-VIII under both static and flow conditions [44]. $\alpha_5\beta_1$ (GPIc-IIa; CD49e/CD29) interacts with RGD sequences in fibronectin and laminin. $\alpha_6\beta_1$ (GPc-IIa; CD49f/CD29) mediates adhesion of platelets to laminin coated surfaces. Two platelet membrane integrins share the β_3-subunit and represent receptors for vitronectin ($\alpha_v\beta_3$; CD51/CD61) and fibrinogen ($\alpha_{IIb}\beta_3$; CD41b/CD61; GPIIb-IIIa). Although $\alpha_v\beta_3$, $\alpha_5\beta_1$ and $\alpha_6\beta_1$ do support adhesion, their roles in hemostasis and thrombosis appear to be minor compared with $\alpha_2\beta_1$ and $\alpha_{IIb}\beta_3$ and the non-integrin glycoprotein (GP) Ib-IX (see below) in normal subjects [45]. However, their roles may be significant in disease states in which $\alpha_{IIb}\beta_3$ or GPIb-IX (CD42) is either not present or cannot function properly.

$\alpha_{IIb}\beta_3$ integrin

$\alpha_{IIb}\beta_3$ is a heterodimeric complex composed of the noncovalent association of two integral membrane proteins, α_{IIb} and β_3. It is the most abundant adhesive receptor on the platelet surface with ~ 50,000 copies per platelet. $\alpha_{IIb}\beta_3$ is also present on the surface canalicular system and in the α-granules of platelets which can be expressed on the cell surface upon activation. $\alpha_{IIb}\beta_3$ on resting platelets has no ligand-binding activity, and it can bind soluble adhesive proteins only after platelet stimulation which induces a conformational change in the receptor. A number of monoclonal antibodies, such as PAC-1, have been produced that recognize activation-dependent determinants on $\alpha_{IIb}\beta_3$ [46, 47]. Moreover, it has been reported that ligand binding by itself induces conformational changes within the complex [47].

$\alpha_{IIb}\beta_3$ serves as the main receptor for fibrinogen on activated platelets, but it can also bind von Willebrand factor (vWf), fibronectin and vitronectin. Recognition of ligands by $\alpha_{IIb}\beta_3$ is mediated, in part, by the RGD peptide sequence [48]. In addition to that, fibrinogen contains a second site that binds to $\alpha_{IIb}\beta_3$, and is a 12-amino acid sequence at the carboxyl terminus of fibrinogen γ-chain [48, 49].

Platelets from patients with Glanzmann's thrombasthenia have qualitative or quantitative abnormalities in the platelet $\alpha_{IIb}\beta_3$ complex [50]. This congenital, hereditary hemorrhagic disorder is characterized by a profound defect of platelet aggregation in response to multiple physiologic agonists such as adenosine diphosphate and thrombin that induce $\alpha_{IIb}\beta_3$-ligand binding.

The platelet glycoprotein (GP) Ib-IX complex

GPIb-IX belongs to the leucine-rich glycoprotein gene family which defines a family of proteins that share a common structural motif composed of a leucine-rich, 24 amino acid consensus sequence. The GPIb-IX complex is present at the level of $\sim 25,000$ copies per platelet, and consists of three transmembrane subunits: GPIbα, GPIbβ and GPIX. GPIbα and GPIbβ are disulfide-linked to each other and noncovalently associated with GPIX in 1:1 stochiometry [43]. In the platelet membrane, GPIb-IX is non covalently associated with another leucine-rich motif protein, GP V, in 1:1 stoichiometry. The primary role of GPIb-IX is to mediate adhesion of platelets to sites of vascular injury by interacting with vWf immobilized in the subendothelium. Although vWf is also present in the plasma, it does not interact spontaneously with platelets in normal subjects, suggesting that the platelet GPIb-IX complex preferentially recognizes vWf bound in the subendothelium [51, 52]. Recent studies have shown that purified vWf enriched in the largest multimers found in plasma binds to platelets in response to shear stress [53, 54]. The vWf binding site appears to reside in the extracellular portion of GPIbα [55].

Platelets from patients with the Bernard–Soulier syndrome have qualitative or quantitative abnormalities in the platelet GPIb/IX complex [50], and as a result, fail to undergo selective vWf-dependent platelet interactions [56]. This congenital platelet disorder is characterized in part by prolonged skin bleeding time, large platelets, and low platelet counts (thrombocytopenia).

2.3 Immunoglobulin superfamily

The immunoglobulin family of receptors is characterized by the presence of the immunoglobulin domain comprising of 70–110 amino acids in a well-defined structure [57]. The members of this family are diverse in nature ranging from soluble and membrane bound immunoglobulins to the multireceptor T-cell antigen receptor complex to monomeric adhesion molecules.

VCAM-1 can be induced on the endothelial surface upon stimulation with cytokines such as IL-1β, TNF-α or IL-4. The message for VCAM-1 is alternatively spliced, resulting in a 6-domain form, which lacks domain 4, and a 7-domain form [58]. The 7-domain form is predominant on cytokine stimulated HUVECs. ICAM-1 is upregulated on endothelial cells after cytokine stimulation and its expression peaks after 24 to 48 hours. ICAM-1 is expressed with five immunoglobulin (Ig) domains. LFA-1, one of the counter-receptors on leukocytes, binds to the first immunoglobulin domain while Mac-1 binds to the third. ICAM-2 is a closely related molecule with two immunoglobulin domains and is another ligand for LFA-1. ICAM-2 is constitutively expressed on endothelial cells and a variety of blood cells. PECAM-1 contains six immunoglobulin homology domains and contains putative proteoglycan binding sequences. PECAM-1 is concentrated in intercellular junctions between endothelial cells and is also expressed on the surfaces of platelets, monocytes, neutrophils and a subpopulation of naive CD8$^+$ T-cells. Since mAb to PECAM-1 on either leukocytes or endothelial cells can block transmigration, it has been proposed that leukocyte PECAM-1 binds in a homophilic manner to endothelial PECAM-1. Recent *in vivo* studies have demonstrated a role of PECAM-1 in the recruitment of neutrophils in a peritonitis model [59]. It has also been reported that PECAM-1 (CD31) binding can induce integrin-mediated adhesion, indicating that CD31 signaling may modulate T-cell adhesion [60].

2.4 Cadherins

Cadherins are transmembrane, calcium dependent family of glycoproteins mediating cell–cell interaction by binding homophilically with adjacent cells. This family of adhesion molecules is responsible for tissue morphogenesis as well as maintaining the integrity of solid tissues. Cadherins form the so called adherens junctions which not only strengthen cell–cell adhesion but also aids signal transfer between neighboring cells.

Members of the cadherin superfamily are type I integral membrane proteins. Their extracellular domain contains a variable number of a repeated domain called the cadherin repeat which is approximately 110 amino acids in length. The cadherin superfamily can be subdivided into 6 gene families based on the number of repeats and sequence homologies in the conserved cytoplasmic domain:

   i)   Classical cadherins type I (E-, N-, P-, R- cadherin)
   ii) Classical cadherins type II (cadherin-6 to -12)
   iii) Cadherins present in desmosomes (desmocollins, desmogliens)

iv) Cadherins with a missing or a very short cytoplasmic domain (LI-, T-cadherin)

v) Protocadherin

vi) Distantly related gene products like the Drosophila *fat* tumor suppress or gene or the dachsous gene [61].

The extracellular domain of mature classical cadherins contains five cadherin repeats (EC1–5). The N-terminal repeat contains the adhesive sequences for mediating cadherin-specific adhesion. Sequencing studies have revealed that all five cadherin-repeats have a common folding topology. The N- and C-termini of each cadherin-repeat are located on opposite ends, such that a tandemly repeated array can form an elongated rod-shaped protein. Within the five cadherin-repeats, four calcium binding domains are formed, each between two successive repeats. Ca^{2+} ions have been shown to be essential in the stabilizing the active conformation of the cadherins.

The adhesive interface or the active site for homophilic recognition of anti-parallel oriented cadherin molecules on adjacent cells is large and complex and includes the histidine-alanine-valine (HAV) motif. High resolution structure determination has revealed that this adhesive binding region resides on the N-terminal cadherin repeat (EC1) unit. X-ray crystallographic studies of the first repeat of N-cadherin have revealed that parallel oriented N-cadherin molecules on the same cell surface dimerize along the strand-dimer interface [62] with their adhesive binding surfaces directed outward from the plasma membrane. In the crystal, each dimer unit interacts with another two dimers in an antiparallel orientation through their adhesive interface, thus forming a continuous linear ribbon structure. The linear cadherin 'zipper' could strengthen adhesion through cooperative effects of relatively weakly interacting monomers. A two dimensional lattice structure can also be formed if the EC1 domains normally interact with one of the other EC domains in the adhesion interface. Such formations could be expected for a junctional contacts [63] or adherens junctions mediated by the cadherins.

The cytoplasmic tail of cadherins interacts with three related proteins of the armadillo family — β-catenin, plakoglobin, and p120. The cadherin–catenin complex is linked to the actin cytoskeleton through α-catenin as shown in Fig. 1.

Another type of adhesive junction is the desmosomal junction which is conspicuous in the epithelia and in cardiac muscle. Desmosomes together with the intermediate filament cytoskeleton not only provide a contiguous network throughout the tissue but also impart a high tensile strength to the tissue. The adhesion receptors of the desmosomes are members of the cadherin superfamily, desmogleins and desmocollins. The adhesive unit important for the formation of permeability barriers in epithelial or endothelial tissues is the tight junction. An integral membrane protein called occludin is involved in the formation of this junction. Tight junctions are diverse structures and their permeability properties vary from tissue to tissue, spanning from the exclusion of whole cells or macromolecules to the selective permeability of protons and other ions.

3 Cell adhesion and signaling mechanisms in the vasculature

3.1 Leukocyte–endothelial cell interactions

The Multistep model of neutrophil extravasation in inflammation
    Upon activation the endothelium begins displaying specific adhesion molecules that tether free-flowing neutrophils (initial contact), slowing them down and causing them to roll in the direction of flow through labile contacts with the vessel wall (rolling). Initial contact and rolling of neutrophils along the endothelium are predominantly mediated by selectins [57, 64, 65]. These transient adhesive interactions bring neutrophils in close apposition with the endothelial cells, allowing them to sample their microenvironment. While rolling, neutrophils encounter activating signals, including soluble substances derived from the endothelium such as interleukin-8 (IL-8), agents presented on the endothelial cell surface such as platelet activating factor (PAF), or direct signal transduction via adhesion receptors themselves. Neutrophil activation upregulates the binding affinity of integrins via both conformational changes and altered interaction with the cytoskeleton. Activation dependent binding of β_2-integrins (Mac-1, LFA-1) on neutrophils to their endothelial ligands (ICAM-1) shifts the transient rolling interactions into firm adhesion [57, 64, 65]. Furthermore, upon activation neutrophils flatten on the endothelium, resulting in increased contact area for integrin-mediated binding and decreased fluid drag and torque on the cell. Subsequently, neutrophils undergo dramatic shape changes allowing them to squeeze through the interendothelial junctions of the vessel wall into the extravascular tissue space and migrate to inflammatory sites [57, 64, 65]. Transmigration does not necessarily accompany neutrophil stable adhesion to the endothelium unless a favorable chemotactic gradient exists across the monolayer [66]. Blocking β_2-integrins has been shown to inhibit both firm adhesion and subsequent transmigration of neutrophils [66]. However, PECAM-1, expressed at the endothelial cell junctions, has recently been reported to mediate transmigration exclusively without influencing firm adhesion, by binding homophilically to PECAM-1 expressed on leukocytes and possibly heterophilically to an as yet unidentified receptor [67]. Each step of this multistage process is essential for the proper function of the immune system in humans as illustrated by leukocyte adhesion deficiency types I

and II syndromes. These patients have impaired immunity due to the inability of neutrophils to either permanently arrest (LAD-I; lack of the β_2-integrin receptors) or roll (LAD-II; absence of the fucose-containing oligosaccharide moiety sialyl Lewis x) on activated endothelium [68].

Research on the effects of flow on leukocyte adhesion to vascular endothelium has revealed that different families of receptors are capable of mediating distinct types of adhesive events such as initial contact/rolling (primary adhesion) or firm adhesion/transmigration (secondary adhesion). This distinction remained unknown until 1987, when Lawrence et al. [69] examined neutrophil adhesion to cytokine-stimulated endothelial cells under well-defined postcapillary venular flow conditions in vitro. These studies showed that neutrophil primary adhesion to cytokine-activated endothelial cells is mediated almost exclusively by β_2-integrin-independent mechanisms, whereas secondary adhesion is β_2-integrin dependent [70]. The initial flow studies were followed by many further studies, both in vivo [68, 71–74] using intravital microscopy and in vitro [75–81] using flow chambers and videomicroscopy that indicate: 1. L-, E- or P-selectins are capable of tethering free-flowing leukocytes; 2. L-, and E- or P-selectins are required for optimal leukocyte rolling; 3. β_2-integrins cannot initiate leukocyte adhesion under flow conditions, except possibly at wall shear stresses less than 1.0 dyn/cm^2 (Figs. 2,3), but can support activation-dependent firm adhesion and subsequent migration.

3.1.1 Extension of the four-step model of neutrophil recruitment
Previous work has shown that neutrophils bound to activated endothelium are capable of supporting tethering and rolling of circulating neutrophils in vitro via L-selectin-dependent mechanisms [82]. It was recently reported that free-flowing neutrophils can form homotypic tethers with neutrophils rolling on purified P-selectin before attaching to the substrate downstream of the primary rolling cells [83] Alon et al. [84] extended these observations to monocytes and activated T cell lines as well as a wide variety of substrates, including E-, P-, L- selectins, peripheral node addressin, VCAM-1 and activated endothelium. Leukocyte–leukocyte tethers are mediated by L-selectin/L-selectin ligand interactions, and can dramatically amplify the rate of initial leukocyte recruitment at sites of inflammation, representing a potential additional pathway in the leukocyte adhesion cascade of events.

Studies of the molecular mechanisms of lymphocyte– or monocyte–endothelial cell adhesive interactions indicate that mononuclear cells may use additional pathways for primary adhesion under conditions of flow. Lymphocytes and monocytes differ from neutrophils in that they express β_1 integrins, including the VLA-4 ($\alpha_4\beta_1$). Jones et al. [85] demonstrated that $\alpha_4\beta_1$ integrins are capable of supporting primary T lymphocyte adhesion to VCAM-1 transfectants under flow (Fig. 3). This finding has been confirmed by others who demonstrated that lymphocytic cell lines [86] or peripheral blood T cells [87] can use α_4 integrins to tether and roll on VCAM-1 expressed on activated endothelium. It was recently shown that domain 1 of VCAM-1 is solely responsible for $\alpha_4\beta_1$ integrin-dependent primary capture under flow [87, 88], whereas both domains 1 and 4 are required for $\alpha_4\beta_1$ integrin-dependent adhesion under static conditions [88]. $\alpha_4\beta_7$ integrins can also support lymphocyte attachment to mucosal addressin under shear forces in the absence of a selectin contribution [89]. In contrast, the β_2/ICAM-1 pathway alone cannot initiate mononuclear cell–endothelial cell adhesive interactions at physiologic flow conditions (Figs. 2 and 3), as shown earlier for neutrophils [85, 90].

Recent studies suggest that monocytes can use multiple receptors to interact with endothelial cells at both primary and secondary adhesion stages. In particular, L-selectin, α_4 integrins and a neuraminidase-sensitive epitope contribute to primary monocyte adhesion to 4-h IL-1β-activated endothelium, while both α_4 and β_2 integrins support firm adhesion (Fig. 3) [90]. These findings provide evidence that the steps of the molecular cascade of events seem to be overlapping and not strictly sequential.

3.1.2 *In vivo* studies of leukocyte adhesion

The multistep adhesion cascade determined by *in vitro* experiments has also been observed *in vivo*. Intravital microscopy has revealed leukocyte rolling and firm adhesion to the wall of postcapillary venules. Spontaneous rolling of leukocytes in rat mesentric venules upon tissue exposure was shown to be mediated by P-selectin induced by agonists like histamine released from mast cell degranulation [91]. In a rabbit model, L-selectin was shown to support the initial rolling of whereas β_2 integrins were solely responsible for mediating firm leukocyte attachment to venular endothelium [73]. Using a rat model of chronic inflammation, Kubes and his co-workers demonstrated that the α_4-integrin was able to mediate both selectin-independent rolling as well as β_2 integrin-independent firm adhesion of periperal blood mononuclear cells at physiological shear rates [92]. Similarly, $\alpha_4\beta_7$ was reported to mediate L-selectin-independent attachment of lymphocytes to lamina propria venules in situ [89].

Recently genetic engineering approaches have been used to further investigate the roles of adhesion molecules in both normal and disease conditions. Gene targeting technology has allowed eliminating the function of a gene and creating the so-called 'knockout' mice. L-selectin deficient mice have been shown to have impaired leukocyte recruitment to inflamed peritoneum after administration of a proinflammatory stimulus, demonstrating the importance of L-selectin in leukocyte recruitment at sites of chronic inflammation [74]. In response to thioglycollate immunologic challenge, E-selectin knockout mice appeared normal with respect to leukocyte emigration. However, blocking P-selectin in these mice almost completely suppressed leukocyte emigration into the peritoneum, while that had no effect in wild-type controls [72]. Correspondingly, blocking E-selectin in P-selectin knockout mice abrogated leukocyte rolling, but had little effect in wild-type control mice. These findings provide evidence that P- and E-selectins are partially redundant and can replace each other as mediators of leukocyte rolling *in vivo*. Ley and his co-workers [93] recently reported that leukocyte rolling is absent in P-selectin/ICAM-1 double knockout mice for a much longer time period than in P-selectin deficient mice, suggesting that ICAM-1 may contribute to leukocyte rolling contact with the vessel wall. Although gene targeting may provide new insight in adhesion molecule research, the results should be interpreted with caution, as the mouse strain may not accurately reflect the physiology of inflammation in humans.

3.1.3 Micromechanics of leukocyte adhesion

The leukocyte receptors involved in the adhesion to endothelial sites of inflammation are not randomly scattered on the cell surface, but are rather localized in or away from the tips of microvilli. β_2 integrins are mainly displayed on the cell body of the leukocytes, whereas L-selectin, PSGL-1, and α_4 integrins are concentrated on microvilli (Fig. 3) [16, 19], a strategic position that seems to be advantageous for initiating leukocyte–endothelial cell interactions under flow. Although microvillous versus cell body epitope presentation does not affect the extent of cell binding under static conditions, it leads to a substantially differential cell capture under flow [16]. The importance of microvillous presentation is markedly evident at higher stresses (~ 3 dyn/cm^2). von Andrian *et al.* [16] suggested that the interaction of endothelial cells with leukocytes through microvillous receptors presumably reduces the area of initial contact and consequently the electrostatic repulsive forces generated by contacts between the glycocalyces of those cells. Furthermore, receptor clustering on microvilli of leukocytes leads to the formation of multiple bonds between the interacting cells that may be required to provide the sufficient adhesive strength to resist the dispersive hydrodynamic forces.

To capture leukocytes moving at high hydrodynamic velocity in the bloodstream a fast rate of bond formation (K_{on}) is required. Furthermore, a fast dissociation rate (K_{off}) coupled with the ability of the bond to withstand considerable strain before breaking (tensile strength) under shear stress will allow the tethered leukocyte to advance forward in the direction of flow while remaining in contact due to rapid K_{on} rates with the vessel wall [94]. The efficiency of an adhesion receptor to mediate rolling is, therefore, dependent on both the K_{on} and K_{off} rates as well as the tensile strength of the bond. However, rolling velocity has been shown to be predominantly determined by the K_{off} rate: the faster the K_{off} rate, the higher the rolling velocity [95]. Finally, resistance to detachment has been suggested to depend on the number of receptor-ligand bonds present at any one time, which is determined by the ratio of K_{on}/K_{off} rates [95] and receptor site densities. Puri *et al.* [95] provided evidence that L-selectin has faster K_{on} and K_{off} rates than E- and P-selectins, but similar strength of binding.

It was recently shown that projection of the ligand-binding domain of P-selectin well above the cell surface is required for optimal neutrophil capture under shear flow conditions. Utilizing chinese ovary cells expressing P-selectin constructs in which various numbers of consensus repeats (CR) were removed, Patel *et al.* [96] demonstrated that deletion of three or four CR minimally affected neutrophil adhesion, whereas deletion of five or more CR significantly impaired or abolished adhesion under flow, but not under static conditions. It is believed that the elongated molecular structure of P-selectin ($\sim$ 40 nm) facilitates contacts with its counter–receptor PSGL-1 ($\sim$ 60 nm long) on flowing neutrophils by overcoming the repulsive forces of the glycocalyx.

Another important property of selectins that clearly differentiates them from $\beta 2$ integrins and makes them specialized in initiating adhesion is that selectin bonds are functional without activation of the leukocyte [65]. This finding makes sense physiologically because circulating leukocytes are normally unactivated but can bind to specific endothelial sites when necessary. It is nonetheless possible that later activation can affect selectin adhesion. Neutrophil activation by signaling molecules such as PAF or IL-8 has been shown to reduce adhesion to P-selectin. Decreased adhesion appears to be the result of redistribution of P-selectin ligands on activated neutrophils mediated by cytoskeletal interactions, and is not related to shedding of L-selectin [97]. These processes may represent ways for the neutrophil to disengage selectin-mediated binding allowing free migration along the endothelial surface.

3.1.4 Signal transduction mechanisms in leukocytes

Neutrophil activation can occur in response to chemotactic factors such as PAF and IL-8, that are recognized by receptors on the neutrophil plasma membrane, including members of the serpentine G-protein-coupled receptor class, receptors that recognize TNF-α and others [57]. Activation of β_2 integrins by chemotactic agents is abolished by blockade of the G-protein-linked receptors on the neutrophil surface. Ligand binding to the G-protein coupled receptors results in activation of phospholipase C which induces hydrolysis of phosphatidylinositol 4,5-bisphosphate and the production of inositol 1,4,5-triphosphate and diacylglycerol. At the same time, there are changes in cytosolic calcium and activation of cellular protein kinases. Several models have been proposed to explain precisely how integrin activation is regulated [98]. One model proposes that the signals (i.e. activation of protein kinases) induce alterations in the integrin cytoplasmic tails (i.e. phosphorylation), or post-translational modification, and/or alterations in the interactions between the α and β tails. A second model proposes the existence of a repressor protein normally bound to the integrin cytoplasmic tails that is removed upon activation. Alternatively, another model proposes the presence of an activator protein that associates with the integrin upon activation. Such changes are transmitted through the cytoplasmic and transmembrane domains to the extracellular domain of the receptor which ultimately undergoes conformational changes and becomes competent to bind ligand.

Several lines of evidence support the notion that selectins play a more complex role than merely capturing free-flowing leukocytes from the bloodstream, and they have been implicated in signal transduction mechanisms. L-selectin cross-linking induces a rise in intracellular calcium, potentiates the oxidative burst of neutrophils and activates MAP kinase as well as other still undefined tyrosine phosphorylated substrates [99, 100]. Furthermore, L-selectin ligation and cross-linking was recently shown to result in a cell signal that upregulates β_2 integrin adhesive responses of neutrophils [101, 102]. In particular, it induces a sufficient level of activation to both LFA-1 and Mac-1, and promotes stable neutrophil adhesion to ICAM-1 at physiologic flow conditions in an *in vitro* model where tethering and rolling are mediated by E-selectin. GlyCAM-1, a physiological ligand for L-selectin, binding to lymphocytes also stimulates adhesion to ICAM-1-bearing monolayers [103].

P-selectin binding to its ligand on monocytes has been reported to activate cells. This interaction induces superoxide anion release [104], enhances PAF synthesis and phagocytosis [105], costimulates secretion of MCP-1 and TNF-α, and is associated with the nuclear translocation of NF-kB [106]. Platelet P-selectin has also been implicated in neutrophil activation [104]. Moreover, binding of purified and recombinant P-selectin to neutrophils enhances phagocytosis mediated by β_2 integrins [107]. However, other studies using P-selectin alone in model membranes or expressed on tranfectants failed to show upregulation of β_2 integrin adhesive activity [78, 108]. Prior work has suggested that binding of neutrophils to E-selectin activates Mac-1-dependent adhesion of C3bi-coated erythrocytes to neutrophils under static conditions [109]. In contrast, recent findings reveal that E-selectin alone is not sufficient to activate β_2 integrin-mediated neutrophil adhesion to ICAM-1 under physiologic conditions of shear [102]. It is possible that under flow, E-selectin does not bind to its ligand on neutrophils for a long enough time to activate the cell. Despite the substantial progress made during the last few years in elucidating the signal transduction pathways in leukocytes, our understanding of this process is far from complete.

3.2 Platelet interactions with thrombogenic surfaces under flow

3.2.1 Molecular mechanisms
Platelet adhesion to exposed subendothelium at sites of vascular injury is a
crucial initiating step in hemostasis and thrombosis. The specific platelet
surface receptors that mediate these processes are determined by the local
fluid dynamic conditions and the extracellular matrix constituents exposed
at the site of a vascular wound. For a platelet to become irreversibly
attached to the blood vessel wall, adhesion bonds should provide sufficient
strength to withstand the dispersive hydrodynamic forces generated by
blood flow. *In vitro* models, utilizing a parallel-plate flow chamber combined with epifluorescence video microscopy and a microphotometric
technique, have been designed to simulate both the mechanical and
biochemical environment encountered in an injured, exposed subendothelial or atherosclerotic plaque surface [44, 51, 110, 111]. Studies on the
molecular mechanisms of arterial thrombus formation reveal that the
initial adhesive (platelet-vessel wall) interactions are mediated by vWf
which forms a bridge between insolubilized type I or type III collagen
fibrils and platelet surface glycoprotein receptors [112]. The source of vWf
involved in the initial stages of adhesion appears to be plasma, since
subendothelial vWf is not uniformly distributed or even absent in many
vessels [113]. vWf from platelet α-granules has been shown to play a key
role in later stages. Platelet aggregation following adhesion requires vWf,
fibrinogen and perhaps other adhesive proteins to bind to the platelet
surface and to bridge the gap between adjacent circulating platelets and
the previously adherent ones [56, 112].

It was recently demonstrated that platelet adhesion to immobilized vWF
requires the sequential involvement of distinct receptors analogous to the
process of leukocyte extravasation in inflammation [111]. Platelet GPIbα
seems to mediate a selectin-like initial adhesive platelet-vWF interaction,
manifested as translocation along the wall (Fig. 4). These activation-inde-
pendent bonds presumably have fast association and dissociation rates and
a high tensile strength, since they appear functional even at 6,000 sec^{-1}.
Following platelet contact with the wall, $\alpha_{IIb}\beta_3$ becomes activated and
binds to vWf leading thus to platelet permanent arrest onto the surface,
and subsequent thrombus formation (Fig. 4). On the other hand, unactivated $\alpha_{IIb}\beta_3$ has been shown to mediate platelet attachment to immobilized fibrinogen only at low shear rates (less than 1500 second^{-1}), a
fact that is attributed to a slow kinetics of bond formation and low tensile
strength [49, 111].

Platelet adhesion and subsequent aggregation on purified type VI collagen are extensive at low shear stresses *in vitro* (4 dyn/cm^2) [110]. In
contrast, little adhesion occurs under high shear conditions (40 dyn/cm^2)

at this substrate. These shear-related events are exactly opposite to those observed with collagen type I [110]. The low shear deposition on type VI collagen requires vWf, GPIbα and $\alpha_{IIb}\beta_3$, and a relatively minor vWf-independent adhesion component [110]. These data suggest a potentially important role for collagen VI in hemostasis and thrombosis in vascular regions characterized by low wall shear rates.

3.2.2 Signal transduction mechanisms in platelets

Platelet activation induced by chemical agonists such as thrombin occurs through G-proteins, and follows the same principles with those outlined for neutrophils in response to chemotactic agents (see Signal transduction mechanisms in leukocytes) [98]. This section will, therefore, focus on the platelet signaling induced by the fluid mechanical shearing stress.

The GPIb-IX complex has been shown to play a direct role in cell signaling. Binding of soluble vWf to platelets under conditions of high shear stimulates Ca^{2+} influx and platelet aggregation [114, 115]. Blockade of vWf binding to GPIbα completely abolishes both of these platelet responses to shear stress. In contrast, blockade of vWf-$\alpha_{IIb}\beta_3$ interactions inhibited aggregation, and only moderately affected shear stress-induced increases in intracellular Ca^{2+}. These results are consistent with the model that shear stress induces the initial binding of vWf to platelet GPIbα [53, 114–116]. Crosslinking of GPIbα by vWf activates platelets by opening a receptor-dependent calcium channel, which may act as an intracellular message and cause activation of $\alpha_{IIb}\beta_3$. vWf binding to $\alpha_{IIb}\beta_3$ mediates irreversible platelet aggregation [115].

Further studies aimed at defining how shear stress-induced vWf binding to GPIbα initiates signals for switching $\alpha_{IIb}\beta_3$ into a ligand-receptive conformation have revealed the activation of a diacylglycerol-independent pathway of protein kinase C (PKC) [116]. The absence of any measurable change in diacylglycerol or hydrolysis of phosphatidylinositol 4,5-bisphosphate in response to shear stress is in contrast to activation by chemical agonists. Platelet tyrosine kinases are also activated in response to pathologic shear stress (90 dyn/cm^2) [116]. Both PKC and tyrosine kinase activation depend on vWf binding to platelet GPIbα and $\alpha_{IIb}\beta_3$.

3.3 Endothelial–endothelial cell interactions

Endothelial cell–cell contact is of paramount importance in maintaining a dynamic interface between circulating blood and the surrounding perivascular tissue space. A number of adhesive proteins are localized in the contact area between endothelial cells giving rise to different kinds of junction components. Endothelial cells are known to express adherens junctions, tight junctions, and gap junctions. However, adherens junctions are more ubiquitous and hence this section will focus on properties of this

junction type regulating endothelial cell permeability and vascular morphogenesis.

The adherens junctions are mainly formed by the cadherin family of
receptors. The major cadherin found in endothelial cells is the VE (vascular endothelial)-cadherin also known as cadherin 5 [117]. Like other
cadherins, VE-cadherin also forms a complex with the catenins and is
linked to the actin cytoskeleton (Fig. 1). However, in endothelial cells, this
complex is very dynamic and varies with the functional state of the
endothelial cells. At early stages of confluency or when detaching and
migrating, endothelial cells have weak junctions. Correspondingly, VE-
cadherin is heavily phosphorylated at tyrosine residues and linked primarily to p120 and β-catenin. Only a small portion of the complex is linked
with plakoglobin and the actin cytoskeleton. When endothelial cells form a
confluent monolayer, VE-cadherin is no longer phophorylated and is
associated with plakoglobin and the actin cytoskeleton. Both p120 and
β-catenin are present in reduced amounts in the VE-cadherin complex.

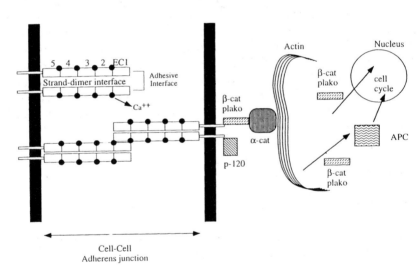

Fig. 1. Representation of Cadherin mediated Adherens Junctions in the endothelium. X-ray crystal studies have suggested that cadherins may aggregate into a tight
parallel strand dimer orientation with the adhesion interface directed outward
from the cell surface and binding to cadherin dimers on the opposite cell. The
cadherin adhesion system is linked to the actin cytoskeletal network through
armadillo proteins (β-cat: β-catenin, plako: plakoglobin and p120). β-catenin and
plakoglobin bind to actin filaments through α-catenin (α-cat) and also mediate
intracellular signaling by binding to transcritional factors within the nucleus.
β-catenin and plakoglobin can bind to oncogene APC which inhibits cell cycle
progression. (Adapted from [117]).

Recently it has been shown that treatment of endothelial cells with thrombin reduces VE-cadherin association with plakoglobin thus weakening the adherens junctions. This could explain the retraction and gap formation of the endothelial monolayer upon thrombin stimulation [118, 119]. It has also been observed that adhesion of polymorphonuclear cells to endothelial cells results in detachment of catenin from VE-cadherin and a breakdown of adherens junctions [120]. This agrees well with *in vivo* observations of increased endothelial permeability accompanying sustained polymorphonuclear leukocyte emigration. Thus results to date support the role of VE-cadherin–catenin complex in regulating endothelial permeability.

Adherens junctions also play an important role during angiogenesis. Angiogenesis is the formation of new capillary blood vessels from existing capillaries and post capillary venules. This phenomenon plays an important role in wound healing and inflammation. Endothelial cell proliferation and migration are critical phases of this event. In early stages of angiogenesis, the junctions loosen to enable endothelial cells to detach from the vascular wall and migrate to the underlying tissues. For this to occur, VE-cadherin–catenin complex has to be inactivated. It has been found that growth factors like epithelial growth factor (EGF) and hepatocyte growth factor (HGF) causes tyrosine phosphorylation of the adherens junction components [121]. It is possible that angiogenic factors could act in a similar way to weaken adherens junction by phosphorylating VE-cadherin and easing endothelial cell migration [117]. Upon forming of the vascular network, adherens junctions could be involved in inhibiting endothelial migration and proliferation.

4 Applications in tissue engineering

4.1 Graft rejection

Tissue rejection could be inhibited by two possible therapies against cell–cell adhesion molecules:

(i) Inhibiting tissue rejection by blocking the adhesion molecules necessary for leukocyte recruitment is the first approach and is widely used for controlling inflammatory and immune events. This in turn would require the determination of the kind of adhesion receptors mediating immunorejection. Recent investigations have revealed ICAM-1, VCAM-1 and E-selectin each may contribute towards rejection [122–124]. In several experimental systems, blocking ICAM-1 and VCAM-1 pathways with monoclonal antibodies following transplant significantly delayed rejection [123,

124]. An anti-ICAM-1 therapy with renal allograft has demonstrated promising results in phase I clinical trials. Recent experimental data from our laboratory [85, 90] using a monocyte as well
as T-cell adhesion model in chronic inflammation has clearly demonstrated the ability of the above mentioned adhesion molecules
based molecular pathways in supporting adhesion of T-cells and
monocytes under physiologic flow conditions. Understanding the
various adhesive mechanisms underlying the different phases of the
multistep adhesion cascade would help in identifying the molecular
target optimal for blocking cell—cell interaction.

(ii)   Induction of tolerance by modulating immune system signals which
require cell—cell adhesion is another mechanism for preventing
tissure rejection. This has been shown in studies using a murine
cardiac allograft model where combination of anti-LFA-1 and
anti-ICAM-1 antibodies produced allospecific long term graft sur-

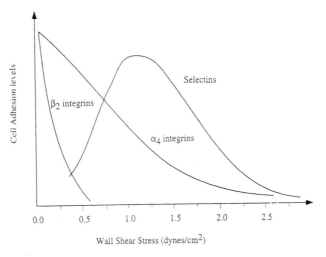

Fig. 2. This figure provides a qualitative description of the roles integrins and
selectins in mediating cell—cell attachment under dynamic shear conditions. β_2
integrins on the leukocyte surface are involved in firm adhesion of under static (no
flow) conditions. With increase in wall shear stresses, the β_2 integrin dependent
component of leukocyte primary adhesion sharply declines. Recent investigations
have revealed that α_4 integrins can also support leukocyte interactions under flow
conditions. The decrease in α_4 integrin dependent component, however is more
gradual than the β_2 integrin component with increase in wall shear stresses.
Selectins, on the other hand are not efficient in mediating primary leukocyte
adhesion at shear stresses below 0.5 dynes/cm². Shear stresses greater than 0.5
dynes/cm² are required for optimal selecitn dependent tethering and rolling of
leukocytes.

vival [125, 126]. The graft survival was a true allospecific tolerance rather than simply immunosuppression, since recipients could tolerate later skin grafts from the same donor strain.

4.2 Platelet–biomaterial interactions

The increasing use of biomaterials in blood-contacting applications underscores the importance of *in vitro* systems used to characterize the molecular mechanisms of platelet adhesion in the assessment of thrombogenecity of potential vascular biomaterials. Recent studies using a variety of artificial surfaces such as polyvinyl alcohol hydrogel indicate extensive material induced platelet activation in whole blood as monitered by microparticle generation and platelet P-selectin expression [127]. Mathematical models of thrombus formation under dynamic flow conditions have been developed to ascertain and predict the concentration of proaggregatory compounds such as thrombin, adenosine diphosphate and thromboxane A_2 [128]. Accurate knowledge of platelet adhesion mechanisms and local proaggregatory substances and fluxes is required for rational development of novel implant systems.

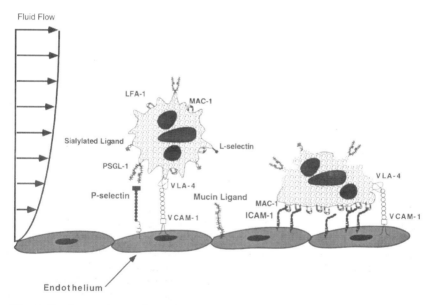

Fig. 3. Leukocyte interaction with activated endothelium. Receptors mediating primary adhesion such as PSGL-1, L-selectin and α_4 integrins are clustered on tips of microvilli, a position particularly advantageous for initial leukocyte-endothelial contact. In contrast, β_2 integrins are displayed on the planar cell surface and are insufficient to initiate adhesion under flow conditions. After initial contact, leukocytes are then able to firmly adhere on the activated endothelium through both β_2 and α_4 integrins dependent pathways.

4.3 Platelet $\alpha_{IIb}\beta_3$ antagonists as therapeutics of coronary artery disease

Percutaneous transluminal angioplasty is now an established approach to improve perfusion to the cardiac muscle. However, this procedure causes injury to the coronary artery that signals the activation of the hemostatic mechanism. Platelet deposition and thrombus formation at the site of injury are associated with acute closure and restenosis. The final common step in platelet aggregation, regardless of the stimulus, involves the interaction of adhesive proteins such as fibrinogen and vWf with the platelet $\alpha_{IIb}\beta_3$ – (Fig. 4). Therefore, use of highly specific agonists of the $\alpha_{IIb}\beta_3$ complex may be beneficial for the prevention of ischemic sequelae after angioplasty. A prospective, randomized, double-blind trial recently reported that thrombotic complications of coronary angioplasty and atherectomy were reduced by an $\alpha_{IIb}\beta_3$ antagonist, ReoPro™ (c7E3 Fab; abcix-

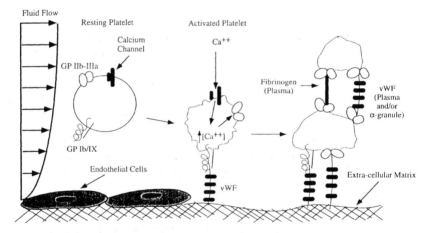

Fig. 4. Schematic of the proposed mechanism of platelet adhesion to the subendothelial matrix and subsequent platelet aggregation under shear conditions. Intact blood vessels are lined with endothelial cells, which under normal conditions prevent the interaction of platelets with the subendothelium. Upon injury to a blood vessel, however, the endothelial cells lining the vessel are damged and/or removed, thereby exposing the extracellular matrix to the formed blood elements. Under high shear conditions, binding of immobilized vWf, which is present in the exposed subendothelium, to GPIb-IX initiates platelet vessel wall adhesive interactions. GPIb-IX signaling and exposure of platelets to locally released chemical agonists such as thrombin and ADP, results in opening of a transmembrane calcium channel leading to Ca^{2+} flux which precedes functional activation of $\alpha_{IIb}\beta_3$ allowing it to bind vWf. Binding of vWf (plasma or α-granules derived) and plasma fibrinogen to $\alpha_{IIb}\beta_3$ results in platelet–platelet aggregation.

imab) [129]. ReoPro, infused to patients undergoing angioplasty, was a potent inhibitor of platelet aggregation induced by shear stress *in vitro*, without interfering with GPIb-IX-vWf interaction that is essential for platelet adhesion and primary hemostasis (Fig. 4) [130, 131]. Furthermore, mild inhibitory effects of ReoPro on direct shear-induced platelet aggregation continued to be observed several days after the termination of ReoPro infusions in patients undergoing angioplasty [130]. Integrelin, a cyclic heptapeptide KGD-antagonist of $\alpha_{IIb}\beta_3$, was reported to be an effective inhibitor of shear-induced platelet aggregation *in vitro* [132]. However, its inhibitory effects were reversible *in vivo* within hours and disappeared after 24 hours of discontinuation of infusion [132]. This difference in the time course of inhibition of direct shear-induced platelet aggregation after discontinuation of antagonist infusion may help explain the relatively better effectiveness of ReoPro in reducing thrombotic complications after angioplasty.

References

1. Johnston, G. I., Kurosky, A. and McEver, R. P., Structural and biosynthetic studies of the granule membrane protein, GMP-140, from human platelets and endothelial cells. *Journal of Biological Chemistry*, 1989, **264**, 1816–1823.
2. Collins, P. W., Macey, M. G., Cahill, M. R. and Newland, A. C., Von willebrand factor release and P-selectin expression is stimulated by thrombin and trypsin but not IL-1 cultured human endothelial cells. *Thrombosis and Haemostasis*, 1993, **70**, 346–350.
3. Khew-Goodall, Y., Butcher, C. M., Litwin, M. S., Newlands, S., Korpelainen, E. I., Noack, L. M., Berndt, M. C., Lopez, A. F., Gamble, J.R. and Vadas, M.A., Chronic expression of P-selectin on endothelial cell stimulated by the T-cell cytokine, interleukin-3. *Blood*, 1996, **87**, 1432–1438.
4. Yao, L., Pan, J., Setiadi, H., Patel, K. D. and McEver, R. P., Interleukin 4 or Oncostatin M induces a prolonged increase in P-selectin mRNA and protein in human endothelial cells. *Journal of Experimental Medicine*, 1996, **184**, 81–92.
5. Barkalow, F. J., Goodman, M. J., Gerritsen, M. E. and Mayadas, T. N., Brain endothelium lack one of two pathways of P-selectin-mediated neutrophil adhesion. *Blood*, 1996, **88**, 4585–4593.
6. Gotsch, U., Jager, U., Dominis, M. and Vestweber, D., Expression of P-selectin on endothelial cells is upregulated by LPS and TNF-α *in vivo*. *Cell Adhesion and Communication*, 1994, **2**, 7–14.
7. Johnson-Tidey, R. R., McGregor, J. L., Taylor and P. R., Poston, R. N., Increase in the adhesion molecule P-selectin in endothelium overlying atherosclerosis plaques. *American Journal of Pathology*, 1994, **144**, 952–961.
8. Patel, K. D., Moore, K. L., Nollert, M. U. and McEver, R. P., Neutrophils use both shared and distinct mechanisms to adhere to selectins under static and flow conditions. *Journal of Clinical Investigation*, 1995, **96**, 1887–1896.

9. Moore, K. L., Stults, N. L., Diaz, S., Smith, D. F., Cummings, R. D., Varki and A., McEver, R. P., Identification of a specific glycoprotein ligand for P-selectin (CD62) on myeloid cells. *Journal of Cell Biology*, 1992, **118**, 445–456.

10. Moore, K. L., Patel, K. D., Bruehl, R. E., Li, F., Johnson, D.A., Lichenstein, H.S., Cummings, R.D., Bainton, D. F. and McEver, R. P., P-selectin glycoprotein ligand-1 mediates rolling of human neutrophils on P-selectin. *Journal of Cell Biology*, 1995, **128**, 661–671.

11. Li, F., Erickson, H. P., James, J. A., Moore, K. L., Cummings, R. D. and McEver, R. P., Visualisation of P-selectin glycoprotein ligand-1 as a highly extended molecule and mapping of protein epitopes for monoclonal antibodies. *Journal of Biological Chemistry*, 1996, **271**, 6342–6348.

12. Wilkins, P. P., McEver, R. P. and Cummings, R. D., Structures of the O-glycans on P-selectin glycoprotein ligand-1 from HL-60 cells. *Journal of Biological Chemistry*, 1996, **271**, 18732–18742.

13. Sako, D., Comess, K. M., Barone, K. M., Camphausen, R. T., Cumming and D. A., Shaw, G. D., A sulfated peptide segment at the amino terminus of PSGL-1 is critical for P-selectin binding. *Cell*, 1995, **83**, 323–331.

14. Kahn, J., Ingraham, R. H., Shirley, F., Migkai, G. I. and Kishimoto, T. K., Membrane proximal cleavage of L-selectin: Identification of the cleavage site and a 6-kD transmembrane peptide fragment of L-selectin. *Journal of Cell Biology*, 1994, **125**, 461.

15. Schleiffenbaum, B. E., Spertini, O. and Tedder, T. F., Soluble L-selectin is present in human plasma at high levels and retains functional activity. *Journal of Cell Biology*, 1992, **119**, 229.

16. von Andrian, U. H., Hasslen, S. R., Nelson, R. D., Erlandsen, S. L. and Butcher, E. C., A central role for microvillous receptor presentation in leukocyte adhesion under flow. *Cell*, 1995, **82**, 989–999.

17. Fuhlbrigge, R. C., Alon, R., Puri, K. D., Lowe, J. B. and Springer, T.A., Sialylated, fucosylated ligands for L-selectin expressed on leukocytes mediate tethering and rolling adhesions in physiologic flow conditions. *Journal of Cell Biology*, 1996, **135**, 837–848.

18. Spertini, O., Cordey, A., Monai, N., Giuffre, L. and Schapira, M., P-selectin glycoprotein ligand-1 is a ligand for L-selectin on neutrophils, monocytes and CD34$^+$ hematopoietic cells. *Journal of Cell Biology*, 1996, **135**, 523–531.

19. Picker, L. J., Warnock, R. A., Burns, A. R, Doerschuk, C. M, Berg, E. L and Butcher, E. C, The neutrophil selectin LECAM-1 presents carbohydrate ligands to the vascular selectins ELAM-1 and GMP-140. *Cell*, 1991, **66**, 921–933.

20. Bevilacqua, M. P, Stenglin, S., Gimbrone , M. A. J. and Seed, B., Endothelial leukocyte adhesion molecule 1 : An inducible receptor for neutrophils related to complement regulatory proteins and lectins. *Science*, 1989, **243**, 1160.

21. Ghersa, P., van Huijsduijnen, R. H, Whelan, J. and DeLamarter, F., Labile proteins play a dual role in the control of endothelial leukocyte adhesion molecule-1 (ELAM-1) gene regulation. *Journal of Biological Chemistry*, 1992, **267**, 19226.

22. Subramaniam, M., Koedam, J. A and Wagner, D. D, Divergent fates of P-and E-selectin after their expression on the plasma membrane. *Molecular Biology of the Cell*, 1993, **4**, 791.

23. Picker, L. J, Kishimoto, T. K, Smith, C. W, Warnock, R. A and Butcher, E. C, ELAM-1 is an adhesion molecule for skin-homing T cells. *Nature*, 1991, **349**, 796.

24. Berg, E. L, Robinson, M. K, Mansson, O., Butcher, E. and Magnani, J. L, A carbohydrate domain common to both Sialyl Le^a and Sialyl Le^x is recognized by the endothelial cell leukocyte adhesion molecule ELAM-1. *Journal of Biological Chemistry*, 1991, **266**, 14869–14872.

25. Varki, A., Selectin ligands. *Proceedings of the National Academy of Sciences of the United States of America*, 1994, **91**, 7390–7397.

26. Steegmaler, M., Levinovitz, A., Isenmann, S., Borges, E., Lenter, M., Kocher, H. P, Kleuser, B. and Vestweber, D., The E-selectin-ligand ESL-1 is a variant of a receptor for fibroblast growth factor. *Nature*, 1995, **373**, 615–620.

27. Zollner, O., Lenter, M. C, Blanks, J. E, Borges, J. E, Steegmaier, M., Zerwes, H. G and Vestweber, D., L-selectin from Human, but not mouse neutrophils binds directly to E-selectin. *Journal of Cell Biology*, 1997, **136**, 707–716.

28. Asa, D., Raycroft, L., Ma, L., Aeed, P. A, Kaytes, P. S, Elhammer, P. and Geng, J., The P-selectin glycoprotein ligand functions as a common human leukocyte ligand for P- and E- selectins. *Journal of Biological Chemistry*, 1995, **270**, 11662–11670.

29. Snapp, K. R, Wagers, A. J, Craig, R., Stoolman, L. M and Kansas, G. S, P-selectin glycoprotein ligand-1 is essential for adhesion to P-selectin but not E-selectin in stably transfected hematopoietic cell lines. *Blood*, 1997, **89**, 896–901.

30. Lawrence, M. B, Kansas, G. S, Kunkel, E. J and Ley, K., Threshold levels of fluid shear promote leukocyte adhesion through selectins (CD62L, P, E). *Journal of Cell Biology*, 1997, **136**, 717–727.

31. Phillips, L., Schwartz, B., Etzioni, A., Bayer, R., Ochs, H., Paulson, J. and Harlan, J., Neutrophil adhesion in leukocyte adhesion deficiency syndrome type 2. *Journal of Clinical Investigation*, 1995, **96**, 2898–2906.

32. Landis, R. C, Bennet, R. I and Hogg, N., A novel LFA-1 activation epitope maps to the I domain. *Journal of Cell Biology*, 1993, **120**, 1519.

33. Yednock, T. A, Cannon, C., Vandevert, C., Goldbach, E. G, Shaw, G., Ellis, D. K, Liaw, C., Fritz, L. C and Tanner, L. I, $\alpha_4 \beta_1$ integrin-dependent cell adhesion is regulated by a low affinity receptor pool that is conformationally responsive to ligand. *Journal of Cell Biology*, 1995, **270**, 28740–28750.

34. Shimizu, Y., van Seventer, G. A, Horgan, K. J and Shaw, S., Regulated expression and binding of three VLA (beta1) integrin receptors on T cells. *Nature*, 1991, **345**, 250–253.

35. Shimizu, Y., Newman, W. and Gopal, T. V, Four molecular pathways of T-cell adhesion to endothelial cells: roles of LFA-1, VCAM-1, and ELAM-1 and changes in pathway hierarchy under different activation conditions. *Journal of Cell Biology*, 1991, **113**, 1203–1212.

36. Miller, L. J, Schwarting, R. and Springer, T. A, Regulated expression of the Mac-1, LFA-1, p150,95 glycoprotein family during leukocyte differentiation. *Journal of Immunology*, 1986, **137**, 2891–2900.

37. Kansas, G. S, Muirhead, M. J and Dailey, M. O, Expression of the CD11/CD18. leukocyte adhesion molecule1, and CD44 adhesion molecules during normal myeloid and erythroid differentiation in humans. *Blood*, 1992, **76**, 2483–2492.

38. Hughes, B. J, Hollers, J. C, Crockett-Torabi, E. and Smith, C. W, Recruitment of CD11b/CD18 to the neutrophil surface and adherence-dependent cell locomotion. *Journal of Clinical Investigation*, 1992, **90**, 1687–1696.

39. Miller, L. J, Bainton, D. F, Borregaard, N. and Springer, T. A, Stimulated mobilization of monocyte Mac-1 and p150,95 adhesion proteins from an intracellular vesicular compartment to the cell surface. *Journal of Clinical Investigation*, 1987, **80**, 535–544.

40. Singer, I. I, Scott, S., Kawka, D. W and Kazazis, D. M, Adhesomes: specific granules containing receptors for laminin, C3bi/fibrinogen, fibronectin, and vitronectin in human polymorphonuclear leukocytes and monocytes. *Journal of Cell Biology*, 1989, **109**, 3169–3182.

41. Anderson, D. C and Springer, T. A, Leukocyte adhesion deficiency: an inherited defect in the Mac-1, LFA-1, and p150,95 glycoproteins. *Annual Review of Medicine*, 1987, **38**, 75–94.

42. Kishimoto, T. K, Hollander, N., Roberts, T. M, Anderson, D. C and Springer, T. A, Heterogeneous mutations in the beta subunit common to the LFA-1, Mac-1, and p150,95 glycoproteins cause leukocyte adhesion deficiency. *Cell*, 1987, **50**, 193–202.

43. Peerschke, E. I. B., Platelet membranes and receptors. In *Thrombosis and Hemorrhage* (J. Loscalzo and A. I. Schafer, eds), Blackwell, Boston, MA 1994,219–245.

44. Saelman, E. U. M., Nieuwenhuis, H. K, Hese, K. M, DeGroot, P. G, Heinjnen, H. F. G., Sage, E. H, Williams, S., McKeown, L., Gralnick, H. R and Sixma, J. J, Platelet adhesion to collagen types I-VIII under conditions of stasis and flow is mediated by GPIa/IIa ($\alpha_2 \beta_1$-integrin). *Blood*, 1994, **85**, 1244–1250.

45. Ross, J. M and McIntire, L. V, Molecular mechanisms of mural thrombosis under dynamic flow conditions. *NIPS*, 1995, **10**, 117–122.

46. Shattil, S. J, Hoxie, J. A, Cunningham, M. C and Brass, L. F, Changes in the platelet membrane glycoprotein IIb-IIIa complex during platelet activation. *Journal of Biological Chemistry*, 1985, **260**, 11107–11114.

47. Frelinger, A. L, III, , Lam, S. C-T., Plow, E. F, Smith, M. A, Loftus, J. C and Ginsberg, M. H, Occupancy of an adhesive glycoprotein receptor modulates expression of an antigenic site involved in cell adhesion. *Journal of Biological Chemistry* 1988, **263**, 12397–12402.

48. Ginsberg, M. H, Loftus, J. C and Plow, E. F, Cytoadhesins, integrins, and platelets. *Thrombosis and Haemostasis*, 1988, **59**, 1–6.

49. Zaidi, T. N, McIntire, L. V, Farrell, D. H and Thiagarajan, P., Adhesion of platelets to surface-bound fibrinogen under flow. *Blood*, 1996, **88**, 2967–2972.

50. Dunlop, L. C, Andrews, R. K and Berndt, M. C, Congenital disorders of platelet function. In *Thrombosis and Hemorrhage* (J. Loscalzo and A. I. Schafer, eds), Blackwell, Boston, MA 1994,615–633.

51. Alevriadou, B. R, Moake, J. L, Turner, N. A, Ruggeri, Z. M, Folie, B. J, Phillips, M. D, Schreiber, A. B, Hrinda, M. E, and McIntire, L. V, Real-time

analysis of shear dependent thrombus formation and its blockade by inhibitors of von Willebrand factor binding to platelets. *Blood,* 1993, **81**, 1263–1276.

52. Turitto, V. T, Weiss, H. J, Zimmerman, T. S and Sussman, I. I, Factor VIII/von Willebrand factor in subendothelium mediates platelet adhesion. *Blood,* 1985, **65**, 823–831.

53. Konstantopoulos, K., Chow, T. W, Turner, N. A, Hellums, J. D and Moake, J. L, Shear stress-induced binding of von Willebrand factor to platelets. *Biorheology,* 1997, **34**, 57–71.

54. McCrary, J. K, Nolasco, L. H, Hellums, J. D, Kroll, M. H, Turner, N. A and Moake, J. L, Direct demonstration of radiolabeled von Willebrand factor binding to platelet glycoprotein Ib and IIb-IIIa in the presence of shear stress. *Annals of Biomedical Engineering,* 1995, **23**, 787–793.

55. Handa, M., Titani, K., Holland, L. Z, Roberts, J. R and Ruggeri, Z. M, The von Willebrand factor-binding domain of platelet membrane glycoprotein Ib. Characterization by monoclonal antibodies and partial amino acid sequence analysis of proteolytic fragments. *Journal of Biological Chemistry,* 1986, **261**, 12579–12585.

56. Weiss, H. J, Turitto, V. T and Baumgartner, H. R, Effect of shear rate on platelet interaction with subendothelium in citrated and native blood. I. Shear rate-dependent decrease of adhesion in von Willebrand disease and the Bernard-Soulier syndrome. *Journal of Laboratory and Clinical Medicine,* 1978, **92**, 750–764.

57. Springer, T. A, Traffic signals for lymphocyte recirculation and leukocyte emigration: The multistep paradigm. *Cell,* 1994, **76**, 301–314.

58. Osborn, L., Vassallo, C. and Benjamin, C. D, Activated endothelium binds lymphocytes through a novel binding site in the alternatively spliced domain of vascular cell adhesion molecule-1. *Journal of Experimental Medicine,* 1992, **176**, 99–107.

59. Vaporciyan, A. A, Delisser, H. M, Yan, H., Mendiguren, I. I, Thorn, S. R, Jones, M. L, Ward, P. A and Albelda, S. M, Involvement of platelet-endothelial adhesion molecule-1 in neutrophil recruitment in vivo. *Science,* 1993, **262**, 1580–1582.

60. Tanaka, Y., Albelda, S. M and Horgan, K. J, CD31 expressed on distinctive T cell subsets is a preferential amplifier of beta 1 integrin-mediated adhesion. *Journal of Experimental Medicine,* 1992, **176**, 245–253.

61. Takeichi, M., The cadherin cell adhesion receptor family: roles in multicellular organization and neurogenesis. *Progress in Clinical and Biological Research,* 1994, **390**, 145–153.

62. Patel, D. J and Gumbiner, B. M, Zipping together a cell adhesion interface. *Nature,* 1995, **374**, 306–307.

63. Hirokawa, N. and Heuser, J. E, Quick-freeze, deep-etch visualization of the cytoskeleton beneath surface differentiation of intestinal epithelial cells. *Journal of Cell Biology,* 1981, **91**, 399–409.

64. Konstantopoulos, K. and McIntire, L. V, Effects of fluid dynamic forces on vascular cell adhesion. *Journal of Clinical Investigation,* 1996, **98**, 2661–2665.

65. Jones, D. A, Smith, C. W and McIntire, L. V, Flow effects on leukocyte adhesion to vascular endothelium. In *Principles of cell adhesion* (P. Richardson and M. Steiner, eds), CRC Press, Inc., Boca Raton, FL. 1995.

66. Smith, C. W, Endothelial adhesion molecules and their roles in inflamma- tion. *Canadian Journal of Physiology and Pharmacology*, 1993, **71**, 76–87.

67. Muller, W. A, Weigl, S. A, Deng, X. and Phillips, D. M, PECAM-1 is required for transendothelial migration of leukocytes. *Journal of Experimen- tal Medicine*, 1993, **178**, 449–460.

68. von Andrian, U. H, Berger, E. M, Chambers, J. D, Ramezani, L., Ochs, H., Harlan, J. M, Paulson, J. D, Etzioni, A. and Arfors, K. E, In vivo behavior of nutrophils from two patients with distinct inherited leukocyte deficiency syndromes. *Journal of Clinical Investigation*, 1993, **91**, 2893–2897.

69. Lawrence, M. B, McIntire, L. V and Eskin, S. G, Effect of flow on polymorphonuclear leukocyte/endothelial cell adhesion. *Blood*, 1987, **70**, 1284–1291.

70. Lawrence, M. B, Smith, C. W, Eskin, S. G and McIntire, L. V, Effect of venous shear stress on CD18-mediated neutrophil adhesion to cultured endothelium. *Blood*, 1990, **75**, 227–236.

71. Ley, K., Bullard, D. C, Arbones, M. L, Bosse, R., Vestweber, D., Tedder, T. F and Beaudet, A. L, Sequential contribution of L- and P- selectin to leukocyte rolling in vivo. *Journal of Experimental Medicine*, 1995, **181**, 669–675.

72. Labow, M. A, Norton, C. R, Rumberger, J. M, Lombard-Gillooly, K. M, Shuster, D. J, Hubbard, J., Bertko, R., Knaack, P. A, Terry, R. W, Harbison, M. L, Kontgen, F., Stewart, C. L, McIntyre, K. W, Will, P. C, Burns, D. K and Wolitzky, B. A, Characterization of E-selectin-deficient mice: demon- stration of overlapping function of the endothelial selectins. *Immunity*, 1994, **1**, 709–720.

73. von Andrian, U. H, Chambers, J. D, McEvoy, L., Bargatze, R. F, Arfors, K. E and Butcher, E. C, Two-step model of leukocyte-endothelial cell interaction in inflammation: distinct roles for LECAM-1 and the leukocyte β_2 integrins in vivo. *Proceedings of the National Academy of Sciences of the United States of America*, 1991, **88**, 7538–7542.

74. Tedder, T. F, Steeber, D. A and Pizcueta, P., L-selectin-deficient mice have impaired leukocyte recruitment into inflammatory sites. *Journal of Experi- mental Medicine*, 1995, **181**, 2259–2264.

75. Abbassi, O., Kishimoto, T. K, McIntire, L. V, Anderson, D. C and Smith, C. W, E-selectin supports neutrophil rolling in vitro under conditions of flow. *Journal of Clinical Investigation*, 1993, **92**, 2719–2730.

76. Jones, D. A, Smith C. W, Picker, L. J and McIntire, L. V, Neutrophil adhesion to 24-hour IL-1-stimulated endothelial cells under flow conditions. *Journal of Immunology*, 1996, **157**, 858–863.

77. Jones, D. A, Abbassi, O., McIntire, L. V, McEver, R. P and Smith, C. W, P-selectin mediates neutrophil rolling on histamine-stimulated endothelial cells. *Biophysical of Journal*, 1993, **65**, 1560–1569.

78. Lawrence, M. B and Springer, T. A, Leukocyte roll on a selectin at physio- logic flow rates: Distinction from and prerequisite for adhesion through integrins. *Cell*, 1991, **65**, 859–873.

79. Lawrence, M. B and Springer, T. A, Neutrophils roll on E-selectin. *Journal of Immunology*, 1993, **151**, 6338–6346.

80. Lawrence, M. B, Berg, E. L, Butcher, E. C and Springer, T. A, Rolling of lymphocytes and neutrophils on peripheral node addressin and subsequent

arrest on ICAM-1 in shear flow. *European of Journal Immunology*, 1995, **25**, 1025–1031.

81. Lawrence, M. B, Bainton, D. F and Springer, T. A, Neutrophil tethering to and rolling on E-selectin are separable by the requirement for L-selectin. *Immunity*, 1994, **1**, 137–145.

82. Bargatze, R. F, Kurk, S., Butcher, E. C and Jutila, M. A, Neutrophils roll on adherent neutrophils bound to cytokine-induced endothelial cells via L-selectin on the rolling cells. *Journal of Experimental Medicine*, 1994, **180**, 1785–1792.

83. Walcheck, B., Moore, K. L, McEver, R. P and Kishimoto, T. K, Neutrophil-neutrophil interactions under hydrodynamic shear stress involve L-selectin and PSGL-1. A mechanism that amplifies initial leukocyte accumulation on P-selectin in vitro. *Journal of Clinical Investigation*, 1996, **98**, 1081–1087.

84. Alon, R., Fuhlbrigge, R. C, Finger, E. B and Springer, T. A, Interactions through L-selectin between leukocytes and adherent leukocytes nucleate rolling adhesions on selectins and VCAM-1 in shear flow. *Journal of Cell Biology*, 1996, **135**, 849–865.

85. Jones, D. A, McIntire, L. V, C. W, S. and Picker, L. J, A two-step adhesion cascade for T cell/endothelial cell interactions under flow conditions. *Journal of Clinical Investigation*, 1994, **94**, 2443–2450.

86. Alon, R., Kassner, P. D, Carr, M. W, Finger, E. B, Helmer, M. E and Springer, T. A, The integrin-VLA-4 supports tethering and rolling in flow on VCAM-1. *Journal of Cell Biology*, 1995, **128**, 1243–1253.

87. Konstantopoulos, K., Kukreti, S., Smith, C. W and McIntire, L. V, Endothelial P-selectin and VCAM-1 each can function as primary adhesive mechanisms for T cells under conditions of flow. *Journal of Leukocyte Biology*, 1997, **61**, 179–187.

88. Abe, Y., Ballantyne, C. M and Smith, C. W, Functions of domain 1 and 4 of vascular cell adhesion molecule-1 in α_4 integrin-dependent adhesion under static and flow conditions are differntially regulated. *Journal of Immunology*, 1996, **157**, 5061–5069.

89. Berlin, C., Bargatze, R. F, Campbell, J. J, von Andrian, U. H, Szabo, M. C, Hasslen, S. R, Nelson, R. D, Berg, E. L, Erlandsen, S. L and Butcher, E. C, α_4 integrins mediate lymphocyte attachment and rolling under physiologic flow. *Cell*, 1995, **80**, 413–422.

90. Kukreti, S., Konstantopoulos, K., Smith, C. W and McIntire, L. V, Molecular mechanisms of monocyte adhesion to IL-1 $-\beta$ – stimulated endothelial cells under physiological flow conditions. *Blood*, 1997, **89**, 4104–4111.

91. Ley, K., Histamine can induce leukocyte rolling in rat mesentric venules. *American Journal of Physiology*, 1994, **267**, H1017–H1023.

92. Johnston, B., Issekutz, T. B and Kubes, P., The α_4-integrin supports leukocyte rolling and adhesion in chronically inflamed postcapillary venules in vivo. *Journal of Experimental Medicine*, 1996, **183**, 1995–2006.

93. Kunkel, E. J, Jung, U., Bullard, D. C, Norman, K. E, Wolitzky, B. A, Vestweber, D., Beaudet, A. L and Ley, K., Absence of trauma-induced leukocyte rolling in mice deficient in both P-selectin and intracellular adhesion molecule 1. *Journal of Experimental Medicine*, 1996, **183**, 57–65.

94. Alon, R., Hammer, D. A and Springer, T. A, Lifetime of the P-selectin-carbohydrate bond and its response to tensile force in hydrodynamic flow. *Nature*, 1995, **374**, 539–542.

95. Puri, K. D, Finger, E. B and Springer, T. A, The faster kinetics of L-selectin than of E-selectin and P-selectin rolling at comparable binding strength. *Journal of Immunology*, 1997, **158**, 405–413.

96. Patel, K. D, Nollert, M. U and McEver, R. P, P-selectin must extend a sufficient length from the plasma membrane to mediate rolling of neutrophil. *Journal of Cell Biology*, 1995, **131**, 1893–1902.

97. Lorant, D. E, McEver, R. P, McIntyre, T. M, Moore, K. L, Prescott, S. M and Zimmerman, G. A, Activation of polymorphonuclear leukocytes reduces their adhesion to P-selectin and causes redistribution of ligands for P-selectin on their surfaces. *Journal of Clinical Investigation*, 1995, **96**, 171–182.

98. Faull, R. J and Ginsberg, M. H, Inside-out signaling through integrins. *Journal of America Society Nephrology*, 1996, **7**, 1091–1097.

99. Waddell, T. K, Fialkow, L., Chan, C. K, Kishimoto, T. K and Downey, G. P, Potentiation of the oxidative burst of human neutrophils. A signalling role for L-selectin. *Journal of Biological Chemistry*, 1994, **269**, 18485–18491.

100. Waddell, T. K, Fialkow, L., Chan, C. K, Kishimoto, T. K and Downey, G. P, Signaling functions of L-selectin: enhancement of tyrosine phosphorylation and activation of MAP kinase. *Journal of Biological Chemistry*, 1995, **270**, 15403–15411.

101. Simon, S. I, Burns, A. R, Taylor, A. D, Gopalan, P. K, Lynam, E. B, Sklar, L. A and Smith, C. W, L-selectin (CD62L) cross-linking signals neutrophil adhesive functions via the Mac-1 (CD11b/CD18) β_2 integrin. *Journal of Immunology*, 1995, **155**, 1502–1514.

102. Gopalan, P. K, Smith, C. W, Lu, H., Berg, E. L, McIntire, L. V and Simon, S. I, Neutrophil CD18-dependent arrest on intercellular adhesion molecule 1 (ICAM-1) in shear flow can be activated through L-selectin. *Journal of Immunology*, 1997, **158**, 367–375.

103. Hwang, S. T, Singer, M. S, Giblin, P. A, Yednock, T., Bacon, K. B, Simon, S. I and Rosen, S. D, GlyCAM-1, a physiologic ligand for L-selectin, activates β_2 integrins on naive peripheral lymphocytes. *Journal of Experimental Medicine*, 1996, **184**, 1343–1348.

104. Nagata, K., Tsuji, T., Todoroki, N., Katagiri, Y., Tanoue, K., Yamazaki, H., Hanai, N. and Irimura, T., Activated platelets induce superoxide anion release by monocytes and neutrophils through P-selectin (CD62). *Journal of Immunology*, 1993, **151**, 3267–3273.

105. Elstad, M. R, La Pine, T. R, Cowley, F. S, McEver, R. P, McIntyre, T. M, Prescott, S. M and Zimmerman, G. A, P-selectin regulates platelet-activating factor synthesis and phagocytosis by monocytes. *Journal of Immunology*, 1995, **155**, 2109–2122.

106. Weyrich, A. S, Elstad, M. R, McEver, R. P, McIntyre, T. M, Moore, K. L, Morrissey, J. H, Prescott, S. M and Zimmerman, G. A, Activated platelets signal chemokine synthesis by human monocytes. *Journal of Clinical Investigation*, 1996, **97**, 1525–1534.

107. Cooper, D., Butcher, C. M, Berndt, M. C and Vadas, M. A, P-selectin interacts with a β2-integrin to enhance phagocytosis. *Journal of Immunology*, 1994, **153**, 3199–3209.

108. Lorant, D. E, Topham, M. K, Whatley, R. E, McEver, R. P, McIntyre, T. M, S. M, P. and Zimmerman, G. A, Inflammatory roles of P-selectin. *Journal of Clinical Investigation*, 1993, **92**, 559–570.

109. Lo, S. K, Lee, S., Ramos, R. A, Lobb, R., Rosa, M., Chi-Ross, G. and Wright, S. D, Endothelial-leukocyte adhesion molecule1 stimulates the adhesive activity of leukocyte integrin CR3 (CD11b/CD18, Mac-1) on human neutrophils. *Journal of Experimental Medicine*, 1991, **173**, 1493–1500.

110. Ross, J. M, McIntire, L. V, Moake, J. L and Rand, J. H, Platelet adhesion and aggregation on human type VI collagen surfaces under physiological flow conditions. *Blood*, 1995, **85**, 1826–1835.

111. Savage, B., Saldivar, E. and Ruggeri, Z. M, Initiation of platelet adhesion by arrest onto fibrinogen or translocation on von Willebrand factor. *Cell*, 1996, **84**, 289–297.

112. McIntire, L. V, 1992 ALZA Distinguished Lecture: bioengineering and vascular biology. *Annals of Biomedical Engineering*, 1994, **22**, 2–13.

113. Ruggeri, Z. M, von Willebrand factor. *Journal of Clinical Investigation*, 1997, **99**, 559–564.

114. Chow, T. W, Hellums, J. D, Moake, J. L and Kroll, M. H, Shear stress-induced von Willebrand factor binding to platelet glycoprotein Ib initiates calcium influx associated with aggregation. *Blood*, 1992, **80**, 113–120.

115. Ikeda, Y., Handa, M., Kamata, T., Kawano, K., Kawai, Y., Watanabe, K., Kawakami, K., Sakai, K., Fukuyama, M., Itagaki, I., Yoshioka, A. and Ruggeri, Z. M, Transmembrane calcium influx associated with von Willebrand factor binding to GPIb in the initiation of shear-induced platelet aggregation. *Thrombosis and Haemostasis*, 1993, **69**, 496–502.

116. Kroll, M. H, Hellums, J. D, McIntire, L. V, Schafer, A. I and Moake, J. L, Platelets and shear stress. *Blood*, 1996, **88**, 1525–1541.

117. Dejana, E., Endothelial adherens junctions: implications in the control of vascular permeability and angiogenesis. *Journal of Clinical Investigation*, 1996, **98**, 1949–1953.

118. Casnocha, S. A, Eskin, S. G, Hall, E. R and McIntire, L. V, Permeability of human endothelial monolayers: effect of vasoactive agonists and cAMP. *Journal of Applied Physiology*, 1989, **67**, 1997–2005.

119. Rabiet, M. J, Plantier, J. L, Rival, Y., Genoux, M. G, Lampugnani, M. G and Dejana, E., Thrombin-induced increase in endothelial permeability is associated with changes in cell to cell junction organization. *Arteriosclerosis Thrombosis Vascular Biology*, 1996, **16**, 488–496.

120. Del Maschio, A., Zanetti, A., Corada, M., Rival, Y., Ruco, L., Lampugnani, M. G and Dejana, E., Polymorphonuclear leukocyte adhesion triggers the disorganization of endothelial cell-to-cell adherens junctions. *Journal of Cell Biology*, 1996, **135**, 497–510.

121. Shibamoto, S., Hayakawa, K. M, Takeuchi, T., Hori, N., Oku, K., Miyazawa, N., Kitamura, M., Takiechi, M. and Ito, F., Tyrosine phosphorylation of β-catenin and plakoglobin is enhanced by hepatocyte growth factor and epidermal growth factor in human carcinoma cells. *Cell Adhesion and Communication*, 1994, **1**, 295–305.

122. Allen, M. D, McDonald, T. O, Himes, V. E, Fishbein, D. P, Aziz, S. and

Reichenbach, D. D, E-selectin expression in human cardiac grafts with cellular rejection. *Circulation*, 1993, **88**, 243–247.

123. Briscoe, D. M and Cotran, R. S, Role of leukocyte-endothelial cell adhesion molecules in renal inflammation: in vitro and in vivo studies. *Kidney International*, 1993, **44**, S27–S34.

124. Rabb, H. A. A., Cell adhesion molecules and the kidney. *American Journal of Kidney Disease*, 1994, **23**, 155–166.

125. Isobe, M., Yagita, H., Okumura, K. and Ihara, A., Specific acceptance of cardiac allograft after treatment woth antibodies to ICAM-1 and LFA-1. *Science*, 1992, **255**, 1125–1127.

126. Isobe, M. and Ihara, A., Tolerance induction against cardiac allograft by anti-ICAM-1 and LFA-1 treatment: T cells respond to in vitro allostimulation. *Transplant Proceedings*, 1993, **25**, 1079–1080.

127. Gemmell, C. H, Ramirez, S. M, Yeo, E. L and Sefton, M.v., Platelet activation in whole blood by artificial surfaces: identification of platelet -derived microparticles and activated platelet binding to leukocytes as material-induced activation events. *Journal of Laboratory and Clinical Medicine*, 1995, **125**, 276–287.

128. Folie, B. J and McIntire, L. V, Mathematical analysis of mural thrombosis: concentration profiles of platelet-activating agents and effects of viscous shear flow. *Biophysical of Journal*, 1989, **56**, 1393–1400.

129. The EPIC investigators., Use of a monoclonal antibody directed against the platelet glycoprotein IIb/IIIa receptor in high-risk coronary angioplasty. *New England Journal of Medicine*, 1994, **330**, 956–961.

130. Konstantopoulos, K., Kamat, S. G, Schafer, A. I, Ba-ez, E. I, Jordan, R., Kleiman, N. S and Hellums, J. D, Shear-induced platelet aggregation is inhibited by in vivo infusion of an anti-glycoprotein IIb/IIIa antibody fragment, c7E3 Fab, in patients undergoing coronary angioplasty. *Circulation*, 1995, **91**. 1427–1431.

131. Turner, N. A, Moake, J. L, Kamat, S. G, Schafer, A. I, Kleiman, N. S, Jordan, R. and McIntire, L. V, Comparative real-time effects on platelet adhesion and aggregation under flowing conditions of in vivo aspirin, heparin, and monoclonal antibody fragment against glycoprotein IIb-IIIa. *Circulation*, 1995, **91**, 1354–1362.

132. Kamat, S. G, Turner, N. A, Konstantopoulos, K., Hellums, J. D, McIntire, L. V, Kleiman, N. S and Moake, J. L, Effects of integrelin on platelet function in flow models of arterial thrombosis. *Journal of Cardiovascular Pharmacology*, 1997, **29**, 156–163.

CHAPTER II.3

Mechanical Forces And Growth Factors Utilized In Tissue Engineering

KEITH J. GOOCH*
TORSTEN BLUNK*
GORDANA VUNJAK-NOVAKOVIC
ROBERT LANGER
LISA E. FREED
Department of Chemical Engineering and Division of Health Sciences and Technology,
Massachusetts Institute of Technology,
77 Massachusetts Ave.,
Cambridge, MA 02139, USA

and

CHRISTOPHER J. TENNANT
University of Maryland,
College Park, MD, USA

1 Introduction

Mechanical and biochemical environments profoundly affect the development, maintenance, and remodeling of tissue *in vivo* and therefore may be important determinants of the quality of engineered tissue grown *in vivo* or *in vitro*. This chapter will discuss three model tissues — cartilage, bone, and blood vessels — to illustrate the important role of mechanical and biochemical environments in tissue engineering. It is not intended as an exhaustive discussion of these tissue types, but instead is meant to highlight general principles and the potential of mechanical forces and growth factors as tools to improve engineered tissue.

The process by which the mechanical and biochemical environments affect tissue can be subdivided into four steps summarizing the conversion of a physical stimulus into biochemical stimuli and culminating in the

---

*The first two authors contributed equally to this chapter.

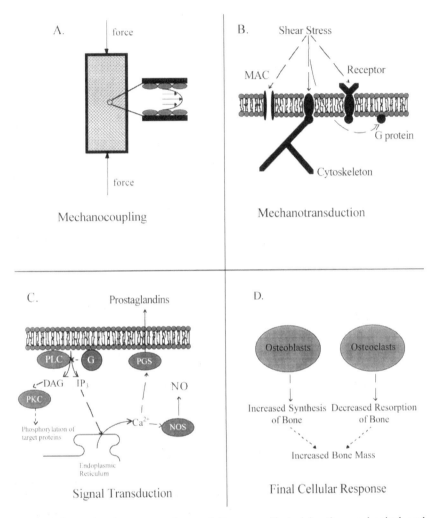

Fig. 1. Tissue development and remodeling are affected by the mechanical and biochemical environments. The sequence of events leading up to the tissue-level response can be subdivided into four steps; mechanocoupling, mechanotransduction, signal transduction and the final cellular response. For a detailed description of these reponses, please see text. Abbreviations used are G, G protein; MAC, mechanically activated ion channels; PGS, prostaglandin synthase; PKC, protein kinase C; PLC, phospholipase C; and NOS, nitric oxide synthase.

physiological alteration of cells and tissue (Fig. 1). These steps were first proposed to describe the responses of bone to mechanical loading but are applicable to mechanically and biochemically induced responses in many other biological systems [1]. Please note that Fig. 1 does not provide full detail regarding the specific response of bone, but rather illustrates the steps by tracing a credible series of events from the application of the mechanical force to a tissue-level response.

The first step, mechanocoupling, is the conversion of the applied physical force to secondary forces or physical phenomena detected by the cells (Fig. 1, Panel A). Here the application of the primary force results in deformation of the bone, which in turn deforms bone cells and elicits pressure gradients and interstitial fluid flow (the secondary forces or phenomena). While bone cells have been shown to respond *in vitro* to deformation, pressure, and fluid flow, the magnitude of the deformation and pressure required to stimulate the cells are much greater than those experienced *in vivo* [1]. Interstitial fluid flow applies a mechanical force called shear stress to osteoblasts and osteocytes. Mathematical models predict that the magnitude of the shear stress applied *in vivo* is comparable to that found to stimulate osteoblasts and osteoclasts *in vitro* [2], supporting the assertion that shear stress is the *in vivo* stimulus to which the cells respond [3].

The second step involves the conversion of either the primary or secondary physical stimulus into an electrical, chemical, or biochemical response (Fig. 1, Panel B). This process, mechanotransduction, is an area of active research and of considerable debate. Primary candidates for mechanotransducers are mechanically activated ion channels (MAC), receptors, G proteins, and the cytoskeleton. The evidence in favor of and against each potential mechanotransducer has recently been reviewed [4].

The third step, signal transduction (Fig. 1, Panel C), entails the conversion of one biochemical signal to another. The signal transduction pathways stimulated by physical forces are in many cases identical to those stimulated by biochemical agonists, the primary difference being that the initial biochemical event (e.g. the activation of a G protein) is mechanically induced. Once the G protein is activated, the resultant downstream biochemical events are similar regardless of whether the protein is activated by a receptor or by shear stress applied to the plasma membrane.

The final step (Fig. 1, Panel D) completes the conversion from initial stimulus to final tissue-level response. The activation of various signal transduction pathways up- or down-regulates cellular activities such as gene expression, protein synthesis, and release of autocrine and paracrine factors, leading to responses on the cellular and tissue levels. For bone, these responses may include increased bone formation by osteoblasts and decreased bone resorption by osteoclasts, resulting in a net increase in bone mass.

The four steps outlined above suggest methods by which the mechanical or biochemical environment may be modified to control the development of engineered tissue. For instance, one might expect tissue-engineered bone to increase its density when subjected to mechanical conditions resembling those known to stimulate net bone formation *in vivo*. Alternatively, application of the appropriate secondary force may be adequate to achieve the desired effect, in this case increasing bone mass. Although mechanotransduction is not clearly understood, a limited opportunity may exist to control tissue development by the use of pharmacological inhibitors and stimulators of potential mechanotransducers. Pharmacological inhibitors and stimulators of most of the components involved in signal transduction also are available and may be employed to control tissue development. Alternatively, biochemicals such as local growth factors, systemic hormones, and steroids, which may initiate the same or different signal transduction pathways as those elicited by mechanical forces, could be used to promote and control tissue development. The remainder of the chapter will discuss approaches to control the development of engineered tissue by modifying its mechanical or biochemical environment.

2 Mechanical forces

The biological effects of mechanical forces have been extensively reviewed in several books [4, 5] and a number of articles. Here, we focus on the implications for the field of tissue engineering. *In vivo*, mechanical forces arise from diverse sources such as muscular contraction resulting in stress and strain of muscle and tendon, and locomotion generating small amplitude cyclic compression of bone and cartilage. Blood flow exerts shear stress on the endothelium, and pressure and cyclic strain is experienced by the endothelium and vessel wall. Other sources of forces are not as readily apparent, such as interstitial fluid flow which applies shear stress to bone cells and growth which stresses skin. Each of these mechanical forces result in acute biological changes on the cellular and subcellular level which culminate in tissue remodeling.

2.1 Cartilage

It has been known for more than 100 years that cartilage responds to mechanical forces and is able to remodel in response to the prevailing stress [8]. More recently, clinical observations and *in vivo* studies have revealed the extent to which this remodeling occurs. Load-bearing surfaces in joints are thicker, stronger, and have higher glycosaminoglycan (GAG) concentrations than nearby nonload-bearing surfaces of the same joint. Decreased loading or immobilization of a joint reduces GAG synthesis and content; however, GAG content is gradually restored once the joint is remobilized [9–11].

Walking generates cyclic compression of articular cartilage at ~ 1 Hz with less than 5% deformation [12]. Small-amplitude cyclic loading of excised cartilage has been extensively employed to study the effects of mechanical stimuli on chondrocyte metabolism. Kim *et al.* indicate that small-amplitude cyclic compression of excised cartilage increased the incorporation rate of [^{35}S]sulfate and [^{3}H]proline (indices of GAG and protein synthesis, respectively) by 20–40% [7]. By comparing predicted values of interstitial fluid velocity and pressure within the excised cartilage to regional GAG synthesis, Kim *et al.* concluded that GAG synthesis co-localized with regions of high velocity but was inversely related to pressure [7]. These results suggest that cartilage, like bone, may respond to mechanical stimulation via loading-induced fluid flow. Regardless of the actual stimulus perceived by cartilage, it is evident that cartilage remodels in response to mechanical loading *in vivo* and *in vitro* by increasing production of extracellular matrix components leading to stronger tissue.

Most of the attempts to culture tissue-engineered cartilage *in vitro* have utilized either stationary cultures or systems generating relatively small mechanical forces. Such systems include spinner flasks, petri dishes on an orbital shaker, and the NASA-developed rotating bioreactors originally designed to simulate microgravity (reviewed in reference [13]). Isolated chondrocytes were seeded on scaffolds made of polyglycolic acid mesh and cultured in petri dishes or spinner flasks under mixed or unmixed conditions. Tissue-engineered cartilage grown under mixed conditions had a larger fraction (based on wet weight) of GAG and collagen than corresponding umixed cultures (1.45 and 0.65% GAG, and 2.92 and 1.10% collagen) [14]. In another study, tissue-engineered cartilage cultured in rotating bioreactors had more GAG and fewer cells than corresponding cartilage cultured in turbulent spinner flasks, suggesting that the mechanical conditions of the rotating bioreactor may favor GAG synthesis or retention [15]. In these mixing studies, however, it is difficult to definitively separate the effects of mixing-induced mechanical forces from those of convection-enhanced transport of nutrients to, and metabolites away from, the tissue.

In vitro cultured tissue-engineered cartilage histologically resembles normal cartilage though its biochemical composition is inferior to native cartilage [14–16]. As suggested by the two studies comparing different bioreactor conditions, the quality of the tissue-engineered cartilage may be enhanced by modifying the mechanical environment. The presence of a more robust mechanical environment than present in these studies may be required to culture cartilage with improved biochemical and mechanical properties. To provide such a mechanical environment, devices currently used to apply small-amplitude cyclic compression to excised cartilage [7] or related systems could be employed. Alternatively, the development of properties identical to native cartilage may be unnecessary, as the tissue-engineered material may continue to remodel following implantation when subject to physiological loading or the *in vivo* biochemical environment.

2.2 Bone

Using a curved wooden beam as an engineering model, Meyer theorized in 1867 that bone density patterns were governed by the distribution of stress [17]. Wolff later hypothesized that bone actively remodels in response to mechanical loading, increasing density and strength in areas exposed to stress while losing density in unstimulated regions [18]. Experimental evidence for mechanically induced remodeling and potential mechanisms regulating this remodeling have been extensively reviewed [1, 19]. It is widely acknowledged that bone mass is rapidly lost under conditions of diminished mechanical loading. It is therefore noteworthy that most attempts to develop tissue-engineered bone *in vitro* (reviewed in reference [20]) from osteogenic cells and appropriate scaffolding have been conducted under conditions of minimal mechanical .forces, conditions that would be expected to result in bone loss *in vivo*.

One study assessing the effects of mechanical forces on the development and differentiation of osteoblasts in culture produced dramatic results [19]. Rat calvarial osteoblasts were cultured on macroporous collagen beads in petri dishes (stationary controls) or in a fluidized-bed bioreactor. The average flow-induced shear stress on the surface of the beads was calculated to be ~ 1 dyn/cm^2 by applying Stoke's Law and ignoring the rotation of the beads. Markers of osteoblast differentiation such as alkaline phosphatase activity and hydroxyapatite (mineral formation) levels were monitored over a period of 1 month; analysis of these markers indicated a more rapid and complete differentiation of osteoblasts cultured in the mechanically active environment. Mineral formation as measured by hydroxyapatite levels was dramatically affected. Osteoblasts maintained under stationary conditions produced undetectable quantities of hydroxyapatite, while those cultured in a mechanically active environment rapidly began to produce mineral and continued for at least one month. The implication of these results for researchers attempting to develop tissue-engineered bone is clear: the mechanical environment is a key determinant of tissue development.

2.3 Blood vessels

Fluctuating blood pressure associated with the cardiac cycle exposes the endothelium and vascular smooth muscle cells to cyclic strain. Moreover, blood flow exerts a shear stress on the endothelium. Due to their constant exposure to a variety of mechanical forces *in vivo* and their significant involvement in vascular disease, the endothelial and vascular smooth muscle cells are two of the cell types whose responses to physical forces have been studied most extensively. Fluid flow, probably acting through shear stress, and cyclic strain regulate many physiologically significant endothelial functions including vasoregulation, control of transvessel transport, and inhibition of thrombogenesis [4–6, 21, 22]. Vascular smooth muscle cells are also dramatically affected by their mechanical environment [4, 23]. In addition to playing an important role in the health and disease state of the vessel, there is compelling evidence that the success of a tissue-engineered blood vessel is influenced by its mechanical environment.

The intimal surface of an artificial blood vessel, like the native endothelium, should be nonthrombogenic. Unfortunately, many of the materials used for vascular prostheses are at least mildly prothrombogenic. While prostheses for large-diameter arteries appear to tolerate limited clotting, thrombosis severely limits the application of vascular prostheses for use in small-diameter (several mm) grafts such as those required for coronary bypass surgery.

One approach to increase the patency of vascular prostheses is to seed the vessel with endothelial cells prior to surgical implantation. The use of this technique has been limited by the detachment of the endothelial cells from the prosthesis upon exposure to fluid flow with a wall shear stress comparable to *in vivo* levels. Cell retention is improved by preconditioning endothelialized grafts by exposing them to flow [24]. Cultured bovine aortic endothelial cells were seeded on 1.5 mm inner diameter spun polyethylene vascular grafts and cultured for 6 days *in vitro* with or without continuous laminar flow. For the first 3 days, flow produced wall shear stress of ~ 2 dyn/cm^2; shear stress was raised to *in vivo* levels of 25 dyn/cm^2 for the next 3 days. Grafts preconditioned with shear and static controls then were exposed to 25 dyn/cm^2 for 25 s. Grafts preconditioned with shear lost ~ 100 times fewer cells and maintained a confluent endothelium, while the continuity of the endothelium in the grafts not pretreated with flow was severely compromised. Furthermore, the pretreated grafts were less thrombogenic than static grafts, suggesting that culturing cells in the appropriate mechanical environment may substantially improve their *in vivo* performance.

A second approach to enhance the non-thrombogenicity of vascular prostheses has been to develop a bioartificial blood vessel more closely resembling a native vessel in the hope of improving patency. Weinberg and Bell constructed a bioartificial blood vessel from collagen and cultured vascular cells [25]. While structurally resembling a native blood vessel, the tissue-engineered blood vessel did not have adequate mechanical strength and needed to be reinforced with Dacron mesh to withstand *in vivo* levels of pressure (~ 150 mmHg) [25].

Vascular smooth muscle cells can exhibit two different phenotypes. The contractile phenotype contributes to the strength of the vessel wall and permits it to contract, while the secretory phenotype is associated with hyperplasia and vessel occlusion. Smooth muscle cells harvested from native vessels and cultured *in vitro* readily transform from the contractile to the secretory phenotype [23, 26]. Clearly the contractile phenotype is desirable, if not required, for a successful bioartificial artery. Therefore, maintaining cultured smooth muscle cells in the contractile phenotype in a developing bioartificial vessel is an important concern. Work by Kanda and Matsuda [27] demonstrates that smooth muscle cells cultured in a type I collagen gel transform to the secretory phenotype, as evidenced by increased number of synthetic organelles. If these collagen gels are stretched isometrically (static stretch), the smooth muscle cells align parallel to the direction of gel elongation but still transform to the secretory phenotype. If these gels are exposed to periodic stretch (1 Hz, 5% stretch), the embedded smooth muscle cells express the contractile phenotype and align parallel to the direction of stretch. These observations are consistent with those of Advanced Tissue Sciences (La Jolla, CA) using human fibroblasts seeded onto a tubular synthetic polymer support; when the supports were radially stretched, the cells aligned circumferentially, i.e., parallel to the direction of elongation [28].

The data of Kanda and Matsuda suggest a mechanism by which two major challenges facing the development of bioartificial blood vessels may be overcome. *In vivo*, arteries are cyclically stretched by the fluctuating blood pressure associated with the cardiac cycle. If the developing bioartificial blood vessel could be exposed to a similar cyclic strain *in vitro*, it would be expected that the contractile phenotype may be preserved while the smooth muscle cells would align in the correct orientation in the vessel wall. Additional implications of mechanical forces on the *in vitro* culture of a bioartificial blood vessel have been reviewed by Ziegler and Nerem [29].

3 Growth factors

Cells may indirectly interact with one another by the release of molecules such as polypeptides or steroids that can act locally in a paracrine/autocrine fashion or systemically in an endocrine manner. The rapidly growing body of knowledge concerning the physiological and pathological functions of these biological molecules may facilitate the integration of their exogenous application into approaches to tissue engineering. This section will focus mainly on the polypeptides commonly called growth factors and cytokines, although the use of specific vitamins as well as hormones such as glucocorticoids and sex steroids may also be considered.

Growth factors can stimulate or inhibit cell division, differentiation, and migration. They up- or down-regulate cellular processes such as gene expression, DNA and protein synthesis, and autocrine and paracrine factor release. By binding to receptors on the cell surface, which are expressed in a cell type-specific manner, the signal transduction cascade is initiated

(Fig. 1, Panel C). Growth factors can interact with one another in an additive, cooperative, synergistic, or antagonistic manner. They may cause dissimilar responses when applied to different cell types or tissues, and their effect on a certain type of cell or tissue may vary according to concentration or time of application. Some are temporarily stored in the extracellular matrix, and some must bind to matrix molecules to become active [30, 31]. For reviews of specifc growth factors and their receptors see [30, 32–41].

Various approaches to the use of growth factors in tissue engineering have been proposed [42, 43]. Growth factors may be added to cell culture media, thus acting on growing tissues *in vitro*. Controlled delivery of growth factors *in vivo* also has been proposed [44]. Controlled-release devices made from biodegradable polymers or gels with incorporated growth factors may be used *in vivo* in order to aid tissue regrowth or regeneration. These devices can be implanted either alone or in conjunction with another structure, such as a polymeric scaffold supporting the regenerating tissue. When the source of growth-factor release is limited to one specific location within the scaffold, the formation of a gradient occurs, thus possibly enabling a preferential cell attachment, assembly and/or a direction of cell migration. Alternatively, the growth factor could be thethered to the polymeric scaffold supporting the tissue, or the growth factor could be released with a controlled time profile if the growth factor is incorporated directly into or adsorbed onto a biodegradable fibrous polymeric scaffold or a highly porous foam on which the tissue grows. Although originally proposed for *in vivo* use, there is no reason why such systems could not also be used in tissue engineering approaches *in vitro*.

3.1 Cartilage

The ability of chondrocytes in articular cartilage to maintain their biochemically and structurally unique extracellular matrix is a prerequisite for proper joint function. In normal articular cartilage this maintenance is accomplished by precisely matching anabolic and catabolic activities with regard to matrix constituents, mainly proteoglycans and collagen type II. Peptide growth factors have been shown to play a major role both *in vivo* and *in vitro* in the preservation of healthy cartilage and its failure in joint disease by influencing growth and metabolism [45–49].

The result of the addition of exogenous growth factors may strongly depend on the experimental conditions used (e.g. age of tissue, concentration of serum, and mechanical environment), which must be taken into consideration when applying this knowledge to tissue engineering. Among the many influential biological molecules, insulin-like growth factor-1 (IGF-1), basic fibroblast growth factor (bFGF), and transforming growth factor-β (TGF-β) have been the most intensively studied. In general,

IGF-1 stimulates articular chondrocyte matrix synthesis and mitotic activity and inhibits chondrocyte-mediated matrix catabolism [49] and markedly increases the synthesis of proteoglycan components [50–52]. bFGF clearly acts as a potent mitogen, but also can promote both anabolic and catabolic matrix processes [49, 52]. IGF-1, but not bFGF, has been shown to preserve mechanical and electromechanical function of adult bovine articular cartilage [53]. Growth factors can thus modulate not only the quantity of matrix components, but also the functional quality of the cartilage. The effect of TGF-β is greatly dependent on its environment; it has opposite effects on chondrocyte proliferation and synthesis of matrix components under different experimental conditions [48, 49]. For example, TGF-β has been reported both to stimulate [54] and inhibit [55] proteoglycan synthesis by articular chondrocytes in excised cartilage. Other factors that stimulate proteoglycan synthesis in articular cartilage explants are the platelet-derived growth factors (PDGF) [56] and bone morphogenetic protein-2B [57].

The factors mentioned thus far act directly on chondrocytes. In contrast, other molecules modulate cartilage metabolism indirectly by interfering with enzymatic degradation cascades. These molecules also may be considered as potential candidates for *in vitro* cartilage tissue engineering. For example, tissue inhibitor of metalloproteinases (TIMP), synthetic matrix metalloproteinases inhibitor [58], and interleukin-4 [59] have been shown to counteract proteoglycan degradation caused by matrix metalloproteinases and other cartilage degradatory systems. Researchers have reported that combinations of the above factors can interact, thereby leading to inhibitory, additive, or synergistic effects on cartilage metabolism [45, 47–49]. The approaches of three research groups, discussed below, illustrate the possible utilization of growth factors in cartilage tissue engineering.

Fujisato *et al.* prepared chondrocyte-collagen composites in order to regenerate cartilage by subcutaneous implantation in the nude mouse [60]. Chondrocytes were isolated from costal cartilage of rats and seeded onto collagen sponges *in vitro*. When a collagen sponge was impregnated with bFGF by a simple adsorption process prior to cell seeding and implantation, regeneration of cartilage tissue was remarkably accelerated. Four weeks after implantation, the composite impregnated with bFGF exhibited distinctly stronger and more widespread histological staining for proteoglycans than the unimpregnated composite. However, quantitative data regarding matrix components are not presented, and the size of the composites decreased significantly depending on the density of seeded cells as compared to the original collagen sponge. The majority (90%) of the bFGF was shown to remain in the sponge even after an extended period of incubation in PBS prior to implantation; it is expected to be released slowly upon enzymatic degradation of the collagen after implantation. The authors attribute the effect of bFGF to the markedly enhanced angiogenesis around the implanted composite, rather than to direct action by the growth factor on the chondrocytes. They point out, however, that bFGF has been reported to promote cartilage repair *in vivo* [61] and inhibit the terminal differentiation of chondrocytes and calcification [62], which otherwise may have been of concern because of transformation of cartilage tissue into bone tissue following vascular invasion.

Zimber *et al.* made a first attempt at growing cartilage tissue *in vitro* on biodegradable polymer scaffolds with the help of a supplemented growth factor, namely TGF-β_1 [16]. Bovine articular chondrocytes obtained from ankle joints of 2- to 3-year-old cows were seeded on a polyglycolic acid mesh (2 mm thick, 10 mm diameter, 3×10^6 cells) and cultured in 6-well plates. The culture medium containing 10% fetal bovine serum was supplemented with TGF-β_1 (20 ng/ml). After 3 weeks of culture, tissues grown in the presence of TGF-β_1 were considerably larger than those receiving no TGF-β_1; their dry weights were more than 2.5-fold greater than those of control tissues. The amounts of matrix components, measured as total collagen and glycosaminoglycan (GAG), increased similarly; this increase was attributed mainly to an increase in cell number. However, the fractions of collagen and GAG in the tissue (0.7% and 1.7% based on dry weight, respectively) did not increase, which would have been desirable to improve the mechanical function of the tissue.

Blunk and co-workers seeded bovine articular chondrocytes obtained from the femoro-patellar groove of 2- to 4-week-old calves on polyglycolic acid meshes (2 mm thick, 5 mm diameter, 3.5×10^6 cells). The ability of IGF-1 (supplement to the regular culture medium containing 10% fetal bovine serum) to promote growth and extracellular matrix production was investigated [63]. Tissue constructs were cultured for 4 weeks in either static or spinner flasks or in NASA-developed rotating bioreactor vessels [13]. In all bioreactor systems studied, the addition of IGF-1 at a concentration of 100 ng/ml led to significant, though differential increases in construct size, cell proliferation, absolute GAG and total collagen amounts, and tissue stiffness as compared to unsupplemented control groups [63]. The most favorable results were achieved in the rotating bioreactors, where IGF-1 distinctly increased the relative GAG fraction in the tissue (3.7% based on wet weight as compared to 2.0% in unsupplemented cultures) and maintained the relative collagen fraction (3% collagen based on wet weight), although the cell number per weight was decreased to two-thirds of the value in unsupplemented cultures (Fig. 2). Apparently the mechanical environment can exhibit a strong influence on the effect of the growth factor. This is potentially not only caused by convection-enhanced molecular transport, but also implies the possibility that a particular cellular response, caused by a growth factor, may be potentiated by the mechanical environment which the cell experiences.

(A)

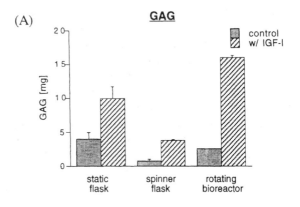

(B)

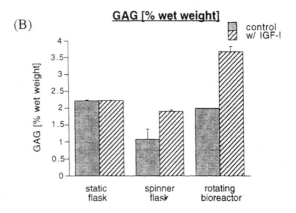

(C)

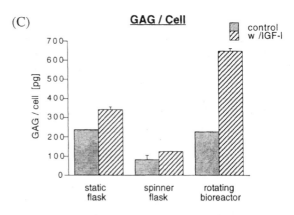

Fig. 2. Bovine articular chondrocytes were seeded on PGA scaffolds (2 mm thick, 5 mm diameter, 3.5×10^6 cells). Tissue constructs were cultured for four weeks in either static or mixed spinner flasks or in rotating bioreactor vessels. The ability of

In each of the studies cited above, one growth factor was applied at a steady concentration for the duration of the experiment. Future studies with combinations of growth factors and application to the developing tissue at specific time points may further improve the engineered cartilage. In the delivery of biologically active molecules to the *in vitro* engineered tissue, one must consider the size of the tissue construct; the pore size within the tissue, which will distinctly decrease as the tissue develops; the size of the growth factor molecule; and mechanically induced convection. Whereas the design of appropriate bioreactors may facilitate improved transport, the development of biodegradable scaffolds that release growth factors with a tailored time-profile to nourish cells in the core of the developing tissue may also be of use. Another potential use for controlled-release devices for growth factors is implantation in conjunction with the engineered cartilage tissue to facilitate integration into the host tissue.

3.2 Bone

The cellular events leading to bone formation [64–69] include chemotaxis of osteoblast precursors, proliferation of committed osteoblast precursors, extracellular matrix maturation, and mineralization. The development of the osteoblast phenotype during these processes is characterized by the ordered expression of genes, e.g. those coding for the structural proteins of bone, such as type I collagen, osteopontin, and osteocalcin. All of these sequential events must be closely regulated. Participating systemic hormones include the calciotropic hormones, parathryroid hormone and 1,25-dihydroxy vitamin D, and others, such as the pituitary and thyroid hormones and sex steroids. However, the predominant influence probably is

---

insulin-like growth factor-1 (IGF-1, 100 ng/ml, supplemented to the regular culture medium containing 10 % fetal bovine serum) to promote growth and extracellular matrix production was investigated [63]. Three constructs were grown per group. The figure shows the effect of IGF-1 on glycosaminoglycan (GAG) content, (A) absolute GAG amount in tissue, (B) relative GAG fraction in tissue (percentage based on wet weight), (C) GAG amount per cell. IGF-1 increased the absolute GAG amount in all three mechanical environments, especially in the rotating bioreactor (6-fold increase) (A). Even more important for the tissue quality, in the spinner flask and the rotating bioreactor, the relative fraction of GAG in the tissue (percentage based on wet weight) was also significantly increased, whereas no change occurred under static conditions (B). The GAG content expressed as amount produced per cell showed slight increases in the static and spinner flask, but a large increase in the rotating bioreactor (C). Thus, the mechanical environment can exhibit a strong influence on the effect of the growth factor. This might not only be due to convection-enhanced molecular transport to the cells within the tissue, but also to flow-induced mechanical forces.

exerted by local growth factors and cytokines. Indeed, bone contains a number of growth factors, including insulin-like growth factors (IGF-1, IGF-2), platelet-derived growth factor (PDGF), acidic and basic fibroblast growth factor (aFGF, bFGF), transforming growth factor (TGF-β_1, TGF-β_2), and bone morphogenetic proteins (BMP). Many of these factors are produced by osteoblasts and show autocrine and paracrine action. The BMPs, members of the TGF-β family, have drawn a lot of attention in recent years, as some of them have the unique capabilities of differentiating mesenchymal stem cells into chondro- and osteoblastic directions, stimulating ectopic bone formation, and inducing endochondral ossification *in vivo*.

Growth factor production can be regulated by both systemic hormones and local mechanical stress. The latter has been demonstrated by increased IGF-1 production after mechanical stimulation of rat osteocytes *in vivo*, followed by increased expression of two matrix proteins, type I collagen and osteocalcin, on the bone surface [70]. In one *in vitro* study, prostacyclin (PGI$_2$), previously shown to exhibit increased production during loading, mimicked early metabolic responses to mechanical loading in osteocytes and cells on the bone surface. PGI$_2$ also led to a greater release of IGF-2 into the cell culture medium, suggesting at least a partial modulation of the mechanical force-induced osteogenesis by PGI$_2$ and IGF-2 [71].

Bone tissue engineering *in vivo* requires the recruitment of appropriate progenitor cells as well as their subsequent differentiation into mature, functioning tissue [72]. Many studies have been conducted utilizing BMP. In a rat femoral defect model, two different dosages of BMP-2 (1.4 and 11 μg) were applied using rat demineralized bone matrix as carrier. Contralateral defects containing demineralized bone matrix without BMP served as control. Whereas in the high-dose group 10 out of 10 defects healed, none of the defects healed in either the low-dose group or the control group [73]. Other experiments employing BMP in animal models have been reviewed [74]. Carriers for BMP included both synthetic polymers and biopolymers. In heterotopic sites, these systems produced new bone, and increasing BMP doses resulted in greater bone formation. When delivered to the site of a skeletal defect, BMP consistently stimulated more bone formation than in the control group receiving only the polymer. In these control groups, the polymer still may act as a scaffold for bone growth without the involvement of any growth factor, but the additional bioactive molecule induces faster bone growth and results in greater bone mass than the scaffold alone [74]. Other growth factors successfully improving osteogenesis in animal studies include osteogenin, TGF-β [74], and bFGF [75]. In the treatment of four human patients with distal tibial metaphyseal non-unions, a biodegradable polymeric delivery system for human allogeneic BMP also was used. The use of BMP resulted in healing of the non-unions in all four patients, enabling them to avoid the amputation that would otherwise have been necessary [76].

Promising *in vitro* approaches to bone tissue engineering include seeding rat stromal or fetal rat calvarial osteoblasts onto three-dimensional scaffolds made from poly(lactic-co-glycolic acid) [20]. Additional bioactive molecules may further improve osteogenesis in a similar experimental system. Mesenchymal stem cells on such a scaffold may be differentiated with the help of BMP, eventually leading to full bone formation. *In vitro* and *in vivo*, scaffolding systems with capacity for both cell seeding and controlled-release of growth factors [77, 78] show great promise for useful application in the future.

3.3 Blood vessels

Normal vascular development, including growth and differentiation, is the result of a delicate balance between stimulators and inhibitors of vascular cell processes [79, 80]. Potent mitogens for the two major cell types of large blood vessels, endothelial cells and smooth muscle cells, have been well characterized. For example, bFGF is mitogenic for both endothelial and smooth muscle cells. Vascular endothelial cell growth factor (VEGF) has been shown to be specific for endothelial cells, whereas platelet-derived growth factor (PDGF) stimulates smooth muscle cells but not endothelial cells [78, 80]. Also, among others, heparin-binding EGF-like growth factor (HB-EGF) and betacellulin [80], two members of the epidermal growth factor (EGF) family, and endothelin-1 (ET-1) [81] are potent mitogens for smooth muscle cells. TGF-β has been reported to either stimulate or inhibit smooth muscle cell proliferation, depending on its concentration and cell density. Furthermore, it can act indirectly by affecting endogenous PDGF production and expression of the PDGF receptor in smooth muscle cells [80]. Inhibitors of endothelial and smooth muscle cell growth also have been reviewed [80]. Examples of growth factors influencing the differentiation state of smooth muscle cells are discussed by Owens [23].

One must bear in mind that each of these growth factors may up- or down-regulate the production of others, thus possibly modulating the originally intended effect. For tissue engineering, it is especially important that the status of the tissue has a huge impact on growth factor action [79]. Whether the cells are quiescent or activated, and whether they are in two-dimensional or three-dimensional culture (including the consequences regarding cell shape), apparently decisively influence endogenous growth factor production and also the response to exogenously applied bioactive molecules [79]. Endothelial and smooth muscle cells cultured together will respond differently to exogenous growth factors than will cultures of each cell type alone, due to subsequent paracrine effects as well as possible interactions of the cells and the growth factor with extracellular matrix proteins [29].

The following example nicely demonstrates the possible involvement of both mechanical forces and growth factors in tissue remodeling. Excised porcine thoracic aortas maintained in organ culture experience smooth muscle cell hyperplasia [82]. In such an aorta, the number of smooth muscle cells increased by 42% after 1 week in culture in the presence of 5% fetal bovine serum. No increase in cell number would be expected *in vivo*, and such rapid proliferation likely would complicate the development of a bioartificial vessel *in vitro*. Removal of the endothelium at the beginning of the experiment substantially inhibited the increase in the number of smooth muscle cells, reducing it to 15%. Incubation of vessels denuded of their endothelium with medium conditioned by intact organ cultures, however, partially restored intimal proliferation (a 30% increase), suggesting that a soluble factor produced by endothelial cells stimulates smooth muscle cell proliferation. Though the authors did not determine which factor was responsible for this increase in intimal proliferation, a likely candidate is ET-1, a smooth muscle cell mitogen whose production by endothelial cells is dramatically increased by the removal of physiological levels of fluid flow [83]. Potentially an exogenous inhibitor or antagonist of the putative biochemical factor added to the organ culture may counteract the intimal proliferation. Alternatively, if ET-1 or any other endothelial-derived compound whose release is controlled by mechanical forces is responsible for the increase in intimal proliferation, then it will be beneficial, perhaps even necessary, to culture any bioartificial blood vessel under the appropriate mechanical environment. This study not only shows that much is still unknown about the regulatory mechanisms that govern tissue remodeling, but it also illustrates the intriguing possibilities of utilizing growth factors and mechanical forces to control the development of engineered tissue.

4 Summary

The experimental work reviewed here clearly demonstrates that both the mechanical and biochemical environments regulate the development and remodeling of engineered tissues. The mechanically induced effects may be the result of convection-enhanced transport or direct mechanical stimulation of the cells. In many cases, mechanically induced stimulation may be mimicked by the application of appropriate biochemical factors. Alternatively, biochemical pathways not stimulated by physical forces but beneficial for tissue development may be initiated by biochemical factors. By judiciously tailoring the mechanical and biochemical environments, the quality and utility of engineered tissues may be substantially improved.

References

1. Duncan, R. L. and Turner, C. H., Mechanotransduction and the functional response of bone to mechanical strain. *Calcified Tissue International*, 1995, **57**, 344–358.

2. Weinbaum, S., Cowin, S. C. and Zeng, Y., A model for the excitation of osteocytes by mechanical loading-induced bone fluid shear stresses. *Journal of Biomechanics*, 1994, **27**(3), 339–360.

3. Reich, K. M. and Frangos, J. A., Effect of flow in prostaglandin E_2 and inositol triphosphate levels in osteoblasts. *American Journal of Physiology*, 1991, **261**, C428–C432.

4. Gooch, K. J. and Tennant, C. J., Mechanical Forces: Their effects on cells and tissues. Springer Verlag, 1997 (in press).

5. Frangos, J. A. (ed), *Physical forces and the mammalian cell*. Academic Press, San Diego, 1993.

6. Davies, P. F., Flow-mediated endothelial mechanotransduction. *Physiological Reviews*, 1995, **75**(3) 519–560.

7. Kim, Y. J., Sah, R. L. Y., Grodzinsky, A. J., Plaas, A. H. K. and Sandy, J. D., Mechanical regulation of cartilage biosynthetic behavior: physical stimuli. *Archives of Biochemistry and Biophysics*, 1994, **311**(1), 1–12.

8. Helminen, H., Jurvelin, J., Kiviranta, I. et al., Joint loading effects on articular cartilage: A historical review, in *Biology and Health of Articular Structures*, eds. H. Helminen, I. Kiviranta and M. Tammi. John Wright, Bristol, 1987, pp. 1–46.

9. Bjelle, A. O., Content and composition of glycosaminoglycans in human knee joint cartilage. Variation with site and age in adults. *Connective Tissue Research*, 1975, **3**, 141–147.

10. Roberts, S., Weightman, B., Urban, J. P. G. and Chappell, D., Mechanical and biochemical properties of human articular cartilage in osteoarthritic femoral heads and in autopsy specimens. *Journal of Bone and Joint Surgery [Br]*, 1986, **68**, 278–288.

11. Slowman, S. D. and Brandt, K. D., Composition and glycosaminoglycan metabolism of articular cartilage from habitually loaded and habitually unloaded sites. *Arthritis and Rheumatism*, 1986, 29, 88–94.

12. Weightman, B. and Kempson, G., Load carriage. In *Adult Articular Cartilage*, ed M. A. R. Freeman, Pitman Medical, London, 1979, pp. 293–341.

13. Freed, L. E. and Vunjak-Novakovic, G. In *The Biomedical Engineering Handbook*, ed J. D. Bronzino. CRC Press, Boca Raton, 1995, pp. 1788–1806.

14. Vunjak-Novakovic, G., Freed, L. E., Biron, R. J. and Langer, R., Effects of mixing on the composition and morphology of tissue-engineered cartilage. *AIChE Journal*, 1996, **42**(3), 850–860.

15. Freed, L. E. and Vunjak-Novakovic, G., Cultivation of cell-polymer tissue constructs in simulated microgravity. *Biotechnology and Bioengineering*, 1995, **46**, 306–313.

16. Zimber, M. P., Tong, B., Dunkelman, N., Pavelec, R., Grande, D., New, L. and Purchio, A. F., TGF-β promotes the growth of bovine chondrocytes in

monolayer culture and the formation of cartilage tissue on three-dimensional scaffolds. *Tissue Engineering*, 1995, **1**, 289–300.

17. Meyer, G. H., Die Architektur der Spongiosa. *Arch. Anat. Physiol. Wiss. Med.*, 1867, **34**, 615–628.

18. Wolff, J., *Das Gesetz der Transformation der Knochen*, Hirschwald, Berlin, 1892.

19. Hillsley, M. V. and Frangos, J. A., Review: Bone tissue engineering: the role of interstitial fluid flow. *Biotechnology and Bioengineering*, 1994, **43**, 573–581.

20. Crane, G. M., Ishaug, S. L. and Mikos, A. G., Bone tissue engineering. *Nature Medicine*, 1995, **1**, 1322–1324.

21. Berthiaume, F. and Frangos, J. A., Effect of flow on anchorage-dependent mammalian cells-secreted products. In *Physical forces and the mammalian cell*, ed JA Frangos, Academic Press, San Diego, 1993, pp. 139–192.

22. Patrick, C. W. and McIntire, L. V., Shear stress and cyclic strain modulation of gene expression in vascular endothelial cells. *Blood Purification*, 1995, **13**, 112–124.

23. Owens, G. K., Regulation of differentiation of vascular smooth muscle cells. *Physiological Reviews*, 1995, **75**, 487–517.

24. Otto, M. J. and Ballermann, B. J., Shear stress-conditioned, endothelial cell seeded vascular grafts: Improved cell adherence in response to in vitro shear stress. *Surgery*, 1995, **117**(3), 334–339.

25. Weinberg, C. B. and Bell, E., A blood vessel model constructed from collagen and cultured vascular cells. *Science*, 1986, **231**, 397–400.

26. Stadler, E., Campbell, J. H. and Campbell, G. R., Do cultured vascular smooth muscle cells resemble those of the artery wall? If not, why not?. *Journal of Cardiovascular Pharmacology*, 1989, **14**(6), S1–S8.

27. Kanda, K. and Matsuda, T., Mechanical stress-induced orientation and ultra-structural change of smooth muscle cells cultured in three dimensional collagen lattices. *Cell Transplantation*, 1994, 3(6), 481–492.

28. Zeltinger, J., Alexander, H. G., Grayson, P., Kidd, I. D., Landeen, L. K., Peterson, A. and Sibanda, B, Tissue engineered vascular grafts. *Keystone Symposium on Tissue Engineering*, 1996.

29. Ziegler, T. and Nerem, R. M., Tissue engineering a blood vessel: Regulation of vascular biology by mechanical stresses. *Journal of Cellular Biochemistry*, 1994, **56**, 204–209.

30. Heath, J. K., *Growth Factors*, Oxford University Press, New York, 1993.

31. McKay, I. A., Types of growth factor activity: Detection and characterization of new growth factor activities. In *Growth Factors — A Practical Approach*, eds. I. A. McKay, and I. Leigh, Oxford University Press, New York, 1993, pp. 1–11.

32. Sporn, M. B. and Roberts, A. B. eds., *Peptide Growth Factors and Their Receptors I, II*, Springer-Verlag, New York, 1991.

33. LeRoith, D., Insulin-like growth factor receptors and binding proteins. *Bailliere's Clinical Endocrinology and Metabolism*, 1996, 10, 49–73.

34. Brown, K. D., The epidermal growth factor: Transforming growth factor-α family and their receptors. *European Journal of Gastroenterology and Hepatology*, 1995, **7**, 914–922.

35. Hart, K. C., Galvin, B. D. and Donoghue, D. J., Structure and function of the platelet-derived growth factor family and their receptors. *Genet. Eng.*, 1995, **17**, 181–208.

36. Vukicevic, S., Stavljenic, A. and Pecina, M., Discovery and clinical applications of bone morphogenetic proteins. *European Journal of Clinical Chemistry and Clinical Biochemistry*, 1995, **33**, 661–671.

37. Wilkie, A. O., Morriss-Kay, G. M., Jones, E. Y. and Heath, J. K., Functions of fibroblast growth factors and their receptors. *Current Biology*, 1995, **5**, 500–507.

38. Fernig, D. G. and Gallagher, J. T., Fibroblast growth factors and their receptors: an information network controlling tissue growth. *Progress in Growth Factor Research*, 1994, **5**, 353–377.

39. Ip, N. Y. and Yancopoulos, G. D., Neurotrophic factor receptors: Just like other growth factor and cytokine receptors? *Current Opinion in Neurobiology*, 1994, **4**, 400–405.

40. Neufeld, G., Tessler, S., Gitay-Goren, H., Cohen, T. and Levi, B. Z., Vascular endothelial growth factor and its receptors. *Progress in Growth Factor Research*, 1994, **5**, 89–97.

41. Malarkey, K., Belham, C. M., Paul, A., Graham, A., McLees, A., Scott, P. H. and Plevin, R., The regulation of tyrosine kinase signalling pathways by growth factor and G-protein-coupled receptors. *Biochemical Journal*, 1995, **309**, 361–375.

42. Langer, R. and Vacanti, J. P., Tissue engineering. *Science*, 1993, **260**, 920–926.

43. Reddi, A. H., Symbiosis of biotechnology and biomaterials: Applications in tissue engineering of bone and cartilage. *Journal of Cellular Biochemistry*, 1994 **56**, 192–195.

44. Saltzman, W. M., Growth-factor delivery in tissue engineering. *MRS Bull.*, Nov. 1996, 62–65.

45. Pfeilschifter, J., Bonewald, L. and Mundy, G. R., Role of growth factors in cartilage and bone metabolism. In *Peptide Growth Factors and Their Receptors II*, eds. M. B. Sporn and A. B. Roberts. Springer-Verlag, New York, 1991, pp. 371–400.

46. Seyedin, S. M. and Rosen, D. M., Cartialge growth and differentiation. In *Cartilage: Molecular Aspects*, eds. B. Hall and S. Newman. CRC Press, Boca Raton, 1991, pp. 131–151.

47. Hill, D. J. and Logan, A., Peptide growth factors and their interactions during chondrogenesis. *Progress in Growth Factor Research*, 1992, 4, 45–68.

48. Malemud, C. J., The role of growth factors in cartilage metabolism. *Rheumatic Disease Clinics of North America*, 1993, **19**, 569–580.

49. Trippel, S. B., Growth factor action on articular cartilage. *Journal of Rheumatology*, 1995, **22**, 129–132.

50. Luyten, F. P., Hascall, V. C., Nissley, S. P., Morales, T. I. and Reddi, A. H., Insulin-like growth factors maintain steady-state metabolism of proteoglycans in bovine articular cartilage explants. *Archives of Biochemistry and Biophysics*, 1988, **267**, 416–425.

51. Curtis, A. J., Ng, C. K., Handley, C. J. and Robinson, H. C., Effect of insulin-like growth factor-I on the synthesis of link protein and hyaluronan in explant cultures of articular cartilage. *Biochimica et Biophysica Acta*, 1992, **1135**, 309–317.

52. Sah, R. L., Chen, A. C., Grodzinsky, A. J. and Trippel, S. B., Differential effects of bFGF and IGF-1 on matrix metabolism in calf and adult bovine cartilage explants. *Archives of Biochemistry and Biophysics*, 1994, **308**, 137–147.

53. Sah, R. L., Trippel, S. B. and Grodzinsky, A. J., Differential effects of serum, insulin-like growth factor-1, and fibroblast growth factor-2 on the maintainance of cartilage physical properties during long-term culture. *Journal of Orthopaedic Research*, 1996, **14**, 44–52.

54. Morales, T. I., Transforming growth factor-β1 stimulates synthesis of proteoglycan aggregates in calf articular cartilage organ cultures. *Archives of Biochemistry and Biophysics*, 1991, **286**, 99–106.

55. van der Kraan, P. M., Vitters, E. L. and van den Berg, W. B., Inhibition of proteoglycan synthesis by transforming growth factor-β in anatomically intact articular cartilage of murine patellae. *Annals of the Rheumatic Diseases*, 1992, **51**, 634–647.

56. Schafer, S. J., Luyten, F. P., Yanagishita, M. and Reddi, A. H., Proteoglycan metabolism is age related and modulated by isoforms of platelet-derived growth factor in bovine articular cartilage explant cultures. *Archives of Biochemistry and Biophysics*, 1993, **302**, 431–438.

57. Luyten, F. P., Yu, Y. M., Yanagishita, M., Vukicevic, S., Hammonds, G. and Reddi, A. H., Natural bovine osteogenin and recombinant human bone morphogenetic protein-2B are equipotent in the maintenance of proteoglycans in bovine articular cartilage explant cultures. *Journal of Biological Chemistry*, 1992, **267**, 3691–3695.

58. Bonassar, L. J., Paguio, C. G., Frank, E. H., Jeffries, K. A., Moore, V. L., Lark, M. W., Caldwell, C. G., Hagmann, W. K. and Grodzinsky, A. J., Effects of matrix metalloproteinases on cartilage biophysical properties in vitro and in vivo. *Annals of the New York Academy Sciences*, 1994, **732**, 439–443.

59. Yeh, L. A., Augustine, A. J., Lee, P., Riviere, L. R. and Sheldon, A., Interleukin-4, an inhibitor of cartilage breakdown in bovine articular cartilage explants. *Journal of Rheumatology*, 1995, **22**, 1740–1746.

60. Fujisato, T., Sajiki, T., Liu, Q. and Ikada, Y., Effect of basic fibroblast growth factor on cartilage regeneration in chondrocyte-seeded collagen sponge scaffold. *Biomaterials*, 1996, 17, 155–162.

61. Cuevas, P., Burgos, J. and Baird, A., Basic fibroblast growth factor (FGF) promotes cartilage repair in vivo. *Biochemical and Biophysics Research Communications*, 1988, **156**, 611–618.

62. Kato, Y. and Iwamoto, M., Fibroblast growth factor is an inhibitor of chondrocyte terminal differentiation. *Journal of Biological Chemistry*, 1990, **265**, 5903–5909.

63. Blunk, T., Sieminski, A. L., Nahir, M., Freed, L. E., Vunjak-Novakovic, G. and Langer R., Insulin-like growth factor-1 (IGF-1) improves tissue engineering of cartilage in vitro. *Keystone Symposium on Bone, Cartilage and Collagen: Growth and Differentiation*, 1997, 19.

64. Baylink, D. J., Finkelman, R. D. and Mohan, S., Growth factors to stimulate bone formation. *Journal of Bone and Mineral Research*, 1993, **8**, Suppl. 2, S565–S572.

65. Price, J. S., Oyajobi, B. O. and Russell, R. G. G., The cell biology of bone growth. *European Journal of Clinical Nutrition*, 1994, **48**, Suppl. 1, S131–S149.

66. Lian, J. B. and Stein, G. S., Development of the osteoblast phenotype: molecular mechanisms mediating osteoblast growth and differentiation. *Iowa Orthopaedics Journal*, 1995, 15, 118–140.

67. Mundy GR, Local control of bone formation by osteoblasts. *Clinical Orthopaedics and Related Research*, 1995, **313**, 19–26.

68. Mundy, G. R., Regulation of bone formation by bone morphogenetic proteins and other growth factors. *Clinical Orthopaedics and Related Research*, 1996, **323**, 24–28.

69. Lind, M., Growth factors: Possible new clinical tools. *Acta Orthopaedics Scandinavica*, 1996, **67**, 407–417.

70. Lean, J. M., Jagger, C. J., Chambers, T. J. and Chow, J. W. M., Increased insulin-like growth factor I mRNA expression in rat osteocytes in response to mechanical stimulation. *American Journal of Physiology*, 1995, **268**, E318-E327.

71. Rawlinson, S. C. F., Mohan, S., Baylink, D. J. and Lanyon, L. E., Exogenous prostacyclin, but not prostaglandin E$_2$, produces similar responses in both G6PD activity and RNA production as mechanical loading, and increases IGF-II release, in adult cancellous bone in culture. *Calcif. Tissue Int.*, 1993, **53**, 324–329.

72. Boyan, B. D., Hummert, T. W., Dean, D. D. and Schwartz, Z., Role of material surfaces in regulating bone and cartilage cell response. *Biomaterials*, 1996, **17**, 137–146.

73. Yasko, A. W., Lane, J. M., Fellinger, E. J., Rosen, V., Wozney, J. M. and Wang, E. A., The healing of segmental bone defects induced by recombinant human bone morphogenetic protein (rhBMP-2): a radiographic, histological, and biomechanical study in rats. *Journal of Bone and Joint Surgery*, 1992, **74A**, 659–670.

74. Yaszemski, M. J., Payne, R. G., Hayes, W. C., Langer, R. and Mikos, A. G., Evolution of bone transplantation: molecular, cellular and tissue strategies to engineer human bone. *Biomaterials* , 1996, **17**, 175–185.

75. Wang, J. S., Basic fibroblast growth factor for stimulation of bone formation in osteoinductive or conductive implants. *Acta Orthopaedica Scandinavia Supplement*, 1996, **269**, 1–33.

76. Johnson, E. E., Urist, M. R. and Finerman, G., Distal metaphyseal tibial nonunion: deformity and bone loss treated by open reduction, internal fixation, and human bone morphogenetic protein (hBMP). *Clinical Orthopaedics and Related Research*, 1990, **250**, 234–240.

77. Lo, H., Kadiyala, S., Guggino, S. E. and Leong, K. W., Poly(L-lactic acid) foams with cell seeding and controlled-release capacity. *Journal of Biomedical Materials Research*, 1996 **30**, 475–484.

78. Brekke, J. H., A rationale for delivery of osteoinductive proteins. *Tissue Engineering*, 1996, **2**, 97–114.

79. D'Amore, P. A. and Smith, S. R., Growth factor effects on cells of the vascular wall: a survey. *Growth Factors*, 1993, **8**, 61–75.

80. Moses, M. A., Klagsbrun, M. and Shing, Y., The role of growth factors in vascular cell development and differentiation. *International Review of Cytology*, 1995, **161**, 1–48.

81. Noll, G., Wenzel, R. R. and Luescher, T. F., Endothelin and endothelin

antagonists: potential role in cardiovascular and renal disease. *Molecular and Cellular Biochemistry*, 1996, **157**, 259–267.

82. Koo, E. W. Y. and Gotlieb, A. I., Endothelial stimulation of intimal cell proliferation in a porcine aortic organ culture. *American Journal of Pathology*, 1989, **134**, 497–503.

83. Kuchan, M. J., Frangos, J. A., Shear stress regulates endothelin-1 release via protein kinase C and cGMP in cultured endothelial cells. *American Journal of Physiology*, 1993, **264**, H150–H156.

Polymer Synthesis

AMARPREET S. SAWHNEY
Focal, Inc., Lexington, MA 02173, USA

and

PAUL D. DRUMHELLER
Gore Hybrid Technologies, Inc., Flagstaff, AZ 86003, USA

1 Introduction

Polymeric biomaterials play numerous critical roles in tissue engineering. In applications of wound healing or tissue regeneration, polymeric biomaterials are used to direct cell ingrowth into specific architectures, or to protect regenerating tissues by acting as a barrier to the undesired growth of scarring tissues or to other phenomena. In combination with the release of bioactive substances immobilized into the bulk or onto the surface, polymeric biomaterials have the ability to influence the physiological regenerative process itself. Finally, in combination with cell culture technology, they become scaffolds or substrates for the growth and differentiation of tissues into functional and therapeutic neo-organs.

Historically, biomaterials were obtained from natural sources, such as purified collagen, gelatin, silk, or cotton. Advances in polymer chemistry supplemented these natural polymers with the first generation medical polymers. The early use of synthetic biomedical polymers, such as polyethylene, polytetrafluoroethylene, polymethacrylates, polyethyleneterephthalate, polyamides, polyurethanes, and silicone rubbers, came about more from adaptations of these commodity polymers to medical applications, rather than from deliberate attempts at rational material design and synthesis. Such synthetic polymers have been used successfully in tissue engineering as tissue regeneration scaffolds; however they are not bioerodible, and the non-absorbed polymer remains an integral part of the tissue. While not necessarily inappropriate for the successful implantation of a tissue engineered product, the desire for polymeric biomaterials that degrade to leave behind only the biological component has been realized in the synthesis of bioerodible materials. Hydrolyzable synthetic polymers were first recognized as potential biomaterials in the 1960's; since then, the poly(α-hydroxy acid) family of biomaterials has been the subject of extensive study. Classical materials of natural origin, such as collagen and hyaluronic acid, still enjoy numerous applications as implantable biomaterials. They have been used in various tissue engineering scaffolds, especially the *ex vivo* regeneration of skin by seeding collagen substrates.

The market for tissue engineering is still in its infancy, but demand is growing at a tremendous rate. Tissue engineered products can be broadly classified as a subcategory of somatic cell/gene therapy, therapies which include combination devices of polymeric biomaterials with mammalian cells for solid organ production, and polymeric scaffolds for tissue repair in periodontal, reconstructive, and regenerative surgery. It has been estimated that the revenues for the cell therapy market alone could exceed $6 billion [1]. In order for the technology to fulfill these expectations, a firm knowledge of biomedical polymer synthesis is needed by those practicing in the field to continue discovery and invention of tissue engineered devices.

The object of this chapter is to present the genre of absorbable and non-absorbable polymers synthesized and used in tissue engineering applications, and to present briefly those next-generation polymers which have potential for use. Often these materials, of either synthetic or biological origin, undergo substantial synthetic modification to the bulk or surface prior to service. The aim of this chapter is not to present the manufacturing steps for these commodity polymers, but rather to illustrate the current state of the art of polymeric biomaterials finding use in tissue engineering applications.

2 Modification of natural polymers

2.1 Proteins

Protein-based polymers, such as collagen, albumin, silk, and gelatin, have been chemically modified to affect mechanical strength or degradation rates. In the simplest manifestation, these proteins have been crosslinked with chemical agents, such as formaldehyde or glutaraldehyde, to form thermoset materials that can be molded into various shapes. Methods have been described that allow for altering the porosity and texture of the crosslinked product to tailor their degradation rate [2]. These crosslinked protein polymers have found many applications, including the lamination product with silicone rubber useful for synthetic skin [3].

In order to prolong the stability of protein-based tissue augmentation materials, their reaction with other crosslinking agents, such as end-activated poly(ethylene glycol) (PEG), has been proposed [4]. The PEG has be functionalized by any one of a variety of techniques that are well known to the art [5] to form activated PEG's that can react with nucleophilic residues along the protein backbone to produce the crosslinked material. This technique has been used to prepare protein polymer material formulations, including crosslinked albumin [6] and crosslinked collagen [4]. Other bioactive species, such as growth factors and cytokines, can be admixed with the protein solution at the time of crosslinking to enable their incorporation into the implant. Upon degradation of the implant, the bioactive species is subsequently released to effect the desired pharmacological response. Proteins, collagen in particular, have been treated with acylating agents, such as acid chlorides, sulfonyl chlorides and anhydrides, to form polymeric substituted adducts that contain ethylenically unsaturated groups which can be polymerized by ultraviolet illumination to form moldable thermoset articles, such as ocular implants [7].

Using genetic engineering and fermentation techniques, silk-like protein polymers expressing selective bioactive peptide sequences have been produced [8]. These materials can be woven to form fibers, used as coatings, or coextruded with plastic resins to produce the final biomaterial article. Additionally, biomimetics based on the repeat peptide unit of elastin have been produced by recombinant techniques [9]. These repeat units can then be crosslinked by gamma irradiation to form crosslinked sheets and hydrogels which can further incorporate cell adhesive peptides or other bioactive species [10].

2.2 Glycosaminoglycans and polysaccharides

Glycosaminoglycans (GAGs), by virtue of their highly hydrated and branched nature, provide the compressive load bearing and shock absorbing component of tissue. GAGs such as hyaluronic acid (HA), chondroitin sulfate, dermatan sulfate, keratan sulfate, and chitin, have been covalently modified to alter mechanical properties and degradation rates. Highly swellable hydrogels have been produced by the covalent crosslinking of these GAGs with PEG [11, 12]. The GAG molecules (first deacetylated or desulfated to provide free amino groups) are reacted with activated PEG to form the hydrogels using techniques similar to those for protein crosslinking. These hydrogels have been used for soft tissue augmentation: in one application, the gels are dried, suspended in a non-aqueous water-miscible vehicle, and injected into the soft tissue of interest [12]; the gel subsequently hydrates with physiologic fluid and can be used to bulk up tissues such as an incompetent sphincter [12]. HA has also been modified by producing interpenetrating polymer networks with other polymers to affect strength and degradation rates [13], or by impregnating HA pastes with polymethacrylate beads to increase putty strength for orthopedic applications [14]. The properties of HA can be modified by esterification of the carboxyl group along the backbone with aliphatic or aromatic alcohols. This modification lowers the water solubility of the HA, so as to make viscous aqueous solutions, or to render the HA soluble in organic solvents. Spun fibers, woven textiles and meshes, films, and shaped articles have been produced from esterified HA [15]. Slow hydrolysis of the ester linkage over time to form the more hydrophilic carboxyl group resolubilizes the HA and promotes bioabsorption.

Acidic polysaccharides, such as carboxymethyl cellulose, carboxymethyl-chitin, and carboxymethylstarch, have been chemically treated using similar techniques. For example, the quaternary ammonium salt of polyanionic polysaccharides has been treated with alkylating or etherifying agents in aprotic solvents, such as dimethyl sulfoxide, to produce chemically unique materials [16]. In a similar manner, a material useful for wound healing and tissue regeneration has been made of sulfated sucrose (via sulfating hydroxyl moieties with sulfur trioxide) [17]. Matsuda *et al.* have synthesized photocurable polysaccharide/gelatin blends based on modification with cinnamate and acrylate groups [18]; photoirradiation forms covalently crosslinked hydrogels which can be used to deliver drugs or to entrap cells.

In other examples, the method to crosslink does not rely upon the formation of covalent bonds, rather they rely upon the coacervation of oppositely charged ionic pairs. Aqueous solutions of sodium alginate, a carbohydrate polymer of mannuronic acid and guluronic acid, can undergo gelation by a complexation mechanism when exposed to divalent cationic species such calcium, magnesium, or barium, or to polycations such as polylysine or spermine. This coacervation phenomenon of sodium alginate has been extensively exploited in the tissue engineering realm, especially for the encapsulation of mammalian cells, and the literature is fertile with reports of chemical modifications or of subtly changed gelation and encapsulation protocols (for example of reviews see [19, 20]). Subsequent crosslinking with a polycation, such as poly(L-lysine), results in the formation of a thin, selectively permeable membrane that aids in the ability of the alginate gel to ultrafilter and be permselective to low molecular weight species. Studies have demonstrated that alginate tissue biocompatibility is enhanced with purified and enriched mannuronic acid content [21]. Combined ionic and covalent crosslinking of natural polymeric biomaterials have also been used to produce hydrogels suitable for cell encapsulation. For example, aqueous blends of sodium alginate/polyvinyl alcohol are crosslinked by the ionic coacervation of the alginate component and the covalent crosslinking of the photosensitized polyvinyl alcohol component [22]. The combination of ionic and covalent crosslinking imparts strength, stability, and the ability to tailor the permselective properties of the blend in manners not readily feasible with a single component system.

3 Synthetic polymers

3.1 Poly(α-hydroxy acids)

The family of poly(α-hydroxyacids), in particular poly(D,L-lactic acid) (also called poly(D,L-lactide) or PLA), poly(glycolic acid) (also known as polyhydroxyacetic acid, poly(glycolide), or PGA), and their copolymers, has been used widely in tissue engineering, in particular as scaffolds for cell seeding and as guides for regenerating soft tissue (such as nerves) or hard tissue (such as bone or cartilage; techniques for fabricating tissue engineering scaffolds are the subject of another chapter in this book). These polymers have an extended history of use in tissue engineering; for example, PLA was used for aortic tissue regeneration [23, 24] and for fixation of fractures [25], and PGA was used to guide bone healing [26] well before tissue engineering was recognized as a discrete discipline. These polymers are also some of the few bioabsorbable materials that have been approved for use in man in various medical devices. A list of medical devices utilizing these materials, in forms that include sutures, clips, staples, pins, fibers, meshes, and drug delivery matrices, has been enumerated by Barrows [27]. One important reason for the widespread use of these polymeric biomaterials is that the constituent units of these polymers are derived from natural metabolites, which generally imparts to them a favorable toxicological profile.

PLA and PGA can be polymerized directly from lactic acid and glycolic acid, respectively, but higher molecular weight polymers are usually obtained by using cyclic monomers such as lactide and glycolide. A ring opening polymerization, typically catalyzed by organometallic catalysts such as stannous octoate, leads to polymerization as illustrated in Fig. 1. Typically, ring opening reactions are carried out from the monomer melt at about 160°C (usually below 200°C) under an inert gas blanket or under vacuum, since air decreases the molecular weight and discolors the polymerization product. Careful drying and purification of the starting materials are needed prior to polymerization, due to competing hydrolysis of monomers from any trace moisture. Ring opening polymerizations can also be performed in the solution phase using solvents such as benzene, toluene, or xylene [28]. Alcohol-containing species initiate the ring opening polymerization through a complexation reaction, and can be used to control the final polymer molecular weight by varying the stochiometry of initiator to monomer. In this fashion, multifunctional hydroxyl-containing compounds, such as glycerol and pentaerythritol, have been used to form branched or multiarmed polymers [29]. In particular, branched polymers exhibit lower viscosities compared to linear polymers of the same molecular weight (due to reduced molar volume) and are less susceptible to melt viscosity degradation. Numerous heavy metal salts or soaps are suitable

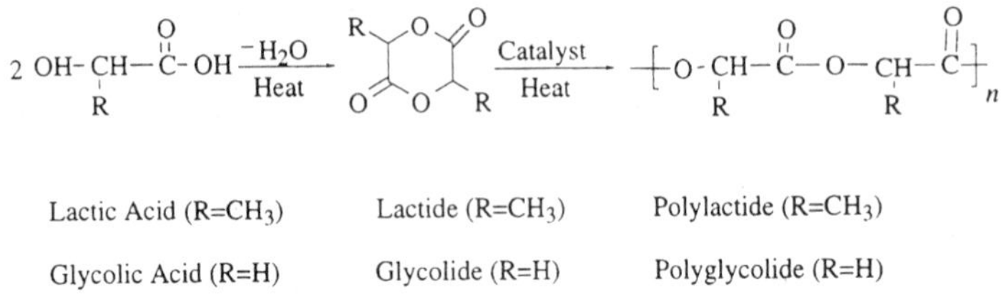

Lactic Acid (R=CH$_3$)   Lactide (R=CH$_3$)   Polylactide (R=CH$_3$)

Glycolic Acid (R=H)   Glycolide (R=H)   Polyglycolide (R=H)

Fig. 1. PLA and PGA polymerization syntheses.

polymerization catalysts, such as antimony trifluoride, stannous chloride, tetraphenyl tin, or zinc oxide; however, stannous octoate at a concentration of 0.02%–0.5% remains the catalyst of choice, primarily due to FDA acceptance (stannous octoate is also approved by the FDA as a food additive).

3.2 Copolymers of poly(α-hydroxy acids)

Copolymerization approaches are considered when the homopolymer does not possess all the required physical and chemical properties by itself for the application at hand. For example, in an attempt to synthesize polymers having a shorter absorption time than PLA and yet having a tough amorphous nature, copolymers of PLA and PGA (PLGA) have been synthesized [30, 31]. The ability to tailor simply the absorption profile makes PLGA polymers popular choices for tissue scaffolding materials. Copolymerization of various α-hydroxy acid monomers has been intensely investigated to alter the mechanical, chemical, or biodegradation properties of the final polymer. Shen *et al.* [32] have synthesized an alternating copolymer of (D,L)-lactic acid and glycolic acid by synthesizing the cyclic monomer (D,L)-3-methyl-1,4-dioxane-2,5-dione, and then carrying out its ring opening polymerization under standard reaction conditions. Copolymers of L-lactide [33] and (D,L)-lactide with ε-caprolactone have also been synthesized [34, 28]. Polymerization rates of comonomer mixtures were noted to be generally lower than the rate of homopolymerization. The dependence of comonomer reactivity ratio upon reaction temperature has been exploited to produce relatively blocky copolymers at lower temperatures but more random copolymers at higher temperatures [33]. In this example, it was observed that more blocky copolymers with longer homopolymeric segments produced polymers with higher modulus and tensile strengths but with lower elongation at break. α-Hydroxy acid monomers have been copolymerized with various other monomers in an effort to reduce the brittleness prevalent in the homopolymer. Poly(ε-caprolactone) (PCL) is a partly crystalline polymer with a very low glass transition temperature. Copolymerization of (D,L)-lactide with ϵ-caprolactone produced polymers which ranged in properties from rigid to elastomeric [35]. Incorporation of the relatively longer aliphatic chains of the ϵ-caprolactone plasticized the hard glassy nature of the PLA and increased its drawability [36]. However, increases in the PCL fraction can produce substantially amorphous PLA/PCL copolymers with very low Tg and which can be gummy and weak at room temperatures. Hence, useful properties are seen when amorphous co-monomers are added in small quantities. Methods to add co-monomers slowly over time to polymerizing glycolide have been described [37]; by adding the co-monomer over time, rather than adding all initially, a block copolymer with increasingly random chain ends is produced [38]. Numerous copolymers of PGA have been prepared since PGA homopolymer is a rigid, crystalline, and brittle material; PGA itself can be formed into braided sutures but the polymer is not suitable for monofilament sutures or large medical devices due to its extreme rigidity. In an effort to enhance the flexibility of PGA products, Casey *et al.* [39] have proposed polymerizing glycolide using polyethylene glycol initiators to form triblock copolymers. Song *et al.* [40] have used similar PGA/PEG/PGA triblock copolymers as initiators for the anionic polymerization of ethylene glycol to form PEG end chains; the high molar content of the hydrophilic PEG produced biodegradable hydrogels. Similar ring opening polymerizations using PEG initiators have been carried out with (D,L)-lactide [41, 42] and L-lactide [43, 44]; multiblock (lactide/glycolide)-co-PEG copolymers have been synthesized by condensation of low molecular weight poly(lactide/glycolide) and PEG precursors [45]. When comparing the rate of degradation of a PLA-PEG copolymer to a PLA homopolymer, the PLA segments in both the homopolymer and the copolymer have been observed to hydrolyze at the same rate, but the presence of the hydrophilic PEG segment promoted PLA water solubility and hence a more rapid onset of overall mass loss for the copolymer [42]. Cohn and Younes [46] have described the synthesis and characterization of PEO/PLA copolymers in detail. Sawhney and Hubbell [47] have synthesized terpolymers of (D,L)-lactide, glycolide, and ϵ-caprolactone, and in an effort to enhance the hydrophilicity of such poly(α-hydroxy acids), they have also initiated α-hydroxy acid monomer polymerization with PEG or triblock copolymers of PEG/polypropylene glycol/PEG (such as the Pluronic series of surfactants manufactured by BASF) [47].

Physical blending, compositing, laminating, and fiber reinforcing are some methods used to engineer desirable properties into a medical device when a single biodegradable material is unable to meet all specifications; a description of these techniques is beyond the scope of this chapter (for examples, see [48, 49]).

3.3 Poly(carbonates) and poly(dioxanones)

Poly(dioxanone) is synthesized by a ring opening reaction of a six membered cyclic lactone monomer under conditions similar to those described for PLA and PGA type of materials [50]. The partly crystalline polymer contains both ether and ester linkages and is suitable for drawing into monofilament fibers which were seen to absorb over a 6 month period [51]. The ability of poly(dioxanone) to retain its strength over a long period of time makes it useful in surgical applications where traditionally nonabsorbable materials were used. The synthesis of poly(dioxanone) is outlined in Fig. 2.

Poly(trimethylene carbonate) by itself is a gummy polymer [52] which when copolymerized with glycolide forms a partly crystalline polymer with useful physical properties [37]. Triblock copolymers of PEG, lactide or glycolide, and trimethylene carbonate have also been synthesized [53]. These materials were useful as coatings for suture materials due to their phase-separated nature that allowed them to swell in the presence of water and provide lubricity. In addition, triblock copolymers of a central segment of (random) poly(lactide/glycolide) with end blocks of trimethylene carbonate have been prepared to form monofilament sutures [54]; the structure of the blocks imparted longer bioabsorption times and greater strength retention compared to random copolymers of poly(glycolide-co-trimethylene carbonate). In a similar fashion, triblock copolymers of a central segment of (random) poly(lactide/dioxanone) with end blocks of glycolide have been synthesized to produce flexible yet strong materials [55]. Random copolymers of L-lactide and polycarbonate have been prepared from dioxanone-containing monomers [56]; these monomers allow the ability to tailor physical and morphological properties by varying the reaction conditions. Triblock copolymers of glycolide/dioxanone/glycolide have been

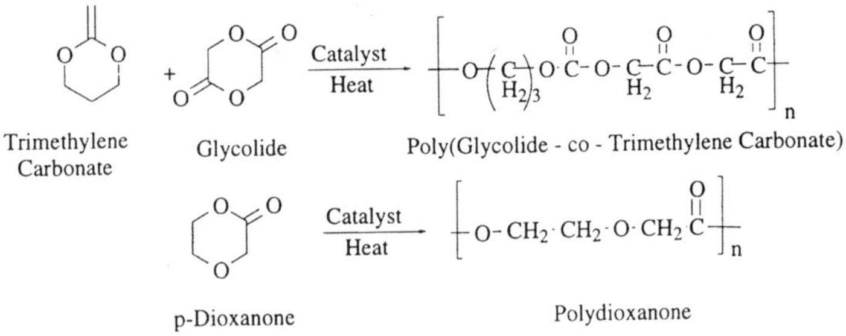

Fig. 2. Synthesis of poly(glycolide-co-trimethylene carbonate) and polydioxanone.

synthesized for spinning into monofilament sutures [57]; the hard/soft/hard segment structure of the polymer produced fibers of good yield strength with good flexibility.

3.4 Poly(phosphazenes)

These polymers are usually prepared in three stages: synthesis of hexachlorocyclotriphosphazene precursors, precursor polymerization to poly(dichlorophosphazene), and substitution of the chlorine atoms along the backbone of this inorganic polymer to form substituted poly(organophosphazenes). Allcock et al. prepared the first soluble and non-crosslinked poly(organozphosphazene) [58], as illustrated in Fig. 3. Preparation of poly(organophosphazenes) from basic chemicals rather than from the synthesis and isolation of monomeric precursors has been reviewed by Potin and Jaeger [59].

A wide range of polyphosphazenes has been synthesized with properties ranging from water solubility to water insolubility, rigid to flexible, crystalline to amorphous, etc. Allcock et al. [60] have produced amorphous and rapidly hydrolyzable poly(phosphazenes) via substitution of glycolic acid ester and lactic acid ester side groups. Photocrosslinkable poly(phosphazenes) that bear chalcone and cinnamate groups have been synthesized which can be crosslinked by ultraviolet irradiation in the absence of exogenous photoinitiators [61]. Poly(phosphazene) that bear phenoxycarboxylate and chalcone groups along the polymer backbone was used to encapsulate cells using a dual crosslinking scheme [61]: the pendant carboxyl groups allowed ionic crosslinking of the polymer in the presence of divalent ions and subsequent photocrosslinking by UV irradiation. The relative transparency of the polymer backbone to UV wavelengths minimized undesirable photoinduced reactions along the backbone.

A variety of nucleophilic agents can replace the chlorine atoms along the polymer backbone to form substituted polyphosphazenes. Moieties bearing hydroxyls or amines are easily substituted along the polymer backbone to produce P–O or P–N linkages. In addition to producing substituted photosensitive or ionic polymer backbones as described above, nucleophilic substitution has been used to immobilize steroids [62], anal-

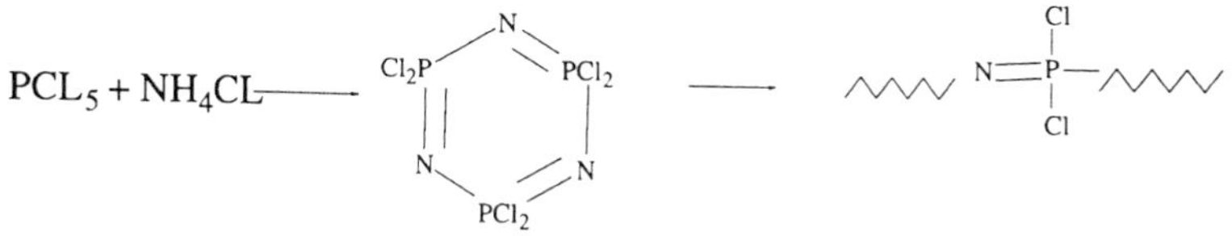

Fig. 3. General synthesis of poly(phosphazenes).

gesics [63], antibiotics [64], and peptides and growth factors [65] to the polyphosphazene backbone. The presence of the alternating P–N double and single bonds gives these materials a unique flexibility [60].

3.5 Poly(anhydrides)

Anhydride bond formation results from the dehydro-acylation of two carboxyl groups. Polyanhydrides are typically prepared by condensation reaction in the solution or melt phase under reduced pressures. Prepolymers can be prepared by reaction of diacids, such as fumaric acid or sebacic acid, with acetic anhydride under reflux conditions [66, 67] as outlined in Fig. 4. The melt polycondensation reaction suffers from reversible thermal depolymerization and thus limits the attainable molecular weight; heat-labile monomers are also difficult to polymerize from the monomer melt. Undesirable polymer cyclization and crosslinking side reactions, especially at high polymerization temperatures, remains unresolved. It has been shown that only aliphatic chain lengths greater than six carbons are stable enough to prevent cyclic anhydride formation [66]. Polyanhydride formation can take place at room temperature through dehydrochlorination by reacting the diacid monomer with an acyl halide, such as acetyl chloride, in the presence of an acid accepting species [68]; however, this reaction is poorly efficient since the carboxylic hydrogen is less reactive than that of more basic groups (such as amines, hydroxyls, or mercaptans). Leong et al. [68] have proposed efficient dehydrative coupling reactions for the high molecular weight synthesis of polyanhydrides. Chiellini et al. have reviewed efforts to synthesize copolymers of anhydrides with glyceric acid, polyethers, and polysaccharides [69].

Hydrolysis and degradation of poly(anhydrides) can be altered by changing the chemical nature of the polymeric backbone; for example, aliphatic polyanhydrides degrade over a few days while aromatic polyanhydrides may require years [70]. Copolymerization of aliphatic and aromatic anhydride moieties produce polymers of intermediate degradation rate. However, since the aliphatic segments hydrolyze faster than the aromatic segments, a gradual increase in the aromaticity of these polymers is seen over time in a physiological environment [70]. In an effort to reduce

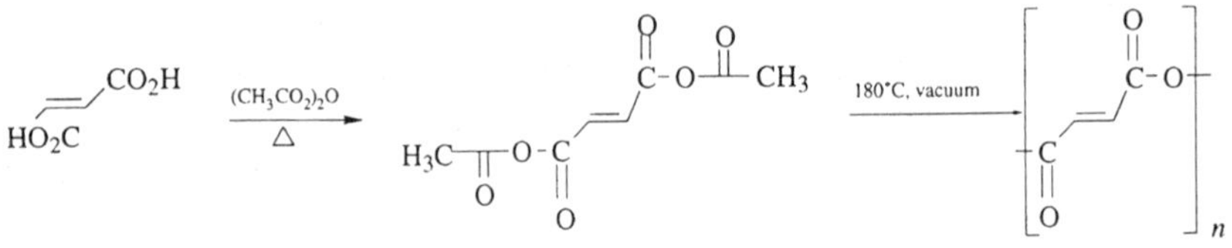

Fig. 4. General synthesis of poly(anhydrides).

increases in polymer aromaticity during degradation, polymers have been synthesized from monomers that incorporate the aliphatic and aromatic moieties into one residue [71]; such polymers show a zero order degradation profile. The stiff mechanical properties and well behaved degradation profile of polyanhydrides have been exploited as drug delivery vehicles/osetoconductive scaffolds for hard tissue regeneration [72].

Highly crystalline polyanhydrides, such as poly(fumaric acid), can be insoluble in most solvents, rendering post-processing a challenge. Solubility properties can be improved by co-polymerization with aliphatic diacids. For example, Yaszemski et al. have synthesized a moldable poly(propylene fumarate)-based material that can be crosslinked in situ via an addition polymerization of N-vinyl pyrrolidone to form an osteoconductive scaffold [73]. Suggs et al. have incorporated poly(ethylene oxide) into the backbone of poly(propylene fumarate) by transesterification to form a solvent-injectable copolymer that can be cured in situ to form shapable implants [74]; transesterification of the poly(propylene fumarate) takes place under elevated temperature and under high vacuum in the presence of antimony trioxide catalyst. In contrast, in an effort to render unsaturated poly(anhydrides) insoluble, Domb et al. have described crosslinkable poly(anhydride) species to form insoluble networks [67].

Albertson and Lundmark reported on a series of thermoplastic degradable elastomers based on block copolymers of polyanhydrides and PEG [75]. These materials exhibited good fiber forming properties and showed an increase in molecular weight and hydrolytic stability with an increase in the aliphatic segment length between acid groups. The surface-eroding properties of hydrophobic polyanhydrides makes them uniquely suited for the controlled release of therapeutic molecules.

3.6 Other polymers

Few bioabsorbable polymers undergo surface erosion absorption. Poly(orthoesters) are one such class of materials that undergoes surface hydrolysis due to its generally hydrophobic nature to yield a gradual degradation profile [76]. However, poly(orthoesters) have low modulus and mechanical strength, and are limited to low load-bearing applications such as drug delivery matrices.

Leong et al. have reported on biodegradable polymers that have a phosphoryl-containing backbone [77]. These poly(phosphoesters) have been synthesized using an interfacial polycondensation reaction that is illustrated in Fig. 5. Porous degradable foams have been developed from these materials and studied as drug delivery matrices and bone regeneration scaffolds [77].

Poly(phosphoester-urethanes) have been synthesized [77] by preparing soft segments of biodegradable bis(2-hydroxyethyl phosphite) and po-

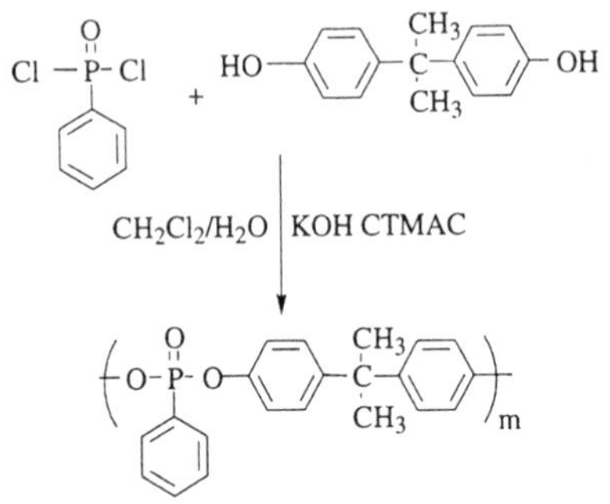

Fig. 5. Synthesis of poly(phosphoesters) (from Leong *et al.* [77]).

lyether, such as poly(ethylene oxide) or poly(tetramethylene oxide), and by preparing hard segments of diisocyanates derived from 1,4-diaminobutane or lysine. A standard two step polymerization scheme resulted in the formation of poly(phosphoester urethanes).

Elastomeric segmented block copolymers of poly(ethylene oxide) and poly(butylene terepthalate) (PEO–PBT) have been synthesized with varying ratios of soft (PEO) and hard (PBT) segments [43, 78]. These materials are sold commercially under the tradename of Polyactive (HC Implants BV, Leiden, Netherlands); these polymers are biodegradable, presumably by the hydrolysis of the ester–ether linkage. PEO–PBT materials have been used for the reconstruction of the tympanic membrane [78], as a ventilation tube [78], and as a skin regeneration template [79].

Tyrosine-derived polycarbonates are high-strength bioabsorbable materials that have potential for use in orthopedic applications [80]. The length of the pendant alkyl side chain can be modified to tailor the physicochemical and mechanical characteristics of the materials. Block copolymers of PEG-co-(tyrosine carbonate) have been prepared from the acylation of end activated PEG upon poly(tyrosine ethyl esters) [81]; the incorporation of the PEG soft segment promoted material flexibility and enhanced *in vitro* degradation rates.

4 Hydrogels

4.1 Poly(ethylene glycol) hydrogels

Poly(ethylene glycol) (PEG)-based materials have been widely used in the medical device, pharmaceutical, and cosmetic industries. PEGs are approved by the FDA for use in foodstuffs and in pharmaceutical products via topical, oral, and parenteral use. Due to their wide acceptance and favorable toxicology, PEGs have formed the basis for the synthesis of a number of crosslinked hydrogels that can have use in tissue engineering applications.

Radiation crosslinked hydrogels based on star shaped poly(ethylene oxide) (used interchangeably with PEG) have been described by Merrill [82]. These gels possess several hydroxyl end groups per divinyl benzene core and can be used to further attach bioactive moieties. PEG chains were activated on both ends using p-nitrophenyl-chloroformate and crosslinked with albumin molecules to form a highly swollen hydrogel [83].

Sawhney et al. [84] have synthesized water soluble macromers based on water soluble PEG chains extended with oligomerized bioabsorbable moieties, such as oligo(D,L-lactic acid), and end-capped with polymerizable groups (Fig. 6). These macromers can be rapidly photopolymerized using long wave ultraviolet or visible light [85], and they present a versatile three-part molecular structure that can be tailored to suit particular applications. For example, increasing the molecular weight of the PEG central segment changes the swollen hydrogels from rigid to highly flexible materials. The identity of the bioabsorbable segment, say glycolate or lactate or trimethylene carbonate, dictates the bioabsorption times of these hydrogels, which can be varied from days, weeks, or months respectively. Elimination of the bioabsorbable segment leads to the formation of hydrogels that do not degrade substantially in a physiological environment [86]. Such tri-block macromers have hydrophobic end groups with hydrophilic central segments; thus they are surface active and form micellar emulsions in aqueous environments which impart rapid rates of polymerization. The viscosity of the starting macromer solution can be raised by increasing the chain length of the bioresorbable comonomer segment as a result of increased micelle aggregation and increased intrinsic viscosity. Since these photopolymerizable hydrogels cure in situ in physiological environments, they are capable of adhering to tissue substrates strongly; this property has been exploited in the use of these materials as tissue sealants [87], as barriers to scar tissue formation [88], and as drug delivery reservoirs [89]. Non-bioabsorbable PEG hydrogels have been used to form immunoprotective barriers around islets of Langerhans [86] using an 'interfacial' photopolymerization technique; using such methods, the thickness of the hydrogel barriers has been accurately controlled [90] to minimize diffusional distances of nutrients and therapeutic agents.

4.2 Other polymeric hydrogels

Other water-swellable polymers have been utilized as polymeric biomedical devices (for example of a review, see [91]), the most commercially successful example being the crosslinked hydroxyethyl methacrylate

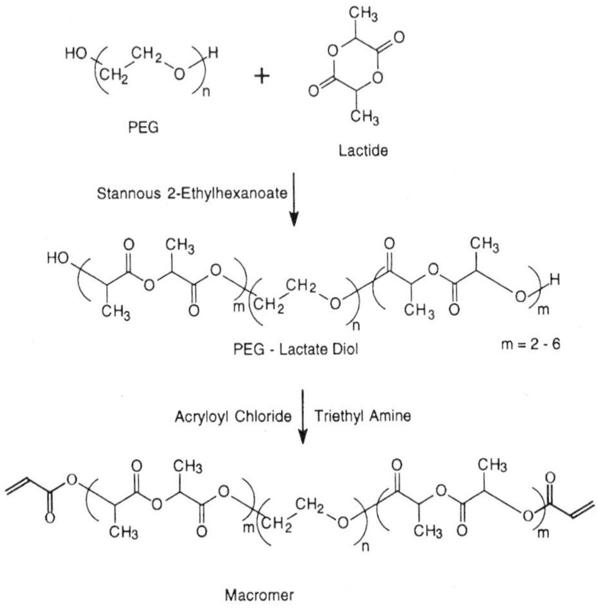

Fig. 6. Macromer synthesis for polymerization of bioabsorbable PEG hydrogels (from Sawhney *et al.* [84]).

(PHEMA) contact lens. Hydrogels of PHEMA have been investigated as intraocular implants, as their inflammatory potential is quite low [92].

Hydrogels of polyvinyl alcohol (PVA) have been used as cervical dilators [93], resorbable surgical sponges [94], drug delivery reservoirs [95], and orthopedic stabilization splints [96]; its use as a blood contacting material has also been investigated [97]. PVA is prepared from the alkaline hydrolysis of polyvinyl acetate; thus the backbone of PVA consists of a 1,3-glycol repeat unit. USP-grade PVA is 85–89% hydrolyzed; the presence of trace acetate groups disrupts hydrogen bonding between adjacent alcohols and allows the polymer to be cold-water soluble. Crosslinking of PVA is effected using physical methods such as freeze-thawing [98] or chemical methods such as acetalization with dialdehydes. Solutions of PVA/PEG blends were crosslinked via complexation with sodium sulfate/potassium hydroxide to produce membranes suitable for cell encapsulation [99]. PVA has also been functionalized using styrylpyridinium groups to form a photocrosslinkable material for the encapsulation of islets of langerhans [100].

Block copolymers of poly(acrylonitrile-co-acrylamide) have been synthesized from polyacrylonitrile homopolymer by the controlled acid hydrolysis of the nitrile group to the amide [101]. In physiological solutions, the hydrophilic acrylamide segment becomes solvated while the hydrophobic acrylonitrile segment undergoes cyano-complexation to form extremely strong, physically crosslinked hydrogels. These hydrogels have been used for surgical augmentation to bulk up tissues, such as vas deferens occlusion [102], and have been proposed for drug delivery [103].

Polyacrylamide hydrogels have found limited use as implantable biomaterials, due to the toxicity of the acrylamide monomer. However, the *in vivo* tissue biocompatibility of the purified crosslinked polymer is quite good [104]. Due to its hydrogel nature, poly(acrylamide-co-methacrylic acid) copolymers have been examined as a possible blood contacting material [105]. N-(2-hydroxypropyl)-methacrylamide hydrogels were polymerized around neuronal cells and were seen to have good biocompatibility and were well tolerated upon implantation in rats [106]. Matsuda *et al.* [107] have used a photosensitive water-soluble acrylamide copolymer partially derivatized with triphenylmethane leucohydroxide which becomes cationically charged on irradiation; this crosslinked hydrogel has been proposed as a potential vehicle for anitsense DNA delivery [107]. A copolymer composed of N-isopropyl acrylamide units and a vinyl monomer which has a cell-adhesion mediating peptide in the side chain, has been synthesized as a thermoresponsive and cell adhesive material. This material was found to induce platelet aggregation on mixing with platelet rich plasma [108].

5 Polymer modification with active fuctional groups

It is well established in the literature that numerous adhesion-dependent cell types exhibit increased longevity and increased cellular function when cultured on extracellular matrix proteins, such as fibronectin and laminin, rather than on virgin plastic substrates. Hence, it is useful to immobilize bioactive molecules or biomimetic species onto polymeric substrate surfaces (reviewed in [109]). Many biomedical polymers possess reactive functional groups that allow the conjugation of bioactive species such as peptides, growth factors, enzymes, proteins, or carbohydrates (reviewed in [110]). Attempts to include functional groups along the backbone of poly(α-hydroxy acids) have been realized by synthesizing cyclic monomers that incorporate lysine/lactic acid [111, 112], aspartic acid/lactic acid [113] or glycine/glycolic acid [114]. The presence of functional groups in the polymeric backbone allows chemical modification of these materials to change their characteristics at the molecular level.

Other techniques have been used to functionalize several commonly used biomedical polymers, such as sodium hydroxide etching (for example, [115]), radiation-induced grafting (for example, [116]), and plasma and ion implantation (reviewed in [117]). Techniques to modify the functional groups that the polymers present to the surrounding environment also include radio frequency glow discharge treatment, protein passivation, and pyrrolitic carbon treatment (reviewed in [118]). Lin *et al.* have functionalized polyurethanes by introducing carbamate salts into the polymer backbone and coupling to the carboxyl end group of a peptide sequence [119]. A convenient photoimmobilization scheme has been described [120] wherein the peptide sequence of interest is coupled to a benzophenone or aryl azide containing moiety and then activated using ultraviolet light on a substrate of interest; the free radical produced during photoactivation abstracts hydrogen from a variety of polymer backbones and forms a covalent attachment. Cell adhesive peptides have been immobilized onto hyaluronic acid materials by the epoxide activation of HA side groups [121]; such a strategy allows for initial cell attachment onto scaffolds composed of such materials, leaving behind only the neotissue as the scaffold and its immobilized peptides degrade.

Covalent attachment of synthetic peptides to substrates has been shown to promote cell adhesion to non-biological substrates such as poly(ethylene terepthalate) and poly(tetrafluoroethylene) [122, 123]. Control of cell adhesion using carbohydrate moieties has also been demonstrated by Kobayashi *et al*: N-acetylactosamine monomer when polymerized demonstrated selective adherence of mammalian hepatocytes known to possess an asialoglycoprotein receptor [114]. Cell adhesive peptides have also been immobilized on PVA hydrogel films using an isocyanate based coupling [124]. Activated cell adhesive peptide sequences have been synthesized using a 4-azidobenxoyloxysuccinamide-based activation of the N-terminal of the peptide sequence [125]: the phenyl azido is easily photolyzed and reacts with surrounding substrates in a non-specific fashion to immobilize the peptide. Immobilization of the cell adhesion signal, such as peptides, carbohydrates, or other bioactive moieties, is probably only part of the process for selectively engineering cellular attachment or tissue regeneration. The complexity of the *in vivo* surroundings can often obscure the selectivity of the biomaterial background. In an attempt to overcome this shortcoming, Drumheller and Hubbell [126] have immobilized PEG into densely crosslinked resins to provide a highly cell adhesion resistant surface that can be rendered selectively cell adhesive by the immobilization of peptide sequences.

6 Conclusions

Considerable progress has been made in the modification of natural polymers and in the preparation of synthetic polymers, and especially in the understanding of their interaction with the biological environment. Increased understanding of cellular mechanisms from molecular biology is in the process of heralding a new era of 'active' biomaterials that go beyond the traditional roles of mechanical support. However, one needs to be careful in extrapolating the understanding of cellular mechanisms from *in vitro* tissue culture experiments to the *in vivo* environment. Tissues in and of themselves are complex composite structures. Most tissue engineering applications have currently relied on either inherent self assembly of cell populations or on a non-specific diffusion of bioactive species for tissue regeneration. Polymeric materials that can selectively guide specific cell phenotypes and support their proper differentiation into functional tissues are needed to meet the challenge of engineered multicellular tissue architectures. One promising area of progress is self assembling macromolecules which have the ability to form highly ordered structure. Self assembly can be initiated by interactions between charged groups, hydrogen bonding and dehydration effects, and through structural stability that arises from an organized structure [127]. Such advances may help generate new levels of organization of tissues within biomaterials. As stated above, most synthetic bioabsorbable polymers degrade by a hydrolytic mechanism. The ability to engineer materials to absorb through cellularly mediated processes would be a significant advance. The field of tissue engineering has fared well with conventional biomaterials to date, but significant future progress can only be made through a multidisciplinary understanding of the issues involved and their solution through creative polymeric biomaterial synthesis.

References

1. World Cell Therapy Markets. Marketing Report No. 5413–43, Frost and Sullivan, San Francisco, 1997.
2. Yannas, I.V., Lee, E. and Ferdman, A., US Patent No. 4,947,840, 1990.
3. Yannis, I.V., Burke, J.F., Gordon, P.L. and Huang, C., US Patent No. 4,060,081, 1977.
4. Rhee, W., Wallace, D.G, Michaels, A.S., Burns Jr, R.A., Fries, L., DeLustro F., and Bentz, H., US Patent No. 5,550,188, 1996.
5. Zalipsky, S. and Lee, C., Use of functionalized poly(ethylene glycols)s for modification of polypeptides. In *Poly(Ethylene Glycol) Chemistry*, ed. J.M. Harris. Plenum Press, New York, 1992, pp. 347–370.
6. Weissleder, R. and Bogdanov, A., US Patent No. 5,514,379, 1996.
7. Kelman, C.D. and DeVore, D.P., US Patent No. 5,480,427, 1996.
8. Anderson, J.P., Cappello, J. and Martin, D.C., Morphology and primary crystal structure of a silk-like protein polymer synthesized by genetically engineered escherichia coli bacteria. *Biopolymers*, 1994, **34**, 1049 1058.
9. McPherson, D.T., Morrow, C., Minehan, D.S., Wu, J.G., Hunter, E. and Urry, D.W., Production and purification of a recombinant elastomeric polypeptide G(VPGVG)^{19}VPGV from *Escherichia coli*. *Biotechnology Progress*, 1992, **8**, 347–352.

10. Radzilowski, L.H. and Strupp, S.I., Nanophase separation in monodisperse rodcoil diblock polymers, *Macromolecules*, 1994, **27**, 7747–7753.

11. Rhee, W.M. and Berg, R.A., US Patent No. 5,510,418, 1996.

12. Rhee, W.M. and Berg, R.A., US Patent No. 5,476,666, 1995.

13. Giusti, P. and Callegaro, L., World Patent Application No. WO 94/01468, 1994.

14. Sander, T.W. and Kaplan, D.S., US Patent No. 5,356,629, 1994.

15. Sung, K.C. and Topp, E.M., Swelling properties of hyaluronic acid ester membranes. *Journal Membrane Science*, 1994, **92**, 157–167.

16. della Valle, F. and Romeo, A., US Patent No. 5,466,461, 1995.

17. Michaeli, D., US Patent No. 4,912,093, 1990.

18. Matsuda, T., Moghaddam, M.J., Miwa, H., Sakurai, K. and Iida, F., Photoinduced prevention of tissue adhesion. *ASAIO Journal*, 1992, **38**, M154–M157.

19. Mikos, A.G., Papadaki, M.G., Louvroukoglou, S., Ishaug, S.L., Thomson, R.C. and Mini-Review: Islet Transplantation to create a bioartificial pancreas. *Biotechnology Bioengineering*, 1994, 43, 673–677.

20. Clayton, H.A., James, R.F.L. and London, N.J.M., Islet microencapsulation: A review; *Acta Diabetologica*, 1993, 30, 181–189.

21. De Vos, P., De Haan, B. and Van Schilfgaarde, R., Effect of the alginate composition on the biocompatibility of alginate-polylysine microcapsules, *Biomaterials*, 1997, **18**, 273–278.

22. Hertzberg, S., Moen, E., Vogelsang, C. and Ostgaard, K., Mixed photo-crosslinked polyvinyl alcohol and calcium alginate gels for cell entrapment, *Applied Microbial Biotechnology*, 1995, **43**, 10–17.

23. Getter, L. and Cutright, D.E., Fracture fixation using biodegradable material. *Journal Oral Surgery*, 1972, **30**, 344–348.

24. Bowald, S., Busch, C. and Eriksson, I., Arterial regeneration following polyglactin 910 suture mesh grafting. *Surgery*, 1979, **86**, 722–729.

25. Ruderman, R.J., Hegyeli, A.F., Hattler, B.G. and Leonard, F., A partially biodegradable vascular prosthesis. *Transactions ASAIO*, 1972, **28**, 30–36.

26. Schmitt, E.E. and Polistima, R.A., US Patent No. 3,463,158, 1969.

27. Barrows, T.H., Synthetic Bioabsorbable Polymers. In *High Performance Biomaterials: A Comprehensive Guide to Medical and Pharmaceutical Applications*, ed. M. Szycher, Technomic, Lancaster, PA, 1991, pp. 243–257.

28. Sinclair, R.G., US Patent No. 4,045,418, 1977.

29. Kim, S.H., Han, Y.-K., Kim, Y.H. and Hong. S.I., Multifunctional initiation of lactide polymerization by stannous octoate/pentaerythritol. *Makromolecular Chemie*, 1992, 193, 1623–1631.

30. Gilding, D.K. and Reed, A.M., Biodegradable polymers for use in surgery - polyglycolic/polylactic acid homo- and copolymers: 1, *Polymer*, 1979, **20**, 1459–1464.

31. Cutright, D.E., Perez, B., Beasley, J.D., Larson, W.J. and Posey, W.R., Degradation rates of polymers and copolymers of polylactic and polyglycolic acids, *Oral Surgery*, 1974, **37**, 142–152.

32. Shen, Z., Zhu, J. and Ma, Z., Synthesis and characterization of poly (DL-lactic acid/glycolic acid). *Makromolecular Chemie Rapid Communications*, 1993, **14**, 457–460.

33. Grijpma, D.W. and Pennings, A.J., Polymerization temperature effects on the properties of l-lactide and ε-caprolactone copolymers. *Polymer Bulletin,* 1991, **25**, 335–341.

34. Hiljanen-Vainio, M.P., Orava, P.A. and Seppälä, J.V., Properties of ε-caprolactone/DL-lactide (ε-CL/DL-LA) copolymers with a minor ε-CL content. *Journal Biomedical Materials Research,* 1997, 34, 39–46.

35. Perego, G., Vercellio, T. and Balbontin, G., Copolymers of L-and D,L-lactide with 6-caprolactone: Synthesis and characterization. *Makromolecular Chemie,* 1993, **194**, 2463–2469.

36. Zhang, X., Wyss, U.P., Pichora, D. and Goosen, M.F.A., Biodegradable polymers for orthopedic applications: synthesis and processability of poly(L-lactide) and poly(lactide-co-ε-caprolactone). *Journal Macromolecular Science Pure Applied Chemistry,* 1993, A30, 933–947.

37. Rosensaft, M.N. and Webb R.L., US Patent No. 4,243,775, 1981.

38. Jamiolkowski, D.D. and Shalaby, S.W., US Patent No. 4,700,704, 1987.

39. Casey, D.J. and Roby, M.S., US Patent No. 4,452,973, 1984.

40. Song, S.S., Kim, H.H. and Yi, Y.W., US Patent No. 5,514,380, 1996.

41. Deng, X.M., Xiong, C.D, Cheng, L.M. and Xu, R.P., Synthesis and characterization of block copolymers from D,L-lactide and poly(ethylene glycol) with stannous chloride. *Journal Polymer Science: Part C: Polymer Letters,* 1990, **28**, 411–416.

42. Shah, S.S., Zhu, K.J. and Pitt, C.G., Poly-DL-lactic acid: polyethylene glycol block copolymers. the influence of polyethylene glycol on the degradation of poly-DL-lactic acid. *Journal Biomaterials Science Polymer Edition,* 1994, **5**, 421–431.

43. Kricheldorf, H.R. and Meier-Haack, J., ABA triblock copolymers of L-lactide and poly(ethylene glycol). *Makromolecular Chemie,* 1993, **194**, 715–725.

44. Cerrai, P. and Tricoli, M., Block copolymers from L-lactide and poly(ethylene glycol) through a non-catalyzed route. *Makromolecular Chemie Rapid Communications,* 1993, **14**, 529–538.

45. Penco, M., Marcioni, S., Ferruti, P., D'Antone, S. and Deghenghi, R., Degradation behavior of block copolymers containing poly(lactic-glycolic acid) and poly(ethylene glycol) segments, *Biomaterials,* 1996, **17**, 1583–1590.

46. Cohn, D. and Younes, H., Biodegradable PEO/PLA block copolymers. *Journal Biomedical Materials Research,* 1988 **22**, 993–1009.

47. Sawhney, A.S. and Hubbell, J.A., Rapidly degraded terpolymers of dl-lactide, glycolide, and ε-caprolactone with increased hydrophilicity by copolymerization with polyethers, *Journal Biomedical Materials Research,* 1990, **24**, 1397–1411.

48. Mooney, D.J., Mazzoni, C.L., Organ, G.M., Puelacher, W.C., Vacanti, J.P. and Langer, R., Stabilizing fiber-based cell delivery device by physically bonding adjacent fibers, *Materials Research Society Symposium Proceedings,* 1994, **331**, 47–52.

49. Tayton, K., Phillips G. and Ralis, Z., Long term effects of carbon fibre on soft tissues. *Journal Bone Joint Surgery.,* 1982, **64 B**, 112–114.

50. Doddi, N., Versfelt, C.C. and Wasserman, D., U.S. Patent No. 4,052,988, 1977.

51. Ray, J.A., Doddi, N., Regula, D., Williams, J.A., Melvegar, A. and Polydioxanone (PDS); a novel monofilament synthetic absorbable suture, *Surgery, Gynecology, and Obstetrics*, 1981, **153**, 497–507.

52. Zhu, K.J., Hendren. R.W., Jenson, K. and Pitt, C.G., Synthesis, Properties, and biodegradation of poly(1,3-trimethylene carbonate), *Macromolecules*, 1991, **24**, 1736–1740.

53. Casey, D.J., Jarrett, P.K. and Rosati, L., U.S. Patent No. 4,716,203, 1987.

54. Muth, R.R., Totakura, N. and Liu, C.-K., US Patent No. 5,322,925, 1994.

55. Kennedy, J., Kaplan, D.S. and Muth, R.R., US Patent No. 5,225,520, 1993.

56. Tang, R.T., Mare, R., Boyle, W.J., Chiu, T.-H. and Patel, K.M., US Patent No. 5,486,593, 1996.

57. Roby, M.S., Bennett, S.L. and Liu, C.-K., US Patent No. 5,403,347, 1995.

58. Allcock, H.R. and Kugel, R.L., US Patent No. 3,370,020, 1968.

59. Potin, P.H. and De Jaeger, R., Polyphosphazenes: synthesis, structures, properties, applications. *European Polymer Journal*, 1991, **27**, 341–348.

60. Allcock, H.R., Pucher, S.R and Scopelianos, A.G., Synthesis of poly(organophosphazenes) with glycolic acid ester and lactic acid ester side groups: prototypes for new bioerodible polymers. *Macromolecules*, 1994, **27**, 1–4.

61. Allcock, H.R., Cameron, C.G. and Smith, D.E., US Patent No. 5,464,932, 1995.

62. Allcock, H.R. and Fuller, T.J., Phosphazene high polymers with steroidal side groups, *Macromolecules* , 1980, **13**, 1338–1345.

63. Allcock, H.R., Austin, P.E. and Neenan, T.X., Phosphazene high polymer with bioactive substituent groups: prospective anesthetic amino phosphazenes, *Macromolecules*, 1982, **15**, 689–683.

64. Allcock, H.R. and Austin, P.E., Schiff base coupling of cyclic and high-polymeric phosphazenes to aldehydes and amines: chemotherapeutic models, *Macromolecules* , 1981, **14**, 1616–1622.

65. Allcock, H.R, Neenan, T.X. and Kossa, W.C., Coupling of cyclic and high-polymeric [(aminoaryl)oxy]phophazenes to carboxylic acids: prototypes for bioactive polymers, *Macromolecules*, 1982, 15, 693–696.

66. Domb, A.J. and Lange:, R., Polyanhydrides: I. Preparation of high molecular weight polyanhydrides, *Journal Polymer Science Polymer Chemistry Edition*, 1987, **25**, 3373–3386.

67. Domb, A.J., Mathiowitz, E., Ron, E., Giannos, S. and Langer, R., Polyanhydrides. IV. Unsaturated and crosslinked polyanhydrides. *Journal Polymer Science Polymer Chemistry Edition.*, 1991, 29, 571–579.

68. Leong, K.W., Simonte, V. and Langer. R., Synthesis of Polyanhydrides: Melt-polycondensation, dehydrochlorination, and dehydrative coupling. *Macromolecules*, 1987, **20**, 705–712.

69. Chiellini E., Solaro, R., Bemporad, L., D'Antone, S., Giannasi, D. and Leonardi, G., New hydrophilic polyesters and related polymers as bioerodible polymeric matrices, *Journal Biomaterials Science Polymer Edition*, 1995, 7, 307–328.

70. Leong, K.W., Brott, B.C. and Langer, R., Bioerodible polyanhydrides as drug-carrier matrices: I. Characterization, degradation and release characteristics, *Journal Biomedical Materials Research*, 1985, **19**, 941–955.

71. Domb, A.J., Gallardo, C.F. and Langer. R., Poly(anhydrides). 3. poly(anhydrides) based on aliphatic-aromatic diacids. *Macromolecules*, 1989, **22**, 3200–3204.

72. Gerhart, T.N., Laurencin, C.T., Domb, A.J., Langer, R.S. and Hayes, W.C., US Patent No. 5,286,763, 1994.

73. Yaszemski, M.J., Payne, R.G., Hayes, W.C., Langer, R. and Mikos, A.G., in vitro degradation of a poly(propylene fumarate)-based composite material. *Biomaterials*, 1996, **17**, 2127–2130.

74. Suggs, L.J., Payne, R.G., Yaszemski, M.J. and Mikos, A.G. US Patent No. 5,527,864, 1996.

75. Albertsson, A.-C. and Lundmark, S., Synthesis, characterization and degradation of aliphatic polyanhydrides. *British Polymer Journal*, 1990, **23**, 205–212.

76. Daniels, A.U., Andriano, K.P., Smutz, W.P., Chang, M.K.O. and Heller, J., Evaluation of absorbable poly(ortho esters) for use in surgical implants. *Journal Applied Biomaterials*, 1994, **5**, 51–64.

77. Leong, K.W., Haiquan, M. and Renxi, Z., Biodegradable polymers with a phosphoryl-containing backbone: tissue engineering and controlled drug delivery applications. *Chinese Journal Polymer Science*, 1995, **13**, 289–314.

78. Grote, J.J., Bakker, D., Hesseling, S.C. and van Blitterswik, C.A., New alloplastic tympanic membrane material, 1991, *Am J Otol*, **12**, 329–334.

79. Beumer, G.J., van Blitterswijk, C.A. and Ponec, M., Degradative behaviour of polymeric matrices in (sub)dermal and muscle tissue of the rat: a quantitative study. *Biomaterials*, 1994, **15**, 551–559.

80. Shieh, L., Tamada, J., Chen, I., Pang, J., Domb, A. and Langer, R., Erosion of a new family of biodegradable polyanhydrides. *Journal Biomedical Materials Research*, 1994, **28**, 1465–1475.

81. Yu, C. and Kohn, J., New poly(ether-carbonate-amide)s containing tyrosine derivatives and PEG exhibit inverse temperature transitions and useful engineering properties. Paper presented at Fifth World Biomaterials Congress, Toronto, May 1996.

82. Merrill, E.W., Poly(ethylene oxide) star molecules: synthesis, characterization, and applications in medicine and biology, *Journal Biomaterials Science Polymer Edi.ion*, 1993, **5**, 1–11.

83. D'Urso, E.M. and Fortier, G., New bioartificial polymeric material: poly(ethylene glycol) cross-linked with albumin. I. Synthesis and swelling properties, *Journal Bioactive Compatible Polymers*, 1994, **9**, 367–387.

84. Sawhney, A.S., Pathak, C.P. and Hubbell, J.A., Bioerodible hydrogels based on photopolymerized poly(ethylene glycol)-co-poly(α-hydroxy acid) diacrylate macromers, *Macromolecules*, 1993, **26**, 581–587.

85. Sawhney, A.S., Pathak, C.P. and Hubbell, J.A., Modification of Langerhans surfaces with immunoprotective poly(ethylene glycol) coatings, *Biotechnology Bioengineering*, 1994, **44**, 383–386.

86. Pathak, C.P., Sawhney, A.S. and Hubbell, J.A., Rapid photopolymerization of gels in contact with cells and tissue, *Journal American Chemical Society*, 1992, **114**, 8311–8312.

87. Lyman, M.L., Pichon, D., Jarrett, P.K. and Sawhney, A.S., Use of a synthetic photopolymerized biodegradable hydrogel as a pneumosealant. Paper presented at Fifth World Biomaterials Congress, Toronto, May 1996.

88. Sawhney, A.S., Pathak, C.P., van Rensburg, J.J., Dunn, R.C. and Hubbell, J.A., Optimization of photopolymerized bioerodible hydrogel properties for adhesion prevention, *Journal Biomedical Materials Research*, 1994, **28**, 831–838.

89. Pathak C.P., Sawhney, A.S. and Hubbell, J.A., In situ photopolymerization and gelation of water soluble monomers: a new approach for local adminis-tration of peptide drugs, *Polymer Preprints*, 1992, **33**, 65–66.

90. Lyman, M.D., Melanson, D. and Sawhney, A.S., Characterization of the formation of interfacially photopolymerized thin hydrogels in contact with arterial tissue, *Biomaterials* , 1996, **17**, 359–364.

91. Andrade, J.D., ed., *Hydrogels for Medical and Related Applications*. ACS Symposium Series, Vol. 31, American Chemical Society, Washington, 1976.

92. Carlson, K.H. and Cameron, J.D., Lindstrom, Assessment of the blood-aque-ous barrier by fluorophotometry following poly(methyl methacrylate), sili-cone, and hydrogel implantation in rabbit eyes, *Journal Cataract Refractive Surgery*, 1993, **19**, 9–15.

93. Bhiwandiwala, P. and Wheeler, R.G., US Patent No. 4,467,806, 1984.

94. Rosenblatt, S., US Patent No. 4,098,728, 1978.

95. Ficek, B.J. and Peppas, N.A., Novel preparation of poly(vinyl alcohol) microparticles with crosslinking agent for controlled drug delivery, *Material Research Society Symposium Proceedings*, 1994, **331**, 223–226.

96. Nelson, T.W., US Patent, No. 5,284, 468, 1994.

97. Llanos and G.R., Sefton, M.V., Immobilization of polyethylene glycol onto a polyvinyl alcohol hydrogel. 2. Evaluation of thrombogenicity, *Journal Biomedical Materials Research*, 1993, **27**, 1383–1391.

98. Hodge, R.M., Bastow, T.J., Edward, G.H., Simon, G.P. and Hill, A.J., Free volume and the mechanism of plasticization in water-swollen poly(vinyl alcohol), *Macromolecules*, 1996, **29**, 8137–8143.

99. Young, T.-H., Yao, N.-Y., Chang, R.-F. and Chen, L.-W., Evaluation of asymmetric poly(vinyl alcohol) membranes for use in artificial islets, *Bioma-terials*, 1996, **17**, 2139–2145.

100. Iwata, H., Amemiya, H., Hayashi, R., Fujii, S. and Akutsu, T., The use of photocrosslinkable polyvinyl alcohol in the immunoisolation of pancreatic islets, *Transplantation Proceedings*, 1990, **22(2)**, 797–799.

101. Stoy, V.A., Stoy, G.P., Lovy, J., US Patent No. 4,943,618.

102. Chvapil, M., Chvapil, T.A., Owen, T.A. and Benson, D.J., Occulusion of the vas derefens in dogs with a biocompatible hydrogel solution, *Journal Repro-ductive Medicine*, 1990, **9**, 905–910.

103. Stoy, V.A., New type of hydrogel for controlled drug delivery, *Journal Biomaterials Applications*, 1989, **3**, 552–604.

104. Gin, H., Dupuy, B., Bonnemaison-Bourignon, D., Bordenave, L., Bareille, R., Latapie, M.J., Baquey, C., Bezian, J.H. and Ducassou, D., Biocompatibility of polyacrylamide microcapsules implanted in peritoneal cavity or spleen of the rat. Effects on various inflammatory reactions in vitro. *Biomaterials Artificial Cell Artificial Organs*, 1990, **18**, 25–42.

105. Yui, N., Suzuki, K., Okano, T., Sakurai, Y., Isikawa, C., Fujimoto, K. and Kawaguchi, H., Mechanism of cytoplasmic calcium changes in platelets in

contact with polystyrene and poly(acrylamide-co-methacrylic acid) surfaces, *Journal Biomaterials Science Polymer Edition*, 1993, **4**, 199–215.

106. Woerly, S., Ulbrich, K., Chytry, V., Smetana, K., Petrovicky, P., Rihova, B. and Morassutti, D.J., Synthetic polymer matrices for neural cell transplantation, *Cell Transplantation*, 1993, **2**, 229–239.

107. Kito, H., Suzuki, F., Nagahara, S., Nakayama, Y., Tsutsui, Y., Tsutsui, N., Nakajima, N. and Matsuda, T., A Total delivery system of genetically engineered drugs or cells for diseased vessels. *ASAIO Journal*, 1994, 40(3), M260–M266.

108. Matsuda, T. and Moghaddam, M.J., Molecular design of artificial fibrin glue, *Materials Science and Engineering, C1*, 1993, **1(1)**, 37–43.

109. Drumheller, P.D., Herbert, C.H. and Hubbell, J.A., Bioactive peptides and surface design. In *Interfacial Phenomena and Bioproducts*, ed. J.L. Brash and P.W. Wojciechowski. Marcel Dekker, New York, 1996, pp 273–310.

110. Drumheller, P.D. and Hubbell, J.A., Surface immobilization of adhesion ligands for investigations for cell-substrate interactions. In *Biomedical Engineering Handbook*, ed. J.D. Bronzino. CRC Press, Boca Raton, FL, 1995, pp. 1583–1596.

111. Barrera, D.A., Langer, R.S., Lansbury, Jr., P.T. and Vacanti, J.P., US Patent No. 5,399,665, 1995.

112. Barrera, D.A., Zylstra, E., Lansbury, P.T. and Langer, R., Copolymerization and degradation of poly(lactic acid-co-lysine). *Macromolecules*, 1995, **28**, 425–432.

113. Hrkach, J.S., Ou, J., Lotan, N. and Langer. R., Poly(L-lactic acid-co-aspartic acid): interactive polymers for tissue engineering. *Materials Research Society Symposium Proceedings*, 1995, **394**, 77–89.

114. Kobayashi, A., Kobayashi, K. and Akaike, T., Control of adhesion and detachment of parenchymal liver cells using lactose-carrying polystyrene as substratum. *Journal Biomaterials Science Polymer Edition*, 1992, **3**, 499–508.

115. Phaneuf, M.D., Quist, W.C., Bide, M.J. and LoGerfo, F.W., Modification of polyethylene terephthalate (Dacron) via Denier reduction: effects on material tensile strength, weight, and protein binding capabilities, *Journal Applied Biomaterials*, 1995, **6**, 289–299.

116. Gupta, B.D., Tyagi, P.K., Ray, A.R. and Singh, H., Radiation-induced grafting of 2-hydroxyethyl methacrylate onto polypropylene for biomedical applications. I. Effect of synthesis conditions, *Journal Macromolecular Science-Chemistry*, 1990, **A27(7)**, 831–841.

117. R. d'Agostino, ed., *Plasma Deposition, Treatment, and Etching of Polymers*. Academic, New York, 1990.

118. Kim, S.W. and Feijen, J., Surface modification of polymers for improved blood compatibility. *CRC Critical Reviews in Biocompatibility*, 1985, **1**, 229–260.

119. Lin, H.-B., Zhao, Z.C., Garcia-Echeverria, C., Rich, D.H. and Cooper, S.L., Synthesis of a novel polyurethane copolymer containing covalently attached rgd peptide. *Journal Biomaterials Science Polymer Edition*, 1992, **3**, 217–227.

120. Clapper, D.L., Kirkham, S.M. and Guire, P.E., ECM proteins coupled to device surfaces improve in vivo tissue integration. *Journal Cellular Biochemistry Supplement*, 1994, **18C**, 283.

121. Dickerson, K.T., Glass, J.T., Liu, L.-S., Polarek, J.W., Craig, W.S., Mullen D.G. and Cheng, S., World Patent Application No. WO 96/20002, 1996.

122. Massia, S.P. and Hubbell, J.A., Covalently grafted RGD and YIGSR containing synthetic peptides support receptor-mediated adhesion of cultured fibroblasts, *Analytical Biochemistry*, 1990, 187, 292–301.

123. Massia, S.P. and Hubbell, J.A., Human endothelial interactions with surface-coupled adhesion peptides on a nonadhesive glass surface and two polymeric biomaterials, *Journal Biomedical Materials Research*, 1991, **25**, 223–242.

124. Kondoh, A., Makino, K. and Matsuda, T., Two-dimensional artificial extracellular matrix: bioadhesive peptide-immobilized surface design, *Journal Applied Polymer Science*, 1993, **47**, 1983–1988.

125. Sugawara, T. and Matsuda, T., Photochemical surface derivatization of a peptide containing Arg-Gly-Asp (RGD), *Journal Biomedical Materials Research*, 1995, **29**, 1047–1052.

126. Drumheller, P.D. and Hubbell, J.A., Polymer networks with grafted cell adhesive peptides for highly biospecific cell adhesive substrates. *Analytical Biochemistry*, 1994, **222**, 380–388.

127. Kossovsky, N., Millett, D., Gelman, A., Sponsler, E. and Hnatyszyn, H.J., Self-assembling nanostructures, *Bio/Technology*, 1993, 11, 1534–1573.

Fabrication of Biodegradable Polymer Scaffolds for Tissue Engineering

MARKUS S. WIDMER,
ANTONIOS G. MIKOS
Institute of Biosciences and Bioengineering,
Rice University,
6100 Main Street,
Houston, TX 77005, USA

1 Introduction

Scaffolds are three-dimensional and highly porous structures with the majority of pores connected to each other. They are mainly used as templates to direct the growth of tissue in the body, or as delivery vehicles for transplanted cells or drugs in an attempt to regenerate structural, and eventually load-bearing tissue. Therefore, these scaffolds may be implanted into a tissue defect without any cells or bioactive compounds previously incorporated, and the tissue regeneration depends on the ingrowth of the surrounding tissue only. Alternatively, the scaffolds may be loaded with cells or compounds, before their implantation, to improve the rate of tissue ingrowth, vascularization, and cell differentiation.

Scaffolds must provide a reproducible microscopic and macroscopic structure with a high surface-area to volume ratio in order to allow a significant amount of cell-surface interactions. Also, a suitable surface chemistry for cell attachment and cell proliferation is essential in order to reorganize a desired tissue, and none of the materials used within the scaffold should provoke an inflammatory response. The average pore size and the macroscopic dimensions of a scaffold are important factors which are associated with cell proliferation and nutrition supply, from tissue culture media *in vitro* and through newly formed blood vessels *in vivo*, to cells and tissue. The pore size of such scaffolds should be sufficient to allow cells to grow in multiple layers in order to form a three-dimensional tissue. The optimal pore size may be highly variable depending also on the intended application of the scaffold. Ideally, *in vivo*, the cell/polymer construct would become vascularized as a result of the cell proliferation into a scaffold. In order to be able to support such new-forming tissue, a scaffold should be strong enough to resist physiological forces within the site of implantation and should prevent pores from collapsing, which would inhibit cell migration and nutrition supply. Further, the flexibility of such a scaffold should be close to that of its surrounding tissue so, once the vascularization starts, no extreme change in the mechanical properties between the host tissue and the scaffold can be experienced by the ingrowing tissue. Those forces could be harmful, not only for the vascularization process, but could also induce the formation of a different tissue than the desired one.

In most applications the support of a scaffold is needed only for a limited time. These temporary scaffolds can not be removed easily because of the tissue grown into the pore structure. Therefore, scaffolds have to be manufactured out of a biodegradable material in which the degradation rate has to be adjusted to match the rate of tissue regeneration. The scaffold should maintain its volume, structure and mechanical stability long enough to allow adequate formation of tissue inside the scaffold. However, none of the degradation products released should provoke inflammation or toxicity.

2 Scaffold materials

There are many natural and synthetic polymers which have a high potential as a biocompatible and biodegradable scaffold material. Many of them have been available for several years and were not originally designed as biomaterials, but were widely used materials which proved useful in medical applications. Among them is the natural polymer collagen and the synthetic polymers of the poly(α-hydroxy ester) family. Poly(α-hydroxy esters) in certain formulations are approved by the FDA as medical implants in the form of sutures [1]. These materials have also been successfully processed to scaffolds for tissue engineering applications [2–4].

Collagen is the major component of mammalian connective tissue that provides strength and flexibility [5]. It provides strength for tensile resistance and rigidity against expansive influence from extracellular matrices by virtue of their ability to attract water [6]. Collagen is a natural polymer that has several unique properties allowing it to be manufactured into three-dimensional scaffolds. There are 14 different types of collagens identified to date, with the most abundant being type I collagen, which is found in high concentrations in tendon, skin, bone and fascia [5]. Collagen used for medical applications in general is derived from bovine submucosa or bovine intestine [7]. Collagen implants degrade by a sequential attack by lysosomal enzymes. The degradation rate can be reduced if the collagen is cross-linked. However, it has been shown that such a cross-linking process reduces the rate of tissue ingrowth into a device manufactured from collagen [8]. Collagen has been used in the form of fibers, sponges and fleeces and has a long established clinical history.

Synthetic polymers offer several advantages over natural materials. They can be prepared with controlled chemical and physical properties and can be made available in almost unlimited quantities. Degradation rate and mechanical properties of synthetic polymers can be altered by simple chemical modification. This results in scaffolds with degradation times starting from several days up to years and in mechanical properties suitable for applications such as skin, cartilage and bone replacement.

Poly(glycolic acid) (PGA) is a highly crystalline, hydrophilic, linear aliphatic polyester with a high melting point (210°C at a molecular weight of 50,000) and a relatively low solubility in most common organic solvents [9]. PGA degrades through hydrolysis and the degradation rate of PGA is affected by the crystallinity which is between 46 and 52% [10, 11]. Typically, it loses most of its mechanical strength over the first 2–4 weeks of degradation in a fluid pH 7 at 37°C *in vitro* [10].

Poly(lactic acids) (PLA) are also polyesters but they are more hydrophobic, have more amorphous character, and an increased solubility to organic solvents compared to PGA [9, 12]. The most common form is Poly(L-lactic acid) (PLLA), which has a lower melting point than PGA (159°C at a molecular weight of 100,000) [9]. During degradation PLLA undergoes hydrolytic de-esterification into lactic acid [13]. If immersed in phosphate-buffered solution at 37°C and pH 6.1, it starts losing its mechanical strength at 8 weeks. *In vivo* it degrades completely within a period of between 10 months and 4 years depending on molecular weight and size of the implant.

Poly(lactic-co-glycolic acid) (PLGA) is a copolymer of lactic and glycolic acid, with the ability to control physical and mechanical properties by changing the copolymer ratio. However, there is no linear relationship between the copolymer ratio and the chemical and mechanical properties of the copolymer. PLGA with a glycolic acid content between 24 and 67 mole% is amorphous, and the degradation rate is highly dependent on the relative amount of each comonomer [12]. Due to a higher degree of crystallinity, high or low copolymer ratios are much more stable against hydrolytic attack than copolymers with a more equimolar ratio [11]. The degradation rate of PLGA also depends on the pH of the solution in which it is immersed. This causes a phenomenon known as autocatalysis, where the degradation products reduce the pH and further induce degradation [14]. For large devices manufactured out of PLGA autocatalysis causes a heterogeneous degradation where the pH decreases in the center of the polymer [15].

3 Processing techniques

A scaffold must provide a suitable substrate for cell attachment, proliferation, differentiated function and cell migration. These critical requirements can not only be met by the choice of an appropriate material, but also by the processing technique. The major requirement of any proposed polymer processing technique to produce scaffolds for tissue engineering purposes is that it should utilize only biocompatible materials, should not affect the biocompatibility of a material, and should alter the degradation behavior only in a controllable manner. It should also allow the manufacture of scaffolds with controlled interconnected pore structure, pore size distribution and pore geometry since these are important factors in organ regeneration. Several techniques were developed which involve either heating the polymers or dissolving them in a solvent. Several of the techniques developed for scaffold fabrication are described in the following sections.

3.1 Fiber bonding

Fibers in the form of sutures were the first commercially available products manufactured from biodegradable synthetic polymers. Such fibers, therefore were used for the fabrication of scaffolds for cell transplantation in the form of tassels and felts [4]. These constructs were useful to demonstrate the feasibility of cell transplantation but lacked the necessary structural stability expected from a scaffold.

To improve the mechanical properties of such structures a fiber bonding technique was developed [16]. In this process, PLLA was dissolved in methylene chloride and cast over a non-woven mesh of PGA fibers. After the solvent was evaporated, the construct was heated above the melting temperature of PGA to weld the fibers at their cross-points. Once the PGA–PLLA construct was cooled down the PLLA was selectively removed by dissolution in methylene chloride. By using this technique the PGA fibers were physically joined without any change in chemical composition and shape compared to untreated fibers. The PLLA was only required to stabilize the PGA fibers and prevent the mesh from collapsing during the heating process. The heating time was identified as a critical parameter for the manufacturing of these scaffolds since increased exposure of the PGA–PLLA composite to heat resulted in the formation of spherical domains at the cross-points of the fibers. However, the necessary immiscibility of the two polymers by dissolution and by melting restricted the application of this technique to a highly limited number of polymer combinations. In addition, the control of the macroscopic shape, porosity and pore size was also limited with this technique.

An alternative method of fiber bonding was developed to prepare tubular scaffolds for the regeneration of intestine, blood vessels, and ureters by coating a non-bonded mesh of PGA fibers with solutions of PLLA or PLGA [17]. For this technique the PGA mesh was wrapped around a rotating cylinder, out of Teflon, and the atomized polymer solution was sprayed over the rotating PGA mesh. After the solvent evaporated the sprayed polymer formed a coating over the PGA fibers and physically bonded adjacent fibers. The thickness of the coating was easily increased by increasing the spraying time. Because of this coating mechanical strength was not only provided by the PGA mesh but also the coating. This was demonstrated by the fact that meshes sprayed with the stronger PLLA resisted higher forces in mechanical tests than the ones sprayed with PLGA. Due to the fact that the entire PGA fiber structure was coated by the sprayed polymer, cell attachment, growth and function, therefore were determined by the coating material rather than the PGA mesh. However successful for the manufacture of porous and biodegradable tubular scaffolds, this method did not allow the manufacture of complex three-dimensional scaffolds.

3.2 Solvent-casting and particulate-leaching

The fiber bonding technique did not allow the manufacturing of scaffolds with a defined pore size and surface-area to volume ratio. To overcome these drawbacks, a solvent-casting and particulate-leaching technique was developed for the production of porous biodegradable PLLA polymer membranes with controlled porosity, pore size, and crystallinity [18]. In this technique, PLLA was dissolved in chloroform and salt particles, sieved for a certain size range, were added. This dispersion was then cast in a glass container and the solvent was allowed to evaporate. Due to the insolubility of the salt particles in chloroform a composite structure containing salt particles of a defined size, bonded together by a PLLA matrix, was formed. For thermal treatment the so-formed composite membranes were then heated above the melting temperature of PLLA. Depending on the desired degree of crystallinity of the PLLA matrix, the membranes were cooled down slowly to form a semicrystalline or fast to yield an amorphous structure within the polymer. The salt particles were leached out by immersing the membranes in water and the resulting salt-free porous PLLA-structures were dried under vacuum. This manufacturing technique is not limited to PLLA and salt, but it can be applied to any polymer–particle combination as long as the solvent does not affect the particles and the polymer is not affected by the leaching process.

Highly porous structures with interconnected pores of a controlled pore size with a porosity up to 93% have been manufactured using this salt-leaching technique. The major drawback of this technique is the fact that it can only be used to produce thin membranes (of thickness up to 2 mm). In order to overcome this problem and produce three-dimensional porous structures with anatomic shapes, a lamination technique was developed [19]. These structures can then be used as scaffolds for the regeneration of tissues such as cartilage and bone whose function is dependent on geometry. Therefore the contour of the desired three-dimensional scaffold was plotted in layers. Porous polymer wafers were manufactured as described above and cut to the contour shape. These layers were then joined in the proper order to form the structure with the desired shape by wetting the attaching surfaces with chloroform prior to lamination. Characterization of the laminated scaffolds showed that the lamination process did not affect the pore structure, no boundary could be seen between the polymer layers, and the bulk properties of the laminated scaffolds were identical to those of the individual membranes.

A similar technique was used to manufacture porous tubular scaffolds for the regeneration of tubular tissues, such as intestine [20]. Porous PLGA membranes were manufactured in a solvent-casting and particulate-leaching method as described above. To form the tubular scaffolds the porous membranes were wrapped around Teflon cylinders, and the overlapping ends were joined together by using a small amount of chloroform. The Teflon tube was then removed from the scaffold. This manufacturing process was limited to scaffolds with a low ratio of wall thickness to inner diameter of the tubular structure due to the relatively brittle behavior of the porous membranes used.

In order to overcome the problems related to the brittle nature of these membranes, a technique was developed to increase the pliability of the porous polymer membranes [21]. Porous membranes were manufactured in the solvent-casting and particulate-leaching process, as described above, by blending the PLGA with poly(ethylene glycol) (PEG). PEG is soluble in water and therefore most of the PEG was removed from the membranes during the leaching process. The incorporation of PEG into the manufacturing process was believed to alter the structure of the walls within the porous membranes by leaving micropores which resulted in the increased pliability of the porous scaffold. In mechanical tests these membranes made from blends of PEG and PLGA were significantly more pliable than the same wafers manufactured from PLGA alone. Due to this additional amount of pliability these membranes could be rolled into tubular scaffolds with a significant higher ratio of wall thickness to inner diameter as described above. The membranes fabricated from the blended polymers did not show any signs of macroscopic damage during rolling as was observed for tubes made out of PLGA alone.

3.3 Melt molding

Melt molding was developed as a process to produce highly complex three-dimensional porous scaffolds out of biodegradable polymers [22]. Porous PLGA scaffolds were produced by loading a cylindrical Teflon mold with a mixture of PLGA powder and pre-sieved gelatin microspheres of a certain diameter range. The Teflon mold was then heated above the glass-transition temperature of PLGA and pressure was applied to the mixture to form it and bond it together. After cooling down, the PLGA-gelatin composites were removed from the mold and the gelatin microspheres were leached out by immersing the composites in water. After drying under vacuum, biodegradable scaffolds, with a shape identical to the shape of the mold, were obtained. The porosity of these scaffolds was easily controlled by varying the amount of gelatin added to the PLGA. The pore size distribution corresponded to the size of the gelatin microspheres. Because this process does not utilize organic solvents and is carried out at relatively low temperatures, it has potential for the incorporation of bioactive molecules for drug delivery applications. The melt molding process can easily be applied to other polymers as well as for PLGA and, in addition to the choice of gelatin, other leachable particles, such as salt can be used.

In an attempt to improve the mechanical strength of these porous biodegradable polymer scaffolds, hydroxyapatite short fibers were incorporated into the melt-molding process [23]. Because of the difficulty in achieving a homogeneous fiber distribution by mixing the three solid phases (polymer, leachable particles, and fibers), membranes were manufactured using the solvent-casting technique, as described above. PLGA was dissolved in methylene chloride, hydroxyapatite short fibers and either gelatin microspheres or salt particles were added, and the mixture was poured into a glass container. The solvent was allowed to evaporate and the resulting membranes, once dry, were cut into pieces. These pieces were then filled into a cylindrical Teflon mold and melted together by a melt-molding process as described above. After removing the composite from the mold, the gelatin or salt was leached out in water and the scaffold was dried under vacuum. This process produced a porous biodegradable and fiber-reinforced scaffold with a porosity up to 84% Yield strength and modulus of such fiber-reinforced scaffolds with a variety of gelatin microsphere and hydroxyapatite fiber ratios were manufactured and characterized by compression testing. The compressive strength of these reinforced scaffolds was higher than for the scaffold manufactured from PLGA alone and maximum yield strength and modulus were measured when the weight ratio of hydroxyapatite fibers to gelatin microspheres was 30/35.

3.4 Extrusion

Porous biodegradable tubular scaffolds for the regeneration of nerves have been manufactured by an extrusion process [24]. Membranes made out of PLGA and salt crystals of size between 150 and 300 μm were manufactured in a solvent-casting process as described above. The resulting dry PLGA–salt composite discs were cut into pieces of less than 5 mm edge

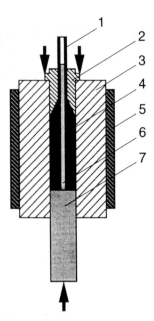

Fig. 1. Piston extrusion tool for the manufacturing of tubular constructs: (1) Extruded tubular PLGA/salt composite construct; (2) Nozzle which forms the outer diameter of the tubular construct; (3) Tool body; (4) Melted PLGA/salt mixture; (5) Heat band with temperature control; (6) Rod defining the inner diameter of the tubular construct; (7) Piston moving the melted PLGA/salt mixture. The arrows indicate the attachment points for the forces involved in the extrusion process.

length, filled into a piston extrusion tool (Fig. 1) and heated to 250°C. First, the temperature within the tool was allowed to equalize and then the PLGA–salt composite was extruded. The extruded structures, tubes with an inner diameter of 1.6 mm and an outer diameter of 3.2 mm, were then cut to their appropriate length on a diamond wheel saw, immersed in water to leach out the salt, and vacuum-dried (Fig. 2). To allow mechanical testing, a short portion of both ends of the tubular scaffolds were embedded into paraffin. The influence of degradation on the mechanical properties was evaluated by maintaining the embedded scaffolds in phosphate buffered saline (PBS) solution at 37°C. The conduits were removed from the PBS solution at various time points for up to 8 weeks and the tensile properties of the conduits were measured by mechanical tension testing. However, while the mechanical strength of the tubular scaffolds decreased with increasing time in solution, there was no observed collapsing of the tubular structure.

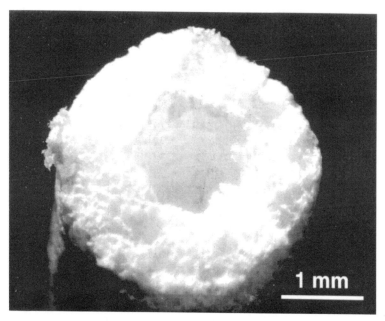

Fig. 2. Photomicrograph of a porous biodegradable tubular scaffold for the regeneration of peripheral nerves manufactured by an extrusion process.

3.5 Freeze-drying

The freeze-drying process was developed as a method to produce porous biodegradable scaffolds out of collagen for use as artificial skin [25]. The collagen used for this study was purified from bovine hide [26]. The hide was treated with lime, in order to depolymerize the collagen, and cut into small, thin pieces. These pieces were then washed for 4 hours with water containing 0.3 wt.% propionic acid and 0.1 wt.% benzoic acid. The water was then separated by filtering and the hide pieces were milled down. The product of this grinding operation was freeze-dried, which yielded loose, shredded, fibrous mats. These mats were then pulverized by a milling process and dispersed in acetic acid by stirring in a refrigerated homogenizer. Chondroitin 6-sulfate was dissolved in acetic acid and added to the collagen dispersion. The homogenate was then formed into membranes and freeze-dried. By removing the fluid from this swollen collagen membrane a highly porous structure resulted. The mean pore size for such scaffolds was measured as 100 μm. These porous collagen scaffolds dried in a freeze-drying process were then compared to scaffolds dried by critical point drying or air drying. The freeze-drying process resulted in the least amount of shrinkage of the membranes, the most porosity, and the highest mean pore-size diameter compared to all other drying processes applied within this study.

3.6 Phase separation

A desirable feature for biodegradable scaffolds used for tissue regeneration would be the ability to load drugs or growth induction factors into them. However, many fabrication techniques do not allow the incorporation of these molecules. The chemical or thermal environment they are exposed to during the scaffold fabrication process, can drastically decrease the activity of such molecules. Therefore, a phase separation process, free of harsh chemical and thermal treatment, was designed to produce porous biodegradable scaffolds with a controlled release function for bioactive molecules [27]. Briefly, a scaffold polymer, such as PLLA, was dissolved in molten phenol or naphthalene and bioactive molecules such as alkaline phosphatase were added. By lowering the temperature of the solution a liquid–liquid phase separation was induced and the resulting bicontinuous polymer and solvent phases were quenched to create a two-phase solid. The solvent was then removed by sublimation to yield a porous, biodegradable scaffold loaded with bioactive molecules. By immersing the scaffolds in PBS at 37°C the *in vitro* alkaline phosphatase release behavior of these scaffolds was measured. According to the resulting release characteristics for a scaffold manufactured with naphthalene as the solvent, approximately 60% of the alkaline phosphatase was merely adsorbed on the surface of the scaffold and about 25% of the enzyme activity was lost due to processing. By using phenol as the solvent in this phase separation process the alkaline phosphatase lost all of its activity. Even though phenol has a lower melting temperature than naphthalene phenol, as a polar solvent, it can interact with proteins and weaken their hydrogen-bonding which can lead to a loss of the protein activity. Therefore phenol as a solvent might be suitable for small drugs or short peptides, which do not have defined conformations.

3.7 High pressure process

Most processes used to fabricate porous biodegradable scaffolds from synthetic polymers typically utilize organic solvents. However, residues of these solvents remaining in the sponges may be harmful for cells or tissues introduced into such scaffolds. To avoid the risk of damaging biological material, a method to produce highly porous scaffolds without the use of organic solvents was developed [28]. In this process solid disks, compression-molded from biodegradable polymers, such as PLGA, were saturated with carbon dioxide (CO_2) at 5.5 MPa and room temperature for 72 hours. The solubility of the gas in the polymer was then decreased rapidly by reducing the CO_2 pressure to atmospheric level. This resulted in the nucleation and growth of gas cells, with sizes between 100 and 500 μm, within the polymer disks. For scaffolds manufactured from PLGA this process yield a porosity of up to 94%. Higher amounts of porosity could be achieved by using solid discs manufactured by a solvent-casting process instead of the compression-molded discs, but at the same time the pore size decreased to values of between 10 and 100 μm. No pore formation was observed by using the same process for discs out of PGA. This leads to the result that the porosity of these scaffolds could be decreased by blending the PLGA with PGA. For compression-molded discs out of 78 wt% PLGA and 22 wt% PGA, the high pressure process yielded scaffolds with a lower porosity (88%) compared to scaffolds manufactured from PLGA alone. In an attempt to improve the mechanical properties of such scaffolds, PGA fibers were incorporated into the compression-molded PLGA discs before the high pressure process was applied. This reinforcement resulted in discs comparable to the ones without fiber reinforcement. For all the scaffolds fabricated by this high pressure process, the pores within these scaffolds were only partially connected and, as a result from rapid diffusion of the gas, a layer without pores was reported close to the surface of the scaffolds. This may initially prevent cells from migrating through the entire scaffold and also provide a barrier for the diffusion of nutrients, until some erosion has occurred. However, the formed skin may be advantageous for guided tissue growth.

3.8 Three-dimensional printing

A three-dimensional printing process was used to manufacture resorbable scaffolds with a complex macroscopic and microscopic structure for drug delivery [29]. Briefly, a layer of polymer powder was laid on top of a piston. This powder layer was then selectively joined by a jet of organic solvent, released from a single printhead nozzle which was mounted on a x–y position control and driven by servo motors. After the joining process of that specific layer was completed, the piston was lowered, a new layer of polymer powder was laid on top of the already processed layer and the powder, again, was selectively joined. This process was repeated until the desired part was completed. To study the drug release profile of such scaffolds, a dye was placed into selected locations inside the scaffolds during the printing process. The release rate of the dye was characterized by placing the scaffolds into water and measuring the amount of dye in the water. Depending on wall thickness and structure of the scaffold, the dye was released within the first 10–30 hours. However, this printing process still needs to be tested for the fabrication of porous scaffolds loaded with bioactive molecules.

4 Conclusions

A large variety of processes have been applied to manufacture scaffolds. However, there is no universal technique to produce scaffolds for the regeneration of all tissues. Often the requirements for a desired scaffold dictate the method of processing and the type of material used. This can be because of the shape of a scaffold, the pore structure, a required degradation rate, drug delivery purposes, or the mechanical properties. Also, different kinds of tissues require different scaffolds. For the repair of hard and brittle tissue, such as bone, scaffolds need to have a high elastic modulus in order to maintain the space they were designated to and provide the tissue with enough space for growth. If scaffolds are used as a temporary load-bearing device, they should be strong enough to maintain that load for the required time without showing any symptoms of fatigue failure. Used in combination with soft tissues the flexibility and the stiffness of the scaffold have to be within the same order of magnitude as the surrounding tissues in order to prevent the scaffold from either breaking or collapsing. The choice of a scaffold-processing technique is therefore a question of assessing critical requirements for each application of such scaffolds. However, many challenges remain in the fabrication of high load-bearing scaffolds, scaffolds with high flexibility and the incorporation and delivery of drugs, bioactive molecules, and cells to stimulate and enhance the growth of a specific tissue.

References

1. Food and Drug Administration, Medical Devices; Reclassification and Codification of Absorbable Poly(Glycolide/L-Lactide) Surgical Suture. In *Federal Register*, Vol. 56, 1991, pp. 47150–47151.
2. Gazdag, A. R., Lane, J. M., Glaser, D. and Foster, R. A., Alternatives to autogenous bone grafts: efficacy and indication. *Journal of American Academic Orthopedic Surgery*, 1995, **3**, 1–8.
3. Hubbell, J. A., Biomaterials in tissue engineering. *BMES Bulletin*, 1994, **18**, 31–34.
4. Cima, L. G., Vacanti, J. P., Vacanti, C., Ingber, D., Mooney, D. and Langer, R., Tissue engineering by cell transplantation using degradable polymer substrates. *Journal of Biomechanical Engineering*, 1991, 113, 143–151.
5. Pachence, J. M., Collagen-based devices for soft tissue repair. *Journal of Biomedical Materials Research*, 1996, **33**, 35–40.
6. McCarthy, J. B., Vachhani, B. and Iida, J., Cell adhesion to collagenous matrices. *Biopolymers*, 1996, **40**, 371–381.
7. Hutmacher, D., Hürzeler, B. and Schliephake, H., A review of material properties of biodegradable and bioresorbable polymers and devices for GTR and GBR applications. *The International Journal of Oral and Maxillofacial Implants*, 1996, **11**, 667–678.

8.   Chvapil, M., Speer, D. P., Holubec, H., Chvapil, T. A. and King, D. H., Collagen fibers as a temporary scaffold for replacement of ACL in goats. *Journal of Biomedical Materials Research*, 1993, **27**, 313–325.

9.   Engelberg, I. and Kohn, J., Physico–mechanical properties of degradable polymers used in medical applications: a comparative study. *Biomaterials*, 1991, **12**, 292–304.

10. Reed, A. M. and Gilding, D. K., Biodegradable polymers for use in surgery — poly(glycolic)/poly(lactic acid) homo and copolymers. *Polymer*, 1981, **22**, 505–509.

11. Miller, R. A., Brady, J. M. and Cutright, D. E., Degradation of oral resorbable implants (polylactates and polyglycolates): rate modification with changes in PLA/PGA copolymer ratio. *Journal of Biomedical Materials Research*, 1977, **11**, 711–719.

12. Gilding, D. K. and Reed, A. M., Biodegradable polymers for use in surgery — polyglycolic/poly(lactic acid) homo- and copolymers: 1. *Polymer*, 1979, **20**, 1459–1464.

13. Laitinen, O., Törmälä, P., Taurio, R., Skutnabb, K., Saarelainen, K., Iivonen, T. and Vainionpää, S., Mechanical properties of biodegradable ligament augmentation devices of poly(L-lactide) in vitro and vivo. *Biomaterials*, 1992, **13**, 1012–1016.

14. von Recum, H. A., Cleek, R. L., Eskin, S. G. and Mikos, A. G., Degradation of polydispersed poly(L-lactic acid) to modulate lactic acid release. *Biomaterials*, 1995, **16**, 441–447.

15. Vert, M., Li, S. M. and Garreau, H., Attempts to map the structure and degradation characteristics of aliphatic polyesters derived from lactic and glycolic acids. *Journal of Biomaterials Science Polymer Edition*, 1994, **6**, 639–649.

16. Mikos, A. G., Bao, Y., Cima, L. G., Ingber, D. E., Vacanti, J. P. and Langer, R., Preparation of poly(glycolic acid) bonded fiber structures for cell attachment and transplantation. *Journal of Biomedical Materials Research*, 1993, **27**, 183–189.

17. Mooney, D. J., Mazzoni, C. L., Organ, G. M., Puelacher, W. C., Vacanti, J. P. and Langer, R., Stabilizing Fiber-Based Cell Delivery Devices by Physically Bonding Adjacent Fibers. In *Biomaterials for Drug and Cell Delivery*, Vol. 331, ed. A. G. Mikos, R. M. Murphy, H. Bernstein and N. A. Peppas. Material Research Society, Pittsburgh, 1994, pp. 47–52.

18. Mikos, A. G., Thorsen, A. J., Czerwonka, L. A., Bao, Y. and Langer, R., Preparation and characterization of poly(L-lactic acid) foams. *Polymer*, 1994, **35**, 1068–1077.

19. Mikos, A. G., Sarakinos, G., Leite, S. M., Vacanti, J. P. and Langer, R., Laminated three-dimensional biodegradable foams for use in tissue engineering. *Biomaterials*, 1993, **14**, 323–330.

20. Mooney, D. J., Organ, G., Vacanti, J. P. and Langer, R., Design and fabrication of biodegradable polymer devices to engineer tubular tissues. *Cell Transplantation*, 1994, **3**, 203–210.

21. Wake, M. C., Gupta, P. K. and Mikos, A. G., Fabrication of pliable biodegradable polymer foams to engineer soft tissues. *Cell Transplantation*, 1996, **5**, 465–473.

22. Thomson, R. C., Yaszemski, M. J., Powers, J. M. and Mikos, A. G., Fabrication of biodegradable polymer scaffolds to engineer trabecular bone. *Journal of Biomaterials Science Polymer Edition*, 1995, **7**, 23–38.

23. Thomson, R. C., Yaszemski, M. J., Powers, J. M., Harrigan, T. P. and Mikos, A. G., Poly(α-Hydroxy Ester)/Short Fiber Hydroxyapatite Composite Foams for Orthopedic Application. In *Polymers in Medicine and Pharmacy*, Vol. 394, ed. A. G. Mikos, K. W. Leong and M. J. Yaszemski. Materials Research Society, Pittsburgh, 1995, pp. 25–30.

24. Widmer, M. S., Evans, G. R. D., Brandt, K., Savel, T., Patrick, C. W. and Mikos, A. G., Porous Biodegradable Polymer Scaffolds for Nerve Regeneration. In *Proceedings of the 1997 Summer Bioengineering Conference*, Vol. 35, ed. K. B. Chandran, R. Vanderby, Jr. and M. S. Hefzy. The American Society for Mechanical Engineers, New York, 1997, pp. 353–354.

25. Dagalakis, N., Flink, J., Stasikelis, P., Burke, J. F. and Yannas, I. V., Design of artificial skin. Part III. control of pore structure. *Journal of Biomedical Materials Research*, 1980, **14**, 511–528.

26. Yannas, I. V., Burke, J. F., Gordon, P. L., Huang, C. and Rubenstein, R. H., Design of an artificial skin. II. Control of chemical composition. *Journal of Biomedical Materials Research*, 1980, **14**, 107–131.

27. Lo, H., Ponticiello, M. S. and Leong, K. W., Fabrication of controlled release biodegradable foams by phase separation. *Tissue Engineering*, 1995, **1**, 15–28.

28. Mooney, D. J., Baldwin, D. F., Suh, N. P., Vacanti, J. P. and Langer, R., Novel approach to fabricate porous sponges of poly(D,L-lactic-co-glycolic acid) without the use of organic solvents. *Biomaterials*, 1996, **17**, 1417–1422.

29. Wu, B. M., Borland, S. W., Giordano, R. A., Cima, L. G., Sachs, E. M. and Cima, M. J., Solid free-form fabrication of drug delivery devices. *Journal of Controlled Release*, 1996, **40**, 77–87.

CHAPTER II.6

Cell–Synthetic Surface Interactions

NINA M. K. LAMBA,
JENNIFER A. BAUMGARTNER,
STUART L. COOPER
Department of Chemical Engineering,
University of Delaware,
Newark, Delaware 19716, USA

1 Introduction

Cell adhesion is an important phenomenon in a wide variety of biological processes, including platelet deposition at sites of vascular injury, recruitment of neutrophils and monocytes at inflammatory or infectious sites, and metastasis of cancer cells. Cell adhesion plays a role in tissue and organ formation, in the generation of traction for the migration of cells, and is also important in determining the biocompatibility of synthetic implant materials. For example, the adhesion of microbial cells to surfaces is believed to be a critical step in the pathogenesis of device-centered infections. In the following sections, mechanisms for the adhesion of both eukaryotic and prokaryotic cells will be discussed, as well as the principles governing protein adsorption and the subsequent influence on cellular interactions with surfaces. Strategies for surface modification of synthetic materials will be addressed, such as the grafting of adhesion peptides or other molecules designed to promote or inhibit the adsorption of specific proteins and cells.

Both animal and microbial cell adhesion is viewed as a two-stage process. In bacterial adhesion, cells initially exhibit Brownian motion as they are deposited near a surface. In this state, the adhesion is reversible [1]. As time passes, the adhesion becomes irreversible, with this phase most likely involving the synthesis of polymers to attach the cell to the surface [2]. In animal cell adhesion, cells first attach to a surface by pseudopodial extensions. Later, the cells spread and form focal contacts. Both stages involve the cell 'probing' the surface for protein ligands [3]. Transmembrane glycoproteins such as integrins may link extracellular matrix proteins to the cytoskeleton inside the cell. The physical nature of cell adhesion is governed by both non-specific and specific interactions. Specific interactions refer to receptor-ligand bond formation, while non-specific interactions do not involve such bond formation, but rather include electrostatic forces, steric stabilization, and van der Waals forces. These interactions are shown in Fig. 1 [4]. Temperature, medium concentration, small ion concentration, and pH of the surrounding medium also affects cell adhesion. In developing models to quantify the physical interactions underlying cell adhesion, Lauffenburger and Linderman have also taken into account the mechanics of cell deformation, the forces on cells such as those imparted by moving fluid, and metabolic effects such as focal adhesions or active movement such as pseudopodial extension [5]. Neu and Marshall and van Loosdrecht *et al.* describe the two predomoninant approaches for describing bacterial adhesion one of which is based on the Gibbs energies of the solid/bacterium, solid/liquid, and bacteria/liquid interfaces, and one based on the Derjaguin, Landau, Verwey, and Overbeek (DLVO) theory for colloidal stability [1, 2]. Loosdrecht *et al.* conclude that DLVO theory more accurately describes experimental observations reported in the literature, but that a more biological approach may be necessary to account for the formation of substances that participate in irreversible adhesion.

Specific cell–surface and cell–cell interactions are mediated by receptors — molecules which have an extracellular domain at the surface of the cell — and adsorbed or soluble proteins or peptide sequence ligands, which bind to these receptors. According to the fluid–mosaic model, the surface of an animal cell is composed of a lipid bilayer composed of phospholipids, cholesterol, and glycolipids [6]. A number of proteins are also associated with the lipid bilayer–instrinsic membrane proteins, transport proteins, ion channels, and receptors, for example. Most cells also possess a polysaccharide-rich outer coat termed a glycocalyx. This glycocalyx is composed of short oligosaccharide chains bound to glycoproteins, glycolipids, and proteoglycans. Four major superfamilies of adhesion receptors have been identified on animal cells–the integrins, the cadherins, the immunoglobulins, and the selectins. Integrins are heterodimers, some of which bind extracellular matrix proteins, and are involved in cell adhesion to surfaces as well as in cell–cell adhesion. In particular, many integrin receptors recognize the specific peptide sequence Arg–Gly–Asp, which is present in many adhesive proteins such as fibronectin, fibrinogen, and vitronectin [7]. The GPIIa–IIIb complex mediating platelet aggregation was the first integrin to be identified [8]. The glycoprotein known as LFA-1 is found on leukocytes and lymphocytes, as well as natural killer cells and monocytes. Mac-1 binds the C3bi component of complement and is found on macrophages, monocytes, and neutrophils. Along with the glycoprotein p150,95, LFA-1 and Mac-1 are believed to contribute to leukocyte adhesion [9, 10]. Cadherins are calcium dependent and promote

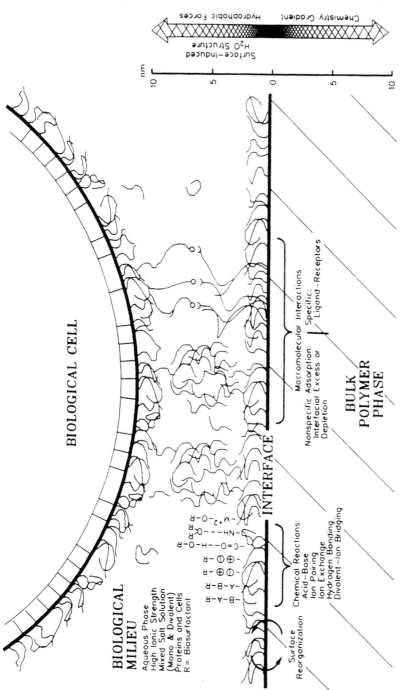

Fig. 1. Diagrammatic view of a polymer interface in contact with biological milieu [4].

adhesion to cells carrying the same cadherin. The selectins are involved in a wide variety of cellular interactions. Endothelial cells express E- selectin, or ELAM-1, which mediates neutrophil adhesion to vascular endothelium. P-selectin is expressed on platelets and endothelial cells following activation, while L-selectin is expressed on the leukocyte. The selectins on the endothelium appear to mediate rolling of cells, while the integrins are responsible for stronger adhesion by binding to immunoglobulin molecules on the endothelium such as ICAM-1 [11].

2 Protein adsorption

Proteins are high molecular weight solutes that can be considered as copolymers of amino acids [12]. Proteins are heterogeneous molecules and contain regions of differing polarity, charge and hydrophilicity; therefore proteins exhibit amphoteric and amphiphilic properties. The precise structure of a protein is not only determined by the specific amino acid sequence, but by the secondary and tertiary structure that determines the conformation of the molecule, and the distribution and orientation of the side groups. Proteins contain anywhere from fifty to over ten thousand residues with a specific chain sequence of amino acids.

Proteins generally prefer an aqueous environment; however, when a protein solution is contacted with another phase, there is a tendency for the protein molecules to accumulate at the interface. At a solid interface, the complex structure of proteins give rise to a number of interactions with the surface, involving hydrophobic, electrostatic and polar forces. The extent of these interactions is not only dependent on the nature of the surface and the nature of the protein, but also on the number and concentration of proteins in the contacting solution, the length of time of contact, the temperature of the system and flow conditions. Thermodynamic driving forces cause protein molecules to condense onto the surface; this adsorption may invoke conformational changes in the protein, or cause the protein to denature. It is generally accepted that the adsorption of plasma proteins onto an artificial surface is the first event to occur when blood contacts a biomaterial, usually within a few seconds [13], preceded only by the adsorption of water and inorganic ions. Blood platelets and other formed elements do not adhere to the surface until after approximately one minute, which corresponds to a protein layer of about 200 Å [14]. On this basis, it is expected that the composition of the adsorbed protein layer influences subsequent events at the solid–liquid interface, as cells must interact with this protein layer. Adhesive proteins may act as bridging molecules between the cell and an artificial surface.

Adsorption is not a static event as adsorbed proteins can undergo conformational changes with time [15, 16], and exchange with other molecules in the contacting solution [14, 17]. The composition of the adsorbed protein layer from plasma or a multi-component protein solution is also time-dependent [18]. The most widely studied compositional change of the protein layer adsorbed from plasma has been termed the 'Vroman Effect'. Vroman and Adams reported that although fibrinogen was adsorbed from plasma onto glass in the initial course of events, it was later 'converted' and could no longer be detected [18]. This 'conversion' was actually the displacement of adsorbed fibrinogen by high molecular weight kininogen (HMWK), a low abundance protein with a higher affinity than fibrinogen for the glass surface [19]. High molecular weight kininogen has been shown to have an anti-adhesive effect on osteosarcoma cells, platelets, monocytes, and endothelial cells [20, 21]. It is likely that the Vroman Effect is only part of a sequence of protein deposition and displacement; the proteins adsorb to the surface according to an order that is determined by their size and relative concentration in the blood, and displaced by lower concentration proteins with higher affinity for the particular surface. The sequence of adsorption and replacement will terminate when a minimum in surface free energy is achieved [17]. The Vroman Effect has been observed in whole blood and plasma, and the limited studies suggest that adsorption and displacement phenomena are similar from the two media. *In vitro* studies have shown that the presence of cells does not influence fibrinogen adsorption from plasma, so it is likely that the Vroman Effect occurs *in vivo*. Thus the dynamics of the adsorbed protein layer are important, as nearly all interactions between mammalian cells and artificial surfaces are mediated by a layer of adsorbed protein [22].

Pierschbacher and Ruoslahti [23] showed that the synthetic fragment arginine–glycine–aspartate (Arg–Gly–Asp) or RGD was the minimal cell-recognizable sequence in many adhesive plasma and extracellular matrix proteins. The RGD sequence has been found in vitronectin, fibronectin, von Willebrand Factor, fibrinogen, and collagen, and has been shown to play a crucial role in mediating cell attachment and subsequent spreading. The RGD sequence has been shown to be adhesive towards platelets, and other cells including fibroblasts and endothelial cells. The Arg–Gly–Asp (RGD) receptor of platelets has been suggested to be the membrane glycoprotein, GPIIb/IIIa [24]. The RGD tripeptide may also act as a ligand for the integrin super-family of receptors. It has been demonstrated that the presence of the synthetic RGD peptide in solution is able to compete with adhesive proteins adsorbed onto a surface for binding receptors on the cell in solution and prevent cellular deposition onto the adsorbed protein [23]. The transport protein albumin does not appear to promote the adhesion of blood platelets to a surface, and can increase the thromboresistance of a synthetic surface. Albumin does not contain the RGD peptide sequence, but when grafted with a RGD residue, shows cellular adhesive activity comparable to fibronectin [25]. Furthermore, albumin does not contain saccharidic residues, which may also contribute to observations of reduced platelet adhesion to albuminated surfaces [26]. Other peptide sequences that are of interest include the Tyr–Ile–Gly–Ser–Arg- (YIGSR) peptide sequence found in the $\beta 1$ subunit of laminin, an important basement membrane glycoprotein. YIGSR is thought to be responsible for promotion of cell adhesion, migration, differentiation and growth. A few studies have also focused on the Arg–Glu–Asp–Val (REDV) sequence present in human fibronectin.

Adsorption of proteins can induce conformational changes within a protein that may expose regions of the protein that are adhesive to cells. Studies of the conformational changes of fibrinogen adsorbed to polyurethane biomaterials support the hypothesis that it is not the amount of the protein that is adsorbed onto a given surface that is important, but the conformation of the adsorbed protein [27, 28]. Fibrinogen has been shown to be less elutable by sodium dodecylsulfate surfactant from polyurethane surfaces as time increases, and this may be due to an increase in the strength of the protein-surface bond [29]. Reduction in the α-helicity of fibrinogen desorbed from glass has been reported [30], and reduced antibody binding to adsorbed fibrinogen has also been observed [31, 32]. Studies of fibrinogen adsorption to sulfonated polyurethanes have shown that these materials adsorb high amounts of fibrinogen when tested *in vitro* [33, 34] and *ex vivo* [35]. The enhanced affinity is believed to be a direct effect of the sulfonate groups. Furthermore, the protein does not appear to be displaced by other blood proteins in a Vroman sequence of events, and despite the high fibrinogen concentration at the surface in an *ex vivo* canine shunt, platelet deposition and thrombus formation onto these materials is reduced [36–38].

3 Targeted cell adhesion

In recent years, there has been an interest in the grafting of synthetic peptide sequences onto polymeric substrates, to promote cell adhesion and attachment for applications such as tissue culture substrates and biomaterials for implantation. By covalently grafting short peptide sequences onto polymeric substrates, it is believed that cellular adhesion can be mediated through receptor–ligand interactions. The long term cellular adhesivity is believed to be superior to surfaces that mediate cell adhesion through protein adsorption, firstly by eliminating the issue of protein desorption, and also by reducing the scope for thermal denaturation and enzyme proteolysis.

RGD peptide sequences have been immobilized onto synthetic surfaces and have been shown to promote cell attachment in a similar manner to fibronectin. Some cell surface receptors have been shown to bind with the

RGD sequence in a specific protein, whereas other receptors may recognize the RGD sequence in more than one protein. The specificity of the RGD protein is believed to be modulated by the conformation of the sequence and this in turn is determined by the amino acids that are immediately adjacent to the RGD sequence [39–41]. Substitution of peptides within the sequence leads to a large reduction in activity [42]. The presence of a peptide at the end of the sequence has been shown to alter the adhesivity of the RGD tripeptide [23, 43]. RGD sequences grafted to polymeric substrates have been shown to increase endothelial cell attachment *in vitro* [44, 45]. RGD grafted materials have been studied with the

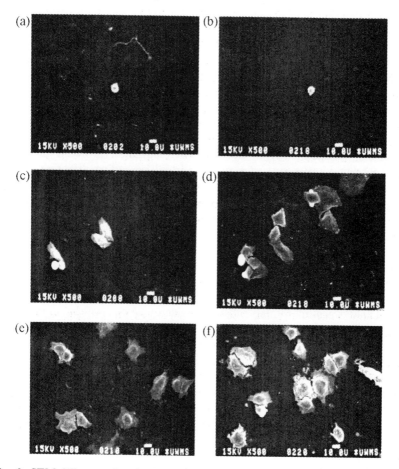

Fig. 2. SEM Micrographs showing endothelial cell adherent on polyurethanes grafted with (a) GRGESY, (b) high density GRGESY, (c) GRGDSY, (d) high density GRGDSY, (e) GRGDVY, and (f) high density GRGDVY [49].

goal of improving the integration of synthetic medical devices into the host organism. RGD has been grafted onto polyethylene terephthalate (PET), polytetrafluoroethylene (PTFE), polyvinyl alcohol (PVA), polyacrylamide and polyurethane [46–48]. Figure 2 shows SEMs of endothelial cells *in vitro*, on RGD grafted polyurethane substrates, including the effect of peptide density [49]. *In vivo*, the presence of this tri-peptide may have undesirable effects, by promoting platelet adhesion and subsequent thrombus formation, and bacterial colonization in the short term. Studies of fibrinogen adhesion to platelets have shown precise peptide-structure property relationships [50, 51]. RGD-protein conjugates have also been synthesized, but problems in defining the precise conformation of the protein once it has adsorbed to the substrate impede interpretation of results. Such an approach to mediate cell-surface interactions may be unsuitable in long term applications due to protein desorption or proteolysis.

Hubbell and co-workers have developed low protein adsorbing surfaces which contain RGD tripeptide sequences. The substrates were prepared from either a glycophase glass [46], or a polyethylene oxide — polyacrylamide semi- interpenetrating network [42]. The low protein adsorbing characteristics of these surfaces allowed the study of cell attachment and spreading independently of protein adsorption. They showed that the RGD and YIGSR peptide sequences directly promoted endothelial cell adhesion independently of adsorbed cell adhesion proteins. They also demonstrated a role for serum proteins in cell spreading, affecting the final extent of spreading, but not the initial rate [46]. Goodman *et al.* showed that the RGD-binding platelet integrin GPIIb/IIIa may have a role in platelet spreading on synthetic surfaces [52].

The structural mechanisms by which microbial cells adhere to surfaces are less well defined, but a number of proteins have been found to bind to bacterial cells. The surfaces of bacterial cells are negatively charged, and are composed of a number of compounds including peptidoglycans, polysaccharides, and proteins. Some strains of microorganisms also produce an extracellular slime substance which may be involved in mediating adhesion to surfaces, but perhaps not in the initial phases of attachment [53]. Adhesion of strains of *Staphylococcus aureus* has been found to be mediated by blood components such as laminin, vitronectin, and fibronectin. Albumin and fibrinogen have also been found to influence the adhesion of *Staphylococcus epidermidis* and *Pseudomonas aeruginosa* [54, 55].

4 Surface modification

One of the greatest features of polymeric biomaterials is their scope for modification. Generally, polymers are easier to modify than other classes of materials such as metals and ceramics, and this has been exploited in order to improve the biocompatibility of polymers. Modification of materials to target cell adhesion via the grafting of specific peptide residues has been discussed above. Other methods that have been employed to influence cellular adhesion are discussed below. The general consensus in the biomaterials field is that protein adsorption mediates cellular interactions with a synthetic substrate. One way, therefore, to improve the performance of polymers for medical applications is to try to control the protein adsorption. This may be achieved through favoring adsorption of specific types or conformations of proteins, or by reducing the amount of protein adsorbed to a surface [56].

It has been shown that the adsorption of the transport protein albumin onto a polymer surface greatly influences the degree of thromboresistance, by reducing platelet interactions with the surface [57-60]. This observation has led to the development of materials that preferentially adsorb albumin, from blood or plasma [60-64]. Generally, these materials have shown preferential adsorption of albumin and improved thromboresistance *ex vivo*. Albumin has also been grafted onto polymeric surfaces, although such materials do not possess long-term stability [60, 65-67].

The grafting of hydrophilic monomers [68] or polymers onto polymeric substrates has been used to improve the interfacial chemistry between polymers and biological fluids. Hydrophilic polymers such as polyvinyl pyrrollidone (PVP), *N*-vinyl pyrrollidone (NVP), polyethylene glycol (PEG), polyethylene oxide (PEO), polyhydroxyethyl methacrylate (HEMA), and polyacrylamide have been grafted onto polymer surfaces. The grafting of polyethylene oxide (PEO) or polyethylene glycol has been widely studied. PEO has been reported to be resistant to the deposition of biological species. It had been proposed that a diffuse hydrophilic biomaterial surface would be blood compatible by exerting steric repulsion to proteins and cells that approach the surface [69]. The motion of PEO chains is thought to contribute to the low deposition of proteins and cells by reducing the contact time between proteins and cells with the artificial surface (Fig. 3 [70]). It has also been postulated that the motion of PEO chains create microflows of water, preventing plasma protein adsorption, although the degree to which PEO chains inhibit deposition may be dependent on the molecular weight of the polymer chains [71]. Theoretical modelling suggests that steric repulsion is the main mechanism by which protein deposition is inhibited by PEO chains bound to a polymer surface.

Materials grafted with phosphorylcholine moieties have been synthesized, based on the observation that the red blood cell membrane, which is largely composed of phosphorylcholine, presents a non-thrombogenic surface. These materials are characterized by low protein adsorption, and enhanced thromboresistance [72-75]. The precise mechanism by which these surfaces resist protein and cellular deposition is not completely

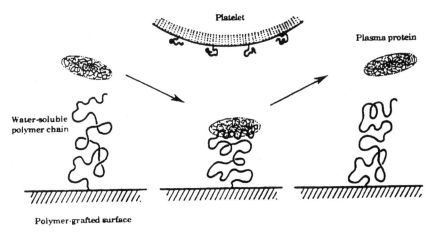

Fig. 3. Steric repulsion of plasma proteins and platelets by surfaces grafted with water-soluble polymers [70].

understood, although it is likely that the reduction in protein adsorption in turn reduces platelet deposition [76].

Other approaches to reduce platelet adhesion includes the grafting of anti-platelet agents onto the surface. Prostaglandin analogues that inhibit platelet aggregation have been grafted onto polyurethanes with favorable results *in vitro* [65, 77]. Heparinized surfaces have been prepared through both covalent and electrostatic binding, with varying reports on the effect of platelet deposition.

Numerous studies have found that the inoculum of bacteria required to initiate infection is dramatically lowered in the presence of a foreign material [78]. Thus, modification of the surface chemistry of synthetic biomaterials to resist bacterial adhesion would be of considerable value in decreasing clinical rates of infection. To this end, a number of surface modifications have been attempted, including modification with surfactants, antimicrobial peptides, quaternary ammonium compounds, and heavy metal ions. Bridgett and coworkers found that polystyrene modified with Pluronic surfactants (block copolymers of PEO and PPO) were capable of significantly decreasing bacterial adhesion compared to polystyrene alone, presumably by steric hindrance [79]. Haynie *et al.* studied a series of polyamide resins possessing bound antimicrobial peptides of the magainin class with respect to their bactericidal activity. Magainins act by disrupting the outer membrane of bacterial cells forming ion channels and upon immobilization to surfaces, can reduce the number of viable cells of a number of strains by several orders of magnitude [80]. The antibacterial effects of quaternary ammonium compounds have been studied for many years, and these compounds are active in many forms [81]. Tertiary amines covalently bound to polystyrene have been found to have bactericidal ability against *Escherichia coli* [82]. Silver has been incorporated into or coated onto polymers such as Nylon, PVC, silicone rubber, and Teflon. Antimicrobial effects against *Pseudomonas aeruginosa, Escherichia coli, Staphylococcus aureus*, and *Staphylococcus epidermidis* were found with all of these materials [83–85].

Another feature of artificial surfaces that can influence cell surface interactions is that of surface texture. Many studies on the influence of surface topography have been performed in the development of synthetic vascular grafts. There have been numerous studies where the texture of the graft has been shown to be more important to tissue ingrowth than polymer surface chemistry [86–90]. Phagocytosis of latex particles has also been shown to be dependent on particle size [91]. Porous surfaces may promote neointima formation on a synthetic biomaterial by improving the anchorage of fibrin and other material, and the pore size can be used to promote different types of cellular ingrowth [92]. Neointima formation is desirable, as the development of a stable lining of endothelial cells within a synthetic vascular graft should improve the thromboresistance and decrease bacterial colonization. Attempts to improve the endothelialization of polyurethanes in blood-contacting applications have also involved seeding device surfaces with endothelial cells prior to implantation [93], pretreating surfaces with growth factors and proteins [94, 95], or through coating with extracellular matrix components prior to seeding [96–98]. Improvements in the extent of endothelialization of implants has been observed in animal models.

References

1. Loosdrecht, M. C. Mv., Physical chemical description of bacterial adhesion. *Journal of Biomaterials Applications*, 1990, **5**(October), 91–106.
2. Neu, T. R. and Marshall, K. C., Bacterial polymers: physicochemical aspects of their interactions at interfaces. *Journal of Biomaterials Applications*, 1990, **5**, 107–133.
3. Mosher, D. F., Adhesive proteins and their cellular receptors. *Cardiovascular Pathology*, 1993, **2**(3), 149S–155S.
4. Vogler, E. A., Interfacial chemistry in biomaterials science. In *Wettability*, ed. J. C. Berg. Marcel Dekker, New York, 1993, pp. 183–250.
5. Lauffenburger, D. A. and Linderman, J. J., Receptors: Models for Binding, Trafficking, and Signaling. Oxford University Press, New York, 1993.
6. Bongrand, P., Intermolecular Forces. In *Physical Basis of Cell-Cell Adhesion*, ed. P. Bongrand. CRC Press, Inc., Boca Raton, 1988, pp. 1–37.
7. Hynes, R. O., Integrins: A family of cell surface receptors. *Cell*, 1987, **48**, 549–554.

8. Phillips, D. R., Charo, I. F. and Scarborough, R. M., GPIIb-IIIa: The responsive integrin. *Cell*, 1991, **65**, 359–362.

9. Springer, T. A., Thompson, W. S., Miller, L., Schmalstieg, F. C. and Anderson, D. C., Inherited deficiency of the Mac-1, LFA-1, p150,95 glycoprotein family and its molecular basis. *Journal of Experimental Medicine*, 1984, **160**(1901).

10. Sanchez-Madrid, F., Nagy, J., Robbins, E., Simon, P. and Springer, T. A., A human leukocyte differentiation antigen family with distinct alpha subunits and a common beta subunit: the lymphocyte function-associated antigen (LFA-1), the C3bi complement receptor (Mac-1), and the p150,95 molecule. *Journal of Experimental Medicine*, 1983, **158**, 1785.

11. Travis, J., Biotech gets a grip on cell adhesion. *Science*, 1993, **260**, 906–908.

12. Norde, W. and Lyklema, J. Why proteins prefer interfaces. In *The Vroman Effect*, ed. C. H. Bamford, S. L. Cooper and T. Tsuruta. VSP, The Netherlands, 1992, pp. 1–20.

13. Vroman, L., Adams, A. L., Klings, M., Fischer, G. C., Munoz, P. C. and Solensky, R. P., Reactions of formed elements of blood with plasma proteins at interfaces. *Annals of the New York Academy of Sciences*, 1977, **283**, 65–76.

14. Baier, R. E., Key events in blood interactions at non-physiologic interfaces — a personal primer. *Artificial Organs*, 1978, **2**(4), 422–426.

15. Gendreau, R. M. and Jacobsen, R. J., Fourier transform infrared techniques for studying complex biological systems. *Applied Spectroscopy*, 1978, **32**(3), 326–328.

16. Sato, H., Kojima, J. and Nakajima, A., Fibrinogen adsorption on artificial surfaces and its effect on platelets. *Journal of Dispersion Science and Technology*, 1993, **14**, 117–128.

17. Brash, J. L., Protein interactions with solid surfaces following contact with plasma and blood. *Makromolecular Chemistry and Macromolecules Symposium*, 1988, 441–452.

18. Vroman, L. and Adams, A. L., Identification of rapid changes at plasma-solid interfaces. *Journal of Biomedical Materials Research*, 1969, **3**, 43–67.

19. Vroman, L., Adams, A. L., Fischer, G. C. and Munoz, P. C., Interaction of high molecular weight kininogen, factor XII and fibrinogen at plasma interfaces. *Blood*, 1980, **55**(1), 156–159.

20. Asakura, S., Hurley, R. W., Skorstengaard, K., Ohkubo, I. and Mosher, D. F., Inhibition of cell adhesion by high molecular weight kininogen. *Journal of Cell Biology*, 1992, **116**(2), 465–476.

21. Ziats, N., Jablonski-Bernasconi, M. and Anderson, J., Inhibition of human cell adhesion by high molecular weight kininogen. *Transactions of the Society for Biomaterials*, 1994, **17**, 74.

22. Massia, S. P. and Hubbell, J. A., Covalently attached GRGD on polymer surfaces promotes biospecific adhesion of mammalian cells. In *Annals of the New York Academy of Science*, eds. W. E. Goldstein, D. DiBiasio and H. Pedersen, 1990, 261–270.

23. Pierschbacher, M. D. and Ruoslahti, E., Cell attachment activity of fibronectin can be duplicated by small synthetic fragments of the molecule. *Nature* 1984, **309**, 30–34.

24. Pytela, R., Pierschbacher, M. D., Ginsberg, M. H., Plow, E. F. and Ruoslahti,

E., Platelet membrane glycoprotein IIb/IIIa: Member of a family of Arg-Gly-Asp specific adhesion receptors. *Science* 1986, **2231**, 1559–1562.

25. Kishida, A., Takatsuka, M. and Matsuda, T. RGD-albumin conjugate: expression of tissue regeneration activity. *Biomaterials* 1992, **13**(13), 924–930.

26. Lee, R. G. and Kim, S. W., The role of carbohydrate in platelet adhesion to foreign surfaces. *Journal of Biomedical Materials Research* 1974, **8**, 393–398.

27. Park, K. D., Mosher, D. F. and Cooper, S. L., Acute surface-induced thrombosis in the canine ex vivo model: importance of protein concentration of the initial monolayer and platelet activation. *Journal of Biomedical Materials Research*, 1986, **20**, 589–612.

28. Collins, W. E., Mosher, D. F. and Diwan, A. R. et al., Ex vivo platelet deposition on fibronectin-preadsorbed surfaces. *Scanning Electron Microscopy* 1987, **1**(4), 1669–167.

29. Rapoza, R. J. and Horbett, T. A., Changes in the SDS elutability of fibrinogen adsorbed from plasma to polymers. *Journal of Biomaterials Science, Polymer Edition*, 1989, **1**(2), 99–110.

30. Chan, B. M. C. and Brash, J. L., Conformational changes in fibrinogen desorbed from glass surface. *Journal of Colloid and Interface Science*, 1981, **84**(1), 263–265.

31. Slack, S. M., Rapoza, R. J. and Horbett, T. A., Changes in the state of fibrinogen adsorbed to biomaterials. The Third World Biomaterials Congress. Kyoto, Japan, 1988, p. 95.

32. Chinn, J. A., Posso, S. E., Horbett, T. A. and Ratner, B. D., Residence time effects in surface bound fibrinogen as indicated by changes in SDS elutability, antibody binding and platelet adhesion. *Transactions of the Society for Biomaterials*, 1990, **16**, 242.

33. Silver, J. H., Hart, A. P. and Williams, E. C. et al., Anticoagulant effects of sulphonated polyurethanes. *Biomaterials*, 1992, **13**(6), 339–344.

34. Santerre, J. P., Hove, Pt., vanderKamp, N. H. and Brash, J. L., Effect of sulfonation of segmented polyurethanes on the transient adsorption of fibrinogen from plasma: possible correlation with anticoagulant behavior. *Journal of Biomedical Materials Research*, 1992, **26**, 39–57.

35. Okkema, A. Z., Yu, X. -H. and Cooper, S. L., Physical and blood contacting properties of propyl sulphonate grafted Biomer. *Biomaterials*, 1991, **12**(1), 3–12.

36. Santerre, J. P., VanderKamp, N. H. and Brash, J. L., Effect of sulfonation of segmented polyurethanes on the transient adsorption of fibrinogen from plasma: possible correlation with anticoagulant behavior. *Transactions of the Society for Biomaterials*, 1989, 15, p. 113.

37. Fabrizius-Homan, D. J. and Cooper, S. L., Competitive adsorption of vitronectin with albumin, fibrinogen, and fibronectin on polymeric biomaterials. *Journal of Biomedical Materials Research*, 1991, **25**, 953–971.

38. Han, D. K., Ryu, G. H., Park, K. D., Jeong, S. Y., Kim, Y. H. and Min, B. G., Adsorption behavior of fibrinogen to sulfonated polyethyleneoxide-grafted polyurethane surfaces. *Journal of Biomaterials Science, Polymer Edition*, 1993, **4**(5), 401–413.

39. Ruoslahti, E. and Pierschbacher, M. D., New perspectives in cell adhesion: RGD and integrins. *Science*, 1987, **238**, 491–497.

40. D'Souza, S. E., Ginsberg, M. H. and Plow, E. F., Arginyl-glycyl-aspartic acid (RGD): a cell adhesion motif. *Trends in Biological Science*, 1991, **16**, 246–250.

41. Hautanen, A., Gailit, J., Mann, M. and Ruoslahti, E., Effects of modification of the RGD sequence and its context on recognition by the fibronectin receptor. *Journal of Biological Chemistry*, 1989, **264**, 1437–1442.

42. Drumheller, P. D. and Hubbell, J. A., Polymer networks with grafted cell adhesion peptides for highly specific cell adhesive substrates. *Annals of Biochemistry*, 1994, **222**, 380–388.

43. Hirano, Y., Okuna, M., Hayashi, T., Goto, K. and Nakajima, A., Cell-attachment activities of surface immobilized oligopeptides RGD, RGDS, RGDV, RGDT, and YIGSR towards five cell lines. *Journal of Biomaterials Science, Polym. Ed.*, 1993, **4**(3), 235–243.

44. Massia, S. P. and Hubbell, J. A., Human endothelial cell interactions with surface coupled adhesion peptides on a nonadhesive glass substrate and two polymeric biomaterials. *Journal of Biomedical Materials Research*, 1991, **25**(2), 223–242.

45. Lin, H. -B., Garcia-Echeverria, C., Asakura, S., Sun, W., Mosher, D. F. and Cooper, S. L., Endothelial cell adhesion on polyurethanes containing covalently attached RGD peptides. *Biomaterials*, 1992, **13**, 905–914.

46. Massia, S. P. and Hubbell, J. A., Covalent surface immobilization of Arg-Gly-Asp and Tyr-Ile-Gly-Ser-Arg containing peptides to obtain well-defined cell adhesive substrates. *Annals of Biochemistry*, 1990, **17**, 292–301.

47. Matsuda, T., Kondo, A., Makino, K. and Akutsu, T., Development of a novel artificial matrix with cell adhesion peptide for cell culture and artificial hybrid organs. *Trans. Am. Soc. Artif. Intern. Organs*, 1989, **35**, 677–679.

48. Brandley, B. K. and Schnaar, R. L., Covalent attachment of an Arg-Gly-Asp sequence peptide to derivatizable polyacrylamide surface: support of fibroblast adhesion and long-term growth. *Annals of Biochemistry*, 1988, **172**, 270–278.

49. Lin, H. -B., Sun, W., Mosher, D. F., Garcia-Echeverria, C., Schaufelberger, K., Lelkes, P. I. and Cooper, S. L., Synthesis, surface and cell adhesion properties of polyurethanes containing covalently grafted RGD-peptides. *Journal of Biomedical Materials Research*, 1994, **28**, 329‑342.

50. Plow, E. F., Pierschbacher, M. D., Rusolahti, E., Marguerie, G. A. and Ginsberg, M. H., The effect of Arg-Gly-Asp containing peptides on fibrinogen and von Willebrand factor binding to platelets. *Proceedings of the National Academy of Science* USA, 1985, **82**, 8057–8061.

51. Plow, E. F., Pierschbacher, M. D., Rusolahti, E, Marguerie, G. A. and Ginsberg, M. H., Arginyl-Glycyl-Aspartic acid sequences and fibrinogen binding to platelets. *Blood*, 1987, **70**, 110–115.

52. Goodman, S. L., Cooper, S. L. and Albrecht, R. M., Integrin receptors and platelet adhesion to synthetic surfaces. *Journal of Biomedical Materials Research* 1993, **27**, 683–695.

53. Peters, G., Locci, R. and Pulverer, G., Adherence and growth of coagulase-negative staphylococci on surfaces of intravenous catheters. *Journal of Infectious Disease* 1982, **146**, 479–482.

54. Herrmann, M., Vaudaux, P. E., Pittet, D., Auckenthaler, R., Lew, P. D., Schumacher-Perdreau, F., Peters, G. and Waldvogel, F. A., Fibronectin,

fibrinogen, and laminin act as mediators of adherence of clinical Staphylococcal isolates to foreign material. *Journal of Infectious Disease*, 1988, **158**(4), 693–701.

55. Mohammad, S. F., Topham, N. S., Burns, G. L. and Olsen, D. B., Enhanced bacterial adhesion on surfaces pretreated with fibrinogen and fibronectin. *Transactions of the American Society for Artificial Internal Organs*, 1988, **34**, 573–577.

56. Brash, J. L., Role of plasma protein adsorption in the response of blood to foreign surfaces. In *Blood Compatible Materials and Devices*, eds. C. P. Sharma and M. Szycher. Technomic, Lancaster, PA, 1991, pp. 3–24.

57. Mustard, J. F., Glynn, M. F., Nishizawa, E. E. and Packham, M. A., Platelet-surface interactions, relationship to thrombosis and haemostasis. *Federation Proceedings*, 1967, **26**, 106–114.

58. Young, B. R., Lambrecht, L. K., Cooper, S. L. and Mosher, D. F., Plasma proteins: their role in initiating platelet and fibrin desposition on biomaterials. In *Biomaterials: Interfacial Phenomena and Applications*, ed. S. L. Cooper and N. A. Peppas. *Adv. Chem. Ser.*, 1982, **199**, 317.

59. Young, B. R., Doyle, M. J. and Collins, W. E. et al., Effect of thrombospondin and other platelet alpha-granule proteins on artificial-surface induced thrombosis. *Trans. Am. Soc. Artif. Intern. Organs*, 1982, **28**, 498.

60. Munro, M. S., Eberhart, R. C., Maki, N. J., Brink, B. E. and Fry, W. J., Thromboresistant alkyl derivatized polyurethanes. *ASAIO J.*, 1983, **6**, 65–75.

61. Eberhart, R. C., Lynch, M. E. and Bilge, F. H. et al., Protein adsorption on polymers, visualization, study of fluid shear and roughness effects, and methods to enhance albumin binding. In *Biomaterials: Interfacial phenomena and applications*, eds. S. L. Cooper, N. A. Peppas. *Adv. Chem. Ser.*, 1982, **199**, 293.

62. Grasel, T. G., Pierce, J. A. and Cooper, S. L., Effects of alkyl grafting on surface properties and blood compatibility of polyurethane block copolymers. *Journal of Biomedical Materials Research*, 1987, **21**, 815–842.

63. Pitt, W. G. and Cooper, S. L., Albumin adsorption on alkyl chain derivatized polyurethanes: I. The effect of C-18 alkylation. *Journal of Biomedical Materials Research*, 1988, **22**, 359–382.

64. Keogh, J. R. and Eaton, J. W., Albumin binding surfaces for biomaterials. *Journal of Laboratory and Clinical Medicine*, 1994, **124**(4), 537–545.

65. Joseph, G. and Sharma, C. P., Prostacyclin immobilized albuminated surfaces. *Journal of Biomedical Materials Research*, 1987, **21**, 937–945.

66. Engbers, G. H. M., An in-vitro study of the adhesion of blood platelets onto vascular catheters. *Journal of Biomedical Materials Research*, 1987, **21**, 613–627.

67. Ryu, G. H., Han, D. K., Kim. Y. H. and Min, B. Albumin immobilized polyurethane and its blood compatibility. *American Society for Artificial Internal Organs Journal*, 1992, **38**(3), M644–M648.

68. Bamford, C. H. and Middleton, I. P., Studies on functionalizing and grafting to poly(ether-urethanes). *European Polymer Journal*, 1983, **19**, 1027–1035.

69. Ikada, Y., Blood compatible polymers. *Advances in Polymer Science*, 1984, (57), 103–140.

70. Amiji, M. and Park, K., Surface modification of polymeric biomaterials with poly (ethylene oxide), albumin and heparin for reduced thrombogenicity. *Journal of Biomaterials Science, Polymer Edition*, 1993, **4**(3), 217–234.

71. Chaikof, E. L., Merrill, E. W., Callow, A. D., Connolly, R. J., Verdon, S. L. and Ramberg, K., PEO enhancement of platelet deposition, fibrinogen deposition and complement C3 activation. *Journal of Biomedical Materials Research* 1992, **26**, 1163–1168.

72. Ishihara, K., Aragaki, R., Ueda, T., Watanabe, A. and Nakabayashi, N., Reduced thrombogenicity of polymers having phospholipid polar groups. *Journal of Biomedical Materials Research*, 1990, **24**, 1069–1077.

73. Ishihara, K., Ziats, N. P., Tierney, B. P., Nakabayashi, N. and Anderson, J. M., Protein adsorption from human plasma is reduced on phospholipid polymers. *Journal of Biomedical Materials Research*, 1991, **25**, 1397–1407.

74. Chapman, D., Biocompatible surfaces based upon the phospholipid asymmetry of biomembranes. *Biochemical Society Transactions*, 1993, **21**, 258–262.

75. Yu, J., Lamba, N. M. K, Courtney, J. M., Whateley, T. L., Gaylor, J. D. S., Ishihara, K., Nakabayashi, N. and Lowe, G. D. O., Polymeric biomaterials: influence of phosphorylcholine polar groups on protein adsorption and complement activation. *International Journal of Artificial Organs*, 1994, **17**(9), 499–504.

76. Iwasaki, Y., Kurita, K., Ishihara, K. and Nakabayashi, N., Effect of reduced protein adsorption on platelet adhesion at the phospholipid polymer surfaces. *Journal of Biomaterals Science, Polymer Edition*, 1996, **8**(2), 151–163.

77. Bamford, C. H., Middleton, I. P. and Satake, Y., Grafting and attachment of anti-platelet agents to poly(ether-urethanes). *Polymer Reprints*, 1984, 24–25.

78. Zimmerli, W., Waldvogel, F. A., Vaudaux, P. and Nydegger, U. E., Pathogenesis of foreign body infection: description and characteristics of an animal model. *The Journal of Infectious Diseases*, 1982, **146**(4), 487–497.

79. Bridgett, M. J., Davies, M. C. and Denyer, S. P., Control of staphylococcal adhesion to polystyrene surfaces by polymer surface modification with surfactants. *Biomaterials* 1992, **13**, 411–416.

80. Haynie, S. L., Crum, G. A. and Doele, B. A., Antimicrobial activities of amphiphilic peptides covalently bonded to a water-insoluble resin. *Antimicrobial Agents & Chemotheropy*, 1995, **39**, 301–307.

81. Petrocci, A. N., Surface active agents: quaternary ammonium compounds. In *Disinfection, Sterilization, and Preservation*, ed. S. S. Block. Lea and Febiger, Philadelphia, 1983.

82. Endo, Y., Tani, T. and Kodama, M., Antimicrobial activity of tertiary amine covalently bonded to a polystyrene fiber. *Applied Environmental Microbiology*, 1987, **53**, 2050–2055.

83. Deitch, E. A., Marino, A. A., Cillespie, T. E. and Albright, J. A., Silver-nylon: a new antimicrobial agent. *Antimicrobial Agents & Chemotheropy* 1983, **23**, 356–359.

84. Farrah, S. R. and Erdos, G. W., The production of antibacterial tubing, sutures, and bandages by in situ precipitation of metallic salts. *Canadian Journal of Microbiology*, 1991, **37**, 445–449.

85. McLean, R. J. C, Hussain, A. A., Sayer, M., Vincent, P. J., Hughes, D. J. and Smith, T. J. N., Antibacterial activity of multilayer silver-copper surface films on catheter material. *Canadian Journal of Microbiology*, 1993, **39**, 895–899.

86. Berhart, W. F., LaFarge, C. G., Liss, R. H., Szycher, M., Berger, R. L. and Poirier, V., An appraisal of blood trauma and the blood-prosthetic interface

during left ventricular bypass in the calf and humans. *Annals of Thoracic Surgery*, 1978, **26**, 427.

87. Hess, F., Jerusalem, C. and Braun, B., The endothelialization process of a fibrous polyurethane microvascular prosthesis after implantation in the abdominal aorta of the rat. *Journal of Cardiovascular Surgery*, 1983, **24**, 516–524.

88. Hess, F., Jerusalem, C., Braun, B. and Grande, P., Three years experience with experimental implantation of fibrous polyurethane microvascular prostheses in the rat aorta. *Microsurgery*, 1985, **6**, 155–162.

89. Whalen, R., Improved textured surfaces for implantable prostheses. *Trans. Am. Soc. Artif. Intern. Organs*, 1988, **34**, 887–892.

90. Kogel, H., Vollmar, J. F. and Proschek, P., New prostheses for venous substitution. *Journal of Cardiovascular Surgery* 1991, **32**, 330–333.

91. Kawaguchi, H., Koiwai, N., Otsuka, Y., Miyamoto, M. and Sasakawa, S., Phagocytosis of latex particles by leukocytes. I. Dependence of phagocytosis on the size and surface potential of particles. *Biomaterials* 1986, 7.

92. Gogolewski, S., Degradable, microporous vascular prosthesis from segmented polyurethane. *Colloid Polymer Science*, 1986, **264**, 854–858.

93. Wachem, P. Bv., Stronck, J. W. S., Koers-Zuideveld, R., Dijk, F. and Wildevuur, C. R. H., Vacuum cell seeding: a new method for the fast application of an evenly distributed cell layer on porous vascular grafts. *Biomaterials* 1990, **11**, 602–605.

94. Soldani, G., Steiner, M., Galletti, P. M., Lelli, L., Palla, M. and Giusti, P., Development of small-diameter vascular prostheses which release bioactive agents. *Clinical Materials* 1991, **8**, 81–88.

95. Zilla, P., Fasol, R. and Grimm, M. et al., Growth properties of cultured human endothelial cells on differently coated artificial heart materials. *Journal of Thoracic Cardiovascular Surgery*, 1991, **101**, 671–680.

96. Bordenave, L., Baquey, C. and Bareille, R. et al., Endothelial cell compatibility testing of three different Pellethanes. *Journal of Biomedical Materials Research* 1993, **27**, 1367–1381.

97. Lee, Y., Park, D. K., Kim, Y. B., Seo, J. W., Lee, K. B. and Min, B., Endothelial cell seeding onto the extracellular matrix of fibroblasts for the development of a small diameter polyurethane vessel. *ASAIO J.* 1993. **39**, M740–M745.

98. Miwa, H., Matsuda, T., Tani, N., Kondo, K. and Iida, F., An *in vitro* endothelialized compliant vascular graft minimizes anastomotic hyperplasia. *ASAIO J.* 1993, **39**, M501–M505.

CHAPTER II.7

Wound Healing

JENNIFER L. WEST
Department of Bioengineering,
Rice University
Houston, TX 77005, USA

Wound healing complications, such as non-healing wounds, excessive scar tissue formation, or loss of tissue function, represent a major clinical issue in today's health care. For instance, approximately 500,000 diabetic patients are treated for chronic ulcers each year in the U.S.: amputation is required in approximately 10% of these cases [1]. In addition, approximately 10% of all major laparotomies in the U.S. are performed to relieve bowel obstruction caused by a type of scar tissue formation known as post-operative adhesion formation [2]. Tissue engineering and biomaterials approaches may be useful for the treatment of many of these wound healing complications. Some of the goals of such approaches may be to increase the rate of tissue regeneration, support and guide tissue regeneration to improve function of the healed tissue, or provide barriers to scar tissue formation.

1 Etiology of wound healing

Wound healing is a complex process involving a number of types of cellular and molecular events over a period of many months. As a result of severe tissue injury, blood vessels are disrupted, leading to coagulation along with platelet aggregation and degranulation. The fibrin-rich clot which results serves to achieve hemostasis over the short-term, and acts as a provisional matrix for cell migration and attachment over the longer-term. Platelet degranulation is also a rich source of growth factors that stimulate the healing process. Over the first several days, the wounded tissue becomes infiltrated with neutrophils, monocytes and fibroblasts. This is the inflammatory phase of wound healing. Macrophages and fibroblasts in the wounded tissue produce growth factors and other cytokines, as well as extracellular matrix proteins, and fibroblast proliferation is marked. During approximately the 2nd week post-wounding, the migration of tissue-specific cell types is observed. This period is the tissue formation phase of wound healing. The tissue remodeling phase, during which there is noticeable collagen turnover and cross-linking, occurs from 2 weeks to 1 year post-wounding. A brief overview of each of the phases of wound healing is provided below.

1.1 Coagulation

The primary function of blood coagulation after wounding is hemostasis. It should be noted, however, that coagulation occurs in the extravascular space even in the absence of frank blood vessel injury. In this case, capillary permeability is increased due to release of histamine and vasoactive kinins from damaged stromal cells, thus allowing plasma to enter the interstitial space where it is converted to a fibrin-rich clot [3]. The clot is composed of cross-linked fibrin, fibronectin, and platelets, with entrapped plasma proteins and blood cells [4]. In addition to achieving hemostasis, the clot serves as a provisional matrix to support the migration of cells involved in wound healing, and some factors involved in or derived from the coagulation cascade may stimulate the inflammatory process [5].

Coagulation begins immediately upon tissue injury, and serves to initiate later events in the wound healing cascade. Coagulation is promoted when blood constituents are exposed to surfaces that support adsorption and activation of coagulant proenzymes (such as Factor XII, an initiator of the intrinsic coagulation pathway) and adhesion of activated platelets. In addition, damaged stromal cells and activated neutrophils and macrophages produce tissue factor, and thus initiate the extrinsic coagulation pathway [6]. Each of these mechanisms causes the production of the prothrombinase complex on the platelet membrane, thus leading to the production of thrombin, the enzyme which produces fibrin from fibrinogen. Fibrin is able to polymerize to form a cross-linked clot matrix. A complex control system exists to regulate the extent of coagulation: the control mechanisms include enzymes that degrade the clot [7–9], protease inhibitors that interact with enzymes in the coagulation cascade [10], and factors that interfere with platelet aggregation [11].

1.2 Inflammation

The inflammation phase of wound healing begins shortly after injury and generally persists for a period of several days, depending on the severity of injury and the degree of wound contamination. Inflammatory cells are recruited to the site of injury by substances originating from platelets (such as platelet-derived growth factor and thrombin) and from damaged cells (such as interleukins and growth factors) and by substances generated as a result of the activation of the coagulation cascade (such as kallikrein and

Table 1. Important mediators of the inflammatory response during wound healing

Classes of inflamamtion mediators	Examples
Growth factors	Platelet-derived growth factor, fibroblast growth factor, transforming growth factor-β
Coagulation factors and products	Factor XIIa, fibrinopeptides
Vasoactive amines	Histamine
Kinins	Bradykinin
Proteases	Collagenase, plasmin
Cytokines	Interleukins, tumor necrosis factor
Complement enzymes	C3a, C5a

fibrinopeptides). Table 1 gives a list of the most important mediators of the inflammatory response during wound healing. The presence of foreign material in the wound site will further stimulate and prolong inflammation.

Neutrophils appear in large numbers very soon after injury and serve to remove foreign materials from the wounded region. Monocytes also appear in the damaged tissue shortly after injury and are then activated to form macrophages. Monocyte accumulation generally persists for a longer duration than that of neutrophils. Macrophages are crucial to tissue repair as they resorb necrotic tissue, phagocytose foreign materials, and release a number of growth factors that facilitate tissue regeneration [12].

1.3 Tissue formation

Tissue repair after wounding is comprised of several distinct processes; repaving of the surface with epithelial, endothelial, or mesothelial cells, formation of a new stroma, and revascularization. Tissue repair begins within the first few days of injury, and may continue for several weeks. Numerous growth factors and cytokines are present in the wound environment, including platelet-derived growth factor, fibroblast growth factor, epidermal growth factor and transforming growth factor. These substances are derived from coagulation and platelet degranulation, the inflammatory process and damaged stromal cells.

Because it is vital to re-establish normal tissue boundaries and barrier properties, repaving of the tissue surface is the first repair process to commence. Repaving involves the migration of epithelial cells (endothelial or mesothelial cells in some tissue types) from the perimeter of the wounded region. Within 48 hours of injury, epithelial cells at the wound margin begin to proliferate [13]. These cells then migrate through the provisional matrix formed during the coagulation process. As they migrate, the matrix proteins in the pathway of the cells are degraded by proteases released at the leading edge of the migrating cells [14]. Newly synthesized basement membrane proteins are deposited into the matrix by epithelial cells just behind the migration front [15].

The formation of new stroma generally begins several days after injury. This newly forming tissue is referred to as granulation tissue due to its granular appearance. It is composed of fibroblasts, macrophages and abundant new capillaries. As these cells migrate into the provisional fibrin-rich matrix [5], they must degrade matrix proteins in their pathway, and then new matrix proteins are synthesized by wound fibroblasts and redeposited in the matrix. These changes in the composition of the extracellular matrix serve as signals to regulate wound healing responses. The stimuli for macrophage and fibroblast migration and angiogenesis include the numerous growth factors present in the injured tissue, the provisional matrix composition, and loss of nearest-neighbor cell contact. Migration of tissue type-specific cells into the wounded region generally lags several days behind that of fibroblasts.

1.4 Tissue remodeling

The processes of tissue remodeling, namely matrix formation and reorganization, may continue for as long as 1 year after injury. During the remodeling process, fibrin, fibronectin and hyaluronic acid (present in the provisional matrix) disappear while collagen is deposited and cross-linked to increase the tensile strength of the tissue. Proteoglycans (predominantly chondroitin sulfate, heparan sulfate, and dermatan sulfate) are also added to the matrix structure during this period. Tissue remodeling restores appropriate mechanical properties, both strength and elasticity, to the damaged tissue. Generally, not more than two-thirds of the initial tissue strength is regained [16].

2 Biomaterials for wound healing applications

Biomaterials may be employed to alter wound healing, either as matrices to support and promote tissue regeneration or as barriers to limit scar tissue formation. Biomaterials of the first type must either be intrinsically conducive to cell attachment or else contain bioactive moieties that support cell adhesion and migration. On the other hand, to prevent scar tissue formation, barrier materials must be highly resistant to protein adsorption and cell adhesion. Biomaterials can also be utilized as barriers to prevent wound contamination and regulate tissue hydration and have been shown to improve wound healing through these mechanisms.

2.1 Scaffolds for tissue growth

A number of polymeric biomaterials, both natural and synthetic, have been investigated for use as scaffolds to support tissue regeneration during wound healing. These materials are being developed to increase the rate or incidence of wound healing and to spatially guide tissue regeneration.

Some of the natural polymers utilized as matrices to facilitate wound healing have included collagen [17, 18], collagen–GAG copolymers [19], and fibrin [20, 21]. In these applications, the implanted material serves as a provisional extracellular matrix that is later resorbed and remodeled by the healing tissue.

In the design of synthetic polymer scaffolds for wound healing, a number of properties have been identified that are crucial. In order to support tissue regeneration, these polymeric materials must be porous so that migrating cells can penetrate into the matrix and so that diffusion of nutrients is not limited. The polymers should also be biodegradable, generating non-toxic degradation products. As in all biomaterials applications, biocompatibility should be a significant concern. Lastly, the polymer material must either readily support adsorption of adhesive proteins or else contain cell-adhesive moieties in order to support biospecific cell adhesion and migration.

The poly(α-hydroxy acid) family of polymers has been utilized in several in vivo wound healing applications. For example, porous D,L-polylactide has been utilized to fill calvarial bone defects in a rabbit model and was shown to support bone ingrowth [22]. In addition, poly-L-lactide/poly-ϵ-caprolactone copolymers have been used to aid in repair and regeneration of severed sciatic nerves in a rat model [23]. These polymers have been used extensively as scaffolds for in vitro seeding of cells, with implantation of a cell/polymer construct to facilitate wound healing. For example, a dermal replacement material composed of polyglycolic acid and cultured human fibroblasts has been evaluated in full thickness skin wounds in athymic mice [24]; these materials were found to be vascularized and epithelialized during wound healing. Polyphosphazenes have also shown promise as scaffold materials for wound healing applications and have been evaluated for both bone [25] and nerve [26] regeneration.

Bioactive polymers, polymer materials that have been enhanced through the incorporation of specific bioactive sequences, show perhaps the greatest degree of promise for use as wound healing scaffolds. Short peptide sequences derived from the cell binding regions of extracellular matrix proteins can be utilized to achieve biospecific adhesion of cells to biomaterials. Table 2 lists a number of these cell adhesion peptides, along with the proteins from which they are derived.

The RGD peptide, present in fibronectin, collagen, and numerous other matrix proteins, is able to support adhesion of most cell types and interacts

Table 2. Cell adhesion sequences derived from extracellular matrix proteins

Peptide sequence	Parent protein
RGD	Fibronectin, collagen, fibrinogen, laminin, vitronectin, von Willebrand's Factor, entactin, tenascin, thrombospondin
YIGSR	Laminin
IKVAV	Laminin
LRE	Laminin
REDV	Fibronectin
DGEA	Collagen
VTXG	Thrombospondin
VGVAPG	Elastin

with several different integrin receptors [27]. Bioactive polymers containing the RGD sequence have been used in several *in vivo* wound healing applications. For example, RGD peptides coupled to *N*-(2-hydroxypropyl) methacrylamide hydrogels encourage ingrowth of glial tissue in the rat [28]. In clinical trials, hydrogels containing RGD peptides were able to significantly promote healing of diabetic ulcers [29] and partial-thickness burn wounds [30]. Other peptide sequences, such as the endothelial cell-selective REDV peptide [31], may impart the biomaterial with cell type selectivity. Laminin-derived peptides, such as YIGSR [32], IKVAV [33], and LRE [34], may be especially suitable for nerve regeneration applications.

Incorporation of other types of bioactive species into biomaterial structures may also improve wound healing characteristics. Biospecific cell adhesion has been achieved through immobilization of carbohydrate moieties such as *N*-acetylglucosamine [35] and lactose [36, 37]. Immobilization of growth factors may also be useful for the stimulation of cell migration and proliferation during wound healing. Insulin has been immobilized onto polyurethanes [38] and onto poly(methyl methacrylate) [39]; in both cases the immobilized growth factor was capable of stimulating cell proliferation. Epidermal growth factor bound to polyethylene oxide surfaces has recently been shown to retain its mitogenic activity [40] and may have a number of applications in wound healing. In addition, several short peptide sequences derived from growth factors have been identified that can bind to the appropriate receptors and elicit the desired biological responses, albeit to a lesser degree than the intact protein. These have included sequences from epidermal growth factor [41] and fibroblast growth factor [42]. The use of short peptides rather than intact proteins may be advantageous in that instability due to protein denaturation and degradation will be reduced.

2.2 Materials as barriers: prevention of scar tissue formation

Post-operative adhesion formation is a significant wound healing complication in many fields of surgery, and may result in infertility, bowel obstruction, and pain [2, 3]. A number of polymeric biomaterials have been evaluated for use as barriers to prevent deposition of a fibrin-rich clot at the site of injury as well as the subsequent infiltration by fibroblasts. In general, such materials should be intrinsically cell non-adhesive, should not generate a marked inflammatory response, and should be biodegradable. Expanded polytetrafluoroethylene membranes have been shown to significantly reduce adhesion formation [43, 44], but are not biodegradable. Oxidized regenerated cellulose mesh has been shown to reduce adhesion formation in some studies [45], but suffers from limitations in its biocompatibility [46, 47]. Photopolymerized polyethylene glycol–co-polylactide hydrogels have been shown to reduce adhesion formation by more than 80% in several animal models [48, 49]. It may be possible to achieve an even greater degree of adhesion prevention through the design of materials that actively interfere with the coagulation or inflammation cascades, for example, through the immobilization of appropriate enzymes within the polymer matrix.

2.3 Materials as barriers: replacement of epithelial functions

Biomaterials and tissue engineering have played a significant role in the improvement of severe wound care through the development of novel wound dressing materials and skin graft substitutes. The focus of the materials discussed in this section is the prevention of wound contamination and maintenance of tissue hydration [50], though such materials may also be engineered to promote and support tissue regeneration. Various polymeric wound dressings have been shown to reduce the incidence of infection, maintain appropriate tissue hydration, and thus increase the rate of wound healing. Some of the polymers utilized as wound dressings have included polyurethanes [51], polyacrylamides [52], poly-ϵ-caprolactone [53], and polyvinyl alcohol [53].

3 Controlled release technologies

As understanding of the biological processes involved in wound healing increases, novel therapeutic modalities will arise for wound treatment. Many of these novel treatments will likely be based on delivery of bioactive macromolecules, such as proteins and oligonucleotides, and thus may require specialized release systems. The goals of controlled release in wound treatment may include increasing the rate of healing, reducing scar tissue formation, or preventing infection.

Local delivery of growth factors has been investigated to increase the rate of wound healing, the ultimate strength of the healed tissue, and the migration of cells into polymer scaffolds. Epidermal growth factor (EGF) released from polymer matrices has been shown to accelerate healing in corneal [54, 55] and skin wounds [56]. Acidic fibroblast growth factor has been shown to increase the rate of wound healing in a diabetic mouse model [57]. Treatment with platelet-derived growth factor after corneal laceration nearly doubled the strength of the wounded tissue, presumably due to the increased fibroblast infiltration and collagen production noted in the treated tissues [58]. Local delivery of transforming growth factor-beta has been shown to accelerate skin wound healing in rats [59]. A number of groups have evaluated the use of bone morphogenetic protein release to stimulate ingrowth of bone into biomaterial matrices [60–62].

A number of studies have examined the effects of administering exogenous plasminogen activators following injury to minimize scar tissue formation by promoting degradation of the fibrin clot, and thus reducing invasion by fibroblasts. Recombinant tissue plasminogen activator (tPA) has been locally delivered from oxidized regenerated cellulose membranes [63], sodium hyaluronate gels [64], and polyethylene glycol–co-polylactide hydrogels [65]. In each of these cases, tPA was shown to significantly reduce adhesion formation, and further, local delivery served to reduce side effects such as bleeding complications. Urokinase plasminogen activator is also able to significantly reduce adhesion formation when locally delivered from an appropriate vehicle [65]. Local delivery of ancrod, a fibrinogenolytic protease derived from snake venom, has also been found to significantly reduce both the extent and the severity of adhesion formation in a rat model [66]. Localization of delivery appears to be more critical than duration of delivery in determining the efficacy of these fibrinolytic and fibrinogenolytic agents.

4 Gene therapy

Gene therapy is the transfer of genetic material into the cells of an organism to treat a disease state. Recent advances in biotechnology have made it possible to transfer genes into mammalian cells and achieve expression of the targeted protein either *ex vivo* (with subsequent implantation of the genetically altered cells) or directly *in vivo*. A number of delivery systems for gene transfer are listed in Table 3.

While transfection rates are generally lower than desired, gene therapy has been successfully utilized to alter wound healing in several animal models. Keratinocytes transfected with the gene for growth hormone have been transplanted into full-thickness skin wounds in rats [67] and pigs [68]. The transfected cells were found to produce growth hormone for approxi-

Table 3. Delivery systems for gene transfer

Type of delivery system	Characteristics
Retrovirus	Enveloped RNA virus, stable integration of transgene into dividing cells
Adenovirus	Non-enveloped DNA virus, transfection of dividing and non-dividing cells, high titers, and efficient transfection
Liposomes	Synthetic lipid vesicles for DNA encapsulation, good efficiency in cell culture but lower *in vivo*
DNA-poly-L-lysine aggregates	Ionic interactions cause PLL to bind and condense DNA into toroid structures. Inefficient *in vivo*.
Electroporation	Naked DNA taken up by cells made permeable by electroporation. Good localization.
Gene Guns	DNA-coated gold particles 'fired' into cells at high velocity. Efficient transfection and good localization.

mately 10 days following implantation, and wound healing was significantly accelerated [68]. *In vivo* transfer of an EGF expression plasmid has been performed in a porcine partial thickness skin wound model [69]. In this case, EGF expression persisted for at least 10 days, and a significant increase in the rate of healing was observed. *In vivo* transfer of genes for both transforming growth factor-beta [70] and acidic fibroblast growth factor [71] has been shown to increase the ultimate strength of the healed tissue. In addition, *in vivo* gene transfer has been utilized to prevent scar tissue formation and restore contractile function in a myocardial infarct model [72]: skeletal muscle differentiation was induced in the healing heart via transfer of the myogenic determination gene. Thus, the advent of gene therapy offers a powerful new tool for the engineering of wound healing responses, as not only can the rate of healing be increased, but the biological properties of the tissue may be drastically altered.

References

1. Centers for Disease Control, The prevention and treatment of complications of diabetes: A guide for primary care practitioners. NIH Publication No. 93-3464, 1991.
2. Rebar, R. W., Drollette, C. M., and Badawy, S. Z. A., Pathophysiology of pelvic adhesions: Modern trends in preventing infertility. *Journal of Reproductive Medicine*, 1992, 37[2], 107–121.

3. Buttram, B. C., Prevention of postoperative adhesions. In *Surgical Treatment of the Infertile Female*, ed. B. C. Buttram and R. C. Reitner. Williams and Wilkins, Baltimore, MD, 1985, pp. 67–88.

4. Mosesson, M. W. The assembly and structure of the fibrin clot. *Journal of Experimental and Clinical Hematology*, 1992, **34**(1), 11–16.

5. Clark, R. A. F., Lanigan, J. M., DellaPelle, P., Manseau, E., Dvorak, H. F., and Colvin, R. B. Fibronectin and fibrin provide a provisional matrix for epidermal cell migration during wound reepithelialization. *Journal of Investigative Dermatology*, 1982, **79**(5), 264–269.

6. Nemerson, Y., Zur, M., Bach, R., and Gentry, R. The mechanisms of action of tissue factor: A provisional model. In *The Regulation of Coagulation*, ed. K. G. Mann and F. B. Taylor. Elsevier, NY, NY, 1980, pp. 193–203.

7. Smariga, P. E. and Marnard, J. R. Platelet effects on tissue factor and fibrinolytic inhibition of cultured human fibroblasts and vascular cells. *Blood*, 1982, **60**(2), 140–147.

8. Loskutoff, D. J. and Edgington, T. S. Synthesis of a fibrinolytic activator and inhibitor by endothelial cells. *Proceedings of the National Academy of Sciences USA*, 1977, **74**(10), 3903–3907.

9. Angles-Cano, E. Overview on fibrinolysis: plasminogen activation pathways on fibrin and cell surfaces. *Chemistry and Physics of Lipids*, 1994, **67**(68), 353–362.

10. Stern, D. M., Naworth, P. P., Marcum, J., Handley, D., Kisiel, D., Rosenberg, R., and Stern, K. Interaction of antithrombin III with bovine aortic segments. *Journal of Clinical Investigations*, 1985, **75**(1), 272–279.

11. Moncada, S., Gryglewski, R., Bunting, S., and Vane, J. R. An enzyme isolated from arteries transforms prostaglandin endoperoxides to an unstable substance that inhibits platelet aggregation. *Nature*, 1976, **263**(8), 663–665.

12. Anderson, J. M. Inflammation, wound healing, and the foreign body response. In *Biomaterials Science*, ed. B. D. Ratner, A. S. Hoffman, F. J. Schoen, and J. E. Lemons. Academic Press, San Diego, CA, 1996, pp. 165–72.

13. Krawczyk, W. S. A pattern of epidermal cell migration during wound healing. *Journal of Cell Biology*, 1971, **49**(4), 247–63.

14. Stricklin, G. P., Li, L., Jancic, V., Wenczak, B. A., and Nanney, L. B. Localization of mRNA representing collagenase and TIMP in sections of healing human burn wounds. *American Journal of Pathology*, 1993, **143**(6), 1657–66.

15. Herman, I. R. Molecular mechanisms regulating the vascular endothelial cell motile response to injury. *Journal of Cardiovascular Pharmacology*, 1993, **22**(Suppl. 4), 25–36.

16. Levenson, S. M., Geever, E. F., Crowley, L. V., Oates, J. F., Berard, C. W., and Rosen, H. The healing of rat skin wounds. *Annals of Surgery*, 1965, **161**(5), 293–308.

17. Christiansen, D., Pins, G., Wang, M. C., Dunn, M. G., Silver, F. H. Collagenous biocomposites for the repair of soft tissue injury. In *Materials Research Society Symposium Proceedings: Tissue-Inducing Biomaterials*, ed. L. G. Cima and E. S. Ron. Materials Research Society, Pittsburgh, PA, 1992, pp. 151–158.

18. Pachence, J. M. Collagen-based devices for soft tissue repair. *Journal of Biomedial Materials Research*, 1996, **33**(1), 35–40.

19. Yannas, I. V. Applications of ECM analogs in surgery. *Journal of Cellular Biochemistry*, 1994, **56**(2), 188–191.

20. Byrne, D. J., Hardy, J., Wood, R. A., McIntosh, R., and Cushieri, A. Effect of fibrin glues on the mechanical properties of healing wounds. *British Journal of Surgery*, 1991, **78**(7), 841–843.

21. Michel, D., and Harmand, M. F. Fibrin seal in wound healing: Effect of thrombin and Ca + 2 on human skin fibroblast growth and collagen production. *Journal of Dermatological Science*, 1990, **1**(5), 325–333.

22. Robinson, B. P., Hollinger, J. O., Szachowicz, E. H., and Brekke, J. Calvarial bone repair with porous D,L-polylactide. *Head and Neck Surgery*, 1995, **112**(6), 707–713.

23. Den Dunnen, W. F., Van der Lei, B., Schakenraad, J. M., Blaauw, E. H., Stokroos, I., Pennings, A. J., and Robinson, P. H. Long-term eavluation of nerve regeneration in a biodegradable nerve guide. *Microsurgery*, 1993, **14**(8), 508–515.

24. Hansbrough, J. F., Cooper, M. L., Cohen, R., Spielvogel, R., Greenleaf, G., Bartel, R. L., and Naughton, G. Evaluation of a biodegradable matrix containing cultured human fibroblasts as a dermal replacement beneath meshed skin grafts on athymic mice. *Surgery*, 1992, **111**(4), 438–446.

25. Laurencin, C. T., El-Amin, S. F., Ibim, S. E., Willoughby, D. A., Attawia, M., Allcock, H. R., and Ambrosio, A. A. A highly porous 3-dimensional polyphosphazene polymer matrix for skeletal tissue regeneration. *Journal of Biomedical Materials Research*, 1996, **30**(2), 133–138.

26. Langone, F., Lora, S., Veronese, F. M., Caliceti, P., parnigotto, P. P., Valenti, F., and Palma, G. Peripheral nerve repair using a polyphosphazene tubular prosthesis. *Biomaterials*, 1995, **16**(5), 347–353.

27. Humphries, M. J. The molecular basis and specificity of integrin-ligand interactions. *Journal of Cell Science*, 1990, **97**(10), 585–592.

28. Woerly, S. Laroche, G., Marchand, R. Pato, J. Subr, V., and Ulbrich, K. Intracerebral implantation of hydrogel-coupled adhesion peptides: Tissue reaction. *Journal of Neural Transplantation*, 1995, **5**(4), 245–255.

29. Steed, D. L.. Ricotta, J. J., Prendergast, J. J., Kaplan, R. J., Webster, M. W., McGill, J. B., and Schwartz, S. L. Promotion and acceleration of diabetic ulcer healing by arginine–glycine–aspartic acid (RGD) peptide matrix. *Diabetes Care*, 1995, **18**(1), 39–46.

30. Hansbrough, J. F., Herndon, D. N., Heimbach, D. M., Solem, L. D., Gamelli, R. L., and Tmpkins, R. G. Accelerated healing and reduced need for grafting in pediatric patients treated with arginine–glycine–aspartic acid peptide matrix. *Journal of Burn Care and Rehabilitation*, 1995, **16**(4), 377–387.

31. Hubbell, J. A., Massia, S. P., Desai, N. P., and Drumheller, P. D. Endothelial cell-selective materials for tissue engineering in the vascular graft via a new receptor. *Bio / Technology*, 1991, **9**, 568–572.

32. Kleinman, H. K., Graf, J., Iwamoto, Y., Sasaki, M., Schasteen, C. S., Yamada, Y., Martin, G. R., and Robey, F. A. Identification of a 2nd active-site in laminin for promotion of cell-adhesion and migration and inhibition of in vivo melanoma lung colonization. *Archives of Biochemistry and Biophysics*, 1989, **272**(1), 39–45.

33. Kanemoto, T. Reich, R., Royce, L., Greatorex, D. Adler, S. H., Shiraishi, N., Martin, G. R., Yamada, Y., and Kleinman, H. K. Identification of an amino

acid sequence from the laminin-A chain that stimulates metastasis and collagen IV production. *Proceedings of the National Academy of Sciences USA*, 1990, **87**(6), 2279–2283.

34. Hunter, D. D., Cashman, N., Morrisvalero, R. Bulock, J. W., Adams, S. P., and Sanes, J. R. An LRE (leucine arginine glutamate)-dependent mechanism for adhesion of neurons to S-laminin. *Journal of Neuroscience*, 1991, **11**(12), 3960–3971.

35. Gutsche, A. T., Parsons-Wigerter, P., Chand, D., Saltzman, W. M., and Leong, K. W. *N*-acetylglucosamine and adenosine derivitized surfaces for cell culture: 3T3 fibroblast and chicken hepatocyte response. *Biotechnology and Bioengineering*, 1994, **43**(11), 801–809.

36. Kobayashi, A. Kobayashi, K., and Akaike, T. Control of adhesion and detachment of parenchymal liver cells using lactose-carrying polystyrene as substratum. *Journal of Biomaterials Science*, 1992, **3**(12), 499–508.

37. Kobayashi, K., Kobayashi, A., and Akaike, T. Culturing hepatocytes on lactose-carrying polystyrene layer via asialoglycoprotein receptor-mediated interactions. *Methods in Enzymology*, 1994, **247**, 409–418.

38. Liu, S. Q., Ito, Y., and Imanishi, Y. Cell growth on immobilized growth factor: covalent immobilization of insulin, transferrin, and collagen to enhance growth of bovine endothelial cells. *Journal of Biomedical Materials Research*, 1993, **27**(7), 909–915.

39. Ito, Y., Inoue, M., Liu, S. Q., and Imanishi, Y. Cell growth on immobilized cell growth factor: Enhancement of fibroblast growth by immobilized insulin and/or fibronectin. *Journal of Biomedical Materials Research*, 1993, **27**(7), 901–907.

40. Kuhl, P. R., and Griffith-Cima, L. G. Tethered epidermal growth factor as a paradigm for growth factor-induced stimulation from the solid phase. *Nature Medicine*, 1996, **2**(9), 1022–1027.

41. Komoriya, A. Hortsch. M. Meyers, C., Smith, M., Kanety, H., and Schlessinger, J. Biologically active synthetic fragments of epidermal growth factor: localization of a major receptor-binding region. *Proceedings of the National Academy of Sciences USA*, 1984, **8**(3), 1351–1355.

42. Baird, A., Schubert, D., Ling, N., and Guillemin, R. Receptor-binding and heparin-binding domains of basic fibroblast growth factor. *Proceedings of the National Academy of Sciences USA*, 1988, **85**(7), 2324–2328.

43. Pagidas, K., and Tulandi, T. Effects of Ringer's lactate, Interceed (TC7), and Gore-Tex surgical membrane on postsurgical adhesion formation. *Fertility and Sterility*, 1992, **57**(1), 199–201.

44. Boyers, S. P., Diamond, M. P., and DeCherney, A. H. Reduction of postoperative pelvic adhesions in the rabbit with Gore-Tex surgical membrane. *Fertility and Sterility*, 1988, **49**(6), 1066–1070.

45. Interceed (TC7) Adhesion Barrier Study Group. Prevention of postsurgical adhesions by Interceed (TC7), an absorbable adhesion barrier: A prospective, randomized multicenter clinical trial. *Fertility and Sterility*, 1989, **51**(6), 933–938.

46. Haney, A. F., and Doty, E. Comparison of the peritoneal cells elicited by oxidized regenerated cellulsoe (Interceed) and expanded polytetrafluoroethylene (Gore-Tex Surgical Membrane) in a murine model. *American Journal of Obstetrics and Gynecology*, 1992, **166**(4), 137–149.

47. Haney, A. F., and Doty, E. Murine peritoneal injury and de novo adhesion formation caused by oxidized regenerated cellulose (Interceed) but not expanded polytetrafluoroethylene (Gore-Tex Surgical Membrane). *Fertility and Sterility*, 1992, **57**(1), 202–208.

48. Hill-West, J. L., Chowdhury, S. M., Sawhney, A. S., Pathak, C. P., Dunn, R. C., and Hubbell, J. A. Prevention of postoperative adhesions in the rat by in situ photopolymerization of bioresorbable hydrogel barriers. *Obstetrics and Gynecology*, 1994, **83**(1), 59–64.

49. Hill-West, J. L., Chowdhury, S. M., Dunn, R. C., and Hubbell, J. A. Efficacy of a resorbable hydrogel barrier, oxidized regenerated cellulose, and hyaluronic acid in the prevention of ovarian adhesions in a rabbit model. *Fertility and Sterility*, 1994, **62**(3), 630–634.

50. Quinn, K. J. Design of a burn dressing. *Burns*, 1987, **13**(5), 377–381.

51. Jonkman, M. F., and Bruin, P. A new high water vapor permeable polyetherurethane film dressing. *Journal of Biomaterials Applications*, 1990, **5**(1), 3–19.

52. Nangia, A., and Hung, T. C. Preclinical evaluation of skin substitutes. *Burns*, 1990, **16**(3), 358–367.

53. Davies, J. W. L. Synthetic materials for covering burn wounds: Longer term substitutes for skin. *Burns*, 1984, **10**(2), 104–108.

54. Gonul, B., Erdogan, D. Ozogul, C., Koz, M., Babul, A., and Celebi, N. Effect of EGF dosage forms on alkali burned corneal wound healing of mice. *Burns*, 1995, **21**(1), 7–10.

55. Raphael, B., Kerr, N. C., Shimizu, R. W., Lass, J. H., Crouthamel, K. C., Glaser, S. R., Stern, G. A., McLaughlin, B. J., Musch, D. C., and Duzman, E. Enhanced healing of cat corneal wounds by epidermal growth factor. *Investigative Opthalmology*, 1993, **34**(7), 2305–2312.

56. Bhora, F. Y., Dunkin, B. J., Batzri, S., Aly, H. M., Bass, B. L., Sidawy, A. N., and Harmon, J. W. Effect of growth factors on cell proliferation and epithelialization in human skin. *Journal of Surgical Research*, 1995, **59**(2), 236–244.

57. Matuszewska, B., Keogan, M., Fisher, D. M., Soper, K. A., Hoe, C. M., Huber, A. C., and Bondi, J. V. Acidic fibroblast growth factor: Evaluation of topical formulations in a diabetic mouse wound healing model. *Pharmaceutical Research*, 1994, **11**(1), 65–71.

58. Murali, S., Hardten, D. R., DeMartelaere, S., Olevsky, O. M., Mindrup, E. A., Karlstad, R., Chan, C. C., Holland, E. J. Effect of topically administered platelet-derived growth factor on corneal wound strength. *Eye Research*, 1994, **13**(12), 857–862.

59. Puolakkainen, P. A., Twardzik, D. R., Ranchalis, J. E., Panket, S. C., Reed, M. J., and Gombotz, W. R. The enhancement in wound healing by transforming growth factor-beta 1 depends on the topical delivery system. *Journal of Surgical Research*, 1995, **58**(3), 321–329.

60. Hollinger, J. O., and Leong, K. Poly(alpha-hydroxy acids): Carriers for bone morphogenetic proteins. *Biomaterials*, 1996, **17**(2), 187–194.

61. Isobe, M., Yamazaki, Y., Oida, S. Ishihara, K., Nakabayashi, N., and Amagasa, T. Bone morphogenetic protein encapsulated with a biodegradable and biocompatible polymer. *Journal of Biomedical Materials Research*, 1996, **32**(3), 433–438.

62. Miki, T., and Imai, Y. Osteoinductive potential of freeze-dried, biodegradable, poly (glycolic acid–co-lactic acid) disks incorporated with bone morphogenetic protein in skull defects in rats. *International Journal of Oral and Maxillofacial Surgery*, 1996, **25**(5), 402–406.

63. Wiseman, D. M., Kamp, L., Linsky, C. B., Jochen, R. F., Pang, R. H. L., and Scholz, P. M. Fibrinolytic drugs prevent pericardial adhesions in the rabbit. *Journal of Surgical Research*, 1992, **53**(4), 362–368.

64. Menzies, D., and Ellis, H. The role of plasminogen activator in adhesion prevention. *Surgery*, 1991, **172**(5), 362–366.

65. Hill-West, J. L., Dunn, R. C., and Hubbell, J. A. Local release of fibrinolytic agents for adhesion prevention. *Journal of Surgical Research*, 1995, **59**(6), 759–763.

66. Chowdhury, S. M., and Hubbell, J. A. Adhesion prevention with ancrod released via a tissue-adherent hydrogel. *Journal of Surgical Research*, 1996, **61**(1), 58–64.

67. Andreatta-vanLeyen, S., Smith, D. J., Bulgrin, J. P., Schafer, I. A., and Eckert, R. L. Delivery of growth factor to wounds using a genetically engineered biological bandage. *Journal of Biomedical Materials Research*, 1993, **27**(9), 1201–1208.

68. Vogt, P. M., Thompson, S., Andree, C., Liu, P., Breuing, K., Hatzis, D., Brown, H., Mulligan, R. C., and Eriksson, E. Genetically modified keratinocytes transplanted to wounds reconstitute the epidermis. *Proceedings of the National Academy of Sciences USA*, 1994, **91**(20), 9307–9311.

69. Andree, C., Swain, W. F., Page, C. P., Macklin, M. D., Slama, J., Hatzis, D., and Eriksson, E. In vivo transfer and expression of a human epidermal growth factor gene accelerates wound repair. *Proceedings of the National Academy of Sciences USA*, 1994, **91**(25), 12188–12192.

70. Benn, S. I., Whitsitt, J. S., Swain, W. F., and Davidson, J. M. Particle-mediated gene transfer with transforming growth factor-beta1 cDNAs enhances wound repair in rat skin. *Journal of Clinical Investigations*, 1996, **98**(12), 2894–2902.

71. Sun, L., Xu, L., Chang, H., Henry, F. A., Miller, R. M., Harmon, J. M., and Nielsen, T. B. Transfection with aFGF cDNA improves wound healing. *Journal of Investigative Dermatology*, 1997, **108**(3), 313–318.

72. Murry, C. E., Kay, M. A., Bartosek, T., Hauschka, S. D., and Schwartz, S. M. Muscle differentiation during repair of myocardial necrosis in rats via genet transfer with MyoD. *Journal of Clinical Investigations*, 1996, **98**(10), 2209–2217.

CHAPTER II.8

Biocompatibility of Tissue Engineered Implants

JAMES M. ANDERSON

Institute of Pathology,
Case Western Reserve University,
Cleveland, Ohio 44106

1 Introduction

The term 'biocompatibility' has been defined as the ability of a material, prosthesis, artificial organ, or biomedical device to perform with an appropriate host response in a specific application [1–5]. The terms 'biocompatibility assessment' and 'safety assessment' have been generally considered to be synonomous. The safety assessment of biomaterials, prostheses, artificial organs and other medical devices generally is considered to be the determination of the biological interactions of the medical device in an *in vivo* environment. The goal of safety testing is to evaluate if a medical device presents potential harm to the patient or user under conditions simulating use. In considering tissue engineering, a tissue engineered implant may be defined as a biologic-biomaterial combination in which some component of tissue has been combined with a biomaterial to create a device for the restoration or modification of tissue or organ function. Thus, biocompatibility assessment pertains to not only the biomaterial component but also the tissue component which is utilized in the creation of the device. A broadened definition of biocompatibility as applied to tissue engineered implants is then the determination of the biological interactions of the device, i.e. biomaterial component, tissue component, and combination of biomaterial and tissue components, in an *in vivo* environment.

Considering the *in vivo* environment, biocompatibility assessment may be considered to be a measure of the degree (magnitude) and extent (duration) of adverse alteration(s) in homeostatic mechanism(s) [1]. The physiological and other biomedical parameters, i.e. homeostatic mechanisms, important in a specific application will determine the host response. In considering the resolution of the alteration in homeostatic mechanism(s), both restitution and reorganization must be considered as endstage events. *In vivo* biocompatibility testing requires introduction of the medical device into the biological environment, i.e. implantation, and this results in an injury to the biological environment. The injury created by the implantation procedure induces alterations in homeostatic mechanisms and the magnitude and duration of these altered homeostatic mechanisms ultimately determine the host response, i.e. biocompatibility of the device. Following injury and alteration in the homeostatic mechanisms, resolution may occur. Resolution is the response of the biological environment to the injury and the presence of the medical device. Restitution or reorganization of the tissues present at the site of application of the medical device may occur in the process of resolution. Restitution is the return of the tissue environment to its normal structure and with biomaterials, prostheses, and other medical devices, this is a rare occurrence. With tissue engineered implants, however, this is commonly the goal or purpose for the device. Reorganization is the result of the wound healing response initiated by the injury and the presence of the medical device. With reorganization, the presence of the foreign body reaction is important in the biocompatibility assessment of the medical device. Reorganization commonly leads to the endstage of the healing response which is fibrous encapsulation of the device. It should be noted that clear definitions are not possible as tissue engineered implants utilizing the concept of immunoisolation may become encapsulated within fibrous tissue but still serve their function by the diffusion of bioactive substances which may function in a paracrine or endocrine manner.

The purpose of this chapter is to provide an overview and perspective on regulatory issues and concerns which pertain to the biocompatibility evaluation of tissue engineered implants. A guiding principle in the regulation of implants and the development of standards for testing the biocompatibility of implants is that unique devices may give unique biological responses or interactions and therefore a unique set of tests are necessary for the biocompatibility assessment. Because of this and other reasons related to the regulation of medical devices, this chapter is divided into sections which address research perspectives on tissue engineered implants, standards for biological response evaluation, and regulatory perspectives on tissue engineered implants. Finally, future perspectives on the biocompatibility assessment of tissue engineered implants is presented.

2 Research perspectives on tissue engineered implants

Approaches and concepts in tissue engineering require an integration of knowledge from engineering, biology, chemistry, materials science, surgery, and medical science disciplines. These approaches and concepts form the basis of device design criteria which will be utilized in device development. Thus, research perspectives utilized in device design and development will also be utilized in the biocompatibility assessment of the tissue engineered implant under consideration.

Table 1 presents approaches and concepts which may be utilized in the development of tissue engineered implants [6]. Investigators in tissue engineering must simultaneously consider the question 'How will we evaluate the biocompatibility and function of our tissue engineered implant?' at the same time that they address the utilization of device design criteria in their device development. Many of the test methods utilized by investigators in evaluating tissue engineered implant prototypes will also be utilized to determine biocompatibility and function of the tissue engineered implant.

Tissue engineered implants may utilize new· biomaterials specifically designed for the tissue engineering application under consideration, biological signals and signal mechanisms considered significant for the application, normal and directed healing mechanisms which may be tissue- or organ-dependent in their magnitude and duration, and delivery and phenotypic expression of cells pertinent to the tissue engineered implant under consideration. Not only do design criteria for tissue engineered implants require a fundamental and applied knowledge of these general areas, but the interactive nature of these areas must be fully appreciated in the design criteria as well as the test methods proposed to adequately and appropriately identify adverse responses in the proposed biocompatibility test methods. Furthermore, the interactive nature of the biomaterials component and the tissue or biological component of a tissue engineered implant must be appreciated in the development of biocompatibility test methods.

3 Standards for biological response evaluation

Over the past seven years, the International Standards Organization (ISO) has made a concerted effort to normalize existing biological evaluation test methods of medical devices in a horizontal international standard focused on the biological evaluation of medical devices. These efforts have led to the development of the standard ISO 10993 'Biological Evaluation of Medical Devices'. The United States' effort within the ISO for this standard is under the guidance of the Association for the Advancement of Medical Instrumentation (AAMI), Arlington, VA [7]. The ISO 10993 standard is composed of 10 parts at the present time with 9 of the parts being American national standards and one part being a AAMI technical information report. These component parts within the ISO 10993 standard have been published by AAMI under the auspices of the American

National Standards Institute (ANSI). The component parts are: Part 1. Guidance on Selection of Tests; Part 2. Animal Welfare Requirements;

Table 1. Research topics in tissue engineering

I. New biomaterials designed for tissue engineering

Biomimetic materials endowed with cell or cell-based signals; synthetic extracellular matrix for enhanced cell interaction, cell polarization, or remodeling; temporal and/or spatial delivery of bioactive agents over short and long time periods

Biomaterials whose chemical, physical or mechanical properties, structure, or form permits active tissue integration of desirable cell types and tissue components

Biointeractive and environmentally responsive materials with controllable biodegradation, cell adhesion, cell activation, and biocompatibility

Computer aided design for material macro-, micro-, and ultrastructure to facilitate tissue engineered implants.

II. Biological signals and signal mechanisms

Mechanisms by which synthetic extracellular matrices induce cell signaling and subsequent cellular responses;

Integrin and other receptor signal transduction by surfaces, tethered ligands, bioactive agents and combinations of these;

Signaling mechanisms which influence or determine biomaterial encapsulation versus integration

The role of apoptosis, ischemia and immune mechanisms in the failure of tissue engineered implants

Role, mechanisms, and regulation of transduced mechanical forces in healing and tissue integration and turnover

III. Normal and directed healing mechanisms in tissue engineered implants

Development and evaluation of resorbable templates that direct tissue formation

Role of nitric oxide, cytokines, growth factors, and other mediators in healing processes

Characteristics of materials designed to modulate healing and angiogenesis

Mechanisms of communication among cell types as well as among cells of the same type in relation to the production and maintenance of extracellular matrix on tissue engineered implants

Induction of matrix development and turnover by tissue engineered implants

Table 1   (*Continued*).

IV. Delivery and phenotypic expression of cells in and on tissue engineered implants

> Novel methods for the delivery of genes and genetic material to cells in or on tissue engineered implants or tissue sites with controlled transcription and targeted cell responses
>
> Enhanced understanding of the factors controlling cell phenotypic expression in implants when tested *in vitro* compared to
> *in vivo*
>
> Optimization of methods for inducing the expression of native or introduced genes in cells in tissue engineered implants
>
> Development of media, synthetic gels, or vehicles for optimized behavior of cells in tissue engineered implants
>
> Development of methods to monitor the metabolism, desired response, or other parameters related to viability and function of cells in tissue engineered implants

---

Part 3. Tests for Genotoxicity, Carcinogenicity and Reproductive Toxicity; Part 4. Selection of Tests for Interactions with Blood; Part 5. Tests for Cytotoxicity: *In vitro* Methods; Part 6. Tests for Local Effects after Implantation; Part 7. Ethylene Oxide Sterilization Residuals; Part 10. Tests for Irritation and Sensitization; and Part 11. Tests for Systemic Toxicity. Part 9, Degradation of Materials Related to Biological Testing, is the AAMI technical information report and while it is not a standards document, it does provide information on the degradation of materials related to biological testing. The ISO Technical Committee 194 (TC 194) continues to work on the development of new component parts for the ISO 10993 standard as well as update and modify existing parts of the standard.

For the purposes of ISO 10993, a medical device is defined as any instrument, apparatus, appliance, material or other article, including software, whether used alone or in combination, intended by the manufacturer to be used for human beings solely or principally for the purpose of diagnosis, prevention, monitoring, treatment or alleviation of disease, injury or handicap; investigation, replacement or modification of the anatomy or of a physiological process; control of conception; and which does not achieve its principal intended action in or on the human body by pharmacological, immunological or metabolic means, but which may be assisted in its function by such means. Two other definitions within ISO 10993 are important for the consideration of the biological evaluation of tissue engineered implants. A material is defined as any synthetic or natural polymer, metal, alloy, ceramic or other non-viable substance, including tissue rendered non-viable, used as a device or any part thereof; and final product is defined as a medical device in its 'as used' state. It is clear from the definitions within ISO 10993 that tissue engineered implants may or may not fall within the guidelines presented in ISO 10993. Nonetheless, ISO 10993 provides a perspective and guidance on the development and utilization of appropriate test methods for the biological evaluation of medical devices. As previously stated, each unique tissue engineered implant will require a unique set of biological response test methods for safety or biocompatibility assessment.

ISO 10993 Biological Evaluation of Medical Devices — Part 1 provides generic guidance on the selection of tests for biological evaluation. It is significant that Part 1 of the ISO 10993 has been accepted by the Food and Drug Administration (FDA) Office of Device Evaluation as a guidance document for biological response evaluation. In 1995, the Office of Device Evaluation of the FDA provided a modified version of Part 1 of ISO 10993 as a guidance document for the selection of tests for biological evaluation of medical devices [8].

Part 1, Guidance on Selection of Tests, of ISO 10993 provides guidance on the fundamental principles governing the biological evaluation of medical devices, the definition of categories of devices based on the nature and duration of contact within the body, and the selection of appropriate tests. Part 1 provides a matrix to assist investigators in determining device categories by body contact and contact duration and suggested biological response test methods.

Table 2 presents ISO 10993-1 categories for selection of biological response test methods. Tissue engineered implants may be surface devices, external communicating devices, or implant devices, and thus fall under any of the three major body contact categories. Regarding contact duration, it is most probable that the prolonged or permanent categories for contact duration will best characterize tissue engineered implants. Details on the appropriate selection of body contact categories and contact duration categories may be found in Vol. 4, Biological Evaluation of Biomedical Devices, in the 1996 edition of the AAMI Standards and Recommended Practices [7].

Appropriate selection of body contact and contact duration of the intended tissue engineered implant permits selection of the biological response tests which should be considered in safety or biocompatibility assessment. Table 3 indicates the biological response tests presented in Part 1 of ISO 10993. Within Part 1 of the ISO 10993 standard, the biological response tests are divided into initial evaluation tests and supplementary evaluation tests. Due to the diversity of medical devices, it is recognized that not all tests identified in a category will be necessary or

Table 2. ISO 10993-1 categories for selection of biological response test methods

Body contact	
Surface devices	Skin
	Mucosal membranes
	Breached or compromised surfaces
External communicating devices	Blood path, indirect
	Tissue/bone/dentin communicating
	Circulating blood
Implant devices	Tissue/bone
	Blood
Contact duration	Limited, less than or equal to 24 hours
	Prolonged, greater than 24 hours and less than 30 days
	Permanent, greater than 30 days

practical for any given device. This would also hold true for tissue engineered implants. Furthermore, it is indispensable for testing that each device shall be considered on its own merits, i.e. unique characteristics, and additional tests not indicated in the matrix tables may be necessary. An important consideration here is the use of biodegradable biomaterials in tissue engineered implants. Knowledge of the time period necessary for complete biodegradation of a biodegradable biomaterial in a tissue engineered implant is important as this information can be used to rationalize or justify time periods for the evaluation of the biomaterials component of the tissue engineered implant.

Table 3. ISO 10993-1 biological response tests

Initial evaluation tests	Cytotoxicity
	Sensitization
	Irritation or intracutaneous reactivity
	Systemic toxicity (acute)
	Subchronic toxicity (subacute toxicity)
	Genotoxicity
	Implantation
	Hemocompatibility
Supplementary evaluation tests	Chronic toxicity
	Carcinogenicity
	Reproductive/developmental
	Biodegradation

In considering the safety or biocompatibility assessment of tissue engineered implants, it is important to appreciate that the range of potential biological hazards is wide. The tissue interaction of a biomaterial cannot be considered in isolation from the overall device design. The best biomaterial with respect to tissue interaction may result in a less functional device, tissue interaction being only one characteristic of a biomaterial. Where the biomaterial is intended to interact with tissue in order for the device to perform its function, i.e. tissue engineered implants, evaluation takes on dimensions not generally addressed in standards and guidelines to date. It must be recognized that biological reactions that are adverse for a biomaterial in one application may not be adverse for the use of the biomaterial in a different application.

4 Regulatory perspectives on tissue engineered implants

The rapid emergence of tissue engineering as a discipline and the development of tissue engineered medical devices has resulted in an interactive atmosphere at the FDA regarding the regulation of tissue engineered implants. The FDA regards tissue engineered implants as combination devices with a biologic component and a biomaterial component. To streamline the review process and resolve the jurisdictional issues for biologic-biomaterial combination products, new regulations and guidance documents have and are being developed by the FDA [9–13].

Three different FDA centers may participate in the regulation of tissue engineered implants. These centers are the Center for Biologics Evaluation and Research (CBER), the Center for Devices and Radiological Health (CDRH), and the Center for Drug Evaluation and Research (CDER). Intercenter agreements have been created for resolving jurisdictional issues for tissue engineered products. The designation of the primary FDA review center for a biologic–biomaterial combination product is based on the primary mode of action of the product. For example, when the primary mode of action of a biologic–biomaterial combination product, such as an artificial hip component coated with a growth factor, is via the device component, CDRH is designated as the primary review center and performs reviews and consultation with CBER. When the primary mode of action of a biologic–biomaterial combination such as a bioartificial liver or pancreas is via the biological component, CBER is designated as the primary review center and evaluates the safety and efficacy of the biologic–biomaterial combination product in consultation with CDRH. Other FDA centers may be involved in the review process. When a combination product is not covered under intercenter agreements, or where the product center jurisdiction is unclear, a request for designation may be made to the

FDA. In either case, sufficient rationale and justification by the sponsor in support of a primary center for review should be made by the sponsor.

The CBER has provided guidance documents called Points to Consider to further assist product developers in identifying regulatory issues of tissue engineered products. In addition, CBER as well as CDRH have published guidance documents to assist in the adequate and appropriate evaluation of the biocompatibility of tissue engineered implants. Table 4 is a list of Points to Consider and guidelines relevant to biologic–biomaterial combinations [9].

Of special concern to CBER is the transmission of adventitious agents to patients via somatic cells, tissue, or cell-derived products which are considered as a major safety concern in the utilization of tissue engineered implants. In addition, *in vivo* inflammatory responses elicited by the biomaterial component have also been identified as a major safety concern regarding tissue engineered implants.

As part of the ongoing effort by centers at the FDA to provide guidance documents, a recent draft guidance document on immunotoxicity testing has been made available for comments by the scientific community. This draft guidance document is entitled 'Immunotoxicity Testing Framework' and has been created by the CDRH Immunotoxicology Working Group. The purpose of this framework is to provide FDA reviewers and manufacturers with a systematic approach for evaluating potential adverse immunological effects of tissue engineered implants which contain biologic–biomaterial combinations. It should be noted that neither the ISO 10993 standard nor the FDA modified (FDA/CDRH G95-1) document provides for guidance on immunotoxicity testing [7, 8]. In this draft guidance document, immunotoxicity refers to any adverse effect mediated by changes in the immune system which is disproportionate to the toxicity manifested in other systems.

As with recent standards and guidance documents, a flowchart for the identification of immunotoxicity testing has been provided. A matrix similar to that provided in the ISO 10993 document indicating where immunological effects should be investigated is presented. Table 5 provides potential immunological effects and responses by tissue engineered implants. Table 6 is a modification from Table 3 of the draft guidance document which is in matrix form. Functional assays, phenotyping, soluble mediators and clinical symptoms are related to specific types of immune responses in the Table 3 matrix in the draft document. Further perspective and an up-to-date guidance document for immunotoxicity testing may be obtained from the Molecular Biology Branch, Division of Life Sciences, Office of Science and Technology, CDRH, FDA, Rockville, MD.

Table 4. Points to consider documents and FDA guidelines relevant to biologic–biomaterial combinations [9]

---

1. Points to consider in human somatic cell therapy and gene therapy; 1991. (P008)
2. Points to consider in the manufacture and clinical evaluation of *in vitro* tests to detect antibodies to the human immunodeficiency virus, type I; 1989 [draft]. (p004)
3. Points to consider in the collection, processing, and testing of *ex vivo*-activated mononuclear leukocytes for administration to humans; 1989. (P003)
4. Points to consider in the characterization of cell lines used to produce biologicals; 1993. (P012)
5. Points to consider in the manufacture of *in vitro* monoclonal antibody products for further manufacturing into blood grouping reagents and anti-human globulin; 1992. (P009)
6. Cytokine and growth factor prepivotal trial information package; 1990. (P007)
7. Points to consider in computer assisted submission for license applications; 1990. (P005)
8. OLEPS advertising and promotional labeling staff procedural guide; 1993
9. Points to consider in the production and testing of interferon Intended for Investigational Use in Humans; 1983. (P014)
10. Points to consider in the manufacture and testing of monoclonal antibody products for human use; 1994. (P013)
11. Points to consider in the production and testing of new drugs and biologicals produced by recombinant DNA technology; 1985. (P002)
12. Supplement to the points to consider in the production and testing of new drugs and biologicals produced by recombinant DNA technology: nucleic acid characterization and genetic stability; 1992. (P011)
13. Points to consider in the manufacture and testing of therapeutic products for human use derived from transgenic animals; 1995. (P015)
14. Guidelines on validation of the limulus ameobocyte lysate test as an end product endotoxin test for human and animal parenteral drugs, biological products, and medical devices; 1987. (G01)
15. Guideline for submitting documentation for the stability of human drugs and biologics; 1987. (G012)
16. Guideline to test residual moisture in biological products; 1990. (G020)
17. Guideline for submitting documentation for packaging for human drugs and biologics; 1987. (G011)
18. Guideline for sterile drug products produced by aseptic processing; 1987. (G013)
19. Guideline on general principles of process validation; 1987. (G014)
20. Draft guidance for reporting-ABR's-adverse reactions to licensed biological products; 1990. (G021)
21. Guideline for adverse experience reporting for licensed biological products; 1993. (G024)

The numbers in the parentheses at the end of each document are the document numbers specified by the Congressional and Consumers Affairs Branch. These documents are available at no charge from the Congressional and Consumers Affairs Branch (HFM-46), Rockwall 1, 6th floor, 11400 Rockville Pike, Rockville, MD 20852-1448.

Table 5. Potential immunological effects and responses by tissue engineered implants

Effects	Hypersensitivity
	Inflammation
	Immunosuppression
	Immunostimulation
	Autoimmunity
Responses	Histopathology
	Humoral response
	Host resistance
	Clinical symptoms
	Cellular responses
	T-Cells
	Natural killer cells
	Macrophages
	Granulocytes

5 Future perspectives on the biocompatibility assessment of tissue engineered implants

The development of biologic–biomaterial combination tissue engineered implants will require new perspectives and approaches to the biocompatibility or safety assessment of these devices. In the past, standards and regulations have considered medical devices on a case by case basis and this will most certainly be true for tissue engineered implants. Moreover, since tissue engineered implants may have their primary mode of action through the biologic component, emphasis in the future will be placed on safety and biocompatibility assessment of the biologic component of the tissue engineered implant. Obviously, safety and biocompatibility assessment of the biomaterial component of a tissue engineered implant also will be necessary. Biocompatibility assessment of the biologic component of the tissue engineered implant will in part be governed by the hypothesis, principles or concepts which are incorporated in the design criteria for the tissue engineered implant. In this regard, investigators and manufacturers will play a major role in creation of biocompatibility or biological response test methods for determining the safety, biocompatibility and function of the tissue engineered implant under consideration. It can be anticipated that increased communication between investigators, manufacturers and regulatory scientists will be necessary for the development of an adequate and appropriate set of test methods to evaluate tissue engineered implants on a case by case basis. In the past, individuals responsible for regulatory issues of medical devices have dealt with standards and guidance docu-

Table 6. Representative tests, indicators, and models for the evaluation of immune responses

Functional assays	Skin testing
	Lymphocyte proliferation
	Plaque-forming cells
	Local lymph node assay
	Mixed lymphocyte reaction
	Tumor cytotoxicity
	Antigen presentation
	Phagocytosis
	Degranulation
	Resistance to bacteria, viruses, and tumors
Phenotyping	Cell surface markers
	MHC markers
Soluble mediators	Antibodies
	Complement
	Immune complexes
	Cytokine patterns (T-cell subsets)
	Cytokines (IL-1, IL-1ra, TNF-α, IL-6, TGF-β, IL-4, IL-13)
	Chemokines
	Bioactive amines
Clinical symptoms	Allergy
	Skin rash
	Urticaria
	Edema
	Lymphadenopathy

ments as a checklist for biocompatibility assessment. This will no longer be possible with the biocompatibility assessment of tissue engineered implants as these will most probably have new performance criteria.

New definitions of medical devices will be necessary, especially for the development of new standards and guidance documents. As previously noted, definitions were for so-called 'non-viable' components of medical devices but that is obviously not the case with tissue engineered implants.

For devices and materials, the application of a rigid set of test methods and pass/fail criteria might result in either unnecessary restriction or a false sense of security in their use. Where a particular application warrants, experts in the product or application area involved may choose to establish specific tests and criteria specified in a product-specific vertical standard. Thus, previously developed standards may have to be expanded to include definitions, issues and concerns relative to tissue engineered implants. On the other hand, new horizontal standards such as ISO 10993 may have to be created to address tissue engineered implants in a specific manner.

Biological response testing relies upon animal models and a device, therefore, cannot be conclusively shown to have the same tissue reaction in humans. In addition, differences between humans suggest that some patients may have adverse reactions even to well-established biomaterials. This places emphasis on the development of appropriate test methods for the biocompatibility assessment of tissue engineered implants. These issues must be considered if tissue engineered implant prototypes utilizing biologic components from non-humans are tested but the final product will depend on the utilization of a human-derived biologic component of the tissue engineered implant. Biological test results are test method- and biological model selection-specific. Therefore, results of tests may be instructive but may lack correlative power and may not adequately predict the clinical safety and efficacy of an implant.

These issues and concerns offer new challenges in safety and biocompatibility assessment of tissue engineered implants. Therefore, it is suggested that investigators simultaneously address biocompatibility issues at the same time that they address the utilization of device design criteria in their device development.

References

1. Anderson, J. M., Perspectives on in vivo testing of biomaterials, prostheses, and artificial organs. *Journal of the American College of Toxicology*, 1988, 7(4), 469–479.
2. Williams, D. F., ed., *Progress in Biomedical Engineering, 4. Definitions in Biomaterials*, Elsevier, Amsterdam, 1987.
3. Ratner, B. D., Hoffman, A. S., Schoen, F. J. and Lemons, J. E., ed., *Biomaterials Science An Introduction to Materials in Medicine*, Academic Press, San Diego, 1996.
4. Greco, R. S., ed., *Implantation Biology The Host Response and Biomedical Devices*, CRC Press, Boca Raton, 1994.
5. Black, J., *Biological Performance of Materials Fundamentals of Biocompatibility*, 2nd edn, Marcel Dekker, Inc., New York, 1992.
6. Tissue engineering in cardiovascular disease: A report. *Journal of Biomedical Materials Research*, 1995, 29, 1473–1475.
7. AAMI Standards and Recommended Practices. Vol. 4: Biological Evaluation of Medical Devices. *Association for the Advancement of Medical Instrumentation*, 3330 Washington Blvd., Suite 400, Arlington, VA 22201, 1996.
8. FDA Modified Matrix of ISO-10993-1. Biological evaluation of medical devices part 1: evaluation and testing. *CDRH Blue Book Memorandum (G95-1)*, May 1, 1995.

9. Chapekar, M. S., Regulatory concerns in the development of biologic-bio-material combinations. *Journal of Biomedical Materials Research (Applied Biomaterials)*, 1996, **33**, 199–203.

10. Hellman, K. B., Picciolo, G. L. and Mueller, E. P., Biomaterials and biotechnology. *Bio / Technology*, 1993, **11**, 1179–1180.

11. Hellman, K. B., Picciolo, G. L. and Fox, C. F., Prospects for application of biotechnology-derived biomaterials. *Journal of Cellular Biochemistry*, 1994, **56**, 210–224.

12. Hellman, K. B., Biomedical applications of tissue engineering technology: Regulatory issues. *Tissue Engineering*, 1995, 1(2), 203–210.

13. Hellman, K. B., Bioartificial organs as outcomes of tissue engineering scientific and regulatory issues. *Annals of New York Academy of Sciences USA*, 1997, in press.

CHAPTER II.9

Tissue Engineered Construct Design Principles

GREGORY P. REECE,
CHARLES W. PATRICK JR.
Laboratory of Reparative Biology and Bioengineering,
Department of Plastic Surgery,
M.D. Anderson Cancer Center,
1515 Holcombe Blvd., Box 62,
Houston, Texas 77030, USA

1 Introduction

The ultimate goals for designing tissue substitutes are: (1) to restore function to and/or to replace tissues lost as a result of congenital abnormalities, trauma, and disease; and (2) to minimize or eliminate the problems associated with obtaining tissues for conventional tissue replacement [1, 2]. At the present time, tissue substitutes range from small groups of encapsulated cells, such as islets of Langerhans [3] and dopamine-secreting cells [4], to tissue engineered constructs (TECs) composed of parenchymal cells seeded on a biomaterial matrix. Bioengineered skin, cartilage, and heart valves are a few examples of TECs under development [5–7]. Ideally, future advances in TEC design may allow the replacement of diseased organs, such as a cirrhotic liver, with a complex, multicellular TEC [8].

The current level of TEC design has been achieved by using several different bioengineering principles and techniques developed over the last several years. The *ex vivo* production of a large parenchymal cell mass required for TEC seeding is now routinely possible with certain types of cells thanks to refinements in bioreactor design. Advances in biomaterials research have produced biocompatibile and biodegradable polymers that are fabricated into desired shapes and used to provide transient structural support and a three-dimensional matrix for the adhesion-dependent parenchymal cells. Moreover, bioengineers can modify cell function and phenotype within and surrounding the TEC by altering biomaterial composition and microstructure, and incorporate controlled-release tissue induction factors to promote growth and differentiation [1, 2, 6–9]. Although these innovations have been crucial to the development of nascent TECs, the fundamental obstacle to the development of larger TECs for clinical use is that TECs are presently design-limited. Specifically, tissue can not be implanted in volumes of greater than 1–3 μl or thicker than a few hundred microns due to limitations of diffusion [9, 10]. Scaling from such insignificant tissue volumes to large, clinically relevant tissue substitutes requires changes in TEC design that accommodates the diffusion limitations and/or facilitates TEC neovascularization at a rate that is fast enough to permit maximal parenchymal cell survival.

The ultimate goal of tissue engineering is to replace missing tissue with a tissue substitute without incurring significant donor site morbidity. Hence, the TEC should be designed to approximate the tissue that is missing or, at least, the tissue that is currently used to repair missing tissue, namely the conventional tissue transplant. The extent that a TEC must model missing tissue or a conventional tissue transplant has largely not been determined. That is, early TECs have primarily focused on tissue structure as a design endpoint and considered tissue function as a secondary, negligible, or unobtainable endpoint. Nevertheless, because the conventional tissue transplant is the 'gold standard' for physicians and surgeons who repair injured or dysfunctional tissue, TECs designed as tissue substitutes naturally will be compared to the conventional tissue transplant. As will be delineated within this chapter, the same requirements that dictate a successful tissue transplant apply to TECs. Generally speaking, physicians/surgeons are 'creatures of habit' when it comes to incorporating new surgical techniques or devices. Thus, to be clinically acceptable, TECs must meet or exceed clinical results realized with current conventional transplants and must employ familiar surgical technology. To be able to contrast a TEC with a conventional tissue transplant, the bioengineer must have a fundamental understanding of how conventional transplants are used, their limitations, and the mechanisms for their survival. In this chapter, we will review the characteristics and limitations of conventional tissue transplants and contrast this to the knowledge base of tissue engineered analogues. The topics reviewed are not all inclusive and will primarily focus on the early survival of the parenchymal and supportive cells of the transplant.

2 Terminology

Before proceeding with a discussion of TEC design principles as related to conventional tissue substitutes, the terminology used by physicians and biologists for conventional tissue transplantation must be understood [11, 12]. The following is a glossary of the most commonly used terms.

Angiogenesis: the growth of capillaries from existing quiescent endothelial cells of capillaries and post-capillary venules present in the host's recipient site. Revascularization, neovascularization, and microvascular remodeling usually occur by angiogenesis.

Donor: the person or animal donating a graft.

Embryonic / Fetal graft: tissues or cells obtained from an embryo or fetus, usually from the same species, and used for transplantation.

Graft: a type of transplant that is generally considered to be any tissue or cells that has been removed from its natural nutritional source and, after transplantation, must derive its nutrition from the surrounding tissues (recipient site).

Grafting: the actual process of transplantation which usually includes harvesting tissue, processing it, and applying the graft to the recipient site.

Graft take: terminology used to convey that a graft successfully obtained a nutrient supply from the recipient site and survived transplantation.

Harvesting: the procedure required to obtain a graft.

Heterotopic transplant: a graft placed in a recipient site that is anatomically and functionally different from the donor site.

Host: the person or animal receiving a graft.

Inosculation: the self-forming anastomoses that occur between capillaries of the graft and those of the host.

Ischemia: the condition in which vascularized tissues do not have sufficient blood flow and avascular tissues do not have access to sufficient nutrition and waste elimination.

Ischemia tolerance: the ability for a tissue to withstand a time interval of ischemia.

Maximum ischemia tolerance: the maximum time interval a tissue will tolerate an ischemic event and still survive.

MHC antigens: the major histocompatibility complex antigens are a group of cell membrane glycoproteins that allow the host's immune system to differentiate the host's cells from the donor's cells. If the donor cells are isogenous in relation to the host, the MHC antigens of the donor cells are identical to those of the host and no immune response forms against the donor cells. If the donor cells are allogenous or xenogenous in relation to the host's cells, the host's MHC antigens are recognized as different from the donor cells and the host's immune system mounts an immune response against the donor cells.

Neovascularization: the growth of new blood vessels (capillaries) from the host's recipient site into a structure, such as a TEC or other porous implant.

Orthotopic transplant: a graft placed in a recipient site that is anatomically similar to the donor site.

Parenchymal cell: the primary cells of a graft. The cells having the desired characteristics for which the graft is being used. For example, the parenchymal cells of bone are osteocytes and osteoblasts.

Recipient site: the wound surface or cavity receiving a graft.

Revascularization: the growth of new blood vessels (capillaries) from the host vessels into a graft. Ingrowing vessels may inosculate with existing graft vessels, invade the graft through 'conduits' provided by the original microvascular network, or invade the extracellular matrix to create a new microvascular network.

Supportive cells: all cells in the transplant that are not parenchymal cells. Examples include, fibroblasts, tissue macrophages, and mast cells.

Transplant: an organ, tissue, or cells transferred from one area of the body to another area on the same or different individual. An organ transplant requires immediate restoration of the microcirculation at the time of transplantation, whereas tissue and cell transplants (grafts) derive their nutrition from the recipient site over a period of several days.

Vascularized: tissues that have a microvasculature and are perfused with blood.

Vasculogenesis: the *in situ* development of blood vessels from islands of endothelial precursor cells (angioblasts and hemangioblasts) in a developing embryo. Vessel development is considered programmed and the endothelial cells forming the microvasculature are considered 'activated', that is, in a proliferative state.

3 Classification and composition of conventional tissue transplants

3.1 Types of conventional grafts

There are many different types of conventional tissue transplants or grafts, all of which may be classified under a variety of schemes. The most commonly used classification scheme subscribed to is based on the source of the donor tissue. Under this classification system, an *autograft* is tissue or cells obtained from one area of the host and transferred to another area of the same host; an *isograft* is tissue or cells obtained from a donor that is genetically identical to the host (an identical clone); an *allograft* is tissue or cells obtained from a donor of the same species but is genetically disparate from the host; and a *xenograft* is tissue or cells obtained from a donor of a species that is different from the host. Xenografts can be further classified by how closely the donor is related to the recipient on a phylogenetic basis [12]. A *concordant transplant* is a xenograft between closely related species, such as humans and non-human primates. A *discordant transplant* is a xenograft between phlogenetically distant species. A skin graft transplanted from a pig to a human is an example of a discordant xenograft.

Other schemes include classification by donor tissue vascularity, donor age, and by the application for which the graft is being used. Examples of classification by vascularity include whether the tissue is avascular (cornea, articular cartilage), relatively avascular (tendon, fat), or vascularized (skin, bone). Classification by age of the donor tissue would include whether the graft was obtained from a fully differentiated adult donor or from a developing embryo or fetus. Examples of classification by application include tissues used for their structural or mechanical properties, for a hormonal product, or for a particular function. Thus, tendon and cartilage are usually grafted for their mechanical properties, parathyroid gland and islets of Langerhans for their hormonal products, and red blood cells for their O_2 carrying properties. Although blood and its cellular products are technically considered a graft, by convention they are generally referred to as a transfusion and given for functional reasons, such as, to replace excessive red blood cell loss during hemorraghe and to correct platelet deficient-coagulation problems.

3.2 Composition of conventional grafts

Almost any tissue can and has been transplanted as a conventional tissue transplant for either clinical or experimental purposes. Examples of commonly used grafts for clinical problems are shown in Table 1. Although this list of transplantable tissues is lengthy and does not include all clinically useable tissues, it does illustrate the broad scope of tissue transplantation available to physicians and some of the reasons they may be used to restore patients to a functional status. Grafts used for experimental purposes consist of many types of tissue and a complete list of grafts used for experimentation would be too lengthy and beyond the scope of this chapter.

All conventional grafts are composed of the tissues from which they are removed. Thus, irrespective of the type of graft, the tissue itself can be separated into its fundamental elements. For example, bone can be separated into its primary parenchymal cells, an extracellular matrix, a microvasculature, nerves, and supportive cellular elements, such as bone marrow, osteoclasts, and mast cells (Fig. 1). Presumably, each component serves a purpose, but it is unknown whether all components of a graft must be present for a graft to function sufficiently.

Avascular grafts obtain their nutrition primarily by diffusion and usually are structurally less complex than grafts that have their own microvasculature. For example, articular cartilage can be up to 3 mm thick and is primarily composed of chondrocytes and surrounded by an extracellular matrix composed of collagen and glycosaminoglycans [52]. The cornea is a 0.5–0.62 mm thick, 5-layered structure composed of an outer layer of epidermal cells, an epidermal basement (Bowman's) membrane, a stromal layer composed of parallel lamellae of collagen fibers and their associated keratocytes surrounded by a glycosaminoglycan ground substance, an

Table 1. Viable grafts commonly used for clinical applications

Tissue	Donor source	Applications	References
Blood			
Whole	Auto/allo	Whole blood replacement	[13, 14]
Packed RBC	Auto/allo	Treatment of anemia	[15, 16]
Platelets	Allo*	Coagulation abnormalites	[17–19]
Leukocytes	Allo*	Treatment of neutropenia	[19, 20]
Skin			
Split thickness	Auto/allo	Wound coverage	[21, 22]
Full thickness	Auto/allo	Wound coverage	[21, 22]
Dermis	Auto	Provide bulk/coverage	[23]
Hair follicle	Auto	Hair replacement	[24, 25]
Epidermis	Auto/allo	Wound coverage	[26]
Bone			
Cortical	Auto/allo†	Long bone replacement	[27, 28]
Cancellous	Auto	Bone fusion/osteoblast source	[29, 30]
Membranous	Auto	Facial bone repair/augmentation	[31]
Cartilage			
Elastic	Auto	Nose/ear/eyelid repair	[32, 33]
Hyaline	Auto/xeno	Nose/ear/eyelid repair	[34, 35]
Nerve	Auto	Repair of nerve gap	[36, 37]
Tendon/fascia	Auto	Tendon replacement, facial paralysis	[38, 39]
Fat (subcutaneous)	Auto	Filling soft tissue defects	[40, 41]
Cornea			
Lamellar	Allo	Partial corneal replacement	[42, 43]
Full thickness	Allo	Corneal replacement	[43, 44]
Blood vessel			
Venous	Auto	Repair vessel gap, renal dialysis fistula	[45–47]
Arterial	Auto	Repair vessel gap	[47, 48]
Parathyroid gland	Auto	Calcium homeostasis	[49]
Bone marrow	Auto/allo	Restore marrow after chemotherapy	[50, 51]

Abbreviations: auto, autograft; allo, allograft; xeno, xenograft; RBC, red blood cells.
* These blood components are pooled from multiple allogenous donors.
† The cells in allografts are generally nonviable secondary to storage techniques.

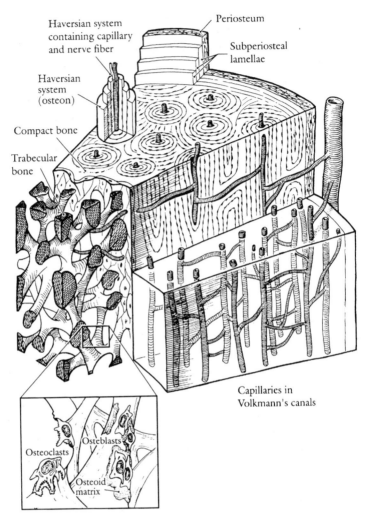

Haversian system
containing capillary
and nerve fiber

Periosteum

Subperiosteal
lamellae

Haversian
system
(osteon)

Compact bone

Trabecular
bone

Capillaries in
Volkmann's canals

Osteoblasts

Osteoclasts

Osteoid
matrix

Fig. 1. Example of a tissue (bone) separated into its fundamental elements: parenchymal cells, extracellular matrix, microvasculature, nerves, and support cells (e.g., marrow, osteoblasts, mast cells).

endothelial (Descement's) membrane, and a unicellular inner layer of endothelial cells [53]. Although both of these grafts are structurally simple in design compared to tissues perfused by a microvascular network, their means of obtaining nutrition and surviving transplantation are unique and specialized.

As a general rule, vascularized tissues *in vivo* are usually arranged such that no parenchymal or supportive cell is greater than 25–50 μm from the nearest blood vessel within the tissue [54]. Because grafts are composed of the tissue from which they are removed, this relationship between the parenchymal cells and their vasculature holds true for grafts as well. Under normal and stressful conditions, the spatial arrangement of the parenchymal cells and the microvasculature in tissue is adaptable. That is, the microvasculature remodels according to the metabolic demands of the surrounding cell mass [55]. Grafting may be the only exception to this rule, at least initially after transplantation, because the metabolic demand of the cell mass is satisfied only to the extent (depth) of diffusion into the graft and by the rate of revascularization from the recipient site.

3.3 Classification and composition of TECs

To date, the majority of TECs consist of a biomaterial matrix seeded with a desired parenchymal cell. However, some TECs are fabricated in stages with more than one parenchymal cell type. Tissue-engineered blood vessels and composite skin substitutes are examples of a multicellular TEC [1, 2, 5–9]. The biomaterial is used to maintain the shape and structural support of the TEC and to provide an attachment site for anchorage-dependent parenchymal cells (Fig. 2) [5–9]. As stated above, the biomaterial may also be used to control cell growth by contact guidance or controlled release of selected tissue-induction factors. Thus, compared to a conventional graft, a TEC can be considered a more selective and sophisticated form of grafting. However, it must be pointed out that the optimal design and natural biological course for transplanted TECs have not been fully elucidated experimentally or clinically.

4 Parallels between conventional and TEC transplants

Similarities between tissues transplanted by conventional and tissue engineering techniques are that they: (1) are usually placed in a wound healing environment, and thus; (2) have the same or similar limitations for obtaining their nutrition from the recipient site after transplantation; (3) must be immunogenetically compatible; (4) usually have the same or similar donor sites; (5) have the potential to transmit infectious disease; and (6) have a limited survival time after harvesting unless special preservation techniques are employed. The two techniques differ in the amount of tissue required for transplantation, the potential for morbidity at the donor site, the economic impact on national health care costs, and the length of time required between harvesting and transplanting the tissue.

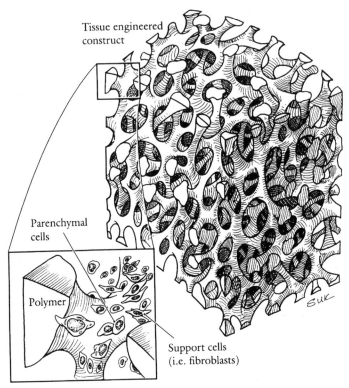

Fig. 2. Illustration depicting a tissue engineered construct. A biodegradable polymer provides a scaffold for parenchymal cells and support cells. As cells proliferate, they generate their own extracellular matrix.

4.1 Conventional grafts

4.1.1 Wound healing environment

The majority of tissues being transplanted for maintenance of function or for repair of diseased or loss tissues are placed in or on a wound, or require wounding to place the transplant in a recipient site, such as the abdominal cavity. Therefore, the graft is usually partially or completely surrounded on all sides by injured tissues. Wound healing is a continuous, dynamic process that is arbitrarily divided into three stages based on the temporal sequence of events that occur after wounding. These three stages are commonly referred to as the inflammatory, proliferative, and tissue remodeling stages [56, 57]. Although these three stages are frequently thought of in a sequential fashion, each stage overlaps the next stage by varying degrees. The first stage of wound healing, the inflammatory stage, begins as soon as the surgeon makes the skin incision and usually lasts for 3–5 days. This stage is characterized by the coagulation of blood, development of a provisional fibrin matrix, vasodilation and edema formation, invasion of inflammatory cells, and early migration of keratinocytes across the outer surface of the wound. When a conventional graft is placed in the wound, the provisional fibrin matrix forms a 'bridge' between the surfaces of the wound and the graft; this event is necessary for invasion of inflammatory and proliferative cells [58, 59]. The proliferative stage of wound healing begins with the arrival of tissue macrophages on the 3rd to 4th day after wounding and ends around the 20th post-wounding day with the arrival of ingrowing nerves. This stage of wound healing is characterized by an orderly, sequential invasion of macrophages, fibroblasts, capillaries, lymphatics and nerves. The final stage of wound healing involves remodeling of the collagen secreted by the first wave of invading fibroblasts, contraction of the wound by myofibroblasts, and a diminution in the number of blood vessels present in the wound as healing comes to a close. A complete review of the events occurring during wound healing are beyond the scope of this chapter and have been exhaustively described in detail elsewhere [56, 57], and has been covered in a previous chapter within this text (see West, Ch II.7).

4.1.2 Immunogenicity

For most situations that require a graft, surgeons prefer to use an autograft as opposed to an allograft or xenograft. The primary reason for this preference is because the patient's (host's) immune system will recognize any graft that is genetically disparate from the host as 'foreign' and will initiate a rejection response against the graft. To obtain tissue from a source other than the host and have the tissue survive, the tissue must have the same histocompatibility antigens, be shielded from the host's immune system, or be grafted to an immunosuppressed or immunodeficient host. Because cartilage is avascular and its cells (chondrocytes) are protected from the wound healing environment by its extracellular matrix, chondrocyte histocompatibility antigens are not typically exposed to the host's immune system [60]. For this reason, the chondrocytes in a cartilage allograft are temporarily able to delay detection by the host's immune system for quite some time. However, the incompatible antigens on the graft chondrocytes and the proteins comprising the ECM are eventually recognized by the host and the graft is slowly absorbed [61]. Because few people have an identical twin with the exact same histocompatible antigens (isogeneic donor), most patients require immunosuppresive medications, such as corticosteroids, azathioprine and cyclosporin, to prevent their immune system from recognizing such a graft and destroying it [62]. While the use of these drugs is warranted for certain types of organ transplants, such as a kidney or a heart transplant, the risks associated with immunosuppression are too great to justify their use for transplantation of tissues that are not integral to maintaining life-sustaining functions. Some of these risks are serious and include organ toxicity, pneumonia, systemic fungal and viral infections, and emergence of malignant tumors [63].

4.1.3 Infectious disease transmission

Another reason that an autograft is preferred to an allograft is to avoid the possibility of transmitting an infectious disease to the host. Although this problem is probably rare, HIV, hepatitis, Creutzfeldt–Jakob and other infectious diseases have been transmitted or have the potential to be transmitted by grafting tissues from any donor other than the host [63, 64]. Prevention of these diseases requires careful screening of tissues for antibodies to HIV, hepatitis, and other organisms, taking a careful medical history from the donor, and careful surveillance of tissue procurement, processing, and preservation by the tissue bank staff [65–68]. Although xenotransplantation is experimental, public health agencies, such as the Food and Drug Administration, Centers for Disease Control and Prevention, and the National Institutes for Health, have collaborated in the development and issuance of guidelines to address infectious disease risks posed by xenotransplantation [12, 64]. The objective of these recommendations is to reduce the risk to the public of human disease that may be transmitted by infectious agents arising from xenotransplantation.

4.1.4 Donor site selection

Conventional grafts are usually obtained from established donor sites that yield high quality tissue without inflicting significant donor site problems. For example, bone grafts may be obtained from the skull, iliac crest of the ilium (pelvic bone), or radius and tibia depending on the type of bone graft required and the application for which the graft is being used. Similarly, skin grafts are removed from areas of the body that are hidden by clothing; tendon and cartilage grafts are removed from muscles and ribs that contribute little in the way of function, respectively. Although these sites are presently considered acceptable locations for graft harvest by most physicians, they still require an operative procedure to obtain the tissue, are painful, are limited by the volume of tissue that can be safely removed, and can result in serious problems, such as visceral injury and contour deformity, even when obtained by highly skilled surgeons [69–74].

4.1.5 Limited transplant survival time

Most conventional grafts are harvested and transferred during the same operative setting. This is done primarily because the survival time for parenchymal and supportive cells within the graft is limited. Some tissues, such as cornea and blood, are transplanted to a recipient in a delayed fashion either because the host can not produce the needed tissue (cornea) or can not produce the tissue fast enough when required (blood). Nevertheless, these tissues are transplanted to recipient environments where the graft obtains its nutrition primarily by diffusion [75–78]. Because diffusion begins as soon as the graft is placed in the recipient site, transplantation in immunogenetically compatible patients is usually successful. Grafting tissues that require perfusion from the microcirculation usually involves creating a wound to place the graft. The cells of grafts placed in such a manner depend on diffusion of nutrients from the wound surface to the graft until the graft can be revascularized. Although tissue preservation techniques for skin, cornea, cartilage, and blood are currently available for maintaining cell viability for limited periods of time, techniques that preserve cell viability for other tissues are very limited or are not successful at all (Table 2).

4.1.6 The recipient site environment

The tissue that makes up a graft is only one key element in repairing a tissue defect. Of equal importance is the condition of the site that receives the graft. The fundamental requirements of the recipient site for conventional grafts are that: (1) the recipient site be well-vascularized; (2) there be no movement between the graft and recipient site; (3) the recipient site provide the proper functional (mechanicochemical) stimuli; and (4) the recipient site be relatively free of bacterial contamination. Each of these requirements will now be discussed in more detail.

To obtain proper nutrition by diffusion and followed by revascularization in a timely fashion (i.e. within the ischemia tolerance of the parenchymal cells) the graft must receive its nutrition over its entire surface area. To accomplish this, the recipient site must provide a large capillary cross-sectional area. Specifically, the number of blood vessels that permit diffusion across a given cross sectional area of the recipient site must be sufficiently high enough to meet the metabolic demands of the transplanted tissue. Likewise, the number of blood vessels over the cross-sectional area of the recipient site that can give rise to angiogenesis must be of a sufficient number to provide adequate neovascularization to support a significant quantity of tissue. Because diffusion of oxygen and other nutrients occurs primarily from the capillary and precapillary arteriole [92–94] and because angiogenesis occurs primarily from capillaries and post-capillary venules [95–97], poorly vascularized recipient sites with low capillary cross-sectional areas, such as subcutaneous fat and cortical bone devoid of periosteum, are unreliable as recipient sites and should be avoided as they do not usually permit graft survival. Tissues with high capillary cross-sectional counts, such as muscle or omentum, generally have a high rate of graft survival.

A corollary to the rule of an adequately vascularized recipient site is that no barrier to diffusion or revascularization should occur between the

Table 2. Maximum ischemia time for viable tissues at different temperatures*

Tissue	Temperature (°C)	Media	MIT	References
Skin	37	Ringer's	5 days	[79, 80]
	25	Saline	11 days	[80]
	4	McCoy	30 days	[81]
	−160	Gly, 15%	180 days	[82]
Cartilage	25	PM	2 days	[83]
	5	Air	56 days	[84]
	−196**	DMSO	153 days	[85]
Blood	25	CPD	24 hours	[86, 87]
	1–6	CPDA-1	35 days	[86, 77]
	1–6	Adsol	42 days	[86]
	−65†	Glycerol 40%	10 years	[87, 78]
	−120†	Glycerol 20%	10 years	[87, 78]
Bone	25	Air	1 hour	[88]
	25	Saline	2–3 hours	[89]
	25	Blood	4–6 hours	[89, 90]
Cornea	25	PM	10 hours‡	[91]
	5	MK	4 days	[75]
	0	Glycerol	Years	[76]

Abbreviations: MIT, maximum ischemia time; d, days; Gly, glycerin; hr, hour; DMSO, dimethylsulphoxide; CPD, citrate-phosphate-dextrose; CPDA- 1, citrate-phosphate-dextrose-adenosine-1; PM, post-mortem; MK, McCarey and Kaufman.

* Maximum time for which a significant portion of the ischemic tissue will survive a clinically revelant transplantation.

** Chondrocytes removed from their matrix. [Chondrocytes in whole cartilage stored in glycerol do not survive cryopreservation. Gibson T. Viability of cartilage after freezing. Proceedings of the Royal Society of Biology 1957, **147**, 528–529].

† Red blood cells.

‡ Time varies from 6 to 12 hours and depends on practices of the eye bank. An endothelial cell count is performed on all potential specimens before being accepted by the eye bank.

graft and the wound surface. Barriers may be anatomical (physical) or functional. An example of a physical barrier is a blood clot or hematoma that may form around the graft at the time of transplantation. If the hematoma is large enough, diffusion and revascularization from the recipient site to the graft would not be possible and the graft would fail [98, 99]. Other anatomical barriers include avascular tissue, such as cartilage, and carbonized or devitalized tissue, such as occurs from excessive use of an electrocautery. Functional barriers occur as the result of a gap (i.e.

deadspace) forming between the graft and recipient site. Such a gap may occur as a result of an air bubble or collection of wound fluid (seroma) [101]. In either case, the distance between the graft and wound surface is too great for diffusion or revascularization to occur in a timely fashion and the cells in the graft die [98–100].

Because blood vessels must proceed from the recipient site into the graft for revascularization to occur, anything that delays the arrival of the vessels from the wound surface into the graft beyond the ischemia tolerance of the parenchymal and supportive cells will lead to cell death and transplant failure. Cancer patients who must be reconstructed using conventional grafts following tumor resection often receive radiation therapy as part of their treatment. However, radiation therapy can delay in blood vessel arrival by inhibiting endothelial cell proliferation. Preoperative radiation therapy to the recipient site is a relative contraindication for placement of a graft. This is because radiation therapy in clinically revelant doses decreases or eliminates the proliferation of blood vessel sprouts from the irradiated recipient site [101–107]. Sholley et al. showed that chemotaxis of endothelial cells from developing vessels still occurs in the limbal blood vessels of irradiated corneas but proliferation was inhibited [101]. Thus, although some aspects of angiogenesis occur, heavily irradiated tissue does not make a reliable recipient site. Like capillary endothelial cells, the endothelium of lymph vessels also appear to be affected by radiation therapy [108, 109]. Van den Brenk noted that 20 Gy of external beam radiation therapy slowed lymphatic growth into the rabbit ear chamber [108]. Radiation also has a direct inhibitory effect on the proliferation of fibroblasts in the irradiated wound [105, 107, 110–112]. This may contribute to poor wound healing.

In addition, to a well vascularized recipient site with no barriers to capillary invasion, the graft must be appropriately secured to the recipient bed such that even micro-shearing movements are eliminated, if possible [99, 113]. Motion between the graft and the recipient site is not well tolerated because it disrupts developing blood vessels and delays revascularization. Moreover, chronic micromotion can exert shearing and tensile forces on tissue that normally does not experience them. This leads to the the third recipient site requirement.

The site must be conducive to graft survival by supplying the appropriate mechanicochemical stimuli to maintain long-term graft viability. Normal intact joints, ligament, tendon, bone, and skeletal muscle require mechanical stresses to aid in growth and healing; without mechanical stress, these tissues atrophy [114–118]. Thus, a bone graft may be adequately nourished if transplanted to a heterotopic recipient site, such as muscle, but graft resorption will occur over a very short period of time if shielded from the appropriate compressional and torsional forces that bone is ordinarily exposed to in an orthotopic grafting site [119]. Similarly, a tendon graft placed as an isolated graft in its tendon sheath or in a knee joint can survive by diffusion of synovial fluid alone, but as much as 50 to 66% of its mass is resorbed in 14 weeks if not subjected to the intermittent tractional forces that tendons normally experience [120, 121].

The final requirement for a recipient site is that it must possess bacteria levels as low as possible. Grafts placed into recipient site environments that are heavily contaminated by bacteria are not conducive to graft survival, even if the recipient site is adequately vascularized [122–124]. Bacteria can significantly alter the recipient site environment. For instance, bacteria compete for the same nutrients as cells in the graft and, as a result, decrease the pO_2 diffusion distance in the dead space [125, 126], increase the concentration of metabolic wastes (lactic acid), increase the acidity of the wound, increase the chance of graft cell injury from exposure to free radicals and lysozomes released during leukocyte–bacterial interactions [127], and facilitate gap formation by producing proteolytic enzymes that lyse the fibrin scaffold between the graft and wound surface [98, 99, 128, 129]. The latter event, allows the graft to pull away from the wound surface and delay graft revascularization. In addition, white blood cells and even serum may not access bacteria that are 'hiding' in small pores or crevices in a graft or devitalized tissue in the wound. Thus, even if prophylactic antibiotics are given prior to transplantation, some bacteria in a heavily contaminated wound may not be exposed to significant antibiotic levels or white blood cells and would continue to proliferate and lead to graft failure. Under these circumstances, the wound must be debrided of all foreign material and devitalized tissue and the graft applied after the bacterial count has decreased to a much lower number, typically less than 10^5 organisms/100 g tissue [130].

4.2 Tissue engineered constructs

4.2.1 Wound healing environment

To the extent that cell-seeded TECs are a selective and sophisticated form of grafting, TECs transplanted into a wound healing environment will be exposed to the same conditions as conventional grafts. However, there are some theoretical differences which must be considered. Some of these differences include: (1) the potential for a biomaterial scaffold to incite a pronounced inflammatory reaction either directly or indirectly which would delay or prevent incorporation of the TEC; (2) parenchymal cell density of the TEC; and (3) the duration of the tissue remodeling phase. In this section, we will briefly go through the wound healing events, but only comment on the similarities and differences in events involved with wound healing in TEC transplants compared to conventional grafts.

The inflammatory events occurring prior to TEC placement should be identical to those occuring before placement of a conventional graft.

Therefore, hemostasis and chemotactic stimulation of neutrophils and macrophages should be no different. Events occuring after TEC placement are similar to a conventional graft except that some biomaterials can elicit a pronounced inflammatory reaction. Residual solvents, monomers, and other agents involved with the polymerization of the TEC biomaterial can stimulate neutrophil and macrophage chemotaxis. Similarly, small particles that are formed as a result of mechanical forces on the TEC are phagocytosed by macrophages and may induce prolonged inflammation. Thus, the TEC must be as biocompatible as possible and, for reasons stated later, should be biodegradable. After TEC placement, a fibrin clot must form between the TEC and the wound surface for the same reasons that it forms during conventional grafting. The fibrin matrix helps anchor the TEC to the wound surface and provides a 'bridge' for neutrophils and the invading front of macrophages, fibroblasts, and endothelial cells.

After transplantation, the wound space will become profoundly hypoxic and thus cells seeded to the TEC will be exposed to an oxygen gradient similar to the cells in a graft [125, 126]. However, the difference between the two types of transplants will be that a TEC will have a hower cell density/mm^3 than a graft and, therefore, a TEC probably will not generate the same lactic acid/cytokine gradient that a graft does under such conditions. Cells located near the outer surfaces of a graft derive their nutrition by diffusion from the wound surface for the first 48–96 hours after transplantation. Because of the hypoxic environment, there is a shift to an anaerobic glycolytic metabolic pathway and a tremendous increase in lactic acid production [131]. Hypoxic cells located deep in the interior of the graft do not receive sufficient nutrients and die. Together, the superficial hypoxic, lactic acid producing cells and the dead/dying cells in the graft's interior create a massive gradient of lactic acid and cytokines that are chemotactic for ingrowing cells. Because most types of cells seeded to the external surface of a TEC and grown in static tissue culture rarely grow deeper than 300 μm from the external surfaces of the TEC, there are very few cells, if any, that are viable in the TEC's interior. Exceptions to this statement are chondrocytes which prefer a relatively anaerobic environment and possibly TECs formed in a bioreactor. Cells can grow further into a TEC formed in a bioreactor because convection currents of the tissue culture media generated in the bioreactor carry nutrients deeper into the TEC. After transplantation, cells seeded along the surface of a TEC would be expected to remain viable and to produce lactic acid as they shift to a glycolytic metabolic pathway in response to the hypoxic environment. However, the magnitude of the lactic acid/cytokine gradient in the TEC will be much smaller than a graft unless a large cell mass is grown or placed in the TEC's interior, tissue-inducing factors are incorporated into the TEC biopolymer, and/or the biopolymer itself is chemotactic for ingrowing cells.

Because TECs do not have an ECM for proteolytic enzymes to attack (unless incorporated in initial TEC fabrication), ECM fragments that are chemoattractants for leukocytes, fibroblasts, and endothelial cells are not present to contribute to that part of the chemoattractant gradient mentioned above. However, the degradation products of some biodegradable polymers used in TEC fabrication are angiogenic and probably contribute to neovascularization. For example, polylactic acid (PLA) is hydrolyzed into lactic acid. Although lactic acid is an angiogenic stimulus and the primary degradation product of PLA, the degradation rate of the PLA polymer must be high enough to generate a significant concentration of lactic acid from the TEC to create a significant lactic acid stimulus. Thus, if the degradation rate is set too low, an insignificant gradient may result; if the degradation rate is too high, the polymer construct may degrade prematurely, losing its shape and mechanical properties. Also, if too much lactic acid is generated, the microenvironment may be too acidic.

During the proliferative phase, the type of matrix being deposited in a TEC is presumably similar to the sequence of matrices described by Kurkinen [132]. Because the provisional fibrin matrix generated from the initial wounding does not penetrate the interior of a TEC to any significant degree, it is assumed that the leukocytes and the first wave of fibroblasts use the polymer itself to migrate into the TEC and deposit cell-derived ECM. The spatial arrangements of the macrophage-fibroblast-capillary unit may be different in a TEC compared to a conventional grafts. That is, one could speculate that macrophages and fibroblasts may be located closer to the capillary sprout than in conventional grafts so that the plasma-derived fibrinogen and fibronectin that leaks out of the sprout may contribute to a provisional matrix as the front invades the TEC.

The tissue remodeling phase of wound healing in a TEC has not been elucidated to any significant degree and, thus, it is unknown whether excess collagen type I or scar contracture will lead to TEC deformity, as is sometimes observed with conventional tissue grafts. Also unknown is whether excess collagen will lead to a decrease in parenchymal cell density due to nutrient limitations, that is, excess collagen acting as a diffusion barrier between the capillary and the parenchymal cells. The answers to these questions require further research.

4.2.2 Immunogenicity

At the present time, all viable tissues obtained for TEC fabrication and transplantation are subject to the same limitations of immunocompatability as for conventional grafts. Thus, tissues obtained from the host or an isogeneic donor would be preferred to tissue obtained from an allogeneic or xenogeneic source. Patients who require a transplantation would need to donate tissue for TEC fabrication prior to surgery. This introduces a logistics problem as far as time is concerned. If the surgery is elective, donating cells for TEC fabrication before surgery would not be a problem. However, if surgery is urgent, there may be insufficient time for TEC fabrication. An allogenous or xenogenous TEC must be available for use at the time of surgery if a TEC is required.

Other circumstances in which allogenous or xenogenous tissue is required include the patient's inability to produce the tissue required or the tissue is present in the patient, but non-functional. Although an allograft is the next most compatible source of cells and tissues, the number of cadavers or aborted fetuses from which cells and tissues could be obtained is restricted by availability and by social and ethical barriers [12]. To get around the problem of allograft availability, scientists are turning to the possibility of xenotransplantation. Xenotransplants are usually tolerated less well than an allograft, especially if the graft is discordant. Xenotransplants are rejected within hours of transplantation by a hyperacute immune reaction which results from a humoral attack of the patient's immune system against the vasculature of the discordant transplant, even if the patient is immunosuppressed. Although hyperacute rejection does not occur in concordant transplants, unless the patient has been previously sensitized, these transplants are rejected through a delayed rejection reaction that occurs within the first few days to weeks after transplantation. Delayed xenograft rejection is mediated by the T-cell component of the patient's cellular immune system and is aimed primarily at endothelial cells in the transplant, but can also react against parenchymal cells. Thus, both rejection phenomena occur in xenotransplants that have a microvasculature, but the delayed rejection reaction may also occur with parenchymal cells that contain discordant or concordant MHC antigens [12].

Because TECs are currently designed with only the parenchymal cells present, the primary immune responses that must be considered are the delayed and chronic rejection reaction. The chronic rejection reaction is poorly defined but is thought to be mediated by both humoral and cellular components of the immune system aimed against foreign antigens on the transplanted cell's surface. This reaction is not controlled by immunosuppressive drugs and is marked by a slow and progressive loss of graft function that begins months to years after transplantation [12]. If a multicellular TEC is created with parenchymal and endothelial cells, a hyperacute rejection reaction like phenomenon may be possible as soon as blood flow is established.

There are currently two strategies that have been developed to avoid xenograft destruction by the host's immune system without using immunosuppressive drugs. They are: (1) to genetically alter the cells prior to transplantation using genetic engineering techniques; and (2) to immunoisolate the cells from the immune system using cell or tissue encapsulation techniques [1, 2, 133]. The genetic engineering techniques revolve around either hiding the MHC antigens from the recipient's immune system or preventing the activation of the complement cascade which is a final effector of graft rejection. Techniques for hiding MHC antigens include antibody masking which uses antibody fragments to conceal MHC class I antigens on the donor cells from the cytotoxic T-cell lymphocytes of the recipient and gene knockout technology to limit or eliminate the expression of the donor MHC antigens and other antigens that lead to rejection [12]. So far, mice are the only gene knockout model available. Techniques that inhibit or block part of the complement cascade include transgenic modification of donor cells to express human decay-accelerating factor which blocks the complement cascade at the C3 stage and human CD59 which blocks the formation attack complex of the complement cascade [12, 134, 135]. All of these techniques have proven feasible in laboratory animals but require further investigation before they are considered reliable for use in humans. All tissues that are genetically altered prior to implantation will require FDA approval and may also require review by the National Institutes of Health Recombinant DNA Advisory Committee [12, 136].

Although encapsulation techniques have been efficacious in some laboratory animal models with transplant survival times of two years or more [137], there are some problems that must be addressed before they can become routinely used in clinical practice. The membrane surrounding the cells is not an absolute barrier and, therefore, will not provide a complete and permanent protection against the host's immune system [3]. The membrane must be thin enough to allow adequate diffusion of nutrients, wastes, and cell products. However, the thinner the barrier and the larger the capsule (macrocapsules), the more fragile the membrane. Micromotion of macrocapsules can lead to fracture of the barrier and exposure and elimination of the encapsulated cells by the host's immune system [138]. Additionally, under the present FDA regulations, encapsulated allogeneic and xenogeneic tissues must be retrievable after transplantation [133]. Although macrocapsules are easier to retrieve, the smaller microcapsules are hard to locate and remove from the patient without removing a significant amount of tissue [139, 140]. Like the genetically altered xenografts, encapsulated cell therapies will require FDA approval before being used in humans [12, 136]. Nevertheless, this technique offers a potential solution for many disease processes.

4.2.3 Potential for infectious disease transmission

Like conventional allografts and xenografts, cells and tissues used in construction of a cell-seeded TEC must be carefully screened for potential infectious disease agents and will be subject to the rules of the same regulatory agencies as conventional non-autologous transplants [12, 136].

4.2.4 Donor site selection

The donor sites suitable for obtaining tissue for TEC fabrication are similar to the donor sites for conventional grafts. However, an advantage of applying tissue engineering principles to this area of medicine is that less tissue is required to accomplish the same goal. Therefore, donor site problems should be less severe which should ultimately lead to a decrease in the costs of 'restoring' the patient. For example, osteoblasts may be removed from a patient using the same technique that is currently used for patients undergoing bone marrow aspiration. That is, osteoblasts are aspirated from the iliac crest of the ilium via a percutaneous approach under a light general anesthetic. Osteoblasts or osteoblast precursor cells would be separated from other cells in the aspirate, placed in tissue culture, and expanded as needed. Although cell harvest may be accomplished more readily, present disadvantages of this approach are the need for a separate procedure, the time required to create the TEC, and TEC availability when needed for transplantation.

Alternatively, allogeneic or xenogeneic parenchymal cells could be genetically modified prior to TEC fabrication, as stated above. If successful, this would eliminate the potential for donor site problems and make tissues readily available at the time of surgery.

4.2.5 Limited transplant survival time

After tissue harvesting and parenchymal cell separation, cells for a TEC can be expanded using standard tissue culture or bioreactor techniques. Furthermore, cells can be preserved at $-80°C$ for prolonged periods of time, although the percentage of viable cells following unthawing is usually less than optimum. Nevertheless, these cells also have a limited survival time once removed from tissue culture. Like a conventional graft, a TEC placed in a wound environment must obtain its initial nutrition by diffusion from the wound surface until neovascularization has occurred. If the biomaterial matrix is prevascularized, the cells seeded into the construct depend on diffusion from the vasculature within the construct until the microvasculature can remodel to meet the metabolic demands of the cell mass.

Thus for both conventional transplants and TECs, there is a biologic clock that starts ticking at the time of tissue harvesting and continues until the transplant either survives or dies. Although this clock may be temporarily slowed or arrested by using tissue preservation techniques, as a general rule, the longer the time interval between harvesting and transplantation, the greater the number of parenchymal cells die and the higher the transplant failure rate.

4.2.6 Recipient site requirements

Theoretically, the requirements for TEC recipient sites are the same as or similar to those stated above in section 4.1.6 for conventional grafts. The recipient site must be well vascularized or TEC parenchymal cells will not survive. There should not be any movement between the TEC and recipient site or ingrowing vessels will be avulsed and wound healing cells can not migrate into the TEC. Theoretically, the recipient site must provide the proper functional (mechanicochemical) stimuli and must be free of contamination and infection. Of these requirements, the last two deserve additional discussion because the ability for the TEC to interface with the recipient site is affected by the presence of the biomaterial in the TEC.

Ideally, a TEC should be placed in an orthotopic recipient site so that it will be exposed to the appropriate functional stimuli that its conventional tissue counter-part would experience. However, this raises some interesting questions. To what extent does the biomaterial transmit functional forces to the parenchymal cells of the TEC? Because the TEC microarchitecture does not exactly mimic the normal tissue microarchitecture and because parenchymal cell phenotype and function can be determined by characteristics of the ECM [141–144], will parenchymal cells of a TEC function *in vivo* the same as parenchymal cells of a conventional graft? For example, do osteoblasts seeded on a biodegradable matrix experience the same or similar compressional forces as osteoblasts in a conventional bone graft do when transplanted as an interpositional bone transplant or does the biomaterial shield the osteoblasts from 'seeing' such compressional forces? Although these questions are probably not important if the TEC is being used like a cancellous bone graft, a source of bone progenitor cells, the questions are pertinent for situations in which the TEC is being used to replicate the structural qualities of a cortical bone graft. At the present time, the answers to these questions remain unknown and require further research.

Wounding frequently introduces bacteria into a wound which are usually eliminated by the patient's defense mechanisms. Even under sterile conditions in the operating room, 35,000–60,000 bacteria fall into an operative wound each hour and yet, for clean cases, such as a hernia repair, the infection rate is less than 4% [122, 123]. The ability for bacteria to cause a wound infection depends on several factors. Most non-immuno-compromised patients will tolerate $\sim 10^5$ bacterial colony-forming counts per gram of tissue before their natural immune defenses are overwhelmed and an infection occurs; patients, who are immunosuppressed, do not have the appropriate active defense mechanisms to remove bacteria and develop wound infections with a smaller quantity of bacteria [122, 123]. The number of bacteria required for a wound infection is markedly decreased by the presence of certain virulent bacterial strains and the presence of necrotic tissue, foreign body, hematoma, and dead space [124, 130]. Bacteria, such as β-hemolytic streptococci, Staphylococcus aureus AM, and Pseudomonas aeruginosa, have evolved mechanisms to evade leukocyte bacteriocidal functions and will predispose the wound to an infection with $< 10^5$ organisms present [122, 145–147]. All foreign bodies, especially those with small pores, crevices, or interstices, interfere with the bacteriocidal activity of the leukocytes and thus, decrease the number of bacteria required for a wound infection [148, 149]. For example, sutures have been shown to increase the risk for infection if placed in the presence of bacteria; sutures that are braided are worse than monofilament suture because braided sutures have an increase in surface area compared to the monofilament suture [145]. Thus, implants, such as a TEC, with pores or crevices that are large enough to accommodate bacteria but too small for the leukocytes to enter and destroy the bacteria increase the risks for a wound infection.

In addition to providing a safe refuge from scavenging leukocytes, necrotic tissue provides nutrients for the bacteria making its presence in a contaminated wound even worse than a foreign body [130]. Subcutaneous hematomas increase the chances for infection by preventing the host's defense mechanisms from localizing bacteria. Krizek and Davis demonstrated that injecting 5×10^7 *Escherichia coli* organisms subcutaneously did not result in sepsis and elevated bacterial tissue counts unless the organisms were injected into a hematoma; animals with injected hematomas had an 80% mortality rate [150]. Although prophylactic antibiotics are helpful, their administration is no guarantee that an infection will not occur. Nevertheless, it is for this reason that surgeons frequently give a patient prophylactic antibiotics before placement of an implant.

As with a graft or an implant, the recipient site for a TEC must be clean, free of necrotic tissue, and placed under sterile conditions. Placement of an implant (TEC) in a contaminated or infected recipient site increases the chance for TEC failure secondary to wound infection. Even with our current technology, an infection occurring around a TEC would require TEC removal for appropriate treatment of the infection.

5 Closing remarks

Building on the knowledge base of conventional tissue grafts and disparate fields of science and engineering, great early strides have been made in the field of tissue engineering with regards to replacement and assembly of functional tissue equivalents. As history has routinely shown, what Nature seems to accomplish simply is often difficult for Science to duplicate. Moreover, what appears simple on the surface typically involves a complex underlying interplay of mechanisms. The development of TECs does not negate this. A case in point is the provision of a microvasculature. What Nature appears to readily provide has become a crux in current TEC design. To be sure, the future design of clinically translatable TECs will continue to involve multidisciplinary research and training in order to address the current and future design limitations realized with TECs. Many current design limitations are not unsurmountable, but only require innovative approaches and/or development of new technologies. For example, once titanium alloys were the implant scaffold of choice. However, today we have biodegradable polymers, 'smart' polymers, and biomimetic materials are just beginning to be realized. The future for the development of clinically translatable TECs is just beginning to be fathomed.

References

1. Langer, R. and Vacanti, J. P., Tissue Engineering. *Science*, 1993, **260**(5110), 920–925.
2. Nerem, R. M. and Sambanis, A., Tissue Engineering: From biology to biological substitutes. *Tissue Engineering* 1995, **1**(1), 3–13.
3. Colton, C. K. and Avgoustiniatos, E. S., Bioengineering in development of the hybrid artificial pancreas. *Journal of Biomechanical Engineering*, 1991, **113**(2), 152–170.
4. Tresco, P. A., Winn, S. R. and Aebischer, P., Polymer encapsulated neurotransmitter secreting cells. Potential treatment for Parkinson's disease. *ASAIO Journal*, 1992, **38**(1), 17–23.
5. Cooper, M. L., Hansbrough, J. F., Spielvogel, R. L., Cohen, R., Bartel, R. L. and Naughton, G., *In vivo* optimization of a living dermal substitute employing cultured human fibroblasts on a biodegradable polyglycolic acid or polygalactin mesh. *Biomaterials* 1991, **12**(2), 243–248.
6. Vacanti, C. A., Langer, R., Schloo, B. and Vacanti, J. P., Synthetic polymers seeded with chondrocytes provide a template for new cartilage formation. *Plastic and Reconstructive Surgery*, 1991, **88**(5), 753–759.
7. Shinoka, T., Breuer, C. K., Tanel, R. E., Zund, G., Miura, T., Ma, P. X., Langer, R., Vacanti, J. P. and Mayer, J. E., Jr., Tissue engineering heart valves: Valve leaflet replacement study in a lamb model. *Annals of Thoracic Surgery*, 1995, **60**(6 Suppl), S513–S516.
8. Davis, M. W., Vacanti, J. P., Toward development of an implantable tissue engineered liver. *Biomaterials*, 1996, **17**(3), 365–372.
9. Vacanti, J. P., Morse, M. A., Saltzman, W. M., Domb, A. J., Perez-Atayde, A. and Langer, R., Selective cell transplantation using bioabsorbable artificial polymers as matrices. *Journal of Pediatric Surgery*, 1988, **23**(1), 3–9.
10. Folkman, J. and Hochberg, M., Self-regulation of growth in three dimensions. *Journal of Experimental Medicine*, 1973, **138**(4), 745–753.
11. Snell, G. D., The terminology of tissue transplantation. *Transplantation*, 1964, **2**(5), 655–657.
12. Institute of Medicine (U.S.). Committee on Xenograft Transplantation. *Xenotransplantation: Science, Ethics, and Public Policy*. National Academy Press, 1996.
13. Mincheff, M. S. and Meryman, H. T., Blood transfusion, blood storage and immunomodulation. *Immunological Investigations*, 1995, **24**, 303–309.
14. Tartter, P. I., Immunologic effect of blood transfusion. *Immunological Investigations*, 1995, **24**, 277–288.

15. Irving, G. A., Perioperative blood and blood component therapy. *Canadian Journal of Anaesthesia*, 1992, **39**, 1105–1115.

16. Toy, P. T., Strauss, R. G., Stehling, L. C., Sears, R., Price, T. H., Rossi, E. C., Collins, M. L., Crowley, J. P., Eisenstaedt, R. S., Goodnaugh, L. T., Predeposited autologous blood for elective surgery: A national multicenter study. *New England Journal of Medicine*, 1987, **316**, 517–520.

17. Seghatchian, M. J. and Brozovic, B., An overview of current trends in platelet preparation, storage and transfusion. *Blood Coagulation Fibrinolysis*, 1992, **3**, 617–620.

18. Herman, J. H. and Kamel, H. T., Platelet transfusion. Current techniques, remaining problems, and future perspectives. *American Journal of Pediatric Hematology-Oncology*, 1987, **9**, 272–286.

19. Luban, N. L., Transfusion therapy with platelets and leukocytes. *Pediatric Annals*, 1983, **12**, 437–444.

20. Wright, D. G., Symposium on infectious complications of neoplastic disease (Part II). Leukocyte transfusions: thinking twice. *American Journal of Medicine*, 1984, **76**, 637–644.

21. Rudolph, R., Fisher, J. C. and Ninnemann, J. L. *Skin Grafting*. Little, Brown and Company, Boston, 1979, pp. 1–205.

22. Reece, G. and Kroll, S., Skin Grafts. In *Basal and Squamous Cell Skin Cancers of the Head and Neck*, eds. R. S. Weber, M. J. Miller and H. Goepfert. Williams and Wilkins, Baltimore, 1996, pp. 225–241.

23. Corso, P. F. and Gerald, F. P., Use of autogenous dermis for protection of the carotid artery and pharyngeal suture lines in radical head and neck surgery. *Surgery Gynecology and Obstetrics*, 1963, **117**, 37–40.

24. Vallis, C., *Hair Transplantation for the Treatment of Male Pattern Baldness*. Charles C Thomas, Springfield, 1982.

25. Griffin, E. I., Hair transplantation. The fourth decade. *Dermatology Clinics*, 1995, **13**(2), 363–387.

26. Gallico, G. G., O'Connor, N. E., Compton, C.C., Permanent coverage of large burn wounds with autologous cultured human epithelium. *New England Journal of Medicine*, 1984, **311**(7), 448–451.

27. Goldberg, V. M. and Stevenson, S., Natural history of autografts and allografts. *Clinical Othopaedics and Related Research*, 1987, **225**, 7–16.

28. Enneking, W. F., Eady, J. L. and Burchardt, H., Autogenous cortical bone grafts in the reconstruction of segmental skeletal defects. *Journal of Bone and Joint Surgery*, 1980, **62**, 1039–1058.

29. Abbott, L. C., Schottstaedt, E. R., Saunders, J. B., The evaluation of cortical and cancellous bone as grafting material. A clinical and experimental study. *Journal of Bone and Joint Surgery*, 1947, **29A**, 381–414.

30. Burwell, R. G., Osteogenesis in cancellous bone grafts: Considered in terms of its cellular changes, basic mechanisms, and the perspective of growth control and its possible abberations. *Clinical Orthopaedics and Related Research*, 1965, **40**, 35–47.

31. Craft, P. D. and Sargent, L. A., Membranous bone healing and techniques in calvarial bone grafting. *Clinics in Plastic Surgery*, 1989, **16**, 11–19.

32. Gorney, M., Murphy, S. and Falces, E., Spliced autogenous conchal cartilage in secondary ear reconstruction. *Plastic and Reconstructive Surgery*, 1971, **47**(5), 432–437.

33. Brent, B., The versatile cartilage autograft: Current trends in clinical transplantation. *Clinics in Plastic Surgery,* 1979, **6**(2), 163–180.

34. Mustardé, J. C., *Repair and reconstruction in the Orbital Region.* E and S Livingstone, Edinburgh, 1969, p. 192.

35. Brent, B., The correction of microtia with autogenous cartilage grafts: I. The classic deformity. *Plastic and Reconstructive Surgery,* 1980, **66**, 1–12.

36. Mackinnon, S. E. and Dellon, A. L., Nerve Grafting. In *Surgery of the Peripheral Nerve,* Thieme Medical Publishers, Inc., New York, 1988, pp. 89–127.

37. Millesi, H., Nerve Grafting. In *Microreconstruction of Nerve Injuries,* ed. J. K. Terzis, W. B. Saunders Company, Philadelphia, 1987, pp.223–237.

38. Wilson, R., Flexor Tendon Grafting. In *Hand Clinics: Symposium on Flexor Tendon Surgery.* ed. J. W. Strickland, W. B. Saunders, Philadelphia, 1985, pp. 97–107.

39. Blair, V. P. Notes on the operative corrections of facial palsy. *Southern Medical Journal,* 1926, **19**(2), 116–123.

40. Billings, E, Jr. and May, J. W., Jr., Historical review and present status of free fat graft autotransplantation in plastic and reconstructive surgery. *Plastic and Reconstructive Surgery,* 1989, **83**, 368–381.

41. Peer, L., Transplantation of fat. In *Reconstructive Plastic Surgery.* 2nd edn, ed. J. M. Converse, W.B. Saunders, Philadelphia, 1977, p. 251.

42. Hovding, G. and Bertelsen, T., Epikeratophakia for keratoconus. Long-term results using fresh, free-hand made lamellar grafts. *Acta Ophthalmologica,* 1992, **70**, 461–469.

43. Anonymous. Report of the organ tansplant panel. Corneal transplantation. Council on Scientific Affairs. *Journal of the American Medical Associatation,* 1988, **259**, 719–722.

44. Dohlman, C. H. and Boruchoff, S. A., Penetrating keratoplasty. *International Ophathalmology Clinics,* 1968, **8**, 655–695.

45. Miller, M. J., Schusterman, M. A., Reece, G. P., Kroll, S. S., Interposition vein grafting in head and neck reconstructive microsurgery. *Journal of Reconstructive Microsurgery,* 1993, **9**, 245–251.

46. Baird, R. N. and Abbott, W. M., Vein grafts: An historical perspective. *American Journal of Surgery,* 1977, **134**(2), 293–296.

47. May, J., Harris, J and Fletcher J., Long-term results of saphenous vein graft arteriovenous fistulas. *American Journal of Surgery,* 1980, **140**(3): 387–390.

48. Canver, C. C. Conduit options in coronary artery bypass surgery. *Chest,* 1995, **108**, 1150–1155.

49. Wells, S. A., Gunnells, J. C., Shelburne, J. D., Schneider, A. B. and Sherwood, L. M., Transplantation of the Parathyroid Glands in Man: Clinical indications and results. *Surgery,* 1975, **78**(1), 34–44.

50. Dicke, K. A., Lotzova, E, Spitzer, G. and McCredie, K. B., Immunobiology of bone marrow transplantation. *Seminars in Hematology,* 1978, **15**(3), 263–282.

51. Long, G. D. and Blume, K. G., Allogeneic and autologous marrow transplantation. In *Williams Hematology,* 5th edn., eds., E. Beutler, M. A. Lichtman, B. S. Colles and T. J. Kipps., McGraw-Hill, Inc., New York, 1995, pp. 172–194.

52. Fulkerson, J. P., Edwards, C. C. and Chrisman, O. D., Articular Cartilage. In *The Scientific Basis of Orthopaedics.* 2nd Edition. eds. J. A. Albright and R. A. Brand, Appleton and Lange, East Norwalk, 1987, pp. 347–371.

53. Pepose, J. S., Ubels, J. L., The Cornea. In *Adler's Physiology of the Eye*, 9th edn., ed. W. M. Hart, Jr., Mosby Year Book, St. Louis, 1992, pp. 29–70.

54. Guyton A. and Hall, J., The microcirculation and the lymphatic system: capillary fluid exchange, interstitial fluid, and lymph flow. In *Textbook of Medical Physiology*, 9th edn, ed. A. C. Guyton and J. E. Hall. W. B. Saunders, Philadelphia, 1996, pp. 183–197.

55. Adair, T. H., Gay, W. J. and Montani, J. P., Growth regulation of the vascular system: Evidence for a metabolic hypothesis. *American Journal of Physiology*, 1990, **259**(3, Pt 2), R393 R404.

56. Cohen, I. K., Diegelmann, R. F. and Lindblad, W. J., *Wound Healing: Biochemical and Clinical Aspects*. W. B. Saunders Company, Philadelphia, 1992.

57. Clark, R. A. F., The Molecular and Cellular Biology of Wound Repair. 2nd edn., Plenum Press, New York, 1996.

58. Tavis, M. J., Thornton, J. W., Harney, J. H., Woodroof, E. A. and Bartlett, R. H., Graft adherence to de- epithelialized surfaces: A comparison study. *Annals of Surgery*, 1976, **184**(5), 594–600.

59. Burleson, R. and Eiseman, B., Nature of the bond between partial thickness skin and wound granulations. *Surgery*, 1972, **72**(2), 315–322.

60. Elves, N. W., Newer knowledge of the immunolgy of bone and cartilage. *Clinical Orthopaedics*, 1976, **120**, 232–259.

61. Moskalewski, S., Kawiak, J. and Rymaszewska, T., Local cellular response evoked by cartilage formed after auto- and allogeneic transplantation of isolate chondrocytes. *Transplantation* 1966, **4**(5), 572–581.

62. Borel, J. F., Feurer, C., Magnee, C. and Stahelin, H. Effects of the new antilymphocytic polypeptide cyclosporin A in animals. *Immunology*, 1977, **32**(6), 1017–1025.

63. Skinner, M. D. and Schwartz, R. S., Immunosuppressive therapy 1. *New England Journal of Medicine*, 1972, **287**(5), 221–227.

64. Public Health Service, Guideline on infectious disease issues in xenotransplantation. *Federal Register*, (September, 23) 1996, **61**(185), 49920–49932.

65. Kakaiya, R., Miller, W. V. and Gudino, M. D., Tissue transplant-transmitted infections. *Transfusion*, 1991, **31**(3), 277–284.

66. Caron, M. J. and Wilson, R., Review of the risk of HIV infection through corneal transplantation in the United States. *Journal of American Optometric Association*, 1994, **65**(3), 173–178.

67. Eastlund, T., Infectious disease transmission through cell, tissue, and organ transplantation: reducing the risk through donor selection. *Cell Transplantation*, 1995, **4**(5), 455–477.

68. Shelby, J., Saffle, J. R. and Kern, E. R., Transmission of cytomegalovirus infection in mice by skin graft. *Journal of Trauma*, 1988, **28**(2), 203–206.

69. Rudolph, R., Skin Grafting. In *The Unfavorable Result in Plastic Surgery*, Vol. 1., 2nd ed., R. M. Goldwyn, Little, Brown and Co., Boston, 1984, pp. 143–149.

70. Younger, E. M. and Chapman, M. W., Morbidity at bone graft donor sites. *Journal of Orthopaedic Trauma*, 1989, **3**(3), 192–195.

71. Laurie, S. W. S., Kaban, L. B., Mulliken, J. B., et al. Donor-site morbidity

after harvesting rib and iliac bone. *Plastic Reconstructive Surgery*, 1984, **73**(6), 933–938.

72. White, W. L. Tendon grafts: A consideration of source, procurement and suitability. *Surgical Clinics of North America*, 1960, **40**(2), 403–413.

73. Thomson, H. G., Ty, K. and Ein, S. H., Residual problems in chest donor sites after microtia reconstruction: A long-term study. *Plastic Reconstructive Surgery*, 1995, **95**(6), 961–968.

74. Tanzer, R. C., Microtia: A long-term follow-up of 44 reconstructed auricles. *Plastic Reconstructive Surgery*, 1978, **61**(2), 161–166.

75. Bourne, W. M., Corneal Preservation. In *The Cornea*, eds. H. E. Kaufman, B. A. Barron, M. B. McDonald and S. R. Waltman, Churchill Livingstone, Inc., New York, 1988, pp. 713–724.

76. Schultz, R. O., Matsuda, M., Yee, R. W., Long-term survival of cryopreserved corneal endothelium. *Ophthalmology*, 1985, **92**(12), 1663–1667.

77. Zuck, T. F., Bensinger, T. A., Peck, R. K., The in vivo survival of red blood cells stored in modified CPD with adenine. *Transfusion*, 1977, **17**(4), 374–382.

78. Young, J. A., Donation, preparation, and storage. In *Textbook of blood banking and transfusion medicine*, ed. S. V. Rudman, Saunders, Philadelphia, 1995, pp. 240–243.

79. Pepper, F. J., Studies on the viability of mammalian skin autografts after storage at different temperatures. *British Journal of Plastic Surgery*, 1954, **6**, 250–256.

80. Smahel, J., Free skin transplantation on a prepared bed. *British Journal of Plastic Surgery*, 1977, **24**(2), 129–132.

81. Hurst, L. N. Brown, D. H. and Murray, K. A., Prolonged life and improved quality for stored skin grafts. *Plastic Reconstructive Surgery*, 1984, **73**(1), 105–110.

82. Bondoc, C. C. and Burke, J. F., Clinical experience with viable frozen human skin and a frozen skin bank. *Annals of Surgery*, 1971, **174**(3), 371–382.

83. Brent, B., Repair and grafting of cartilage and perichondrium. In: *Plastic Surgery*, Vol. 1, ed. J. G. McCarthy, WB Saunders, Philadelphia, 1990, pp 559–582.

84. Hagerty, R. F., Calhoon, T. B., Lee, W. H. and Cuttino, J. T., Human cartilage grafts stored in air. *Surgery, Gynecology, and Obstetrics*, 1960, **110**, 433–436.

85. Schachar N. S. and McGann L. E., Investigations of low-temperature storage of articular cartilage for transplantation. *Clinical Orthopaedics and Related Research*, 1986, **208**, 146–150.

86. Young, J. A. and Rudmann, S. V., Blood component preservation and storage. In *Textbook of blood banking and transfusion medicine*, ed. S. V. Rudman, Saunders, Philadelphia, 1995, pp. 269–271.

87. American Association of Blood Banks, *Standards for Blood Banks and Transfusion Services*, 17th edn., Bethesda, MD, 1996, pp. 22–24

88. Bassett, C. A. L., Clinical implications of cell function in bone grafting. *Clinical Orthopaedics and Related Research*, 1972, **87**, 49–59.

89. Puranen, J., Reorganization of fresh and preserved bone transplants. *Acta Orthopaedica Scandanavia*, 1966, **92**(Suppl), 9–95.

90. Haas, S. L., A study of the viability of bone after removal from the body. *Archives of Surgery*, 1923, **7**, 213–226.

91. Eye Bank Association of America: *Medical standards for member eye banks.* Houston, June, 1996, pp. 12–13.

92. Popel, A. S. and Gross, J. F., Analysis of oxygen diffusion from arteriolar networks. *American Journal of Physiology*, 1979, **237**(6), H681–H689.

93. Popel, A. S., Pittman, R. N. and Ellsworth, M. L., Rate of oxygen loss from arterioles is an order of magnitude higher than expected. *American Journal of Physiology*, 1991, **256**(3 Pt. 2), H921–H924.

94. Fihlo, I. P., Kerger, H. and Intaglietta, M. pO_2 measurements in arteriolar networks. *Microvascular Research*, 1996, **51**(2), 202–212.

95. Cliff, W. J., Observations on healing tissue: A combined light and electron microscopic investigation. *Philosophical Transactions of the Royal Society (British)*, 1963, **246**, 305–325.

96. Schoefl, G. I., Studies on inflammation. III. Growing capillaries: Their structure and permeability. *Virchows Archives of Pathology, Anatomy, and Physiology*, 1963, **337**, 97–141.

97. Ausprunk, D. H. and Folkman, J., Migration and proliferation of endothelial cells in preformed and newly formed blood vessels during tumor angiogenesis. *Microvascular Research*, 1977, **14**(1), 53–65.

98. Tavis, M. J., Thornton, J. W., Harney, J. H., Danet, R. T., Woodruf, E. A. and Bartlett, R. H., Mechanism of skin graft adherence: Collagen elastin and fibrin interactions. *Surgical Forum*, 1977, **28**, 522–524.

99. Teh, B. T., Why do skin grafts fail? *Plastic and Reconstructive Surgery*, 1979, **63**(3), 323–332.

100. Littlewood, A. H. M., Seroma: An unrecognized cause of failure of split-thickness skin grafts. *British Journal of Plastic Surgery*, 1960, **13**, 42–46.

101. Sholley, M. M., Ferguson, G. P., Seibel, H. R., Montour, J. L. and Wilson, J. D., Mechanisms of neovascularization. Vascular sprouting can occur without proliferation of endothelial cells. *Laboratory Investigations*, 1984, **51**(6), 624–634.

102. Shamberger, R. Effect of chemotherapy and radiotherapy on wound healing: Experimental studies. *Recent Results in Cancer Research*, 1985, **98**, 17–34.

103. Stearner, S. P., Sanderson, M. H., Mechanisms of acute injury in the gamma-irradiated chicken: Effects of a protracted or split-dose exposure on the fine structure of the microvasculature. *Radiation Research*, 1972, **49**(2), 328–352.

104. Dewhirst, M. W., Gustafson, C., Gross, J. F. and Tso, C. Y., Temporal effects of 5.0 Gy radiation in healing subcutaneous microvasculature of a dorsal flap window chamber. *Radiation Research*, 1987, **112**(3), 581–591.

105. Prionas, S. D., Kowalski, J., Fajardo, L. F., Kaplan, I., Kwan, H. H. and Allison, A. C., Effects of X irradiation on angiogenesis. *Radiation Research*, 1990, **124**(1), 43–49.

106. Dimitrievich, G. S., Fischer-Dzoga, K. and Griem, M. L., Radiosensitivity of vascular tissue. I. Differential radiosensitivity of capillaries: A quantitative in vivo study. *Radiation Research*, 1984, **99**(3), 511–535.

107. Fischer-Dzoga, K., Dimitrievich, G. S.and Griem, M. L., Radiosensitivity of

vascular tissue. II. Differential radiosensitivity of aortic cells *in vitro*. *Radiation Research*, 1984, **99**(3), 536–546.

108. Van den Brenk, W. A. S., The effect of ionizing radiation on the regeneration and behaviour of mammalian lymphatics. *American Journal of Roentgenology, Radiation Therapy, and Nuclear Medicine*, 1957, **78**(5), 837–849.

109. Malek, P. and Vrubel, J., Lymphatic system and organ transplantation. *Lymphology*, 1968, **1**(1), 4–22.

110. Macierira-Coelho, A., Diatloff, C., Billard, M., Fertil, B., Malaise, E. and Fries, D., Effect of low dose irradiation on the division of cells in vitro. IV. Embryonic and adult human lung fibroblast-like cells. *Journal of Cellular Physiology*, 1978, **95**(2), 235–238.

111. Rudolph, R., Vande Berge, J., Schneider, J. A., Fischer, J. C. and Poolman, W. L., Slowed growth of cultured fibroblasts from human radiation wounds. *Plastic Reconstructive Surgery*, 1988, **82**(4), 669–677.

112. Smith, K. C., Hahn, G. M., Hoppe, R. T. and Earle, J. D., Radio sensitivity in vitro of human fibroblasts derived from patients with a severe skin reaction to radiation therapy. *International Journal of Radiation Oncology, Biology, and Physics*, 1980, **6**(11), 1573–1575.

113. Bassett, C. A. L., Clinical implications of cell function in bone grafting. *Clinical Orthopaedics and Related Research*, 1972, **87**, 49–59.

114. Akeson, W. H., Amiel, D. and Woo, S. L.-Y., Immobility effects on synovial joints: the pathomechanics of joint contracture. *Biorheology*, 1980, **17**(1–2), 95–110.

115. Amiel, D., Akeson W. H. Harwood, F. L. and Frank, C. B., Stress deprivation effect on metabolic turnover of the medial collateral ligament collagen: a comparison between nine and 12 week immobilization. *Clinical Orthopaedics and Related Research*, 1983, **172**, 265–270.

116. Gelberman, R. H., Amiel, D., Gonsalves, M., Woo, S. L.-Y., and Akeson, W. H., The influence of protected passive mobilization on the healing of flexor tendons: a biochemical and microangiographic study. *Hand*, 1981, **13**(2), 120–128.

117. Woo, S. L.-Y., Kuei, S. C., Amiel, D., Gomez, M. A., Hayes, W. C., White, F. C. and Akeson, W. H., The effect of prolonged physical training on the properties of long bone: a study of Wolff's Law. *Journal of Bone and Joint Surgery*, 1981, **63A**, 780–787.

118. Booth, F. W., Time course of muscle atrophy during immobilization of hindlimbs in rats. *Journal of Applied Physiology:Respiratory Environmental and Exercise Physiology*,1977, **43**(4), 56 661.

119. Wolff, J. *The Law of Bone Remodeling*. (Translated Maquet, P. and Furlong, R.), Springer-Verlag, Berlin, 1986.

120. Matthews, P., The fate of isolated segments of flexor tendon within the digital sheath – a study in synovial nutrition. *British Journal of Plastic Surgery*, 1976, **29**(2), 216–224.

121. Lundborg, G. and Rank, F., Experimental intrinsic healing of flexor tendons based upon synovial fluid nutrition. *Journal of Hand Surgery*, 1978, **3**(1), 21–31.

122. Robson, M. C., Krizek, T. J. and Heggars, J. P., Biology of surgical infections. In *Current Problems in Surgery*, ed. M. M. Ravitch, Year Book, Chicago, 1973, pp. 1–62.

123. Krizek, T. J. and Robson, M. C., Evolution of quantitative bacteriology in wound management. *American Journal of Surgery*, 1975, **130**(5), 579–584.

124. Rodeheaver, G. T., Pettry, D., Turnbull, B., Edgerton, M. T. and Edlich, R. F., Identification of the wound infection – potentiating factors in soil. *American Journal of Surgery*, 1974, 128(1), 8–14.

125. Silver, I. A., The measurement of oxygen tension in healing tissue. *Progress in Respiratory Research*, 1969, **3**, 124–135.

126. Hunt, T. K. and Hutchinson, J. G. P., Studies on the oxygen tension in healing wounds. In *Wound Healing*, ed. C. F. Illingsworth, Lister Symposium, Churchill, London, 1966, pp. 257–266.

127. Heggars, J. P. and Robson, M. C., Prostaglandins and Thromboxanes. In *Traumatic Injury – Infection and Other Immunologic Sequelae*, ed. J. Ninnemann, University Park Press, Baltimore, 1983, Chpt. 5.

128. Liedburg, N. C. F., Reiss, E. and Artz, C. P., The effect of bacteria on the take of split thickness skin grafts in rabbits. *Annals of Surgery*, 1955, **142**(1), 92–96.

129. Perry, A. W., Sutkin, H. S., Gottlieb, L. J., Stadelmann, W. K. and Krizek, T. J., Skin graft survival – the bacterial answer. *Annals of Plastic Surgery*, 1989, **22**(6), 479–483.

130. Edlich, R. F., Rodeheaver, G., Thacker, J. G. and Edgerton, M. T., Technical Factors in Wound Management. In *Fundamentals of Wound Management*, ed. by T. K. Hunt and J. E. Dunphy, Appleton-Century-Crofts, New York, 1979, pp. 416–423.

131. Hira, M. and Tajima, S., Biochemical study on the process of skin graft take. *Annals of Plastic Surgery*, 1992, **29**, 47–54.

132. Kurkinen, M., Vaheri, A., Roberts, P. J. and Stenman, S., Sequential appearance of fibronectin and collagen in experimental granulation tissue. *Laboratory Investigation*, 1980, **43**(1), 47–51.

133. Christenson, L., Dionne, K. E. and Lysaght, M. J., Biomedical applications of immobilized cells. In *Fundamentals of Animal Cell Encapsulation and Immobilization*, ed. M. F. A. Goosen, CRC Press, Boca Raton, 1993, pp. 7–41.

134. Bach, F. H., Turman, M. A., Vercellotti, G. M., Platt, J. L. and Dalmasso, A. P., Accomodation: A working paradigm for progressing toward discordant xenografting. *Transplantation Proceedings*, 1991, **23**(1, Pt. 1), 205–208.

135. Dalmasso, A. P., Vercellotti, G. M., Platt, J. L. and Bach, F. H., Inhibition of complement-mediated endothelial cell cytotoxicity by decay-accelerating factor. Potential for prevention of xenograft hyperacute rejection. *Transplantation*, 1991, **52**(3), 530–533.

136. U.S. Food and Drug Administration (FDA). Application of Current Statutory authorities to human somatic cell therapy products and gene therapy products. *Federal Register Notice*, October 14, 1993.

137. Soon-Shiong, P., Feldman, E., Nelson, R., Heintz, R., Merideth, N., Sandford, P., Zheng, T. and Komtebedde, J., Long-term reversal of diabetes in the large animal model by encapsulated islet transplantation. *Transplantation Proceedings*, 1992, **24**(6), 2946–2947.

138. Lanza, R. P., Sullivan, S. J. and Chick, W. L., Islet transplantation with immunoisolation. *Diabetes*, 1992, **41**(12), 1503–1510.

139. Lum, Z. P., Tai, I. T., Krestow, M., Norton, J., Vacek, I. and Sun, A. M., Xenografts of rat islets into diabetic mice. An evaluation of new smaller capsules. *Transplantation*, 1992, **53**(6), 1180–1183.

140. Sun, A. M., Vacek, I., Sun, Y. L., Ma, X. and Zhou, D., In vitro and in vivo evaluation of microencapsulated porcine islets. *ASAIO Journal*, 1992, **38**(2), 125–127.

141. Damsky, C. H. and Werb, Z., Signal transduction by integrin receptors for extracellular matrix: Cooperative processing of extracellular information. *Current Opinion in Cell Biology*, 1992, **4**(5), 772–781.

142. Ingber, D. E. and Folkman, J., Mechanochemical switching between growth and differentiation during fibroblast growth factor-stimulated angiogenesis *in vitro*: Role of extracellular matrix. *Journal of Cell Biology*, 1989, **109**(1), 317–330.

143. Norton, L. A., Rodan, G. A. and Bourrett, L. A., Epiphseal cartilage cAMP changes produced by mechanical and electrical pertubations. *Clinical Orthopaedics and Related Research*, 1977, **124**, 59–68.

144. Grodzinsky, A. J., Electromechanical and physicochemical properties of connective tissue. *CRC Critical Review of Biomedical Engineering*, 1983, **9**, 133–199.

145. Elek, S. D., Conen, P. E., The virulence of staphylococcus pyogenes for man: A study of the problems of wound infection. *British Journal of Experimental Pathology*, 1957, **38**, 573–586.

146. Melly, M. A., Duke, J., Liau, D. F. and Hash, J. H., Biological properties of the encapsulated *Staphylococcus aureus* M. *Infections and Immunity*, 1974, **10**(2), 389–397.

147. Dimitracopoulos, G. and Bartell, P. F., Slime glycolipoproteins and the pathogenicity of various strains of *Pseudomonas aeruginosa* in experimental infection. *Infections and Immunity*, 1980, **30**(2), 389–397.

148. Roettinger, W., Edgerton, M. T., Kurtz, L. D., Prusak, M. and Edlich, R. F., Role of inoculation site as a determinant of infection in soft tissue wounds. *American Journal of Surgery*, 1973, **126**(3), 354–350.

149. Cruse, P. J. and Foord, R., A five year prospective study of 23,649 surgical wounds. *Archives of Surgery*, 1973, **107**(2), 206–210.

150. Krizek, T. J. and Davis, J. H., The role of the red cell in subcutaneous infection. *Journal of Trauma*, 1965, **5**(1), 85–95.

Musculoskeletal Tissue Engineering for Orthopedic Surgical Applications

MICHAEL J. YASZEMSKI
Orthopedic Surgery,
Adult Reconstruction and Spine Surgery,
Mayo Clinic,
200 First Street SW,
Rochester, MN 55905, USA

ALAN W. YASKO
Orthopaedic Surgical Oncology,
M.D. Anderson Cancer Center,
1515 Holcombe Blvd.,
Houston, TX 77030, USA

1 Introduction

The practice of orthopedic surgery involves caring for patients who have diseases of the musculoskeletal system. The orthopedic surgeon manages these diseases via non-operative and operative measures that target the tissues of the musculoskeletal system: bone, cartilage, muscle, ligament, intervertebral disc, and tendon [6, 47]. There have been tissue engineering efforts to regenerate most of these tissues. This chapter will present and discuss those efforts that involve the regeneration of bone and cartilage. These two tissues, although they are by no means the only musculoskeletal tissues which present unsolved problems to the treating orthopedic surgeon, are the two tissues of the musculoskeletal system which have received the greatest attention from a tissue engineering perspective. The first part of this chapter will discuss skeletal repair and regeneration, and will present tissue engineering strategies to reconstruct portions of the skeleton where and when the natural repair and regeneration processes have failed. The second part of the chapter will discuss cartilage biology, biomechanics, and tissue engineering strategies for cartilage regeneration. Each of the two sections will discuss the relevant clinical needs, current options for their solution, and potential means by which the proposed tissue engineering strategies might address these needs. In addition, there will be a discussion of the *in vitro* and *in vivo* methods by which the tissue engineering applications have been evaluated. We will discuss the potential limitations of biologic regeneration techniques from a surgical perspective.

2 Skeletal repair and regeneration

The regeneration of bone via tissue engineering strategies will likely exploit several key properties of normal bone. Bone tissue has the capacity to regenerate in response to injury. The new tissue, called woven bone, consists of randomly oriented collagen fibers that mineralize, and then remodel into an organized, anisotropic bone structure that optimizes itself mechanically for the local stress environment it experiences. The remodeled bone is called lamellar bone. This transformation of woven bone into lamellar bone that occurs in fracture healing also occurs in embryogenesis of the skeleton.

This section will first discuss the anatomy of bone, which determines in part the mechanical properties of bone tissue in different body locations. The process of fracture healing, which is the natural bone regeneration process that tissue engineering strategies attempt to mimic, will be described next. Following this will be a discussion of currently available bone transplantation methods, and a presentation of their strengths and weaknesses. The section will conclude with a discussion of tissue engineering techniques that are currently being explored as alternatives to the currently used bone transplantation techniques.

There are several descriptors of bone that are pertinent to this discussion. Bone is either cortical (also called compact or osteonal) or trabecular (also called cancellous or spongy). Cortical bone occurs in the walls of the shafts of long bones and as the outer shell of flat bones such as the pelvis. It is dense with respect to trabecular bone, and contains closely spaced groups of concentric lamellar bone rings called osteons. Cortical bone functions mechanically in tension, compression, and torsion. Trabecular bone occurs in the interior of flat bones, and near the ends of long bones in the region of the joint surface. It consists of a porous array of interconnected lamellar bone rods and plates (trabeculae). Trabecular bone functions mainly in compression. The spaces between the trabeculae contain the bone marrow. The bone marrow houses, among others, the blood forming cells of the hematopoietic system and the infection fighting cells of the immune system. The marrow also contains growth factors that direct the development of several cell types, including bone and cartilage. The outer surface of the shafts of long bones is covered by a tissue called periosteum. The periosteum contains two distinct layers: an outer, fibrous layer, and an inner, cellular layer called the cambium layer. This cambium layer of the periosteum is the source of cells that become bone forming cells in the process of fracture repair. The periosteum has been used in several tissue engineering strategies as a tissue that can provide uncommitted cells capable of becoming cartilage or bone forming cells.

The body's skeleton, in addition to providing mechanical support for the organs and a site of attachment for muscles used in locomotion, provides several other functions. It is the largest reservoir of calcium and phosphorus in the body, and constantly exchanges these substances with the bloodstream to provide for a close control of their concentrations in the body fluids. In addition, the marrow function of cellular regeneration has been described above. Tissue engineering strategies for skeletal regeneration need replace only the mechanical function of the bone; the other functions can be adequately provided by bone elsewhere in the skeleton. However, the mechanical function in the area of bone deficiency can only be restored by bone regeneration in that area that is in mechanical continuity with host bone at the margins of the defect.

Fracture healing occurs in a well defined sequence, and is usually successful in healing skeletal injury. The sequence can be viewed as three overlapping physiologic stages. A second description of the process is that of four biomechanical stages. The physiologic stages are inflammation, repair, and remodeling. Inflammation begins at the moment of injury. The body recruits acute inflammatory cells to the area of the fracture, and begins to organize a fibrous network of tissue that begins as the blood clot at the fractured bone surfaces. The area becomes populated with cells from the periosteum, and cells brought via the neovasculature of the organizing fibrous tissue. These cells first make a cartilaginous framework which mineralizes in an orderly process called ossification. The mineralized cartilage framework has become woven bone. This stage of fracture healing is the repair stage, and overlaps in time with the inflammatory stage. As soon as woven bone is formed, it is subject to remodeling along lines of stress by the removal and redeposition of its mineral and organic phases. The biomechanical classification of fracture healing begins with the first stage, in which the repair tissue is fibrous, and not capable of performing bone's mechanical function. The second stage is the soft callus stage. This corresponds to the formation of a cartilaginous framework, which is not as strong or stiff as normal bone, and will be the location where a refracture will occur should the healing bone experience further injury. The third stage is the hard callus stage. At this time, the healing bone has increased its strength and stiffness. It is not yet as strong as normal bone, but a refracture at this time would likely pass partially through the original, healing fracture, and partially through previously intact, uninjured bone. The fourth stage, remodeling, is a stage in which a new fracture is just as likely to pass through previously uninjured bone as it is to pass through the original fracture.

These stages of fracture healing include the recruitment of cells and bioactive molecules to a specific body area, the elaboration of a matrix which first becomes a cartilaginous scaffold and then ossifies, and then proceeds with the remodeling of this newly formed woven bone into strong lamellar bone. They represent the natural process that skeletal tissue engineering strategies seek to reproduce at a time and in a body location that a specific patient requires.

3 Current options for the treatment of bone defects

Bone transplantation is a commonly required surgical procedure. Skeletal loss occurs as a result of trauma and removal of tumors. Skeletal augmentation is needed for arthrodeses (joint fusions) and during some joint replacement procedures. The currently available options are autograft, allograft, bone cement, and several synthetic materials that have recently come to the market. Autograft is currently the gold standard against which other bone transplantation choices are measured. Autograft is bone harvested from one location in a person, and transplanted to a different location in the same person. Autograft has the benefit of being immunologically identical to the patient, and hence there are no rejection concerns with its use. The cells and bioactive molecules theoretically have an opportunity to survive the journey from harvest site to recipient site. This potential transplantation of bone forming cells and bioactive growth factors is not possible with the other currently available bone transplantation options. The disadvantages of autograft include morbidity at the donor site (described below), a limited supply, and limitations on the size and shape that the graft can take. The donor site, often the anterior or posterior iliac crest, can be a long term source of pain for the patient, fracture after the graft is taken, or become infected. Allograft is bone harvested from a deceased donor, processed, and transplanted into a recipient of the same species. The processing removes all cells and bioactive molecules, and decreases the antigenicity of the graft. The benefits of allograft include an abundant supply of both cortical and trabecular bone in a variety of shapes and sizes. The disadvantages of allograft are that it does occasionally incite an immune response in the recipient, has a somewhat lower rate than autograft for incorporation with host bone at the margins of the defect, and can potentially transmit viral diseases.

There are several synthetic choices available to fill bone defects. Bone cement is poly(methyl methacrylate). It is a non-degradable material. Its advantages include a ready, abundant supply, the ability to fill defects of varying shapes and sizes, and the provision of immediate mechanical strength to the reconstructed region, since it does not need to become incorporated as autograft and allograft do. This property of permanence is also one of bone cement's major disadvantages. Since it does not allow the reconstructed skeletal region to regenerate bone, the repetitive stresses of daily activities eventually will cause its failure. These very stresses would have caused natural bone to remodel and become more capable of resisting them. The region which bone cement occupies is also more liable than natural bone to become infected, since there is no vascular supply or infection fighting cellular and molecular capability there.

There are various synthetic materials that use mineral (calcium phosphate) and xenograft collagen in different forms and combinations. Xenograft is material from a member of one species transplanted into a member of another species. The collagen in these synthetic bone void fillers is usually of bovine origin. The calcium phosphate is usually in the form of hydroxyapatite or β-tricalcium phosphate. An injectable form of calcium phosphate that hardens into dahllite has been used clinically.

4 Tissue engineering strategies as alternatives to the current bone defect options

The natural skeletal repair process of fracture healing utilizes bone forming cells, a cartilaginous scaffold upon which the new woven bone forms, and bioactive molecules to direct the repair sequence. Tissue engineering strategies endeavor to provide some or all of these conditions to the area of desired bone formation. Let us review some features of these three factors, and then describe several current strategies to duplicate them.

The body recognizes the cartilaginous scaffold in some physical or chemical manner, and ossifies it. The scaffold is porous, and allows vascular and cellular ingrowth. A majority of the tissue engineering approaches utilize a porous scaffold. The scaffolds have often been degradable polymers, but ceramics, pyrolyzed carbon, and non-degradable polymers have also been used. The cells that ossify the cartilaginous fracture scaffold are osteoblasts (bone forming cells). Several bone regeneration strategies deliver osteoblasts, harvested from the host, expanded in tissue culture on degradable polymers, and delivered to the desired site of bone regeneration. There are several bioactive molecules reported to have a role in the direction of the bone regeneration process [16]. These are the bone morphogenic proteins (BMP's), transforming growth factor-β (TGF-β), and basic fibroblast growth factor (b-FGF). Finally, prior to discussion of specific strategies, the terms osteoconduction and osteoinduction need definition. Osteoconduction is the provision of a scaffold to enable bone growth at a time and in a location where it will occur because of the presence of cells to make osteoid (the bone organic matrix), and bioactive molecules to direct the process. Osteoinduction is the process of effecting bone growth in a location and at a time where and when it otherwise would not occur by providing the scaffold, the bone forming cells, and the bioactive molecules. We will now discuss several specific strategies that use combinations of scaffolds, bone forming cells, and bioactive molecules to engineer new bone [2, 10].

Bovine periosteum was grown on a non-woven poly (glycolic acid) (PGA) mesh, wrapped in a cylindrical poly (glycolic acid)- poly (lactic acid) (PGA-PLA) tube, and placed in an athymic (nude) rat femoral segmental defect [14]. The femur was stabilized with a titanium plate. This model represents a xenograft. The femoral defect, 9 mm long, is a critical size defect in the rat. This means that it will reliably not heal without intervention in the lifetime of the animal. The cylindrical tube is an example of guided tissue regeneration, in which fibrous tissue is prevented from filling the defect by the cylindrical barrier. This allows the cells of the cell–polymer construct time to populate the defect and express bone matrix. The PGA mesh was 97% porous. Most of the polymer scaffold had resorbed at 12 weeks post-implantation, and the new tissue was tested for osteocalcin and type-1 collagen, both of which were present. There was no evidence of inflammation noted at the sites of the defect. Seven of 10 rats with the polymer–cell construct bridged the gap with bone at 12 weeks. Specimens harvested at 6 weeks had primarily cartilage in the gap. None of the control animals, who received either no treatment or polymer without cells, bridged the gap with new bone in the 12 weeks of the study.

New bone formed in cell–polymer constructs passes through a cartilaginous scaffold phase prior to ossification. We will discuss neo-cartilage formation in the second section of this chapter, but a brief mention of the cartilaginous phase merits description here. Bovine osteoblasts cultured on a degradable polymer and implanted into rats form cartilage, then bone, as described above. However, bovine chondrocytes cultured on the same degradable polymer and implanted into rats form cartilage that does not ossify [17]. Composite tissues containing both bone and cartilage have been produced by seeding chondrocytes and periosteal derived cells (presumably of osteoblast lineage) separately on degradable polymers, and then joining the polymers together via a suture [18]. After implantation, bone tissue grows on the side seeded with osteoblasts (after passing through a cartilage phase), and cartilage grows on the side seeded with chondrocytes. At one time point, both areas are populated by cartilage tissue, but the part initially seeded with osteoblasts subsequently vascularizes and ossifies, while the other part does not. Vascularization of bone tissue is necessary for its growth and survival, while mature articular (hyaline) cartilage is avascular, aneural, and alymphatic. There is a putative angiogenesis inhibiting factor present in cartilage (and hence expressed by the chondrocytes) that prevents vascular ingrowth and subsequent ossification.

Polyphosphazines have been investigated as bone regeneration scaffolds [24]. These polymers have a phosphorus–nitrogen backbone, and are rendered hydrolytically degradable by the addition of an imidazolyl or amino acid alkyl ester side chain. The nature of the side chain and percent substitution can affect the degradation rate and cell attachment rate [8]. These polymers degrade, in the presence of water, to ammonia, phosphate, and the organic side group. Osteoblast-like cells have grown on three dimensional matrices made of these polymers [1, 7, 9].

A polyetherester copolymer, poly (ethylene oxide)/poly (butylene terephthalate) (PEO/PBT) was processed into a cylinder via sintering of the particulate polymer, and implanted in a goat unicortical femoral drill hole. Bone grew into the scaffold in the cortical region, but not into the part of the device which was intramedullary [14]. This strategy did not use a degradable polymer, and depended upon cells and growth factors from the local host environment to populate the scaffold. Another study explored the piezoelectric properties of drawn poly-L-lactic acid (PLLA) fibers in bone regeneration [2]. It is known that electrokinetic phenomena are associated with healing fractures, and that electrical stimulation appears to be effective in treating some fracture non-unions clinically. In the study using drawn PLLA, increasing the draw ratio increased the piezoelectric constants of the fibers. These fibers were placed in a cat tibia fracture model, which was stabilized with an intramedullary rod. Control specimens were treated with undrawn fibers or polyethylene fibers. There was more bone callus present at 8 weeks in the drawn fiber groups than in the other groups, and the amount of bone present increased linearly with an increase in the fiber draw ratio.

Neonatal rat calvarial osteoblasts have expressed their phenotypic function on degradable poly (lactic-co-glycolic acid) (PLGA) matrices. This was determined by measuring the production of type 1 collagen and alkaline phosphatase [5]. The same feasibility was demonstrated for human osteoblasts harvested during total hip replacement surgery, transported to the lab, and grown on tissue culture polystyrene [15]. The demonstration of this feasibility was followed by studies that showed rat calvarial osteoblasts would migrate *in vitro* on poly (α-hydroxy esters) and express their differentiated function on both two dimensional foams [4] and in three dimensional foams [3]. Osteoblasts are attachment dependent cells and require a vascular supply for their nutrient and waste exchange. Foam pre-vascularization in an ectopic site, which had been demonstrated previously [9], was applied to osteoblast transplantation in the rat mesentery and yielded bone formation in the PLGA scaffold [6]. The concept explored in this study is that the vascularized scaffold, with new bone formation in it and a transplantable vascular supply attached to it, could then be moved to the area of bone deficiency. The concept of a transplantable, vascularized newly formed bone graft was demonstrated as feasible in an ovine rib periosteal bed model [10]. This experiment did not use a degradable polymer scaffold, but used instead trabecular bone graft in preformed poly (methyl methacrylate) chambers. These had an open face sutured to the periosteal bed, and the new bone in them had a vascular supply based on the intercostal vessels. The new bone and its vessels were then successfully transplanted to a defect in the mandible, and incorporated well. This process would find clinical utility in poorly vascularized recipient sites, such as a mandible which had experienced osteoradionecrosis after radiation for cancer.

Poly (propylene fumarate) has been fabricated into two and three dimensional scaffolds for skeletal regeneration [12, 16]. It has demonstrated degradation *in vitro* [11, 20], and has supported bone growth into it in an *in vivo* rat model [19]. This material has propylene glycol and fumaric acid as its monomeric units, both of which the body can excrete via normal metabolic pathways. Fumaric acid is a component of the Krebs cycle, and propylene glycol has a long history of use as an intravenous fluid component.

The above descriptions are attempts to deliver osteoblasts on temporary biodegradable scaffolds to skeletal regions where bone regeneration is desired. Issues that are involved are those of mechanical stability during bone regeneration, the source and rapidity of neovascularization, the degradation rate of the temporary scaffold, and maintenance of sterility of the polymer–cell construct during *in vitro* processing. The mechanical stability of several scaffold-cell devices is supplied by additional fixation in the form of rods, plates, and screws, while others rely on the intrinsic mechanical properties of the scaffold-cell construct. The requirements for additional fixation will likely be both device and application specific. We will now move to a discussion of novel cartilage regeneration using tissue engineering techniques.

5 Articular cartilage biology, biomechanics, and injury

Articular cartilage is a specialized tissue that covers the bone ends at synovial joints and forms the low friction mating surface at which joint motion occurs. Articular cartilage is also called hyaline (glassy) cartilage, a descriptive term that emphasizes its smooth appearance. It consists of chondrocytes and organic matrix called chondroid. The articular cartilage is avascular, aneural, and alymphatic. The cells obtain their nutrition via diffusion from the synovial fluid. The tissue is relatively hypocellular with respect to bone tissue. The collagen fibers of the matrix are oriented in arches which result in the collagen further from the surface being oriented nearly perpendicular to the surface, and collagen near the surface being oriented nearly parallel to the surface. The matrix contains proteoglycans, which render it hydrophilic. Hence, when the cartilage is loaded, some of the bound water is forced out. When the load is removed, the hydrophilic matrix causes synovial fluid, with its nutrients, to move into the cartilage. This cyclic loading and unloading is in part responsible for the nutrition of the cartilage as the cartilage performs its mechanical functions of load transmission across the joint and low friction relative motion of the joint surfaces.

The mechanical functions mentioned above are enhanced by lubrication of the mating cartilage surfaces. These surfaces function as deformable bearings that can imbibe and release fluid. They also have a layer of adsorbed lubricating glycoproteins on them which separate them and provide a boundary lubrication mechanism. The lubrication provided by the deformable surfaces which sequentially imbibe and release fluid is termed elastohydrodynamic lubrication. Thus, several modes of lubrication act simultaneously at the cartilage surfaces of joints.

As cartilages ages, it loses both proteoglycan content and water content. It becomes stiffer, and these age related changes lead to the clinical condition of osteoarthritis. The cartilage thickness decreases, osteophytes form at the joint margins, the subchondral bone becomes more dense (sclerotic), and the joint geometry and mechanics are altered. The aneural cartilage eventually wears through to well innervated bone, which functions poorly as a bearing surface and results in painful joint motion. Cartilage that proceeds through this sequence of degradation will not regenerate. A second pathway of cartilage degeneration is an acute injury in a young person. This will typically produce a discrete area of cartilage loss with normal cartilage elsewhere in the joint. This area of cartilage loss will likely be a source of pain for the person, and the altered joint congruity it causes will also likely result in decreased functional ability. The abnormal forces at the margins of such a defect will often lead to earlier degeneration of the surrounding normal cartilage than would otherwise happen.

6 Current options available to treat articular cartilage loss

Persons who have developed osteoarthritis have several treatment options available to them. Treatment will include oral anti-inflammatory medication, weight loss attempts toward ideal body weight to decrease loading on the joints, use of ambulatory aids such as a cane, and injection of anti-inflammatory medication into the arthritic joint. Surgical options include arthrodesis (fusion) of the joint, total joint replacement, and allograft osteoarticular replacement of part of the joint surface with attached bone. These treatments find use in a variety of clinical situations, but none is ideal, and none is universally applicable to persons with articular cartilage loss.

The problem of localized cartilage loss in a young person whose remaining joint surface is normal presents a particularly difficult management dilemma. Techniques such as arthroscopic abrasion of the injured cartilage or drilling of the subchondral bone have been used, but neither has provided good results. The goal of abrasion is to smooth the area of injury into the surrounding normal area so that there is not a sudden discontinuity of the joint surface, and also that the process of abrading the area might stimulate the chondrocytes to become active and repair the defect. The goal of subchondral drilling is to expose the defect to the vascular subchondral bone in the hope, again, of stimulating a repair process. When these techniques do result in repair, the new tissue is usually a less mechanically attractive fibrocartilage rather than hyaline cartilage.

The successful harvest, *in vitro* expansion in tissue culture, and transplantation of chondrocytes into a cartilage defect was reported several years ago [4]. The defect is covered with a flap of periosteum, which serves to contain the cartilage cell suspension. This technique has shown promise in clinical trials. Another technique undergoing clinical trial involves periosteal transplantation into the base of the localized defect rather than at the defect surface. The periosteum, in this case, is the source of the undifferentiated cells that will repopulate the defect and regenerate cartilage. This method has resulted in the regeneration of hyaline cartilage in some cases, and fibrocartilage in others. It represents a potential solution for the problem of localized cartilage loss in a young person.

The articular cartilage, once injured, has a limited ability to repair itself. In this respect, it is very different from bone tissue, in which repair and remodeling usually occur with great efficiency. The current strategies to treat articular cartilage injury have varying degrees of success, but are particularly problematic in the young person. We will now discuss several tissue engineering approaches that attempt to address this need.

7 Tissue engineering strategies as alternatives to the current cartilage defect options

Chondrocyte attachment to and growth on biodegradable polymer scaffolds has received much attention. There has been the suggestion that the polymers function as synthetic basement membranes for cell attachment [28]. Good growth results were achieved using non-woven PGA mesh 100 μm thick, having a fiber diameter of 15 μm, and a fiber separation of 150–200 μm [28]. The same study also demonstrated that bovine chondrocytes were capable of maintaining their function after storage at 4°C for 30 days, and that the optimal cell seeding density for cartilage growth was

50–100 million cells per milliliter of polymer mesh. The age of the person from whom cartilage cells are harvested is thought to be important in the ability of the cells to regenerate cartilage matrix, in that cells from younger donors are expected to be more metabolically active. However, cartilage cells harvested from a 100-year-old donor, when expanded *in vitro* and attached to PGA mesh, formed neocartilage *in vivo* [44]. Many of the scaffolds have been PGA, but chondrocytes grown on three-dimensional PLA matrices have survived and expressed their phenotypic function [8]. Combinations of PGA and PLA have been used for matrices [12]. In a study that used non-woven PGA immersed in a 5% solution of PLA to physically cross-link the PGA intersections, 96% of nude (athymic) mice grew neocartilage when they had 2–100 million cells per milliliter of polymer implanted subcutaneously [36]. This appeared to be the optimum cell seeding density to produce hyaline cartilage. Lesser densities produced varying amounts of fibrocartilage and hyaline cartilage. Some applications, such as the temporomandibular joint disk, require fibrocartilage, and will benefit from lower cell seeding densities to effect its formation [39]. The degradation rate of the PGA and PLA meshes differs. The PLA exists in a racemic, optically inactive form, and in optically active L- and D-forms. The scaffold degradation rate depends upon both the crystallinity of the polymer and its hydrophobicity. Well ordered, highly crystalline domains degrade more slowly than amorphous domains, and hydrophobic polymers degrade more slowly than hydrophilic ones. When considering both crystallinity and hydrophilicity, the rate of poly α-hydroxy ester scaffold degradation, from fastest to slowest, is PGA > racemic PLA > PLLA (poly L-lactic acid) [15, 27, 50].

The passage of chondrocytes has been found to not affect their ability to form cartilage on polymer scaffolds. Primary cultures were compared to chondrocytes cultured after three passages, both on non-woven PGA mesh [11]. The passaged cells performed as well as the primary cultured cells. In addition to the effect of passaging, the effect of culture technique has been addressed. Cells cultured in polymer microcarriers and mixed well in a simulated microgravity counterrotating bioreactor demonstrated decreased doubling time compared to control cells grown under static conditions [13]. The uniform mixing allowed for more efficient exchange of nutrients and wastes.

Several *in vivo* studies have demonstrated the feasibility of cartilage regeneration using biodegradable scaffolds. Rabbit chondrocytes grown on PGA, transplanted with the polymer mesh into 1–2 mm defects of the rabbit distal femur, and covered with a PGA mesh, formed hyaline cartilage [45]. Rabbit perichondral cells grown to confluence on porous D-PLA, racemic D,L-PLA, and L-PLA and implanted into knee cartilage defects grew neocartilage in 15 of 16 specimens [7]. Comparison of isogeneic with allogeneic rabbit articular chondrocytes cultured on a collagen gel and implanted into a rabbit distal femoral defect demonstrated that the isogeneic cells were better at forming neocartilage [31]. The PGA scaffolds, sprayed with a 5% solution of PLLA, seeded with bovine chondrocytes and implanted into nude mice, demonstrated shape stability in the form of nasoseptal cartilage during the formation of the neocartilage [37]. There is a hypothesis that only those cells which firmly attach to the scaffold and thus survive the physical transfer from the culture medium to the implantation site will produce cartilage matrix. This accounts for the observation that new cartilage maintains the shape of the implanted scaffold [23]. The phenotypic expression of chondrocytes has been assessed relative to cells that possess bone forming ability. Fetal calf chondrocytes grown on PGA mesh and implanted into calvarial defects produced only cartilage, but not bone [22]. PGA meshes immersed in a 2% PLLA solution and seeded either with bovine chondrocytes or with bovine periosteum and implanted into nude mice produced only cartilage when seeded with chondrocytes. However, those same meshes, seeded with periosteum, which contains cells that possess bone forming ability, formed cartilage that became bone [48].

Chondrocytes, suspended in a calcium alginate gel, were injected subcutaneously in the presence of calcium chloride, which polymerized the gel [32, 33]. The polymerized gel degrades enzymatically into guluronic and mannuronic acids. These injectable preparations would add clinical flexibility with respect to the mode of cell/polymer delivery. Concerns regarding the possible immunogenicity of calcium alginate led to demonstration of the injectable concept using a poly (ethylene oxide) (PEO) gel [42]. Carbon fibers have been used in both laboratory *in vivo* studies and in clinical trials [14]. Rabbit distal femoral defects, covered with a periosteal flap, had either chondrocytes or chondrocytes plus a carbon fiber scaffold injected beneath them [5]. The cells alone did better than either the cells plus the scaffold or the periosteal flap without cells. A clinical study used woven carbon fiber plugs in knee chondral defects [3]. The cylindrical plugs were pressed into drill holes into the subchondral bone and left protruding at a level even with the uninjured cartilage surface. The defects filled in with fibrocartilage rather than hyaline cartilage. This was not a tissue engineering method in that no transplantation of cells or growth factors was involved. However, the anchorage of the carbon fiber scaffold to subchondral bone demonstrates a technique for addressing the problem of scaffold immobilization.

We have now discussed the clinical needs for both skeletal deficiency and cartilage loss, current options to address each of these needs, and the limitations of those options. We have presented several strategies that attempt to address these current clinical limitations, and have seen that a wide variety of approaches exists. The methods that ultimately find their way into clinical practice will likely be flexible enough to accommodate the inevitable differences in clinical presentation of the same problem in different patients. These methods will likely involve a degradable polymeric scaffold that supports cells, growth factors, or combinations of both components. It would certainly be desirable to find universal solutions to the clinical problems of musculoskeletal tissue deficiency, but the more likely situation of initial solutions that address some of the current clinical deficiencies will benefit many persons.

References

1. Attawia, M. A., Devin, J. E. and Laurencin, C. T., Immunofluorescence and confocal laser scanning microscopy studies of osteoblast growth and phenotypic expression in three-dimensional degradable synthetic matrices. *Journal of Biomedical Materials Research*, 1995, **29**, 843–848.

2. Bennett, S., Connolly, K., Lee, D. R., Jiang, Y., Buck, D., Hollinger, J. O. and Gruskin, E. A., Initial biocompatibility studies of a novel degradable polymeric bone substitute that hardens in situ. *Bone*, 1996, **19**(1 Suppl), 101S–107S.

3. Brittberg, M., Faxen, E. and Peterson, L., Carbon fiber scaffolds in the treatment of early knee osteoarthritis. A prospective 4-year follow-up of 37 patients. *Clinical Orthopaedics and Related Research*, 1994, **307**, 155–164.

4. Brittberg, M., Lindahl, A., Nilsson, A., Ohlsson, C., Isaksson, O. and Peterson, L., Treatment of deep cartilage defects in the knee with autologous chondrocyte transplantation. *New England Journal of Medicine*, 1994, **331**, 889–895.

5. Brittberg, M., Nilsson, A., Lindahl, A., Ohlsson, C. and Peterson, L., Rabbit articular cartilage defects treated with autologous cultured chondrocytes. *Clinical Orthopaedics and Related Research*, 1996, **326**, 270–283.

6. Cao, Y., Vacanti, J. P., Ma, X., Paige, K. T., Upton, J., Chowanski, Z., Schloo, B., Langer, R. and Vacanti, C. A., Generation of neo-tendon using synthetic polymers seeded with tenocytes. *Transplantation Proceedings*, 1994, **26**, 3390–3392.

7. Chu, C. R., Coutts, R. D., Yoshioka,M., Harwood, F. L., Monosov, A. Z. and Amiel, D., Articular cartilage repair using allogeneic perichondrocyte-seeded biodegradable porous polylactic acid (PLA): a tissue-engineering study. *Journal of Biomedical Materials Research*, 1995, **29**, 1147–1154.

8. Chu, C. R., Monosov, A. Z. and Amiel, D., In situ assessment of cell viability within biodegradable polylactic acid polymer matrices. *Biomaterials*, 1995, **16**(18), 1381–1384.

9. Devin, J. E., Attawia, M. A. and Laurencin, C. T., Three-dimensional degradable porous polymer-ceramic matrices for use in bone repair. *Journal of Biomaterials Science, Polymer Edition*, 1996, **7**(8), 661–669.

10. Flahiff, C. M., Blackwell, A. S., Hollis, J. M. and Feldman, D. S., Analysis of a biodegradable composite for bone healing. *Journal of Biomedical Materials Research*, 1996, **32**(3), 419–424.

11. Freed, L. E., Grande, D. A., Lingbin, Z., Emmanual, J., Marquis, J. C. and Langer, R., Joint resurfacing using allograft chondrocytes and synthetic biodegradable polymer scaffolds. *Journal of Biomedical Materials Research*, 1994, **28**(8), 891–899.

12. Freed, L. E., Marquis, J. C., Nohria, A., Emmanual, J., Mikos, A. G. and Langer, R., Neocartilage formation in vitro and in vivo using cells cultured on synthetic biodegradable polymers. *Journal of Biomedical Materials Research*, 1993, **27**(1), 11–23.

13. Freed, L. E., Vunjak-Novakovic, G. and Langer, R., Cultivation of cell-polymer cartilage implants in bioreactors. *Journal of Cellular Biochemistry*, 1993, **51**(3), 257–264.

14. Hendrickson, D. A., Nixon, A. J., Grande, D. A., Todhunter, R. J., Minor, R. M., Erb, H. and Lust, G., Chondrocyte-fibrin matrix transplants for resurfacing extensive articular cartilage defects. *Journal of Orthopaedic Research*, 1994, **12**(4), 485–497.

15. Hollinger, J. O., Brekke, J., Gruskin, E. and Lee, D., Role of bone substitutes. *Clinical Orthopaedics and Related Research*, 1996, (324), 55–65.

16. Hollinger, J. O. and Leong, K., Poly(alpha-hydroxy acids): carriers for bone morphogenetic proteins. *Biomaterials*, 1996, **17**(2), 187–194.

17. Ikada, Y., Shikinami, Y., Hara, Y., Tagawa, M. and Fukada, E., Enhancement of bone formation by drawn poly(L-lactide). *Journal of Biomedical Materials Research*, 1996, **30**(4), 553–558.

18. Ishaug, S. L., Crane, G. M., Miller, M. J., Yasko, A. W., Yaszemski, M. J. and Mikos, A. G., Bone formation by three-dimensional stromal osteoblast culture in biodegradable polymer scaffolds. *Journal of Biomedical Materials Research*, In Press.

19. Ishaug, S. L., Payne, R. G., Yaszemski, M. J., Aufdemorte, T. B., Bizios, R. and Mikos, A. G., Osteoblast migration on poly(α-Hydroxy Esters). *Biotechnology and Bioengineering*, 1996, **50**, 443–451.

20. Ishaug, S. L., Yaszemski, M. J., Bizios, R. and Mikos, A. G., Osteoblast function on synthetic biodegradable polymers. *Journal of Biomedical Materials Research*, 1994, **28**, 1445–1453.

21. Ishaug-Riley, S. L., Crane, G. M., Gurlek, A., Miller, M. J., Yaszemski, M. J., Yasko, A. J. and Mikos, A. G., Ectopic bone formation by marrow stromal osteoblast transplantation using poly(DL-lactic-co-glycolic acid) foams implanted into the rat mesentery. *Journal of Biomedical Materials Research*, In Press.

22. Kim, W. S., Vacanti, C. A., Upton, J. and Vacanti, J. P., Bone defect repair with tissue-engineered cartilage. *Plastic and Reconstructive Surgery*, 1994, **94**(5), 580–584.

23. Kim, W. S., Vacanti, J. P., Cima, L., Mooney, D., Upton, J., Puelacher, W. C. and Vacanti, C. A., Cartilage engineered in predetermined shapes employing cell transplantation on synthetic biodegradable polymers. *Plastic and Reconstructive Surgery*, 1994, **94**(2), 233–237.

24. Laurencin, C. T., Attawia, M. A., Elgendy, H. E. and Herbert, K. M., Tissue engineered bone-regeneration using degradable polymers: the formation of mineralized matrices. *Bone*, 1996, **19**(1 Suppl), 93S–99S.

25. Laurencin, C. T., El-Amin, S. F., Ibim, S. E., Willoughby, D. A., Attawia, M., Allcock, H. R. and Ambrosio, A. A., A highly porous 3-dimensional polyphosphazene polymer matrix for skeletal tissue regeneration. *Journal of Biomedical Materials Research*, 1996, **30**(2), 133–138.

26. Laurencin, C. T., Norman, M. E., Elgendy, H. M., el-Amin, S. F., Allcock, H. R., Pucher, S. R. and Ambrosio, A. A., Use of polyphosphazenes for skeletal tissue regeneration. *Journal of Biomedical Materials Research*, 1993, **27**(7), 963–973.

27. Li, S., Vert, M., Hydrolytic degradation of the coral/poly(DL-lactic acid)bioresorbable material. *Journal of Biomaterials Science, Polymer Edition*, 1996, 7(9), 817–827.

28. Mikos, A. G., Sarakinos, G., Leite, S. M., Vacanti and J. P., Langer, R., Laminated three-dimensional biodegradable foams for use in tissue engineering. *Biomaterials*, 1993, **14**, 323–330.

29. Mikos, A. G., Sarakinos, G., Lyman, M. D., Ingber, D. E., Vacanti, J. P. and Langer, R., Prevascularization of porous biodegradable polymers. *Biotechnology and Bioengineering*, 1993, **42**, 716–723.

30. Miller, M. J., Goldberg, D. P., Yasko, A. W., Lemon, J. C., Satterfield, W. C., Wake, M. C. and Mikos, A. G., Guided bone growth in sheep: A model for tissue-engineered bone flaps. *Tissue Engineering*, 1996, **2**, 51–59.

31. Noguchi, T., Oka, M., Fujino, M., Neo, M. and Yamamuro, T., Repair of osteochondral defects with grafts of cultured chondrocytes. Comparison of allografts and isografts. *Clinical Orthopaedics and Related Research*, 1994, (302), 251–258.

32. Paige, K. T., Cima, L. G., Yaremchuk, M. J., Schloo, B. L., Vacanti, J. P. and Vacanti, C. A., De novo cartilage generation using calcium alginate-chondrocyte constructs. *Plastic and Reconstructive Surgery*, 1996, **97**(1), 168–180.

33. Paige, K. T., Cima, L. G., Yaremchuk, M. J., Vacanti, J. P. and Vacanti, C. A., Injectable cartilage. *Plastic and Reconstructive Surgery*, 1995, **96**(6), 1390–1400.

34. Peter, S. J., Nolley, J. A., Widmer, M. S., Merwin, J. E., Yaszemski, M. J., Yasko, A. W., Engel, P. S., Mikos and A. G., In vitro degradation of a poly(propylene fumarate)/β-Tricalcium phosphate composite orthopaedic scaffold. *Tissue Engineering*, In Press.

35. Peter, S. J., Yaszemski, M. J., Suggs, L. J., Payne, P. G., Langer, R., Hayes, W. C., Unroe, M. R., Alemany, L. B., Engel, P. S. and Mikos, A. G., Characterization of partially saturated poly(propylene fumarate) for orthopaedic application. *Journal of Biomaterials Science, Polymer Edition*, Submitted.

36. Puelacher, W. C., Kim, S. W., Vacanti, J. P., Schloo, B., Mooney, D. and Vacanti, C. A., Tissue-engineered growth of cartilage: The effect of varying the concentration of chondrocytes seeded onto synthetic polymer matrices. *International Journal of Oral and Maxillofacial Surgery*, 1994, **23**(1), 49–53.

37. Puelacher, W. C., Mooney, D., Langer, R., Upton, J., Vacanti, J. P. and Vacanti, C. A., Design of nasoseptal cartilage replacements synthesized from biodegradable polymers and chondrocytes. *Biomaterials*, 1994, **15**(10), 774–778.

38. Puelacher, W. C., Vacanti, J. P., Ferraro, N. F., Schloo, B., Vacanti and C. A., Femoral shaft reconstruction using tissue-engineered growth of bone. *International Journal of Oral and Maxillofacial Surgery*, 1996, **25**(3), 223–228.

39. Puelacher, W. C., Wisser, J., Vacanti, C. A., Ferraro, N. F., Jaramillo, D. and Vacanti, J. P., Temporomandibular joint disc replacement made by tissue-engineered growth of cartilage. *Journal of Oral and Maxillofacial Surgery*, 1994, **52**(11), 1172–1177.

40. Radder, A. M., Leenders, H. and van Blitterswijk, C. A., Application of porous PEO/PBT copolymers for bone replacement. *Journal of Biomedical Materials Research*, 1996, **30**(3), 341–351.

41. Ruder, C. R., Dixon, P., Mikos, A. G., Yaszemski, M. J., The growth and phenotypic expression of human osteoblasts. *Cytotechnology*, 1996, **22**, 263–267.

42. Sims, C. D., Butler, P. E., Casanova, R., Lee, B. T., Randolph, M. A., Lee, W. P., Vacanti, C. A. and Yaremchuk, M. J., Injectable cartilage using polyethylene oxide polymer substrates. *Plastic and Reconstructive Surgery*, 1996, **98**(5), 843–850.

43. Thomson, R. C., Yaszemski, M. J., Powers, J. M. and Mikos, A. G., Fabrication of biodegradable polymer scaffolds to engineer trabecular bone. *Journal of Biomaterials Science, Polymer Edition*, 1995, **7**, 23–38.

44. Vacanti, C. A., Cao, Y. L., Upton, J. and Vacanti, J. P., Neo-cartilage generated from chondrocytes isolated from 100-year-old human cartilage. *Transplantation Proceedings*, 1994, **26**(6), 3434–3435.

45. Vacanti, C. A., Kim, W., Schloo, B., Upton, J. and Vacanti, J. P., Joint resurfacing with cartilage grown in situ from cell-polymer structures. *American Journal of Sports Medicine*, 1994, **22**(4), 485–488.

46. Vacanti, C. A., Kim, W., Upton, J., Vacanti, M. P., Mooney, D., Schloo, B. and Vacanti, J. P., Tissue-engineered growth of bone and cartilage. *Transplantation Proceedings*, 1993, **25**(1 Pt 2), 1019–1021.

47. Vacanti, C. A., Paige, K. T., Kim, W. S., Sakata, J., Upton, J. and Vacanti, J. P., Experimental tracheal replacement using tissue-engineered cartilage. *Journal of Pediatric Surgery*, 1994, **29**(2), 201–204.

48. Vacanti, C. A. and Upton, J., Tissue-engineered morphogenesis of cartilage and bone by means of cell transplantation using synthetic biodegradable polymer matrices. *Clinics in Plastic Surgery*, 1994, **21**(3), 445–462.

49. Vacanti, C. A. and Vacanti, J. P., Bone and cartilage reconstruction with tissue engineering approaches. *Otolaryngologic Clinics of North America*, 1994, **27**(1), 263–276.

50. Vert, M., Mauduit, J. and Li, S., Biodegradation of PLA/GA polymers: increasing complexity. *Biomaterials*, 1994, **15**(15), 1209–1213.

51. Yaszemski, M. J., Payne, R. G., Hayes, W. C., Langer, R., Aufdemorte, T. B. and Mikos, A. G., The ingrowth of new bone tissue and initial mechanical properties of a degrading polymeric composite scaffold. *Tissue Engineering*, 1995, **1**, 45–52.

52. Yaszemski, M. J., Payne, R. G., Hayes, W. C., Langer, R. and Mikos, A. G., The in vitro degradation of a poly(propylene fumarate)-based composite material. *Biomaterials*, 1996, **17**, 2127–2130.

Tissue Engineering and Plastic Surgery

MICHAEL J. MILLER,
GREGORY R. D. EVANS
Laboratory of Reparative Biology and Bioengineering,
Department of Plastic Surgery,
The University of Texas M.D. Anderson Cancer Center,
1515 Holcombe Blvd,
Houston, Texas 77030, USA

Tissue engineering will change the practice of plastic and reconstructive surgery perhaps more than any other clinical speciality. Both fields share the same purpose: to fashion replacements for lost or damaged tissues. Plastic surgery has a long tradition of devising new methods for tissue replacement, and tissue engineering is a natural 'next step'.

1 Historical Perspective

'Not for self, not for the fulfillment of any earthly desire or gain, but solely for the salvage of suffering humanity.'

Sushutra Samhita (ca. 600 B.C.)

'We restore, repair and make whole, those parts ... which nature has given but which fortune has taken away...'

Gaspare Tagliacozzi (1545–1599)

The name of the speciality is derived from the Greek word *plastikos,* meaning 'to form.' A plastic surgeon is one who performs operations that change the shape of tissues. The French surgeon Desault first used the term 'plastique' to describe this kind of surgery by in 1798. Zeis published his classic text entitled *Handbuch der Plastischen Chirurgie* in 1838, and 'plastic' surgery began to be established in medical terminology [1, 2].

Plastic Surgery has been recognized as a speciality in the United States only since 1938, but it has been practiced throughout recorded history. The Ebers Papyrus (ca. 1600 BC) describes procedures to restore missing noses and ears in the Egyptian Middle Kingdom. Sushutra Samhita (ca. 600 BC), a member of a caste of pottery craftsmen in ancient India, recorded surgical techniques to restore nasal deformities that are still useful today

[3]. Celsus (25 BC–50 AD) of Rome described in detailed procedures on the nose, lips, eyelids and ears. Galen (138–201 AD) sought to remodel tissue by using elliptical incisions for skin tumor removal and wrote about repair of tendons, nerves and muscle. During the Italian Renaissance, Tagliacozzi laid the cornerstone of modern plastic surgery by describing techniques for tissue transfer that endured well into the 20th century. As professor at the University of Bologna and Chief Surgeon to the Grand Duke of Tuscany, he provided the first comprehensive report of plastic surgery techniques in his treatise *De Curtorum Chirurgia per Insitionem* [4]. His approach might be considered the first attempts to 'engineer' tissues by performing procedures intended to improve their blood supply and render them more reliable for transfer and reshaping.

Over the intervening years, progress was slow but a tradition of innovation was built that has allowed plastic surgery to make important contributions to every other surgical specialty and even help establish new ones. Examples are found in hand surgery, transplantation surgery and craniofacial surgery. Joseph E. Murray, a Boston plastic surgeon, reported the first successful human kidney transplantation with colleagues Merrill and Harrison in 1955 [5]. Murray was later awarded the Nobel Prize in medicine for his pioneering work in this area. Another plastic surgeon, John M. Converse, was a founding member and first president of the Transplantation Society in the United States. Principals of craniofacial surgery were introduced by Paul Tessier *et al.* at the Fourth International Congress of Plastic and Reconstructive Surgery in 1967. These techniques have since been adopted by a variety of specialities treating complex surgical disorders of the head and neck.

Plastic surgery experienced rapid changes in the 1970's and 80's. Discoveries related to the anatomy of regional blood supply in humans resulted in the rapid introduction of many new techniques. Refinements in operating microscopes and instrumentation lead to the advent of reconstructive microsurgery, the most sophisticated method for tissue transfer. Plastic surgery is poised now for the next major advance in tissue replacement: Tissue Engineering.

2 Plastic surgery principles

The process of surgical reconstruction is essentially the same for all defects with a tissue deficit. There is a two-step process involving (a) *transfer* of new tissue from an uninjured location and (b) *modifying* that tissue to replace or simulate that which has been lost (Fig. 1).

2.1 Tissue transfer

Tissue transfer methods are available to the modern surgeon may be

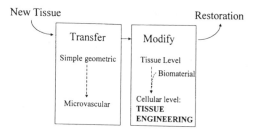

Fig. 1. Steps for surgical reconstruction of deformities associated with a tissue deficit.

classified as either tissue grafts or flaps. A graft is any tissue transferred without its blood supply. Graft healing depends upon nutritive support passively available in the tissues surrounding the defect. Skin, dermis, bone, cartilage, and fascia may be transferred in this way. These tissues are obtained from an uninjured location on the same patient (i.e. autologous) or, in special cases, from another individual (i.e. allogeneic). Allograft tissue is usually obtained from cadaver donors and must be properly treated to reduce the risks of immune reactions and disease transmission.

When a large amount of tissue is required or the recipient site is compromised, a tissue graft will not adequately restore a defect. Trauma, bacterial contamination, or radiation exposure (as with cancer treatment) impairs the ability of tissue to incorporate grafts. In these circumstances, tissue must be transferred with a blood supply that originates from outside the zone of injury. This is the definition of a surgical 'flap,' the traditional term for a unit of tissue moved to another location with preservation of its blood supply. Tissue transferred as flaps may be moved into a compromised area because they will not depend on the ability of the surrounding tissues to supply nutritive support. A great variety of surgical flaps have been described that provide skin, fat, muscle, bone, nerve, and viscera. Flaps are defined based upon the type of tissue they provide and the anatomic pattern of blood supply. They are composed of autologous tissue obtained from a location on the patient away from the defect. The most advanced transfer technique is a microvascular transfer. This method involves isolating a tissue unit on its primary vascular supply and temporarily dividing the blood vessels, cutting it 'free' from the patient. Tissues transferred in this way are often called 'free flaps.' The vessels supplying the flap, usually 1–3 mm in diameter, are sewn to other vessels near the defect using an operating microscope and extremely fine suture materials. Usually, two microvascular anastomoses are required, one for the artery and one for the vein. Tissues transferred in this way heal in the normal way with little contracture or loss of substance, except for denervation atrophy in flaps containing muscle. Most of the history of plastic surgery consists of advances in techniques to transfer tissue.

2.2 Tissue modification

After transfer, tissues must be reshaped to simulate missing structures. In contrast to tissue transfer techniques, tissue modification methods have changed little through the centuries. Surgeons still learn to manually alter tissues at the time of surgery, an often difficult and time-consuming process. A skin flap used for a nasal reconstruction must be folded and redraped to establish multiple surfaces. Muscle flaps must be tailored to fill deep spaces or adipose tissue flaps sculpted to simulate the shape of a formed structure like a breast. Long, straight bones harvested from an extremity must be fashioned to simulate the curved appearance of a mandible or maxillary bone. The results are never exact. Original structures are approximated to a degree that varies with each patient depending on the nature of the tissues and the personal skill of the surgeon.

The primary factors limiting these methods are (a) the need to preserve adequate blood supply to all portions of the tissue and (b) the complex anatomy of the structures to be replaced. Efforts to overcome these limitations have focused improving the blood supply and 'prefabricating' structures prior to transfer [6–13]. An example is tissue expansion. This involves placing an inflatable device called a tissue expander beneath the tissues. The expander is gradually inflated with physiologic saline solution injected over several weeks. When the expansion is complete, the device is removed and the tissue is ready to use. This process has been shown to increase the amount of tissue and improve the blood supply.

More elaborate methods have been described experimentally and used in selected patients. As early as 1963 attempts were made to revascularize tissues by direct transfer of blood vessels [14, 15]. In 1971 Orticochea used vascular induction from direct blood vessel transfer to fabricate a composite flap containing skin and cartilage [16]. Erol and Spira [17] demonstrated that the greater omentum, a large structure composed primarily of fat found inside the abdominal cavity, could be wrapped around cartilage or bone grafts to create a prefabricated flap that could then be transferred to a distant location in laboratory animals. Tzuyano has performed similar studies using an entire hind limb [18]. Khouri and coworkers successfully transferred flaps created by wrapping isolated knee joints in rats with vascularized fascia. They later went on to use this approach to successfully reconstruct a damaged finger joint in a child [19]. Others have applied this strategy to create a variety of flaps with, for example, multiple epithelial linings [20]. Such techniques are based upon rearranging mature tissues elements into useful configurations prior to transfer.

A more advanced strategy than simple rearrangement is direct modification of the tissue elements by molding and transformation using implants and induction factors. Titanium molds filled with morcellized bone graft yield formed bone segments when implanted in laboratory animals

[21]. Silicon rubber molds have been used to create soft tissue flaps of different shapes [22]. Khouri et al. added osteogenin (i.e. Bone Morphogenetic Protein-3, or BMP-3), a potent bone induction factor, to soft tissue molds and transformed muscle into bone flaps [23]. Despite these laboratory studies that demonstrate the feasibility of fabricating surgical flaps into different shapes, clinically useful techniques have yet to emerge. Only small amounts of tissue have been produced that have failed to retain proper shape after removing the mold.

The goal of these studies is to improve our ability to modify tissues by shifting from working with whole tissues to more fundamental levels. From a surgeons viewpoint, tissue engineering is modification of existing tissues at the cellular or molecular level to fabricate new tissues for reconstructive surgery.

3 Clinical problems

Plastic surgery operations may be broadly classified as either primarily reconstructive or aesthetic depending on the nature of the deformity. This distinction is primarily one of degree. Cosmetic procedures address deformities that are anatomically within normal limits but nevertheless present an appearance that is unsatisfactory to the patient. Reconstructive operations correct more extreme problems that usually associated with a significant functional disturbance and are clearly 'abnormal' or life threatening. Even the most complex reconstructive procedure, however, must still follow proper aesthetic principals. After all, the most sophisticated nasal reconstruction that does not look like a human nose will not be well accepted. Both reconstructive and aesthetic operations may require tissue replacement and therefore may be influenced by developments in tissue engineering.

3.1 Aesthetic surgery

Most cosmetic operations modify available tissues and do not require significant tissue replacement. An exception is aesthetic nasal surgery, or rhinoplasty, in which additional small pieces of cartilage or bone may be utilized to alter the shape of the nose. The malar (i.e. cheek) bones, chin, and posterior angle of the mandible are other areas that may be altered by inserting additional tissue or tissue substitutes to modify the shape of the underlying skeletal support. Breast enlargement surgery, or augmentation mammoplasty, is one of the most common aesthetic procedures. Several hundred cubic centimeters of additional soft tissue or tissue equivalent may be required. Usually this is supplied by breast implants consisting of an envelope made of silicone elastomer filled with either saline solution or silicone gel. The ability to engineer additional fat would potentially eliminate the need for artificial breast implants.

3.2 Reconstructive surgery

Most reconstructive operations involve some form of tissue replacement. The exact type will vary depending on the anatomic structures to be restored. This section will review the variety of problems encountered by region. The specific tissue engineering strategies under development for these different tissues are considered in depth elsewhere in this text.

3.2.1 Skin
Skin is the largest organ in the body and serves primarily as a barrier and in thermoregulation. Skin may be lost due to trauma, malignancy, and destruction from certain kinds of inherited skin disorders. Large burns pose the most challenging problem. Ultimate survival of a patient with a major burn depends in large part on how rapidly the burned skin can be replaced. Current therapy is based on autologous skin grafting, a procedure that actually increases the size of the injury for a short period until the graft donor sites heals (Fig. 2). Long term healing is affected by the amount of scarring that occurs especially around joints and other areas of normally mobile skin cover. A permanent full-thickness bioartificial skin replacement would be a great advance in caring for these patients.

3.2.2 Head and neck
This region is anatomically the most complex in the entire body. Large defects usually involve multiple kinds of tissues in complicated three-dimensional relationships. Functional and aesthetic deformities may severely impair normal living and even threaten survival.

Reconstructive requirements by location. Priorities for the scalp and forehead are to protection the brain and to provide thin soft tissue that affords a natural-appearing contour that is durable enough to support a wig to restore the hair-bearing scalp. Coverage for the forehead should be thin and match the color of other parts of the face. The conventional surgical options include skin grafting and local rotation flaps. For large defects, free tissue transfers using the greater omentum [24] or a variety of skin and muscle flaps [25] have proven effective. In addition to soft tissue, significant defects of the cranial bone require replacement using alloplastic materials like methyl methacrylate [26] or autogenous bone as a graft or flap [27–29].

Specialized features in the midface are especially difficult to reconstruct with tissue. The eyes, ears and nose are focal points of personality expression and individual social recognition. Even subtle imperfections are difficult to hide. A variety of tissue flaps have been described for advanced defects, including 'prefabricated' [30] tissues and delicate composite trans-

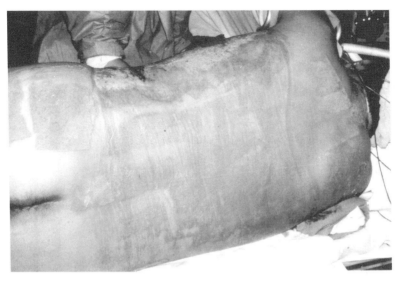

Fig. 2. Healing skin graft donor sites in a patient treated for a major burn. Harvesting skin for burn surgery temporarily increases the injury to the patient, causing the physiologic equivalent of a controlled partial thickness burn.

fers of skin and cartilage from the anterior superior helical rim [31, 32]. Nevertheless, it is difficult to obtain consistently good results. An artificial prosthesis usually provides the best restoration, with tissue reconstruction designed primarily to afford a durable foundation (Fig. 3).

The cheeks, lips, and mandible are the principle components of the lower third of the face. Each has unique reconstructive considerations. Cheek defects require restoration of two epithelial surfaces with minimal intervening soft tissue. Extensive defects may also require suspension of the oral commissure or simultaneous facial reanimation procedures. The lips participate in mastication and speech and ideal restoration is mobile and sensate. The mandible provides the underlying structure in this part of the face. The functional objective of mandible reconstruction is to restore mastication and speech. The aesthetic objective is to restore a normal-appearing contour to the lower face. Loss of the anterior mandible is particularly debilitating, and reconstruction requires a free flap containing bone in most cases. A variety of bone flaps have been described to accomplish these goals, but the most commonly used are the fibula [33], iliac crest [34], and scapular [35] flaps. Each of these options provides soft tissue as well as bone that may be used for cutaneous reconstruction after mandible restoration (Fig. 4). A tissue engineered alternative to bone flap harvest in these cases would be especially beneficial.

(a)

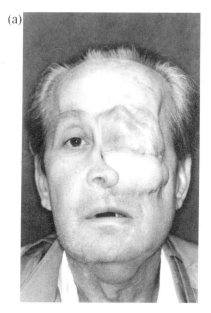

(b)

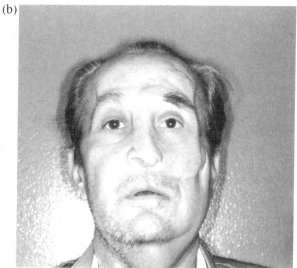

Fig. 3. This patient has undergone complex midface reconstruction using two microvascular flap transfers for treatment of a large tumor of the paranasal sinuses. (A) Appearance after surgery to restore the palate, provide bone for placement of maxillary dental implants, and restore the soft tissue protecting base of the skull. (B) Appearance after fitting with a facial prosthesis (Courtesy James Lemon, D.D.S.).

3.2.3 Chest wall and breast

The chest wall is a structure that is flexible to allow respiratory function yet rigid enough to afford protection of vital organs and resist deformation due to changes in air pressure associated with breathing. Defects of the chest wall can occur following major trauma, resection of advanced cancer, or radiation injury. The priorities for chest wall reconstruction are skeletal stabilization, stable soft tissue coverage, obliteration of empty spaces inside the thoracic cavity (i.e. 'dead space'), and restoration of normal appearance. Current methods to restore skeletal stability mostly rely on rigid biomaterials, such as acrylic, or semirigid replacements like synthetic mesh or sheets [36] (Fig. 5), but procedures using rib transfer have been described [37]. These implanted materials are permanent and generally work well, but there is still the risk of extrusion and exposure, particularly in cases of irradiated chest walls following tumor therapy. Obliteration of intrathoracic 'dead space' and external soft tissue coverage are typically provided by transfer of soft tissue flaps. Flaps for these purposes may be taken from the adjacent chest wall or upper abdomen based on the large muscles such as the pectoralis major, latissimus dorsi, or rectus abdominis muscles [38]. They necessarily result in scarring and the possibility of secondary deformities. There is a role for tissue engineered substitutes for each of these needs.

Breast reconstruction is one of the most common reconstructive procedures. The usual indication is to restore the breast following complete removal (i.e. total mastectomy) performed for cancer treatment. It has been shown that women with breast cancer must deal with two separate emotional issues, the reality of a life-threatening disease and the possibility of losing a breast. Loss of the breast is not life-threatening, but many women find the deformity emotionally and psychologically disturbing [39] (Fig. 6). A mastectomy causes a significant functional and cosmetic deformity, replacing the soft, projecting breast with a long, flat scar. The breast is a significant part of female body image and sense of femininity. The patient is reminded of her cancer experience every time she looks into a mirror. It can be difficult to find clothing. For some women, an external prosthesis may be satisfactory, but many find them cumbersome and unacceptable. Breast reconstruction is intended to overcome these problems and enhance the quality of life for women following mastectomy for breast cancer.

Breast reconstruction requires restoring missing breast skin and the underlying soft tissue volume. Current methods rely on breast implants [40], soft tissue flaps [41], or a combination of these [42]. Each technique offers certain advantages depending on the patient.

Reconstruction based on implants usually begins with a period of tissue expansion. When adequate skin and soft tissue is created, a process requiring usually 4–6 months, then the expander is removed and replaced by a permanent breast implant. Permanent implants are filled with silicone

(a)

(b)

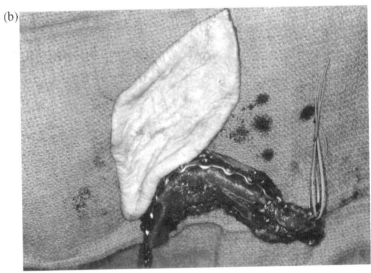

gel or physiologic saline solution. The major advantages of breast implants are that they are easy to place, require minimal additional surgery, and pose the lowest risk to the patient. The main disadvantage is that, with an implanted device, it is more difficult to control the shape and contour of the breast. Breast implants are not a good option in patients who have been treated with radiation because of a tendency for firm scars to form around the implant.

(c)

(d)

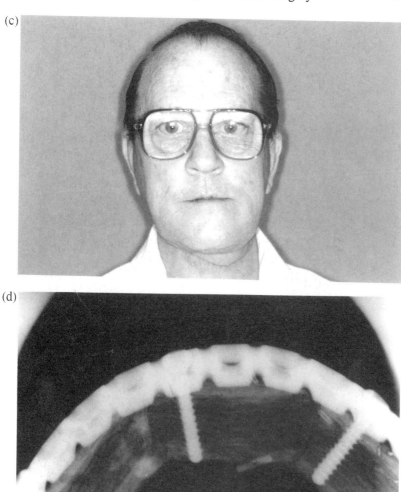

Fig. 4. This patient had cancer involving the oral cavity and required removal of the anterior floor of the mouth and mandible. (A) Pre-operative appearance. (B) View of a fibula osteocutaneous free flap the contoured to fit a titanium plate. Note the latex loop on the right encircling the blood vessels (peroneal artery and vein) that will be reattached to vessels in the neck by microvascular surgery. (C) Appearance of the patient six weeks following surgery. (D) Post-operative radiograph demonstrating the appearance of a healed fibula free flap secured by a reconstruction plate.

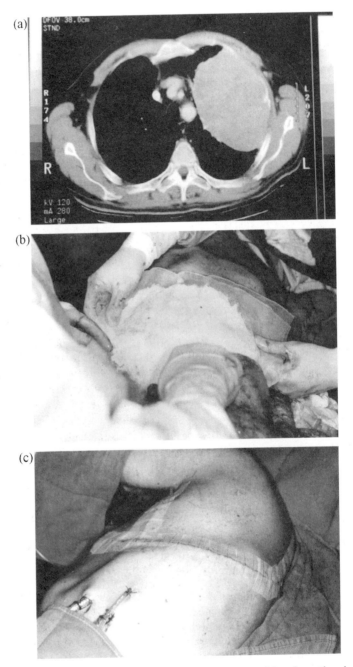

Fig. 5. Computerized tomography scan of a large tumor arising from the chest wall and nearly filling the left chest cavity (A). Surgical removal of the tumor required resection of 75% of the left chest wall. Stability of the chest wall was restored using a composite formed of poly (methyl methacrylate) and polypropylene mesh (B). Appearance of after coverage of the implant with local muscle flaps and skin (C).

Autologous tissue reconstruction is most often performed using skin and fat obtained from the lower anterior abdominal wall as a flap called a transverse rectus abdominis musculocutaneous (TRAM) flap (Fig. 7). The tissue may be transferred by either keeping the rectus abdominis muscle attached superiorly or by performing a microvascular transfer [43]. The skin and fat may then be shaped to simulate the appearance of the breast. The tissue has a similar consistency to breast tissue and provides a reconstructed breast that looks and feels the most natural. The transverse scar at the donor site, located midway between the umbilicus and the pubic area, is easily hidden by clothing. In cases when there is inadequate tissue on the lower abdomen, the procedure can be combined with placement of a breast implant. In such cases, tissue may be harvested either from the back or the abdomen. If tissue is used, however, the results tend to be more natural and long-lasting. The disadvantages of these operations are that they are more time-consuming, have greater risks, and cause more scarring than other techniques. A tissue engineered soft tissue alternative would have wide application in post-mastectomy breast reconstruction.

3.2.4 Genitourinary

Genitourinary reconstruction addresses deformities of the genitalia and perineum, including the anus. Reconstructive problems in the region are usually the result of congenital deformities or cancer treatment. Radiation tissue injury and bacterial contamination are often concerns when performing reconstructive surgery in this region. A variety of regional soft tissue flaps may be rotated into this area from the lower abdomen and thighs. Free tissue transfers are also possible in special cases.

Total penile reconstruction is accomplished using regional muscle flaps [44] or free tissue transfer [45, 46]. Tissues are fashioned into a neophallus surrounding an epithelial-lined tube that serves as a neourethra. Ideally, innervation is provided by coapting sensory nerves of the surgical flap to the transected local nerves [47]. Sexual function can be restored using either a rigid or a more complicated inflatable prosthesis implanted in the shaft of the neophallus [48].

Congenital absence of the vagina is an unusual deformity that is frequently not identified until puberty. This is often corrected using with skin grafts [49]. Vaginal deformities are more often are more often the result of surgery for advanced anorectal tumors. In these cases, reconstruction is usually accomplished using rotation flaps harvested from the medial thighs [50] or brought through the abdomen from the anterior abdominal wall [51]. These techniques yield a satisfactory function in most cases but cause additional scarring, altered donor muscle function, and provide vaginal lining formed by cutaneous tissue rather than vaginal mucosa with the normal lubricating glands.

Other reconstructive problems in the perineum include restoration of support in the pelvic floor, anal sphincter reconstruction, and replacement

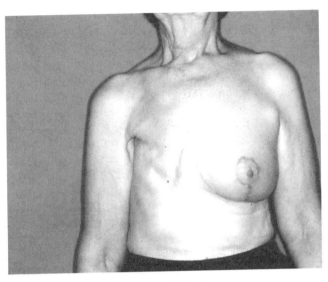

Fig. 6. Unreconstructed post-mastectomy deformity.

of cutaneous coverage. Tissue engineering may contribute to improved methods to for address each of these perineal and genitourinary problems. Fabricating tubular structures lined with uroepithelium, devising replacements for urethral and anal sphincter mechanisms, creating specialized soft tissues and epithelial coverings able to withstand the unique local environment of the perineum will prove a challenging area for engineering tissue replacements.

3.2.5 Extremities

Extremity reconstruction involves restoring osseous support, sensory and motor function, and durable soft tissue coverage. The largest bones in the body are the long bones of the extremities. Replacement is challenging because of the large bone mass required to provide dynamic support and stable fixation points for muscle action. Current reconstructive methods are often limited by tissue availability. Significant segmental bone loss may be restored using a metal prostheses. The disadvantage of these devices is that they tend to loosen over time and cause adjacent bone resorption [52]. They resist infection poorly and may not be used in contaminated wounds. Cadaver allografts may be used in the extremities, but they are associated with the potential for fracture and poor resistence to infection, and disease transmission [53, 54]. The most reliable method to major long bone defects is to use vascularized autologous bone flaps (Fig. 8).

Motor and sensory function mandate intact peripheral nerves and functioning target organs. Tumor resection may require ablation of adjcent peripheral nerves resulting in a segmental nerve defect and loss of distal motor and sensory function. If functional units can be preserved distal to the resection, then it is sometimes possible to provide reinervation by nerve grafting. This technique involves harvesting a sensory nerve from a location where the resulting neurologic deficit is expected to cause the least morbidity (e.g. the sural nerve) [55]. The graft is then microsurgically placed between the cut nerve ends and serves as a bridging conduit. Regenerating axons cross the gap into the distal nerve segment and reach the end organ at a rate of approximately 1 mm/day. The axons of transected peripheral nerves will regenerate if the nerve cell body, located in the central nervous system, is not injured. Major peripheral nerves contain a mixture of motor and sensory fibers organzied in groups of fascicles with a very specific intraneural topography. Individual grafts placed between matching fascicles are thought to yield the best results. The most precise way to identify matching groups is by intraoperative nerve mapping prior to division of the nerve. This is performed by dissecting the proximal nerve under magnification and electrically stimulating individual groups of fascicles. Matching fascicles distal to the anticipated defect are marked. The tumor resection is completed including the segment of involved peripheral nerve. Each fasicular group is then individually restored by an interposing graft. Important factors influencing outcome include the timing and technical precision of the microneural repair, level of defect, and patient age. Immediate reconstruction is superior because of the opportunity to identify matching fascicles and less time for atrophy of target organs. Defects located more distally yield better results than proximal lesions because of shorter distances to target organs and greater intraneural homogeneity. Younger patients achieve better function than older ones for reasons that are not completely understood but may be related to slower target organ atrophy and greater cortical plasticity. If motor units are lost, then reconstruction can be accomplished using local tendon transfers or free muscle transfers that are reinervated [56]. The disadvantage of these is that motor function of the donor site must be sacrificed. Sensory unit are more difficult to replace, but if there is an associated cutaneous loss, then a sensory enervated skin flap may be transferred to the defect [57, 58].

Soft tissue coverage depends upon adequate peripheral blood supply. If critical major blood vessels are injured, they must be replaced. This may be done using a synthetic vascular prosthesis or an autologous vein graft. Large areas of soft tissue loss often require a free tissue transfer, especially on the distal extremities. Tissue engineering may yield a variety of solutions to these problems. Bioartificial bone substitutes would provide a more available source of tissue replacement with less risk and morbidity. Biodegradable nerve conduits may make it possible to avoid the neurologic deficits associated with harvesting autologous nerve grafts. Soft tissue substitutes would perhaps make it easier to large deformities.

(a)

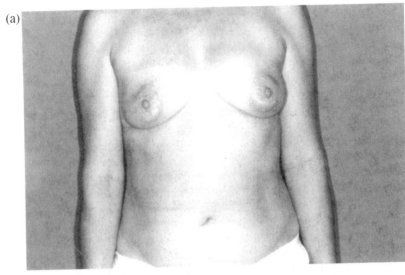

(b)

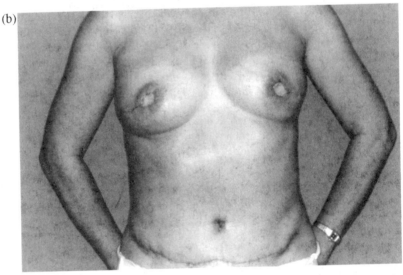

Fig. 7. This patient underwent bilateral mastectomies for breast cancer. (A) Appearance before surgery. (B) Appearance after bilateral breast reconstruction with microvascular transverse rectus abdominis musculocutaneous (TRAM) flaps. The donor site is represented by the long transverse scar on the lower abdomen.

4 Summary

This overview of the history and scope of clinical plastic surgery illustrates several important things. Plastic surgery is a clinical speciality with a long

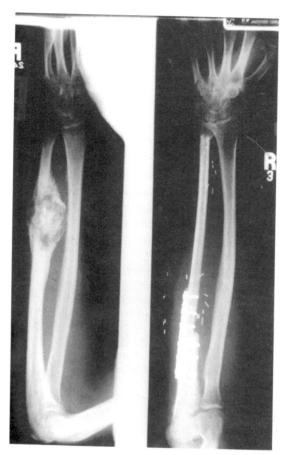

Fig. 8. Radiographs from a patient treated for a bone tumor of the upper extremity. The pre-operative image (left) shows the tumor in the distal ulna and the post-operative view demonstrates the reconstruction using a free fibula flap.

traditional developing and applying tissue replacement methods to patient care. Tissue engineering is the logical next frontier. Clinician scientists in plastic surgery are becoming increasingly committed to investigations in this field. Secondly, the speciality occupies a strategic position in surgery. Reconstructive surgeons consult regularly with colleagues from every other speciality. They develop experience with full spectrum of tissue replacement problems and understand the benefits and limitations of current therapies. These insights can serve to guide the direction of tissue engineering research and help to efficiently translate experimental therapies into clinical practice. The goal of tissue engineering is to change clinical practice. This will be realized through the collaborative efforts of bioengineers, bioscientists, and clinician scientists. What lies ahead? Certainly, surgeons and their patients in the next century will look back and consider our reconstructive surgery of the twentieth century as quaint, perhaps even 'barbaric,' thanks to contributions from tissue engineering.

References

1. Desault, P. J., Oeuvres Chirurgicales ou Exposé do la Doctrine et de la Plastique, Vol 2. Megegnon, Paris, 1798.
2. Zeis, E., Handbuch der plastischen Chirurgie (nebsteiner Vorrede von J. F. Dieffenbach). Reimer, Berlin, 1838.
3. Bhishagratna, K. K., An English translation of the Sushruta Samhita based on original Sanskrit text. 3 Vols. Bose, Calcutta, 1916.
4. Tagliacozzi, G., De Curtorum Chirurgia per Insitionem. Gaspare Bindoni, Venice, 1597.
5. Murray, J. E., Merrill J. P. and Harrison J. H., Renal homotransplantation in identical twins. *Surgical Forum*, 1955, **6**, 432.
6. Özgenta, H. E., Shenaq, S. and Spira, M., Prefabrication of a secondary TRAM flap. *Plastic and Reconstructive Surgery*, 1995, **95**, 441–449.
7. Itoh, Y., An experimental study of prefabricated flaps using silicone sheets, with reference to the vascular patternization process. *Annals of Plastic Surgery*, 1992, **28**, 140–146.
8. Mulliken, J. B. and Glowacki, J., Induced osteogenesis for repair and construction in the craniofacial region. *Plastic and Reconstructive Surgery*, 1980, **65**, 553–560.
9. Hirase, Y., Valauri, F. A. and Buncke, H. J., Prefabricated sensate myocutaneous and osteomyocutaneous free flaps: An experimental model. Preliminary report. *Plastic and Reconstructive Surgery*, 1988, **82**, 440–445.
10. Stark, G. B., Hong, C. and Futrell, J. W., Enhanced neovascularization of rat tubed pedicle flaps with low perfusion of the wound margin. *Plastic and Reconstructive Surgery*, 1987, **80**, 814–824.
11. Hussl, H., Russell, R. C., Zook, E. G. and Eriksson, E., Experimental evaluation of tissue Revascularization using a transferred muscular-vascular pedicle. *Annals of Plastic Surgery*, 1986, **17**, 299–305.
12. Hyakusoku, H, Okubo, M, Umeda, Ta and Fumiiri, M. A., prefabricated hair-bearing island flap for lip reconstruction. *British Journal of Plastic Surgery*, 1987, **40**, 37–39.
13. Khouri, R. K., Tark, K. C. and Shaw, W. W., Prefabrication of flaps using an arteriovenous bundle and angiogenesis factors. *Surgical Forum*, 1992, 597–599.
14. Dickerson, R. C. and Duthie, R. B., The diversion of arterial blood flow to bone. *Journal of Bone Joint Surgery*, 1963, **45**(A), 356.
15. Woodhouse, C. F., The transplantation of patent arteries into bone. *Journal of the International College of Surgeons*, 1963, **39**, 437.
16. Orticochea, M., A new method for total reconstruction on the nose: the ears of donor areas. *British Journal of Plastic Surgery*, 1971, **24**, 225.
17. Erol Ö.O. and Spira, M., Utilization of a composite island flap employing omentum in organ reconstruction: An experimental investigation. *Plastic and Reconstructive Surgery*, 1981, **68**, 561–570.

18. Tzuyao, S., Experimental study of tissue graft vascularization by means of vascular implantation and subcutaneous burying. *Plastic and Reconstructive Surgery*, 1984, **73**, 403–410.

19. Khouri, R. K., Upton, J. and Shaw, W. W., Prefabrication of composite free flaps through staged microvascular transfer: An experimental and clinical study. *Plastic and Reconstructive Surgery*, 1991, **87**, 108–115.

20. Upton, J., Ferraro, N., Healy, G, Khouri, R and Merrell, C. The use of prefabricated fascial flaps for lining of the oral and nasal cavities. *Plastic and Reconstructive Surgery*, 1994, **94**, 573–579.

21. Albrektsson, T., Branemark, P. A., Erikson, A. and Lindstrom, J., *Scand J Plastic and Reconstructive Surgery*, 1978, **12**, 215–223.

22. Fisher, J. and Yang, W. Y., Experimental tissue molding for soft tissue reconstruction: a preliminary report. *Plast Reconstr Surg* 1988, **82**, 857–864.

23. Khouri, R. K., Koudsi, B and Reddi, H., Tissue transformation into bone in vivo. *Journal of the American Medical Association*, 1991, **266**, 1953–1955.

24. McLean, D. H. and Bunke, H. J., Autotransplant of the omentum to a large scalp defect with microsurgical revascularization. *Plastic and Reconstructive Surgery*, 1972, **49**, 268.

25. Swartz, W. M. and Banis J. C., eds. *Head and Neck Microsurgery*. Williams and Wilkins, Philadelphia, 1992.

26. Argenta, L. C. and Newman, M. H., The use of methylmethacrylate cranioplasty in forehead reconstruction. *European Journal of Plastic Surgery*, 1986, **9**, 94.

27. McCarthy, J. G. and Zide, B. M., The spectrum of calvarial bone grafting: the introduction of the vascularized calvarial bone flap. *Plastic and Reconstructive Surgery*, 1984, **74**, 10.

28. Munro, I. R. and Guyuron, B., Split-rib cranioplasty. *Annals of Plastic Surgery*, 1981, **7**, 341.

29. Robson, M. C. et al., Reconstruction of large cranial defects in the presence of heavy radiation damage and infection utilizing tissue transferred by microvascular anastomosis. *Plastic and Reconstructive Surgery*, 1989, **83**, 438.

30. Baudet, J., Martin, D. and Cullen, K., Maximizing donor tissue while minimizing donor morbidity. In *Problems in Plastic and Reconstructive Surgery*. J. B. Lippincott, Philadelphia, 1991, pp. 64–124.

31. Pribaz, J. J. and Falco, N., Nasal reconstruction with auricular microvascular transplant. *Annals of Plastic Surgery*, 1993, **31**, 289.

32. Shenaq, S., Dinh, T. and Spira, M., Nasal alar reconstruction with ear helix free flap. *Journal of Reconstructive Microsurgery*, 1989, **1**, 63.

33. Hidalgo, D. A., Fibula free flap mandible reconstruction. *Clinical Plastic Surgery*, 1994, **21**, 25.

34. Shenaq, S. M. and Klebuc, M. J. A., The iliac crest microvascular free flap in mandibular reconstruction. *Clinical Plastic Surgery*, 1994, **21**, 37.

35. Robb, G. L., Free scapular flap reconstruction of the head and neck. *Clinical Plastic Surgery*, 1994, **21**, 45.

36. McCormack, P., Bains, M. S., Beattie, E. J. and Martini N., New trends in chest wall reconstruction after resection of chest wall tumors. *Annals of Thoracic Surgery*, 1981, **31**, 45.

37. Pers, M. and Medgyesi, S., Pedicle muscle flaps and their applications in the surgery of repair. *British Journal of Plastic Surgery*, 1973, **26**, 313.

38. Pairolero, P. C. and Arnold, P. G., Thoracic wall defects: surgical management of 205 consecutive patients. *Mayo Clinic Proceedings*, 1986, **61**, 557.

39. Gilboa, D., Borenstein, A. and Floro, S. et al., Emotional and psychological adjustment of women to breast reconstruction and detection of subgroups at risk for psychological morbidity. *Annals of Plastic Surgery*, 1990, **25**, 397–401.

40. Cohen, B. E., Casso, D. and Whetstone, M., Analysis of risks and aesthetics in a consecutive series of tissue expansion breast reconstructions. *Plastic and Reconstructive Surgery*, 1992, **89**, 840–843.

41. Bostwick, J. and Jones, G., Why I choose autogenous tissue in breast reconstruction. *Clinical Plastic Surgery*, 1994, **21**, 165.

42. Fisher, J. and Hammond, D., The combination of expanders with autogenous tissue in breast reconstruction. *Clinical Plastic Surgery*, 1994, **21**, 309.

43. Schusterman, M. A., Kroll S. S. and Weldon M. E., Immediate breast reconstruction: Why the free TRAM over the conventional TRAM? *Plastic and Reconstructive Surgery*, 1992, **90**, 255–261.

44. Santi, P., Berrino, P. and Canavese, G. et al., Immediate reconstruction of the penis using an inferiorly based rectus abdominis myocutaneous flap. *Plastic and Reconstructive Surgery*, 1988, **81**, 961.

45. Beimer, E., Penile reconstruction by the radial forearm flap. *Clinical Plastic Surgery*, 1988, **15**, 425.

46. Chang, T. and Hwang, W., Forearm flap in one-stage reconstruction of the penis. *Plastic and Reconstructive Surger*, 1984, **74**, 251.

47. Gilbert, D. A., Williams, M. W. and Horton, C. E. et al., Phallic reinnervation via the pudendal nerve. *Journal of Urology*, 1988, **140**, 295.

48. Levine, L. A., Zachary, L. S. and Gottlieb, L. J., Prosthesis placement after total phallic reconstruction. *Journal of Urology*, 1993, **149**, 593.

49. Sadove, R. C. and Horton, C. E., Utilizing full-thickness grafts for vaginal reconstruction. *Clinical Plastic Surgery*, 1988, **15**, 443.

50. McCraw, J., Massey, F., Shanklin, K. and Horton, C., Vaginal reconstruction using gracilis myocutaneous flaps. *Plastic and Reconstructive Surgery*, 1976, **58**, 176.

51. DeHaas, W. G., Miller, M. J. and Temple, W. J. et al., Perineal wound closure with the rectus abdominis musculocutaneous flap following tumor ablation. *Annals of Surgery Oncology*, 1995 (Sept.), 2(5), 400–406.

52. Jaffe, W. L. and Scott, D. F., Total hip arthroplasty with hydroxyapatite-coated prostheses. *Journal of Bone and Joint Surgery*, 1996, **78**A, 1918.

53. Tomford, W. W., Transmission of disease through transplantation of musculoskeletal allografts. *Journal of Bone Joint Surgery*, 77A, 1742.

54. Mankin, H. J. Springfield, D. S. Gebhardt, M. C. and Tomford, W. W., Current status of allografting for bone tumors. *Orthopedics*, 1992, **15**, 1147.

55. Millesi, H., Fascicular nerve repair and interfascicular nerve grafting. In *Reconstructive microsurgery*, eds. R. K. Daniel and J. K. Terzis. Little, Brown Company, Boston, 1977, pp. 430–422.

56. Hidalgo D. A., Free muscle transplantation. In *Microsurgery in trauma*, eds. S. S. Shaw and D. A. Hidalgo. Futura Publishing Company, Mount Kisko, New York, 1987, pp. 389–396.

57. Lee, W. P. and May, J. W. Jr., Neurosensory free flaps to the hand. Indications and donor selection. *Hand Clinics*, 1992, **8**, 465.

58. Goldberg, J. A., Trabulsy, P., Lineaweaver, W. C. and Buncke H. J., Sensory reinnervation of muscle free flaps for foot reconstruction. *Journal of Reconstructive Microsurgery*, 1994, **10**, 7.

CHAPTER II.12

Visceral Surgery

ROSA S. CHOI,
JOSEPH P. VACANTI
Department of Surgery,
Harvard Medical School, Children's Hospital
300 Longwood Ave.,
Boston, MA 02115, USA

1 Introduction

The purpose of this chapter is to provide a brief history of the origins of visceral transplantation surgery (liver, intestine, pancreas). Pertinent general issues that are applicable to tissue engineering of visceral organs will be discussed such as epithelial cell biology, vascularization, polymer properties for cell transplantation, and innervation. A complete discussion of anatomy, physiology, and issues involved in tissue engineering a specific organ can be found in their respective chapters.

2 History

2.1 Liver

Although the history of liver transplantation (LTx) has been short, spanning four decades, it has undergone explosive growth as a result of innovative surgical techniques and improved drugs for immunosuppression and infection control. The field of vascular surgery, fathered by Alexis Carrel and his description of vascular anatomosis, facilitated the surgical development of LTx [1]. Fifty years after Carrel's pioneering work, the first experimental LTx was described in a canine model by C.S. Welch of Albany, N.Y., and J.A. Cannon of the University of California, Los Angeles in 1955 and 1956, respectively. Welch described placing an auxiliary liver heterotopically into the pelvis or right paravertebral gutter [2]. Cannon described an orthotopic LTx which was later proven by Starzl *et al.*, to be more advantageous and represents the current standard for LTx [3, 4] (Fig. 1).

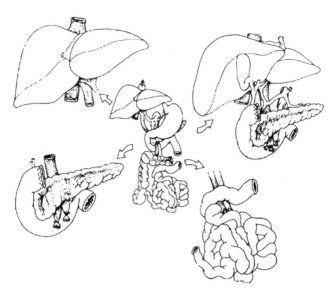

Fig. 1. Schematic diagram of visceral organs from which come liver, intestine, pancreas and liver-pancreas grafts (from Starzl, T. E. et al., transplantation of gastrointestinal organs. Gastroenterology, 1993, 104, 674).

Two groups primarily contributed to the development of clinical LTx. In Boston, the team at the Peter Bent Brigham Hospital headed by Francis D. Moore focused on the immunological aspects of organ transplantation. In 1954, the first successful organ (renal) transplantation between identical twins was performed by Joseph E. Murray (Nobel prize laureate in 1990) at the Brigham Hospital. In Chicago, the team at the Northwestern University headed by Thomas E. Starzl focused on the physiology of liver regeneration and hepatotrophic growth factors. In 1963, the first clinical attempt in LTx was in a 3-year-old child with biliary atresia who bled to death [5]. Four years later, the first successful LTx was performed in a 1.5-year-old child with a hepatocellular carcinoma [6]. She survived for 400 days but died of recurrent cancer. For the next decade and a half, intraoperative mortality was 5–10% with 1-year survival rates of less than 30% [7, 8]. In 1987, advances in organ preservation occurred with the development of the University of Wisconsin solution by Belzer and Southard, which increased organ preservation time up to 24 hours [9]. Other advances were made in operative techniques with development of the venovenous bypass and heparin-bonded Gott cannulas [10] which allowed splanchnic portal venous decompression during the anhepatic phase, avoidance of systemic anticoagulation and bleeding, and improved cadiovascular stability.

Clinical LTx could not have been possible without advances in immuno-suppressive therapies. Sir Peter Medawar was the first to observe that graft rejection had an immunological basis in his work with skin grafts in the rabbit [11, 12]. Initial immunosuppressive drug therapies included use of azathioprine and prednisone. In 1979, the era of cyclosporine began when Roy Calne of Cambridge, England, introduced cyclosporine to the transplant community [13]. Reports of improved 1-year survival rates up to 70% were reported from multiple institutions (Pittsburgh [14], Cambridge [15], Hannover [16], and Groningen [17]) at the National Institutes of Health consensus conference in 1983. It was decided that LTx was no longer an experimental but a therapeutic treatment for end-stage liver failure. The numbers of LTx have since increased steadily. In 1995, 3,926 LTx were performed in contrast to 1,713 in 1988 [18]. There were 5,701 patients waiting to receive LTx in 1995. Organ shortage continues as a significant obstacle for patients with end-stage liver failure.

2.2 Small intestine

Small bowel transplantation (SBTx) also dates back to the work of Alexis Carrel when he described the technique of segmental intestinal transplantation into the neck of dogs [19, 20]. In 1959, Lillehei et al., at the University of Minnesota began their work with autotransplantation of intestine after cold ischemia in dogs [21]. Unfortunately, clinical small bowel transplantation has not met with the same success as LTx due to several inherent properties of the small intestine. It functions to provide a barrier between the host and its external luminal environment and thus contains the largest mass of lymphoid tissue transplanted within an organ. The large immune load of the small bowel can result in graft versus host disease. The mucosal barrier can be compromised due to injury from ischemia or rejection leading to problems of infection. Intestinal epithelial cells express class II MHC antigens which make them more immunogenic compared to other epithelial cell types. Animals models continue to be developed to better understand its physiologic properties. In 1971, heterotopic SBTx was described by Monchik and Russell [22] and in 1973, orthotopic SBTx by Kort [23] in rats. In 1993, Zhong described the technique in mice [24].

Over two decades, more than a dozen patients have undergone SBTx but died within 3 months due to rejection [25]. Interest waned with poor clinical results and with the development of total parenteral nutrition to sustain patients with short bowel syndrome. Cyclosporine renewed interest with promising results of successful SBTx in dogs [26] and pigs [27] but proved to be disappointing in humans [19]. Current success can be attributed to two main discoveries: multivisceral transplantation and development of the immunosuppressant FK506. Several reports found that LTx could provide some immunologic protection to other transplanted organs in dogs and rats [28–30]. The first patient to receive a multivisceral transplantation with cyclosporine immunosuppression survived for 6 months but died of B-cell lymphoma, a complication of excessive immuno-suppressive therapy [31]. Starzl and his team in Pittsburgh described the success of isolated SBTx with FK506 treatment. Using the protocols established in Pittsburgh, centers worldwide have reported 70% 1-year patient survival rate for combined liver-small bowel transplantation and 70% 1-year graft survival rate for isolated SBTx using FK506 [19].

2.3 Pancreas

Insulin was discovered in 1921 by Banting and Best [32]. This discovery extended the lives of millions of people with diabetes mellitus. Unfortu-nately, control of serum glucose levels with insulin therapy did not prevent its complications. Microvascular disease of diabetes is the leading cause of blindness, renal failure, and limb losses. The first description of pancreatic transplantation was in the late 1800's, when an English surgeon trans-planted pancreatic extracts from a sheep to the abdominal wall of a comatose diabetic patient [33]. In 1959, a large animal model of pancreatic transplantation proved that it was a feasible technique and revealed that complications of surgery were related to the exocrine portion of the gland [34]. The first clinical whole organ pancreatic transplantation was per-formed by Kelly and Lillehei at the University of Minnesota in 1966 [35]. Over the years, the technique of pancreatic transplantation has undergone various modifications regarding the drainage of the exocrine duct. Cur-rently, European centers use either enteric drainage or the technique of duct injection and obliteration with a polymer described by Dubernard [36]. Surgeons in North America use bladder drainage, initially described by Cook and Sollinger [37] and later modified by Ngheim [38]. Bladder drainage of the exocrine gland has reduced complications of peripancreatic infection tracking from the bowel; however, detection of early rejection still remains a clinical obstacle. It is reported that 90% to 95% of the islet cells have died by the time hyperglycemia is clinically detected, making it a poor marker for rejection surveillance. There has been significant improve-ment with higher patient and graft survival rates with simultaneous pan-creas-kidney transplantation (SPK) (Fig. 2). Reports of 5-year survival rates for SPK are 62% versus 28% for isolated pancreas transplantation and 23% for those who had a previous kidney transplant [18]. Successful pancreas transplantation also appears to halt the progression of complica-tions seen with diabetes. There is improvement of neuropathy, stabilization of retinopathy, and prevention of diabetic nephropathy in the simultane-ously or previously transplanted kidney.

An alternative to whole organ pancreas transplant is the transplantation of pure islets of Langerhans. The adult pancreas is comprised of 95%

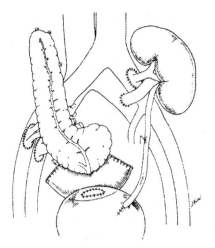

Fig. 2. Simultaneous pancreas-kidney transplant with bladder drainage (from Sollinger, H. W. and Geffner, S. R., Pancrease transplantation. Horizons in Organ Transplantation, 1994, 74, 1188).

exocrine and 5% endocrine gland tissue, approximately one million islet cells. It is estimated that approximately 500,000 islet cells are necessary for normal carbohydrate metabolism. Much work has been done to isolate the islet cells by various perfusion techniques in rodents and large animal models. Others have explored the possibility of using fetal islet cells, which comprise 50% of the pancreas in a 16–20-week-old fetus and may also provide some immunological advantages. Currently, human islet cell transplantations have been autotransplants to prevent diabetes in patients who required total pancreatectomy [39]. Islet cells are delivered into the liver bed through a portal vein injection. Rodent studies have shown that islets do best at a well vascularized site with portal blood flow [40]. The viability of the islets and their contribution to glucose homeostasis is questionable and disappointing, with graft failure after 6–8 months [41]. Much work still needs to be done to realize islet cell transplantation as a treatment for diabetes. New investigations of islet cell transplantation using immuno-isolation devices may allow xenotransplantation, and holds promise for the future.

Although the field or organ transplantation has improved dramatically with better patient and graft survival, the demand for new organs continues to exceed supply. This demand has pushed science to new frontiers of cell transplantation and tissue engineering. Tissue engineering involves the collaborative efforts of cell biologists, chemical engineers and medical/surgical clinicians to better understand the organ systems in order to design new tissue. Issues such as mechanism of epithelial cell survival and proliferation, vascularization and innervation of tissue, and polymer properties optimal for tissue engineering will be discussed.

3 Epithelial cell isolation, growth and survival

One common theme in working with epithelial cells derived from the fetal endoderm such as hepatocytes, enterocytes, and islets of Langerhams, are the difficulties of isolating pure populations of epithelial cells, maintaining these cells in tissue culture for extended periods of time, and the inability to expand epithelial cultures. The following discussion will focus on what is known and has been studied pertaining to the small bowel, but the general principles can be applicable to other epithelial cell types.

During embryogenesis, normal organogenesis depends upon interactions between the three layers of endoderm, mesoderm, and ectoderm. A review of the developmental biology studies of the small bowel demonstrates that the cells of the endoderm and mesoderm or mesenchyme can have permissive or inductive properties on one another [42]. A permissive effect triggers previously determined developmental programs in contrast to an inductive effect, which may reprogram the developmental schedule. Kedinger showed the mesenchymal permissive effect on endoderm by grafting chick intestinal mesenchyme and rat intestinal endoderm to form intestinal mucosa with phenotypic expressions of rat brush border enzymes (lactase, sucrase, maltase, and alkaline phosphatase) [43]. Mesenchyme can also have an inductive effect on endoderm. Combined grafts of chick intestinal mesenchyme and chick stomach endoderm resulted in intestinal mucosa [44–46]. LeDouarin's work showed the inductive properties of the endoderm by relocating chick embryonic endoderm to somatopleural mesenchyme. Not only did the endoderm demonstrate its self-differentiating properties to form various visceral organs (liver, intestine, pancreas, stomach, and esophagus), it also reprogrammed the mesenchyme to differentiate into smooth muscle rather than skeletal muscle [47]. Lacroix reconfirmed LeDouarin's work demonstrating the inductive properties of chick endoderm to change skin or lung fibroblasts into smooth muscle [48]. Others have demonstrated the importance of mesenchyme for morphogenesis and cytodifferentiation in organ cultures of isolated undifferentiated intestinal endoderm [49, 50]. These cultures differentiated only into absorptive enterocytes lacking the other three cell types (paneth, globlet, and neuroendocrine cells) and lacked any evidence of villus morphogenesis. Progenitor crypt cells do retain the ability to differentiate when provided the mesenchymal support, as demonstrated between combined grafts of IEC-17 cells and fetal mesenchyme, forming intestinal mucosa with all four types of differentiated cells [51, 52]. These types of epithelial-mesenchymal cell-cell interactions in embryogenesis are analogous to adult small bowel self-maintenance. The epithelial-mesenchymal unit, comprised of a pericryptal fibroblastic sheath that demonstrates proliferative and migrating properties in association with epithelial cells along the crypt-villus axis, has been described in several animal models [53–58].

Based on these principles, our laboratory's effort to tissue engineer small intestine has focused on isolating epithelial-mesenchymal units to serve as the cellular components of our transplantation work. Our technique, adapted from a technique described by Evans *et al.*, isolates intestinal epithelial organoid units from young rats by enzyme digestion and mechanical dissociation [59, 60]. Epithelial organoid units can be grown in tissue culture up to 3–4 weeks. The primary culture consists of > 90% epithelial cells and the remaining < 10% are comprised of myofriboblast-like cells and endothelium. These cultures contain the four types of differentiated epithelium derived from the progenitor crypt cell. These organoid units, seeded onto biodegradable scaffolds of polyglycolic acid (PGA) and transplanted into syngeneic adult rat, can generate complex tissue resembling small bowel. Formation of neomucosa with differentiated epithelium, crypt-villus morphogenesis, and underlying smooth muscle-like cells is evident [60]. Expression of brush border enzymes, sucrase and lactase, and formation of basement membrane identified by immunofluorescent stains for laminin and collagen suggest phenotypic maturation of the epithelial cells (unpublished work). Previous work from our laboratory transplanting pure populations of progenitor crypt cells onto biodegradable scaffolds resulted in stratified epithelium of 8–12 cell layers at 72 days post-implantation [61–63]. This morphology is reminiscent of fetal gut development and again suggests the importance of mesenchymal support. Other experimental work involving cotransplantation of hepatocytes with non-parenchymal cells (biliary epithelial cells, sinusoidal endothelial cells, Ito cells, and Kuppfer cells) has shown improved hepatocyte engraftment [64], suggesting again the importance of mesenchymal support for epithelial cells.

The role of cell-matrix interaction is also important and should not be underestimated. Several studies have defined the major structural components of basement membrane in adult and fetal intestine as laminin, fibronectin, collagen type IV, nidogen, heparin sulfate proteoglycans, and various glycoproteins [65]. It is possible that the mesenchymal signal to epithelial cells may be also mediated through the ECM, i.e. basement membrane. Hahn *et al.*, demonstrated that fetal rat intestinal epithelial cells can differentiate *in vitro* on various basement membrane components in absence of mesenchymal cell support [66]. The exact cell origin of the various basement membrane components is not well understood, but one chimeric study between a chick and rat demonstrated that collagen type IV was derived from the mesencymal cell [67]. More recently, others have tried to elucidate this mesenchymal signal, of which there may be more than one. One such factor may be transforming growth factor-β (TGF-β) [68].

4 Vascularization

One of the most critical issues in cell transplantation and tissue engineering is vascularization. The ability to transplant the maximum cell mass required to replace whole organ function is rate limited by the ability of those cells to survive via diffusion only until neovascularization from the hose occurs within 48–72 hours. For example, the liver is the largest solid organ containing 5×10^8 number of cells [69]. Ten to 15% of total cell mass is necesary for replacement of hepatic function [70]. Our experiments with hepatocyte transplantation have revealed that hepatocyte survival occurs along the periphery of the polymer discs where vascularization is most prominent. We have also shown that prevascularization of the polymers *in vivo* improves hepatocyte engraftment [71]. Of interest, the most potent hepatotrophic factor, hepatocyte growth factor (HGF), if produced by sinusoidal endothelium along with other non-parenchymal cell types. It may be more advantageous to co-culture or co-seed hepatocytes with other cell types, more specifically, endothelium; therefore, one approach to solving the problem of cell mass transfer is to build into the tissue engineered organ (liver) its own microvascular system which can be connected directly to the host upon transplantation. Our laboratory is actively involved with cotransplantation of hepatocytes and with sinusoidal and vascular endothelium on three-dimensional scaffolds in a bioreactor system. Preliminary work of hepatocytes and vascular endothelial cells in static culture on our three-dimensional scaffold suggest that these cells are able to self-organize, forming endothelium-lined channels surrounded by hepatocytes mimicking he normal architecture of liver [72]. A bioreactor system functions like an artificial vascular system, providing flow for media and allowing exchange of oxygen, nutrients, and metabolic wastes for large number of cells.

5 Polymer

It is a major challenge to design and to develop the optimal polymer to meet the requirements of a specific organ to be engineered. The polymer scaffold functions as a cell carrier device and provides a three-dimensional framework for cell organization and regeneration. It must be biocompatible, biogradable, and reproducible into various shapes and sizes. It must have high porosity with large surface/volume ratio to allow diffusion and exchange of oxygen, metabolic nutrients and waste products for cell survival prior to neovascularization from the host. It must also resist compressional forces during fibrovascular ingrowth from the host.

Our laboratory's initial efforts focused on the use of sheets of a non-woven array of fibers made of PGA (Albany International, Mansfield, MA). The PGA fiber diameter is 15 μm with a 250 μm pore size within the felt. This polymer has the desirable characteristics described previously and provides a large surface area for cell attachment. However, it lacks structural integrity. New techniques of improving polymer strength have been developed. Co-polymers of PGA and poly-lactic acid (PLA) have been made into tubes capable of resisting compressive forces. This is applicable to engineering tissues such as vessels, intestine, trachea, and ureter [73, 74]. PGA fibers can also be stabilized by bonding with solutions of PLA or poly (D,L-lactic-co-glycolic acid) (PLGA) (Fig. 3) [75, 76]. A novel technique of creating three-dimensional scaffolds has been developed. A complex solid structure with varying channel sizes can be made from liquid PLGA by spraying a binder from an ink-jet printer head onto a powder base of polymer [72]. It is especially attractive for tissue engineering of the liver which contains two types of vascular (microvascular and sinusoidal) beds in direct contact with hepatocytes. This three-dimensional scaffold allows simultaneous seeding and culturing of endothelial cells and parenchymal cells to establish a microvasculature needed to allow mass cell transplantation without the limits of cell diffusion for nutrient delivery and waste removal. Current investigation in our laboratory continues with this polymer.

Many agree that one of the advantages to synthetic polymer scaffolds is the ability to change or manipulate the surface properties with the addition of various macromolecules. Hrkach et al., has attached lysine to the backbone of PLA, which creates a functional group for attachment of macromolecules such as arginine-glycine-aspartate (RGD) or basement membrane components (laminin, collagen, or fibronectin) to optimize cell attachment [77]. Delivery of growth factors such as EGF incorporated into microspheres made of PGA and PLGA has been shown to improve

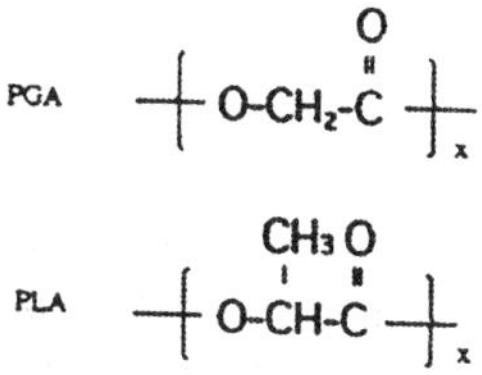

Fig. 3. Chemical backbone of Polyglycolic Acid (PGA) and Polylactic Acid (PLA) from Lee, H. Vacanti, J. P., Tissue Engineering of the Liver. In Synthetic Biodegradable Polymer Scaffolds. ed. A. Atala and D. Mooney, Birkhauser, Boston, 1997, 238.

hepatocyte engraftment [78]. Delivery of other growth factors of interest and angiogenic factors such as basic-fibroblast growth factor is feasible.

6 Innervation

All visceral organs are innervated by the parasympathetic and sympathetic nervous systems. Aside from the vascular innervation and regulation of blood flow into the solid organ, the degree of parenchymal innervation may vary from organ to organ and from one species to another [79]. Parenchymal innervation and its role and significance are incompletely understood. Whole organ transplantation has demonstrated that organs can function without their extrinsic innervation. Studies performed in LTx and SBTx biopsy specimens demonstrate that some degree of reinnerva-tion does occur in a limited fashion but its significance is unclear [80, 81].

The enteric nervous system is impressive in size and function. It is important for motor coordination and propulsion of luminal chyme; and it also plays a vital role in regulation of enterocyte absorption and secretion and immunomodultion of lymphoid cells in the bowel. Quantitative studies of the enteric nervous system demonstrate that there are nine submucosal neurons per intestinal villus [82]. Each villus is innervated by at least 70 submucosal neurons, and each neuron supplies about eight villi. Consider-ing the complexity of the enteric nervous system, attempts to approach and incorporate neurons into tissue engineered small intestine will not be a trivial problem.

7 Summary

Much progress has been made in the field of transplantation but with an ever increasing problem of organ donor shortage, tissue engineering may become one possible solution to many. In our combined efforts to tissue engineer various organs and tissues, it is critical that we understand normal anatomy and physiology and also understand the abnormal or diseased state that requires tissue replacement. The issues of cell isolation, survival, and growth, as well as vascularization and innvervation combined with the optimal polymer scaffold make the field both challenging and interesting.

References

1. Starzl, T. E., The French heritage of clinical kidney transplantation. *Trans-plantation Review*, 1993, 7, 65–71.

2. Welch, C. S., A note on transplantation of the whole liver in dogs. *Transplant Bulletin*, 1955, **2**, 54.

3. Cannon, J. A., Brief report. *Transplant Bulletin*, 1956, **3**, 7.

4. Starzl, T. E., Kaupp, H. A. Jr., Brock, D. R., Lazarus, R. E. and Johnson, R. V., Reconstructive problems in canine liver homotransplantation with special reference to the postoperative role of hepatic venous flow. *Surgery Gynecology and Obstetrics*, 1960, **111**, 733–743.

5. Starzl, T. E., Marchioro, T. L., VonKaulla, K. N., Hermann, G., Brittain, R. S. and Waddell, W. R., Homotransplantation of the liver in humans. *Surgery Gynecology and Obstetrics*, 1963, **117**, 659–676.

6. Starzl, T. E., Groth, C. G., Brettschneider, L., Penn, I., Fulginiti, V. A., Moon, J. B., Blanchard, H., Martin, A. J. Jr. and Porter, K. A., Orthotopic homotransplantation of the human liver. *Annuals of Surgery*, 1968, **168**, 392–415.

7. Starzl, T. E., Groth, C. G. and Makowka, L., In *Liver transplantation*. Austin, TX., Silvergirl, 1988, 164–169.

8. Starzl, T. E., Porter, L. A. and Putnam, C. W., Liver replacement in children. In *Liver Diseases in Infancy and Childhood*, ed. S. R. Berenberg, Williams and Wilkins, Baltimore, MD., 1976, 97–128.

9. Belzer, F. O. and Southard, J. H., Principles of solid-organ preservation by cold storage. *Transplantation*, 1988, **45**, 673–676.

10. Griffith, B. P., Shaw, B. W., Hardesty, R. L., Iwatsuki, S., Bahnson, H. T. and Starzl, T. E., Veno-venous bypass without systemic anticoagulation for transplantation of the human liver. *Surgery Gynecology and Obstetrics.*, 1985, **160**, 270–272.

11. Medawar, P. B., The behavior and fate of skin autografts and skin homografts in rabbits. *Journal of Anatomy.*, 1944, **78**, 276–299.

12. Medawar, P. B., Second study of behavior and fate of skin homografts in rabbits. *Journal of Anatomy.*, 1945, **79**, 157.

13. Calne, R. Y., Rolles, K., White, D. J., Thiru, S., Evans, D. B., McMaster, P., Dunn, D. C., Croddock, G. N., Henderson, R. G., Aziz, S. and Lewis, P., Cyclosporin A initially as the only immunosuppressant in 34 recipients of cadaveric organs: 32 kidneys, 2 pancreases and 2 livers. *Lancet*, 1979, **2**, 1033–1036.

14. Starzl, T. E., Iwatsuki, S., Shaw, B. W. Jr., Van Thiel, D. H., Gartner, J. C., Zitelli, B. J., Malatack, J. J. and Schade, R. R., Analysis of liver transplantation. *Hepatology*, 1984, **4**, 47S–49S.

15. Rolles, K., Williams, R., Neuberger, J. and Calnes, R., The Cambridge and King's College Hospital experience of liver transplantation, 1968–1983. *Hepatology*, 1984, **4**, 50S–55S.

16. Pichlmayr, R., Brolsch, C. H., Wonigeit, K., Neuhaus, P., Siegismund, S., Schmidt, F. W. and Burdelski, M., Experiences with liver transplantation in Hanover. *Hepatology*, 1984, **4**, 56S–60S.

17. Krom, R. A. F., Gips, C. H., Houthoff, H. J., Newton, D., van der Waaij, D., Beelen, J., Haagsma, E. B. and Slooff, M. J., Orthotopic liver transplantation in Groningen. The Netherlands (1979–1983). *Hepatology*, 1984, **4**, 61S–65S.

18. 1966 annual report of the US Scientific Registry of Transplant Recipients and The Organ Procurement and Transplant Network. Transplant Data: 1988–1955. Richmond VA: UNOS. Under contract to the US Department of

Health and Human Services. Health Resources and Services Administration. Bureau of Health Resources Development. Division of Transplantation. 1966.

19. Asfar, S., Zhong, R. and Grant, D., Small bowel transplantation. *Surgery Clinics of North America*, 1944, **74**, 1197–1210.

20. Carrel, A., La technique operatoire des anastomoses vasculaires et la transplantation des visceres. *Lyon MEO*, 1902, **98**, 859.

21. Lillehei, R. C., Goott, B. and Miller, F. A., The physiological response of the small bowel of the dog to ischemia including prolonged *in vitro* preservation of the bowel with successful replacement and survival. *Annuals of Surgery*, 1959, **150**, 543.

22. Monchik, G. J. and Russell, P. S., Transplantation of small bowel in the rat: Technical and immunological considerations. *Surgery*, 1971, **70**, 693.

23. Kort, W. J., Westbroeck, K. L., MacDicken, I. and Lameijer, L. D., Orthotopic total small-bowel transplantation in the rat. *European Surgical Research.*, 1973, **5**, 81.

24. Zhong, R., Zhang, Z., Quan, D., Duff, J., Stiller, C. and Grant, D., Development of a mouse intestinal transplantation model. *Microsurgery*, 1993, **14**, 141.

25. Pritchard, T. J. and Kirkman, R. L., Small bowel transplantation. *World Journal of Surgery*, 1985, **9**, 860–867.

26. Reznick, R. K., Craddock, G. N., Langer, B., Gilas, T., and Cullen, J. B., Structure and function of small bowel allografts in the dog: Immunosuppression with cyclosporin A. *Canadian Journal of Surgery*, 1982, **25**, 51.

27. Grant, D., Duff, J., Zhong, R., Garcia, B., Lipohar, C., Keown, P. and Stiller, C., Successful intestinal transplantation in pigs treated with cyclosporin. *Transplantation*, 1988, **45**, 279.

28. Calne, R. Y., Sells, R. A., Pena J. R., Davis, D. R., Millard, P. R., Herbertson, B. M., Binns, R. M. and Davies, D. A., Induction of immunological tolerance by porcine liver allografts. *Nature*, 1969, **223**, 472.

29. Kamada, N. and Wight, D. G. D., Antigen-specific immunosuppression induced by liver transplantation in the rat. *Transplantation*, 1984, **38**, 217.

30. Zhong, R., He, G., Sakai, Y., Li, X. C., Garcia, B., Wall, W., Duff, J., Stiller, C. and Grant, D., Combined small bowel and liver transplantation in the rat: Possible role of the liver in preventing intestinal allograft rejection. *Transplantation*, 1991, **52**, 550.

31. Starzl, T. E., Rowe, M., Todo, S., Jaffe, R., Tzakis, A., Hoffman, A. L., Esquivel, C., Porter, K. A., Venkataramanan, R. and Makowka, L. et al., Transplantation of multiple abdominal viscera. iJournal of the American Medical Association, 1989, **26**, 1449–1457.

32. Banting, F. G., Best, C. H., Collip, J. B., Pancreatic extracts in the treatment of diabetes mellitus: A preliminary report. *Canadian Medical Association Journal*, 1922, **12**, 141–146.

33. Williams, P. W., Notes on diabetes treated with extract and by grafts of sheep's pancreas. *British Medical Journal*, 1984, **2**, 1301.

34. Brooks, J. R. and Gifford, G. H., Pancreatic homotransplantation. *Transplant Bulletin.*, 1959, **6**, 100.

35. Kelly, W. D., Lillehei, R. C., Merkel, F. K., Idezuki, Y. and Goetz, F. C., Allotransplantation of the pancreas and duodenum along with the kidney in diabetic nephropathy. *Surgery*, 197, **61**, 827–837.

36. Dubernard, J. M., Traeger, J., Neyra, P., Touraine, J. L., Tranchant, D. and

Blanc-Brunat, N., New method of preparation of a segmental pancreatic graft for transplantation: Trials in dogs and in man. *Surgery,* 1978, **84**, 634.

37. Sollinger, H. W., Cook. K., Kamps, D., Glass, N. R. and Belzer, F. O., Clinical and experimental experience with pancreaticocystostomy for exocrine pancreatic drainage in pancrease transplantation. *Transplant Proceedings,* 1984, **16**, 749–751.

38. Ngheim, D. D. and Bentel, W. D., Doudenocystostomy for exocrine drainage in total pancreatic transplantation. *Transplant Proceedings,* 1986, **18**, 1753.

39. Sutherland, D. E. R. and Najarian, J. S., In *Pancreas and Islet Transplantation in Surgery of the Pancreas,* ed. J. R. Brooks, WB Saunders and Co., Philadelphia, 1983.

40. Kemp, C. B., Knight, M. J., Sharp, D. W., Ballinger, W. F. and Lacy, P. E., The effect of transplantation site on the results of pancreatic islet isographs in diabetic rats. *Diabetologia,* 1973, **9**, 486.

41. Cameron, J. L., Mehigan, D. G., Harrington, D. P. and Zuidema, G. D., Metabolic studies following intra-hepatic autotransplantation of pancreatic islet grafts. *Surgery,* 1980, **87**, 397.

42. Haffen, K., Kedinger, M. and Simon-Assmann, P., Mesenchyme-dependent differentiation of epithelial progenitor cells in the gut. *Journal of Pediatric Gastroenterology and Nutrition,* 1987, **6**, 14–23.

43. Kedinger, M., Simon, P. M., Grenier, J. F. and Haffen, K., Role of epithelial-mesenchymal interactions in the ontogenesis of intestinal brush-border enzymes. *Developmental Biology,* 1981, **86**, 339–347.

44. Haffen, K., Kedinger, M., Simon-Assmann, P. and Lacroix B., Mesenchyme-dependent differentiation of intestinal brush-border enzymes in the gizzard endoderm of the chick embryo. In *Embryonic Development; Part b: Cellular Aspects,* ed. A. R. Liss, New York, 1982, 261–270.

45. Ishizuya-Oka, A. and Muzuno, T., Intestinal cytodifferentiation in vitro of chick stomach endoderm induced by the duodenal mesenchyme. *Journal of Embryology and Experimental Morphology* 1984, **82**, 163–176.

46. Ishizuya-Oka, A. and Mizuno, T., Chronological analysis of the intestinalization of chick stomach endoderm induced in vitro by duodenal mesenchyme. *Willhelm Roux's Archives of Developmental Biology,* 1985, **194**, 301–305.

47. Le Douarin, N., Bussonnet, C. and Chaumont, F., Etude des capacités de différenciation et du role morphogene de l'endoderme pharyngien chez l'embryon d'oiseau. *Annals of Embryology Morphology,* 1968, **1**, 29–39.

48. Lacroix, B., Kedinger, M., Simon-Assmann, P. and Haffen, K., Enzymatic response to glucocorticoids of the chick intestinal endoderm associated with various mesenchymal cell types. *Biology Cell,* 1985, **54**, 235–240.

49. Minzuno, T. and Ischizuya, A., Etude au microscope electronique de la différenciation in vitro de l'endolderme d'intestin grele de jeuna embryon d'oiseau, associe ou non a son mesenchyme. *CR Society of Biology,* 1982, **176**, 580–584.

50. Ishizuya, A., Epithelial differentiation of the intestine of young quail embryos cultures organotypically in vitro. *Journal of Faculty Science Tokyo,* 1981, **15**, 71–79.

51. Quaroni, A., Wands, J., Trelstad, R. L. and Isselbacher, K. J., Epithelioid cell cultures from rat small intestine: characterization by morphologic and immunologic criteria. *Journal of Cell Biology,* 1979, **80**, 248–265.

52. Kedinger, M., Simon-Assmann, P. M., Lacroix, B., Marxer, A., Hauri, H. P. and Haffen, K., Fetal gut mesenchyme induces differentiation of cultured intestinal endodermal and crypt cells. *Developmental Biology*, 1986, **113**, 474–483.

53. Pascal, R. R., Kaye, G. I. and Lane, N., Colonic pericryptal fibroblast sheath: replication, migration, and cytodifferentiation of a mesenchymal cell system in adult tissue. I. Autoradiographic studies of normal rabbit colon. *Gastroenterology*, 1968, **54**, 835–851.

54. Kaye, G. I., Lane, N. and Pascal, R. R., Colonic pericryptal fibroblast sheath: replication, migration, and cytodifferentiation of mesenchymal cell system in adult tissue. II. Fine structural aspects of normal rabbit colon. *Gastroenterology*, 1968, **54**, 853–865.

55. Marsh, M. N. and Trier, J. S., Morhpology and cell proliferation of subepithelial fibroblasts in adult mouse jejunum. I. Structural features. *Gastroenterology*, 1974, **67**, 622–635.

56. Marxh, M. N. and Trier, J. S., Morphology and cell proliferation of subepithelial fibroblasts in adult mouse jejunum. II. Radioautographic studies. *Gastroenterology*, 1974, **67**, 636–645.

57. Parker, F. G., Barnes, E. N. and Kaye, G. I., The pericryptal fibroblast sheath. IV. Replication, migration, and differentiation of the subepithelial fibroblasts of the cryp and villus of the rabbit jejunum. *Gastroenterology*, 1974, **67**, 607–621.

58. Maskens, A. P., Rahier, J. R., Meersseman, F. P., Dujardin-Loita, R. M. and Hoot, J. G., Cell proliferation of pericryptal fibroblasts in the rat colon mucosa. *Gut*, 1979, **20**, 775–779.

59. Evans, G. S., Flint, N., Somers, A. S., Eyden, B. and Potten, C. S., The development of a method for the preparation of rat intestinal epithelial cell primary cultures. *Journal of Cell Science*, 1992, **101**, 219–231.

60. Choi, R. S. and Vacanti, J. P., Preliminary studies of tissue-engineered intestine using isolated epithelial organoid units on tubular synthetic biodegradable scaffolds. *Transplant Proceedings*, 1997, **29**, 848–851.

61. Organ, G. M., Mooney, D. J., Hansen, L. K., Schloo, B. and Vacanti, J. P., Transplantation of enterocytes utilizing polymer-cell constructs to produce a new intestine. *Transplantation Proceedings*, 1992, **24**, 3009–3011.

62. Organ, G. M., Mooney, D. J. and Hansen, L. K., Enterocyte transplantation using cell-polymer devices to create intestinal epithelial-lined tubes. *Transplantation Proceedings*, 1993, **25**, 998–1001.

63. Organ, G. M. and Vacanti, J. P., Tissue engineering neointestine. In *Principles of Tissue Engineering*, ed. R. Lanza, R. Langer and W. Chick. R. G. Landes Company, 1997, 441–462.

64. Sano, K., Cusick, R. A., Lee H., Pollok, J. M., Kaufmann, P. M., Uyama, S., Mooney, D., Langer, R. and Vacanti, J. P., Regenerative signals for heterotopic hepatocyte transplantation. *Transplant Proceedings*, 1996, **28**, 1859–1860.

65. Simon-Assmann, P., Kedinger, M. and Haffen, K. Immunocytochemical localization of extracellular-matrix proteins in relation to rat intestinal morphogenesis. *Differentiation*, 1986, **32**, 59–66.

66. Hahn, U., Stallmach, A., Hahn, E. G. and Riecken, E. O., Basement membrane components are potent promoters of rat intestinal epithelial cell differentiation in vitro. *Gastroenterology*, 1990, **98**, 323–335.

67. Simon-Assmann, P., Bouziges, F. and Arnold C., Epithelial-mesenchymal interactions in the production of basement membrane components in the gut. *Development*, 1988, **102**, 339–347.

68. Halttunen, T., Marttinen, A. and Rantala, I. et al., Fibroblasts and transforming growth factor B induce organization and differentiation of T84 human epithelial cells. *Gastroenterology*, 1996, **111**, 1252–1262.

69. Uyama, S., Kaufmann, P. M., Takeda, T. and Vacanti, J. P., Delivery of whole liver-equivalent hepatocyte mass using polymer devices and hepatotrophic stimulation. *Transplantation*, 1993, **55**, 932–935.

70. Asonuma, K., Gilbert, J. C., Stein, J. E., Takeda, T. and Vacanti, J. P., Quantitation of transplanted hepatic mass necessary to cure the Gunn rat model of hyperbilirubinemia. *Journal of Pediatric Surgery*, 1992, **27**, 298.

71. Stein, J. E., Asonuma, J., Gilbert, J. C., Schloo, B., Ingber, D. and Vacanti, J. P., Hepatoctype transplantation using prevascularized porous polymers and hepatotrophic stimulation. Presented at the American Pediatric Surgical Associations meeting. Lake Buena Vista, Fl., May 1990.

72. Griffith, L. G., Wu, B., Cima M. J., Chaignaud, B. and Vacanti, J. P., In vitro organogenesis of liver tissue. *Annals of the New York Academy of Science*, 1997, in press.

73. Mooney, D. J., Organ, G. M., Vacanti, J. P. and Langer, R., Design and fabrication of biodegradable polymer devices to engineer tubular tissues. *Cell Transplantation*, 1994, **3**, 203–210.

74. Mooney, D. J., Breuer, C., McNamara K., Vacanti, J. P. and Langer, R., Frabricating tubular devices from polymers of lactic and glycolic acid for tissue engineering. *Tissue Engineering*, 1995, **1**, 107–118.

75. Mikos, A. G., Bao, Y., Cima, L. G., Ingber, D. E., Vacanti, J. P. and Langer, R., Preparation of poly(glycolic acid) bonded fiber structures for cell attachment and transplantation. *Journal of Biomedical Materials Research*, 1993, **27**, 183–189.

76. Mooney, D. J., Mazzoni, C. L. and Breuer, C., Stabilized polyglycolic acid fibre-based tubes for tissue engineering. *Biomaterials*, 1996, **17**, 115–124.

77. Hrkach, J. S., Ou, J., Lotan, N. and Langer, R., Synthesis of poly (l-lactic acid-co-lysine) graft copolymers. *Macromolecules*, 1995, **28**, 4736–4739.

78. Mooney, D. J., Kaufmann, P. M., Sano, K., Schwendeman, S., McNamara, K., Schloo, B., Vacanti, J. P. and Langer, R., Localized delivery of EGF improves the survival of transplanted hepatocytes. *Biotechnical Bioengineering*, 1996, **50**, 422–429.

79. Lautt, W. W. Hepatic nerves: a review of their functions and effects. *Canadian Journal of Physiology Pharmacology*, 1980, **58**, 105–123.

80. Dhillon, A. P., Sankey, E. A. and Wang, J. H., Immunohistochemical studies on the innervation of human transplanted liver. *Journal of Pathology*, 1992, **167**, 211–216.

81. Hirose, R., Taguchi, T., Hirata, Y., Yamada, T., Nada, O. and Suita, S., Immunochistochemical demonstration of enteric nervous distribution after syngeneic small bowel transplantation in rats. *Surgery*, 1995, **117**, 560–569.

82. Song, Z. M., Brookes, S. J. H., Steel, P. A. and Costa, M., Projections and pathways of submucous neurons to the mucosa of the guinea-pig small intestine. *Cell and Tissue Research*, 1992, 87–98.

CHAPTER II.13

Immunoisolation

MICHAEL V. SEFTON,
JULIA E. BABENSEE,
MICHAEL H. MAY
Department of Chemical Engineering and Applied Chemistry
and Centre for Biomaterials,
University of Toronto,
Toronto, Ontario, M5S 3E5, Canada

Immunoisolation devices such as for the treatment of insulin dependent diabetes must provide for the appropriate behaviour of an adequate cell mass for extended periods of time. This requires control of device size and shape, membrane permeability and permselectivity, intracapsule environment and device 'biocompatibility'. The principles underlying the mass transport limitations to immunoisolation devices and especially microencapsulated cell size, the role of the internal capsule environment on immunoisolated cell behaviour and the consequences of implantation are discussed.

1 Introduction

Immunoisolation is a means of implementing cell transplantation using selectively permeable membranes to isolate the cells from the host's immune system. The encapsulated cells, from either the same species (allogeneic) or a different species (xenogeneic), produce the desired therapeutic product while the membrane prevents their rejection. Cells of primary (e.g. islets of Langerhans), immortalized (e.g. pheochromocytoma, PC12 cells) or engineered (e.g. baby hamster kidney, BHK, cells engineered to produce nerve growth factor, NGF) origins can be used in immunoisolation devices because the membrane protects the cells against immune attack.

The membrane material must ensure diffusion of both nutrients (e.g. oxygen, glucose, etc.) to the encapsulated cells and the therapeutic product to the host system. The membrane may be of natural or synthetic origins; however, all must have the appropriate transport and geometry for optimal functioning and survival of cells. Biocompatibility and biostability of the membrane material impact on its transport and immunoisolatory properties. Encapsulated cells are often suspended or immobilized within a hydrogel matrix material within the membrane. The matrix material defines the environment within the membrane with nutrients and cell products that diffuse across it. The matrix material provides a distribution of cells within the membrane, thereby inhibiting the formation of large cell clusters and cell necrosis. The matrix may be tailored to provide the most suitable environment for continued cell viability and cell secretory function. For example, the matrix may provide an attachment site for anchorage-dependent cells.

Immunoisolation may be achieved either by filling a pre-formed membrane device (e.g. a tube or hollow fibre, 'macroencapsulation') with cells and matrix or by forming the membrane around the cells from a polymer solution [1] or from the monomer [2] (e.g. in the form of spheres, 'microencapsulation', Table 1).

Successful implementation of an immunoisolation device, like any tissue engineering construct, requires (1) an adequate, viable cell mass (2) the appropriate behaviour of the cells and (3) sufficient durability of the function, *in vivo*. The specific requirements are determined by the application, the nature of the cells, the implantation site and the biocompatibility of the device. In turn, these depend on the device geometry, the membrane permeability and associated mass transfer characteristics; the phenotype of the cells and their sensitivity to nutrient limitations, cytokines and extracellular matrix; the cell-type and source and the inflammatory and immune response to the device or its components. Some aspects of these three requirements are highlighted here.

2 Mass transport limitations

There are three main designs for diffusion-based immunoisolation devices: spherical microcapsules, cylindrical, hollow fibre macrocapsules and rectangular diffusion chambers (Fig. 1). The first concern is total surface area for a given volume of transplanted tissue. An aliquot of microcapsules will have a higher surface area than the corresponding tube or flat sheet but the larger surface area for the spherical geometry also means a greater volume. The implant volume is related to the amount of tissue that is needed (i.e. how productive the cells are relative to the required dose of therapeutic agent) and the immunoisolation chamber(s) shape and wall thickness:

$$\text{volume} = \frac{\text{\# cells needed}}{\text{cell density}} \times \frac{\text{outside volume}}{\text{core volume}}$$

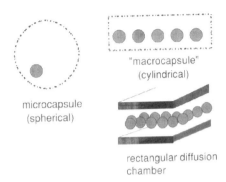

microcapsule
(spherical)

"macrocapsule"
(cylindrical)

rectangular diffusion
chamber

Fig. 1. Types of immunoisolation 'capsules'.

For spherical microcapsules, this is equivalent to:

$$\text{volume} = \frac{\#\text{ cells needed}}{\#\text{ cells/capsule}} \times \text{capsule volume}$$

The different geometries are compared in Table 2, for a particular set of dimensions.

The volume advantage of the other geometries, however, is diminished somewhat due to other practical considerations. For instance if $5*10^5$ islets (typically 150 μm in diameter) are needed to control glycaemia [3], then a cylindrical device would have a total length of 75 metres and a rectangular chamber would be 11cm square. The only way to reduce these dimensions to ones realistic for implantation would be to 'stack' the islets, which would introduce further resistances to nutrient and cell product diffusion. A promising means to balance surface area, and thus diffusion, considerations with total implant volume constraints is to use a device that is virtually the same size as the immunoisolated islet. For this reason we have developed a method to conformally coat cell aggregates in a thin HEMA–MMA membrane less than 5 μm thick [4].

Minimizing the implant volume also requires the use of the minimum amount of tissue. Mass transport limitations increase the required cell volume because of limited nutrient supply (or product buildup) leading to central necrosis or because of the effect of critical parameters (e.g. pO_2) on the secretion behaviour of the cells (e.g. islets, [5]). The nutrient supply issue is amenable to modelling to account for the effects of nutrient concentrations, nutrient consumption rate, membrane permeability and thickness, device size, the number of cells/capsule, and the diffusivity of nutrients within the cell mass.

Nutrient consumption in the presence of capsule wall and intra-tissue diffusion can be estimated through a variant of the well-known Thiele Modulus/Biot Number problem in the cell immobilization [6] and chemi-

Table 1. Microencapsulation methods

Microencapsulation technique	Capsule membrane formation principle
Agarose	Thermally-induced gelation
Agarose and Poly(styrene sulfonic acid) (PSSa) [and polybrene + chondroitin sulfate A, carboxymethyl cellulose, poly(acrylic acid)]	Thermally-induced gelation and polyionic complexation
Agarose/polyacrylamide	Interfacial photopolymerization
Alginate	Ca^{2+}-induced gelation
Alginate	Ba^{2+}-induced gelation
Alginate-polylysine	Interfacial polyelectrolyte complexation of anionic alginate and cationic polylysine
Alginate-polylysine/polyethylene-imine/protamine sulfate/heparin	Interfacial polyelectrolyte complexation of anionic alginate and cationic polyethyleneimine and protamine sulfate/heparin
Poly(ethylene glycol) surface modification of alginate polylysine capsules	Interfacial polyelectrolyte complexation of anionic alginate and cationic polylysine
Carboxymethyl cellulose and chitosan	Interfacial polyelectrolyte complexation
Collagen/chitosan and alginate	Interfacial polyelectrolyte complexation
Matrigel/carboxymethyl cellulose/chondroitin sulfate A and chitosan/polygalacturonate	Interfacial polyelectrolyte complexation
Cellulose sulfate and poly(dimethyldiallylammonium chloride)	Interfacial polyelectrolyte complexation
Polyacrylates (e.g. polyHEMA–MMA)	Interfacial precipitation
Polyacrylate polyelectrolyte polymers [e.g. polydimethylaminoethyl methacrylate (DMAEMA) – methacrylic acid (MAA)]	Interfacial polyelectrolyte complexation
Polyethylene glycol — diacrylate	Photopolymerization
Polyacrylonitrile-sodium methylsulphonate	Interfacial precipitation
Polyphosphazenes and poly(lysine)	Interfacial polyelectrolyte complexation
Polyvinyl alcohol (PVA-SbQ)	Photopolymerization

Table 2. Effect of immunoisolation shape on surface to volume ratio and implant volume

	Surface/volume (cm^{-1})	Outside/inside volume ratio	Total volume (ml)
Rectangular	17	1.33	13.3
Cylindrical	67	1.77	17.7
Spherical	100	2.37	23.7

Critical dimension: 600 μm inside diameter or thickness; Membrane thickness: 100 μm; 10 ml of cell suspension = 10^8 cells/10^7 cells/ml

cal engineering literatures. This results in the calculation of an 'effectiveness factor' for the encapsulated tissue which is the ratio of the consumption rate of oxygen in the presence of mass transfer resistance to the theoretical consumption if mass transfer resistance were absent. It is then possible to assume that steady state function (secretion of insulin, for example) is proportional to the fraction of tissue that is adequately nourished with some critical nutrient, such as oxygen. Then the problem of calculating product delivery rate is simplified to estimating the fraction of tissue that is 'satisfied' with its nutrient supply, which is the same as the effectiveness factor. Additional cells are needed to make up for an effectiveness factor which is less than unity, so that the total cell requirement becomes:

$$\frac{theoretical\ cell\ number}{effectiveness\ factor}$$

Effectiveness factors for the three geometries with membrane thicknesses of 50 μm and various device sizes are listed in Table 3, for a representative set of parameters that are specifically relevant to HEMA–MMA encapsulated pancreatic islets [7]. Also shown is the total implant volume, which includes the extra tissue required to make up for the fraction of tissue that is producing insulin at sub-maximal levels. Although large capsules with thin membranes have the largest effectiveness factors, they result in unrealistic total implant volumes. The smaller, thinner spherical capsules provide the best compromise between function and volume considerations. Indeed, the effectiveness and total implant volume for islets conformally coated in a 5 μm HEMA–MMA membrane are 0.75 and 2.5 ml, respectively. With such a small implant volume, $5 \times 10^5/(0.75)$ of these coated islets could be easily implanted at a variety of sites. Moreover, the ultra-thin membrane results in shorter lag times for insulin response to glucose.

Table 3. Effectiveness factors and total implant volumes for different geometries

Size	Spherical		Cylindrical		Rectangular	
(diameter or thickness, μm)	Effectiveness factor	Total implant volume (ml)	Effectiveness factor	Total implant volume (ml)	Effectiveness factor	Total implant volume (ml)
300	0.38	19	0.28	19	0.19	18
600	0.69	82	0.42	51	0.18	38
900	0.77	246	0.48	99	0.17	62

Based on oxygen diffusion to a single islet in a microcapsule of given size or to closely packed islets in a cylindrical microcapsule or to a single layer of closely packed islets between two flat membrane sheets

The Thiele modulus based effectiveness factor accounts only for nutrient limited transport effects. There are other features of the microcapsule that can result in the need for extra tissue (i.e. an even lower effectiveness factor) and increase the demands on the source of tissue (a special problem with primary tissues such as islets) and increase the device volume. These include various downregulatory effects caused by product inhibition, a lowered pO_2 [5], by the three-dimensional nature of the cell aggregate or by the simple diffusion limiting effect of the capsule wall on the cell product.

3 Cell phenotype

It is well known that the extracellular matrix, through binding to cell surface integrins, has a strong influence on cell behaviour. Accordingly adding ECM components to cells prior to encapsulation or otherwise altering the substrate on or within which the cells are grown provides a means to control the behaviour of the transplanted cells. For example, Matrigel grown hepatocytes have elevated mRNA synthesis as compared to cells grown on plastic or in collagen gel [8]. P-450 activities were also induced in Matrigel but not collagen I [9]. Alteration of the ECM or other features of the cell microenvironment thus provides another means for controlling the performance of the encapsulated cell.

The ECM is a dynamic environment undergoing constant remodeling (e.g. through the indirect effect of cytokines and growth factors) [10]. The capsule wall will likely affect the extent of remodelling, e.g. by impeding the inward transport of interstitial proteases that otherwise cause ECM turnover and consequently stabilizing the intracapsule ECM. On the other hand, necrotic cells may release significant amounts of lysosomal enzymes which can degrade biomolecules. The capsule wall will impair their outward diffusion and could cause a build-up of degrading enzymes or of the acidic lysosomal environment.

Minimization of the diffusion limitations associated with microencapsulation is also critical to the success of encapsulation. For example, depletion of a particular nutrient (e.g. oxygen) may cause cells to grow at a lower, diffusion limited rate or the centre of a cell cluster may become necrotic. This has been observed [11] and is similar to what occurs with tumour spheroids [12]. Because the cells in spheroids are at a distance from the surrounding medium gradients of critical nutrients and growth factors and metabolites (and also pH) are set up [13]. As a spheroid grows, the number of proliferating cells decreases, the proportion of quiescent cells increases and eventually, due to nutrient deprivation and waste product accumulation, necrosis develops at the centre of spheroids [13]. The diffusion of larger molecular weight species (such as growth factors, hormones, and cytokines) into spheroid regions, their interaction with the receptors of cells in the outer few layers of an aggregate [14] and cell contact with the ECM are also important.

The microenvironment of spheroidal aggregates appears to maintain the *in vivo* differentiated cell characteristics for longer times than in monolayer culture [15,16], presumably, due to the three dimensional arrangement and the corresponding diffusion gradients and ECM–cell and high cell–cell contact. With the capsule wall controlling the aggregate size and strongly influencing the diffusion gradients, such differentiated cells with prolonged lifespans would presumably be ideally suited for microencapsulated cell applications. Microcapsules are also characterized by a locally high cell density and this too is expected to influence intracapsule cell behaviour.

4 Biocompatibility: immune and inflammatory response

The two main components of microencapsulated cells are the polymer membrane and the transplanted cells. The latter may be of isogeneic, allogeneic, or xenogeneic origin. Microcapsules containing living cells may be transplanted into a variety of tissue sites such as the peritoneal cavity, subcutaneously, or in the brain. Regardless of the site, the implantation of polymer encapsulated cells combines concepts of biomaterials and cell transplantation. Implantation of a biomaterial (without transplanted cells) initiates a sequence of events akin to a foreign body reaction starting with an acute inflammatory response and leading in some cases to a chronic inflammatory response and/or granulation tissue development. The duration and intensity of each of these is dependent upon the extent of injury created in the implantation and the physical and chemical characteristics of the biomaterial [17] and involving biomaterial associated protein adsorption, complement activation and leukocyte adhesion and activation. A central consequence of the inflammatory response is the activation of macrophages resulting in the release of cytokines, growth factors, proteolytic enzymes, and reactive oxygen and nitrogen intermediates as mediators of the inflammatory response. The nature of the inflammatory response determines the degree of fibrosis and vascularization of the tissue reaction. A fibrous tissue reaction surrounding the implanted microcapsule may act as a barrier to nutrient and product diffusion [18–20], to an extent depending on its thickness. In contrast, a granular tissue reaction would include vascular structures to facilitate the delivery of nutrients and the absorption of cell-derived therapeutic products and has been induced with particular membrane architectures [21–23].

Furthermore, the polymer capsule, by promoting a non-specific inflammatory reaction, recruiting antigen-presenting cells (APC's) (e.g. dendritic cells or macrophages) and inducing their activation, may act as an adjuvant in the immune response to antigens originating from encapsulated cells (Fig. 2). This may therefore intensify an immune response to shed antigens or shorten the time until its onset. Taken together, the host response to microcapsules containing live cells may be viewed as an inflammatory response to the biomaterial and an immune response to whatever portion of the transplanted cells that can be 'seen' by the host. The interaction and interconnections of these two capsule components with the host are fundamental to understanding the *in vivo* response to encapsulated cells.

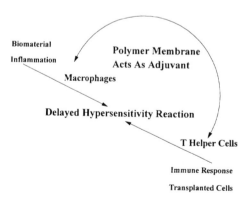

Fig. 2. Schematic of the merging of inflammation induced by the implantation of a biomaterial and the immune response to the transplanted cells within the polymer capsule; the delayed hypersensitivity reaction. Macrophage activation by the adjuvant polymer membrane would enhance a T helper cell-mediated immune response towards shed cellular antigens.

Originally, the idea of immunoisolation, using a polymer membrane, was to physically separate allogeneic and even xenogeneic tissue from the host immune system by preventing contact with immunoglobulin, complement components and immune and inflammatory cells. In this way, the presence of the polymer membrane would prevent an immune response originating from the outside of the capsule acting on cells inside the capsule. However, an immune response is potentially also possible towards antigens shed by encapsulated cells, such as foreign proteins secreted by the cells including the therapeutic agent, cell surface molecules (e.g. donor MHC molecules), or cell components (e.g. proteins, phospholipids, DNA), released upon cell death, which then diffuse through the polymer membrane to be recognized as foreign by the host's immune system. Degradation of capsule components such as the extracellular matrix could also generate immunologic or proinflammatory products. Antigen shedding from microcapsules sensitizes the host to the transplanted cells and could initiate potentially a humoral or molecular cytotoxic tissue response around the implant. This would enhance the tissue reaction to the implant which in turn would further affect encapsulated cell viability and function.

These shed antigens may be internalized, processed and presented in association with the host Class II Major Histocompatibility Complex (MHC) molecules, most effectively by macrophages or dendritic cells, to host $CD4^+$ helper T cells: the indirect pathway of antigen recognition [24]. This pathway is more likely since the direct presentation of graft antigens by donor APC's (e.g. carrier leukocytes) to host helper or cytotoxic T cells [25], would not be expected because the polymer membrane should prevent the required cell contact. The indirect pathway may lead to the activation of $CD4^+$ T helper cells, which secrete cytokines and provide the necessary signals for the growth (autocrine, IL-2-mediated T cell proliferation), maturation and activation of effector $CD8^+$ cytotoxic T cells, B cells, macrophages, inflammatory leukocytes and endothelial cells. In this way, they would promote and regulate humoural and cell-mediated immune responses and inflammation. Immunologists term this type of cell-mediated immune response and the associated inflammation, a delay-type hypersensitivity reaction [26].

Cytokines, such as IL-1β, IFN-γ, IL-6, and TNF-β released by inflammatory and immune cells in the tissue reaction to polymer microcapsules containing cells [27], may be present in sufficiently high concentrations intracapsularly [28], to affect adversely islet [29] or hepatocyte [30] viability and function, possibly resulting in graft failure [31]. Interestingly, some microcapsules appear to prevent IL-1β inhibition of insulin secretion [32], suggesting that the intracapsular microenvironment is such that the sensitivity of islets to IL-1 is lessened after encapsulation. This lowered sensitivity is unlikely to be because of permeation limitations since the membrane molecular weight cut-off is far greater than the ~ 20 kDa that would otherwise be required. Also, encapsulated cells may secrete cytokines or growth factors that influence the inflammatory and immune response to these capsules [33, 34] or induce angiogenesis surrounding the cell-containing implant [35–37]. Using a membrane to prevent adverse cytokine effects is not a very promising approach, since such membranes would likely compromise cell viability due to nutrient diffusion restrictions.

It is conceivable that non-specific lysis of encapsulated cells may be mediated by lysosomal enzymes or reactive oxygen (e.g. O_2^-, H_2O_2) or nitrogen (e.g. NO) intermediates released by activated leukocytes [38, 39]. These free radicals are small but highly reactive with a short lifetime until inactivation by chemical reaction (half life of NO in H_2O is ~ 3–50 seconds [40]); hence, their function is usually confined locally to their site of secretion. There has not yet been any demonstration that these very small radicals need to be excluded from the encapsulated cells.

5 Antibody / complement exclusion

Antibody exclusion and molecular weight cut-offs of 50–100 kDa were the original hallmarks of immunoisolation. Concerns over antigen shedding and cytokines has raised doubts about the sufficiency of such a cut-off. It may be preferable to have a much higher cut-off (one that allows for IgG permeation) and a corresponding higher permeability to nutrients and growth factors with a view to reducing the dose of shed antigens. Generally, antibodies which permeate the immunoisolating polymer membrane may not on their own destroy antibody-targeted cells if the required immunocellular and complement components are excluded. An exception would be antibodies directed against an essential ligand (e.g. transferrin).

Potential antibodies, specific to encapsulated cells, would be those recently produced by the host in an immunological response to shed antigens, or preformed antibodies due to an immunologically-based disease etiology (e.g. Type I diabetes) or those present in discordant xenogeneic transplantation. To cause a problem, it is necessary for C1 to become activated through the binding of the C1q component to an aggregate of IgG molecules or an individual IgM molecule. This would initiate the classical pathway of complement activation that, with the addition of the subsequent complement components, (of lower molecular weight) would potentially result in the assembly of the membrane attack complex (MAC) and cell lysis [41]. Alternatively, complement activation by the alternative pathway may occur if not inhibited by regulatory membrane proteins on the surface of the encapsulated mammalian cells. Depending on what is the predominant mechanism of complement activation resulting in lysis of encapsulated cells may determine the required molecular weight cut-off of the membrane.

Deciphering the relative importance of inflammatory reaction to the biomaterial and immune reaction to the encapsulated cells and the effects of cell source (xenogeneic vs. allogeneic) and capsule permeability and molecular weight cut-off have become crucial issues in translating this technology into a clinically practical reality.

References

1. Sefton, M. V. and Broughton, R. L., Microencapsulation of erythrocytes. *Biochimica et Biochema Acta*, 1983, **717**, 473–477.
2. Dupuy, B., Gin, H., Baquey, C. and Ducassou D., In situ polymerization of a microencapsulating medium around living cells. *Journal of Biomedical Materials Research*, 1988, **22**, 1061–1070.
3. Colton, C. K. and Avgoustiniatos, E. S., Bioengineering in development of the hybrid artificial pancreas. *Transactions ASME* 1991, **113**, 152–170.
4. May, M. H. and Sefton, M. V., Conformal coating of small particles and cell aggregates at a liquid–liquid interface, (submitted).
5. Dionne, K., Colton, C. and Yarmush, M., Effect of hypoxia on insulin secretion by isolated rat and canine islets of Langerhans. *Diabetes*, 1993, **42**, 12–21.
6. Karel, S. F., Libicki, S. B. and Robertson, C. R., Immobilization of whole cells. *Chemical Engineering Science*, 1985, **40**, 1321–1354.
7. May, M., Conformal coating of mammalian cells. Ph.D. thesis, University of Toronto, 1997.
8. Bissell, D. M., Caron, J. M., Babiss, L. E. and Friedman, J. M., Transcriptional regulation of the albumin gene in cultured rat hepatocytes role of basement-membrane matrix. *Molecular Biology Medicine*, 1990, **7**, 187.
9. Schuetz, E. G., Li, D., Omecinski, C. J., Muller-Eberhard, U., Kleinman, H. K., Elswick, B. and Guzelian, P. S., Regulation of gene expression in adult rate hepatocytes cultured on a basement membrane matrix, *Journal of Cellular Physics*, 1988, **134**, 309.
10. Alexander, C. M. and Werb, Z., Proteinases and extracellular matrix remodeling. *Current Opinion in Cell Biology*, 1989, **1**, 974.
11. Babensee, J. E., De Boni, U. and Sefton, M. V., Morphological assessment of hepatoma cells (HepG2) microencapsulated in a HEMA–MMA copolymer with and without Matrigel. *Journal of Biomedical Materials Research*, 1992, **26**, 1401.
12. Sutherland, R. M., Cell microregions: the multicell spheroid model. *Science*, 1988, **240**, 177–184.
13. Sutherland, R. M., Importance of critical metabolites and cellular interactions in the biology of microregions of tumors. *Cancer*, 1986, **58**, 1668–1680.
14. Erlanson, M., Daniel-Szolgay, E. and Carlsson, J., Relations between the penetration, binding and average concentration of cytostatic drugs in human tumour spheroids. *Cancer Chemotherapy and Pharmacology*, 1992, **29**, 343.

15. Landry, J., Bernier, D., Ouellet, C., Goyette, R. and Marceau, N., Spheroidal aggregate culture of rat liver cells: histotypic reorganization, biomatrix deposition, and maintenance of functional activities. *Journal of Cell Biology*, 1985, **101**, 914.

16. Van Der Schueren, B., Denef, C. and Cassiman, J-J., Ultrastructural and functional characteristics of rat pituitary cell aggregates. *Endocrinology*, 1982, **110**, 513.

17. Anderson, J. M., Inflammatory response to implants. *Transactions, American Society for Internal Organs*, 1988, **24**, 101–107.

18. Anderson, J. M., Niven, H., Pelagalli, J., Olanoff, L. S. and Jones, R. D., The role of the fibrous capsule in the function of implanted drug-polymer sustained release systems. *J. Biomed. Mater. Res.*, 1981, **15**, 889–902.

19. Wood, R. C., LeCluyse, E. L. and Fix, J. A., Assessment of a model for measuring drug diffusion through implant-generated fibrous capsule membranes. *Biomaterials*, 1995, **16**, 957–959.

20. Bodziony, J., Bioartificial endocrine pancreas: foreign-body reaction and effectiveness of diffusional transport of insulin and oxygen after long-term implantation of hollow fibers into rats. *Research in Experimental Medicine*, 1992, **192**, 305–316.

21. Siebers, U., Sturm, R., Renardy, M., Planck, H., Zschocke, P., Bretzel, R. G., Zekorn, T. and Federlin, K., Morphological studies on biocompatibility of artificial membranes for immunoisolated islet transplantation. In *Methods in Islet Transplantation*, eds. K. Federlin, R. G. Bretzel and B. J. Hering. Georg Thieme Verlag, Stuttgart, 1990, pp. 206–208.

22. Brauker, J. H., Carr-Brendel, V. E., Martinson, L. A., Crudele, J., Johnston, W. D. and Johnson, R. C., Neovascularization of synthetic membranes directed by membrane microarchitecture. *Journal of Biomedical Materials Research*, 1995, **29**, 1517–1524.

23. Picha, G. J. and Drake, R. F., Pillared-surface microstructure and soft-tissue implants: Effect of implant site and fixation. *Journal of Biomedical Materials Research*, 1996, **30**, 305–312.

24. Sayegh, M. H., Watschinger, B. and Carpenter, C. B., Mechanisms of T cell recognition of alloantigen. The role of peptides. *Transplantation*, 1994, **57**, 1295–1302.

25. Wolf, L. A. Coulombe, M. and Gill, R. G., Donor antigen-presenting cell-independent rejection of islet xenografts. *Transplantation*, 1995, **60**, 1164–1170.

26. Abbas, A. K., Lichtman, A. H. and Pober, J. S., *Cellular and Molecular Immunology*. W. B. Saunders Company, Harcourt Brace Jovanovich, Philadelphia, PN, USA, 1991.

27. Weber, C., Ayres-Price, J., Costanzo, M., Becker, A. and Stall, A., NOD mouse peritoneal cellular response to poly-L-lysine-alginate microencapsulated rat islets. *Transplantation Proceedings*, 1994, **26**(3), 1116–1119.

28. Bergmann, L., Kroncke, K-D., Suschek, C., Kolb, H. and Kolb-Bachofern, V., Cytotoxic action of IL-1β against pancreatic islets is mediated via nitric oxide formation and is inhibited by NG-monomethyl-L-arginine. *Federation of European Biochemical Societies*, 1992, **299**, 103–106.

29. Rabinovitch, A., Sumoski, W., Rajotte, R. V. and Warnock, G. L., Cytotoxic

effects of cytokines on human pancreatic islet cells in monolayer culture. *Journal of Clinical Endocrinology Metabolism*, 1990, **71**, 152–156.

30. Perlmutter, D. H., Colten, H. R., Adams, S. P., May, L. T., Sehgal, P. B. and Fallon, R. J., A cytokine-selective defect in interleukin-1 β-mediated acute phase gene expression in a subclone of the human hepatoma cell line (HepG2). *Journal of Biological Chemistry*, 1989, **264**, 7669–7674.

31. Cole, D. R., Waterfall, M., McIntyre, M and Baird, J. D., Microencapsulated islet grafts in the BB/E rat: a possible role for cytokines in graft failure. *Diabetologia*, 1992, **35**, 231–237.

32. Zekorn, T., Siebers, U., Bretzel, R. G., Renardy, M., Planck, H., Zschocke, P. and Federlin, K., Protection of islets of Langerhans from interleukin-1 toxicity by artificial membranes. *Transplantation*, 1990, **50**, 391–394.

33. Fraser, R. B., MacAulay, M. A., Wright, J. R., Sun, A. M. and Rowden, G., Migration of macrophage-like cells within encapsulated islets of Langerhans maintained in tissue culture. *Cell Transplantation*, 1995, **5**, 529–534.

34. Tsukui, T., Kikuchi, K., Mabuchi, A., Sudo, T., Sakamoto, T., Asano, G. and Yokomuro, K., Production of interleukin-1 by primary cultured parenchymal liver cells (hepatocytes). *Experimental Cell Research*, 1994, **210**, 172–176.

35. Brauker, J., Martinson, L. A., Hill, R. S., Young, S. K., Carr-Brendel, V. E. and Johnson, R. C., Neovascularization of immuno-isolation membranes: The effect of membrane architecture and encapsulated tissue. In First International Congress of Cell Transplantation Society, Pittsburgh, PA, June 13–15, [*Cell Transplantation*, 1992, **1**(2/3), 163].

36. Kan, M., Huang, J., Mansson, P. E., Yasumito, H., Carr, B. and McKeehan, W. L., Heparin-binding growth factor 1 (acidic fibroblast growth factor): A potential biphasic and paracrine regulator of hepatocyte regeneration. *Proceedings of the National Academy of Science*, 1989, **86**, 7432–7436.

37. Folkman, J. and Klagsbrun, M., Angiogenic factors. *Science*, 1987, **23**, 442–447.

38. Kroncke, K. D., Funda, J., Berschick, B., Kolb, H. and Kolb-Bachofen, V., Macrophage cytotoxicity towards isolated rat islet cells: Neither lysis nor its protection by nicotinamide are Beta-cell specific. *Diabetologia*, 1991, **34**, 232–238.

39. Wiegard, F., Kroncke, K.-D. and Kolb-Bachofen, V., Macrophage-generated nitric oxide cytotoxic factor in destruction of alginate-encapsulated islets. *Transplantation*, 1993, **56**, 1206–1212.

40. Moncada, S., Palmer, R. M.J. and Higgs, E. A., Nitric oxide: physiology pathophysiology, and pharmacology. *Pharmacology Review*, 1991, **43**, 109.

41. Cooper, N. R. and Cochrane, C. G., The biochemistry and biologic activities of the complement and contact systems. In *Hematology*, 3rd edn, eds. W. J. Williams, E. Beutler, A. J. Erslev and M. A. Lichtman. McGraw-Hill Book Company, 1983.

Drug Delivery

YOU HAN BAE,

Department of Materials Science and Engineering,
Kwangju Institute of Science and Technology,
Kwangju, 506-712, Korea

and

SUNG WAN KIM

Center for Controlled Chemical Delivery,
University of Utah,
Salt Lake City, Utah 84112, USA

1 Controlled release technology

The technology for controlled or sustained release has been developed in many different ways and its aim is to supply therapeutic agents in a desired release patterns. Delivering agents to specific cells, tissues, or organs is termed 'drug targeting' and is related to the local delivery of cell regulatory molecules for new tissue formation in tissue engineering or for preventing tissue from over-growth after injury.

Typical release patterns include first- and zero-order kinetics, pulsatile release, and timely release, as schematically presented in Fig. 1. First-order kinetics can be obtained from a slab-shaped monolithic device where a drug is solubilized or dispersed throughout a polymeric matrix and passively diffuses out of the matrix, when it is exposed to a release medium, with increasing diffusional length with time. Drug release based on first-order kinetics has been described by Higuchi [1]. The release kinetics often, however, deviates from the expected and is strongly influenced by a variety of parameters. These parameters include device geometry, the physicochemical properties of the polymer matrix (molecular weight, swelling, crystallinity, and glass transition temperature), initial drug concentration profile, drug–matrix interactions, and osmotic pressure associated with drug solubility and its loading content. Diffusion of a lipophilic drug in hydrophobic polymers can be explained by the partition mechanism and diffusion of a water soluble drug through swellable matrices by the free-volume (or pore) mechanism [2]. However, both mechanisms are commonly involved in a system. When a bulk-degradable polymer is used,

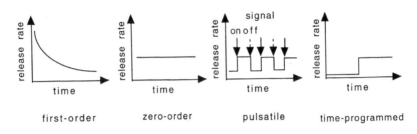

Fig. 1. Typical drug release patterns from drug delivery systems.

a more complex release pattern is observed due to the combined effects of drug diffusion and bulk-degradation of the matrix.

Historically much effort in developing drug delivery systems has centered on achieving zero-order release kinetics, anticipating maximum pharmacological effect, while minimizing undesirable side effects that can be caused by large fluctuations in the plasma drug concentration beyond the therapeutic index of a drug. This type of release kinetics is achieved by several approaches. One representative example has been obtained by passive diffusion of a drug through a rate determining barrier enveloping a core of drug formulation where the thermodynamic activity of the drug remains constant for a given period of time [3]. This reservoir-type drug delivery system can provide long-term constant release rate with a potential risk of a drug overdose due to defect or failure of the barrier membrane which is typically fabricated from a non-degradable polymeric material. Other approaches to zero-order release kinetics include systems of swelling- [4], surface erosion-controlled release [5, 6] and osmotic pump technology [7]. The constant release rate based on a swelling-controlled mechanism can be obtained with drug loaded hydrogel monolithic devices when the diffusional rate of a dissolved drug through the swollen gel phase is faster than the speed of the swelling front movement in an initially glassy hydrogel. The swelling front is defined as the interface of a swollen gel phase and a drug containing glassy phase, where the polymer chain is relaxed by interacting with water molecules supplied from the swollen phase. A constant release rate also requires the constant movement of the swelling front [4]. With bioerodible matrices, zero-order release is typically obtained with a hydrophobic matrix which erodes from the surface by chain cleavage or by solubilzation of the polymer by hydrolysis of the backbone or polymer side chains [5, 6]. Osmotic pump technology uses constant permeation of water through a semipermeable membrane into a compartment containing an osmogen, resulting in the squeezing of a drug containing compartment and thus pumping the drug formulation out of the device [7].

There is increasing evidence that constant release may not be the way to maximize drug effects and minimize side effects, because most organ functions demonstrate circadian rhythms or other physiological cycles. For the maximum pharmacodynamic effects, drug targeting to a specific time scheme should be taken into account in developing more sophisticated drug delivery systems, depending on the variation of pharmacokinetic parameters and pharmacodynamic effects with time [8]. There are also increasing findings of an oscillatory pattern in cellular communication based on hormonal release and subsequent cellular response [9]. This oscillatory pattern may be a natural process to compensate for the down regulation induced by receptor internalization or phosphorylation when the receptors are constantly exposed to binding ligands for a long period of time.

The release of hormones from endocrine glands or secretory cells (pituitary thyroid, parathyroid, hypothalamus, adrenal cortex, ovaries, testes, kidney, pancreas, and gastrointestinal tract) is regulated by physiological cycles, by circadian cycles, or by specific input signals such as messenger hormones, neurotransmitters, electric potential and metabolic concentrations. As a result, self-regulating drug delivery systems [10, 11] and oscillating release patterns have become an emerging issue in drug delivery technology in an attempt to mimic nature.

When a drug delivery system designed with passive control is applied to the body, this can not meet the unpredictable or irregular drug requirement in the body. This has prompted researchers to investigate responsive drug delivery systems from which the drug release rate can be controlled by internal or applied external signals. The delivery technology based on time programmed release and stimulus activated release could be particularly relevant to tissue engineering and cell therapy because of the potential of down-regulation of growth factor receptors caused by a continuous supply of growth factor. In the design of modulated or time programmed delivery systems, stimuli-responsive polymers play a key role because such polymers can interact with input signals, change their own structure or properties, and subsequently influence drug release rate. The signals that are most often employed in stimuli-responsive drug delivery systems include pH [12], ions [13], temperature [14–16], electric current [17, 18], photoirradiation [19], magnetic field [20], and ultrasound [21]. The property changes resulting from these stimuli, in turn, affect the permeability or diffusional rate of the incorporated drug. In addition to the bulk property change of the polymers, the surface property can also dramatically influence the solute release characteristics. An example is 'on–off' release obtained when a dense skin layer was reversibly formed on a matrix surface [15]. When external energy, such as thermal and ultrasonic energy, is applied to a drug delivery system responsive to these forms of energy, the energy changes the polymer degradation rate [22] and its physical properties associated to glass transition temperature [23] and polymer melting temperature [24]. These property changes influence mass transport rates through the polymer matrix, thus resulting in modulated drug release.

2 Polymers for drug delivery

2.1 Non-degradable polymers

Polymers are widely investigated materials, among others, for the controlled delivery of therapeutics and these materials can be classified depending on biodegradability and hydrophilicity [2]. Hydrophobic non-degradable polymers including polydimethylsiloxane (PDMS), polyurethanes (PU), and poly(ethylene-co-vinyl acetate) (EVAc copolymers) have been used since the development of polymeric drug delivery systems. PDMS has been attractive in designing the diffusional barrier in reservoir-type delivery systems and is still a material of choice for long-term implantable systems. PDMS is characterized as an amorphous rubbery material and shows high permeability for water-insoluble drugs and gases. It has been commercially available in biomedical grade for implantation and with low surface energy, it minimizes interactions with host components. This property allows long-term application in the body without deteriorating diffusional characteristics.

PU have been applied to many biomedical implantable devices. Their mechanical performance in the body coupled with their capability of drug loading and release and variety of surface modifications have made them useful materials for the local delivery of cardiovascular drugs [25] and for non-thrombogenic surfaces [26].

EVAc copolymers are known to have good biocompatibility, mechanical properties with flexibility, and selective permeability for the design of various reservoir-type drug delivery systems [2]. These polymers have also been used as monolithic devices carrying macromolecular drugs [27].

Other useful materials for controlled delivery technology are those that swell, but do not dissolve in water. These materials are classified as hydrogels [28]. The first example of a hydrogel is poly(2-hydroxy ethylmethacrylate) (poly(HEMA)) introduced by Wichterle and Lim in 1960. Poly(HEMA) was used as the first material in a soft contact lens due to its excellent optical properties and processability [29]. Poly(HEMA) is an amorphous glassy polymer at room temperature and is soluble in alcohol but swells in water even without chemical crosslinking [30]. By virtue of such characteristics, poly(HEMA) had been investigated as a polymeric carrier for water soluble drugs. Poly(HEMA) can be fabricated into either a dense or a porous matrix depending on water content during polymerization. Polymerization can be accomplished by various conventional radical polymerization techniques with or without crosslinking agents. The degree of swelling in dense poly(HEMA) is fairly constant around 40–50% water content regardless of crosslinking density. A broader control of swelling can be obtained by copolymerization. Extensive efforts had been made to find an appropriate hydrogel system that, when loaded with a drug, has the potential of zero-order release kinetics by swelling-controlled mechanism as briefly described before. A variety of hydrogels have been synthesized by copolymerization of water soluble monomers with other comonomers and crosslinkers. However, true zero-order release kinetics has not entirely been obtained in aqueous systems with hydrogels.

For modulated or pulsatile delivery of drugs in response to external signals, a new class of stimuli-sensitive polymeric materials has been developed. These materials can be sorted according to the stimuli which modulate their drug delivery. Among the different stimuli, pH, temperature, and electric current have been frequently used stimuli in modulated drug delivery applications.

pH-sensitive hydrogels contain ionizable groups in the main chain or on side chains which are weakly acidic [31] or basic moieties [32], such as carboxylic acids and primary or substituted amines. Polymers containing carboxylic acid in pendent side chains are utilized as pH responsive hydrogels. The ionization state of these groups changes in response to pH changes, leading to changes in the swelling property. When the medium pH increases, these gels show an increase in swelling due to more ionization of the carboxylic acid. No practical systems are available even though the swelling properties and permeability of pH-sensitive hydrogels have been extensively studied.

Electric field or current is another useful signal which can be applied to polymeric systems for modulated release, because the signal strength can be easily adjusted using relatively small and simple electric circuit, such as an electrophoretic device for transdermal application. The electrical stimulus influences local pH via electrochemical reactions around the electrodes, inducing various electro-kinetic phenomena, such as electro–diffusion process, electro-osmosis, and electrophoretic augmentation of solute flux, and electro-static partitioning of a charged solute into the charged polymeric materials [33]. Drug release caused by an applied electric stimulus can be affected by the accompanying electro-kinetic phenomena, electro-chemical reactions, as well as electroactivity. Macromolecule release in an on–off manner from electroerodible polymer complexes was also demonstrated [18, 34].

Drug delivery systems which are responsive to temperature utilize various polymer properties, including the thermally reversible coil–globule transition of polymer molecules, swelling change of crosslinked polymers, glass transition, and crystalline melting. The swelling of different polymer networks in an aqueous environment is influenced by temperature in different ways, increased (positive thermosensitivity) or decreased (negative thermosensitivity) swelling with increasing temperature. Polymers with a lower critical solution temperature (LCST) exhibit negative thermosensitivity in an aqueous solution. The LCST is defined as the critical temperature at which a polymer solution undergoes phase transition from a soluble to an insoluble state when the temperature is raised. A range of hydrophilic/hydrophobic (HPL/HPB) balanced water soluble polymers exhibit LCST phenomenon and the HPL/HPB balance is either from the repeating unit in homopolymer or from the polymer composition in random copolymers. Accordingly, the LCST and the swelling transition temperature of the hydrogel can be manipulated by copolymerization. Environmental temperature affects the molecular interactions, such as hydrogen bonding and hydrophobic interaction between polymer chains or between polymer chains and solvent molecules. In particular, when the temperature rises above or lowers below the critical temperature, such interactions are dramatically influenced, resulting in large changes in polymer properties, such as degree of swelling or solubility. Recently, N-isopropylacrylamide(NiPAAm) polymers and its copolymers have been studied due to their pronounced negative thermosensitivity in solubility and swelling as crosslinked gels. These gels have been studied for applications in solute separation [35], concentrating enzyme solution [36], immobilization of enzymes [37], and drug delivery in an 'on–off' pulsatile manner [15]. One advantage of the use of thermosensitive polymers is the possibility of polypeptide and protein drug loading in aqueous phase without use of organic solvents. A drug can be mixed into an aqueous polymer solution at low temperature and a gel or solid matrix containing the drug can be formed as the temperature increases. Based on this concept, polymeric beads were fabricated from linear pH/temperature-sensitive polymers using terpolymers of NiPAAm, butylmethacrylate, and acrylic acid. The beads were evaluated as a potential polypeptide drug carrier by studying their loading efficiency, pH-dependent release, and preservation of bioactivity of the loaded drug [38].

2.2 Degradable polymers

Biodegradable polymers are high molecular weight substances that can decompose into non-toxic monomers or low molecular weight substances in a living organism. These polymers have been extensively researched over the past decades and some of them are currently available as commercial resorbable sutures and as drug delivery microspheres for human applications [39]. Biodegradable polymers include both natural polymers of polysaccharides and proteins and a variety of synthetic polymers. Because the synthetic biodegradable polymers can be designed with various monomers in different combinations, a wide spectrum of synthetic biodegradable polymers are available either commercially or in research laboratories. Toxicity concerns about the degradation products of the polymers have lead to the design of biodegradable polymers whose degradation products are safe and naturally occurring in the body. Biodegradable polymers, although limited in application due to government regulatory affairs, have strong potential as polypeptide and protein drug delivery systems in the future. Injectable or implantable delivery systems using biodegradable polymers are becoming more feasible for polypeptide drugs requiring longer sustained action.

Among various biodegradable polymers, aliphatic polyesters of polylactide, polyglycolide, and poly(lactide-co-glycolide) are the most popular biodegradable carriers because of their long clinical experience as resorbable sutures and their favorable regulatory status with the Food and Drug Administration. Polylactide (PLA) with high molecular weight has typically been synthesized by ring-opening melt polymerization. The low molecular weight polymers of PLA are gaining more popularity in controlling the degradation rate and have favorable processability. The degradation products of PLA are natural metabolites in the body and can be controlled by many parameters including the copolymer composition, end group functionality, molecular weight, and stereochemistry of lactide [40]. The major degradation process occurs by bulk hydrolysis and is autocatalized by the carboxylic groups produced by hydrolysis, resulting in multi phase release characteristics of incorporated drugs [41].

Favorable zero-order release kinetics from biodegradable polymers was obtained from hydrophobic, surface erodible polymers including poly(ortho esters) and polyanhydrides [42]. The surface degradation rate is controlled by the polymers' structure and acidic or basic adjuvants since the degradation is catalyzed by acid for poly(ortho esters) and base for polyanhydrides. Incorporated low and high molecular weight drugs are successfully released from the matrix with almost zero-order kinetics.

All these degradable polymers require organic solvents during formulation of drug containing delivery systems. Despite technological advances in the delivery of polypeptide and protein drugs, the use of organic solvents results in loss of protein bioactivity and in low protein loading efficiency. For safe loading and release of labile drugs, biodegradable hydrogels with sol–gel transitions have become important in protein delivery technology. A typical example is alginate which is a natural polysaccharide extracted from a seaweed, algae. The polymer is a copolymer of α-L-guluronic acid and β-D-mannuronic acid and instantly crosslinks in the presence of divalent cations, such as calcium ions, as modeled by the 'egg box' mechanism [43]. The thermally reversible sol–gel transition of agarose and gelatin is another example utilized in protein loading and controlled release. These polymers can go into an aqueous solution at high temperature and turn into a gel upon cooling with hysteresis between gelation and gel dissolution temperatures. Some other natural and synthetic water soluble polymers form a gel in the presence of hydrolyzable chemical crosslinkers, but this process often involves chemical reactions unfavorable for polypeptides.

2.3 Future development of polymers for drug delivery

Functional polymers are useful for more sophisticated modulated drug delivery but most known functional polymers, such as temperature sensitive polymers and polyelectrolytes, are not biodegradable, while biodegradable polymers usually lack functionality. In the future development of polymeric carrier of polypeptide drugs for injectable and implantable devices, the combined properties of biodegradability and functionality will provide more versatility in pulsatile or modulated delivery of polypeptide drugs. Polymers with combined properties of functionality, biodegradability, and processability would open a new avenue in polypeptide and protein drug delivery technology. Another future direction in polypeptide delivery is the development of polymers which do not require the use of organic solvents in the protein delivery formulation. Using the phase transition of biodegradable polymers in aqueous solutions caused by mild change in environmental conditions, such as temperature and pH, will be a useful approach in achieving this.

3 Protein delivery system

Most molecules which regulate cell proliferation and differentiation are polypeptides and controlled local delivery technology may be important in tissue regeneration or mammalian cell culture for cell therapy.

The first *in vitro* and *in vivo* demonstrations of macromolecular drugs loading and release from non-degradable polymeric matrices, such as poly(2-hydroxyethyl methacrylate), poly(vinyl alcohol), and copolymers of ethylene vinyl acetate, were made by Langer *et al.* [44]. The macromolecules were lysozyme (MW 14,400), soybean trypsin inhibitor (MW 21,000), alkaline phosphatase (MW 88,000), catalase (MW 250,000) and tumor angiogenesis factor. Since then, the same group had extended their research to the sustained-release of insulin, bovine serum albumin, methyl β-lactoglobulin A, and inulin from ethylene-vinyl acetate copolymers. The polymer and protein were mixed in an organic solution, casted in a mold to make drug containing matrix, and cold-dried, and if necessary, the dried matrix was coated with the pure polymer to retard the release rate. The release pattern of these macromolecules was characterized by an initial big burst effect followed by sustained release. By improving the solvent-casting method and developing a sintering technique, a more controlled and reproducible release pattern was obtained and they were able to show almost zero-order release of insulin from a slab with polymer coating and a hole drilled in it.

This type of delivery system requires surgical removal of the exhausted device after a certain period of time, because the polymers used are not degradable. Recently, great efforts using microparticulate systems made of biodegradable polyesters, typically poly(α-hydroxy acids), have been devoted to the development of a long-acting parenteral preparation of polypeptide and protein drugs. Naferelin was among the first peptides to be successfully incorporated with biodegradable poly(lactide-co-glycolide) (PLGA) microspheres [45]. A single administration of Naferelin microspheres suppresses the secretion of leuteinizing hormone and follicle-stimulating hormone for 30 days, with subsequent abolition of the secretion of progesterone in females and testosterone in males [46]. Leuprolide, another analogue of leuteinizing hormone releasing hormone (LHRH), has been incorporated into PLGA microspheres to obtain a controlled-release dosage form [47]. After evaluating microspheres, a formulation derived from PLGA (lactide:glycolide = 75:25) of molecular weight 14,000 was reported to be ideal for a one month delivery system.

Microparticulate formulations were also applied to large and labile macromolecules in the early 1970s. The first example of the application of microspheres to protein delivery is credited to Chang [48], who reported insulin and asparaginase release from microspheres. Kwong et al. [49] have also reported on insulin-loaded PLA microspheres, where the insulin was loaded up to 20% by weight. However, large quantities of insulin crystals on the surface and loosely bound insulin led to considerable burst effect in vivo. This investigation demonstrated that insulin-loaded PLA formulations were effective in lowering blood glucose levels in diabetic rats for about 18 days. For the same purpose, insulin-loaded microspheres have been developed with serum albumin.

Recombinant human interferon has also been incorporated into PLGA by microemulsion of precipitated protein in the polymer solution, followed by rapid solvent removal by spray drying [50]. Vaccines offer an excellent opportunity for controlled delivery because many vaccines require repeated booster regimens after a primary treatment. A programmed vaccine delivery system would be possible by using polymers which degrade at different time intervals. Antigen- or antibody-containing formulations have been incorporated into microparticles for programmed delivery [51]. Although some success has been achieved, none of the above methods offers a general solution applicable to most polypeptides.

In general, the design of biodegradable microparticulate systems for polypeptides is always confronted by problems related to the properties of the specific polypeptide, such as high molecular weight, easy denaturation and chemical modification. There is an approximate log–log correlation between molecular weight (M) and diffusion coefficient (D), that is, log D = $a - b$ log M, such that D decreases as the molecular weight increases [52]. For polypeptide where M is large and the diffusion coefficient becomes extremely low, diffusing polypeptide cannot be accommodated by the free-volume of polymer arising from the segmental mobility. Hydrophobic polyesters, such as PLA, PGA, and PCL are not likely to allow partition-dependent diffusion of hydrophilic polypeptides through the polyester phase to occur. Most polypeptide delivery systems based on degradable polymers resulted in a multiphase release pattern [41]. Kent et al. [53] reported the feasibility of controlled release of LHRH analogues from PLGA (lactide:glycolide = 50:50), representing a typical multiphase release pattern from the microspheres. The initial rapid release over the first few days was attributed to the diffusion of the drug from superficial areas of microspheres. The second phase of lower release rate then occurred and continued until the onset of the third, major phase of release. The third phase is due to the biodegradation of the polymer matrix. Similar complications, such as a long induction period, drug burst, and discontinuity, have been frequently reported with polyester microspheres. All these undesirable release patterns come from the bulk erosion, which is due to the rapid water penetration into the polymer matrix. Although there are complications in formulating protein drugs with biodegradable polymers and subsequent release from them, they are still actively researched for the development of predictable release rates and duration. Custom-tailoring polymers with special properties for safe protein drug formulation, desired release characteristics, and appropriate degradation profile should be continuously researched in the future.

4 Delivery of therapeutics to cells and tissues

Drugs relevant to tissue engineering are polypeptides known as growth factors which induce cell chemotaxis, proliferation, differentiation and migration, and are associated with neovascular formation (angiogenesis). Growth factors have widely been applied in hepatocyte culture, in the formation of cartilage and bone, in nerve regeneration, and in wound healing. Growth factors can be supplied either by adding them in culture media, or by coculturing and cotransplantation of growth factor producing natural or genetically engineered cells along with tissue forming cells, or by controlled local delivery technologies. Recently, several investigators have reported the effect of growth factors on tissue engineering as summarized in Table 1. Although these delivery systems include non-degradable polymeric carriers, such as EVAc copolymers and poly(vinyl alcohol)

Table 1. Recent examples of growth factor delivery in tissue engineering

Growth factor	Carrier	Target cell/ tissue	Remarks	Refs.
EGF, aFGF, bFGF, TGF-α	Ca^{2+}-alginate beads	angiogenesis	subQ, mice	[54]
NGF	hyaluronane derivative microspheres	(symphathetic and certain sensory neurons)	*In vitro*	[55]
NGF	Ca^{2+}-alginate microspheres	Central cholinergic nerve	CNS, rat	[56]
NGF	atelocollagen mini-pellet	Hippocampal neuron Mongolian gerbil	Intrahippocampal,	[57]
NGF	NGF-producing fibroblasts	Adrenal chromaffin cells	Striatum, rat	[58]
NGF	Poly(ethylene-co-vinyl acetate) NGF-producing cell line	Fetal septal cell	Brain, adult rat	[59]
bFGF	Heparin-Sepharose/ Ca^{2+}-alginate beads	Arterial smooth muscle cell	Perivascular system, rat	[60]
aFGF	Heparin/ethylcellulose solution	Wound healing	Topical, diabetic mouse	[61]
FGF	Collagen sponge	Chondrocyte	SubQ implant, mouse	[62]
bFGF	Alzet osmotic pump	Articular cartilage cavity, rabbit	Articular knee	[63]
FGF, EGF	Elwax 40 (poly(ethylene-co-vinyl acetate)	Angiogenesis	Cornea, rabbit	[64]
EGF	PLGA w/o/w emulsion/PLA sponge	Hepatocyte	Mesentery, rat	[65]
BMP	PLA-PEO block (650–200 d), viscous semiliquid	Bone with hematopoitic marrow	Fasciae of the dorsal muscle, mouse	[66]
BMP	PHAs	Bone	Review	[40]

BMP = bone morphogenetic proteins; EGF = epidermal growth factor, FGF = fibroblast growth factor; aFGF = acidic fibroblast growth factor; bFGF = basic fibroblast growth factor; NGF = Nerve growth factor; TGF = transforming growth factor.

sponge, and the growth factor release rate was not fully investigated, the growth of seeded cells in a scaffold in the presence of growth factor delivery system was well contrasted with control experiment without the delivery system. The results indicate the release of appropriate growth factors from biodegradable polymer microspheres with well-defined release rates will be beneficial in various animal cell culture systems. More research on the delivery of growth factors should be performed and detailed parameters including dose size, duration, and release pattern (constant, pulsatile, and time programmed) should be customized for each tissue engineering system.

Restenosis result from reobstruction of peripheral arteries, such as carotid, femoral, and coronary artery, by neointimal proliferation after catheter based balloon angioplasty combined with expandable stent, which is used to treat artery obstruction. One potential approach to prevent restenosis, which has attracted great attention, is to deliver antiproliferatives, antithrombogenic agents, antiplatelet agents, gene or antisense to the injured site [67]. This could be a typical example of in situ manipulation or inhibition of cell growth, contrasting to tissue regeneration or cell therapy.

Several nitric oxide donor molecules were proven to be effective *in vitro* in reducing restenosis risk. By incorporating the $N_2O_2^-$ functional group as a part of polymer pendent groups or in the polymer backbone, NO released for a prolonged period of time (over 800 hours) and effectively inhibited vascular smooth muscle cell proliferation *in vitro* and reduced thrombogenecity *in vivo* when a commercial vascular implant coated with the polymer was grafted in an artery-to-vein (A–V shunt) [68]. Since the over-growth of smooth muscle cells (SMC) after vessel wall injury is partially associated with release of growth factors, particularly bFGF, and upregulation of bFGF receptor expression on the SMC, bFGF conjugated with saporin, a ribosome-inactivating protein, selectively kill proliferating cells that express bFGF receptor [69]. The concept of direct delivery of the conjugate via a local infusion system has proven maximum therapeutic effects without apparent side effect on non-proliferating artery cell in a model animal [70].

References

1. Higuchi, T., Mechanism of sustained-action medication. Theoretical analysis of rate of release of solid drug dispersed in solid matrix. *Journal of Pharmaceutical Science*, 1963, **52**, 1145–1149.

2. Kim, S. W., Peterson, R. V. and Feijen, J., Polymeric drug delivery system. In *Drug Design*, Vol. 10, ed. E. J. Ariens. Academic Press, New York, 1980, pp. 193–250.

3. Sivan, I., Sanchez, F. A., Diaz, S., Holma, P., Coutinho, E., McDonald, O., Robertson, D. N. and Stern., J., Three year experience with Norplant subdermal contraception. *Fertility and Sterility*, 1989, **39**, 799–808.

4. Korsmeyer, R. W. and Peppas, N. A., Macromolecular and modeling aspects of swelling-controlled system. In *Controlled Release Delivery Systems*, eds. T. J. Roseman and S. Z. Mansdorf. Marcel Dekker, New York, 1983, 77–90.

5. Heller, J., Controlled release of biologically active compounds from bioerodible polymers. *Biomaterials*, 1980, **1**, 51–57.

6. Hsieh, D. S. T., Rhine, W. D. and Langer, R., Zero-order controlled-release polymer matrices for micro- and macromolecules. *Journal of Pharmaceutical Science*, 1983, **72**, 17–22.

7. Eckenhoff, B. and Yum, S. I., The osmotic pump: novel research tool for optimizing drug regimens. *Biomaterials*, 1981, **2**, 89–97.

8. Lemmer, B., Circadian rhythm and drug delivery. *Journal of Controlled Release*, 1991, **16**, 63–74.

9. Li, Y. -X. and Goldbeter, A., Frequency specificity in intercellular communication. Influence of pattern of periodic signaling on target cell responsiveness. *Biophysics Journal*, 1989, **55**, 125–145.

10. Kim, S. W. and Bae, Y. H., Smart drug delivery systems. In *Innovations in Drug Delivery. Impact on Pharmacotherapy*, ed. A. P. Sam and J. G. Fokkens. The Anselmus Foundation, Houten, The Netherlands, 1995, pp. 112–121.

11. Heller, J. Chemically self-regulated drug delivery system. *Journal of Controlled Release*, 1988, **8**, 111–125.

12. Bronstead, H. and Kopecek, J., pH-Sensitive hydrogels: Characteristics and potential in drug delivery. In *Polyelectrolyte Gels*, ed. R. S. Harland and R. Prud'homme. ACS Symp. Ser., 1992, pp. 285–304.

13. Bartolini, A., Gliozzi, A. and Richadson, I. W., Electrolytes control flows of water and sucrose permeability through collagen membranes. *Journal of Membrane Biology*, 1973, **13**, 283–298.

14. Hoffman, A. S., Afrassiabi, A. and Dong, L. C., Thermally reversible hydrogels: II. Delivery and selective removal of substances from aqueous solution. *Journal of Controlled Release*, 1986, **4**, 213–222.

15. Okano, T., Bae, Y. H., Jacobs, H. and Kim, S. W., Thermally on-off switching polymers for drug permeation and release. *Journal of Controlled Release*, 1990, **11**, 255–265.

16. Katono, H., Maruyama, A., Sanui, K., Ogata, N., Okano, T. and Sakurai, Y., Thermo-responsive swelling and drug release switching of interpenetrating polymer networks composed of poly(acrylamide-co-butyl methacrylate) and poly(acrylic acid). *Journal of Controlled Release*, 1991, **16**, 215–228.

17. Weiss, A. M., Grodzinsky, A. J. and Yarmush, M. L., Chemically and electrically controlled membranes: Size specific transport of fluorescent solute through PMAA membrane. *AIChE Symposium Series*, 1986, **82**, 85–98.

18. Kwon, I. C., Bae, Y. H. and Kim, S. W., Electrically erodible polymer gel for controlled release of drug. *Nature*, 1991, **254**, 291–293.

19. Ishihara, K., Hamada, N., Kato, S. and Shinohara, I., Photoresponse of the release behavior of an organic compound by an azoaromatic polymer device. *Journal of Polymer Science: Polymer Chemistry Edition*, 1984, **22**, 881–884.

20. Kost. J., Wolfrum, J. and Langer, R., Magnetically enhanced insulin release in diabetic rats. *Journal of Biomedical Materials Research*, 1987, **21**, 1367–1373.

21. Kost. J., Leong, K. and Langer, R., Ultrasound-enhanced polymer degradation and release of incorporated substances. *Proceedings of the National Academy of Sciences USA*, 1989, **86**, 7663–7666.

22. Liu, L. -S., Kost, J., D'Emanuele, A. and Langer, R., Experimental approach to elucidate the mechanism of ultrasound-enhanced polymer erosion and release of incorporated substances. *Macromolecules*, 1992, **25**, 123–128.

23. Kydonieus, A. F., Decker, S. C. and Shah, K. R., Temperature activated controlled release. *Proceedings of the International Symposium on the Controlled Release of Bioactive Materials*, 1991, **18**, 417–418.

24. Mohr, J. M., Schmitt, E. E. and Stewart, R. F., Drug delivery with side chain crystallizable polymer blends. *Proceedings of the International Symposium on the Controlled Release of Bioactive Materials*, 1991, **18**, 409–410.

25. Labhasetwar, V. and Levy, R. J., Polymer systems for cardiovascular drug delivery. *Polymer News*, 1992, **17**, 336–342.

26. Kim, S. W. and Jacobs, H., Design of nonthrombogenic polymer surfaces for blood-contacting medical devices. *Blood Purification*, 1996, **14**, 357–372.

27. Langer, R. and Folkman, J., Polymers for the sustained release of proteins and other macromolecules. *Nature*, 1976, **263**, 797–800.

28. Ratner, B. D. and Hoffman, A. S., Synthetic hydrogels for biomedical applications. In *Hydrogels for Medical and Related Applications*, ed. J. D. Andrade. ACS, Washington, D.C., 1976, pp. 1–36.

29. Wichterle, O. and Lim, D., Hydrophilic gel for biological use. *Nature*, 1960, **185**, 117–118.

30. Peppas, N. A. and Moynihan, H. J., Structure and physical properties of poly(2-hydroxyethyl methacrylate) hydrogels. In *Hydrogels in Medicine and Pharmacy*, Vol. II., ed. N. A. Peppas. CRC Press, Boca Raton, 1987, pp. 49–64.

31. Kou, J. H., Amidon, G. L. and Lee, P. I., pH-Dependent swelling and solute diffusion characteristics of poly(hydroxyethyl methacrylate-co-methacrylic acid) hydrogels. *Pharmaceutical Research*, 1988, **5**, 592–597.

32. Siegel, R. A., Falamarzian, M., Firestone, B. A. and Moxley, B. C., pH-Controlled release from hydrophobic/polyelectrolyte copolymer hydrogels. *Journal of Controlled Release*, 1988, **8**, 179–182.

33. Grimshaw, P. E., Grodzinsky, A. J., Yarmush, M. L. and Yarmush, D. M., Selective augmentation of macromolecular transport in gels by electrodiffusion and electrokinetics. *Chemical Engineering Science*, 1990, **45**, 2917–2929.

34. Kwon, I. C., Bae, Y. H. and Kim, S. W., Heparin release from polymer complex. *Journal of Controlled Release*, 1994, **30**, 155–159.

35. Feil, H., Bae, Y. H., Feijen, J. and Kim, S. W., Molecular separation by thermosensitive hydrogel membranes. *Journal of Membrane Science*, 1991, **64**, 283–294.

36. Freitas, R. F. S. and Cussler, E. L., Temperature sensitive gels as extraction solvents. *Chemical Engineering Science*, 1987, **42**, 97–103.

37. Park, T. G. and Hoffman, A. S., Effect of temperature cycling on the activity and productivity of immobilized β-galactosidase in a thermally reversible hydrogel bead reactor. *Applied Biochemistry and Biotechnology*, 1988, **19**, 1–9.

38. Kim, Y. H., Bae, Y. H. and Kim, S. W., pH/Temperature sensitive polymers for macromolecular drug loading and release, *Journal of Controlled Release*, 1994, **28**, 143–152.

39. Gombotz, W. R. and Pettit, D. K., Biodegradable polymers for protein and peptide drug delivery. *Biocon. Journal Chem.*, 1995, **6**, 332–351.

40. Hollinger, J. and Leong, K., Poly(α-hydroxy acids): carriers for bone morphogenetic proteins. *Biomaterials*, 1996, **17**, 187–194.

41. Lewis, D. H., Controlled release of bioactive agents from lactide/glycolide polymers. In *Biodegradable Polymers as Drug Delivery*, eds. M. Chasin and L. Langer. Marcel Dekker, New York, 1990, pp. 1–41.

42. Heller, J., Controlled release of biologically active compounds from bioerodible polymers. *Biomaterials*, 1980, **1**, 51–57.

43. Morris, E. R., Molecular interactions in polysaccharide gelation. *British Polymer Journal*, 1986, **18**, 14–21.

44. Folkman, J. and Langer, R., Polymers for the sustained release of proteins and other macromolecules. *Nature*, 1976, **263**, 797–800.

45. Sanders, L. M., McRae, G. I., Vitale, K. M., Vickery, B. H. and Kent, J. S., Controlled release of an LHRH analogue from biodegradable injectable microspheres. *Journal of Controlled Release*, 1985, **2**, 187–195.

46. Tice, T. R., Mason, D. W. and Gilley, R. M., Clinical use and future of parenteral microsphere delivery systems. In *Novel Drug Delivery and its Therapeutic Application*, eds. L. F. Prescott and W. S. Nimmo. John Wiley, New York, pp. 223–235.

47. Okada, H., Heya, T., Ogawa, Y. and Shimamoto, T., One month release injectable microcapsules of a leuteinizing hormone-releasing hormone agonist (leuprolide acetate) for treating experimental endometriosis in rats. *Journal of Pharmacology and Experimental Therapeutics*, 1988, **244**, 744–750.

48. Chang, T. M. S., Biodegradable semipermeable microcapsules containing enzymes, hormones, vaccines, and other biologicals, *Journal of Bioengineering*, 1975, **1**, 25.

49. Kwong, A. K., Chou, S., Sun, A. M., Sefton, M. V. and Goosen, M. F. A., In vitro and in vivo release of insulin from poly(lactic acid) microbeads and pellets. *Journal of Controlled Release*, 1986, **4**, 47–62.

50. Eppstein, D. A., van der Pas, M. A., Schryver, B. B., Felgner, P. L., Gloff, C. A. and Soike, K. F., Controlled-release and localized targetting of interferons. In *Delivery Systems for Peptide Drugs*, eds. S. S. Davis, L. Illum and E. Tomlinson. Plenum Press, New York, 1986, p. 277.

51. Singh, M., Singh, A and Talwar, G. P., Controlled delivery of diphtheria toxoid using biodegradable poly(D,L-lactide) microcapsules. *Pharmaceutical Research*, 1991, **8**, 958–961.

52. Hutchison, F. G. and Furr, B. J. A., Biodegradable polymer systems for the sustained release of polypeptides. *Journal of Controlled Release*, 1990, **13**, 279–294.

53. Kent, J. S., Sanders, L. M., Tice, T. R. and Lewis, D. H., Microencapsulation of the peptide nafarelin acetate for controlled release. In *Long-acting Contraceptive Delivery Systems*, eds. G. I. Zatuchni, A. Goldsmith, J. D. Shelton and J. J. Sciarra. Harper and Row, Philadelphia, 1984, pp. 169–179.

54. Downs, E. C., Robertson, N. E., Riss, T. L. and Plunkett, M. L., Calcium alginate beads as a slow release system for delivering angiogenic molecule in vivo and in vitro. *Journal of Cell Biology*, 1992, **152**, 422–429.

55. Ghezzo, E., Benedetti, L., Rochira, M., Biviano, F. and Callegaro, L., Hyaluronate derivative microspheres as NGF delivery devices: Preparation methods and in vitro release characterization. *International Journal of Pharmacology*, 1992, **87**, 21–29.

56. Maysinger, D., Jalsenjak, I. and Cuello, A. C., Microencapsulated nerve growth factor: effects on the forebrain neurons following devascularizing cortical lesions. *NeuroScience Letters*, 1992, **140**, 71–74.

57. Yamamoto, S., Yoshimine, T., Fujita, T., Kuroda, R., Irie, T., Fujioka, K. and Hayakawa, T., Protective effect of NGF atelocollagen mini-pellet on the hippocampal delayed neuronal death in gerbils. *NeuroScience Letters*, 1992, **141**, 161–165.

58. Niijima, K., Chalmers, G. R., Peterson, D. A., Fisher, L. J., Patterson, P. H. and Gage, F. H., Enhanced survival and neuronal differentiation of adrenal chromaffin cells cografted into the striatum with NGF-producing fibroblasts. *Journal of NeuroScience*, 1995, **15**, 1180–1194.

59. Krewson, C. E. and Saltzman, W. M., Nerve growth factor delivery and cell aggregation enhance choline acetyltransferase activity after neural transplantation. *Tissue Engineering*, 1996, **2**, 183–196.

60. Edelman, E. R., Nugent, M. A. and Karnovsky, M. J., Perivascular and intravenous administration of basic fibroblast growth factor: Vascular and solid organ deposition. *Proceedings of the National Academy of Science USA*, 1993, **90**, 1513–1517.

61. Matuszewska, B., Keogan, M., Fisher, D. M., Soper, K. A. Hoe, C. -M., Huber, A. C. and Bondi, J. V., Acidic fibroblast factor: Evaluation of topical formulations in a diabetic mouse wound healing model. *Pharmaceutical Research*, 1994, **11**, 65–71.

62. Fujisato, T., Sajiki, T., Liu, Q. and Ikada, Y., Effect of basic fibroblast growth factor on cartilage regeneration in chondrocyte-seeded collagen sponge scaffold. *Biomaterials*, 1996, **17**, 155–162.

63. Cuevas, P., Burgos, J. and Baird, A., Basic fibroblast growth factor (FGF) promotes cartilage repair in vivo. *Biochemistry and Biophysics Research Communication*, 1988, **156**, 611–618.

64. Gospodarowicz, D., Bialecki, H. and Thakral, K., The angiogenic activity of the fibroblast and epidermal growth factor. *Experimental Eye Research*, 1979, **28**, 501–514.

65. Mooney, D. J., Kaumann, P. M., Sano, K., Schwendeman, S. P., Majahod, K., Schloo, B., Vacanti, J. P., and Langer, R., Localized delivery of epidermal growth factor improves the survival of transplanted hepatocytes. *Biotechnology and Bioengineering*, 1996, **50**, 422–429.

66. Miyamoto, S., Takaoka, K., Okada, T., Yoshikawa, H., Hashimoto, J., Suzuki, S. and Ono, K., Polylactic acid-polyethylene glycol block copolymer. A new biodegradable synthetic carrier for bone morphogenic protein. *Clinical Orthopaedics and Related Research*, 1993, **294**, 333–343.

67. Levy, R. J., Labhasetwar, V., Song, C., Lerner, E., Chen, W., Vyavahare, N.

and Qu, X., Polymeric drug delivery systems for treatment of cardiovascular calcification, arrhythmias and restenosis. *Journal of Controlled Release*, 1995, **36**, 137–147.

68. Smith, D. J., Chakravarthy, D., Pulfer, S., Simmons, M. L., Hrabie, J. A., Citro, M. L., Saavedra, J. F., Davies, K. M., Hutsell, T. C., Moorradian, D. L., Hanson, S. R. and Keefer, L. K., Nitric oxide-releasing polymers containing the [N(O)N(O)]$^-$ group. *Journal of Medicinal Chemistry*, 1996, **39**, 1148–1156.

69. Mattar, S. G., Hanson, S. R., Pierce, F., Chen, C., Hughes, J. D., Cook, J. E., Shen, C., Noe, B. Suwyn, A. C. R., Scott, J. R. and Lumsden, A. B., Local infusion of FGF-saporin reduces intimal hyperplasia. *Journal of Surgical Research*, 1996, **60**, 339–344.

70. Chen, C., Mattar, S. G., Hughes, J. D., Pierce, G. F., Cook, J. E., Ku, D. N., Hanson, S. R. and Lumsden, A. B., Recombinant mitotoxin basic fibroblast growth factor-saporin reduces venous anastomotic intimal hyperplasia in the arteriovenous graft. *Circulation*, 1996, **94**, 1989–1995.

Gene Therapy in Tissue Engineering

JEFFREY R. MORGAN,
MARTIN L. YARMUSH

Center for Engineering in Medicine,
Massachusetts General Hospital,
Shriners Burns Institute and Harvard Medical School,
Boston, MA 02115, USA

Gene therapy, the transfer of genes to achieve a therapeutic effect, has numerous applications in many areas of medicine, including tissue engineering. In this chapter, we review the properties and attributes of the various gene transfer technologies and the gene delivery strategies in which these technologies are being applied. In addition, we review the current research efforts and in some cases, clinical efforts, which are using gene therapy technologies to help achieve the goals of tissue engineering. This merger of gene therapy and tissue engineering is being applied in multiple organ / tissue / cell systems for a growing list of applications. The prominence of gene therapy technologies in tissue engineering will continue to increase as gene transfer technologies are improved and more genes are identified and characterized.

1 Introduction

Recent advances in molecular biology have resulted in the development of new technologies for the introduction and expression of genes in human somatic cells. This emerging field, known as gene therapy, is broadly defined as the transfer of genetic material to cells/tissues/organs in order to achieve a therapeutic effect. Although this technology was originally developed as a potential means for the treatment of inherited diseases, gene therapy technologies are now under consideration for a variety of acquired diseases including cancer and infectious diseases. To date, over 160 gene transfer clinical protocols, using a variety of gene transfer technologies to treat numerous diseases, have been approved, are in progress or completed (Table 1) [1].

Table 1. This table shows the current clinical experience using gene transfer (as of 1997)

Human gene transfer protocols	Number	Percentage
Gene marking studies	28	21
Gene therapy protocols	132	79
Infectious diseases	17	12
Inherited diseases	28	21
Cancer	87	65
Other diseases	3	2
Strategies for gene delivery		
Ex vivo protocols	95	59
In vivo protocols	65	41
Gene transfer technologies		
Permanent genetic modification		
Recombinant retrovirus	101	63
Recombinant adeno associated virus	2	1
Temporary genetic modification		
Recombinant adenovirus	27	17
Cationic liposome	23	15
Plasmid DNA	5	3
Particle mediated	2	1

Another potential area for the application of gene therapy technologies lies in the emerging field of tissue engineering which seeks to create substitutes or replacements for defective tissues or organs. In many, but not all tissue engineering scenarios, cells are an integral part of the organ/tissue substitute. Gene therapy and its technologies to transfer genes into cells can be used to enhance the function of the cells in an organ/tissue substitute. Such genetic modifications could be used to impart new functions to the cells or enhance existing cellular activities. Either way, genetic modification should be thought of as an additional tool which can aid in the performance of an organ/tissue substitute. Listed in Table 1 under the heading, 'Other diseases/disorders', are three protocols which address goals in tissue engineering namely, arterial restenosis, rheumatoid arthritis, and peripheral artery disease.

In the first part of this chapter, we will review the strategies being employed for genetic modification of cells and tissues, the basic technologies which have been developed for the transfer of genes into cells and the important properties of these technologies. In the second part, we will review the most recent research efforts that utilize a gene therapy strategy to achieve the goals of tissue engineering. Lastly, we will summarize and speculate how the importance of gene therapy technologies in tissue engineering may increase as more and more genes are isolated and their functions characterized.

2 Strategies for gene delivery

Before we discuss the numerous technologies which have been developed for the delivery of genes it is important to understand the two principal settings in which gene transfer is performed. Genes can be delivered to cells either *ex vivo* or *in vivo*, and each strategy has distinct advantages and limitations. For *ex vivo* gene therapy, a biopsy of the desired tissue is obtained, its component cells cultured, the cells genetically modified *ex vivo* and transplanted to the patient. For *in vivo* gene therapy, the therapeutic gene is delivered directly to the desired organ/tissue *in vivo*. Illustrated in Fig. 1 are a variety of *ex vivo* and *in vivo* gene transfer approaches to multiple tissue types which have been tested experimentally and, in some cases, tested clinically. As we will discuss later in the chapter, some of the gene transfer technologies are best suited for exclusive use in either *ex vivo* or *in vivo* gene therapy, whereas some gene transfer technologies can be used for both.

2.1 Ex vivo gene delivery

The clinical success of these *ex vivo* and *in vivo* strategies will ultimately depend on their suitability for the treatment of specific diseases as well as the inherent advantages and limitations of each gene transfer technology. Some of the first experimental successes which demonstrated the feasibility of gene therapy, as well as the first clinical trials, utilized an *ex vivo* approach. Two primary considerations must be kept in mind for all *ex vivo* approaches. First, the target cell must be amenable to culturing *ex vivo*, and adequate culture conditions must be available to produce sufficient numbers of cells. Second, after the cells are genetically modified there must be an effective means for the delivery of the cells to the patient. Cell delivery can be achieved in a variety ways including classical transplantation techniques as well as other methods still under development; alternative transplantation methods are discussed in detail in other chapters of this book, and include techniques such as encapsulated implants or extracorporeal devices.

The first experiments to demonstrate the feasibility of *ex vivo* gene therapy used bone marrow stem cells as a target [2]. Cells of the bone marrow were obtained from mice, cultured *ex vivo*, genetically modified and transplanted by simple reinfusion into an irradiated recipient. The first

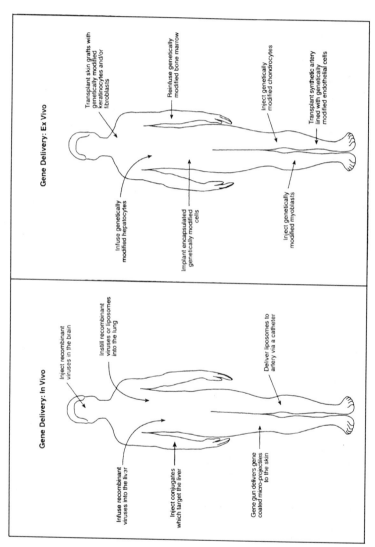

Fig. 1. This figure shows the two prinicpal strategies for gene delivery.

clinical trial was also an *ex vivo* approach whereby lymphocytes were obtained from a patient, cultured *ex vivo*, genetically modified and returned to the patient by reinfusion [3]. Other cells/tissues for which experimental *ex vivo* strategies have been successfully used include keratinocytes and fibroblasts of the skin, endothelial cells, myoblasts, chondrocytes, and hepatocytes [4–9].

Typically the cells used in an *ex vivo* strategy are **autologous** (obtained from the same patient who will receive the cells after genetic modification). The use of autologous cells ensures that problems of immune rejection after transplantation are minimized. While this may be crucial for many applications in gene therapy, especially those applications which seek to attain permanent engraftment of the genetically modified cells, the use of autologous cells is labor intensive as it requires that cells be obtained from each patient, cultured, genetically modified, and returned to the same patient. There is also a significant time lag in preparing genetically modified autologous cells, so this approach is not applicable to acute diseases which require immediate treatment.

In some instances, *ex vivo* gene therapy strategies can utilize **allogeneic** (from unrelated donors) or even **xenogeneic** (from a different species) cells. In these scenarios, the cells are cultured and genetically modified using the same methods. However, because the cells are obtained from either an allogeneic or xenogeneic source, it is possible to prepare large numbers of genetically modified cells, test their effectiveness, cryopreserve the cells, and use this cell bank to treat large numbers of patients on demand. Because of the obvious immunological mismatch, the issues of graft rejection must be addressed. Efforts are underway to develop devices and membranes which isolate and protect the cells from the immune response. As discussed later in this chapter, gene therapy strategies may also be useful for suppressing the local immune response, thus prolonging the survival of allogeneic or xenogeneic cell transplants.

2.2 In vivo gene delivery

The *in vivo* gene therapy approach offers a number of distinct advantages over the *ex vivo* approach. The primary advantage is that since the genes are delivered directly to the tissue/organ *in vivo*, the time and expense of culturing cells is avoided. Moreover, genes can be delivered to cells which are not readily cultivated *ex vivo* or to cells for which there is no effective means of transplantation. In many respects, the *in vivo* gene therapy approach has similarities to classical pharmaceuticals which are also delivered *in vivo*. Although, as we'll discuss in more detail in the next section, there are multiple gene transfer technologies, whose routes of administration are similar to those of pharmaceuticals (i.e. intravenous, injection, aerosol, etc). Thus, the issues relevant to pharmaceuticals such as half-life, biodistribution, and toxicity are also relevant to the gene transfer technologies used *in vivo*. However, because all of the *in vivo* gene transfer technologies are either macromolecules or macromolecular assemblies (e.g. DNA-liposome complexes, virus particles), there are additional obstacles such as transport limitations and immunogenicity which must be overcome. Another significant challenge is the targeting of the genes to a desired set of cells. In organs, which are typically a mixture of cell types and often contain membrane barriers, the selective and efficient delivery of macromolecular gene therapeutics to the desired cells can be problematic.

3 Technologies of gene transfer

Multiple methods, agents and devices have been employed for the purposes of gene delivery. These include physicochemical means for gene delivery such as DNA liposome complexes and DNA coated microprojectiles, as well as virus particle-mediated gene transfer such as recombinant retroviruses and recombinant adenoviruses. Although each gene transfer technology fundamentally seeks to attain the same goal, namely the introduction and expression of a therapeutic gene into a target cell or tissue, the success of the clinical application of these technologies will ultimately depend on the advantages and limitations of each technology and its suitability to its medical application. Although each of these technologies is widely used, most of them are still under active investigation in the hopes of improving their effectiveness.

3.1 Permanent versus temporary genetic modification

Before we discuss each of the gene transfer technologies, its important to point out a few fundamental differences in the manner in which the technologies deliver genes. The most significant difference is whether or not the genetic modification of the cells is **temporary** or **permanent**. In temporary genetic modification, the delivered genes are not integrated into the genome of the target cell, nor do they have the correct genetic elements which would allow them to be replicated as the cell divides. These genes will persist and be expressed in the cell for a finite time which can be quite significant as well as therapeutically relevant. Over time, however, the genes will be lost and gene expression will subside. As we will discuss later in this chapter, for some therapeutic applications, temporary genetic modification may be ideal. For some applications, the medical need is for long term genetic modification, thus, either technologies which mediate permanent genetic modification must be employed, or strategies must be devised for repeated dosing of those gene therapeutics which mediate temporary genetic modification.

In permanent genetic modification the delivered genes are integrated into the genome of the target cell or the genes contain additional genetic elements which facilitate the replication of the genes as the cell divides. These genes will persist and be expressed for the lifetime of the cell. Moreover, as the cell divides, the introduced genes will be replicated and expressed by each subsequent generation of daughter cells. For some therapeutic applications, especially some inherited diseases, permanent genetic modification is the ideal modality since it can potentially provide a long lasting cure. In applications where it is possible to grow large numbers of cells *ex vivo*, permanent genetic modification is preferred because after the initial genetic modification, large numbers of genetically modified cells can be expanded and banked.

There are obvious exceptions and qualifications to these categories of temporary and permanent genetic modification. For example, some of the gene transfer technologies which mediate temporary genetic modification can in fact produce permanent genetic modification of some types of cells. The frequency of these permanent events, however, is usually exceedingly low ($\sim 10^{-5}-10^{-6}$) and insignificant when compared to the numbers of cells which are temporarily modified (100% of the cells for some technologies). These rare events of permanent genetic modification can be extremely valuable for some applications and in these cases the cells can be isolated; selection schemes which use genes that act as dominant selectable markers can be employed to achieve this goal. For example, after gene delivery using DNA liposomes, large numbers of cells will take up the complex and express the genes of interest. The vast majority of the cells will contain and express the genes only transiently. However, by a process of illegitimate recombination a few cells in the population will have an integrated copy of the introduced gene which is stable. To find these rare cells, a dominant selectable marker gene, such the neo gene, is introduced into the cells during transfer of the gene of interest. The neo gene encodes the neomycin phosphotransferase, and confers resistance to the antibiotic, G418. When grown in medium containing G418, only those few rare cells which contain a stable integrated copy of the neo gene will be able to grow, all other cells will die.

There are qualifications to permanent genetic modification as well. Although the technologies which mediate permanent genetic modification can do so at fairly high frequencies ($\sim 10-100\%$), and these genes are inherited by all daughter cells, gene expression from these introduced genes can sometimes be variable, especially *in vivo*. After introduction of the genes *in vivo* or transplantation of *ex vivo* modified cells, expression of the genes can decline with time. This can be due to either down regulation of the genetic elements which control gene expression or loss (apoptosis, immune recognition) of the genetically modified cells. The net result is that the patient's genetic modification was only temporary. Allogeneic cells can be temporarily engrafted and made to secrete growth factors useful for tissue repair; this will be discussed later in the chapter when we discuss applications of gene therapy in the skin.

3.2 Technologies for permanent genetic modification

3.2.1 Recombinant retrovirus

Retroviruses are single-strand RNA-containing viruses with a genome approximately 8 kilobases in size [10, 11]. Murine retroviruses adapted for gene transfer encode only three viral genes: pol, which encodes the enzyme, reverse transcriptase; env, which encodes a glycoprotein which directs the binding of virus particles to specific receptors on the cell surface; and gag, which encodes the major capsid protein [10, 11]. Retroviral-mediated gene transfer begins by insertion of a gene of interest into the DNA of a recombinant retroviral vector. This vector, derived from the Moloney murine leukemia virus, contains no viral genes but retains the genetic elements necessary for the transcription, packaging, reverse transcription, integration and expression of the recombinant vector and its inserted gene. To generate virions capable of transmitting the recombinant vector with its gene of interest, it is introduced into a specialized *packaging cell line* derived from 3T3 fibroblasts. These packaging cell lines have been genetically engineered to provide the retroviral proteins (gag, pol, env) necessary to produce virions containing the recombinant vector which encodes the gene of interest [12]. The packaging cell lines, however, are incapable of producing wild-type replication competent virus. Recombinant retroviral particles are continuously shed into the medium by the packaging cell lines and these stocks of recombinant retrovirus are free of wild-type virus and are unable to replicate. They can, however, efficiently introduce genes into target cells. Useful stocks of recombinant retroviruses contain between 10^6 and 10^7 active retroviral particles/ml.

When recombinant retrovirus is added to the medium of a target cell, the virion is absorbed on the cell surface by a specific receptor, internalized and uncoated. Its packaged RNA genome is then reverse transcribed and the resulting DNA is transported to the nucleus and is integrated into the genome of the target cell. This integrated copy of the recombinant vector, with its gene of interest, is stable and inherited by daughter cells as any other autosomal gene. Moreover, this newly introduced copy of the gene is actively expressed so the cells may begin to synthesize high levels of a gene product they normally might not express.

As with all the gene transfer technologies, recombinant retroviruses have inherent advantages as well as limitations. The principle advantages are that recombinant retroviruses are capable of genetically modifying numerous cell types; including many primary cells; the genes are faithfully transmitted without rearrangements, the transferred genes are stably integrated into the chromosomal DNA and recombinant retroviruses can be used to simultaneously modify large numbers of cells. For many *ex vivo* protocols, recombinant retroviruses are ideal because of these properties. Since the genes are stably integrated, it is possible to produce large numbers of genetically modified cells.

One limitation is that recombinant retroviruses are limited with respect to the size of the gene(s) that can be packaged and transmitted by recombinant virions (7–8 kilobases of inserted sequence). In addition, the viral particle is unstable and decays with a half-life of approximately 6–8 hours at 37°C, and recombinant retroviruses can only transfer genes to cells which are actively dividing. Thus, their use for *in vivo* gene transfer is significantly limited since few tissues are actively proliferating. However, in some cancer applications, this last property has been used to selectively target 'suicide' genes to tumor cells which are actively proliferating [13].

3.2.2 Recombinant lentivirus

Recently, a new class of retroviral vectors has been developed based on human lentiviruses (HIV, human immunodeficiency virus) which can be used to transfer genes to non-dividing cells *in vitro* as well as *in vivo* [14]. Like the recombinant vectors based on murine retroviruses, these recombinant lentivirus vectors are stably integrated into the genome of the target cells. In fact, the same recombinant murine based vectors encoding the genes of interest are introduced into a new lentiviral particle containing HIV proteins.

The packaging and production of recombinant lentivirus particles is accomplished via a transient transfection procedure in which three plasmids are introduced into the same cell [14]. The first plasmid encodes a disabled derivative of the HIV genome which is unable to be packaged and lacks the normal HIV env gene. This HIV derivative supplies all the other proteins necessary for the generation of HIV lentivirus particles. The second plasmid encodes a heterologous envelope protein which can be either the amphotropic env protein used in murine recombinant retroviruses or an envelope protein from the vesicular stomatitis virus (VSV G protein). These envelope proteins are incorporated into the virions and endow the particles with a broad host range making it possible to transfer genes to a wide variety of tissues, including human cells. The third plasmid, which encodes the genes to be transmitted, is based on a murine recombinant retroviral vector and also encodes the necessary cis acting sequences for packaging, reverse transcription and integration of the recombinant genome. These three plasmids are co-transfected into 293-cells (an epithelial cell line) and recombinant lentivirus containing medium is harvested

48–60 hours after transfection. The concentration of recombinant lentivirus particles is somewhat low ($\sim 10^5$/ml). However, these recombinant lentivirus particles are able to mediate stable gene transfer to growth arrested cells *in vitro* (fibroblasts) as well as non-dividing cells *in vivo* (terminally differentiated neurons).

3.2.3 Recombinant adeno associated virus

Adeno associated viruses (AAV) are single-strand DNA-containing viruses with a genome which is 4675 base pairs in size [15]. Each virion contains a single DNA molecule of either polarity. AAV viruses encode only two genes, rep and cap which encode for replication and capsid proteins, respectively. This native AAV genome is replication defective and can not replicate unless the cell is infected with either an adenovirus or a herpes virus. When wild type AAV infects a cell in the absence of one of these helper viruses, the AAV genome becomes integrated into the genome of the target cell. If in the future the cell becomes infected with a helper virus, the integrated wild type AAV genome is able to replicate. It is this property of stable integration which makes AAV attractive as the basis for a gene transfer technology.

To develop a recombinant AAV vector, the rep and cap genes are removed and replaced with foreign genes of interest. These AAV recombinant genomes retain those AAV sequences which are required for packaging of the recombinant DNA into AAV particles. To produce recombinant AAV virus, the AAV vector, containing the genes of interest as well as the necessary packaging sequences, is transfected into cells along with another plasmid which encodes the rep and cap genes, but lacks the packaging sequences. Shortly after transfection, these cells are infected with an adenovirus. Two days later, the cells are lysed to produce a stock of mixed viruses which contain recombinant AAV as well as adenovirus. The helper adenovirus can be removed by inactivation and purification strategies and the result is a stock of recombinant AAV ($\sim 10^4 - 10^5$/ml) which are free of helper virus.

A major advantage of recombinant AAV is that the virus mediates stable integration of the gene of interest into quiescent non-dividing cells. For wild-type AAV, integration occurs at a preferred site on chromosome 19, although incorporation of foreign DNA into the vector may alter this specificity [16]. This property makes AAV vectors ideal for permanent genetic modification during *in vivo* gene transfer into tissues such as the lung which have few dividing cells [17]. The present limitations of recombinant AAV are that the packaging cell systems for producing recombinant AAV are cumbersome resulting in low yields, the vector can accommodate only 4.6–5.0 kilobases of foreign DNA and steps must be taken to eliminate wild-type adenovirus from the final stock.

3.3 Technologies for temporary genetic modification

3.3.1 Recombinant adenovirus

Adenoviruses are DNA-containing viruses with a double-stranded genome size of 35 kilobases [18]. The wild-type adenovirus genome has two groups of transcriptional units depending on whether they are expressed before or after viral DNA replication (early or late). In general, early transcriptional units encode regulatory proteins which activate the transcription of other viral genes, whereas late transcriptional units encode predominantly structural proteins which comprise the icosahedral viral capsid.

The first generation of recombinant adenoviral vectors were based on a mutant of adenovirus type 5 (Ad5) in which the early region 1 (E1 region) was deleted. Replication of E1 minus mutants on normal cells is severely impaired. However, E1 minus mutants can be grown to very high titers on a specialized packaging cell line (293-cells) which provide the missing viral E1 function [19]. Second generation adenovirus vectors with deletions and mutations in other viral genes, as well as new packaging cell lines, have been developed to minimize expression of viral proteins after gene transfer [20].

To generate a recombinant adenovirus, a transcriptional unit encoding the gene of interest is inserted into the deleted E1 region of a mutant virus. This is accomplished by homologous recombination in which the transcriptional unit of the gene of interest is first constructed on a standard bacterial plasmid cloning vector. Moving from left to right, this transcriptional unit contains the following features: the inverted terminal repeat found at the left end of the normal adenovirus genome, the promoter/enhancer of the E1 early region with the packaging sequence, the gene of interest, and flanking adenovirus sequence which promotes homologous recombination. Once the construct is complete, the DNA of this plasmid is transfected into 293 cells which are infected with an E1 minus mutant adenovirus. Homologous recombination occurs between the transfected DNA and the replicating E1 minus mutant adenovirus. The resulting viral stock is diluted and screened for rare homologous recombination events in which the transcriptional unit has correctly recombined with the E1 minus adenovirus. This recombinant virus is plaque purified and grown in quantity on the 293-packaging cell line. Recombinant adenoviruses can be grown to high titers on 293-cells, and the stable viral particles can be purified to produce stocks of virus with titers of 10^{10} to 10^{12} particles/ml.

Shortly after recombinant adenoviruses are mixed with cells, the viral DNA enters the nucleus and early genes are expressed including the gene of interest which is linked to an early gene promoter. Late viral genes (capsid proteins) are not expressed or are expressed at very low levels because the deletion in the vector removes the major early viral proteins (E1) responsible for activation of other viral genes. Other deletions and mutations have been introduced to minimize expression of late viral genes.

A major advantage of recombinant adenoviruses is that gene transfer is not limited to dividing cells and can be used to transfer genes to tissues *in vivo*. For some tissues, gene transfer is very efficient and large numbers of cells will become genetically modified and sustain the expression of the genes for as long as 2–4 weeks [21]. The size of the genes which can be accommodated by recombinant adenoviruses is finite, but significantly larger than either recombinant retroviruses or recombinant AAV. The present limitations of recombinant adenoviruses are that the proteins of the virus particles can sometimes elicit an inflammatory response *in vivo*, and evoke lasting immune reactivity which will neutralize repeated doses of the virus, unless steps are taken to suppress the immune system during administration [22].

3.3.2 Cationic liposomes

Liposomes made with cationic lipids can be used to deliver genes to cells *in vitro* as well as *in vivo* [23–25]. Small unilamellar liposomes approximately 100 nm in diameter are prepared by sonication of a mixture of cationic lipids such as DOTMA, *N*-[1-(2,3-dioleyloxy)propyl]-*N*,*N*,*N*-triethylammonium and neutral lipids. These cationic liposomes are mixed with plasmid DNA to form a non-covalent complex between the positively charged cationic lipids and the negatively charged DNA molecules. Recently, a new class of cationic lipids have been developed which are more effective at mediating gene transfer, especially *in vivo* [26].

Gene delivery to cells *in vitro* is achieved by simply mixing the liposome/DNA complex with cells. This process is also called lipofection, and soon after treatment many cells *in vitro* take up and express the genes. For gene delivery *in vivo*, several methods of administration have been used. For example, the arterial wall can be genetically modified by infusion of cationic liposomes into the empty space created by a double-balloon catheter [27]. Genes have also been delivered to the lung by either aerosolization or direct instillation of cationic liposomes [28].

One advantage of cationic liposomes is that genes can be delivered to dividing, as well as non-dividing cells. Moreover, since there is no need to package the genes into a virus particle, larger genes and more genetic information can be delivered. When administered *in vivo*, cationic liposomes do not evoke a strong immune response as do viral particles, however, the efficiency of gene transfer *in vivo* is well below the level attained by recombinant adenoviruses.

3.3.3 Plasmid DNA

Successful gene transfer and expression has been accomplished by the direct injection of DNA into tissues *in vivo*. Several reports have shown that a simple injection of DNA into skeletal and cardiac muscles of rodents with a syringe can genetically modify these tissues [29, 30]. Similar results have been reported with non-human primates [31].

Muscle cells appear to have the ability to take up and express injected DNA. Successful gene transfer was achieved in 10% to 30% of the cells at the injection site, after a single injection into the quadricep muscle of the mouse, or approximately 1.5% of the cells of the entire quadricep [29]. Implantation of DNA pellets is also an effective means of gene delivery to muscle cells. Injected DNA is not stably integrated, although expression can persist for as long as 2 months [29].

Like cationic liposomes, there are few constraints on the size of the genes which can be delivered via DNA injection. For example, a normal copy of the gene which is defective in Duchenne's muscular dystrophy (dystrophin) is 12 kilobases in size. After intramuscular injection into a mouse strain deficient for dystrophin, approximately 1% of the myofibers expressed the dystrophin protein [32]. However, a major limitation of direct plasmid injection is the efficiency of gene transfer.

3.3.4 DNA coated microprojectiles

The use of DNA coated microprojectiles is a physical means by which genes have been delivered to cells *in vitro* and *in vivo*. The first report of this 'gene gun' used biologically inert tungsten particles approximately 1 mm in size which were coated with a DNA calcium phosphate precipitate [33]. DNA-coated microprojectiles were loaded onto a macroprojectile inside the gene gun. Using the force from a gunpowder discharge, the macroprojectile was accelerated at high velocity through a vacuum chamber until it hit a stopping plate which caused the release of the DNA-coated microprojectiles through a hole in the stopping plate. The resulting spray of microprojectiles was able to penetrate the target tissue and the bombarded cells were shown to express the encoded genes. Numerous modifications have been made to the hardware of this system, including the development of a hand-held device, the use of gold microprojectiles of various sizes, and the acceleration of the macro-projectile using either a burst of helium or a voltage discharge [34, 35]. These improvements have increased gene-transfer efficiencies and have decreased the number of cells killed during bombardment.

A hand-held version of the gene gun has been developed and used to genetically modify tissues *in vivo*. Genes have been introduced and expressed in skin, spleen, liver, muscle, and intestine. Each tissue differed with respect to gene-transfer efficiency. For example, an estimated 20% of bombarded epidermal cells of the skin expressed the gene, whereas only 1–3% of muscle cells expressed the gene [34, 35]. Expression of the genes in all tissues was transient (e.g. skin 14 days, liver 23 days).

Unlike recombinant viruses, the size of the genes delivered by microprojectiles is not constrained by packaging limits. However, efficiency of gene delivery *in vivo* by microprojectiles can be low compared to some viruses (i.e. adenovirus), and there is some tissue injury associated with the gene delivery process. Although it will be difficult to target genes to specific cell types using microprojectiles, it may be possible to restrict gene expression to specific cell types by using genetic control elements which facilitate expression of the gene of interest in only the desired cells. This approach, to restrict gene expression of the transferred genes, is also relevant to other methods which deliver plasmid DNA (e.g. direct injection, cationic liposomes).

3.3.5 Other gene transfer methods

In addition to the technologies discussed, there are other methods for the transfer of genes to cells and tissues. These include recombinant viruses such as vaccinia virus and herpes virus, as well as methods which conjugate plasmid DNA to various ligands (i.e. asialoglycoproteins, transferrin), and 'synthetic' viruses [36–40]. These methods, although capable of mediating gene transfer, have not been discussed in detail because they have not been tested in the clinic, they are still under development and their competitive advantage not yet firmly established, or they are not likely to be useful for tissue engineering purposes (e.g. recombinant vaccinia virus).

3.4 Summary of gene transfer technologies

In summary, gene transfer technologies, which were originally developed to treat inherited diseases, are now being applied to numerous medical applications. Gene transfer is performed in two principal settings, either *ex vivo*, (to genetically modify cultured cells which are subsequently transplanted) or *in vivo*, (to transfer genes to tissue/organs directly). Numerous technologies for the delivery of genes have been developed, each with advantages and disadvantages and improvements to these technologies is an active area of investigation. A primary difference between the different gene transfer technologies is whether or not the genetic modification is permanent or temporary. No single gene transfer strategy or technology is ideal for all medical applications, rather various gene transfer strategies/technologies will be employed depending on their specific attributes and suitability for the treatment of specific diseases.

4 Applications of gene therapy technologies to tissue engineering

As stated earlier, tissue engineering seeks to create substitutes or replacements for defective tissues or organs which are needed in various disease states. Often these diseases include trauma, cancer, and congenital defects where there is an insufficient supply of suitable tissue available to transplant and reconstructive surgeons. However, tissue engineering clearly has applications in other diseases which can benefit from tissue/organ substitutes such as autoimmune diseases (i.e. juvenile diabetes, arthritis), as well as inherited diseases.

Inherited diseases are responsible for a wide spectrum of diseases which afflict nearly all organ systems, with varying degrees of severity. Over 5,000 different inherited diseases have been described, and a growing number of defective genes have been isolated and characterized [41]. All inherited diseases are due to DNA alterations which affect protein structure/function which can adversely influence cell and organ function. Multiple modalities have been developed to treat some, but not all inherited diseases. These include dietary interventions, surgical interventions, metabolic inhibitors, replacement of the defective protein, and organ transplantation. Like the clinical examples cited above, the use of organ transplantation to treat inherited disease is also limited by the availability of suitable tissue. Thus, tissue engineering can also be expected to impact the treatment of some inherited diseases. Conversely, gene therapy technologies, which were originally developed for inherited diseases with no treatment, are now being applied to other diseases.

In this section, we will review applications of gene therapy technologies to the goals of tissue engineering. Gene transfer technologies can be useful in a number of potential tissue engineering applications in multiple organ systems. Genetic modification can help to enhance or augment the performance of organs, tissues, cells, or cell/biomaterial composites (Table 2). In these examples, genetic modification is helping to achieve the goals of tissue engineering by providing cells with new or enhanced function that serves to overcome one or more of the limitations of the performance of an organ/tissue substitute.

There is little limitation to the types of genes which can be introduced into cells, thus it is possible to genetically engineer a cell to overexpress a protein it normally expresses at only very low levels, and it is also possible to genetically engineer a cell to express a protein which it normally does not express.

The list of isolated and characterized human genes is steadily growing and will continue to grow until the completion of the Human Genome Project. These genes encode proteins involved in all aspects of cell/tissue/organ function. A small fraction of these examples include: diffusible growth factors which control cell proliferation, proteins of the extracellular matrix and cell surface which control cell shape and function, and enzymes which control key metabolic pathways. In addition to this collection of normal genes, there is a growing list of altered or mutated genes which encode proteins with improved function such as: longer half-lives, in-

Table 2. This table shows potential applications of gene therapy in tissue engineering

Genetically enhanced function	Representative genes	Applications
Synthesis and delivery of proteins	Growth hormone, clotting factors, insulin	Systemic delivery of therapeutic proteins
Promote angiogenesis	PDGF, VEGF, aFGF	Enhance the take of grafts and cellular implants, and bio-compatibility of materials
Promote matrix synthesis	PDGF, TGF-β, collagenase and protease inhibitors, extracellular matrix molecules	Dermal repair, plastic surgery, non-healing ulcers, bone and cartilage repair
Inhibit immune reactions	IL-1 RA, IL-10, FAS ligand, TGF-β	Promote allograft survival, promote bio-compatability of cell transplant devices
Promote cell growth	TGF-α, EGF, IGF-1, KGF, NGF-β	Re-epithelialization of wounds, liver regeneration, nerve regeneration
Inhibit matrix synthesis	Collagenases, metalloproteinases	Inhibit scar formation, surgical adhesions, and fibrosis
Alter cell surface	TPA, UPA, cyclooxygenase	Decrease thrombogenicity of vascular grafts and implantable devices
Enhanced metabolic functions	Cytochrome P450, adenosine deaminase	Detoxification, prodrug metabolism, metabolic engineering

creased enzymatic activities, and improved biodistribution. This list of mutated genes is the byproduct of a large research effort in molecular biology that uses site-directed mutagenesis techniques to probe protein structure and function. In some applications of tissue engineering, these altered proteins may outperform the normal protein.

4.1 Gene therapy in tissue engineering of the skin

The skin is particularly well suited for applications in both tissue engineering and gene therapy. Skin biopsies are easy to obtain and methods exist for the *ex vivo* cultivation of large numbers of keratinocytes of the epidermis, as well as fibroblasts of the dermis. Estimates are that the keratinocytes of a 1 cm^2 biopsy can be expanded to 1 m^2 of cultured epithelium within 30 days (average cell doubling time of 18–20 hours) [42]. As discussed in other chapters, these cultured skin cells have been used to generate various skin equivalents which have been used clinically to repair defects of the skin including severe burns and ulcers [44–47].

Cultured human keratinocytes have also been genetically modified *ex vivo* and tested in animals for multiple applications. In one of the first studies, human keratinocytes were genetically modified with a recombinant retrovirus encoding human growth hormone (hGH), a protein normally secreted by the pituitary gland [4]. Modified keratinocytes secreted biologically active growth hormone, and when transplanted to athymic mice, the modified cells formed an epidermal structure containing hGH. This study demonstrated that therapeutically relevant genes could be delivered to the skin and illustrated the concept that genetic modification could be used to transfer an important function of one organ (pituitary) to a different organ (skin). In this way, gene therapy technologies can provide a tissue/organ substitute simply by transferring the key gene(s) of the diseased organ to a different organ/tissue which is healthy and amenable to manipulation. As we will discuss later in this chapter, this paradigm of using gene therapy to create an ectopic site of gene function is relevant to other tissues/organs. Subsequent studies have shown that genetically modified human keratinocytes secreting apoE [48] and clotting factor IX [49] can deliver these proteins to the systemic circulation of grafted animals.

Similar types of studies have been performed with fibroblasts of the dermis. Transplants of genetically modified fibroblasts have been shown to deliver transferrin, factor IX, factor VIII, granulocyte colony-stimulating factor, β-glucuronidase, α_1-antitrypsin and erythropoeitin to the systemic circulation [50–56]. In each of these examples, the genetically modified skin cells are acting as an organ/tissue substitute because they are providing a function (e.g. secreted protein) which is normally performed by a different organ/tissue. This type of organ substitution, through genetic modification, is not necessarily restricted to secreted proteins, and could potentially be extended to other metabolic functions.

Patients suffering from inherited diseases of the skin, particularly the epidermis (i.e. some forms of epidermolysis bullosa and lamellar ichthyosis), are often in desperate need of a skin replacement or substitute which can function long term on sizable areas of the body surface. Unfortunately, none of the available skin substitutes or synthetic dressings are capable of adequately addressing this problem. Thus, gene therapy methods are being tested as a potential means of correcting the genetic defect, thereby re-establishing normal skin function. Lamellar ichthyosis, a disfiguring skin disease characterized by abnormal epidermal differentiation and defective cutaneous barrier function, is due to mutations in the transglutaminase 1 gene [57]. Recently, it was shown that when cultured keratinocytes from patients with lamellar ichthyosis were genetically modified with a recombinant retrovirus encoding normal transglutaminase 1 and transplanted to athymic mice, the modified cells formed an epidermis in which barrier function and epidermal differentiation were both normalized [58]. Thus, for patients with inherited diseases of the skin which have very few therapeutic options, gene therapy technologies might some day be able to provide a suitable alternative by correcting the underlying genetic defective.

A primary goal of tissue engineering of the skin is to correct acquired defects in the skin such as burns and ulcers. Recently it was demonstrated how gene therapy technologies might be useful for this application. In these studies, human keratinocytes were genetically modified using recombinant retroviruses expressing various wound healing growth factors. Modified keratinocytes were shown to synthesize and secrete high levels of platelet derived growth factor-A chain (PDGF-A) or insulin-like growth factor (IGF-1) *in vitro* [59, 60]. PDGF, which is normally synthesized at low levels by keratinocytes, is mitogenic and chemotactic for fibroblasts and mononuclear cells, and topical application of recombinant PDGF protein accelerates granulation tissue formation. IGF-1, which is not normally synthesized by keratinocytes, is a mediator of keratinocyte proliferation and migration and stimulates the proliferation of fibroblasts and promotes angiogenesis. When modified keratinocytes were transplanted as an epithelial sheet to athymic mice, PDGF-A expressing cells formed a differentiated epidermal structure, comparable to unmodified cells. The newly synthesized connective tissue layer subjacent to the PDGF-A modified grafts was significantly thicker and showed an increase in cellularity, vascularity and fibronectin deposition, when compared to control grafts of unmodified cells or grafts expressing IGF-1 [59, 60]. IGF-1 expressing keratinocytes also formed a stratified epidermis but showed an increase in the proliferation of modified cells, demonstrating that genetic modification can be used to modify the autocrine control of epidermal cell proliferation. These data demonstrate the feasibility of genetically modifying the cells of a skin substitute to secrete high levels of wound healing growth factors and the ability of this genetically modified skin substitute to affect the tissue formation process *in vivo*.

Such skin substitutes, which use genetically modified **allogeneic** cells, could be applied to a wound to act as a temporary dressing that sustains the local synthesis and secretion of a wound healing growth factor. Clinical experience with normal allogeneic keratinocytes has indicated that the cells are not permanently engrafted, but are gradually lost over a period of 2 weeks without overt rejection [61, 62]. Thus, although the genetic modification of the cells is permanent, the graft and genetic treatment is temporary. Permanent genetic modification with a recombinant retrovirus does, however, make it possible to produce large numbers of modified keratinocytes *ex vivo*.

Local growth factor delivery by these temporary grafts of allogeneic cells might be expected to enhance the efficacy of a growth factor by providing sustained delivery, avoiding high toxic doses, and maximizing the bioavailability of the growth factor by cell-based delivery. In addition, when used for either local or systemic delivery, such a cell-based method for the combined production and delivery of a therapeutic protein obviates the need for the large scale production and purification of a recombinant protein. Moreover, since these are mammalian cells, they are competent to perform the majority of post-translational modifications required for producing a bioactive protein.

In vivo gene delivery strategies have also been applied to the skin. Recently, several groups have investigated the use of DNA coated microprojectiles to affect cutaneous wound healing *in vivo*. One group demonstrated that the *in vivo* transfection of porcine partial-thickness wounds with a vector expressing epidermal growth factor (EGF) resulted in a modest increase in the rate of reepithelialization and shortened the time of wound closure [63]. Another group demonstrated an increase in tensile strength of healing rat tissue following particle mediated gene transfer of transforming growth factor beta-1 (TGF-β_1) [64]. These studies demonstrate that microprojectile-mediated transfection can be a simple and effective means of enhancing the rate of wound repair.

Other *in vivo* gene transfer methods have been used to deliver genes to the skin. Topical application to mouse skin of cationic liposomes complexed with a reporter gene (lacZ), resulted in gene expression in the cells of the hair follicle [65]. Injection of plasmid DNA with a standard syringe or a tattooing device resulted in transient expression of reporter genes (lacZ, luciferase) in cells of the epidermis and the dermis in animal models (mouse and pig), as well as human skin grafted to athymic mice [66–68].

4.2 Gene therapy in tissue engineering of the cardiovascular system

Numerous applications of gene therapy have been tested in the cardiovascular system. As discussed in other chapters, there is an important medical need for a small diameter vascular graft which remains patent. Although numerous synthetic materials have been tested, none have been found to be adequate. One approach to improving the patency of these synthetic grafts is to line the lumen of the grafts with endothelial cells in hopes of reducing the thrombogenicity of the graft surface, and inhibiting the neointimal hyperplasia which occurs at the anastomosis. While this approach is somewhat encouraging, others have suggested that the anti-thrombogenicity of the endothelial cells could be enhanced by genetic modification through an *ex vivo* gene therapy strategy. The first studies demonstrated the feasibility of transplanting genetically modified endothelial cells as part of a synthetic graft. Endothelial cells, genetically modified with a retroviral vector encoding the lacZ reporter gene, were seeded onto Dacron grafts and transplanted as carotid interposition grafts [69]. Modified endothelial cells persisted on the graft and expressed the lacZ gene for 5–6 weeks after transplantation.

Others have used recombinant retroviruses to modify endothelial cells to over express tissue plasminogen activator (TPA) or glycosylphosphatidylinositol-anchored urokinase-type plasminogen activator (a-UPA), both anti-thrombotic enzymes [70, 71]. Modified cells were seeded at subconfluent densities onto the lumenal surface of Gore-Tex (4 mm internal diameter) which had been coated with collagen. These thrombogenic grafts, containing either modified cells or control unmodified cells, were interposed in exteriorized arteriovenous shunts in baboons and observed for 60 minutes. The accumulation of platelets and fibrin on the collagen surface of the grafts was significantly reduced in grafts containing modified endothelial cells expressing either TPA or a-UPA [71]. This type of *ex vivo* approach to enhancing the anti-thrombogenicity of the endothelium may be useful for fabricating small-diameter grafts resistant to graft failure, or for improving the performance of intravascular stents used for the treatment of iliac and coronary stenosis [71, 72].

Alternatively, *in vivo* gene therapy strategies have been used to intervene directly in the vascular wall. The *in vivo* approach may be useful for addressing the events which lead to restenosis after balloon angioplasty. The initial study demonstrated that the lacZ reporter gene (recombinant retrovirus or cationic liposomes) could be delivered to endothelial cells and vascular smooth muscle cells of the porcine iliofemoral artery via a double balloon catheter [27]. Others have shown that recombinant adenoviruses are also effective in this system for gene delivery [73–79]. In one study, a recombinant adenovirus encoding the herpesvirus thymidine kinase gene (*tk*) was delivered to porcine arteries after balloon angioplasty induced injury [73]. Cells expressing the herpesvirus *tk* are sensitive to the drug, ganciclovir, and when the transfected animals were treated with ganciclovir, the intimal hyperplasia of the smooth muscle cells of the injured artery was delayed.

Others have used the *in vivo* approach to alter the thrombogenicity of the vessel wall. A recombinant adenovirus, encoding cyclooxygenase-1, which catalyzes the rate-limiting step in prostacyclin synthesis, has been delivered directly to the vessel wall (carotid artery) after an angioplasty induced injury [76]. High doses of the recombinant adenovirus inhibited thrombus formation. Likewise, delivery to injured rat carotid arteries of a recombinant adenovirus encoding hirudin, a thrombin inhibitor, resulted in a 35% reduction in neointima formation in the injured artery [79]. However, one study has raised questions regarding the inflammation caused by some of the first generation recombinant adenovirus vectors [75].

In vivo gene therapy is also being used to stimulate angiogeneisis in hopes of treating peripheral artery disease. Plasmid DNA encoding vascular endothelial growth factor (VEGF) was applied to the hydrogel polymer coating of an angioplasty balloon and used to deliver the DNA to ischemic arteries *in vivo*. VEGF gene delivery enhanced the development of collateral vessels and improved blood flow [80]. This method has since been advanced to clinical trials [81, 82]. Recombinant adenoviruses encoding either basic fibroblast growth factor (bFGF) or VEGF have also been shown to stimulate angiogenesis in animal models [83, 84].

An *ex vivo* gene therapy treatment of whole organs may also be useful for modulating the local immune response to help prolong cardiac allograft survival. Several cytokines are known to have immunosuppressive properties such as TGF-β, IL-10 and viral IL-10 (vIL-10), a homologue of IL-10 encoded by the Epstein Barr virus. When plasmids encoding vIL-10 or TGF-β were injected into murine cardiac allografts, survival of treated versus control allografts was prolonged (36 or 26 days, respectively vs 13 days for grafts injected with a control plasmid) [85]. In another study, cardiac allografts were injected with recombinant retrovirus encoding vIL-10 and the survival of these allografts was also prolonged (39 days vs 12 days for controls injected with a retrovirus encoding lacZ) [86].

The use of cells of the cardiovascular system as a site of ectopic expression of selected gene functions has also been demonstrated. Endothelial cells modified *ex vivo* with a recombinant retrovirus expressing the lacZ gene were injected into the muscle microvasculature and demonstrated to adhere and spontaneously become incorporated into the capillary wall [87]. Such cells could be modified to express and secrete into the bloodstream therapeutically relevant proteins. Likewise, vascular smooth muscle cells which had been modified *ex vivo* with a recombinant retrovirus expressing the adenosine deaminase gene were seeded onto denuded arteries and became incorporated into the artery [88]. Adenosine deaminase activity is missing in some inherited forms of severe combined immunodeficiency. The modified smooth muscle cells expressed significant levels of adenosine deaminase for as long as six months.

4.3 Gene therapy in tissue engineering of the musculoskeletal system

Loss or degeneration of cartilage is the consequence of various disease states, including trauma, osteoarthritis, inflammatory diseases such as rheumatoid arthritis, and invading tumor cells. As discussed in other chapters, cartilage regeneration is slow and limited, and new methods are needed to repair these defects. One approach is the transplantation of cultured chondrocytes. Recently a series of human studies have evaluated this procedure [89]. As discussed in other chapters of this book, porous polymer matrices such as collagen matrices and synthetic polymers (poly-L-lactide and polyglycolide) have also been evaluated as scaffolds for transplantation of cultured chondrocytes.

Genetic modification of chondrocytes as well as synovial cells of the joint has been tested as a means to enhance the performance and thera-peutic benefit of these cells. In an *ex vivo* approach, human chondrocytes were genetically modified with a recombinant adenovirus encoding inter-leukin-1 receptor antagonist (IL-1RA) [8]. IL-1, has direct and indirect catabolic effects on articular cartilage, and is thought to be one of the mediators of the degradation of cartilage extracellular matrix associated with osteoarthritis. Accordingly, IL-1RA, which can block the activities of IL-1, has been shown to have anti-arthritic potential. Modified chondro-cytes expressing IL-1RA were seeded onto the surface of osteoarthritic cartilage in organ culture, and were able to protect the osteoarthritic cartilage from IL-1 induced matrix degradation [8].

In another study, the synovial cells, which line the joint, were genetically modified *ex vivo* with a recombinant retrovirus encoding IL-1RA [90, 91]. After transplantation by intra-articular injection, the modified cells recolo-nized the synovium and synthesized IL-1RA. When these joints were challenged with an IL-1 injection, the joints with modified cells were resistant to IL-1 induced pathologies. Currently a human gene therapy protocol to assess the safety and efficacy of transferring a potentially anti-arthritic cytokine gene to human joints with rheumatoid arthritis is in progress [92].

Ligaments, specialized connective tissues linking bone to bone, consist of parallel bundles of collagen fibers containing dermatan-sulfate, in-terspersed with spindle-shaped cells. A variety of injuries can affect liga-ments, including laceration, rupture, and cumulative stress. If primary healing of the ligament does not occur, it may be necessary to replace it with an autologous graft. Synthetic substitutes of Dacron, PTEE (Gore-tex), or carbon fiber of woven nylon have been tested, but do not perform as well as grafts. Clearly, if the injury is not great, healing of the injured ligament is preferred, but this can be problematic due to the hypocellular-ity of the ligament. Methods to stimulate ligament repair may help provide for more rapid and stronger ligament healing.

Using an *in vivo* gene therapy approach it was demonstrated that genes could be delivered to the cells of the healing ligament. A liposome/virus complex containing the lacZ gene was injected directly into a rat patellar ligament 3 days after injury [93]. Fibroblasts and macrophages/monocytes expressed the lacZ gene up to 56 days after gene transfer. The maximum numbers of cells expressing the lacZ gene occurred at 7 days and comprised approximately 7% of the cells of the wound site. Although the efficiency of gene transfer was modest, this study demonstrates the feasibility of genetically modifying the cells of a healing ligament and might be useful for introducing genes which stimulate ligament repair.

Muscle myoblasts have been used in an *ex vivo* gene therapy strategy for the treatment of inherited diseases of the muscle and for the ectopic expression of therapeutic genes [7]. For ectopic expression, myoblasts were genetically modified *ex vivo* to express hGH, erythropoietin, factor IX, or factor VIII [7, 94–98]. The modified myoblasts were transplanted by several methods including injection and incorporation into existing muscle fibers, implantation of microencapsulated cells, and implantation of cells into the liver or spleen. In each case, systemic levels of the therapeutic proteins were detected, which in some cases were high enough to correct some animal models of disease (hGH, erythropoietin).

4.4 Gene therapy in tissue engineering for diabetes

Treatment of diabetes is an important area of investigation for tissue engineering, and gene therapy approaches have been used as part of three general strategies. In the first strategy, gene transfer technologies are being used in a stepwise fashion to create a genetically modified cell line which synthesizes, processes, and releases insulin in a regulated manner [99, 100]. Such a cell line would be encapsulated to provide immune protection prior to implantation.

In the second strategy, gene therapy technologies are being used to engineer ectopic sites of insulin production. A recombinant retrovirus encoding rat insulin was used to genetically modify the cells of the liver of severely diabetic rats [101]. The low level production of insulin prevented ketoacidosis and the animals exhibited normoglycemia during a 24 hour fast.

In the third strategy, gene transfer technologies are being used to suppress the local immune response and help prolong the survival of pancreatic allografts. For this approach, a composite 'organ' of cultured myoblasts mixed with isolated pancreatic islets was created. The syngeneic myoblasts were genetically modified to express the ligand for the Fas receptor (FasL) which induces apoptosis of Fas + T-cells. These modified myoblasts were mixed with allogeneic pancreatic islets to form the composite graft [102]. Composite grafts with myoblasts expressing FasL survived and maintained normoglycemia after transplantation (> 84 days), whereas composite grafts with unmodified myoblasts did not (< 10 days). Survival of the allogeneic pancreatic islets was mediated by the ability of the syngeneic myoblasts expressing FasL to deliver a local death signal that killed allo-reactive Fas + T-cells infiltrating the graft.

4.5 Gene therapy in tissue engineering of the nervous system

One area of application in which gene therapy can aid tissue engineering in the nervous system is in helping to overcome some of the limitations of delivering substances across the blood–brain barrier [103]. Solutions to this problem may help address some neurodegenerative diseases. In one study, fibroblasts were genetically modified to express nerve growth factor (NGF) and these modified cells were implanted into the brains of rats which had received surgically created lesions of the fimbria-fornix [104]. Local NGF production and delivery by the modified cells helped prevent the degeneration of cholinergic neurons. The presence and accurate placement of the NGF secreting fibroblasts has also been shown to prevent neuronal degeneration in monkeys and this type of approach may be useful for treating diseases such as Alzheimer's disease [105]. In another study, NGF expressing fibroblasts, implanted into the nucleus basalis magnocellularis of rats with age impaired memory, persisted for 6 weeks and reversed some memory impairment [106]. Using a similar approach, other neurotrophic growth factors such as, brain-derived neurotrophic factor (BDNF) and neurotropin 3, have been delivered to the adult brain [107, 108]. Likewise, xenogeneic cells expressing ciliary neurotrophic factor (CNTF) have been encapsulated and implanted into the lumbar intrathecal space and CNTF could be measured in the cerebrospinal fluid of these patients for up to 17 weeks [109].

Fibroblasts have also been modified to express tyrosine hydroxylase which converts tyrosine to L-DOPA. L-DOPA is subsequently converted by endogenous decarboxylase activity into the neurotransmitter dopamine. Intracerebral implantation of these modified cells into a rat model of Parkinson's disease resulted in significant improvement [110]. Fibroblasts have also been modified to express choline acetyltransferase. Modified cells produced and released acetylcholine at levels which varied with the level of choline in the media as well as choline infused in the brain [111].

The feasibility of *in vivo* delivery of genes directly to the nervous system has also been demonstrated. Recombinant adenoviruses encoding either the lacZ reporter gene or the α_1-antitrypsin gene have been delivered to the lateral ventricle of rats [112]. Modified lacZ positive cells were found among the ependymal cells lining the ventricles, and α_1-antitrypsin was detected in the cerebrospinal fluid for about 1 week. The lacZ adenovirus was also successfully delivered to the globus pallidus and substantia nigra after stereotactic injection [112]. The potential for stable permanent genetic modification of the terminally differentiated neurons has also been demonstrated using a lentivirus vector based on the human immunodeficiency virus [14].

5 Summary

Central to the success of gene therapy are methods to introduce genes into cells. Numerous technologies, based on a variety of methods (i.e. viral-mediated, physical/ chemical) have been developed to achieve gene transfer. A fundamental distinction between these gene transfer technologies, which often has an important impact on their applicability, is whether or not the genetic modification is permanent or temporary. These gene transfer technologies are used in two primary settings, either for *ex vivo* or *in vivo* genetic modification. Both strategies, as well as each of the gene transfer technologies, have inherent advantages and disadvantages which must be evaluated for specific medical applications.

Gene therapy strategies and gene transfer technologies are currently being applied to the field of tissue engineering, and genetic modification should be thought of as an additional technology which can aid in the performance of organ/tissue substitutes. Gene transfer technologies have been applied to tissue engineering of multiple organ/tissue systems including the skin, the cardiovascular system, the endocrine system, the musculoskeletal system, and the central nervous system. Through genetic modification the cells/tissues of these organ systems or substitutes can be modified to express genes they normally do not express, or the normal level of expression of endogenous genes can be augmented or in some cases downregulated. In either case, the goal of genetic modification is to enhance the performance of these cell/tissue/organ substitutes so they are better prepared to address their medical need. The majority of the effort is currently evaluating the feasibility of this approach in animal models. A few cases have advanced to clinical trials. As more genes are identified and characterized, and the *in vivo* role of their gene products elaborated, the importance of gene therapy strategies in tissue engineering is expected to increase.

References

1. Office of Recombinant DNA Activity ORDA, NIH, 31/4B11, Bethesda, MD.
2. Williams, D. A., Lemischka, I. R., Nathan, D. G. and Mulligan, R. C., Introduction of new genetic material into pluripotent haematopoietic stem cells of the mouse. *Nature*, 1992, **310**, 476–480.
3. Rosenberg, S. A., Immunization of cancer patients using autologous cancer cells modified by insertion of the gene for tumor necrosis factor. *Human Gene Therapy*, 1992, **3**, 57–73.

4. Morgan, J. R., Barrandon, Y., Green, H. and Mulligan, R. C., Transfer and expression of foreign genes in transplantable human epidermal cells. *Science*, 1987, **237**, 1476–1479.

5. Louis, D. S. T. and Verma, I. M., An alternative approach to somatic cell gene therapy. *Proceedings of the National Academy of Science*, 1988, **85**, 3150–3154.

6. Wilson, J. M., Birinyi, L. K., Salomon, R. N., Libby, P., Callow, A. D. and Mulligan, R. C., Implantation of vascular grafts lined with genetically modified endothelial cells. *Science*, 1989, **244**, 1344–1346.

7. Dhawan, J., Pan, L. C., Pavlath, G. K., Travis, M. A., Lanctot, A. M. and Blau, H. M., Systemic delivery of human growth hormone by injection of genetically engineered myoblasts. *Science*, 1991, **254**, 1509–1512.

8. Baragi, V. M., Renkiewicz, R. R., Jordan, H., Bonadio, J., Hartman, J. W. and Roessler, B. J., Transplantation of transduced chondrocytes protects articular cartilage from interleukin 1 induced extracellular matrix degradation. *Journal of Clinical Investigation*, 1995, **96**, 2454–2460.

9. Wilson, J. M., Jefferson, D. M., Roy Chowdhury, J., Novikoff, P. M., Johnston, D. E. and Mulligan, R. C. *Proceedings of the National Academy of Sciences of the United States of America*, 1988, **85**, 3014–3018.

10. Weiss, R. N., Teich and Varmus, H., Molecular biology of tumor viruses. In *RNA Tumor Viruses*, 2nd ed., Cold Spring Harbor Laboratory, Cold Spring Harbor, New York, 1982.

11. Morgan, J. R., Tompkins, R. G. and Yarmush, M. L., Advances in recombinant retroviruses for gene delivery, *Advanced drug delivery reviews*, 1993, **12**, 143–158.

12. Danos, O. and Mulligan, R. C., Safe and efficient generation of recombinant retroviruses with amphotropic and ecotropic host ranges. *Proceedings of the National Academy of Sciences of the United States of America*, 1988, **85**, 6460.

13. Culver, K. W., Ram, Z., Wallbridge, S., Ishii, H., Oldfield, E. H. and Baese, R. M. In vivo gene transfer with retrovidral vector-producer cells for treatment of experimental brain tumors. *Science*, 1992, **256**, 1550–1552.

14. Naldini, L., Blömer, U., Gallay. P., Ory, D., Mulligan, R., Gage, F. H., Verma, I. M. and Trono, D., In vivo delivery and stable transduction of nondividing cells by a lentiviral vector. *Science*, 1996, **272**, 263–267.

15. Carter, B. J., Adeno-associated virus vectors. *Current Opinion in Biotechnology*, 1992, **3**, 533–539.

16. Dong, J-Y., Fan, P-D. and Frizzell, R. A., Quantitative analysis of the packaging capacity of recombinant adeno-associated virus. *Human Gene Therapy*, 1996, **7**, 2101–2112.

17. Flotte, T. R., Afione, S. A., Conrad, C., McGrath, S. A., Solow, R., Oka, H., Zeitlin, P. L., Guggino, W. B. and Carter, B. J., Stabe in vivo expression of the cystic fibrosis transmembrane conductance regulator with an adeno-associated virus vector. *Proceedings of the National Academy of Sciences of the United States of America*, 1993, **90**, 10613–10617.

18. DNA Tumor Viruses, J. Tooze, ed., Cold Spring Harbor Laboratory, Cold Spring Harbor, New York, 1981.

19. Wilson, J. M., Adenovirus as gene-delivery vehicles. *New England Journal of Medicine*, 1996, 334, 1185–1187.

20. Wang, Q. and Finer, M. H., Second-generation adenovirus vectors. *Nature Medicine*, 1996, **2**, 714–716.

21. Wilson, J. M., Adenovirus as gene-delivery vehicles. *New England Journal of Medicine*, 1996, **334**, 1185–1187.

22. Yang, Y., Trinchieri, G. and Wilson, J. M., Recombinant IL-12 prevents formation of blocking IgA antibodies to recombinant adenovirus and allows repeated gene therapy to mouse lung. *Nature Medicine*, 1995, **1**(9), 890–893.

23. Felgner, P. L., Gadek, T. R., Holm, M., Roman, R., Chan, H. W., Wenz, M., Northrop, J. P., Ringold, G. M. and Danielsen, M., Lipofection: a highly efficient, lipid-mediated DNA-transfection procedure. *Proceedings of the National Academy of Sciences of the United States of America*, 1987, **84**, 7413–7417.

24. Felgner, P. L. and Ringold, G. M., Cationic liposome-mediated transfection, *Nature*, 1989, **337**, 387–388.

25. Felgner, P. L. and Rhodes, G., Gene therapeutics. *Nature*, 1991, **349**, 351–352.

26. Wheeler, C. J., Felgner P. L., Tsai, Y. J., Marshall, J., Sukhu, L., Doh, G. S., Hartikka, J., Nietupski, J., Manthorpe, M., Nichols, M., Plewe, M., Liang, X., Norman, J., Smith, A. and Cheng, S. H., A novel cationic lipid greatly enhances plasmid DNA delivery and expression in mouse lung. *Proceedings of the National Academy of Sciences of the United States of America*, 1996, **93**, 11454–11459.

27. Nabel, E. G., Plautz, G. and Nabel, G. J., Site-specific gene expression in vivo by direct gene transfer into the arterial wall. *Science*, 1990, **249**, 1285–1288.

28. Stribling, R., Brunette, E., Liggitt, D., Gaensler, K. and Debs, R., Aerosol gene delivery in vivo. *Proceedings of the National Academy of Sciences of the United States of America*, 1992, **89**, 11277–11281.

29. Wolff, J. A., Malone, R. W., Williams, P., Chong, W., Acsadi, G., Jani, A., and Felgner, P. L., Direct gene transfer into mouse muscle in vivo, 1990, *Science*, **247**, 1465.

30. Lin, H., Parmacek, M. S., Morle, G., Bolling, S. and Leiden, J. M., Expression of recombinant genes in myocardium in vivo after direct injection of DNA. *Circulation*, 1990, **82**, 2217–2221.

31. Jiao, S., Williams, P., Berg, R. K., Hodgeman, B. A., Liu, L., Repetto, G. and Wolff, J. A., Direct gene transfer into nonhuman primate myofibers in vivo. *Human Gene Therapy*, 1992, **3**, 21–33.

32. Acsadi, G., Dickson, G., Love, D. R., Jani, A., Walsh, F. S., Gurusinghe, A., Wolff, J. A. and Davies, K. E., Human dystrophin expression in mdx mice after intramuscular injection of DNA constructs. *Nature*, 1991, **352**, 815–818.

33. Klein, T. M., Wolf, E. D., Wu, R. and Sanford, J. C., High-velocity microprojectiles for delivering nucleic acids into living cells. *Nature*, 1987, **327**, 70–73.

34. Yang, N.-S., Burkholder, J., Roberts, B., Martinell, B. and McCabe, D., In vivo and in vitro gene transfer to mammalian somatic cells by particle bombardment. *Proceedings of the National Academy of Sciences of the United States of America*, 1990, **87**, 9568–9572.

35. Williams, R. S., Johnston, S. A., Riedy, M., DeVit, M., J., McElligott, S. G. and Sanford, J. C., Introduction of foreign genes into tissues of living mice

by DNA-coated microprojectiles. *Proceedings of the National Academy of Sciences of the United States of America*, 1991, **88**, 2726–2730.

36. Moss, B., Poxvirus vectors: cytoplasmic expression of transferred genes. *Current Opinion in Genetics and Development*, 1993, 3(1), 86–90.

37. Geller, A. I., Keyomarsi, K., Bryan, J. and Pardee, A. B., An efficient deletion mutant packaging system for defective herpes simplex virus vectors: potential applications to human gene therapy and neuronal physiology. *Proceedings of the National Academy of Sciences of the United States of America*, 1990, **87**, 8950–8954.

38. Wu, G. Y. and Wu, C. H., Receptor-mediated gene delivery and expression in vivo. *Journal of Biological Chemistry*, 1988, **263**, 14621–14624.

39. Lozier, J. N., Thompson, A. R., Hu, P. C., Read, M., Brinkhous, K. M., High, K. A. and Curiel, D. T., Efficient transfection of primary cells in a canine hemophilia B model using adenovirus-polylysine-DNA complexes. *Human Gene Therapy*, 1994, 5(3), 313–322.

40. Curiel, D. T., High-efficiency gene transfer employing adenovirus-polylysine-DNA complexes. *Natural Immunity*, 1994, 13(2–3), 141–164.

41. Scriver, C. R., Beaudet, A. L., Sly, W. S. and Valle, D. The Metabolic Basis of Inherited Disease, sixth edition, McGraw-Hill Information Services Company, Health Profession Division, 1989, pp. 1–164.

42. Green, H., Kehinde, O. and Thomas, J., Growth of cultured human epidermal cells into multiple epithelia suitable for grafting. *Proceeding of the National Academy of Sciences of the United States of America*, 1979, **76**(11), 5665–5668.

43. Leigh, I. M., E. PP, Navsaria, H.A. and Phillips, T.J., Treatment of chronic venous ulcers with sheets of cultured allogenic keratinocytes. *British Journal of Dermatology*, 1987, **117**(5), 591–597.

44. Phillips, T. J., Kehinde, O., Green, H. and Gilchrest, B. A., Treatment of skin ulcers with cultured epidermal allografts. *Journal of the American Academy of Dermatology*, 1989, **21**, 191–199.

45. Cuono, C., Langdon, R. and McGuire, J., Use of cultured epidermal autografts and dermal allografts as skin replacement after burn injury. *Lancet*, 1986, 1(8490), 1123–1124.

46. De Luca, M., Albanese, E., Bondanza, S., Megna, M., Ugozzoli, L., Molina, F., Cancedda, R., Santi, P. L., Bormioli, M., Stella, M. and Magliacani, G., Multicentre experience in the treatment of burns with autologous and allogenic cultured epithelium, fresh or preserved in a frozen state. *Burns* 1989, **15**(5), 303–309.

47. Madden, M. R., Finkelstein, J. L., Staiano-Coico, L., Goodwin, C. W., Shires, G. T., Nolan, E. E. and Hefton, J. M., Grafting of cultured allogeneic epidermis on second- and third-degree burn wound on 26 patients. *Journal of Trauma*, 1986, **26**(11), 955–962.

48. Fenjves, E. S., Smith, J., Zaradic, S. and Taichman, L. B., Systemic delivery of secreted protein by grafts of epidemal keratinocytes: Prospects for keratinocyte gene therapy. *Human Gene Therapy*, 1994, **5**, 1241–1248.

49. Gerrard, A. J., Hudson, D. J., Brownlee, G. G. and Watt, F. M., Gene therapy for haemophilia B using primary human keratinocytes. *Nature Genetics*, 1993, **3**, 180–183.

50. Tani, K., Ozawa, K., Ogura, H., Takahashi, T., Okano, A., Watari, K., Matsudiaira, T., Tajika, K., Karasuyama, H., Nagata, S., Asano, S. and Takaku, F., Implantation of fibroblasts transfected with human granulocyte colony-stimulating factor cDNA into mice as a model of cytokine-supplement gene therapy. *Blood*, 1989, **74**, 1274–1280.

51. Scharfmann, Axelrod, J. H. and Verma, I. M., Long-term in vivo expression of retrovirus-mediated gene transfer in mouse fibroblast implants. *Proceeding of the National Academy of Sciences*, 1991, **88**, 4626–4630.

52. Hoeben, R. C., Fallaux, F. J., van Tilburg, N. H., Cramer, S. J., van Ormondt, H., Briet, E. and van der Eb, A. J., Toward gene therapy for hemophilia A: long-term persistence of factor VIII-secreting fibroblasts after transplantation into immunodeficient mice. *Human Gene Therapy*, 1993, **4**, 179–186.

53. Petersen, M. J., Kaplan, J., Jorgensen, C. M., Schmidt, L. A., Li, L., Morgan, J. R., Kwan, M. K. and Krueger, G. G., Sustained production of human transferrin by transduced fibroblasts implanted into athymic mice: A model for somatic gene therapy. *Journal of Investigative Dermatology*, 1995, **104**, 171–176.

54. Moullier, P., Bohl, D., Heard, J. M. and Danos, O., Correction of the lysosomal storage in the liver and spleen of MPS VII mice by implantation of genetically modified skin fibroblasts, *Nature Genetics*, 1993, **4**, 154–159.

55. Garver, R. I., Chytil, A., Courtney, M. and Crystal, R. G., Clonal gene therapy: transplanted mouse fibroblast clones express alpha 1-trypsin gene in vivo. *Science*, 1987, 714–718.

56. Naffakh, N., Henri, A., Villeval, J. L., Rouyer-Fessard, P., Moullier, P., Blumenfeld, N., Danos, O., Vainchenker, W., Heard, J. M. and Beuzard, Y., Sustained delivery of erythropoietin in mice by genetically modified skin fibroblasts. *Proceedings of the National Academy of Sciences of the United States of America*, 1995, **92**(8), 3194–3198.

57. Huber, M., Rettler, T., Bernasloni, K., Frenk, E., Lavrijsen, S. P., Ponee, M., Bon, A., Lautenschlager, S., Schorderet, D. F. and Hohl D., Mutations of keratinocyte transglutaminase in lamellar ichthyosis. *Science*, 1995, **267**, 525–528.

58. Choate, K. A., Medalie, D. A., Morgan, J. R. and Khavari, P. A., Corrective gene transfer in the human skin disorder lamellar ichthyosis. *Nature Medicine*, 1996, **2**(11), 1263–1267.

59. Eming, S. A., Lee, J., Snow, R. G., Tompkins, R. G., Yarmush, M. L. and Morgan, J. R., Genetically modified human epidermis overexpressing PDGF-A directs the development of a cellular and vascular connective tissue stroma when transplanted to athymic mice — Implications for the use of genetically modified keratinocytes to modulate dermal regeneration, *Journal of Investigative Dermatology*, 1995, **105**, 756–763.

60. Eming, S. A., Snow, R. G., Yarmush, M. L. and Morgan, J. R., Targeted expression of IGF-1 to human keratinocytes: Modification of the autocrine control of keratinocyte proliferation, *Journal of Investigative Dermatology*, 1996, **106**, 113–120.

61. Burt, A. M., Pallett, C. D., Sloane, J. P., O'Hare, M. J., Schafler, K. F., Yardeni, P., Eldad, A., Clarke, J. A. and Gusterson, B. A., Survival of

cultured allografts in patients with burns assessed with a probe specific for Y chromosome. *British Medical Journal*, 1989, **298**, 915–917.

62. Aubock, J. E. I., Romani, N., Kompatscher, P., Hopfl, R., Herold, M., Schuler, G., Bauer, M., Huber, C. and Fritsch, P., Rejection, after a slightly prolonged survival time, of Langerhans cell-free allogeneic cultured epidermis used for wound coverage in humans. *Transplantation*, 1988, **45**(4), 730–737.

63. Andree, C., Swain, W. F., Page, C. P., Macklin, M. D., Slama, J., Hatzis, D. and Eriksson, E., In vivo transfer and expression of a human epidermal growth factor gene accelerates wound reapair. *Proceedings of the National Academy of Sciences*, 1994, **91**, 12188–12192.

64. Benn, S. I., Whistitt, J. S., Broadley, K. N., Nanney, L. B., Perkins, D., He, L., Patel, M., Morgan, J. R., Swain, W. F. and Davidson, J. R. Particle mediated gene transfer with TGF-B1 cDNAs enhances wound repair in rat skin. *Journal of Clinical Investigation*, 1996, **98**, 2894–2902.

65. Alexander, M. Y. and Akhurst, R. J., Liposome-mediated gene transfer and expression via the skin. *Human Molecular Genetics*, 1995, **4**(12), 2279–2285.

66. Li, L. and Hoffman, R. M., The feasibility of targeted selective gene therapy of the hair follicle. *Nature Genetics*, 1995, **1**, 705–706.

67. Hengge, U. R., Walker, P. S. and Vogel, J. C., Expression of naked DNA in human, pig, and mouse skin. *The Journal of Clinical Investigation*, 1996, **97**(12), 2911–2916.

68. Ciernik, I. F., KrayenbYhl, B. H. and Carbone, D. P., Puncture-mediated gene transfer to the skin. *Human Gene Therapy*, 1996, **7**, 893–899.

69. Wilson, J. M., Birinyi, L. K., Salomon, R. N., Libby, P. L., Callow, A. D. and Mulligan, R. C., Implantation of vacular grafts lined with genetically modified endothelial cells. *Science*, 1989, **244**, 1344–1346.

70. Zwiebel, J. A., Freeman, S. M., Kantoff, P. W., Cornetta, K., Ryan, U. S. and Anderson, F., High-level recombinant gene expression in rabbit endothelial cells transduced by retroviral vectors. *Science*, 1989, **243**, 220–222.

71. Dichek, D. A., Anderson, J., Kelly, A. B., Hanson, S. R. and Harker, L. A., Enhanced in vivo antithrombotic effects of endothelial cells expressing recombinant plasminogen activators transduced with retroviral vectors. *Circulation*, 1996, **93**, 301–309.

72. Dichek, D. A., Neville, R. F., Zwiebel, J. A., Freeman, S. M., Leon, M. B. and Anderson W. F., Seeding of intravascular stents with genetically engineered endothelial cells. *Circulation*, 1989, **80**, 1347–1353.

73. Ohno, T., Gordon, D., San, H., Pompili, V. J., Imperiale, M. J., Nabel, G. J. and Nabel, E. G., Gene therapy for vascular smooth muscle cell proliferation after arterial injury. *Science*, 1994, **265**, 781–784.

74. Steg, P. G., Feldman, L. J., Scoazec, J. Y., Tahlil, O., Barry, J. J., Boulechfar, S., Ragot, T., Isner, J. M. and Perricaudet, M., Arterial gene transfer to rabbit endothelial and smooth muscle cells using percutaneous delivery of an adenoviral vector. *Circulation*, 1994, **90**, 1648–1656.

75. Newman, K. D., Dunn, P. F., Owens, J. W., Schulick, A. H., Virmani, R., Sukhava, G., Libby, P. and Dichek, D.A., Adenovirus-mediated gene transfer into normal rabbit arteries results in prolonged vascular cell activation,

inflammation, and neointimal hyperplasia. *Journal of Clinical Investigation*, 1995, **96**, 2955–2965.

76. Zoldhelyi, P., McNatt, J., Xu, X. M., Loose-Mitchell, D., Meidell, R. S., Clubb, F. J., Buja, L. M., Willerson, J. I. and Wu, K. K., Prevention of arterial thrombosis by adenovirus-mediated transfer of cyclooxygenase gene. *Circulation*, 1996, **93**, 10–17.

77. Muhlhauser, J., Pili, R., Merrill, M. J., Maeda, H., Passaniti, A., Crystal, R. G. and Capogrossi, M. C., In vivo angiogenesis induced by recombinant adenovirus vectors coding either for secreted or nonsecreted forms of acidic fibroblast growth factor. *Human Gene Therapy*,1995, **6**, 1457–1465.

78. Mulhauser, J., Merrill, M. J., Pili, R., Maeda, H., Bacic, M., Bewig, B., Passaniti, A., Edwards, N. A., Crystal, R. G. and Capogrossi, M.C., VEGF165 expressed by a replication-deficient recombinant adenovirus vector induces angiogenesis in vivo. *Circulation Research*, 1995, **77**, 1077–1086.

79. Rade, J. J., Schulick, A. H., Vidrmani, R. and Dichek, D. A., Local adenoviral-mediated expression of recombinant hirudin reduces neointima formation after arterial injury. *Nature Medicine*, 1996, 2(3), 293–298.

80. Takeshita, S., Tsurumi, Y., Couffinahl, T., Asahara, T., Bauters, C., Symes, J., Ferrara, N. and Isner, J. M., Gene transfer of naked DNA encoding for three isoforms of vascular endothelial growth factor stimulates collateral development in vivo. *Laboratory Investigation*, 1996, **75**(4), 487–501.

81. Isner, J. M., Walsh, K., Rosenfield, K., Schainfeld, R., Asahara, T., Hogan, K. and Pieszek, A., Arterial gene therapy for restenosis. *Human Gene Therapy*, 1996, 7(8), 989–1011.

82. Isner, J. M., Pieczek, A., Schainfeld, R., Blair, R., Haley, L., Asahara, T., Rosenfield, K., Razvi, S., Walsh, K. and Symes, J. F., Clinical evidence of angiogenesis after arterial gene transfer of phVEGF165 in patient with ischaemic limb. *Lancet*, 1996, **348**(9024), 370–374.

83. Mühlhauser, J., Merrill, M. J., Pili, R., Maeda, H., Bacic, M., Bewig, B., Passaniti, A., Edwards, N. A., Crystal, R. G. and Capogrossi, M. C., VEGF$_{165}$ expressed by a replication-deficient recombinant adenovirus vector induces angiogenesis in vivo. *Circulation Research*. 1995, **77**, 1077–1086.

84. Mühlhauser, J., Pili, R., Merrill, M. J., Maeda, H., Passaniti, A., Crystal, R. G. and Capogrossi, M. C., In vivo angiogenesis induced by recombinant adenovirus vectors coding either for secreted or nonsecreted forms of acidic fibroblast growth factor. *Human Gene Therapy*, 1995, **6**, 1457–1465.

85. Qin, L., Chavin, K. D., Ding, Y., Tahara, H., Favaro, J. P., Woodward, J. E., Suzuki, T., Robbins, P. D., Lotze, M. T. and Brombert, J. S. Retrovirus-mediated transfer of viral IL-10 gene prolongs murine cardiac allograft survival. *Journal of Immunology*, 1996, **156**, 2316–2323.

86. Qin, L., Chavin, K. D., Ding, Y., Favaro, J. P., Woodward, J. E., Lin, J., Tahara, H., Robbins, P., Shaked, A., Ho. D. Y. et al., Multiple vectors effectively achieve gene transfer in a murine cardiac transplantation model. Immunosuppression with TGF-beta 1 or vIL-10. *Transplantation*, 1995, **59**(6), 809–816.

87. Messina, L. M., Podrazik, R. M., Whitehill, T. A., Ekhterae, D., Brothers, T. E., Wilson, J. M., Burkel, W. E. and Stanley, J. C., Adhesion and incorporation of lacZ-transduced endothelial cells into the intact capillary wall in the

rat. *Proeceedings of the National Academy of Sciences of the United States of America*, 1992, **89**(24), 12018–12022.

88. Lynch, C. M., Clowes, M. M., Osborne, W. R., Clowes, A. W. and Miller, A. D., Long-term expression of human adenosine deaminase in vascular smooth muscle cells of rats: a model for gene therapy. *Proceedings of the National Academy of Sciences of the United States of America*, 1992, **89**(3), 1138–1142.

89. Brittberg, M., Lindahl, A., Nilsson, A., Ohlsson, C., Isaksson, O., Peterson, L., Treatment of deep cartilage defects in the knee with autologous chondrocyte transplantation. *New England Journal of Medicine*, 1994, **331**, 889–895.

90. Bandara, G., Mueller, G. M., Galea-Lauri, J., Tindal, M. H., Georgescu, H. I., Suchanek, M. K., Hung, G. L., Glorioso, J. C., Robbins, P. D. and Evans C. H., Intraarticular expression of biologically active interleukin 1-receptor-antagonist protein by ex vivo gene transfer. *Proceedings of the National Academy of Sciences*, 1993, **90**, 10764–10768.

91. Hung, G. L., Galea-lauri, J., Mueller, G. M., Georgescu, H. I., Larkin, L. A., Suchanek, M. K., Tindal, M. H., Robbins, P. D. and Evans, C. H., Supression of intra-articular responses to IL-1 by transfer of the IL-1 receptor antagonist to synovium. *Gene Therapy*, 1994, **1**, 64–69.

92. Evans, C. H., Robbins, P. D., Ghivizzani, S. C., Herndon, J. H., Kang, R., Bahnson, A. B., Barranger, J. A., Elders, E. M., Gay, S., Tomaino, M. M., Wasko, M. C., Watkins, S. C., Whiteside, T. L., Glorioso, J. C., Lotze, M. T. and Wrighe, T. M., Clinical trial to assess the safety, feasibility, and efficacy of transferring a potentially anti-arthritic cytokine gene to human joints with rheumatoid arthritis. *Human Gene Therapy*, 1996, **7**(10), 1261–1280.

93. Nakamura, N., Horibe, S., Matsumoto, N., Tomita, T., Natsuume, T., Kaneda, Y., Shino, K. and Ochi, T., Transient Introduction of a foreign gene into healing rat patellar ligament. *Journal of Clinical Investigation*, 1996, **97**, 226–231.

94. Al-Hendy, A., Hortelano, G., Tannenbaum, G. S. and Chang, P. L., Correction of the growth defect in dwarf mice with nonautologous microencapsulated myoblasts — an alternative approach to somatic gene therapy. *Human Gene Therapy*, 1995, **6**(2), 165–175.

95. Hamamori, Y., Samal, B., Tian, J. and Kedes, L., Myoblast transfer of human erythropoietin gene in a mouse model of renal failure. *Journal of Clinical Investigation*, 1995, **95**(4), 1808–1813.

96. Zatloukal, K., Cotten, M., Berger, M., Schmidt, W., Wagner, E. and Birnstiel, M. L., In vivo production of human factor VII in mice after intrasplenic implantation of primary fibroblasts transfected by receptor-mediated, adenovirus-augmented gene delivery. *Proceedings of the National Academy of Sciences of the United States of America*. 1994, **91**(11), 5148–5152.

97. Dai, Y., Roman, M., Naviaux, R. K. and Verma, I. M., Gene therapy via primary myoblasts: long-term expression of factor IX protein following transplantation in vivo. *Proceedings of the National Academy of Sciences of the United States of America*, 1992, **89**(22), 10892–10895.

98. Hortelano, G., Al-Hendy, A., Ofosu, F. A. and Chang, P. L., Delivery of human factor IX in mice by encapsulated recombinant myoblasts: a novel approach towards allogeneic gene therapy of hemophilia B. *Blood*, 1996, **87**(12), 5095–5103.

99. Newgard, C. B., Cellular engineering and gene therapy strategies for insulin replacement in diabetes. *Diabetes*, 1994, **43**(3), 341–350.

100. Stewart, C., Taylor, N. A., Green, I. C., Docherty, K. and Bailey, C. J., Insulin-releasing pituitary cells as a model for somatic cell gene therapy in diabetes mellitus. *Journal of Endocrinology*, 1994, **142**(2), 339–343.

101. Kolodka, T. M., Finegold, M., Moss, L. and Woo, S. L., Gene therapy for diabetes mellitus in rats by hepatic expression of insulin. *Proceedings of the National Academy of Sciences of the United States of America*, **92**(8), 3293–3297.

102. Lau, H. T., Yu, M., Fontana, A. and Stoeckert, Jr., C. J., Prevention of islet allograft rejection with engineered myoblasts expressing fasL in mice. *Science*, 1996, **273**, 109–112.

103. Karpati, G., Lochmuller, H., Nalbantoglu, J. and Durham H., The principles of gene therapy for the nervous system. *Trends in Neurosciences*, 1996, **19**(2), 49–54.

104. Rosenberg, M. B., Friedmann, T., Robertson, R. C., Tuszynski, M., Wolff, J. A., Breakefield, X. O. and Gage, F. H., Grafing genetically modified cells to the damaged brain: restorative effects of NGF expression. *Science*, 1988, **242**(4885), 1575–1578.

105. Fisher, L. J., Raymon, H. K. and Gage, F. H., Cells engineered to produce acetylcholine: therapeutic potential for Alzheimer's disease. *Annals of the New York Academy of Sciences*, 1993, **695**, 278–284.

106. Chen, K. S. and Gage, F. H., Somatic gene transfer of NGF to the aged brain: behavioral and morphological amelioration. *Journal of Neuroscience*, 1995, **15**(4), 2819–2825.

107. Licidi-Phillipi, C. A., Gage, F. H., Shults, C. W., Jones, K. R., Reichardt, L. F. and Kang, U. J., Brain-derived neurotrophic factor-transduced fibroblasts: produciotn of BDNF and effects of grafting to the adult brain. *Journal of Comparative Neurology*, 1995, **354**(3), 361–376

108. Senut, M. C., Tuszynski, M. H., Raymon, H. K., Suhr, S. T., Liou, N. H., Jones, K. R., Reichardt, L. F. and Gage, F. H., Regional differences in responsiveness of adult CNS axons to grafts of cells expressing human neurotrophin 3. *Experimental Neurology*, 1995, **135**(1), 36–55.

109. Aebischer, P., Schluep, M., Deglon, N., Joseph, J. M., Hirt, L., Heyd, B., Goddard, M., Hammang, J. P., Zurn, A. D., Kato, A. C., Regli, F. and Baetge, E. E., Intrathecal delivery of CNTF using encapsulated genetically modified xenogeneic cells in amyotrophic lateral sclerosis patients. *Nature Medicine*, 1996, **2**(6), 696–699.

110. Suhr, S. T. and Gage, F. H., Gene therapy for neurologic disease. *Archives of Neurology*, 1993, **50**(11), 1252–1268.

111. Tuszynski M. H., Roberts, J., Senut, M. C., U H. S. and Gage, F. H., Gene therapy in the adult primate brain: intraparenchymal grafts of cells genetically modified to produce nerve growth factor prevent cholinergic neuronal degeneration. *Gene Therapy*, 1996, **3**(4), 305–314.

112. Bajocchi, G., Feldman, S. H., Crystal, R. G. and Mastrangeli, A., Direct in vivo gene transfer to ependymal cells in the central nervous system using recombinant adenovirus vectors. *Nature Genetics*, 1993, **3**(3), 187–189.

CHAPTER II.16

Ethical Considerations of Tissue Engineering on Society

TIMOTHY W. KING,
CHARLES W. PATRICK, JR.
Laboratory of Reparative Biology and Bioengineering,
Department of Plastic Surgery,
M.D. Anderson Cancer Center,
1515 Holcombe Blvd., Box 62,
Houston, TX 77030, USA

1 Introduction

We, as human beings, are planners, designers, builders, and achievers. We tinker, explore, invent, and create. This is part of our genetic heritage. We are the survivors of the natural selection process of human biological evolution. Our recent brain development has propelled us into a cultural evolution that has outstripped the capacities of our physical selves. Our vital organs break down, but our brains compel us to fix them, to renovate them, and to replace them. Technology and innovation are part of our inner drives. It should come as no surprise to anyone that, sooner or later, an artificial heart would be implanted in the chest of one of our fellow human beings [1].

In this statement, Margery W. Shaw eloquently expresses the force that drives scientists and engineers to create and develop new and innovative devices that enhance and improve our lives. In October 1995, the media reported that researchers at MIT had successfully grown a human ear on the back of a mouse using tissue engineering techniques [2–10]. In February 1997, Scottish scientists announced they cloned a sheep naming her 'Dolly' [11]. These stories created a frenzy of controversy within the national and international media, but at the same time, brought the field of tissue engineering to the forefront of mainstream society. While the wide-spread public debate diminished after several weeks, the important underlying ethical, legal, and social issues of using tissues in research and as implants still remain. This chapter is designed to introduce the reader to some of these issues. To be sure, this chapter illustrates the point that novel therapeutics and medical modalities are not only dictated by innovative experimental thinking, but are also dictated by the culture within which they are developed. That is, tissue engineering will not become a viable clinical solution unless it is accepted by society. The aim of this chapter is to sensitize the reader to these issues and stimulate the reader to formulate a basic understanding and personal philosophy regarding tissue engineering and its use.

2 The role of tissue engineering in healthcare and society

Since the beginning of recorded time, the concept of taking tissues from one person, modifying them and transplanting them either back into the same person or into another person, has been a topic that has captivated science, medicine, and society [12]. Examples include the use of Adam's rib to create Eve in the book of *Genesis* in the Bible and the novel Frankenstein, written by Mary Shelly in 1831. These examples, although not scientifically based, emphasize the positive and negative potential of tissue engineering. Recent scientific discovery and quantitative methodologies have made it possible for this potential to be developed further. Therefore, as with any new or emerging biomedical discipline, it is important to consider the impact tissue engineering will have on the healthcare industry and society as a whole. While the full impact of tissue engineering can not be predicted, there is great potential for the field to decrease human suffering by providing engineered tissues for both functional and cosmetic replacement or repair.

The elaboration of tissue engineered constructs that recapitulate the structure and function of *in vivo* tissues is conceptually simple but developmentally complex. Because of the complexity of problems to be solved, tissue engineering necessitates a multidisciplinary approach to be effective. In the past, many other biomedical disciplines have brought together research scientists from several disparate fields to focus on a common medical problem or goal. One example of the multidisciplinary approach is the 'team' that, by combining individual talents, elucidated the structure of DNA. Although James Watson and Francis Crick, a biologist and physicist, respectively, received the Nobel Prize for this discovery, several other scientists, from fields such as X-ray crystallography, protein chemistry, nucleic-acid chemistry, biochemistry, and mathematics aided in this landmark revelation [13]. Another example is the development of the artificial heart for which cardiovascular surgeons, engineers, and biologists focused their efforts and expertise on a singular goal: the production of an artificial, implantable pump with long term life-sustaining capabilities. Similarly, tissue engineering combines several different fields, although over a much broader spectrum of individual disciplines. As depicted in Table 1, tissue engineering currently combines professionals from engineering, clinical medicine, biological science, physical science, and com-

Table 1. Scientific fields involved in tissue engineering

Engineering
  Chemical
  Biomedical
  Biomaterials
  Electrical
  Mechanical

Biological science
  Cell biology
  Molecular biology
  Genetics
  Biochemistry
  Immunology
  Histology

Clinical medicine
  Surgery
    Plastic and reconstructive
    Orthopedics
    Organ transplantation/artificial organs
    Oral
    Urology
    Otolaryngology
  Dermatology
  Pathology
  Internal medicine
    Endocrinology
    Immunology
    Hematology/oncology
    Nephrology
    Hepatology
  Anesthesiology
  Neurology
  Dentistry

Miscellaneous
  Microscopy
  Digital image analysis
  Computer modeling
  Magnetic resonance imaging and spectroscopy

---

puter science creating an environment where the potential for progress is synergistic rather than merely additive. 'Tissue engineering is like an orchestra,' states Ron Cohen, Advanced Tissue Science's (ATS) Vice President for Medical Affairs. 'Everyone has been off studying their own instrument for 12 years, and now they're coming together to play Mozart' [14].

The field of tissue engineering has begun to encourage a broad cooperation through tissue engineering initiatives such as the Pittsburgh Tissue Engineering Initiative (PTEI) [15–20]. These partnerships bring together research institutions and clinical medical centers with businesses, entrepreneurs, investors, manufacturing facilities, and local governments in order to provide an opportunity to maximize the possibilities of the field. Through these consortiums, local communities are able to diversify their economic and manufacturing base in one step [21].

The ultimate goal of tissue engineering is to improve the quality of lives of people. Due to the inherent diversity of tissue engineering, the potential impact of the field on society is vast. However, while tissue engineering has the potential to impact over 39 million patients or procedures per year [22], this reality may not occur for several years. Once the development of functional tissues and organs is complete, these engineered tissues must still be approved by regulatory agencies (e.g. the Food and Drug Administration (FDA) in the U.S.), a process that is measured in years rather than months. In the interim, tissue engineering can have a viable and significant impact through the use of the tissue engineered constructs for diagnostic and research purposes. Some of the possible applications include toxicity testing and macromolecule (i.e. antibodies, hormones, etc.) synthesis for clinical use [21].

3 The ethics of tissue engineering

As the field of tissue engineering advances, we must be dedicated to the pursuit and enforcement of ethical standards. The importance of adhering to ethical standards in medicine and biomedical research is illustrated in the tragedy of the Tuskegee syphilis study. In the study conducted by the U.S. Public Health Service (PHS) on U.S. citizens, 400 African–American men with syphilis were actively denied penicillin, a curative therapy, for 40 years so the investigators could study the natural course and long-term effects of the disease [23–27]. The United States government still pays $18 million per year to the survivors and their families of this tragic event. Unfortunately, this was not an isolated incident. During World War II, Nazi physicians and scientists conducted inhumane experiments on imprisoned Jews against their will. These historical tragedies reaffirm the need for the establishment of ethical standards for biomedical research endeavors such as tissue engineering. Prior to discussing ethical issues associated with tissue engineering, a brief synopsis of biomedical ethics fundamentals is presented.

3.1 Teleological approach to ethics

Teleological ethics are consequence oriented and emphasize the ends, or results, of an action, rather than the action itself. Furthermore, the determination of an act being 'right' or 'wrong' depends upon the results of the act itself. The importance of inherent traits of an act, such as whether the act is in accord with a moral rule against the act are downplayed, in order to concentrate on the end results or consequences of the act. For example, one might claim that physical abuse is wrong because of the consequences (i.e. the psychological, emotional, and physical damage) of the act. Utilitarianism is the best known example of the teleological approach to ethics [28, 29].

3.1.1 Act utilitarianism
In act utilitarianism, the person determines the utility of each act individually, and then the act with the greatest utility is implemented. Act utilitarianism emphasizes a particular act in a particular situation [28].

3.1.2 Rule utilitarianism
When a person applies rule utilitarianism, the person determines the utility of executing one kind of act in a certain type of situation. In this case, the utility of following one rule, or set of rules, is compared with that of other rules [28].

3.2 Deontological approach to ethics

Deontological ethics is not consequence or goal oriented. Rather, deontological ethics is rule oriented and emphasizes the acts in and of themselves. The act itself, and not the consequences of that act, possess some inherent quality of 'right' or 'wrong'. The act is 'right' by its virtue of simply being that act. For example, one might claim that physical abuse is 'wrong', not because of the consequences of the act, but due to the very nature of the act because it violates dignity, autonomy, and respect [28, 29].

3.3 Principles of ethics in health care

There are certain principles of ethics that apply directly to the field of health care. Childress and Self describe six of these principles [28, 29]:

- Principle of Non-maleficience: the duty or obligation to cause no harm. This duty is sometimes included in beneficence.
- Principle of Beneficence: the duty or obligation to benefit others, to do that which is good, or to help others further their own legitimate interests.

- **Principle of Autonomy**: the duty or obligation to promote the self-determination of others.
- **Principle of Justice**: the duty or obligation to allocate social burdens and benefits fairly among all persons.
- **Principle of Universalizability**: the duty or obligation to do only those acts which you would be willing for anyone else to do under relevantly similar circumstances.
- **Principle of Rationality**: the duty or obligation to do only those acts which you can give reasons and justifications for doing.

3.4 Criteria for a sound ethical decision

The criteria for a sound ethical decision are [28]:

- It must be rationally arrived at (i.e. not based on just 'gut feelings'). Reasons and justifications, sometimes better or worse, for choosing one alternative over others can be given.
- All relevant data must be fairly considered with an open mind and the decision must not be biased (i.e. all data is objectively considered).
- It must be universalizable. You would be willing for everyone to do likewise under relevantly similar conditions. This is thought to be the cornerstone of morality by most people. Examples of this include Kant's Categorical Imperative and the Golden Rule.
- It must be coherent and consistent (i.e. it fits in well with the moral beliefs of society).

These four criteria define the concepts needed to make a sound ethical decision. The specific codes regulating human experimentation will now be considered.

3.5 The Nuremberg Code

Following World War II, Nazi physicians and scientists who performed unethical research on Jews, Gypsies, prisoners, and the physically and mentally disabled were placed on trial for their actions. The Nuremberg Code, guidelines on human research, was written by the Nuremberg tribunal as a result of these trials. This code is the basis for current ethical standards in human trials. The Nuremberg code reads [30]:

The proof as to war crimes and crimes against humanity

Judged by any standard of proof the record clearly shows the commission of war crimes and crimes against humanity substantially as alleged in counts two and three of the indictment. Beginning with the outbreak of World War II criminal medical experiments on non-German nationals, both prisoners of war and civilians, including Jews and 'asocial' persons, were carried out on a large scale in Germany and the occupied countries. These experiments were not the isolated and casual acts of individual doctors and scientists working solely on their own responsibility, but were the product of coordinated policy-making and planning at high governmental, military, and Nazi Party levels, conducted as an integral part of the total war effort. They were ordered, sanctioned, permitted, or approved by persons in positions of authority who under all principles of law were under the duty to know about these things and to take steps to terminate or prevent them.

Permissible medical experiments

The great weight of evidence before us is to the effect that certain types of medical experiments on human beings, when kept within reasonably well-defined bounds, conform to the ethics of the medical profession generally. The protagonists of the practice of human experimentation justify their views on the basis that such experiments yield results for the good of society that are unprocurable by other methods or means of study. All agree, however, that certain basic principles must be observed in order to satisfy moral, ethical and legal concepts:

1. The voluntary consent of the human subject is absolutely essential. This means that the person involved should have legal capacity to give consent; should be so situated as to be able to exercise free power of choice, without the intervention of any element of force, fraud, deceit, duress, over-reaching, or other ulterior form of constraint or coercion; and should have sufficient knowledge and comprehension of the elements of the subject matter involved as to enable him to make an understanding and enlightened decision. This latter element requires that before the acceptance of an affirmative decision by the experimental subject there should be made known to him the nature, duration, and purpose of the experiment; the method and means by which it is to be conducted; all inconveniences and hazards reasonably to be expected; and the effects upon his health or person which may possibly come from his participation in the experiment.

   The duty and responsibility for ascertaining the quality of the consent rests upon each individual who initiates, directs or engages in the experiment. It is a personal duty and responsibility which may not be delegated to another with impunity.

2. The experiment should be such as to yield fruitful results for the good of society, unprocurable by other methods or means of study, and not random and unnecessary in nature.

3. The experiment should be so designed and based on the results of animal experimentation and a knowledge of the natural history of the disease or other problem under study that the anticipated results will justify the performance of the experiment.

4. The experiment should be so conducted as to avoid all unnecessary physical and mental suffering and injury.

5. No experiment should be conducted where there is an *a priori* reason to believe that death or disabling injury will occur; except, perhaps, in those experiments where the experimental physicians also serve as subjects.

6. The degree of risk to be taken should never exceed that determined by the humanitarian importance of the problem to be solved by the experiment.

7. Proper preparations should be made and adequate facilities provided to protect the experimental subject against even remote possibilities of injury, disability, or death.

8. The experiment should be conducted only by scientifically qualified persons. The highest degree of skill and care should be required through all stages of the experiment of those who conduct or engage in the experiment.

9. During the course of the experiment the human subject should be at liberty to bring the experiment to an end if he has reached the physical or mental state where continuation of the experiment seems to him to be impossible.

10. During the course of the experiment the scientist in charge must be prepared to terminate the experiment at any stage, if he has probable cause to believe, in the exercise of the good faith, superior skill and careful judgment required of him that a continuation of the experiment is likely to result in injury, disability, or death to the experimental subject.

In 1966, 15 years after the Nuremberg Code was composed, the World Medical Association adopted the guidelines outlined in the Helsinki Declaration, recommendations prepared at the 18th World Medical Assembly [31, 32]. Unfortunately, 2 years later, research scientists were still not considering the rights of their subjects. In a sentinel article published in the *New England Journal of Medicine* in 1966, Dr. Henry Beecher outlined ethical violations occurring in twenty-two research studies [33]. Shortly after this article was published, the U.S. PHS mandated that any institution receiving PHS funding must create an Institutional Review Board (IRB). The IRB's purpose is to review all research (not just PHS funded) involving human subjects and ensure the rights and welfare of these subjects are not violated. In addition, the IRB is responsible for ensuring that proper informed consent is obtained from each subject prior to their participation in the investigation.

The ethical guidelines for human subjects were further expanded in 1979 by the Belmont Report, the final three volume document which resulted as an outgrowth of an intensive 4-day period of discussions held at the Smithsonian Institution's Belmont Conference Center and the monthly deliberations of the National Commission for the Protection of Human Subjects of Biomedical and Behavioral Research [34–36]. Together with the World Medical Association's Declaration of Helsinki and the Belmont Report, the Nuremberg Code and IRB's strive to ensure ethical principles are consistently maintained in clinical research.

With this introduction to ethical principles and codes, the ethical issues with respect to tissue engineering may be approached. For a more detailed discussion of bioethics the reader is referred to books edited by J. Howell and W. Sale [37], R. Veatch [38], R. E. Bulger [39], and R. Gillon [40], and an article by Albert Johnsen [41].

3.6 General ethical considerations for tissue engineering

The assumption will be made that every tissue engineering research project will follow the ethical principles previously discussed. Specifically, no research project should begin which is not justifiable or rational in its scientific method, approach, and application. Further, the importance of the principle of universalizability should always be considered when designing a tissue engineered construct or clinical trial. This principle requires that all ethnic groups, races, and sexes be included in the investigation. If one of these groups will be excluded, justification for this exclusion is required. The principle of beneficence is difficult to balance in any new, innovative clinical research project. Of course, the entire goal of tissue engineering is to benefit mankind. However, when considering these principles, the individual must be the primary concern.

4 The use of animals and xenotransplants in tissue engineering constructs

4.1 Historical perspective of xenotransplantation

Organ transplantation became a viable therapy in 1954 when Drs. Joseph Murray and John Merrill successfully performed the first kidney transplant between monozygotic twins [12, 42, 43]. The field quickly grew and the number and type of organs being transplanted increased dramatically. Presently human heart, lung, liver, kidney, pancreas, islet cell, cornea, bone, bone marrow, skin, ovary, testicle, nerve, middle ear, and small intestine transplantation procedures are clinically available [12, 44–46]. The number of transplantation procedures performed is currently limited by the small number of available donor organs, as shown in Table 2.

The procurement and distribution of organs in the U.S. is based on a weighted formula of medical criteria. Each organ possesses its own allocation policy approved by the United Network for Organ Sharing (UNOS). Among the 11 UNOS regions in the United States, there are 277 transplantation centers with 866 organ transplant programs divided between these centers [44, 49]. In 1988, the number of patients on the waiting list was 25% higher than the number of transplants performed. This percentage rose to greater than 50% by 1994 [49, 50]. The number of reported deaths has doubled since 1988. However, because the number of registrants has also doubled, the overall death rate has remained constant at 5.5% [44]. Due to the shortage of available organs, patients must meet

Table 2. Societal need of organ transplantation in the United States [12, 44–48]

Organ	First viable transplant (worldwide)	Transplants/year‡	People on the waiting list†	People/year who die waiting for a transplant‡
Kidney	1954	10,892	34,599	1,520
Liver	1963	3,925	7,576	804
Heart	1967	2,361	3,695	770
Lung	1981	871	2,328	342
Other	—	1,096	2,090	12
Total	—	19,145	50,288	3,448

‡Totals are for 1995.
†As of January 22, 1997.

stringent requirements in order to be placed on the waiting lists. These strict rules prevent many patients that could benefit from organ transplantation from being placed on the waiting list until their medical condition becomes 'serious enough'. For example, there are only 3,700 people on the waiting list for heart transplants. However, over 40,000 people with heart failure could benefit from a heart transplant [51]. In fact, some estimate 100,000 Americans die annually who could have benefited from an organ transplant [52].

Because of the shortage of available donor organs, surgeons have begun to develop alternative procedures. Xenotransplantation (cross-species transplantation) is one proposed treatment modality to supplement heterologous transplants. During the 1960s and the 1970s several xenotransplantations between chimpanzees or baboons and humans were performed [53]. These transplants were unsuccessful because of the poor understanding of the immune system and lack of the proper surgical technique. In 1984, the nation watched the plight of Baby Fae, a 15-day-old baby with a lethal congenital heart defect, as she received a baboon heart. Twenty days later, at the age of 35 days, Baby Fae died. Since then, with few notable exceptions, surgeons have maintained a voluntary moratorium on human xenotransplants [54]. In 1992, after receiving a baboon heart transplant at the University of Pittsburgh School of Medicine, a 35-year-old recipient survived for 2.5 months. The patient's death was not due to rejection, but rather, an infection that developed secondarily [54]. Although this individual xenotransplant was somewhat 'successful', continued research into this area has been limited to primate, porcine (pig), and ovine (sheep) models. Recent developments in the field of tissue engineering have contributed to the advancement of the field of transplantation surgery. In fact, combined with the increased knowledge-base in immunology, transplantation surgery is once again actively considering human xenotransplantations [55].

4.2 The ethics of xenotransplantation

Originally, non-human primates, specifically baboons, were thought to be ideal candidates for xenotransplantation. However, concerns over the risk of infectious disease, the length of required to breed primates, the fact that several species of primates are endangered, and the belief that primates are 'too close' to humans to justify breeding them as organ donors has led scientists to look for alternative organ sources. The most promising source of organs for human implantation appears to be the pig. Porcine models have been successfully used as tissue donors for over 25 years. As will be discussed in detail elsewhere in this book, transgenic pigs with human DNA have been developed to decrease, or even eliminate, rejection [47, 55–57]. The issue for this chapter is the ethical impact of this developing technology.

Many questions have been raised concerning xenotransplantation. Some of the issues include [53, 58–61]:

- Prevention of the introduction of dangerous animal pathogens into the human population.
- Will society find the idea of transplanting animal organs into humans acceptable?
- Protection of the rights of the first patients receiving this therapy.
- Should animals be bred for the sole purpose of donating implantable organs?
- Is there any difference between breeding animals for food and breeding animals for saving human lives? If so, are the medical purposes of higher cultural value?
- Should animals be used at all in research?
- Is xenotransplantation scientifically justified?
- Should xenotransplantation be avoided until all other possibilities of human organ donation and artificial man-made devices are exhausted?
- Is xenotransplantation or its associated research 'playing God'?
- Does the patient actually benefit from xenotransplantation?

These questions were considered so important that in January 1997 clinical trials in pig-human xenotransplants were barred in the United Kingdom (U.K.) until some of these issues could be appropriately addressed [62]. A U.K. panel of government advisors, consisting of scientists, lawyers, and lay representatives and headed by Ian Kennedy, professor of law and medicine at King's College London, was convened to address the safety and efficacy of xenotransplantation as well as the welfare of the animal models being considered. In their report entitled 'Animal Tissue Into Humans', the panel stated that transplanting pig organs would be ethically acceptable. However, they felt that the use of primates as a source of donor organs was not acceptable. The panel recommended that primates should only be used in xenotransplantation research when there was no alternative method of obtaining critical information [62]. The report also recommended further investigation into the risk of introducing new human diseases via pig pathogens. Transplant advocates downplay this risk by pointing out that humans have been eating pigs for thousands of years and diabetic patients have used pig extracted insulin for decades without any infectious disease consequences [52]. Additionally, fluids derived from other animals, such as antibodies from rabbits and horses, and hormones from horses and sheep (e.g. Premarin, an oral estrogen replacement pill used by millions of post-menopausal women, is derived from female horse urine), have been used for decades [63]. However, tissue culture experiments conducted by Robin Weiss at the Institute for Cancer Research in London have shown that a porcine retrovirus, carried in the genome and passed on to offspring by the parent, can infect human cells [62]. The panel's recommendation was not disregarded. In fact the U.K. government, following the advice of the panel, has, created a new national body entitled the U.K. Xenotransplantation Interim Regulatory Authority, chaired by biologist and former Arch-bishop of York, Lord Habgood. This regulatory agency oversees the developments in xenotransplantation while legal guidelines are established [62].

According to the principles of non-maleficience and autonomy, the rights of patients undergoing xenotransplantation must be protected. While the patient has the right to participate in a research investigation (autonomy), the investigators should not needlessly endanger their life (non-maleficience). Article 5 of the Nuremberg Code states, 'No experiment should be conducted where there is *a priori* reason to believe death or disabling injury will occur; except perhaps in those experiments where the experimental physicians also serve as subjects' [30]. Although current knowledge indicates human recipient xenotransplants are ineffective, Paul D. Simmons states that, with respect to surgery, a morally justifiable act is one that: '(a) holds the prospect for meaningful patient recovery or the advance of medical knowledge with the knowledge and consent of the patient, and (b) that does not extend suffering unnecessarily or without compensatory factors such as the prolongation of meaningful living' [64]. This belief was eloquently communicated in a *New York Times* editorial, 'the purpose of medicine is to improve life's quality...To prolong life beyond its natural span is no favor unless reasonable quality is also provided. Without it, the physician has only succeeded in prolonging death' [65]. Unless the quality of life of the patient has been sustained or improved, the patient is done a disservice. However, to develop a viable alternative to homologous transplantation the research and surgeries must occur. Further, through the development and refinement of tissue engineered products, there is the possibility to increase the number of organs available for transplant. Thus, by creating a biocompatible xenotransplant, a patient may not have to wait until they were close to death to receive their new organ. Rather, they will have the opportunity to undergo transplantation prior to extreme morbidity and improve both their quantity and quality of life [66].

5 The use of fetal tissue in tissue engineering

Fetal tissue engineered constructs have the potential to help relieve the symptoms, if not cure, many (at least 20), previously, incurable chronic diseases, including Parkinson's disease, Alzheimer's disease, diabetes mellitus, hepatic enzyme/factor deficiencies, and hematopoetic disorders (thallesemias, sickle cell anemia, etc.). In fact, the use of neonatal tissues in tissue engineered constructs has already become a success. Companies such as Advanced Tissue Sciences, Organogensis, LifeCell, and BioSurface Technology are using neonatal foreskins, a normally discarded piece of tissue from circumcisions, to isolate fibroblasts used to create bioartificial skin grafts [14, 22, 67–73]. The use of fetal tissues offers several advantages over tissue from an adult or even an infant. Fetal tissues demonstrate a remarkable ability to grow, divide, and differentiate. Moreover, fetal tissues have been shown to have the ability to withstand hypoxia and also have decreased immunogenic effects [74].

5.1 Historical perspective of fetal research

The use of fetal tissue in biomedical research dates back to 1928 when Fichera unsuccessfully attempted to transplant human fetal pancreas tissue in patients with type I diabetes [75, 76]. The NIH began supporting fetal tissue research in the 1950s and has funded a fetal tissue procurement center for more than 30 years. The fetal tissue research field has received millions of dollars in grants from the NIH, including a research project involving the transplantation of fetal tissue into human subjects [77]. By the late 1960s, the transplantation of fetal thymus cells to treat DiGeorge's syndrome (congenital thymic hypoplasia) was recognized as an effective therapy [78, 79]. Commercial fetal cell lines have been used in laboratory and clinical medicine for years. In fact, the research and preparation of vaccines against polio, rabies, and other viruses has utilized fetal kidney, lung, and brain cell lines [80, 81].

In 1987, an intramural NIH research proposal to transplant fetal neural tissue, derived from an induced abortion, into patients suffering from Parkinson's disease was submitted for review. In October 1987, after the proposal had cleared the internal review, the director of the NIH, Dr. James Wyngaarden, voluntarily sought the advice of Dr. Robert Windom, Assistant Secretary of Health and Human Services (HHS), concerning the propriety of funding fetal tissue research. In March 1988, after consulting with officials in the Reagan administration, Dr. Windom declared a moratorium on federally funded fetal tissue research and asked the NIH to form a committee to determine the appropriateness of, and create guidelines for, federal funding of fetal tissue research. Volumes have been written on what has transpired since that time [77, 80, 82–88]. For detailed information see the excellent reviews by N. Bell, D. Vawter, R. Hurd, and N. Rojansky [77, 80, 86, 87]. Briefly, the NIH formed the Human Fetal Tissue Transplantation Research (HFTTR) Panel, a special advisory panel composed of medical researchers, lawyers, physicians, politicians, clergy, and ethicists, which, in December 1988, sent its recommendations to the standing advisory committee of the NIH. Both committees agreed that, assuming certain guidelines were followed, federal funding should be provided to fetal tissue transplantation and research. In November 1989, Dr. Louis Sullivan, the Secretary of HHS rejected the panel's conclusions and announced the moratorium would continue indefinitely. This ban continued, despite unsuccessful attempts by Congress to end it, through the Bush administration. However, on January 22, 1993, just 2 days after his inauguration, President Clinton instructed Donna Shalala, the new Secretary of HHS to end the moratorium on fetal tissue transplantation research. The Republican Congress of 1996 responded to this action by placing a clause to the HHS Appropriations Bill. This clause states, 'None of the funds made available in this act may be used for...research in which a human embryo or embryos are destroyed, discarded, or knowingly subjected to risk of injury or death greater than allowed for research on fetuses in utero...' [89]. This policy was enforced in October 1996 when the NIH terminated its relationship with Mark Hughes, a geneticist who developed a technique for extracting DNA from single cells for genetic diagnosis of embryos [90]. Ironically, Hughes had been aggressively recruited by the NIH and upon his appointment in 1994 was asked by Harold Varmus, NIH director, to serve on a blue-ribbon NIH panel writing guidelines for a new policy on human embryo research [91]. Roger Pedersen, a fertility researcher at the University of California-San Francisco and Alan Handyside, a British *in vitro* fertilization pioneer, believe that Hughes' treatment will discourage knowledgeable researchers from entering the field and push them into the unregulated private research laboratories [89].

The announcement of the successful cloning of a sheep in February 1997, and the speculation that humans might also be cloned, sparked a global ethical debate and calls throughout the western world for greater governmental control of cloning experiments [92–97]. Ethicists state that science can offer opportunities for change long before society has wrestled with the ethical, scientific, and evolutionary implications [98]. Two days following the announcement, the Clinton administration asked a national bioethics advisory commission to review what the White House called the 'troubling' implications of cloning for human beings [99]. The European Union, which already bans human cloning, requested a scientific committee to recommend whether other forms of genetic manipulation should be regulated. Just one week after the announcement, the British government decided to end funding of the project [100]. A *Time / CNN* survey conducted shortly after the announcement showed 89% of Americans felt cloning humans is morally unacceptable. In addition, of those surveyed, 74% believed cloning humans was 'against the will of God', 66% felt cloning animals was morally unacceptable, 49% said they would eat cloned fruits and vegetables while only 33% said they would eat meat from cloned animals, and 65% felt the government should regulate the cloning of animals. Finally, only 7% of those surveyed said they would clone themselves if given the chance [97, 100–102]. In March 1997, the Clinton administration asked the National Bioethics Advisory Committee (NBAC) to write an opinion by May 1997. However, while waiting for the report, the Clinton administration banned all federally funded human-cloning research [91, 97]. At the time of publication, the issue has not been resolved and it appears that a consensus may be difficult to reach [91].

5.2 The ethics of using fetal tissue in research

Fetal tissues are derived from electively aborted fetuses. Spontaneously aborted fetuses and ectopic pregnancies are undesirable because of the high incidence of infection, genetic/chromosomal anomalies, and necrotic tissue secondary to anoxia [82]. The fundamental ethical issue surrounding the use of fetal tissue is whether its use can be separated from the act of elective abortion. The supporters of the moratorium claim that abortion is so immoral that any use of the tissue is immoral as well. In addition, they fear that [77, 80, 103]:

- Women will be encouraged, or pressured, to abort fetuses they would otherwise carry to term.
- Elective abortion will become even more legitimized, leading to the relaxation of abortion policies.
- Abortion procedures will be modified to yield the maximum amount of fetal tissue samples.
- Women may begin to conceive children for the sole purpose of creating a transplant donor.

- Science and society would become dependent on abortions for medical therapies.
- The mother or the physician might receive potential economic or academic benefits based upon the acquisition of fetal tissue.

There is no empirical evidence that any of these scenarios would occur. After at least thirty years of fetal tissue transplantation and research around the world, there is not a single documented report of a woman aborting her fetus for the purpose of donating tissue. Further, in existing surveys of women who had received abortions, there was no evidence that the welfare of anonymous third parties played a role in choosing an elective abortion [77]. In fact, in a U.K. survey, 94% of women were in favor of research using fetal tissue. There was no difference in the opinion of women who had received an abortion and those who had not [104].

However, since fetal tissue research and transplantation has such broad social and political implications, and because the moral issues are so complex, it is imperative that ethical principles are followed. According to Lee Sanders and Thomas Raffin, the principles of autonomy and non-maleficience assume special meaning in the context of fetal tissue research. Autonomy must be considered for two individuals, the mother and the fetus. The principle of non-maleficience may find conflicting definitions among physicians involved in fetal tissue research. The notion of abortion might be considered by some physicians to be inherently maleficient, while other physicians, see the oath to 'do no harm' residing specifically between themselves and their patient, the mother [83]. Further, the principle of beneficence could also play a role in the ethical discussion. Science can now claim that something 'good' (fetal tissue transplants) may arise out of something 'distasteful' (elective abortion) [83].

In order to prevent the problems listed above, several countries and organizations have created guidelines to be implemented [80, 87, 88, 105]. A compilation of some of these guidelines follows.

- The commercial use or sale of the tissues is prohibited.
- The tissue recipient can not be designated by the donor.
- The tissue is not provided by an exchange for financial gain.
- The physicians performing the abortion would not be allowed to benefit from the subsequent use.
- There must be a clear separation between abortion counseling and fetal tissue procurement.
- Fetal tissue procurement should be discussed only after a firm decision to have an abortion is made.
- The plaintiff must complete an informed consent form.

These guidelines may aid in ensuring that the subsequent use of the tissue does not influence a woman's decision to have an abortion. If these guidelines are followed, fetal tissue can be ethically obtained for transplantation research. This conclusion is supported by a survey conducted on Canadian physicians. The survey revealed that 75% of the physicians agreed with the third and fourth points above. Further 90% and 94% believe in the fifth and sixth points, respectively [74]. The remaining issues mentioned above were not addressed by the Canadian survey.

Finally, the impact of tissue engineering on fetal research could be tremendous. It is conceivable that, through tissue engineering, the need for freshly harvested fetal tissue would eventually be eliminated through the establishment of cell lines developed from these samples.

6 Patents in tissue engineering

'The patent system added the fuel of interest to the fire of genius.'
— *Abraham Lincoln* [106, 107]

One of the outcomes of tissue engineering will be inventions requiring patents. However, biotechnology has raised some legal and ethical issues regarding both submitted and potential patent applications [106, 108–116]. Since patents are a vital aspect of, and the financial security for, many innovations, it is necessary to understand fundamentals of the patent process. This discussion will be limited to the U.S. patent process, although many of the concepts are amenable to international patent processes.

6.1 Historical perspective and general patent information

The power to enact laws relating to patents is granted to the Congress by Article I, section 8, clause 8 of the U.S. Constitution which reads: 'Congress shall have power...to promote the progress of science and useful arts, by securing for limited times to authors and inventors the exclusive right to their respective writings and discoveries.' Utilizing this authority, Congress enacted the first patent law in 1790. The original law has been revised four times. The Plant Patent Act was ratified in 1930 and revised in 1970. In 1952, further revision of the general patent law occurred and is known as Title 35, United States Code. Specifically, section 101 of this law states, 'Whoever invents or discovers any new and useful process, machine, manufacture, or composition of matter, or any new and useful improvements thereof, may obtain a patent therefor [sic], subject to the conditions and requirements of this title.' Finally, in 1988, the Process Patent Act was passed by Congress [106, 110, 117]. It should be noted that in the United States a patent is awarded to the first person to derive the concept, rather than the first person to file a patent application. On the other hand, in many other countries the first person to file a patent application receives the patent. This means that if a person can prove, through legal documentation (i.e. notarized, bound notebooks, etc.), that they are the originator of an invention, the patent will be awarded to them, even if another individual has previously filed for the invention.

According to guidelines published by the U.S. PTO, a patent is a 'grant of property right by the government to the inventor to exclude others from making, using, or selling the invention without the consent of the inventor.' The inventor has the right to sell all or part of his interest in the patent. Further, the U.S. patent process only protects the invention in the United States. If filing for a patent in other countries within six months of the U.S. application filing, a license must be obtained from the Commissioner of Patents and Trademarks. Any product manufactured outside the U.S. that uses a process patented in the U.S. can not be imported into the U.S. This protection is provided by the Process Patent Act of 1988.

Only the true inventor can receive a patent. A machine, a process or method of doing something, any device that is manufactured, or a composition of matter (a new chemical or a new mixture of old elements) are all patentable. Conversely, methods of doing business, printed matter, mere ideas or suggestions, algorithms, principles of nature or products of nature and matters against public policy are not patentable. For example, while a natural gene is not patentable, its cDNA may be patentable [118]. The Atomic Energy Act of 1954 excludes patents which are solely useful for the utilization of special nuclear material or atomic energy for atomic weapons [119]. The U.S. Patent and Trademark Office (PTO), requires an invention to be [118]:

- New — not known or used by others in the U.S. or described anywhere in the world in a publication.
- Useful — something that can be used in society.
- Not obvious to someone skilled in the art — something that would not be known or apparent to anyone with knowledge in the science or particular area of the invention.

With the exception of design patents, which have a 14-year term, patents are granted for a term of 17 years. Once the patent has expired, with the exception of some pharmaceuticals, it can only be renewed through a special act of Congress and the inventor loses the exclusive rights to the invention. If an invention is in public use or sale for more than one year prior to filing the patent application, a patent may not be obtained on that invention [120]. Further, in the U.S., once an invention is publicly disclosed, the inventor has one year to file a patent. However, internationally, the right to patent is forfieted if public disclosure occurs prior to filing for the patent.

There are three major types of patents: utility, design, and plant. Utility patents are granted to a person who meets the criteria outlined in Title 35, Section 101 (see above). Design patents are granted to any person who has 'invented a new, original, and ornamental design for an article of manufacture. The appearance of the article is protected' [121]. Plant patents are granted to anyone who has 'invented or discovered and asexually reproduced any distinct and new variety of plant, including cultivated sports, mutants, hybrids, and newly found seedlings, other than a tuber-propagated plant or a plant found in an uncultivated state' [120].

Other important issues in the patent process are the cost to obtain a patent and the length of time it takes to receive a patent. In general, most patent applications cost between $5,000 and $10,000. However, in specialized fields, such as Tissue Engineering, the costs may exceed $20,000. Once the patent application has been filed, it will take, on average, approximately two years before the patent is issued. In some cases it can take as long as five years before the patent will be awarded [118].

6.2 Patents in biotechnology

The first protection of 'biotechnology' by the patent process can be traced to plant breeders in the 1930's [106]. More recently, in 1971 Anand Chakrabarty, an employee of General Electric Corporation, filed a patent application for a genetically altered bacteria that had the ability to digest oil hydrocarbons. This was the first patent of a non-plant 'living' organism. The patent office initially rejected the patent on the grounds that 'products of nature' were not patentable. In a 1980 landmark decision of the case Diamond v. Chakrabarty, the U.S. Supreme Court, by a vote of 5–4, upheld the patent granted to Dr. Chakrabarty [106, 110, 122]. In the affirmative argument, Chief Justice Warren Burger stated that the microbe was a 'human-made invention'.

The decision was supported by committee reports accompanying the Patent Act of 1952. These reports state that Congress intended for patents to 'include anything under the sun that is made by man.' Being alive or inanimate was not a major factor in the court's decision. Since that time, the PTO has further expanded this ruling. In 1985, the PTO ruled that plants, seeds, and plant tissues could be patentable. In 1987, all 'multicellular living organisms, including animals' (but excluding humans) were held patentable, provided the organism complies with the normal requirements of patentability (i.e. it is novel, unobvious, useful, enables description, and especially relevant to biotechnology, man-made). In 1988, the PTO granted the first patent for a genetically altered animal [114]. Since that time, patents for human cells, other animals, gene therapy protocols, and even medical treatments have been submitted [106, 108, 109, 113, 115, 122, 123]. Patents for medical treatments are not granted in the United States. The decision by the PTO to grant patents for human cells became very relevant in the 1990 sentinel case Moore v. Regents of the University of California.

In 1976, John Moore, a patient at the University of California-Los Angeles (UCLA) Medical Center, was diagnosed with hairy-cell leukemia. At the time of diagnosis, Dr. David Golde, Mr. Moore's attending physician, determined that Mr. Moore's cells had unique properties and were potentially of 'great value in a number of commercial and scientific efforts' and would give researchers 'competitive commercial and scientific advantages'. However, although Dr. Golde knew of the potential profitability of Mr. Moore's cells, he did not inform Mr. Moore of this fact. Following the diagnosis, Dr. Golde recommended a splenectomy. Prior to the surgery, Mr. Moore signed a consent form which included a clause allowing research to occur on his discarded tissues. Dr. Golde and his research assistant Ms. Shirley Quan, obtained some of Mr. Moore's spleen cells with the intention of creating a immortalized cell line. Over the next 8 years, Mr. Moore returned to UCLA Medical Center several times for follow-up. This is a common practice in cancer patients to ensure their well-being and health. During each of these visits, additional tissue samples, including blood, skin, bone marrow, and sperm, were obtained and used in the development of the cell line. The creation of the cell line was completed in 1979. Listing Dr. Golde and Ms. Quan as the inventors, the 'Mo' cell line was patented by the Regents of the University of California (UC). This action was in accordance with the policy of the university that a patent should be obtained for any invention of its research staff. Once the patent was approved, with the cell line having an estimated value exceeding several billion dollars, Dr. Golde and UC sought collaboration with Genetics Institute and Sandoz Pharmaceuticals. Mr. Moore returned to UCLA Medical Center in 1983 and was asked to sign an additional consent form as a mere formality. After he asked about the profitability of his tissues, Mr. Moore refused to sign the consent form and, instead, filed a lawsuit against Golde, Quan, UC, Sandoz, and Genetics Institute, with the claim that he was entitled to a share of the profits from the cell line derived from his tissues. Specifically, Mr. Moore claimed that the use of his cells was unauthorized, that there was no informed consent, and that Dr. Golde breached his fiduciary duty. The California Superior Court dismissed Moore's claims. However, on appeal, the California Court of Appeals ruled in favor of Mr. Moore. In 1990, the California Supreme Court, in a 5–2 decision, reversed the lower court's ruling. Specifically, the court determined that Mr. Moore did not have property rights over tissues excised from his body. They did, however, rule that Mr. Moore could sue Dr. Golde for breech of fiduciary duty because he prospectively knew of the potential commercial value of the cells and did not provide Mr. Moore with that information prior to his surgery. The court concluded that, in order for a physician to satisfy their fiduciary duty, they must 'disclose personal interests unrelated to the patient's health, whether research or economic, that may affect his judgment' [111, 122, 124]. In summary, Moore v. Regents of the University of California broadened the ability of the PTO to grant biotechnology patents.

While the ethical and legal controversy continues, without further legislative guidance from Congress, the Supreme Court will defer to the Patent Act of 1952. Congress is free to amend this statue, if they so desire, to exclude living organisms from patent protection. However, there is no indication that this will occur [106]. As mentioned previously, only US patent law has been discussed. Space does not permit a discussion of foreign patent law. However, in most European and Asian countries, the patent laws are much more restrictive and do not allow the patenting of even plants, much less living organisms [106]. For excellent reviews of patents in biotechnology see references [106, 110–112, 125].

7 Society's response to tissue engineering

When the professional ethics of any field are being considered, it is important to also reflect on society's impression of that field. In the field of tissue engineering, this issue is even more important because of the wide variety of ethical issues associated with the field. Because tissue engineering is such a young discipline, it is possible that misperceptions and misunderstanding can occur. If they are not addressed immediately, they could perpetuate and severely impair the reputation of tissue engineering. It would be a tragedy if there was a decrease in tissue engineering support because the public perceived the field to be 'playing God' or 'building Frankensteins'. In fact, as shown in the case of 'Dolly' the sheep and the mouse with the human ear on its back, many lay people believe that is exactly what scientists are doing. However, if the time and effort is made to educate the public about how tissue engineering may improve the lives of their loved ones, people will be more likely to support tissue engineering research programs. Previous sections of this chapter (sections 4 and 5) have included some of society's impressions of the issues discussed. Similarly, animal rights issues were mentioned in the xenotransplantation portion of this chapter. In this segment, we will discuss a few areas of society that are important but have not yet been mentioned.

7.1 Religious leaders view of tissue engineering

Society often turns to their religious leaders for guidance concerning complex issues such as biomedical science. It is therefore important to understand the philosophical and ethical positions of religious leaders.

Because of its infancy, theologians have not addressed the field of tissue engineering. However, we must look to related fields for guidance.

Most religions support organ and tissue donation and transplantation. In general, religious leaders feel that, as long as the donor is not in danger and the transplant would offer real medical hope for the recipient, organ transplantation is acceptable [126]. However, few religions have formally addressed the issue of xenotransplantation. It should be noted that, while there have not been any formal statements, the majority of ethics committees have reserved seats for the clergy. This allows religions to be proactive in their opinions and aids in educating the committee on the complex issues between religion and bioscience.

Religious leaders have united over the issue of patent rights. On May 18, 1995, approximately 200 American clergy from over 80 different religions, including Roman Catholic, Jewish, Protestant, Buddhist, Hindu, and Muslim, assembled in Washington D.C. to oppose the patenting of life [127, 128]. Under the banner of the 'Joint Appeal Against Human and Animal Patenting', the coalition said: 'We are opposed to the patenting of human embryos, genes, cells, and the patenting of animals including those with human genes engineered into their permanent genetic code.' Religious leaders maintain that patenting animals or humans, in any form, violates divine law. United Methodist Bishop Kenneth Carder points out that throughout history scientists have identified elements such as oxygen, but did not claim ownership of the elements. He states that genes should be treated the same way the elements are treated. Using this philosophy, patenting products or processes that use the elements (or genes) is acceptable [127, 128]. For example, as mentioned earlier, the cDNA for a gene may be patentable while the original, natural gene is not.

7.2 The public's view of tissue engineering

It is difficult to know how the general public feels about tissue engineering. No detailed surveys on the subject have been conducted. However, there have been several small surveys that can aid in elucidating society's perspective. In a broad based survey of Americans conducted by Princeton Research Associates for *Newsweek*, 70% of those surveyed believe scientists will be able to replace damaged or diseased human organs with laboratory made substitutes [129]. In a Gallup poll conducted in 1993, 75% of Americans said they would be willing to donate an organ at the time of their death [130]. In a similar survey of the general population, conducted in the same year by the Partnership for Organ Donation in the USA, 80% of those surveyed said they would accept an organ transplant if it would save their life and 50% said they would accept a xenotransplant if no human organ was available [63]. From these limited surveys it appears that society supports and agrees with tissue engineering research goals. It should be mentioned that the *Time/CNN* poll discussed earlier in this chapter revealed a lack of support for at least some aspects of genetic engineering. However, public opinion can change quickly. This possibility was expressed in March 1997 by NIH director Harold Varmus in a U.S. Congressional subcommittee evaluating the policy on human cloning. Varmus stated that it could take just one infertile couple, arguing that cloning provides their only chance to bear a child, to turn public opinion [97]. Therefore, it is possible that, through education and public awareness, the lay public will garner an understanding of tissue engineering and thereby become more supportive of tissue engineering principles and projects.

8 Summary

The ultimate objective of tissue engineering is to create tissue constructs that can be used to restore, repair, or replace damaged tissues. Ideally, using a small biopsy sample from the patient and *ex vivo* tissue engineering techniques, a tissue construct could be produced which would serve as an implantable autograft. If tissue engineers want to reach this goal, present research must continue. It is likely that the public focus and debate will wax and wane as new discoveries occur. However, the significant ethical, legal, and social issues of using tissues in research and as implants will remain. Therefore, it is important for each individual actively involved in tissue engineering to develop a set of ethical standards based upon ethical principles. As C. Krauthammer stated in a *Time* magazine article, 'you cannot repeal biology'. No amount of regulation will prevent the progression of science or technology [131]. The human mind is a creative vessel that continually invents ways of improving the life it sees around it. As with any scientific or engineering venture, the development of tissue engineered products is a natural instinct. Dr. Lewis Thomas summarizes this well in his 'Notes of a Biology Watcher':

> Is there something fundamentally unnatural, or intrinsically wrong, or hazardous for the species, in the ambition that drives us all to reach a comprehensive understanding of nature, including ourselves? I cannot believe it. It would seem to me a more unnatural thing, and more of an offense against nature, for us to come on the scene endowed as we are with curiosity, and naturally talented as we are for the asking of clear questions, and then for us to do nothing about it, or worse, to try to suppress the questions [132].

Hopefully, by reading this chapter, the reader will think about these issues, reflect on their importance, and start to formulate a basic understanding and personal philosophy regarding the ethics of tissue engineering and its use.

References

1. Shaw, M. W., *After Barney Clark*. UT Press, Austin, TX, 1984, pp. ix–4.
2. Pabst, P. L., Ear no evil: The controversy over the mouse with the human ear. *Tissue Engineering*, 1996, **2**(1), 83–84.
3. Toufexis, A., An eary tale. *Time*, November 6, 1995, **146**(19), 60.
4. Cowley, G., Underwood, A. and Murr, A., Replacement parts. *Newsweek*, January 27, 1997, **129**(4), 66–69.
5. Hirshberg, C., Altered states. *Life*, December, 1995, **18**(14), 80–85.
6. Maddry, L., How would you like your ear back? In *The Virginian-Pilot*, November 22, 1995, E1.
7. Steinmetz, G. and Hirshberg, C., Altered states. *Life*, December, 1995, **18**(14), 80–85.
8. Clark, C., Human ear grown on mouse. *Blood Weekly*, November 6, 1995, 16–17.
9. Toufexis, A., An eary tale. *Time Australia*, November 6, 1995, (44), 83.
10. Durst, W., Drunken tree torture. *Progressive*, January, 1996, **60**(1), 12.
11. Wilmut, I., Schnieke, A. E., McWhir, J., Kind, A. J. and Campbell, K. H. S., Viable offspring derived from fetal and adult mammalian cells. *Nature*, 1997, **385**, 810–813.
12. Lamb, D., *Organ Transplants and Ethics*. Routledge, New York, 1990, pp. 7–23.
13. Watson, J. D., *The Double Helix*. Mentor Books, New York, 1968, pp. 143.
14. Erickson, D., Material help — Bioengineers produce versions of body tissues. *Scientific American*, 1992, **267**(2), 114–116.
15. Initiative launches program. *Pittsburgh Business Times*, June 12, 1995, **14**(44), 28.
16. Cash is the connective tissue. *Pittsburgh Business Times*, August 12, 1996, **16**(2), 10.
17. Strohl, L., Big game fishing. *Executive Report*, December, 1995, **14**(4), 62–63.
18. Robinet, J. E., Medical initiative grows tissue, cultures an industry. *Pittsburgh Business Times*, January 6, 1997, **16**(23), 4.
19. Robinet, J. E., Growth culture. *Pittsburgh Business Times*, August 12, 1996, **16**(2), 1–2.
20. Dickerson, L. A., Tommy fell asleep. *Executive Report*, May, 1996, **14**(9), 3.
21. The Pittsburgh Tissue Engineering Initiative. Center for Biotechnology and Bioengineering, Pittsburgh, PA, 1996, http://www.pittsburgh-tissue.net/.
22. Langer, R. and Vacanti, J. P., Tissue engineering. *Science*, 1993, **260**(5110), 920–926.
23. United States. Department of Health and Welfare. Public Health Service, Final report of the Tuskegee Syphilis Study Ad Hoc Advisory Panel. Washington D.C., 1973.
24. Kampmeier, R. H., Final report on the 'Tuskegee syphilis study'. *Southern Medical Journal*, 1974, **67**(11), 1349–1353.
25. Curran, W. J., The Tuskegee Syphilis Study. *New England Journal of Medicine*, 1973, **289**(14), 730–731.

26. Cobb, W. M., The Tuskegee Syphilis Study. *Journal of the National Medical Association,* 1973, **65**(4), 345–348.

27. Nishimi, R. Y., From the congressional office of technology assessment: Biomedical ethics in US public policy. *Journal of the American Medical Association,* 1993, **270**(24), 2911.

28. Self, D. J., Medical ethics (PHIL 480) course syllabus. Texas A & M University, College Station, TX, Spring, 1990.

29. Childress, J. F., The normative principles of medical ethics. In *Medical Ethics,* ed. R. M. Veatch. Jones and Bartlett, Boston, 1989, pp. 27–48.

30. Trials of War Criminals before the Nuremberg Military Tribunals under Control Council Law. U.S. Government Printing Office, 1949, **2**(10), 181–182.

31. World Medical Association, Declaration of Helsinki: Recommendations Guiding Medical Doctors in Biomedical Research Involving Human Subjects. Helsinki, Finland, June 1964.

32. United States Food and Drug Administration, IRB Operations and Clinical Investigation Requirements; Appendix G: World Medical Association Declaration of Helsinki. FDA, World Wide Web, Downloaded January 22, 1997, IRB Operations: http://www.fda.gov/oc/oha/toc.html. Appendix G: http://www.fda.gov/oc/oha/appendg.html.

33. Beecher, H. K., Ethics and clinical research. *New England Journal of Medicine,* 1966, **274**(24), 1354–1360.

34. United States. The National Commission for the Protection of Human Subjects of Biomedical and Behavioral Research, The Belmont Report: Ethical Principles and Guidelines for the Protection of Human Subjects of Research. Department of Health, Education, and Welfare (DHEW), U.S. Government Printing Office, Washington, D.C., 1978. Publications (OS) 78-0012 — 78-0014.

35. United States National Institutes of Health, The Belmont Report. NIH, World Wide Web, April 18. 1979. Converted to HTML April 1996. Downloaded January 22, 1997, http://www.nih.gov/grants/oprr/belmont.htm.

36. United States Food and Drug Administration, IRB Operations and Clinical Investigation Requirements; Appendix F: The Belmont Report. FDA, World Wide Web, Downloaded January 22, 1997, IRB Operations...: http://www.fda.gov/oc/oha/toc.html. Appendix F: http://www.fda.gov/oc/oha/appendf.html.

37. Howell, J. H. and Sale, W. F., eds. *Life Choices: A Hastings Center Introduction to Bioethics.* Georgetown University Press, Washington, D.C., 1995, pp. 537.

38. Veatch, R. M., ed. *Medical Ethics.* Jones and Bartlett, Boston, 1989.

39. Bulger, R. E., Heitman, E. and Reiser, S. J., eds. *The Ethical Dimensions of the Biological Sciences.* Cambridge University Press, New York, 1993, pp. 294.

40. Gillon, R. and Lloyd, A., eds. *Principles of Health Care Ethics.* John Wiley, New York, 1994, pp. 1118.

41. Johnsen, A. R., *The birth of bioethics.* Hastings Center Report, 1993, **23**(6), S1–S15.

42. Cooper, T., Survey of development, current status, and future prospects for organ transplantation. In *Human Organ Transplantation: Societal, Medical-Legal, Regulatory, and Reimbursement Issues,* ed. D. H. Cowan et al. Health Administration Press, Ann Arbor, Michigan, 1987, pp. 18–51.

43. Murray, J. E., Merrill, J. P. and Harrison, J. H., Renal homotransplantation in identical twins. *Surgical Forum*, 1955, **6**, 432–436.
44. United Network for Organ Sharing, 1995 Annual Report of the U.S. Scientific Registry for Transplant Recipients and the Organ Procurement and Transplantation Network-Transplantation Data: 1988–1994. United Network for Organ Sharing, Richmond, VA, and the Division of Transplantation, Bureau of Health Resources Development, Health Resources and Services Administration, U.S. Department of Health and Human Services, Rockville, MD, 1995. http://www.unos.org/UNOS – annual – report/.
45. United Network for Organ Sharing, United Network for Organ Sharing U.S. Waiting List Statistics. In Weekly U.S. Waiting List Statistics, January 22, 1997. Downloaded January 29, 1997, http://www.unos.org/usd1-22-97.htm.
46. United Network for Organ Sharing, Reported Deaths on the Waiting List. Downloaded January 29, 1997, http://www.unos.org/sta – dol.htm.
47. Bisbee, C. A., Bioethics and law issues stand at the genetic frontier of biotechnology. *Genetic Engineering News*, February 1, 1994, 28.
48. Fox, R. C. and Swazey, J. P., *Spare parts: Organ Replacement in American Society*. Oxford University Press, New York, 1992, pp. 3–244.
49. Hauptman, P. J. and O'Connor, K. J., Medical progress: Procurement and allocation of solid organs for transplantation. *New England Journal of Medicine*, 1997, **336**(6), 422–431.
50. Pierce, G. A., Graham, W. K., Kauffman, H. M., Jr. and Wolf, J. M., The United Network for Organ Sharing: 1984 to 1994. *Transplantation Proceedings*, 1996, **28**(1), 12–15.
51. Costanzo, M. R., Augustine, S., Bourge, R., Bristow, M., O'Connell, J. B., Driscoll, D. and Rose, E., Selection and treatment of candidates for heart transplantation. A statement for health professionals from the Committee on Heart Failure and Cardiac Transplantation of the Council on Clinical Cardiology, American Heart Association. *Circulation*, 1995, **92**(12), 3593–3612.
52. Stipp, D. and Moore, A. H., Replaceable you. *Fortune*, November 25, 1996, **134**(10), 131–135.
53. Chiche, L., Adam, R., Caillat-Zucman, S., Castaing, D., Bach, J. F. and Bismuth, H., Xenotransplantation: Baboons as potential liver donors? *Transplantation*, 1993, **55**(6), 1418–1421.
54. Nowak, R., Xenotransplants set to resume. *Science*, 1994, **266**, 1148–1151.
55. Dillner, L., Pig organs approved for human transplants. *British Medical Journal*, 1996, **312**(7032), 657.
56. Concar, D., The organ factory of the future. *New Scientist*, 1994, **142**(1930), 24–29.
57. Pierson, I. R. N., White, D. J. G. and Wallwork, J., Ethical considerations in clinical cardiac xenografting. *The Journal of Heart and Lung Transplantation*, 1993, **12**(5), 876–878.
58. Koshal, A., Ethics issues in xenotransplantation. Bioethics Bulletin, University of Alberta, Canada, 1996, pp. 1–3. http://gpu.srv.ualberta.ca/~ethics/bb6-3xen.htm.
59. Kielstein, R. and Sass, H.-M., From wooden limbs to biomaterial organs: The ethics of organ replacement and artificial organs. *Artificial Organs*, 1995, **19**(5), 475–480.

60. Palca, J., Animal organs for human patients? *Hastings Center Report*, 1995, **25**(5), 4.

61. Platt, J. L., A perspective on xenograft rejection and accommodation. *Immunological Reviews*, 1994, **141**, 127–149.

62. Williams, N., Pig-human transplants barred for now. *Science*, 1997, **275**(5299), 473.

63. Cooper, D. K. C., Xenotransplantation: Benefits, risks and regulation. *Annals of the Royal College of Surgery England*, 1996, **78**, 92–96.

64. Simmons, P. D., Ethical considerations of the artificial heart implantations. *Annals of Clinical and Laboratory Science*, 1986, **16**, 1–12.

65. Quoted by Kaukas, D., Artificial hearts. In *Scene Magazine, The Louisville Times*. Louisville, KY, September 22, 1984, 5.

66. Asonuma, K. and Vacanti, J. P., Cell transplantation as replacement therapy for the future. *Critical Care Nursing of North America*, 1992, **4**(2), 249–254.

67. Hamilton, J. O. C., Miracle cures may be in your skin cells. *Business Week*, December 6, 1993, (3349), 76–81.

68. Cima, L. G. and Langer, R., Engineering human tissue. *Chemical Engineering Progress*, 1993, **89**(6), 46–54.

69. Ezzell, C., Tissue engineering and the human body shop: designing 'bio-artificial' organs. *The Journal of NIH Research*, 1995, **7**, 49–53.

70. Hubbell, J. A. and Langer, R., Tissue engineering. *Chemical and Engineering News*, March 13, 1995, **73**(11), 42–54.

71. Lipkin, R., Tissue engineering: Replacing damaged organs with new tissues. *Science News*, July 8, 1995, **148**(2), 24–26.

72. Pitta, J., Biosynthetics. *Forbes*, May 10, 1993, **151**(10), 170–171.

73. Nerem, R. M., Tissue engineering in the USA. *Medical and Biological Engineering and Computing*, 1992, **30**(4), CE8–CE12.

74. Mullen, M. A., Williams, J. I. and Lowy, F. H., Transplantation of electively aborted human fetal tissue: physicians' attitudes. *Canadian Medical Association Journal*, 1994, **151**(3), 325–330.

75. Downing, R., Historical view of pancreatic islet transplantation. *World Journal of Surgery*, 1984, **8**(2), 137–142.

76. Fichera, G., Implanti omoplastici feto-umani nei cancro e nel diabete. *Tumori*, 1928, **14**, 434.

77. Vawter, D. E. and Caplan, A., Strange brew: The politics and ethics of fetal tissue transplant research in the United States. *The Journal of Laboratory and Clinical Medicine*, 1992, **120**(1), 30–34.

78. Goldsobel, A. B., Hass, A. and Stiehm, E. R., Bone marrow transplantation in DiGeorge syndrome. *Journal of Pediatrics*, 1987, **111**(1), 40–44.

79. Hayflick, L., Fetal tissue banned...and used. *Science*, 1992, **257**(5073), 1027.

80. Rojansky, N. and Schenker, J. G., The use of fetal tissue for therapeutic applications. *International Journal of Gynecology and Obstetrics*, 1993, **41**(3), 233–240.

81. Lehrman, D., *A summary: Fetal research and fetal tissue research.* Association of American Medical Colleges, Washington D.C., 1988, pp. 34.

82. Garry, D. J., Caplan, A. L., Vawter, D. E. and Kearney, W., Are there really alternatives to the use of fetal tissue from elective abortions in transplantation research? *New England Journal of Medicine*, 1992, **327**(22), 1592–1595.

83. Sanders, L. M. and Raffin, T. A., Ethical ground rules for fetal tissue research in the post-moratorium era. *Chest,* 1994, **106**(1), 2–4.

84. Auerbach, A. D., Umbilical cord blood transplants for genetic disease: Diagnostic and ethical issues in fetal studies. **Blood Cells,** 1994, **20**(2–3), 303–309.

85. Rosner, F., Risemberg, H. M., Kark, P. R., Sordillo, P. P., Bennett, A. J., Buscaglia A., Cassell, E. J., Farnsworth, P. B., Halpern, A. L., Henry, J. B., Landolt, A. B., Loeb, L., Rogatz, P., Numann, P. J., Ona, F. V., Lowenstein, R. E., Sechzer, P. H. and Wolpaw, J. R., Fetal tissue and transplantation. *New York State Journal of Medicine,* 1993, **93**(3), 174–177.

86. Bell, N. M. C., Regulating transfer and use of fetal tissue in transplantation procedures: The ethical dimensions. *American Journal of Law and Medicine,* 1994, **20**(3), 277–294.

87. Hurd, R. E., Ethical issues surrounding the transplantation of human fetal cells. *Clinical Research,* 1992, **40**(4), 661–666.

88. Hoffer, B. J. and Olson, L., Ethical issues in brain-cell transplantation. *Trends in Neurosciences,* 1991, **14**(8), 384–388.

89. Marshall, E., Embryologists dismayed by sanctions against geneticist. *Science,* 1997, **275**(5299), 472.

90. Handyside, A. H., Lesko, J. G., Tarin, J. J., Winston R. M., and Hughes, M. R., Birth of a normal girl after in vitro fertilization and preimplantation diagnostic testing for cystic fibrosis. *New England Journal of Medicine,* 1992, **327**(13), 951–953.

91. Marshall, E., Rules on embryo research due out. *Science,* 1994, **265**(5175), 1024–1026.

92. Marshall, E., Mammalian cloning debate heats up. *Science,* 1997, **275**(5307), 1733.

93. Buck, C., McGuire, S., Underwood, A., Rogers, A. and Hager, M., Little lamb, who made thee? *Newsweek,* March 10, 1997, **129**(10), 53–59.

94. Woodward, K. L. and Underwood, A., Today the sheep... *Newsweek,* March 10, 1997, **129**(10), 60.

95. Williams, N., Cloning sparks calls for new laws. *Science,* 1997, **275**(5305), 1415.

96. Pennisi, E. and Williams, N., Will Dolly send in the clones? *Science,* 1997, **275**(5305), 1415–1416.

97. Gorman, C., Neti and ditto. *Time,* March 17, 1997, **149**(11), 60.

98. Kotulak, R., First mammal, a sheep, is cloned: One of nature's taboos broken. In *Houston Chronicle,* Houston, TX, February 24, 1997, 1A, 6A.

99. Carey, J., The biotech century. *Business Week,* March 10, 1997, 78–90.

100. Lederer, E. M., Britain will terminate funds for project that cloned lamb. In *Houston Chronicle,* Houston, TX, March 2, 1997, 26A.

101. Cable News Network Inc., CNN — Poll: Most Americans say cloning is wrong. CNN Interactive, March 1, 1997, http://cnn.com/TECH/9703/01/clone.poll/index.html.

102. Kluger, J. and Thompson, D., Will we follow the sheep? *Time,* March 10, 1997, **149**(10), 66–72.

103. Anand, K. J. S. and Hickey, P. R., Pain and its effects in the human neonate and fetus. *New England Journal of Medicine,* 1987, **317**(21), 1321–1329.

104. Anderson, F., Glasier, A., Ross, J. and Baird, B. T., Attitudes of women to fetal tissue. *Journal of Medical Ethics*, 1994, **20**, 36–40.

105. Tuch, B. E., Human fetal tissue for medical research. *The Medical Journal of Australia*, 1993, **158**(9), 637–639.

106. Hueni, A., Patents in biotechnology. *Medicinal Research News*, 1992, **12**(1), 41–53.

107. Lipscomb, E. B., *Lipscomb's Walker on Patents*. 3rd edn., Vol. I. Lawyers Co-Operative Publishing Co., Rochester, NY, 1984, pp. 58.

108. Loughlan, P. L., The patenting of medical treatment. *The Medical Journal of Australia*, 1995, **162**(7), 376–380.

109. Nisselle, P., The ethics of patenting medical treatment. *The Medical Journal of Australia*, 1995, **162**(7), 341.

110. McCoy, T. J., Biomedical process patents: Should they be restricted by ethical limitations? *The Journal of Legal Medicine*, 1992, **13**(4), 501–519.

111. Hartman, R. G., Beyond Moore: Issues of law and policy impacting human cell and genetic research in the age of biotechnology. *The Journal of Legal Medicine*, 1993, **14**(3), 463–477.

112. Fleising, U. and Smart, A., The development of property rights in biotechnology. *Culture, Medicine and Psychiatry*, 1993, **17**(1), 43–57.

113. Butler, D., US company comes under fire over patent on umbilical cord cells. *Nature*, 1996, **382**(6587), 99.

114. Kimbrell, A., A question of ethics. *Modern Maturity*, 1995, **38**(3), 28–29.

115. Swinbanks, D., Gene therapists face double check. *Nature*, 1994, **369**(6475), 5.

116. ACOG Committee Opinion: Committee on Ethics, Commercial ventures in medicine: Concerns about the patenting of procedures. *International Journal of Gynecology and Obstetrics*, 1994, **44**(1), 87.

117. U.S. Patent and Trademark Office, Patent Laws. July 1995, Downloaded January 8, 1997, http://www.uspto.gov/web/offices/pac/doc/general/laws.html.

118. Office of Technology Transfer and Intellectual Property, Patent Issues: Questions and Answers. University of Pittsburgh, Pittsburgh, PA, 1996.

119. U.S. Patent and Trademark Office, What can be patented? July 1995, Downloaded January 8, 1997, http://www.uspto.gov/web/offices/pac/doc/general/what − can − be − patented.html.

120. U.S. Patent and Trademark Office, General Information. July 1995, Downloaded January 8, 1997, http://www.uspto.gov/web/offices/pac/doc/basic/geninfo.html.

121. U.S. Patent and Trademark Office, A guide to filing a design patent application. Downloaded January 8, 1997, http://www.uspto.gov/web/offices/pac/design/.

122. Bereano, P. L., Body and soul: The price of biotech. In *The Seattle Times*, Seattle, August 20, 1995, B5.

123. Bereano, P. L., Patent pending: The race to own DNA. In *The Seattle Times*, Seattle, August 27, 1995, B5.

124. Greenberg, W. and Kamin, D., Property rights and payment to patients for cell lines derived from human tissues: An economic analysis. *Social Science and Medicine*, 1993, **36**(8), 1071–1076.

125. Eisenberg, R. S., Structure and function in gene patenting. *Nature Genetics*, 1997, **15**, 125–130.

126. Committee on Medicine and Religion, Religious views on organ/tissue donation and transplant. Texas Medical Association, Austin, TX, 1996, pp. 1–4. http://www.donor-network.org/pubedu/Relig.html.

127. Religious leaders oppose patenting life. Los Angeles Times News Service, May 18, 1995, http://weber.u.washington.edu/~radin/stand.htm.

128. Charatan, F. B., US religious groups oppose gene patents. *British Medical Journal*, 1995, **310**(6991), 1351.

129. Living in the 21st century. *Newsweek*, January 27, 1997, **129**(4), 57.

130. Berry, P. H., Ethics in transplantation. *Texas Medicine*, 1997, **93**(2), 37–42.

131. Krauthammer, C., The age of cloning. *Time*, March 10, 1997, **149**(10), 60–61.

132. Thomas, L., Notes of a biology-watcher the hazards of science. *New England Journal of Medicine*, 1977, **296**(6), 324–328.

CHAPTER II.17

Tissue Engineering: Product Applications and Regulatory Issues

KIKI B. HELLMAN

Center for Devices and Radiological Health,
United States Food and Drug Administration,
5600 Fishers Lane, Rockville,
Maryland 20857, USA

EMMA KNIGHT

Center for Biologics Evaluation and Research,
United States Food and Drug Administration,
1401 Rockville Pike,
Rockville, Maryland 20857, USA

and

CHARLES DURFOR

Center for Devices and Radiological Health,
United States Food and Drug Administration,
5600 Fishers Lane,
Rockville, Maryland 20857, USA

The mission of the United States Food and Drug Administration (FDA), a science-based regulatory agency in the United States Public Health Service (PHS), is to promote and protect the public health by assuring the safety of foods, cosmetics, and radiation-emitting electronic products, as well as the safety and effectiveness of human and veterinary pharmaceuticals, biologicals, and medical devices, as assessed by scientific principles and methods. To accomplish its mission over the wide range of products in its regulatory purview, the FDA has six Centers, each staffed with the scientific and regulatory expertise to evaluate the products in the Center's jurisdiction. The FDA recognizes that an important segment of the products that it regulates arises from new technological achievements and innovations. One example is the products developed through tissue engineering technology which applies life sciences and engineering principles to the restoration, maintenance, modifica-

tion, improvement, or replacement of human tissue or organ function. Tissue engineered medical products span a range of products including: transplanted human tissues or organs (i.e. autologous or allogeneic tissue); animal tissues or organs (e.g. transgenic animals or xenotransplants); processed, selected, or expanded mammalian cells (e.g. somatic and genetic cellular therapies) with or without biomaterials; and totally synthetic materials of biomimetic design. Representatives of these product classes are in different stages of development: some have been approved for use in the marketplace, while others are under regulatory evaluation or experimental investigation. Regulatory evaluation of products is conducted on a case-by-case basis. The FDA has adopted a cooperative approach across the appropriate FDA Centers in developing science-based regulatory approaches for tissue engineered products (TEPs). These include rulemaking, InterCenter Agreements, a Proposed Approach to Regulation of Cellular and Tissue-Based Products, and initiatives of cross-cutting working groups such as the Wound Healing Clinical Focus Group, the FDA InterCenter Tissue Engineering Working Group, and the Tissue Reference Group. Legislative and regulatory changes simplified and facilitated the administrative process for evaluating novel combination products emanating from such interdisciplinary technology as tissue engineering and to resolve questions of product regulatory jurisdiction. Under these procedures, the FDA may designate a lead FDA Center for product review based on the primary mode of action of the combination product, with additional Center(s) designated to assist in the evaluation in a collaborative or consultant capacity. In addition, a proposed regulatory framework is provided as a measure to define the criteria for product characterization for the regulation of human cellular and tissue-based products. It recommends the level of regulatory oversight based on the degree of public health risk and protection deemed appropriate. For those TEPs requiring premarket review, the assessments of safety and effectiveness constitute the basic elements of the evaluation. Issues of product manufacture, such as product consistency and stability, preclinical safety such as materials sourcing, adventitious agents, and toxicity testing, as well as preclinical activity, such as materials characterization, biomaterials compatibility testing, and in vitro / animal models are considered in this evaluation. Important elements in clinical investigations are clinical trial design and efficacy endpoints.

Postmarket studies may be necessary when all questions of product safety and effectiveness cannot reasonably be determined during premarket clinical trials. The critical elements that will continue to be important in the regulatory review for all products are: the manufacturers' claim of intended use; the product quality control procedures; and the product's performance. As tissue engineering technology evolves, additional issues of product safety and effectiveness will be addressed by the FDA as part of its continuing review and assessment of TEPs. The FDA Centers will continue to use multiple approaches in the science-based premarket review and postmarket surveillance of TEPs, including research, data and information monitoring, regulatory guid-

ance, training and education, and cooperation with public and private groups. These efforts will continue to address the scientific and regulatory issues of TEPs and, thus, will contribute to establishing the proper niche for these products in the armamentarium of clinical medicine.

1 Introduction

1.1 The FDA — a scientific regulatory agency: mission and background

The U.S. Food and Drug Administration (FDA) is a science-based regulatory agency in the U.S. Public Health Service (PHS). The FDA's mission is to promote and protect the public health by ensuring the safety of foods, cosmetics, and radiation-emitting electronic products as well as the safety and effectiveness of human and veterinary pharmaceuticals, biologicals and medical devices as assessed by scientific principles and methods. The authority of the FDA includes premarket approval and postmarket surveillance of products covered by its mandate as derived from the following legislation: Food, Drug, and Cosmetic (FD & C) Act of 1938, Section 351 and Section 361 of the United States Public Health Service (PHS) Act (1944), 1976 Medical Device Amendments to the FD & C Act, the Safe Medical Devices Act (SMDA) of 1990 and the Medical Device Amendments of 1992. The FDA evaluates and approves or clears new medical products for the marketplace, inspects manufacturing facilities before and during commercial distribution, and takes corrective action to remove products from commerce when they are unsafe, ineffective, misbranded, or adulterated. The burden of proof with regard to the product's safety and effectiveness is the responsibility of the manufacturer. The premarket review of all products is based on the manufacturer's stated intended use of each product and the information submitted.

To accomplish its mission over the wide range of products in its regulatory purview, the FDA's structure provides for six centers, each staffed with the scientific and regulatory expertise to deal with the products in the centers' jurisdiction. Regulations for human medical products are based on whether the product is a biological, device, or drug with products reviewed by the FDA Center with the lead responsibility and jurisdiction for the particular class of products as follows:

- Center for Biologics Evaluation and Research (CBER) — Biologics
- Center for Devices and Radiological Health (CDRH) — Devices
- Center for Drug Evaluation and Research (CDER) — Drugs

The three additional Centers and their areas of responsibility are:

- Center for Food Safety and Applied Nutrition (CFSAN) — Foods and Cosmetics
- Center for Veterinary Medicine (CVM) — Animal Drugs and Feed
- National Center for Toxicological Research (NCTR) — Product-related toxicology research

In addition, the following FDA Offices cooperate with the individual centers by providing assistance on regulatory procedures and processes, as necessary:

- Office of Chief Council (OCC)
- Office of the Chief Mediator and Ombudsman
- Office of Health Affairs (OHA)
- Office of International Affairs (OIA)
- Office of Legislative Affairs (OLA)
- Office of Orphan Products (OOP)
- Office of Policy
- Office of Regulatory Affairs (ORA)

In conjunction with premarket approval and postmarket surveillance, the FDA conducts research to support the scientific evaluation of products. This includes studies of biological effects for information on product safety and effectiveness, product analysis and testing to assess quality, and development of test methods for product assessment [1, 2]. For example, FDA research has provided the basis for the regulatory evaluation of transgenic animal systems for biopharmaceutical production, vaccines developed through nucleic acid recombinant techniques, and biomaterial-containing medical devices for implantation.

1.2 Tissue engineered products: a continuum

The FDA recognizes that an important segment of the products that it regulates arises from new technological achievements and innovations. Products developed through tissue engineering (TE) technology, i.e. tissue engineered medical products, are one example. Tissue engineering is the application of the principles of life sciences and engineering to develop biological substitutes for the restoration, maintenance, modification, improvement, or replacement of tissue or organ function [3]. In the broadest sense, tissue engineered medical products span a spectrum of products including: transplanted human tissues or organs (i.e. autologous or allogeneic tissue); animal tissues or organs (e.g. transgenic animals or xenotransplants); processed, selected or expanded mammalian cells (e.g. somatic and genetic cellular therapies) in combination with or without biomaterials; and totally synthetic materials of biomimetic design. Both the range of products and the rapid evolution in product design have generated a certain degree of ambiguity in the definition of tissue engineered medical products. There is no precise and globally accepted definition for tissue engineered products (TEPs). Skalak and Fox [4] and Langer and Vacanti [5] define TEPs not as a type of product, but rather as the result of applying engineering and life science methods and principles to create biological substitutes to restore, maintain or improve functions. Presentations from the Australian, Canadian, European, Japanese and United States regulatory authorities at the May 1996 Toronto Workshop on Tissue Engineering [6], focused on a narrower range of products, including human or animal cells or tissues in combination with or without a supportive natural scaffold (e.g. collagen or hyaluronic acid) or a biopolymer support (e.g. copolymers of polyglycolic acid (PGA), polyurethane).

1.3 Evolving technologies and product regulatory strategies

Since tissue engineered therapies do not constitute a single class of medical products but rather a broad continuum of treatment modalities, it is not surprising that the responsibility for overseeing their development and commercialization within the United States (U.S.) government has been divided among many different regulatory agencies, centers, and programs. The Health Resources Services Administration (HRSA) is charged with the oversight of the National Organ Transplant Program and the National Marrow Donor Program. Other minimally processed tissue products (e.g. musculoskeletal tissue, skin, cornea) are the regulatory responsibility of the Human Tissue Program, Office of Blood Research and Review, CBER, FDA. Premarket review of tissue-engineered medical devices (e.g. cellular wound healing products incorporating a supportive matrix) is performed by the CDRH, FDA. Biological products (e.g. xeno-transplant, somatic cell, encapsulated cell and gene therapies) are regulated by the Office of Therapeutics Research and Review, CBER, FDA.

While generalizing the regulatory considerations for all TEPs and the Federal government programs cited above is probably not possible and certainly not within the scope of this chapter, the level of FDA premarket review for TEPs is based on the concept of a continuum of risk to the public health associated with the following factors: transmission of communicable disease; the degree of cell/tissue manipulation and control of processing; clinical safety and effectiveness, including whether the effect is local and structural or systemic and metabolic; promotion and labeling, including the specific claims for any given product and the need for long term monitoring; and identifying the scope of the industry and its products.

To many who view the promise of TEPs, the possibility of FDA premarket review raises certain questions. Will the premarket review significantly delay the availability of needed products derived from new technologies? Will the agency's requests for extensive testing escalate the already high costs of product development? Are the current FDA regulatory authorities (i.e. FD & C Act, the PHS Act and Title 21 of the Code of Federal Regulations (CFR)) appropriate and sufficiently flexible for this new field of medical products?

Previous experience often provides the best answer. A discussion of the FDA's response to the first recombinant DNA products may assist in predicting how FDA may respond to future dramatic changes in medical product technology. In general, there are four phases to this response: (1) recognition of the development of a new technology; (2) review of the first few premarket applications for products employing the new technology; (3) determination of the common issues for these products which can lead to the development of initial guidance documents for FDA review staff and the industry; and (4) the continual updating of these guidance documents as well as the modification of existing FDA policies and regulations.

The first recombinant DNA product was experimentally expressed in 1973 [7]. This was followed by the expression of recombinant human insulin in September 1978 [8], with its eventual FDA approval in 1982. As an agency composed of research and review staff in clinical medicine and the biological, chemical, and engineering sciences, the initial FDA response to recombinant DNA technology was two-fold: (1) FDA researchers began studying this new area as demonstrated by the publication of numerous articles on recombinant DNA technology between 1973 and 1982; and (2) FDA scientific and clinical reviewers developed an understanding of the ramifications associated with recombinant DNA technology, through information exchange with research colleagues, literature review, conference attendance, and sponsoring of, or participation in, scientific workshops.

The results of research and product reviews and several FDA-sponsored workshops led to the publication of the CBER document, 'Points to Consider in the Characterization of Cell Lines to Produce Biologicals' in 1984. Updates of this 'Points to Consider' document were released in 1987 and 1993. Table 1 lists examples of other recent modifications in FDA policies and regulations indicative of the agency's continual refinement of its policies based on new information and the eventual final response to recombinant DNA technology.

An approach for considering how the FDA will respond to the challenges posed by TEPs is to revisit the questions posed previously. The following discussion considers each of these questions.

Will the premarket review significantly delay the availability of new technologies? It is difficult to predict the length of time required for the FDA premarket review of products since the complexity of the preclinical, clinical, and product manufacturing data can vary considerably among product applications. However, the 21 CFR 814.40 statute requires that the FDA, within 180 days after receipt of a premarket approval application

Table 1. Examples of FDA policies/regulations applicable to recombinant or tissue engineered products

---

Human tissue intended for transplantation, Final Rule. *Federal Register*, July 29, 1997, Vol. 62, p. 40429.

Application of current statutory authorities to human somatic cell therapy products and gene therapy products, Notice. *Federal Register*, October 14, 1993, Vol. 58, p. 53248.

FDA guidance document concerning use of pilot manufacturing facilities for the development and manufacturing of biological products. *Federal Register*, July 11, 1995, Vol. 60, p. 35750.

Elimination of establishment license application for specified biotechnology and specified synthetic biological products. *Federal Register*, May 14, 1996, Vol. 61, p. 24227.

FDA guidance concerning demonstration of comparability of human biological products, Notice. *Federal Register*, April 1996, Vol. 61, p. 18612.

Current good manufacturing practice: amendment of certain requirements for finished pharmaceuticals; proposed rule, *Federal Register*, May 3, 1996, Vol. 61, p. 20103.

Announcement of the availability of regulations for the humanitarian device exemption, Final Rule. *Federal Register*, June 26, 1996, Vol. 61, p. 33231.

Proposed approach to cellular and tissue-based products, notice of availability. *Federal Register*, March 4, 1997, Vol. 62, p. 9721.

InterCenter agreement between the Center for Drug Evaluation and Research and the Center for Devices and Radiologic Health, 1991.

InterCenter agreement between the Center for Biologics Evaluation and Research and the Center for Drugs Evaluation and Research, 1991.

InterCenter agreement between the Center for Biologics Evaluation and Research and the Center for Devices and Radiological Health, 1991.

(PMA), send the applicant an approval order (paragraph 814.44(d), 314), an approvable letter (814.44(e), 314), a not approvable letter (814.44(f), 314), or an order denying approval (814.45 decision on the approvability of a PMA). This statute requires that the application be complete and that the applicant not submit a major amendment to the PMA during this time frame. Similarly, the CFR and the Prescription Drug User Fee Act of 1992 require actions on a biologic license application (BLA) or product license application (PLA) within 180 days. The FDA is committed to expediting the review of products that offer a significant benefit to patients with serious or life threatening illnesses [9]. For example, human insulin was approved in less than 18 months even though it was the first recombinant DNA product to be commercialized. It is also clear that, as the FDA gains further insight into a technology, its product application(s) review times are shortened significantly. For example, DNAase for the treatment of cystic fibrosis was approved by the CBER in 9 months and Dermagraft TC for the temporary covering of burn wounds was approved by the CDRH in under 12 months.

Will the Agency's requests for extensive testing escalate the already high costs of product development? There are many unknowns associated with the safety of any new technology. For example, during the premarket review of recombinant human insulin, concerns were raised regarding the: (1) clinical significance of an anti-insulin immune response; (2) DNA concentrations in excess of 10 pg per injected dose; (3) stability of the expression system; and (4) possible presence of oncogenic material in the final product. These were not hypothetical questions raised by regulatory officials, but rather the unknown questions shared by all scientists and clinicians attempting to understand the implications of this new technology. Fifteen years after the approval of insulin and with many other recombinant products approved during this period, some of the concerns were appropriate, some were unnecessary, and some required specific refinement. It is reasonable to assume that a similar growth in understanding will occur during the review of the first few TEP premarket applications. Thus, the testing involved in evaluating these products will address questions that may in the future appear appropriate, unnecessary, or valid after modification by improvements in an understanding of TEPs, as a result of the normal growth and evolutionary process characteristic of progress in scientific experimentation and clinical investigation.

Are the current FDA regulatory authorities appropriate for this new field of medical products? The FDA policies and regulations will be modified and revised whenever appropriate to foster innovation in medical products while achieving the primary mission of the FDA, i.e. to ensure the public health and safety.

A final question for consideration is: where is the FDA in the process of assessing TEP technology? Several actions suggest that the FDA is entering the later phases of this response. For example, the FDA researchers have authored manuscripts on transgenic animal models, bone morphogenic proteins, and the host's response to implanted biomaterials. Review of the scientific and medical literature suggests that many first generation TEPs are undergoing either clinical evaluation or premarket review. As discussed previously, the human dermal fibroblast skin substitute, Dermagraft-TC, was approved in March 1997 and several FDA guidance documents applicable to TEPs have been released (Table 2). The May 1996 FDA-sponsored Workshop on 'Biotechnology Biomaterials: A

Global Regulatory Perspective for Tissue-Engineered Products' at the 5th World Congress for Biomaterials provides an additional example of the FDA's response to TEP development.

Table 2. A partial list of FDA guidance documents appropriate for products developed by tissue engineering technology*

Characterization of cell and tissue sources available from CBER

Points to consider in human somatic cell therapy and gene therapy, 1991.

Points to consider in the characterization of cell lines to produce biologicals, 1993.

Morbidity and mortality weekly report, HIV screening of tissue donors, Vol. 43, No. RR-8, May 20, 1994.

Draft points to consider in the manufacture and testing of monoclonal antibody products for human use, 1994.

Memorandum to all establishments engaged in manufacturing plasma derivatives, disposition of products derived from donors diagnosed with, or at known high risk for, Creutzfeldt–Jakob disease, August 1995.

Draft document concerning the screening and testing of donors of human tissue intended for transplantation, June 1995.

Draft points to consider in manufacture and testing of therapeutic products for human use derived from genetically altered transgenic animals, 1995.

Draft public health service guideline on infectious disease issues in xenotransplantation, September 1996.

Characterization of cell and tissue sources available from CDRH

Guide for 510(k) review of processed human dura mater, 1990.

'Blue Book Memorandum #G95-1, Use of ISO-10993, Biological Evaluation of Medical Devices Part 1: Evaluation and Testing,' 1992.

Product specific guidances available from CBER

Points to consider in the collection, processing and testing of *ex vivo* — activated mononuclear leukocytes for administration to humans, 1989.

Points to consider on efficacy evaluation of hemoglobin- and perfluorocarbon-based oxygen carriers, 1994.

Draft document concerning the regulation of peripheral blood hematopoietic stem cell products intended for transplantation or further manufacturing into injectable products, 1996.

Guidance on applications for products composed of living autologous cells manipulated *ex vivo* and intended for structural repair or reconstruction, 1996.

Table 2 (*Continued*).

Guidance for the submission of chemistry, manufacturing, and controls information and establishment description for autologous somatic cell therapy products, 1997.

Product specific guidances available from CDRH

Guidance document for the preparation of IDE and PMA applications for bone growth stimulator devices, August 12, 1988.

Guide for 510(k) review of processed human dura mater, June 26, 1990.

Draft guidance: human heart valve allografts, June 21, 1991.

Guidance document for testing orthopedic implants with modified metallic surfaces apposing bone or bone cement, April 28, 1994.

Draft replacement heart valve guidance, 1994.

Draft guidance for the preparation of an IDE submission for a interactive wound and burn dressing, April 4, 1995.

Guideline on validation of the limulus amebocyte lysate test as an end-product endotoxin test for human and animal parenteral drugs, biological products and medical devices, 1987.

---

* While not comprehensive, this is a list of guidance documents pertinent to tissue engineered products. The documents may be obtained by contacting individual centers of the Food and Drug Administration as follows: Center for Biologics Evaluation and Research — (301)827-3844 or (888)223-7329; Center for Devices and Radiological Health — (301)827-0111 or (800)899-0381; Center for Drug Evaluation and Research — (301)827-0577 or (800)342-2722; or the FDA Home Page at http://www.fda.gov.

In addition, a Proposed Approach to Regulation of Cellular and Tissue-Based Products was released for public comment by the FDA in February 1997 [10] as a measure to define clearly the criteria for product characterization in the regulation of human cellular and tissue-based products. This initiative was designated as a Reinvention of Government (REGO) Initiative, identifying it as a high priority in the FDA as well as in other areas of government where input is necessary. The FDA held a public meeting on March 17, 1997 to outline the approach and to offer an opportunity for public input. The proposed regulatory framework provides a unified, consistent approach to the regulation of both conventional and new cellular and tissue-based products. It specifies criteria for regulation(s) and provides for harmonized review of product premarket applications by the appropriate Centers within the Agency. Additionally, the framework provides for Federal government oversight based on the degree of public health protection deemed appropriate. For example, the framework provides flexibility and innovation without an application review process for products with limited public health concerns. It is believed that this new system would provide a rational and comprehensive framework under which manufacturers of cellular and tissue based products, which include some TEPs, could develop and market their products. It strives to ensure that innovation and product development in this rapidly growing field can proceed unhindered by unduly burdensome regulation. At the same time, it would provide physicians and patients with the assurance of safety that the public has come to expect from drugs, biologicals, and medical devices approved by the FDA [10].

In summary, the FDA has established many different approaches in its consideration of the new technological issues posed by products developed through TE technology. In some cases, the lessons learned from previous products may be applicable (e.g. recombinant DNA engineering, facility control for products with great lot-to-lot variability, screening of allogeneic donors, propagation of cell lines). In other cases, the FDA will look to the best scientific minds and methods to determine innovative and appropriate resolution.

2 Science-based regulation and tissue engineered products

The development of any medical product begins with inspiration and need, followed by design, laboratory and animal testing, clinical studies, premarket review, product approval, commercial use, and product refinement (e.g. new indications).

Progress in cell and tissue culture technology, biomaterials science, engineering, computer sciences, and surgical approaches have contributed to the development of tissue engineering as a source of products for ameliorating different medical conditions. Interdisciplinary research teams, focusing on imaginative approaches, and industries, that are translating the research into products, are responsible for the development of TEPs over the last few years to address pathologies of virtually every organ system. There are many types of products in different stages of development; some have been approved and are in use, while others are under regulatory evaluation. They include: wound covering and repair systems, e.g. artificial skins; combination products of bone morphogenic proteins, collagen, and bone for surgery; bone graft substitutes; blood substitutes; cardiovascular products, e.g. replacement heart valves and endothelialized vascular grafts; human tissue products; encapsulated cells for restoration of tissue and organ function, used either as implants, i.e. secretory tissue 'organoids', or *ex vivo* as metabolic support systems; and combinations with drugs or genes, such as in drug delivery or as vehicles for gene therapy. In addition, cultures of structural, secretory, and other cells, tissues, or organs developed for TEPs can be utilized as *in vitro* model systems for evaluating modulation of structure and/or function by toxicological or other agents [2, 3, 11].

Tissue engineered products, like biotechnology-derived and conventional products, are subject to evaluation by the national regulatory agencies responsible for overseeing the approval of medical products before release into the marketplace. The process of product review is basically an assessment of safety and efficacy. Issues of product manufacture such as product consistency and stability, preclinical evaluation such as product preclinical safety and activity, and clinical investigation are considered in this assessment [3, 11].

The 1997 FDA Proposed Approach to Cellular and Tissue-Based Products [10], discussed previously, is an effort to define the characteristics related to tissue and tissue engineered products. Because of the diversity of these products and their intended uses, it is difficult to address all the products through one mechanism. While factors such as the risk of transmission of communicable disease, the degree of manipulation and others outlined in the approach represent an element considered in the overall regulatory framework, the product itself is likely to dictate how each factor determines the necessary degree of regulation.

2.1 Human tissue intended for transplantation

Transmission of communicable disease is addressed as a minimal level of safety for tissues in the Interim Rule for Human Tissues Intended for Transplantation [12] implemented under the authority of Part 361 of the PHS Act. It was published in response to a recognized public health threat and provides for donor screening and testing for hepatitis and human immunodeficiency viruses. In addition to basic testing requirements for 'conventional tissues' including corneas, musculoskeletal and bone tissue, it also requires basic recordkeeping and written procedures and establishes the authority for inspection of tissue establishments. While the interim rule is intended primarily for these conventional tissues, it is clear that such basic requirements also apply to other products, including TEPs, which have characteristically required greater FDA oversight. Thus, these basic requirements represent the minimum level of regulatory oversight required for all cellular and tissue-based products.

2.2 Tissue engineered products requiring premarket review

Those TEPs which require greater oversight under 351 of the PHS Act or under the FD & C Act as either biological products or medical devices should also meet the basic requirements for human tissues intended for transplantation. Products subject to premarket review will be expected to meet these requirements in part through current Good Manufacturing Practices (GMPs), through submission of information to the product application, or through certification of existing standards. For a product consisting of allogeneic cells or tissues, the risk of transmission of infectious diseases must be controlled through adequate donor screening and testing, especially for such life threatening agents as human immunodeficiency virus (HIV). Although certain processing methods may decrease the survival of viruses and, thus, decrease the risk of exposure, no one process is known to be effective for complete inactivation of all viruses without inactivation of the cellular or tissue product as well. The extrinsic risk of introducing adventitious agents during the procurement or sourcing of cells and tissues or during processing also must be considered. The product sensitivity, that is, the effect of the processing procedure or parts thereof on a given product, should also be considered as a risk. The risk to the health care worker as well as to those involved in the processing procedure must also be considered and necessitates adherence to universal precautions. This also raises questions of quarantine and labeling to protect the safety of those involved during procurement or processing until the infectious status of the source cells or tissues is known. Although autologous cells or tissues would be expected to present a lower level of risk of disease transmission, it must be recognized that manipulation and processing can activate latent viruses in some instances. Recognition of this possibility is important so that the recipient does not receive an increased infectious load which, in turn, may present an increased risk. Such concerns may also relate to products incorporating animal tissue(s) other than human [13].

The degree of manipulation or processing of cells or tissues, whether minimal or extensive, is a difficult factor to define when determining the necessary level of regulatory oversight by the FDA. Tissues which are processed only to prevent disease transmission and to maintain tissue integrity generally fall under the Interim Rule and require no submission for efficacy to the FDA. However, most TEPs do not meet this criterion and, thus, require submission of safety and effectiveness data. It must be noted that minimal manipulation itself is difficult to define and the definition is subject to change as processing procedures become standardized. It is recognized that, as new processing methods are developed and encompassed by industry, they gradually become accepted as the norm and may be accepted as standards. As a result, what was considered extensive manipulation several years ago may evolve to represent the norm. Cell selection and separation is one area where existing data and processing may require the submission of less information. If this is the case and a standard industry practice exists and is followed, less information may need to be submitted. Further, in those cases where standards exist or could be adopted, the need for the premarket review of clinical data could decrease or be eliminated. However, processing procedures, such as cellular expansion, activation, encapsulation, or modification are known to change or potentially change the biological characteristics of cells and, as such, may continue require that such cells or tissues be subject to premarket review and the submission of data. Examples of such applications include manipulated autologous cells for structural repair, and the use of such substances as growth hormones and activated lymphocytes for the treatment of cancer, and gene therapies. The addition of biomaterials or mechanical components, as with encapsulated pancreatic islet cells, also raises a greater level of concern due to the greater risks associated with systemic or life-threatening adverse events, concerns over compatibility of the additional component, and the effect of the component at a local or systemic level. In general, the greater the possibility of systemic interactions the greater the concern.

The degree to which processing may affect the end product is also relevant to the degree of regulatory oversight required. For example, certain blood components may be especially sensitive to heat or chemicals, and even though clinical data may not be required, chemistry, manufacturing and controls, product specifications and/or certification with a defined standard must be met. Further, although the requirements for clinical data are not increased, the assurance that precise processing methods are employed may require review.

Reproducible processing methods must be assured for all products. The processing must provide a reproducible end product which is characterized by identified parameters. These include cell viability, cell number, or any other measurable parameter. These parameters should be established based upon their known clinical significance so that they may serve as surrogates in establishing effectiveness.

2.3 Preclinical evaluation of tissue engineered products

The following topics of preclinical safety and efficacy may be appropriate for those TEPs undergoing premarket review. Issues of preclinical product safety include: sourcing for both the biological and biomaterial product components; adherence to safety precautions associated with the use of autologous tissue; appropriate donor sourcing and screening procedures for allogeneic and xenogeneic cells or tissues; testing and process validation to assure that the product is as free as possible from adventitious agents such as potentially infectious endogenous viruses or zoonotic pathogens; toxicity testing (local and systemic, acute or chronic); testing for potential carcinogenicity and immunogenicity via allo- and xenoantigens or immune reactivities to biological or biomaterial combinations; and sterility and effects of sterilization on products [2, 3, 11, 14, 15]. It should also be considered that some biomaterials may produce adverse effects only in the immediate post-implantation period, while others produce only chronic effects with few or no acute effects from biodegradable breakdown products.

Characterization of both the structure and functional activity of the biological and biomaterial components, as well as appropriate tests for biomaterial compatibility, are elements of consideration in assessing the product's preclinical activity. It has been emphasized that biocompatibility testing is only of value and meaningful when the product tested has first been characterized to enable lot-to-lot consistency thereby permitting the repetition of studies and assurance that the clinical products will be manufactured from materials representative of those tested [16]. Moreover, it would be difficult to evaluate analytically any adverse events arising from postmarket surveillance of the products in the absence of adequate biomaterial characterization [16].

In vitro and animal models utilized as assessments of a product's preclinical efficacy must be appropriate, not only for modeling products, but also for assessing therapeutic performance [2, 11, 15]. Cross-species issues of immunoreactivity and pharmacological incompatibility are important issues to consider when designing animal model studies [15]. With the increased use of allogeneic and xenogeneic cell and tissue sources for TEPs comes the need to assure immunocompatibility, graft and host immunocompatibility, or ways to minimize or eliminate the host's untoward immunological reaction or inflammatory response through immunoisolation or other means [17]. In addition, there may be subpopulations of individuals for whom the cells or tissues may have altered safety, such as immunosuppressed individuals. Further, the inflammatory response at the implant site, which is possible for all implants, must be considered [2, 11]. Cross-species immunoreactivity may require evaluating analogous biological components, i.e. canine cells in a dog to model human cells in a human [15]. Moreover, cross-species pharmacological incompatibility, i.e. species specific effect(s), may necessitate separate animal models for addressing the surgical or engineering aspects of the product and its pharmacological activity [15].

Certain scientific issues regarding the biological and biomaterials components of TEPs have been identified as important for the continued progress of product development [2, 11]. For the molecular or cellular component of TEPs, the control of cellular proliferation, differentiation, and modulation for the desired phenotype must be considered together with the development of adequate test methods to monitor cellular activity. When products consist of more than one cell type, it may be important to understand how different cells communicate with one another and how this can be optimized in the desired setting. In addition to the potential inflammatory and immunological considerations described above, the cell's mutagenic and carcinogenic potential must be considered and, if genetically modified cells are utilized, the potential for cell transformation by the vector, vector stability, and optimal functioning of the inserted gene must be considered. For the biomaterials component of TEPs, it is important to determine what structural or mechanical characteristics are dictated by the *in situ* biological environment and whether a naturally-derived or a synthetic material is the appropriate choice for a particular product. To a certain extent, the choice of the biomaterial is dependent on the biomaterial's intended function. Will it persist indefinitely or will it be absorbed over a certain time period? In addition, it may be important to minimize or eliminate reactions at the biomaterial-host interface for both implantable products and those designed for extracorporeal use. For biodegradable materials, it may be important to understand, control, and measure the rate of degradation, as well as to identify and characterize the mechanisms of degradation and its byproducts. Since some inflammation and fibrotic response may accompany any soft tissue implant or surgical procedure, it will be important to determine the extent of fibrosis at the host–material interface and whether the intensity of the response represents a safety risk to the host or interferes with the product's effectiveness [2, 11]. Since a product's functional activity must be consistent and predictable from product to product, attempts should be made to minimize any lot-to-lot variation by consistent processing methods and testing parameters [3].

2.4 Clinical investigations of tissue engineered products

2.4.1 Premarket considerations

The premarket requirement for clinical data depends on at least two factors, e.g. the amount of clinical information known and the impact a therapy may have on a particular disease or disease process. The type and amount of clinical data required will also reflect the sponsor's or manufacturer's claims of intended use for the product. For example, a conventional tissue such as a tendon used for replacement of a tendon may not raise additional clinical questions and would probably not require clinical data submission. If, however, a synthetic material is added to surround the tendon, other questions of biocompatibility and local tissue reactions may be raised. Each escalation of product complexity and marketing claim increases the need for additional clinical data.

In general, the clinical data requirements and the establishment of safety and effectiveness is less intricate for products with local and structural modes of action than those with systemic and metabolic actions. Local effects tend to be more easily measured or monitored with an overall lower clinical risk, while metabolic and systemic actions require more sophisticated monitoring and increased patient risk. The concerns may also overlap. For example, encapsulated pancreatic islet cells may need to be evaluated for local reactions to synthetic capsules, while the greater focus is associated with the overall metabolic action of the cellular secretions. Further, the potential life-threatening and long term effects of a product can increase the concern for its clinical use. In addition, as the size of the population exposed increases so does the concern for potential side effects.

Since both the FD & C Act and the PHS Act require demonstration of product safety and effectiveness prior to commercial distribution, an exemption from these laws is required before the initiation of human clinical trials. Thus, most manufacturers or clinical investigators first interact with the FDA during the process of preparing an exemption application , either an Investigational New Drug Application (IND), or an Investigational Device Exemption (IDE). In general, the contents of both applications are similar and will include a sufficient description of manufacturing methods and prior studies, both preclinical and clinical, if available, to permit an accurate evaluation of product safety and potential benefit. The INDs and IDEs also contain a clinical protocol which describes the proposed patient population, treatment regimen, and follow-up methods. Since there are several differences in the specific requirements of an IND or IDE application (e.g. cost-recovery, device risk assessment), it is suggested that sponsors (e.g. product manufacturer or clinical investigator) review the detailed description of the contents of an application presented in Title 21 of the CFR Part 312.20 (IND) and Part 812.20 (IDE) before preparing a FDA submission.

Once the FDA concurs that human studies may begin, a series of studies designed to investigate product performance in humans are initiated. Clinical investigations are generally performed in different phases of product development to assess the safety and efficacy in actual use situations for the intended patient population. Phase 1 and Phase 2 studies assess safety and efficacy in a limited number of patients. These studies seek to obtain preliminary information concerning product safety, product activity, optimal product dosages, treatment regimens, product pharmacokinetics, systemic availability and/or decomposition mechanisms. They often lead to a better understanding of how product structure correlates with clinical performance, and sponsors often use this information to refine product design or formulation. Consequently, changes in product manufacture or final product composition often result from these preliminary clinical evaluations [18]. Phase 3 studies assess efficacy in expanded clinical trials. These Phase 3 studies usually enroll a sufficient number of patients to permit a predefined level of clinical benefit (i.e. experimental versus control therapies) to be observed with a significance of 0.05 (i.e. a 5% probability that one treatment will be perceived as superior, when both therapies are equivalent) and a power of 0.80 (i.e. a 20% probability that the two treatments will appear equal, when one therapy is superior).

Will full scale Phase 3 studies be required for each new tissue engineered medical product? The answer is, not necessarily. For example, in situations where the cellular components are minimally processed and are used in a manner consistent with physiological function, requirements for clinical testing may be less than that required for a product containing extensively manipulated, expanded, or selected cells used in a non-physiological manner.

This flexible approach to clinical evaluations is illustrated in the CBER document, 'Guidance on Applications for Products Comprised of Living Autologous Cells Manipulated *Ex Vivo* and Intended for Structural Repair or Reconstruction' which states that premarket clinical trials may involve studies that are of short term (< 1 year.) In such cases, subsequent post-approval studies may provide all long term safety and effectiveness data. The FDA also believes that extensive screening of large numbers of patients for systemic toxicity may not be required for products containing *ex vivo* manipulated autologous cells if the product is implanted locally and product action does not involve systemic effects.

The state of knowledge of the natural history of the disease or disease process and alternate treatment interventions at any given point in the disease process must be considered in clinical study design. Considerable thought should be given to the design such as: the selection of appropriate control groups, including historical and sham controls; identification of meaningful endpoints for clinical efficacy and the adequacy of surrogate endpoints; methods for monitoring studies to assure safety; and the appropriate target population. Attention to age, gender, and ethnic subgroup responses should be considered. While clinical efficacy studies are optimally randomized and controlled, cellular implants and other TEPs requiring surgery may not be amenable to the use of surgical sham controls for ethical reasons. Masking of control groups or, if not possible, then masking of evaluators should be attempted, although this is not always possible [2]. In addition, variations among individuals, sensitive subpopulations of individuals, and an understanding of existing underlying pathologies are issues for consideration in actual use situations, since these factors may pose certain safety risks as well as affect the desired end results [3]. Further, since clinical trials are often conducted in patients with serious illnesses or life-threatening diseases, there may be a high degree of variability in response or outcome [15]. Existing FDA regulatory mechanisms, such as accelerated or expedited review, may be appropriate for life threatening or novel therapies.

Product monitoring occurs in two forms, i.e. premarket requirements for clinical safety and effectiveness, and postmarket requirements where long term outcomes cannot reasonably be expected for years. For example, encapsulated pancreatic islet cells may prove to be beneficial in a clinical study of two years duration by establishing less fluctuation in blood glucose levels and, thus, adequate control of blood glucose. However, it is unlikely that long term effects, such as life long freedom from insulin regimens, will be established without long term monitoring and data collection.

As noted previously, these premarket studies are conducted under IND or IDE prior to the submission of a marketing application, either a BLA or PMA.

2.4.2 Postmarket studies and surveillance

The role of postmarketing vigilance and of monitoring product performance during use and commercial distribution is an important consideration. The extent of postmarketing surveillance required for TEPs may vary considerably. At a minimum, product manufacturers and users will need to provide adverse event information through the FDA MedWatch process. In addition, it will be incumbent on manufacturers to develop methods for maintaining records which correlate the original donor for each cell or tissue production lot with the final product recipient when appropriate. Databases which track adverse event reports and permit detection of emerging trends in product toxicity or operator misuse can provide significant information. These methods and databases will be valuable when emerging data suggest potential safety concerns requiring product recalls or rapid communication with physicians and patients.

Additional postmarketing studies may be necessary when the sponsor seeks a change in product labeling, e.g. long term patient outcomes or new patient populations. Postmarketing studies may also be required as a condition of FDA approval. This can occur when questions of long term safety (e.g. implant durability) or efficacy (e.g. validation of surrogate study endpoints) cannot be determined reasonably during the premarket clinical studies. Thus the premarket review process and the extent of postmarketing surveillance will be determined on a case-by-case basis with factors such as clinical indication, patient need, product manufacture and final composition, and premarketing study results being important in such considerations.

2.5 Overall considerations

In summary, the critical elements to be considered in the development of TEPs are: (1) an understanding of the organ or tissue that is being replaced or restored, i.e. its structure, organization, function, physiology, and metabolic requirements; (2) methods to optimize performance *in situ*; (3) meaningful functional assays of performance; and (4) long term monitoring of ongoing safety [3].

A number of regulations and guidance documents have been established by the Federal government which have, in large measure, contributed to a comprehensive approach for TEPs (Table 2). In addition there is increased interaction and cooperation among those Federal agencies responsible for health care in the preparation of such guidance documents. For example, the guidance on the application of xenotransplantation was developed with the active collaboration of the Center for Disease Control (CDC), FDA and National Institutes of Health (NIH), with HRSA providing the lead in the areas of human organ transplantation and the National Donor Marrow Program [13].

The articulation of policy, regulation, and guidance has had an overall beneficial effect on the creation of regulatory pathways leading to premarket approvals in the diverse areas constituted by TEPs. As exemplified by its successes, this evolving regulatory framework has also accommodated the rapid expansion of technologies and therapies. The concept of regulation based on the continuum of risk has been advocated and implemented. As more information is gained in these novel areas and, as standards are developed, the FDA's premarket regulatory requirements and general oversight will adjust accordingly. Tissue engineered products will require innovative regulatory strategies since, among others, they may include novel biomaterials requiring further safety characterization, and the biological components may contribute to product variability and testing complexity. Since they are combination products in many cases, they will require review and cooperation by different centers in the FDA [11, 15].

2.6 Combination products and their evaluation

A subset of products requiring premarket review may include combinations of a biologic, drug, or device. As discussed previously, the FDA regulatory approaches for medical products are based on whether the product is a biologic, drug, or device with products reviewed by the FDA Center with the lead responsibility for the particular class of products — biologicals, CBER; devices, CDRH; and drugs, CDER. The FDA, realizing the complex nature of TEPs which often are combination products, i.e. combinations of a biologic, drug, or device, has taken a cooperative approach across the FDA Centers in developing regulatory approaches for these products. These include rulemaking, InterCenter Agreements, and the initiatives of cross-cutting working groups.

The rule on combination products, 21 CFR Part 3.7, defines these products and discusses the appropriate regulatory approaches. The FDA InterCenter Agreements clarify product jurisdictional issues and describe the agreements entered into by the CBER, CDER, and CDRH; and guidance documents between the Centers describe the allocation of responsibility for certain products and product classes. Thus, Section 16 of the SMDA of 1990 [21 U.S.C. section 353(g)] describes how the FDA will determine which Center within FDA will have the primary jurisdiction for the premarket review and regulation of any combination product comprised of drugs, devices, or biologicals.

The use of synthetic or mechanical components in conjunction with cellular or tissue components usually results in a combination product. Designation of the lead FDA Center for product review is based on the primary mode of action of the combination product. The SMDA states explicitly that the FDA can use any Agency resources necessary to ensure the appropriate review of the combination product.

In November 1991, the FDA adopted a final rule implementing the SMDA mandate. To enhance the efficiency of Agency operations, the scope of the rule was extended to apply to any drug, device, or biological product where the jurisdiction is unclear or in dispute. The regulation specifies how a sponsor can obtain an Agency determination early in the process before any required filing, and which FDA Center will have primary jurisdiction for the premarket review and regulation of the product. The regulation also establishes a 'product jurisdiction officer' within the FDA's Office of the Commissioner to oversee these procedures and processes. The FDA Ombudsman is the Product Jurisdiction Officer.

The regulation applies to two categories of products: (1) any product that constitutes a combination of a drug, device, or biological product; and (2) any drug, device, or biological product where the FDA Center with primary jurisdicion of the premarket review and regulation is unclear or in dispute. The FDA designation assigns a Center to take the administrative lead and also designates the responsibilities of any consulting or collaborating FDA Centers [2].

In addition, the Agency has established a Tissue Reference Group (TRG) to assist in making jurisdictional decisions and applying consistent policy to individual products. The TRG consists of members from the CBER and CDRH and provides a single reference point for all tissue-related questions received by the Centers or the Office of the Chief Mediator and Ombudsman [10] or posed directly from sponsors.

2.7 FDA InterCenter agreements

The FDA InterCenter Agreements, noted in Table 1, were established to clarify product jurisdictional issues for combination products. The CBER, CDER, and CDRH have entered into these agreements and, there are guidance documents between the Centers that describe the allocation of responsibility for certain products and product classes. Currently, tissue engineered products are under the purview of both the CBER and CDRH. Both Centers are involved in the evaluation of these products, with the lead Center for product review currently identified by the primary mode of action of the product; the other Center acts as a consultant in the product evaluation [2, 3].

The sponsor of a premarket application or required investigational filing for a combination or other product covered by these guidance documents may contact the designated Agency Center identified in the InterCenter Agreement before submitting an application for premarket review or to confirm coverage and to discuss the application process. For a combination product not covered by a guidance document or for a product where the Center with primary jurisdiction is unclear or in dispute, the sponsor of an application for premarket review should follow outlined procedures to request a designation of the Center with the primary jurisdiction before submitting the applications. For example, the InterCenter Agreement between the CBER and CDRH considers jurisdictional issues related to biologicals–biopolymer combinations, important for TE applications.

2.8 FDA working groups

Finally, there are the initiatives of working groups for tissue and TEP-related scientific and regulatory issues that cut-across FDA Centers. These include the: Clinical Wound Healing Focus Group; the FDA InterCenter Tissue Engineering Working Group (TEWG); and the Tissue Reference Group (TRG). These groups do not work in isolation but, rather, are engaged in a cooperative, complementary way in addressing issues and providing input/recommendations regarding the scientific issues and the regulatory oversight of TEPs.

The TEWG is composed of scientific research and review staff from five participating FDA Centers, the CBER, CDER, CDRH, CFSAN, and CVM. It was established in July 1994 to identify and address the emerging scientific and science-based regulatory issues of cell and tissue engineered products. The Working Group has facilitated communication, information exchange on scientific and regulatory developments in TE, enhanced cooperation among research, review, and administrative staff across FDA Centers in the review and regulation of TEPs through its different regulatory strategies, and explores global perspectives on TEPs [3, 19]. An important goal is to support and strengthen product review and it works to promote regulatory consistency for TEPs across FDA through networking mechanisms on science and review issues.

The first accomplishment of the TEWG was to focus the Agency's attention on TEPs and to analyze the reviewed products, identifying safety and efficacy issues and providing an overview of research and development. The result of this analysis is the TE White Paper, an internal Agency resource document on product and review issues. It has been converted into an electronic format and entered into the FDA Tissue Engineering Knowledge Base (TEKB). The TEKB also includes: citations and abstracts of the pertinent scientific literature; applicable guidance documents; access to other Agency/Centers databases; pertinent regulations; and a listing of FDA staff with scientific and regulatory expertise for TEPs. The TEKB is currently accessible to all of FDA through the FDA Intranet and has hyperlinks to TE information on the Internet, Agency expertise, and to a conferencing area. It's benefits to the Agency are many — from rapid dissemination of TE developments to an educational resource for new reviewers.

Since the support of product review is an important goal of the Working Group, it has begun an inventory of TEPs that have been approved by FDA or are under investigation in order for FDA reviewers and other staff to identify rapidly the products by category, to monitor trends in product development, and, by entering the inventory into the TEKB, to provide Cross-Centers availability for reviewers of these products.

Recognizing that TEPs are being developed for a potential world-wide marketplace, the TEWG thought it important to understand how entities other than the FDA were regulating these products in order to acquire a global regulatory perspective for the evaluation of TEPs. Together with representatives from Canada and Australia, the TEWG, together with other FDA staff, organized the May 1996 Toronto Workshop for the purpose of understanding the regulatory approaches for TEPs by different national regulatory bodies. The discussions indicated that, while there are certain differences in procedures, there are also similarities in the requirements/standards for product manufacture and performance of TEPs.

Since communication is a very important function for all activities of the TEWG, different approaches are being used to disseminate information on TEPs both within and outside the Agency. These include an education/training program in the Agency and symposia, workshops, and short courses in cooperation with scientific societies, institutes, and academe.

3 The FDA and future perspectives for tissue engineered products

Among the future challenges for the FDA with regard to its regulatory oversight of TEPs are the continued development of science-based rationales for regulatory decision-making that will not only provide a road map for FDA reviewers but also for the industry. Such an approach will continue to enhance product review and address questions for manufacturers early on in product development. Standards and guidance may be very important. Whether a standard or a guidance is the more appropriate mechanism for achieving product consistency and reproducibility may depend, to a certain extent, on the level of regulatory oversight, i.e. the less oversight there may be in place, the greater the need for guidance and the converse for cases of greater established regulatory oversight. Material sourcing continues to be an issue of concern, primarily from the potential for contamination of TEPs with adventitious agents. Control and documentation of source material and manufacturing processes will continue to be important for development of all TEPs containing a biological component. Further, it will not be possible to develop pertinent and realistic standards or guidance unless approaches to adequately characterize the biomaterials used in TEPs are addressed and their reproducibility is assured. Finally, meaningful functional assays of performance for TEPs are critical in order to adequately evaluate their effectiveness and safety.

The FDA Centers will continue to use different approaches in the science-based, premarket review and postmarket surveillance of TEPs. As indicated, these approaches include:

- research (bioeffects analysis and testing; test method development);
- data and information monitoring (databases; literature review); post-market surveillance;
- regulatory guidance (generic Points to Consider and product specific guidance) and standards;
- training and education for both FDA staff and the research and development community (publications, FDA Staff Colleges, industry and FDA reviewer training, workshops and conferences); and
- cooperation with public and private groups (active participation and input are encouraged to develop consensus on important issues and considerations for the development of policy).

Just as the FDA has come together in a cross-cutting approach to study TE technology and its products, the continuing cooperation and communication with the public and private sector are encouraged so that the input of private and other public groups is assured and that safe and effective products can reach the public as quickly as possible.

The FDA regulatory process for all products, including TEPs will continue to be based on product-by-product review and, the Agency's science-based approach will continue to provide the information necessary for decision-making. There are essentially three critical elements that will continue to be important in the regulatory review for all products: (1) the manufacturer's claim of intended use; (2) the product quality control procedures; and (3) the product's performance. As TE technology evolves, additional specific issues of product safety and effectiveness will be addressed by the Agency as part of its continuing review and assessment of TEPs.

References

1. Hellman, K.B., Biotechnology regulatory policy for biomedical products: The United States perspective. *Current Science*, 1992, **63**(3), 123–126.
2. Hellman, K.B., Biomedical applications of tissue engineering technology: Regulatory issues. *Tissue Engineering*, 1995, **1**(2), 203–210.
3. Hellman, K.B., Picciolo, G.L., Durfor, C.N., Goldman, D.W., Knight, E., Black, L.E., Chapekar, M.S. and Tripathi, S.C., Directions in tissue engineering: Technologies, applications, and regulatory issues. *Tissue Engineering*, in preparation.
4. Shalak, R. and Fox, C.F., eds., Tissue Engineering. *Proceedings for a Workshop held at Granlibakken, Lake Tahoe, California*, February 26–29, 1988, Alan Liss, New York.
5. Langer, R. and Vacanti, J.P., Tissue engineering. *Science*, 1993, **260**(5110), 920–926.
6. Durfor, C.N., Biotechnology biomaterials: A global regulatory perspective for tissue engineered products: Summary report and future directions. *Tissue Engineering*, 1997, **3**(1), 115–120.
7. Cohen, S.N., Chang, A.C.Y., Boyer, H.W. and Helling, RB, Construction of biologically functional bacterial plasmids *in vitro*. *Proceedings of the National Academy of Sciences, USA*, 1973, **70**, 3240–3244.
8. Johnson, I.S., Human insulin from recombinant DNA technology. *American Association for the Advancement of Science, Washington, D.C.*, 1983, **219**, 632–637.
9. G94-2 Blue Book — PMA/510(k) 'Expedited Review', A Subpart E 21 CFR 601.40, 'Accelerated Approval of Biological Products for Serious or Life-Threatening Illness.'
10. Proposed approach to cellular and tissue-based products. Notice of availability. *Federal Register*, March 4, 1997, Vol. 62, No. 42.
11. Hellman, K.B., Bioartifical organs as outcomes of tissue engineering: Scientific and regulatory issues. *Annals of the New York Academy of Sciences*, in press.
12. Human Tissue Intended for Transplantation; Interim Rule. *Federal Register*, December 14, 1993, Vol. 58, No. 238.
13. Draft Public Health Service Guideline on Infectious Disease Issues in Xenotransplantation. *Federal Register*, September 23, 1996, Vol. 61, No. 185.
14. Hellman, K.B., Adventitious agents from animal-derived raw materials and production systems. *Developments in Biological Standardization*, in press.
15. Goldman, D., Black, L., Chapekar, M., Tripathi, S., Picciolo, G.L. and the FDA InterCenter Tissue Engineering Working Group (TEWG). *Horizons for Tissue Engineering Products: Outcomes-Based Clinical Research, Society for Biomaterials Short Course, Baltimore, Maryland*, December 4–5, 1996.
16. Wallin, R.F., Improving biocompatibility standards for the global market. *Medical Devices and Diagnostic Industry*, 1996, **18**(12), 36–40.
17. Galletti, P.M., Hellman, K.B. and Nerem, R.M., Tissue engineering: From basic science to products: A preface. *Tissue Engineering*, 1995, **1**(2), 147–149.

18. Durfor, C.N. and Scribner, C.L., An FDA perspective of manufacturing changes for products in human use. *Annals of the New York Academy of Sciences*, 1992, **665**, 356–363.
19. Hellman, K.B., Tissue engineered products and regulatory challenges: An update, Who's Doing What in Tissue Engineering: Reality, Regulation and Research Panel. *Society for Biomaterials Annual Meeting, New Orleans, Louisiana*, April 29–May 4, 1997.

SECTION III
TISSUE ENGINEERING APPLIED TO SPECIALIZED TISSUES

CHAPTER III.1

Tissue Engineered Adipose Tissue

CHARLES W. PATRICK JR.,
PRISCILLA B. CHAUVIN,
GEOFFREY L. ROBB

Laboratory of Reparative Biology and Bioengineering,
Department of Plastic Surgery,
M.D. Anderson Cancer Center,
1515 Holcombe Blvd., Box 62,
Houston, Texas 77030, USA

1 Perspective

With tissue engineering as a broad multidisciplinary field dedicated to the advancement of the human biological environment in healing, correction of deformity, as well as overall tissue function, current research application efforts are strongly directed at the successful engineering of soft tissue, such as human skin, which will have important implications for greatly improved survivals of extensively burned individuals, and, with progressive development of knowledge attained for further clinical applications, even add to the potential engineering beyond single tissues to that of compound tissues for the future. The structurally important bone and cartilage basic tissue equivalents are now being intensely concentrated on as the next focus for engineering application efforts.

A natural next step for soft tissue structuring efforts lies in the area of adipose tissue. As anticipated, the multidisciplinary efforts of engineering and materials science, cell biology, and surgical science will interact to help produce viable fat tissue solutions for presently limited reconstructive applications in soft tissue augmentation and, ultimately, for incorporation into compound flap tissue for clinical use to increase soft tissue bulk and help create or repair appropriate superficial body contour and shape where well-vascularized soft tissue is needed. There are extensive indications for the use of such tissue-engineered fat, primarily for reconstructive purposes in congenital deformities, post-traumatic repair, and cancer rehabilitation, especially for women desiring breast augmentation or undergoing mastectomy for breast cancer. In 1995, 182,000 American women were diagnosed with breast cancer and 85,000 of these women underwent a mastectomy as part of their treatment [1].

Clearly, a common potential application for tissue-engineered fat would be aesthetic breast augmentation or reconstruction following mastectomy, considering the fact that almost 26,000 breast reconstructions were done in the United States in 1994 [1]. However, controversy will always surround the use of silicone implants for breast surgery because of the silicone foreign body casing that is used to hold the saline or other filler liquid. Using an autologous tissue source that does not require a significant donor site and which avoids the use of an implant would revolutionize the cosmetic approach to the breast. There are also many other areas of asymmetric physical shape or actual tissue deficiency that would benefit from autologous tissue engineered fat. These would include the calf and buttock areas and pectoral area in the male, in which at present, firm silicone implants are currently used for augmentation and shape. The biggest drawback for the use of implants, in general, is the development of a scar capsule around the implant that contains myoepithelial cells that can actively contract and tighten the skin and subcutaneous tissues permanently, producing unnatural external contours and chronic local pain secondary to the contraction process. The breast implant position, as an example, tends to shift asymmetrically high on the chest wall, which is very difficult to correct aesthetically for the long term. Moreover, the patient is often more concerned about the increasingly unnaturally firm feel of the breast as the contracture process matures around the implant. A surgical procedure is then usually necessary to open the chest wall skin, muscle, and underlying scar capsule with a release of the capsule itself to create more space for the implant and reposition it in a more symmetric location.

Ordinarily, for a patient's own tissue to be used rather than an implant in breast reconstruction, a large donor area from the back or abdomen would be required to provide adequate vascularized fascial and fat substance for a potential augmentation and shaping using a pedicle or free tissue transfer, which often requires the addition of muscle in the flap tissue to keep the fat organized together and adequately well-vascularized [2]. If not sufficiently vascularized, the fat will heal as fibrous scar which tends to be lumpy and can mimic a residual or recurrent cancerous mass, which can be especially distressing to a cancer patient. Another factor to consider is that the donor site for these compound flaps can develop significant hypertrophic scarring which is not uncommon on the back, which can then display a contour deformity which translates into 'robbing Peter to pay Paul' for the patient. Although uncommonly severe, functional deficiencies from pain in shoulder activity or limits in range of motion of the shoulder or arm can also occur from the donor procedure to harvest an adequate amount of fatty soft tissue for a face or torso defect.

Another present option for providing available fat for reconstruction is using the omentum from the abdomen. However, the harvest of this tissue requires a laparotomy, which entails some potential morbidity in the surgical procedure as well as in the post-operative healing. This procedure can also entail an extensive midline abdominal scar, unless the omentum is extracted endoscopically, which may limit the actual amount of omental fat available. A limited amount of fat for reconstruction can also be a problem for patients who are slender or who have an overall low body fat percentage. Other tissue options, such as muscle, must then be used for most more extensive soft tissue reconstruction requirements [3]. Ideally, in the future, the reconstructive goal would be to completely avoid using functional tissues, such as muscle, for soft tissue restructuring, since other tissue-engineered sources would potentially be available at minimal donor cost to the patient. And considering the fact that the general cost of reconstruction is high, in both the monetary as well as the physical sense, an economic impetus exists to reduce costs. At present, muscle flaps cost almost $5,000 and free muscle flaps used for breast reconstruction are about $17,000. Many health management organizations and insurance companies attempt to restrict breast reconstructions or at least make the option for reconstruction difficult to get approved. In 1995, 84% of plastic surgeons reported having up to 10 patients who were denied insurance for breast reconstruction following mastectomy [1].

On the other hand, with tissue-engineering facilitated reconstructions, it may be possible to curtail physical as well as monetary costs in several ways. Operative morbidity from the various flap donor sites could be diminished significantly, with actual hospitalization times greatly reduced. Maximum reconstructive cost efficiency could then be realized, with improved rehabilitation results possible as well. Hopefully, the reconstructive outcome would also positively impact the key element for the majority of patients undergoing soft tissue reconstructions today, which is the resulting quality of life which the patients either enjoy or endure. And since the quality of life is the ultimate reconstructive goal for all plastic surgical patients, we recognize that we have now evolved well beyond the earlier challenge of conceptualizing what reconstructive tissue approach will best solve the patient's defect dilemma and then successfully completing that task. We are now at the crossroads of the need to definitively fine tune and improve that reconstructive process so that functional and cosmetic outcomes can be a quantum level better and more predictable, more efficient in cost and rehabilitation, and less morbid for the patient overall.

The key to fine tuning our present, still somewhat primitive, reconstructive finesse lies in refining and modifying (even generating) our flaps of autologous tissue on a cellular level, which is the evolutionary future of tissue engineering. As is well-documented, autologous tissue used, for example, as an alternative to foreign body implants in breast reconstruction following mastectomy provides more predictable long term aesthetic results with a similar breast tissue consistency and a more natural-appearing breast shape [4]. The present reconstructive approach of taking skin, fat, and muscle from the back or the abdomen, requiring its own blood supply in an attached pedicle or as a free microvascular anastomotic transfer, to replace the breast in the appropriate location, can in the future be supplanted by the placement of an appropriate absorbable scaffold construct on the chest wall onto which fat cells obtained from minimally invasive liposuction can be integrated. More difficult reconstructions could be completed as free tissue transfers of prefabricated skin/scaffold/fat complexes, perhaps allowed to develop and mature in the lower abdomen as the least visible donor site.

2 Adipose transplantation

The idea of using autologous fat for soft tissue repair and augmentation is not new. Fat grafts have been used in plastic and reconstructive surgery for over a century to repair tissue defects [5]. The first use of human fat autotransplantation was described by Van der Meulen in 1889 [5]. The procedure involved grafting a free omentum and fat autograft between the liver and diaphragm. The first use of free fat autografts was performed by Neuber in 1893, in which multiple small fat grafts from the upper arm were used to fill facial soft tissue depressions [6]. Lexer in 1910 described the first use of fat plasty applied for support [7]. Specifically, autogenous fat graft from the abdomen was used to treat a depression of the molar infraorbital area and build out a receding chin. The first reported use of fat in breast reconstruction was performed by Czerny in 1895 [8]. A large lipoma was used to fill a breast defect. However, innovative as these procedures were, free grafting of adipose tissue is known to give poor results, as a considerable reduction (40–60%) in the volume of the graft is observed [9]. Because of this reduction in volume following fat implantation, surgeons use hypercorrection to offset [10]. This difficulty with autologous fat transplantation was summarized in a 1987 report by the Ad Hoc Committee on New Procedures of the American Society of Plastic and Reconstructive Surgeons [11]. The report concluded that only 30% of injected fat can be expected to survive for 1 year, and further stated that overcorrection is necessary when performing fat transplants. The reader can appreciate that a fair amount of 'guess work' is required to ascertain the volume of fat needed to produce the ultimate contour of the recipient site over an extended period of time.

Peer and Smahel have stressed that survival of fat grafts is dependent on early neovascularization [12–14]. Bartynski demonstrated that neovascularization begins along the graft periphery by 5 days [15]. Before this time, the grafts live solely by diffusion and inosculation [16]. This neovascularization never proceeds past a peak at 10 days, and only occurs along the graft edges. Thus the graft never achieves an adequate blood supply at its center to maintain its long-term viability [15]. Fat transplantation of relatively small fragments is feasible since plasmatic nutrition is more effective for small grafts and they are more rapidly revascularized, giving a fast recirculation of the transplant after spontaneous anastomosis with the recipient bed. However, the isolated implantation of small fragments is a considerable procedure and the volume of tissue that can be implanted (at least in a single session) is typically negligible compared to the augmentation area or tissue defect [9]. Moreover, because of the decrease in tissue volume post-transplantation fat transplants seem to be less predictable that other transplants, although fat is the most natural filling material for soft tissue defects [17].

Due in part to the limitations associated with autologous fat transplantation, investigators naturally contemplated using single cell suspensions of mature adipocytes. Single cells are more apt to be supported by diffusion until neovascularization initiates. In 1977 Illouz introduced the technique known as liposuction. Bircoll in 1984 first presented his technique of injecting autologous adipose tissue removed by liposuction [18]. However, fat from aspiration appears to be too traumatized, since capillary structure is broken up and cells are torn apart during the process [17]. Although adipose tissue has been considered to be poorly vascularized, each adipocyte is actually in contact with at least one capillary [19]. The blood supply is adequate to support the very active metabolism of the thin rim of cytoplasm surrounding the large lipid inclusions. For the best graft survival, the capillary structure of the transplant should be preserved. Nguyen demonstrated that suctioned adipose tissue examined with light microscopy was a mixture of fragments containing approximately 90% elongated, irregularly shaped, and ruptured adipocytes and only 10% unchanged, normal-appearing adipocytes [20]. The reticulin network in Wilder's staining was irregular and shattered. Two months after injections of aspirated adipose tissue, there was focal necrosis of adipocytes, irregular size and shape of lipid cysts, and only a few unchanged adipocytes. There was also a large foreign-body reaction and inflammation. Four and 6 months post-injection, only a few adipocytes were found. There was still evidence of a foreign-body reaction and fibrosis around adipocytes and lipid cysts. Nine months post-injection, no adipocytes were evident. The consensus to date is that aspirated fat is inferior as autologous grafting material.

Although the clinical utility of fat grafts and aspirated adipose tissue has not been realized, research in these areas led to the demonstration that mature adipocytes differentiate from mesenchymal adipose cell precursors. Wasserman in 1926 and 1929 described the preadipocyte [21, 22]. He studied the transplantation of undifferentiated connective-tissue cells from within immature fat bodies. He concluded that the immature connective-tissue cells were already specialized cells which at the time of transplantation could not be distinguished from fasciae latae fibroblasts, but which could accumulate lipids after transplantation. In a landmark article, Wertheimer and Shapiro in 1948 reported that fat developed from very primitive adipose cells and the structure of the cells and tissue was much like that of the fibroblasts of connective tissue [23]. In 1954, Hausberger demonstrated that the final size of the autotransplant was dependent on, among other factors, the amount of preadipocytes transplanted and that these young mesenchymal cells had special potential to become mature adipose cells [24]. In 1955 he presented the theory that it is the preadipocyte, or mesenchymal adipose cell precursor, that is destined to become a mature adipose cell [25].

The study of the preadipocyte was facilitated by the development of adipose tissue culture techniques. Smith in 1971 first described 'fibroblast-like' cells grown in tissue cultures of human adipose tissue [26]. Poznanski et al. concluded in a study that the stromal vascular fraction contained cells that were adipocyte precursors with the potential for multiplication and differentiation into mature adipose cells [27]. They observed significant differences between preadipocytes and fibroblasts with preadipocytes having increased lipoprotein lipase activity, $15 \times$ more triglyceride synthetase activity, and $2 \times$ more fatty acid synthetase activity. Green and Kehinde in 1979 used tissue culture techniques to select an established preadipocyte cell line that was susceptible to adipose differentiation [28]. These cells were injected subcutaneously in the lower chest and upper abdomen of nude mice. Over 6 weeks, a soft fat pad of mature adipose cells that molded to body contour developed at the site. Hence, this established cell line of adipocyte precursors was shown to differentiate in vivo, resulting in soft tissue augmentation. With recent work delineating the dedifferentiation of mature autograft fat cells into 'fibroblast-like' preadipocytes, the fate of autologous fat transplants can be clarified. The conditions of inadequate oxygen and nutrients at the graft site during the initial period of ischemia and during the first 8 weeks after transplantation may stimulate the graft fat cell dedifferentiation to a preadipocyte state or necrose. When the blood supply adequately supplies oxygen and nutrients to the graft, the adipocyte precursor pool of immature preadipocytes could differentiate into mature adipose tissue, albeit of less volume [5]. The preadipocyte has a better chance to survive the initial hypoxic period after transplantation than the mature adipocyte [5, 17].

3 Tissue engineered adipose

Although fat is the most natural filling material for soft tissue defects, autologous fat grafts remain minimally effective [17]. Consequently, various synthetics, including paraffin, liquid silicone, encapsulated silicone,

silicone rubber, plastic, and various human and bovine collagen extracts have been tried as alternatives for soft tissue repair and augmentation [29]. As stated before, the controversy surrounding silicone implants, potential for scar capsule formation, and the need for hypercorrection with current fat transplantation has sparked a new interest to develop new strategies for soft tissue repair. Table 1 lists some criteria for tissue engineered adipose tissue. Of particular importance is the need to provide a microvascular network for the adipose tissue equivalent. This is a prerequisite if volumes required for breast reconstruction are to be attained. Conceptually, tissue engineered adipose tissue would first involve isolating preadipocytes from fat depots within the patient. There is certainly no problem finding

Table 1. Engineering criteria for the design of adipose tissue equivalents

Use autologous adipose tissue if possible
   Biopsy
   Excised fat
   Aspirated fat

End product must result in mature adipocytes

Tissue engineered equivalent must structurally and functionally resemble autologous adipose tissue
   Lipid deposition and mobilization
   Equivalent lipid accumulation
   Equivalent lipoprotein lipase, triglyceride synthetase, and fatty acid synthetase activity
   Specific gravity of 0.74

Adipocyte scaffold or end product must be able to be contoured into complex, asymmetric shapes
   Ultra-thin sheets for orbital floor and rim repair
   Spheres for enucleation and evisceration procedures
   Blocks for surgeon's custom shaping
   Eccentric hemispheres for breast reconstruction

The end product must be of sufficient volume to be clinically useful
   0.5 ml for facial augmentation
   1,000 ml for breast reconstruction

Adipose equivalent must be well-vascularized and rapidly promote anastomoses with recipient bed

If a polymer scaffold is used, must provide interconnecting pore structure to encourage tissue growth and capillary invasion

Adipose equivalent must not significantly resorb over time

sufficient fat tissue from which to isolate these cells. Humans can grow fat in most areas of the body [20]. Next, the cells would be expanded *ex vivo*, seeded on a bioactive scaffold prefabricated into a desired anatomical shape, and the entire construct sutured *in vivo* to repair or augment a tissue defect.

To date, tissue engineered equivalents of adipose tissue have not been developed, despite the overwhelming clinical need. As a first step, we have begun to develop tissue engineered constructs using rat adipose tissue as an initial model. We hypothesize that isolated preadipocytes can be seeded onto biodegradable polymer scaffolds. Polymer scaffolds have advantages that they can be milled into complex shapes, their surfaces can be derivatized to increase adipocyte adhesion and alter function, and they degrade into cytocompatible products concomitant with proliferation of seeded cells. Conventionally, differentiated cells are seeded onto polymer scaffolds. However, isolation and maintenance in culture of mature adipocytes poses unique problems due to the buoyancy and fragility of the adipocytes themselves, and maintaining viable mature adipocytes requires some special considerations. A number of methods have been described for the maintenance in culture of isolated mature adipocytes. Novakofski describes a double layer of collagen gels with cells in the top layer [30]. Briefly, a layer of collagen is spread on the bottom of T-25 flasks, 1.5 ml per flask, minimum concentration of 2 mg/ml of collagen, and allowed to gel overnight at 37°C. After isolation the cells are plated by mixing the cooled cells in a second collagen solution, then spreading the solution containing cells over the previously gelled collagen layer. This method proved ineffective in our laboratory because the amount of lactic acid released by the adipocytes was sufficient to drop the pH of the medium enough to dissolve the collagen gels [31]. The adipocytes were then free floating in the culture flask and lost when the medium was aspirated off to change the feeding solution.

Shillabeer *et al.* addressed the buoyancy problem by plating the floating adipocytes on a culture flask in a very small amount of medium supplemented with 4% BSA (AAM, Adipocyte Adhesion Medium) and quickly inverting the flasks to allow adherence of cells to the flask [31]. When the flasks are inverted, the floating adipocytes contact the surface of the flask and adhere. Inverted flasks are incubated for 90 minutes at 30°C after which time they are turned over and medium is added to bring up the volume in the flask. This method also allows for loss of a significant number of the initially plated cells which do not adhere to the plastic surface during the inverted incubation time. We have modified this inverted flask method in several ways. The flasks have been coated with a collagen gel (Vitrogen) at 2 mg/ml. The collagen is allowed to gel at 37°C overnight. Isolated mature adipocytes suspended in AAM are seeded onto the flasks. The flasks (T-25's) are then filled to capacity with medium (about 75 ml/flask). Caps are tightened and the flasks are incubated inverted at 37°C overnight. The following day, flasks can be righted and excess medium can be removed, leaving the usual amount on each flask. However, very few of the mature adipocytes survive this process, as evidenced by large pools of free lipid floating on the medium. Hence, isolating mature adipocytes is not feasible for tissue engineering applications due to the fact the cells are too fragile (they would never survive a polymer-seeding process) and mature adipocytes do not proliferate. Preadipocytes, however, can be isolated, and they are more resistant to trauma (mechanical and ischemia) and proliferate.

To isolate preadipocytes, epididymal fat pads are excised aseptically from CO_2 euthanized Sprague–Dawley rats weighing from 400 to 900 grams. Figure 1 depicts an excised fat pad. The fat pads are placed directly into sterile 50 ml screw-cap tubes containing 30 ml of DMEM supplemented with 500 U/ml penicillin and 500 μg/ml streptomycin (Harvest Medium), 5 × the normal amount of antibiotic present in Growth Medium. Growth Medium consist of DMEM supplemented with 10% FBS, 100 U/ml penicillin, and 100 μg/ml streptomycin. The pads are placed in 100 ml plastic culture dishes with just enough harvest medium to keep them moist. Using a dissecting microscope, tissue containing blood vessels is dissected free of the surrounding fat and removed. This minimizes connective tissue fibroblast contamination of the *ex vivo* cultures.

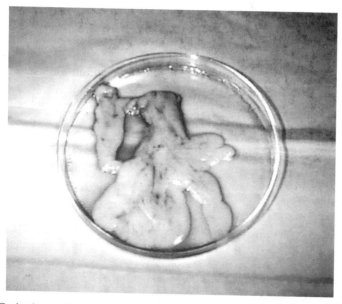

Fig. 1. Excised and filleted rat epididymal fat pad. A single fat pad has an average mass of 2.5 g.

Isolated fat tissue are rinsed with an additional 10 ml of Harvest Medium and minced into small pieces using two #10 scalpel blades. The minced tissue is then transferred into a siliconized, sterile 100 ml glass screw-cap bottle. Enzymatic dissociation is carried out in Type I collagenase (5% w/v) supplemented with bovine serum albumin (BSA, 5% w/v) in Ca^{2+}/Mg^{2+}-free Hank's Balanced Salt Solution (HBSS) at 37°C on a shaker for 20 minutes. For the tissue from four fat pads, 5 ml of the dissociating medium is used.

After 20 minutes incubation, the undissociated fat is removed with a large bore pipet to a separate 50 ml tube, and the slurry of cells at the bottom of the bottle is transferred to a 15 ml screw cap centrifuge tube. An additional 5 ml of HBSS/BSA are added to the tube to abrogate the enzymatic dissociation. The slurry is spun down at $200 \times g$ for 3 minutes at room temperature. The supernatant is removed and placed in a separate tube on ice to cool. The pellet is resuspended in 5 ml growth medium and set aside.

The undissociated tissue is triturated gently with a 5-ml pipet to further break up tissue clumps and filtered through a 250 μm stainless screen. The filtrate is transferred to another 15 ml tube, and 5 ml of the HBSS/BSA are added. This is spun down at $200 \times g$ for 3 minutes at room temperature. The supernatant is removed and pooled with the previously iced supernatant from above. This pellet is resuspended in Growth Medium and combined with the pellet from the cell slurry. The combined suspension of cells is filtered through a 90 μm nylon filter to remove cellular aggregates and microvascular endothelial cells (from capillary network innervating the fat tissue). This filtrate consists mainly of the preadipocyte stromal fraction of the fat tissue. The cells are cultured in T-25 flasks and fed Growth Medium. The preadipocytes differentiate from a fibroblast spindle shape with small lipid inclusions, to a multilocular stage with numerous large inclusions, to a state of a single, large lipid inclusion. As lipid accumulates, the plasma membrane shows several micropinocytotic vesicular areas, and the external laminae elaborates. Figure 2 depicts an early preadipocyte with many lipid inclusions, a preadipocyte with a single, large lipid inclusion, and mature adipocytes. The mature adipocytes closely resemble *in vivo* mature adipocytes (Fig. 3). Preliminary experiments are presently being conducted in which preadipocytes are seeded onto polymer scaffolds and implanted subcutaneously in rats to assess *in vivo* differentiation and survival of preadipocytes.

4 Concluding remarks

Tissue engineering is an interdisciplinary field that applies the principles of engineering and life sciences towards the development of biological substi-

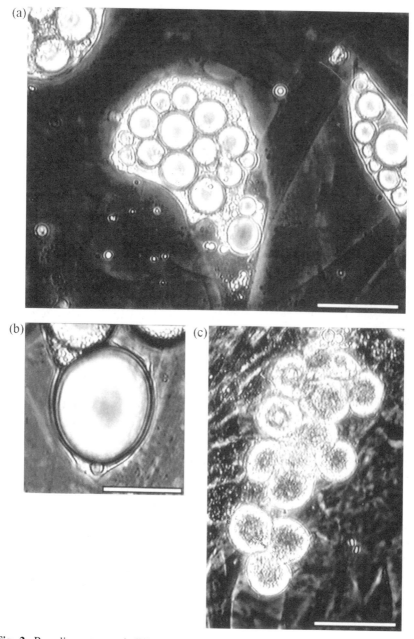

Fig. 2. Preadipocytes and differentiating adipocytes. (A) Initially, differentiating adipocytes are multilocular, with many small lipid inclusions. (B) Over time, the lipid inclusions coalesce to form a unilocular adipocyte. (C) Mature adipocytes in culture. Bars represent 50 μm.

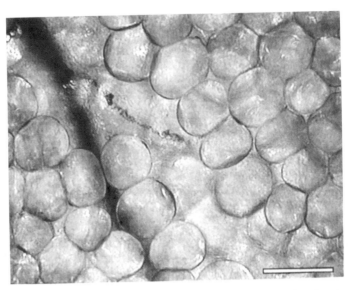

Fig. 3. Adipose tissue from rat epididymal fat pads. The tissue is a loose association of lipid filled adipocytes innvervated with capillaries and held together with collagen fibers. Adipocytes are 60–85% lipid by weight and the lipid is 90–99% triglyceride [19]. Bars represent 50 μm.

tutes that restore, maintain, or improve tissue function. To date, tissue engineering has not been applied to developing clinically translatable adipose tissue equivalents. The scope of this research couples the disciplines of cell biology, engineering, and material science to address the critical problem of supplying autologous adipose tissue for reconstructive surgery applications. Although tissue engineered adipose tissue could be used for soft tissue augmentation, such as in facial augmentation to correct congenital problems, it would have immediate and widespread use in breast reconstruction following tumor resection. The emotional distress which accompanies a lump found in the breast is compounded by the apparent problems noted with reconstructive procedures. Without a trusted reconstruction procedure, the specter of a disfiguring operation adds to the fear of the cancer. This fear can lead to patient denial and, inevitably, to avoidance of breast cancer screenings and delayed treatment. Tissue engineered adipose tissue can provide patients with breast tissue comprised of the patient's own cells, negating donor site morbidity and complications associated with flap surgery. New breast reconstruction methods that are reliable, safe, and improve cosmesis can have a positive impact on women, leading to appropriate cancer screening and prompt treatment. Moreover, they can provide a positive psychological benefit for women, giving them a renewed sense of wholeness and a return to a normal state. The use of autologous tissue source that does not require a significant donor site and avoids the use of a conventional implant would revolutionize reconstructive cancer surgery.

References

1. American Society of Plastic and Reconstructive Surgeons and Plastic Surgery Educational Foundation.
2. Hartram, Jr., C. R., Scheflan, M. and Black, P. W., Breast reconstruction with a tranverse abdominal island flap. *Plastic and Reconstructive Surgery*, 1982, **69**, 216–224.
3. Mathes, J. S. and Nahai, F., *Clinical Applications for Muscle and Musculocutaneous Flaps*. Mosby-Year Book, St. Louis, 1982.
4. Kroll, S. S. and Marchi, M., Immediate reconstruction: Current status in cancer management. *Texas Medicine*, 1991, **87**, 67–72.
5. Billings Jr., E. and May, J. W., Historical review and present status of free fat graft autotransplantation in plastic and reconstructive surgery. *Plastic and Reconstructive Surgery*, 1989, **83**, 368–381.
6. Neuber, G. A., Fetttransplantation (Fat transplantation). *Deutsche Gesellschaft Chir.*, 1893, **22**, 66.
7. Lexer, E., Freie fetttransplantation (Free fat transplantation). *Deutsche Medizinische Wochenschrift*, 1910, **36**, 640 (abstract).
8. Czerny, A., Plastischer ersatz der brustdruse durch ein lipoma (Reconstruction of the breast with a lipoma). *Chir. Kongr. Verhandl.*, 1895, **2**, 216.
9. Smahel, J., Experimental implantation of adipose tissue fragments. *British Journal of Plastic Surgery*, 1989, **42**, 207–211.
10. Kononas, T. C., Bucky, L. P., Hurley, C. and May Jr., J. W., The fate of suctioned and surgically removed fat after reimplantation for soft-tissue augmentation: A volumetric and histologic study in the rabbit. *Plastic and Reconstructive Surgery*, 1993, **91**, 763–768.
11. American Society of Plastic and Reconstructive Surgeons, Report on autologous fat transplantation by the ASPRS ad hoc committee on new procedures. ASPRS, Chicago, 1987.
12. Peer, L. A., The neglected 'free fat graft', its behavior and clinical use. *Plastic and Reconstructive Surgery*, 1956, **11**, 40–47.
13. Peer, L. A., Transplantation of fat. In *Reconstructive Plastic Surgery: Principles and Procedures in Correction, Reconstruction, and Transplantation*, Vol. 1, ed. J. M. Converse. Saunders, Philadelphia, 1977, p. 251.
14. Smahel, J., Failure of adipose tissue to heal in the capsule preformed by a silicone implant. *Chirurgia Plastica*, 1985, **8**, 109–115.
15. Bartynski, J., Marion, M. S. and Wang, T. D., Histopathologic evaluation of adipose autografts in a rabbit ear model. *Otolaryngology Head and Neck Surgery*, 1990, **102**, 314–321.
16. Ersek, R. A., Transplantation of purified autologous fat: A 3-year follow-up is disappointing. *Plastic and Reconstructive Surgery*, 1991, **87**, 219–227.

17. Fagrell, D., Enestrsm, S., Berggren, A. and Kniola, B., Fat cylinder transplantation: An experimental comparative study of three different kinds of fat transplants. *Plastic and Reconstructive Surgery*, 1996, **98**, 90–96.

18. Bircoll, M., Autologous fat transplantation. In *California Society of Plastic Surgery*, 1985.

19. Greenwood, M. R. C. and Johnson, P. R., The adipose tissue. In *Histology: Cell and Tissue Biology*, ed. L. Weiss. Elsevier Biomedical, New York, 1983, pp. 178–199.

20. Nguyen, A., Pasyk, K. A., Bouvier, T. N., Hassett, C. A. and Argenta, L. C., Comparative study of survival of autologous adipose tissue taken and transplanted by different techniques. *Plastic and Reconstructive Surgery*, 1990, **85**, 378–386.

21. Wassermann, F., Über speicherung, entspeicherung und wiederspeicherung der fettorgane. *Verhandlugnen der Anatomischen Gesellschaft*, 1929, **38**, 181–194.

22. Wassermann, F., Die fettorgane des menschen: Entwicklung, bau und systematische stellung des sogenannten fettgewebes. *Ztschr. f Zellforsch. u Mikr. Anat.*, 1926, **3**, 235–328.

23. Wertheimer, E. and Shapiro, B., The physiology of adipose tissue. *Phsyiological Reviews*, 1948, **28**, 451–464.

24. Hausberger, F. X., Development of autotransplants of immature adipose tissue of rats. *Anatamical Record*, 1954, **118**, 389 (abstract).

25. Hausberger, F. X., Quantitative studies on development of autotransplants of immature adipose tissue of rats. *Anatamical Record*, 1955, **122**, 507–575.

26. Smith, U., Morphological studies of human subcutaneous adipose tissue in vitro. *Anatamical Record*, 1971, **169**, 97–104.

27. Poznanski, W. J., Waheed, I. and Van, R., Human fat cell precursors: Morphologic and metabolic differentiation in culture. *Laboratory Investigation*, 1973, **29**, 570–576.

28. Green, H. and Kehinde, O., Formation of normally differentiated subcutaneous fat pads in an established preadipocyte cell line. *Journal of Cell Physiology*, 1979, **101–171**, 169.

29. Rubin, L., *Biomaterials in Reconstructive Surgery*. C.V. Mosby, St. Louis, 1983.

30. Novakofski, J. E., Primary cell culture of adipose tissue. In *Biology of the Adipocyte: Research Applications*, eds. G. J. Hausman and R. J. Martin. Van Norstrand Company, New York, 1987, pp. 160–197.

31. Shillabeer, G., Li, Z. H., Hatch, G., Kumar, V. and Lau, D. C. W., A novel method for studying preadipocyte differentiation *in vitro*. *International Journal of Obesity*, 1996, **20**(Suppl 3), S77–S83.

CHAPTER III.2

Options For Engineering Bone

ELISA A. BURGESS,
JEFFREY O. HOLLINGER
Oregon Health Sciences University,
Department of Surgery,
3181 SW Sam Jackson Park Rd L352A,
Portland, OR 97201-3098, USA

1 Introduction: An operational framework

The operational framework for this chapter on options for engineering bone evolves from definitions of *engineering, engineer, tissue engineering,* and *bone.* *Tissue engineers* can exploit applied technologies to design products if the *engineer* understands structure and function of the recipient of that product. Therefore, to accomplish the objective of this chapter (i.e. to present options for engineering bone) the authors provide working definitions, a concise review of bone basics, and selected design considerations for bone products.

2 Starting at the beginning: Consensus definitions

An operational framework is braced by consensus definitions that unambiguously define scope, thus providing a comprehensive, sound, and comfortable foundation. Therefore, a likely starting point for the relatively new discipline of *tissue engineering* (Reviewed by Nerem and Sambanis [1]) is to provide the definition for the old discipline, *engineering*: 'Engineering is the profession in which knowledge of the mathematical and natural sciences gained by study, experience, and practice is applied with judgment to develop ways to utilize, economically, the materials and forces of nature for the benefit of mankind [2]'. The next logical move is to define *engineer*: 'Engineer shall mean a person who, by reason of his special knowledge and use of mathematical, physical, and engineering sciences and the principles and methods of engineering analysis and design, acquired by education and experience, is qualified to practice engineering [2].' Clarifying the definition a bit more is the fact an *engineer* often conducts research with the goal of *solving a problem* [2]. (Tissue engineers in bone are challenged with serious problems involving either regeneration or augmentation of an extremely complex *organ*.) Moreover, the ' ... end result of an engineering effort — generally referred to as *design* — is a device, structure, or process which satisfies a need' [2]. Tissue engineers in bone are compelled to *design a physiological, functionally dynamic device* that should integrate non-immunologically with the body; that should be biocompatible (*Biocompatibility*: 'ability of a material to perform with an appropriate host response to a specific situation'[3].); and that should be physiologically and biomechanically equivalent to bone (i.e. restore form and function through either regeneration or augmentation). To fulfill this formidable array of design criteria, the bone tissue engineer will use a number of basic tools from a tool kit to construct a device that may include combinations of natural components (e.g. cells, soluble factors, extracellular matrix), laboratory-derived products (e.g. synthetic extracellular matrix), and recombinantly engineered components (e.g. recombinantly engineered soluble factors). The *tools* are *technologies* such as cell and molecular biology, ceramic chemistry, polymer chemistry, immunology, and physiology.

Another definition is needed to buttress our operational framework, that of *tissue engineering*. A reiteration from Langer and Vacanti [4]) and Hubbell (in 1995 [5]) is the following: '*Tissue engineering* combines the principles and methods of the life sciences with those of engineering to elucidate fundamental understanding of structure-function relationships in normal and diseased tissues, and to create entire tissue replacements' [5]. A concise, yet remarkably encompassing definition for *tissue engineering* was offered in the inaugural edition of the journal Tissue Engineering: '*Tissue engineering* is an interdisciplinary field in which the principles of engineering and the life sciences are applied toward the generation of biologic substitutes aimed at the creation, preservation, or restoration of lost organ function' [6].

3 Bone basics: A concise focus on structure and function

Successful application of *components* to a build a device (i.e. a product to restore lost organ function) requires an understanding of *structural and functional basics* about the recipient that receives the device. Therefore, a focused, abridged review of salient *bone basics* will provide instructive guidance to engineer therapies.

The definition of an *organ* is ' ... a part of the body ... arranged according to a characteristic structural plan (that) performs a special function or functions; (and) ... is composed of various tissues, one of which is primary in function' [7]. Therefore, bone is an *organ*; it consists of two *specialized* tissues: *vascular* and *calcified*.

3.1 Vascular tissue

Bone houses the *marrow*: located in the medullary zones of tubular bones, vertebrae, ribs, sternum, and flat bones of the pelvis and craniofacial complex and is the habitat for hematopoietic cells of erythropoietic and granulocytic lineage, and the undifferentiated mesenchymal stem cells that can replenish selected phenotypes [8, 9]. In addition to *marrow*, an extensive weave of blood vessels constitutes the periosteal-diaphyseal-metaphyseal and epiphyseal-physeal vascular tapestries needed for metabolic exchange and trafficking of molecular signals to sustain osseous physiology dynamics. Moreover, a lush vascular web of Haversian systems and Volkmanns' canals penetrates virtually every aspect of bone (reviewed by Buckwalter and colleagues [10, 11]). (These will be mentioned later in the chapter.) The vascular design supports mineral homeostasis and provides indispensable, life-sustaining enrichment to bone cells (i.e. osteocytes and osteoblasts). Bone cells will perish without blood vessels, a partnership recognized in the insightful work of Trueta over thirty years ago [12].

3.2 Calcified tissue

Bone as a 'hard tissue' may be partitioned conveniently into four components: cells, organic and inorganic matrices, and soluble factors. The *microstructural components* of bone — the cells, the molecules comprising the organic matrix, and the ions assembled as a mineralized superstructure — are integrated and yield a distinctive *macrostructure*: *cortical (compact) and cancellous (trabecular) bone*. Macrostructural form is correlated with functional properties (e.g. protection, locomotion) and physiologic properties (e.g. responsiveness to metabolic needs, such as calcium balance).

3.2.1 Cortical bone

On gross appearance (i.e. *macroscopically*), cortical bone *looks* dense and compact. It comprises about 80% of the skeleton [10], forming the outer and inner layers of tubular-appearing appendicular bone, inner and outer tables of flat bones, and the inner and outer surfaces of vertebral bodies and pelvic structures. Cortical bone is engineered to protect vital organs, to resist biofunctional challenges, to enable locomotion by muscle-lever arm activities, and to provide safe haven for hematopoiesis. Microscopic inspection of cortical bone reveals the *Haversian systems* (*osteons*), sets of 4–20 circumferential rings of concentrically arranged *lamellae* encompassing a central canal 22–110 μm in diameter (the *Haversian canal*) that contains blood vessels, lymphatics, and sometimes nerves (reviewed by Hollinger and McAllister [13]). (The average Haversian system diameter in man is about 200 μm [14].) Each lamella ring is populated by a variable number of *osteocytes*, with each osteocyte linked to sibling osteocytes in the same lamella and to osteocytes in adjacent lamella through a gossamery net of cell processes traversing *canaliculi*. Furthermore, *Volkmann canals* penetrate cortical bone in an oblique direction, anastomizing with the Haversian systems, providing additional vascular-lymphatic channels for metabolic exchange and trafficking of soluble signals, such as hormones. Vascular dependency of bone is emphasized by the rich networking of passageways penetrating its structure, ensuring no cell lies more than 300 μm from a blood vessel [15]. Moreover, the extensive vascular freeways enable *cross-talk* (e.g. exchanging of signaling molecules) among osteocytes and endosteal and periosteal lining osteoblasts, thereby facilitating and promoting physiologic and functional responses.

Aside from roles as osteocyte condominiums and vascular highways, Haversian systems function as buttresses, cleverly designed to optimize responsiveness to biofunctional challenges; they are parallel to the diaphysis in the appendicular bones. Moreover, navigating from the central canal in a centrifugal direction, each concentric lamella component of the system spirals in an opposite direction: clockwise, then counter clockwise, and so on [16], thereby ensuring for maximal response to compressive and torsional loading.

The structural format of cortical bone (with its Haversian systems and inorganic and organic matrices) yields a product with remarkable physical and physiologic properties that may be beyond the design capacities of the tissue engineer.

3.2.2 Cancellous bone

Interposed between cortical bone is the *spongy-looking* cancellous bone. A three-dimensional lattice of *trabeculae* is the hallmark of this structure. In general, spatial orientation of trabeculae is random (i.e. *isotropic*). Rather than being engineered for loading, cancellous bone has been designed to respond rapidly to physiologic requirements. This is underscored by the roughly 20 times greater surface area and concomitant cell density per unit volume of cancellous bone than cortical bone [10].

3.3 Cells

Osteoblasts are bone forming cells derived from mesenchymal lineage cells [17]. Osteoblasts are metabolically active secretory cells that express soluble signaling factors (e.g. bone morphogenetic proteins, transforming growth factor-β, insulin-like growth factors I and II, interleukin-1, and platelet derived growth factors [18]) and *osteoid*, a product whose extracellular matrix modification yields an organic insoluble substratum consisting mostly of type I collagen. Osteoid is produced at a rate of about 2–3

μm/day, and when reaching a width of approximately 20 μm, it begins to mineralize at a rate of 1–2 μm/day [19]. The life span for human osteoblasts may range from one to ten weeks, at which time these cells may disappear (by apoptosis), become lining cells, and about 15% turn into *osteocytes*. The osteocyte is a relatively inactive cell, yet this subdued activity is crucial to maintain bone. Dynamic interactions among osteocytes and osteoblasts are accomplished by an elegantly designed web of inter-connecting tunnels that course through bone. Tunnels (that is, *canaliculi*) contain a thread-like web of cytoplasmic tendrils emanating from osteo-cytes, enabling osteocyte correspondence through gap junctions, thus permitting signal transmission among osteocytes and even to osteoblasts. Appositional craftsmanship of bone through osteoid deposition, mineral-ization, and maintenance involves a neatly choreographed interplay among the osteoblasts and osteocytes and the trafficking of soluble signaling molecules that switch apposition 'on and off' [18, 20].

Balancing bone apposition is bone resorption, a physiologic act per-formed by a multinucleated giant cell known as the *osteoclast* [21]. This cell originates from a circulating monocyte and through asynchronous fusion yields a multinucleated giant cell up to 100 μm in diameter with an average of 10–12 nuclei [22]. Interleukins-1, -3, -6, -11, and probably tumor necrosis factor alpha along with transforming growth factor alpha appear to be significant modulators for osteoclast development [23]. A dynamic, reciprocal relationship between osteoblasts and osteoclasts affects resorp-tive-appositional activities [24, 25]. For example, when osteoblasts disperse as local lining cells on a bone surface (in response to parathyroid hor-mone), an exposed runway is provided for osteoclasts to attach [24], and following expression of a soluble factor or second messenger from the osteoblasts, osteoclasts are prompted to resorb bone [23, 26].

That bone is a vital organ continuously in motion is underscored by the rich vascularity and cellular activity of its cells. The perception that bone, once formed, remains static unless challenged by an acute biofunctional challenge, such as a fracture, fortunately has been dispelled. Therefore, if the *tissue engineer* is to design a relevant device suitable for bone, basic physiologic dynamics of the recipient must be addressed: *modeling* and *remodeling*. In the simplest terms, *modeling* may be viewed as the cellular, functional, and biochemical events that shape maturing bone until epiphy-seal plate closure [27]. *Remodeling* consists of similar events responsible for the repair of mature bone to restore form and function [27]. Human bone (e.g. cortical bone) remodels at a rate of about 2–10 per cent per year, depending on the site and will remodel at even higher rates, up to 20%, around loaded endosseous implants [28]. (Basic principles of model-ing and remodeling have been reviewed extensively by several authors [21, 29–32].) Packets of cells (i.e. osteoblasts, osteocytes, osteoclasts) known as the *basic multicellular unit* (BMU) function as instruments for remodeling and are responsible for the sum activities of cell activation, resorption, and formation [33]. The quantum of bone formed by a BMU is referred to as the *basic structural unit* and the period from BMU activation through completion of the cycle is known as *sigma*, which lasts for 3–6 months in humans [33].

3.4 Matrices

Approximately 35% of the dry weight of bone is the *organic matrix* (reviewed by Hollinger [13]). Type I collagen is the principal component (approximately 90%) of the organic matrix with a relatively small (about 10%) non collagenous component [34]. The structure, expression, and regulation of the major non collagenous proteins have been discussed capably in several noteworthy publications [34–38].

The collagenous organic matrix of bone may be viewed as a solid state *extracellular matrix* (ECM) providing important directional cues linking structure and function of cells with soluble signaling factors, thereby affecting cell differentiation, and ultimately, phenotype expression [39–44]. For example, the ECM provides a docking site for cell-expressed soluble factors (e.g. bone morphogenetic proteins), positioning them to bind to *serine / threonine kinase cell receptors* [45]. Moreover, binding of molecular factors to ECM may protect and facilitate controlled release in register with local requirements, a property to be exploited in engineering therapeutics. Cell attachment (i.e. *anchorage*) through *integrin* binding is another crucial role for ECM [42]. The attachment of cells to ECM initiates a cascade of intracellular mechanochemical switches regulating differentiation, mitogenesis, and competence [46–48]. The ECM is a *substratum* for cell attachment and it orients the configuration of soluble factors for cooperative cell interaction [42]. These pivotal roles underscore the need for a suitable substratum component to a tissue engineered bone product.

The *inorganic matrix* accounts for around 60–70% of the dry weight of bone and holds about 99% of the calcium found in the body, around 85% of the phosphorous, and approximately 40–60% of the sodium and magnesium (reviewed by Buckwalter *et al.* [10, 11]). The inorganic matrix fulfills life-sustaining *physiological roles* by providing appropriate aliquots of minerals for homeostasis and functioning as a cathedral for hematopoiesis.

Regulation of skeletal mineral homeostasis generally focuses on three ions: calcium, phosphate, and magnesium, with concentration modulation by vitamin D_3, parathyroid hormone (PTH), and calcitonin (reviewed by Guyton [49]). Calcium is a key divalent cation required for physiologic processes, such as action potential dynamics and cell membrane permeability. Phosphorous is associated with purine nucleotide synthesis, phosphorylation of regulatory proteins, and cell membrane assembly. Magnesium is an important cofactor for certain enzymatic and neuromuscular reactions.

PTH promotes calcium reabsorption in the kidney and activates osteoblasts, that in turn express an as yet unidentified factor prompting osteoclast bone resorption, thus releasing calcium [50]. Calcitonin mutes osteoclast activity (mechanism unknown) and facilitates restoration of the basal calcium level.

3.5 Soluble factors

Probably the most spirited interests in bone focus on the soluble factors due to their compelling therapeutic merit. Consequently, a wealth of information is being generated on their actual and hypothesized physiologic roles. Clearly, the marquis player attracting a frenzy of attention is the clan of molecules known as the *bone morphogenetic proteins* (BMPs). BMPs are a member of the transforming growth factor beta 'super family', and at the time of this writing, there are probably more than 40 molecules within the TGF-β *superfamily*. The major subdivisions within the TGF-β classification have been detailed in several comprehensive reviews and include TGF-β_{1-5} and BMPs 2–13 (BMP-1 is not part of the TGF-β superfamily), the growth/differentiating factors (GDF)1–10 (a subclass of BMPs [51]), the inhibins, activins, Vg-related genes, nodal-related genes, Drosophila genes (e.g. dpp, 60A), and glial-derived neurotropic factor [52–60].

4 Tissue engineered bone: Is it possible?

Bone is a highly complex organ consisting of two distinctive, integrated tissue types speckled with multiple cell phenotypes. Bone is dynamic; it is richly vascularized; it fulfills rigorous biomechanical roles; and it supports challenging physiologic efforts. In light of these awesome capabilities and responsibilities, is the tissue engineer naively arrogant in assuming deficient bone can be augmented or regenerated? Of course not! Surgeons have been exploiting the capacity for bone to regenerate for over 300 years [61]. With the contemporary boom in cell and molecular biology, incredible new technologies are available for bone tissue engineers. It is the opinion of the authors that mimicking the functional and physiological capabilities of bone with a laboratory-engineered product is in the future; however, the promotion of bone formation with a tissue engineered composition is possible. There may be several design considerations with components that include *cells, soluble factors,* and *extracellular matrix.*

4.1 Cells

4.1.1 Mesenchymal stem cells

It is a well-known clinical fact that *bone marrow* is effective in promoting bone regeneration [62]. We mentioned that marrow contains a mixture of several types of lineage cells (e.g. erythropoietic, granulocytic, stromal [17]). Caplan *et al.* have shown that *mesenchymal stem cells* (MSCs) found in marrow can be isolated, separated, their numbers expanded, and an osteoblast phenotype can be derived [9, 63]. This is a remarkable accomplishment towards engineering a strategically crucial component of bone. However, unless a suitable substratum (i.e. ECM) is included in the engineered package, it is highly unlikely that cells alone will be effective. Caplan and Bruder acknowledge there is a 'lack of detailed understanding about the key parameters involved in cell–cell, cell–matrix, and/or vascular relationships involved in osteogenic activity of secretory osteoblasts' [64]. This is underscored by the authors when they emphasize that ' ... it is important to construct a delivery system which has the potential to place sheets of osteoprogenitor cells in a configuration where *host vasculature* (emphasis added) can quickly position itself to provide the appropriate cuing' [64]. Caplan and Bruder point out that if MSCs aggregate and vascularity is absent, cartilage is produced [64].

4.1.2 Osteoblast-like cells

Both primary osteoblast-like cells (isolated from periosteum [65–67]) and transformed cell lines (MC3T3-E1 [68, 69]) have been investigated as candidates for tissue engineered bone. Cells were cultured on *polymer substrata*. It is doubtful whether *in vitro* technology will achieve the level of sophistication to promote and to sustain human osteoblast phenotype and function to the extent practical for clinical applications. A significant problem with human cells in tissue culture is the phenomenon of *dedifferentiation*. Therefore, novel, clever *in vitro* methodologies must be developed to sustain osteoblast phenotypes, perhaps after those noted in a perceptive paper by Sittinger *et al.* [70]. However, engineering bone from combinations of substrata and osteoblast-like cells derived through *in vitro* technology may not evolve into a practical clinical therapeutic, retaining value only as a test bed for cell–material interactions.

4.2 Soluble factors

Earlier in the chapter, we mentioned there is a wealth of soluble factors associated with bone and we chose to showcase the bone morphogenetic proteins (BMPs), now available in sufficient therapeutic quantity through genetic engineering.

BMPS direct the progression of cells and their organizational format to tissues and organs in the embryo [71–74]; influence body patterning, limb development, size and number of bones [75–79]; and modulate post-fetal chondro-osteogenetic maintenance [57, 58]. No currently identified soluble factor has the impact of BMP for regenerating bone; consequently, the allure of harnessing the chondro-osteogenic capacity of BMPs has prompted spirited investigations of its applications in animal studies [80–91] with the goal to develop human therapies. Despite remarkable clinical promise, it is not practical to use BMP without a carrier: a delivery system. A delivery system may function as a solid state extracellular matrix (ECM) to position, localize, and actively deploy BMP for cell interactions. Candidate ECMs may be either polymers (e.g. collagen, poly-α hydroxy acids [92]) or ceramics (Reviewed by Ripamonti and Vukicevic [93] and Yaszemski *et al.* [93]).

4.3 The tissue engineer's equivalent to the extracellular matrix: The delivery system

First: An Opinion. We are unashamedly biased in our engineering protocol for preferring a design composition consisting of *BMP and a delivery system* either to augment or to regenerate bone. Based on criteria of clinical convenience and ramping up for manufacturing, cell-laden devices have draw backs. We mentioned dedifferentiation with cell lines and primary cells. Moreover, MSC therapy (i.e. MSC with some carrier) does not appear suitable as an acute bone remedy: time is needed to acquire a patient's marrow, to isolate, identify, and amplify MSCs, and to push a critical mass number of these cells to an osteoblast phenotype that can be loaded onto a carrier for administration to a patient. In addition, it has not been proven how durable MSCs are within a carrier. How susceptible are these cells to variables such as storage, handling, operating room environments, and surgeons? A powerful virtue of MSC therapy is application in cell-deficient patients. For example, Haynesworth *et al.* (1994) have reported that MSCs plummet with advancing age [94]. Consequently, the effectiveness of BMPs, a differentiating factor promoting osteoblast phenotype from an undifferentiated mesenchymal cell, should be less robust in an elderly patient than in a teenager. Supplemental MSC therapy to the BMP plus carrier could off-set local cell decrement.

4.3.1 Candidate delivery systems

There appear to be a number of strong delivery system candidates that can be engineered with BMP. The major candidates are ceramics, collagen (a natural polymer), and synthetic polymers (e.g. the poly-α hydroxy acids). Competent reviews detailing ceramics for cell [64] and BMP delivery [95] should be consulted. Moreover, synthetic polymers for similar applications have been capably reviewed and should be sought by the inquisitive [92, 95–97]. A clever strategy recently reported by Fang *et al.* (1996) involved genetically manipulating fibroblasts to produce plasmid-coded proteins with BMP-like activity [98]. The intriguing report by Fang *et al.* described how long bone ostectomies in rats were regenerated with this therapy; however, only efficacy in higher order mammals can validate potential clinical relevance.

4.3.2 Are there design criteria for BMP delivery systems?

In the section on bone basics, the authors focused on specific organizational and structural patterns, with emphasis on vascularity, Haversian system–Volkmann canal design, and dynamic reciprocity among bone cells. Moreover, the ECM was highlighted for its indispensable roles to position soluble factors, to provide attachment for cells, and to promote a cooperative interaction between positioned soluble factors and ECM anchorage domains. While space constraints limit the details we can pursue in this chapter, there are design parameters for a BMP delivery system that can be derived from the reports about the regulatory roles provided by ECM [39, 40, 43, 44, 46, 99]. Unfortunately, specifications for particular characteristics of the design remain in large part a mystery. Some design criteria that may be important include: surface topography (i.e. texture), surface charge, hydrophobicity/hydrophilicity, and porosity. Moreover, the dynamic nature of bone, especially during regeneration, requires a product designed with distinctively engineered BMP release kinetics. It remains a challenge for the bold to develop investigative protocols to determine specific values for the numerous design criteria offered.

5 Conclusion — and — where do we go from here?

The authors presented basic definitions to serve as an operational framework for this chapter. We reviewed salient structural and functional components of bone and highlighted potential design considerations for the enterprising tissue engineer. Design options were not inclusive, space constraints precluded discussions of *guided bone regeneration* [100, 101] and *cell surface modification* (e.g. altering cell epitopes) [102]. We purposefully were cursory on design specifications for candidate delivery systems for cells and BMPs because not much is known on these details. Therefore, the investigative spirit of the inquiring tissue engineer should provide the inspirational vigor to pursue and identify fundamental design characteristics and ensure compliance with the dynamic structure and function of bone.

References

1. Nerem, R. M. and Sambanis, A., Tissue engineering: From biology to biological substitutes. *Tissue Engineering*, 1995, **1**(1): 3–13.
2. Arvide, R. E., Jenison, R. D., Mashaw, L. H. and Northup, L. L., *Engineering Fundamentals and Problem Solving*. McGraw Hill, New York, 1996.
3. Black, J., *Biological Performance of Materials. Fundamentals of Biocompatibility*. New York: Marcel Dekker, 1992, pp. 1–390.
4. Langer, R. and Vacanti, J. P., Tissue engineering. *Science*, 1993, **260**, 920–926.
5. Hubbell, J. A., Biomaterials in tissue engineering. *Biotechnology*, 1995, **13**, 565–576.
6. Vacanti, C. A. and Mikos, A. G., Letters from the editors. *Tissue Engineering*, 1995, **1**(1): 1–2.
7. Dorland, *Dorland's Illustrated Medical Dictionary*. W.B. Saunders Company, Philadelphia, 1988.
8. Urist, M. R., Bone morphogenetic protein induced bone formation and the bone-bone marrow consortium. In *Bone transplantation*, eds. M. Aebi and P. Regazzoni. Springer-Verlag, Berlin, 1989, pp. 185–197.
9. Bruder, S. P., Fink, D. J. and Caplan, A. I., Mesenchymal stem cells in bone development, bone repair, and skeletal regeneration therapy. *Journal of Cell Biochemistry*, 1994, **56**, 283–294.
10. Buckwalter, J. A., Glimcher, M. J., Cooper, R. R. and Recker, R., Bone biology. Part I: Structure, blood supply, cells, matrix, and mineralization. *Journal of Bone and Joint Surgery*, 1995, **77**-A(8), 1256–1275.
11. Buckwalter, J. A., Glimcher, M. J., Cooper, R. R. and Recker, R., Bone biology. Part II: Formation, modeling, remodeling, and regulation of cell function. *Journal of Bone and Joint Surgery*, 1995, **77**-A(8), 1276–1289.
12. Trueta, J., The role of blood vessels in osteogenesis. *Journal of Bone and Joint Surgery*, 1963, **45**B, 402–418.
13. Hollinger, J. O. and McAllister, B. M., Bone and its repair. In *Bioceramics*, eds. L. H. J. Hench and D. Greenspan. Pergamon-Elsevier Science Ltd., London, 1995, pp. 3–11.
14. Jowsey, J., Studies of haversian systems in man and some animals. *Journal of Anatomy*, 1966, **100**(4), 857–864.
15. Kelly, P. J., Anatomy, physiology, and pathology of the blood supply of bones. *Journal of Bone and Joint Surgery*, 1968, **50**-A, 766–783.
16. Kessel, R. G. and Kardon, R. H., *Tissues and Organs: A Text-Atlas of Scanning Microscopy*. W.H. Freeman, New York, 1979, p. 25.
17. Owen, M., *Lineage of osteogenic cells and their relationship to the stromal system*. Elsevier Science Publishers, Amsterdam, 1985, pp. 1–25.
18. Delmas, P. D. and Malaval, L., The proteins of bone. In *Physiology and Pharmacology of Bone*, ed. G. R. Mundy. Springer-Verlag, New York, 1993, pp. 673–724.
19. Parfitt, M. A., The cellular basis of bone remodeling: The quantum concept reexamined in light of recent advances in the cell biology of bone. *Calcified Tissue International*, 1984, **36**, 37–45.

20. Wlodarski, K. H., Properties and origin of osteoblasts. *Clinical Orthopaedics and Related Research*, 1990, **252**, 276–293.

21. Sorensen, M. S., Temporal bone dynamics, the hard way. *Acta Otolaryngolica*, 1994, **5**, 5–22.

22. Alvarez, J., Ross, P., Athanasou, N., Blair, H., Greenfield, E. and Teitelbaum, S., Osteoclast precursors circulate in avian blood. *Calcified Tissue International*, 1992, **51**, 48–53.

23. Athanasou, N. A., Current concepts review: Cellular biology of bone resorbing cells. *Journal of Bone and Joint Surgery*, 1996, **78**-A(7), 1096–1112.

24. Baron, R., The cellular basis of bone resorption: Cell biology of the osteoclast. In *Bone Formation and Repair*, eds. C. T. Brighton, G. Friedlaender and J. M. Lane. American Academy of Orthopaedic Surgeons, Rosemont, 1994, pp. 247–252.

25. Raisz, L. G., Recent advances in bone cell biology Interactions of vitamin D with other local and systemic factors. *Bone and Mineral*, 1990, **9**, 191–197.

26. Fuller, K., Gallager, A. C. and Chambers, T. J., Osteoclast resorption-stimulating activity is associated with the osteoblast cell surface and/or the extracellular matrix. *Biochemical and Biophysical Research Communications*, 1991, **181**, 67–73.

27. Hollinger, J. O. and Wong, M. E. K., The integrated processes of hard tissue regeneration with special emphasis on fracture healing. *Oral Surgery, Oral Medicine and Oral Pathology*, in press.

28. Roberts, W. E., Helm, F. R., Marshall, K. J. and Gongloff, R. K., Rigid endosseous implants for orthodontic and orthopedic anchorage. *Angle Orthodontics*, 1990, **59**(4), 247–256.

29. Frost, H. M. *Intermediary Organization of the Skeleton*. CRC Press, Boca Raton, FL, 1986, pp. 1–365.

30. Frost, H. M. *Intermediary Organization of the Skeleton*. Boca Raton, FL: CRC Press, 1986, pp. 1–331.

31. Kimmel, D., A paradigm for skeletal strength homeostasis. *Journal of Bone and Joint Mineral Research*, 1993, **8**(2), 515–522.

32. Recker, R. R. Bone remodeling and metabolic bone disease. Portland Bone Symposium. Portland, OR, 1993

33. Parfitt, A. M., Pharmacologic manipulation of bone remodeling and calcium homeostasis. In *Calcium Metabolism*, ed. A. J. Kanis. Karger, Basel, 1990, pp. 1–27.

34. Young, M. F., Kierr, J. M., Ibaraka, K., Heegaard, A. and Robey, P. G., Structure, expression, and regulation of the major noncollagenous matrix proteins of bone. *Clinical Orthopaedics and Related Research*, 1991, **281**, 275–294.

35. Watrous, D. A. and Andrews, B., The metabolism and immunology of bone. *Seminars in Arthritis and Rheumatism*, 1989, **19**(1), 45–65.

36. Hauschka, P. V., Growth factor effects in bone. In *The Osteoblast and Osteocycte*, ed. B. K. Hall. The Telfor Press, 1990, pp. 103–170.

37. Sandberg, M. M., Aro, H. T. and Vuorio, E. I., Gene expression during bone repair. *Clinical Orthopaedics and Related Research*, 1993, **289**, 292–312.

38. Robey, P. G., Bone matrix proteoglycans and glycoproteins. In *Principles of*

Bone Biology, eds. J. P. Bilezikian, L. C. Raisz and G. A. Rodan. Academic Press, New York, 1996, pp. 155–166.

39. Reddi, A., Extracellular matrix and development. In *Extracellular Matrix Biochemistry*, eds. K. A. Piez and A. H. Reddi. Elsevier, New York, 1984, pp. 375–412.

40. Sage, E. H. and Bornstein, P., Extracellular matrix proteins that modulate cell-matrix interactions. *Journal of Biological Chemistry*, 1991, **266**(23), 14831–14834.

41. Scutt, A., Mayer, H. and Wingender, E., New perspectives in the differentiation of bone-forming cells. *BioFactors*, 1992, **4**(1), 1–13.

42. Lin, C. Q. and Bissel, M. J., Multi-faceted regulation of cell differentiation by extracellular matrix. *FASEB Journal*, 1993, **7**, 737–743.

43. Haralson, M. A., Extracellular matrix and growth factors: An integrated interplay controlling tissue repair and progression to disease. *Laboratory Investigation*, 1993, **69**(4), 369–372.

44. Raghow, R., The role of extracellular matrix in postinflammatory wound healing and fibrosis. *FASEB Journal*, 1994, **8**, 823–831.

45. Attisano, L., Wrana, J. L., Lopez-Castillas, F. and Massagué, J., Transforming growth factor-beta receptors and actions. *Biochimica et Biophysica Acta*, 1994, **1222**, 71–80.

46. Ingber, D. E. and Folkman, J., Mechanochemical switching between growth and differentiation during fibroblast growth factor-stimulated angiogenesis in vitro: Role of extracellular matrix. *Journal of Cell Biology*, 1989, **109**, 317–330.

47. Ingber, D. E., Deepa, P., Sun, Z., Betensky, H. and Wang, N., Cell shape, cytoskeletal mechanics, and cell cycle control in angiogenesis. *Journal of Biomechanics*, 1995, **28**(12), 1471–1484.

48. Nagahara, S. and Matsuda, T., Cell-substrate and cell-cell interactions differently regulate cytoskeletal and extraceullar matrix potein gene expression. *Journal of Biomedical Materials Research*, 1996, **32**, 677–686.

49. Guyton, A. C. *Textbook of Medical Physiology*. W.B. Saunders, Philadephia, 1991, pp. 2–23.

50. Fitzpatrick, L. A. and Bilezkian, J. P., Actions of parathyroid hormone. In: *Principles of Bone Biology*, eds. J. P. Bilezkian, L. G. Raisz and G. A. Rodan. Academic Press, New York, 1996, pp. 339–346.

51. Nishitoh, H., Ichijo, H., Kimura, M., Matsumoto, T., Makishima, F., Yamaguchi, A., Yamashita, H., Enomoto, S. and Miyazono, K., Identification of type I and type II serine/threonine kinase receptors for growth/differentiation factor-5. *Journal of Biological Chemistry*, 1996, **271**(35), 21345–21352.

52. Massagué, J., The transforming growth factor-beta family. *Annual Review of Cell Biology*, 1990, **6**, 597–641.

53. Massagué, J., Cheifetz, S., Laiho, M., Ralph, D. A., Weis, F. M. B. and Zente, A., Transforming growth factor-β. *Cancer Surveys*, 1992, **12**, 81–103.

54. Reddi, A. H., Regulation of cartilage and bone differentiation by bone morphogenetic protein. *Current Opinion on Cell Biology*, 1992, **4**, 850–855.

55. Sporn, M. and Roberts, A., Transforming growth factor beta: Recent progress and new challenges. *Journal of Cell Biology*, 1992, **119**, 1017–1021.

56. Wozney, J. M., The bone morphogenetic protein family and osteogenesis. *Molecular Reproduction and Development*, 1992, **32**, 160–167.

57. Kingsley, D., What do BMPs do in mammals? Clues from the mouse short-ear mutation. *Trends in Genetics*, 1994, **10**(1), 16–21.

58. Kingsley, D. M., The TGF-beta superfamily: new members, new receptors, and new genetic tests of function in different organisms. *Genes and Development*, 1994, **8**, 133–146.

59. Reddi, A. H., Bone and cartilage morphogenesis: Cell biology to clinical applications. *Current Opinion in Genetics and Development*, 1994, **4**, 737–744.

60. Hogan, L. B., Bone morphogenetic proteins in development. *Current Opinion in Genetics and Development*, 1996, **6**, 432–438.

61. Meek'ren, J. J. V., Observations. *Medico-Chirurgicaae*, 1668.

62. Connolly, J., Guse, R., Lippiello, L. and Dehne, R., Development of an osteogenic bone marrow preparation. *Journal of Bone and Joint Surgery*, 1989, **71**-A(5 (June)), 684–691.

63. Caplan, A. I., Mesenchymal stem cells. *Journal of Orthopedic Research*, 1991, **9**(5), 641–650.

64. Caplan, A. I. and Bruder, S. P., Cell and molecular engineering of bone regeneration. In: *Principles of Tissue Engineering*, eds. R. Lanza, R. Langer and W. Chick. Academic Press, San Diego, 1997, pp. 604–618.

65. Ishaug, S. L., Yaszemski, M. J., Bizios, R. and Mikos, A. G., Osteoblast function on synthetic biodegradable polymers. *Journal of Biomedical Materials Research*, 1994, **28**, 1445–1453.

66. Ishaug, S. L., Payne, R., Yaszemski, M. J., Aufdemorte, T., Bizios, R. and Mikos, A. G., Osteoblast migration on poly(alpha-hydroxy esters). *Biotechnology Bioengineering*, 1996, **50**, 443–451.

67. Puelacher, W. C., Vacanti, J., Ferraro, N. F., Schloo, B. and Vacanti, C., Femoral shaft reconstruction using tissue-engineered growth of bone. *International Journal of Oral Maxillofacial Surgery*, 1996, **25**, 223–228.

68. Elgendy, H. M., Norman, M. E., Keaton, A. R. and Laurencin, C. T., Osteoblast-like cell (MC3T3-E1) proliferation on bioerodible polymers: an approach towards development of a bone-erodible polymer composite material. *Biomaterials*, 1993, **14**(4), 263–269.

69. Laurencin, C. T., Norman, M. E., Elgendy, H. M., El-Amin, S. F., Allcock, H. R., Pucher, S. R. and Ambrosio, A. A., Use of polyphosphazenes for skeletal tissue regeneration. *Journal of Biomaterials Research*, 1993, **27**, 963–973.

70. Sittinger, M., Bujia, J., Rotter, N., Reitzel, D., Minuth, W. W. and Burmester, G. R., Tissue engineering and autologous transplant formation: practical approaches with resorbable biomaterials and new cell culture techniques. *Biomaterials*, 1996, **17**, 237–242.

71. Tomizawa, K., Matsui, H., Kondo, E., Miyamoto, K., Tokuda, M., Itano, T., Nagahata, S., Akagi, T. and Hatase, O., Developmental alteration and neuron-specific expression of bone morphogenetic protein-6 (BMP-6) mRNA in rodent brain. *Molecular Brain Research*, 1995, **28**, 122–128.

72. Iwasaki, S., Hattori, A., Sato, M., Tsujimoto, M. and Kohno, M., Characterization of the bone morphogenetic protein-2 as a neurotropic factor. *Journal of Biological Chemistry*, 1996, **271**(29), 17360–17365.

73. Reissmann, E., Ernsberger, U., Francis-West, P. H., Rueger, D., Brickell, P. M. and Rohrer, H., Involvement of bone morphogenetic protein-4 and bone morphogenetic protein-7 in the differentiation of the adrenergic phentoype in developing sympathetic neurons. *Development*, 1996, **122**, 2079–2088.

74. Zhao, G. Q., Deng, K., Labosky, P. A., Liaw, L. and Hogan, B. L., The gene encoding bone morphogenetic protein 8b is required for the initiation and maintenance of spermatogenesis in the mouse. *Genes and Development*, 1996, **10**, 1657–1669.

75. Duboule, D., The vertebrate limb: A model system to study the Hox/HOM gene network during development and evolution. *Bioessays*, 1992, **14**, 375–384.

76. Duboule, D., How to make a limb? *Science*, 1994, **266**, 575–576.

77. Laufer, E., Nelson, C. E., Johnson, R. L., Morgan, B. A. and Tabin, C., Sonic hedgehog and Fgf-4 act through a signaling cascade and feedback loop to integrate growth and patterning of the developing limb bud. *Cell*, 1994, **79**, 993–1003.

78. Tickle, C., On making a skeleton. *Nature*, 1994, **368**, 587–588.

79. Winnier, G., Blessing, M., Labosky, P. and Hogan, B. L. M., Bone morphogenetic protein-4 is required for mesoderm formation and patterning in the mouse. *Genes and Development*, 1995, **9**, 2105–2116.

80. Toriumi, D. M., Kotler, H. S., Luxunberg, D. P., Holtrop, M. E. and Wang, E. A., Mandibular reconstruction with a recombinant bone-inducing factor. Functional, histologic, and biomechanical evaluation. *Archives of Otolaryngology Head and Neck Surgery*, 1991, **117**(10), 1101–1112.

81. Yasko, A. W., Lane, J. M., Fellinger, E. J., Rosen, V., Wozney, J. M. and Wang, E. A., The healing of segmental bone defects induced by recombinant human bone morphogenetic protein (rhBMP-2). *Journal of Bone and Joint Surgery*, 1992, **74**-A(5), 659–671.

82. Gerhart, T. N., Kirker-Head, C. A., Kriz, M. J., Holtrop, M. E., Hennig, G. E. and Wang, E. A., Healing segmental femoral defects in sheep using recombinant human bone morphogenetic protein (rhBMP-2). *Clinical Orthopaedics and Related Research*, 1993, **293**, 317–326.

83. Kenley, R., Yim, K., Abrams, J., Ron, E., Turek, T., Marden, L. and Hollinger, J., Biotechnology and bone graft substitutes. *Pharmacological Research*, 1993, **10**(10), 1393–1401.

84. Cook, S. D., Baffes, G. C., Wolfe, M. W., Sampath, T. K. and Rueger, D. C., Recombinant human bone morphogenetic protein-7 induces healing in a canine long-bone segmental defect model. *Clinical Orthopaedics*, 1994, **301**, 302–312.

85. Kenley, R., Marden, L., Turek, T., Jin, L., Ron, E. and Hollinger, J., Osseous regeneration in the rat calvarium using novel delivery systems for recombinant human bone morphogenetic protein-2 (rhBMP-2). *Journal of Biomedical Materials Research*, 1994, **28**(10), 1139–1147.

86. Marden, L. J., Hollinger, J. O., Chaudhari, A., Turek, T., Schaub, R. and Ron, E., Recombinant bone morphogenetic protein-2 is superior to demineralized bone matrix in repairing craniotomies defects in rat. *Journal of Biomedical Materials Research*, 1994, **28**, 1127–1138.

87. Cook, S. D., Wolfe, M. W., Salkeld, S. L. and Rueger, D. C., Effect of recombinant human osteogenic protein-1 on healing of segmental defects in non-human primates. *Journal of Bone and Joint Surgery*, 1995, **77**-A(5), 734–750.

88. Sandhu, H. S., Kanim, L. E., Kabo, J. M., Toth, J., Zeegan, E., Liu, D., Seeger, L. J. and Dawson, E. G., Evaluation of rhBMP-2 with an OPLA carrier in a canine posterolateral (transverse process) spinal fusion model. *Spine*, 1995, **20**(24), 2669–2683.

89. Schimandle, J. H., Boden, S. D. and Hutton, W. C., Experimental spinal fusion with recombinant human bone morphogenetic protein-2. *Spine*, 1995, **20**(12), 1326–1337.

90. Bostrom, M., Lane, J., Tomin, E., Browne, M., Berberian, W., Turek, T., Smith, J., Wozney, J. and Schildhauer, T., Use of bone morphogenetic protein-2 in the rabbit ulnar nonunion model. *Clinical Orthopaedics and Related Research*, 1996, **327**, 272–282.

91. Mayer, M. H., Hollinger, J. O., Ron, E. and Wozney, J., Repair of alveolar clefts in dogs with recombinant bone morphogenetic protein and poly(a-hydroxy acid). *Plastic Reconstructive Surgery*, 1996, **98**(2), 247–259.

92. Hollinger, J. O. and Leong, K., Poly(α-hydroxy acids): Carriers for bone morphogenetic proteins. *Biomaterials*, 1996, **17**(2), 187–194.

93. Ripamonti, U. and Vukicevic, S., Bone morphogenetic proteins: from developmental biology to molecular therapeutics. *South African Journal of Science*, 1995, **91**, 277–280.

94. Haynesworth, S., Goldberg, V. and Caplan, A. I., Diminution of the number of mesenchymal stem cells as a cause for skeletal aging. In: *Musculoskeletal Soft Tissue Aging: Impact on Mobility*, eds. J. A. Buckwalter, V. Goldberg and S. Soo. Am. Academy of Orthopedic Surgeons, Rosemont, 1994, pp. 80–86.

95. Yaszemski, M. J., Payne, R. G., Hayes, W. C., Langer, R. and Mikos, A. G., Evolution of bone transplantation: Molecular, cellular and tissue strategies to engineer human bone. *Biomaterials*, 1996, **17**, 175–185.

96. Cima, L., Vacanti, J. P., Vacanti, C., Ingber, D., Mooney, D. and Langer, R., Tissue engineering by cell transplantation using biodegradable polymer substrates. *Journal of Biomechanical Engineering*, 1991, **113**, 143–151.

97. Hollinger, J. *Biomedical Applications of Synthetic Biodegradable Polymers.* CRC Press, Boca Raton, FL, 1995, pp. 1–257.

98. Fang, J., Zhu, Y., Smiley, E., Bonadio, J., Rouleau, J., Goldstein, S., McCauley, L., Davidson, B. and Roessler, B., Stimulation of new bone formation by direct transfer of osteogenic plasmid genes. *PNAS, USA*, 1996, **93**(June), 5753–5758.

99. Madri, J. A. and Basson, M. D., Extracellular matrix-cell interactions: Dynamic modulators of cell, tissues and organisms structure and function. *Laboratory Investigation*, 1992, **66**, 519–521.

100. Wachtel, H. C., Langford, A., Bernimoulin, J. P. and Reichart, P., Guided bone regeneration next to osseointegrated implants in humans. *International Journal of Oral Maxillofacial Implants*, 1991, **6**, 127–135.

101. Robinson, B., Hollinger, J. O., Szachowicz, E. and Brekke, J., Calvarial bone repair with porous D,L-polylactide. *Archives of Otolaryngology Head and Neck Surgery*, 1995, **112**(6), 707–713.

102. Sandrin, M. S., Fodor, W. L., Mouhtouris, E., Osman, N., Choney, S., Rollins, S. A., Guilmette, E. R., Setter, E., Squinto, S. P. and McKenzie, I. F., Enzymatic remodeling of the carbohydrate surface of a xenogeneic cell substantially reduces human antibody binding and complement-mediated cytolysis. *National Medical*, 1995, **1**(12), 1261–12567.

CHAPTER III.3

Tissue Engineering of Cartilage

ANGELA M. RODRIGUEZ,
CHARLES A. VACANTI
University of Massachusetts Medical Center,
S-2, Room 751, 55 Lake Avenue North,
Worcester, MA 01655, USA

1 Introduction

Cartilage has tremendous potential in tissue engineering because it has a broad spectrum of use in multiple areas of the body and has a small capacity for self-repair [1–5]. This mesodermal derivative possesses specific characteristics that make it a unique material in terms of reconstructive capabilities [6–7]. Cartilage loss can be due to multiple conditions such as arthritic, congenital or traumatic diseases. The use of isolated chondrocytes as therapy for loss of tissue has started to evolve to a clinical phase [8]. Engineered cartilage has been shown to be useful as a experimental model in such fields as reconstructive plastic surgery, orthopedics surgery. otolaryngology and other medical fields [9–14]. These issues led tissue engineering researchers to focus on cartilage regeneration.

We will present an overview of the tissue engineering research related to cartilage, its most significant advances as well as current concepts and future clinical applications. Since tissue engineering is dependent of advances related to tissue culture, material sciences and transplantation, we will review each one of these fields and its relation to the engineering of cartilage [15].

2 The need for cartilage

Many clinical conditions lead to loss of cartilage. Osteoarthritic, traumatic and congenital diseases account for most of them. Current therapies include autogenic or allogenic transplantation and the use of synthetic prosthetic devices which are limited either because of scarce supply and difficulties in shaping it once tissue is harvested (the former) or because of host-device interaction (the latter) [9, 16–18]. Autogenous cartilage has the most desirable characteristics as a graft material including, low absorption rate, low blood supply requirement, capability to retain bulk and others [6, 7].

Several methods to engineer cartilage grafts have been attempted. Low proliferative rate of the chondrocytes and the need of an appropriate matrix onto which to seed and shape the cells and resulting tissue, are major limiting factors. Research efforts have responded partially to these needs by developing a method of tissue regeneration in specific sizes and shapes [19–21]. Tissue engineering offers the potential to use autologous cells in the attempt to create self custom-designed replacements, providing the cells with an appropriate synthetic biodegradable and bioabsorbable matrix which allows them to recreate the shape and size needed [10].

Our laboratory has developed a technique permitting the engineering of cartilage. By seeding high density solution of chondrocytes onto biocompatible and biodegradable polymer scaffolds (with specific chemical and physical properties that can be manipulated) [21–23]. This support structure permits anchorage of the cells, guides their reorganization and growth and enables nutrition and gas exchange. After *in vivo* implantation, polymer scaffolds guide chondrocytes to tissue formation by providing a three dimensional environment which allows matrix production with the ultimately formation of cartilage in the shape of the polymer material [24–28].

3 Background

Several reasons led us to believe that cartilage would be an 'ideal tissue to engineer' To name a few, limited regeneration, one cell type, low oxygen requirement and the capability of being stored for relatively long periods of time [2–5, 29]. These characteristics suggested that cartilage was a good candidate to engineer for transplantation [24–26].

For years scientists had attempted to regenerate cartilage from isolated chondrocytes. Results were not encouraging because once cells were delivered (without an appropriate matrix), they disperse into the surrounding tissue. Moskalewski reported cartilage nodules formation after transplantation, which on gross examination were similar to the original tissue (nasal septal cartilage) [32]. Kawiak was able to generate small amounts of elastic cartilage from transplantation of isolated cells in serum, by intramuscular injection of the suspension, but such cartilage displayed an irregular arragement of cells. In addition the quality of the newly regenerated cartilage did not seem to mimic the properties of the original tissue [31]. Wakitani reported regeneration of defective tissue when attempting to repair rabbit articular cartilage from bone marrow and periostium-derived mesenchymal cells [33]. These results helped researchers to realize the need for (a) a vehicle for adequate support of the cell suspension and (b) the isolation of the proper cellular type.

Greene, in the 1970s, had already predicted that the development of synthetic biocompatible materials would be helpful in the regeneration of functional tissue [26]. In our laboratory initial attempts involved seeding cells in a monolayer onto a small wafer of biodegradable polyanhydride polymer and then placing them *in vivo*. Cell density and cell number were inadequate for reliable, successful engraftment when using this design [15]. The next step was to increase the surface area available for attachments to improve the cell anchorage and density by using branching systems as polymer scaffolds for transplantation. Results have been consistently successful using a nonwoven mesh of polyglycolic acid (PGA) 15 μm in diameter and various hydrogels [23, 48, 49, 52].

4 Chondrocytes *in vitro*

Chondrocyte culture constitutes the method which allows cell proliferation. High cellular density in the chondrocyte-polymer constructs has been shown to improve the quality of the tissue formed. It provides enough cells for generation of a successful engraftment [34].

Bearing in mind that one of the main problems of cartilage as a grafting tissue is its limited source. *In vitro* culture offers the possibility to induce chondrocyte proliferation. Our studies have primarily involved static Petri dishes to date. Other methods currently available are rotating bioreactors, spinner flasks, and perfused vessels. Some of them have been shown to demonstrate certain advantages over the others as was seen when continuous mixing in the tissue culture environment resulted in a better distribution of cells along the polymer and improved biochemical properties [20, 35]. Recently, Bujia reported *de novo* generation of cartilage after 50 days of *in vitro* perfusion culture of human nasal chondrocytes seeded onto bioresorbable polymer fleeces [36].

Several studies have shown the dedifferentiation effects of a two dimensional culture in chondrocytes of multiple animal species, including humans. Perhaps the most studied phenomena, is the switch of collagen synthesis from type II to type I, associated with a change in the cell configuration, which acquires a fibroblast-like appearance [37–40]. Although this concept is widely accepted, some recent studies using molecular biology techniques postulate that morphological changes are not related at all to functional properties, and that chondrocytes *in vitro* continue expressing phenotipically distinguished proteins such as collagen type II for long periods of time [41, 42]. It is also the believe that once chondrocytes are returned to a three dimensional environment, they recover their normal matrix production [37, 38, 41]. Three questions regarding the utility of this knowledge for tissue engineering can arise: (1) How long can chondrocytes be allowed to proliferate avoiding the undesirable effects of a two dimensional culture? (2) Will seeding of chondrocytes onto a three dimensional device permit redifferentiation of cells and allow proper cartilage formation? (3) Are other factors needed on the way to optimize the engineered tissue?

We have studied the *in vitro* behavior of cell-polymer constructs, and consistently demonstrated that chondrocytes appear to readily adhere to polymer fibers 14–15 μm in diameter, in multiple layers retaining and regaining their round configuration which is related with a differentiated function. Histological analysis of these constructs showed matrix formation and the presence of strongly acidic sulfate mucopolysaccharides in this novel formed matrix [21].

5 Chondrocytes *in vivo*

Several studies have shown that the implantation of chondrocyte-polymer constructs leads to cartilage formation [10–15]. Initial studies concentrated on seeding bovine chondrocytes onto unwoven synthetic biodegradable polymer fibers of polyglactin 910, a 90:10 copolymer of glycolide and lactide, coated with polyglactin 370 and calcium stearate (0-0 Vicryl suture material, Ethicon Co., Somerville, NJ). These fibers provided a temporary scaffold onto which cells attached until they created their own support matrix. These chondrocyte-polymer constructs were then implanted *in vivo* for different periods of time for up to 6 months [21]. Results showed a reliable, consistent way to deliver chondrocytes for implantation. By increasing the surface area with the branching characteristics of the polymer, chondrocytes were able to engraft and form cartilage that increased in weight and had progressively more mature characteristics as shown by hematoxylin and eosin (HandE) and specific stains for cartilage. Polymer was subsequently replaced by cartilage. This engineered tissue also showed collagen type II content (a representative component of mature cartilage matrix). Controls, done by implanting *in vivo* polymer fiber with no cells or by injecting cells in suspension, formed no cartilage.

After the use of polyglactin fibers other polymers and configurations were used with similar results. Sheets of a non-woven mesh of fibers of polyglycolic acid (PGA) polymer 100 μm thick, with interfiber spaces of 150–200 μm and a fiber diameter of 15 μm (PGA, Davis and Geck, Danbury, CT) were configured in specific shapes and immersed in a 2% solution of Poly-L-lactic acid (PLLA, Polysciences). In this way the crosslinkages of individual polymer fibers are chemically bonded and can retain the shape. Prior to immersion in the PLLA solution, the PGA consisted of a mesh similarly in consistency to a cotton ball, unable to retain a specific shape when wet. Successful cartilage formation in prede-termined shapes was demonstrated using this approach [19]. In another experiment, seeded cells were exposed *in vitro* to BrdU (thymidine ana-logue 5-bromo 2'-deoxyuridine) before implantation. After harvesting BrdU was detected in the lacunae of the new cartilage, which indicated that the new tissue was produced at least partially by the implanted chondrocytes that had been labeled *in vitro*.

Experience with polyglycolic acid polymer fibers allowed our laboratory to determine the ideal cell concentrations to generate new tissue, and to optimize and test the survival capacity of these cell during storage [34, 43]. Naturally occurring matrices as collagen have also been studied as a method to engineer cartilage, but the ability to manipulate mechanical properties is an important limiting factor, as well as a potential inmuno-logic recipient reaction [28]. Other authors have reported the use of fibrin glue and perichondrium as vehicles to grow new cartilage with varying rates of success [44–46]. Several polymers with different chemical and physical properties have been used. Injectable gels present an interesting alternative with potential advantages (Fig. 1). Cells can be evenly sus-pended within a hydrogel in known quantities and pre-determined shapes. Alginate, a naturally occurring polymer derived from seaweed, can be ionically crosslinked in the presence of calcium [48]. Calcium alginate seeded with chondrocytes and then injected subcutaneously on the *dorsum* of the nude mouse, resulted in cartilage formation within several weeks. This demonstrates that hydrogels can provide a three dimensional scaffold for the delivery of chondrocytes [49]. In further experiments ideal concen-tration of cells and their relation to calcium and alginate have been identified [50].

6 Toward clinical applications

Three issues are of critical importance for the future application of tissue engineering as an alternative for replacing lost tissue in human beings. First, is the behavior of human cells, second is the development of experimental models using autologous cells and last is the design of experiments in animal models with specific clinical applications.

6.1 Human cell studies

The ultimate goal in the engineering of cartilage is the use of human chondrocytes. In our laboratory we have begun to generate tissue from human cells. Chondrocytes from elastic (auricular), hyaline (articular) and costal cartilage have been isolated and resulted in cartilage formation after seeding onto appropriate scaffoldings or hydrogels and implanting the

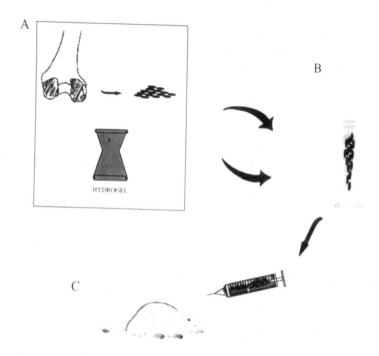

Fig. 1. Injectable polymer model system. Experimental model. (A) bovine chondrocytes are harvested from the articular surface of the distal femur, isolated by enzymatic digestion, and combined with hydrogel (B). This chondrocyte-hydrogel solution is then injected into the dorsum of athymic mice (C) to form neocartilage.

constructs *in vivo* in nude mice. A constant limiting factor has been the period of time that cells have to be maintained in culture to allow for sufficient cell replication. Our recent experience suggests that human chondrocytes maintained *in vitro* for ten weeks, at low plating density and after 4 passages generate cartilage of less than ideal quality after being seeded onto scaffolds. Actually, tissue formed from constructs made from cells after only one passage is decreased in weight and size, with progressively increasing amounts of fibrous tissue being found in the engineered cartilage.

Specific proteins such as elastin have been demonstrated in the matrix of the engineered human tissue (elastic cartilage) (Fig. 2). The average number of cell doubling for human auricular chondrocytes has be determined to be approximately 2 doubling per week the first 3 weeks in culture, and 1 doubling per week the following 10 weeks. Recent reports suggest in the case of hyaline chondrocytes, that cells have a limited lifespan of 34 to 37 doubling. We have not seen a significant difference

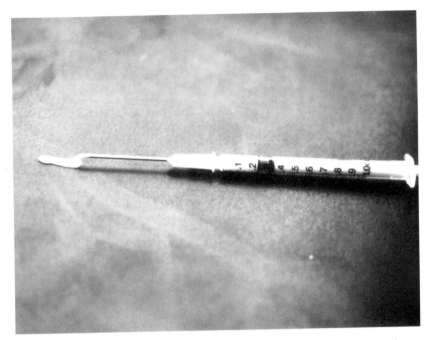

Fig. 2. Injectable gels. Hydrogels are an interesting alternative in tissue engineering. Their physical properties allow them to be delivered by injection and manipulated for different requirements. This photograph shows calcium alginate gel coming out of a needle. Alginic acid is a polysaccharide isolated from seaweed and ionically cross-linked in the presence of calcium.

among donors between 4 and 20 years of age, regarding cellular popullation doubling or the ability of cells to generate engineered elastic cartilage [41, 51].

6.2 Autologous cells

The use of autologous cells is perhaps one of the most important and attractive characteristics of tissue engineering. In 1994 we demonstrated the feasibility of joint resurfacing with autologous chondrocytes and PGA scaffolds in rabbits [10]. Ongoing studies in our laboratory have shown successful engraftment of autologous chondrocytes in larger animal models [52].

6.3 Clinical applications

Very specific experiments of engineering cartilage have been performed in route to clinical applications, using experimental models. Nasal cartilaginous implants have been recreated in the shape of human nasal cartilages by using bovine cells, demonstrating that an appropriate three dimensional shape can be generated [12]. Temporomandibular joint disc has also been mimicked using engineered cartilage [13]. Complex structures such as the human ear have been developed, by seeding cells onto synthetic polymer molds [11]. Cartilaginous tubes have been engineered for tracheal replacement (Fig. 3) [14]. Cartilage has been used also to cover cranial bone defects using nude rats as an animal model [53, 54].

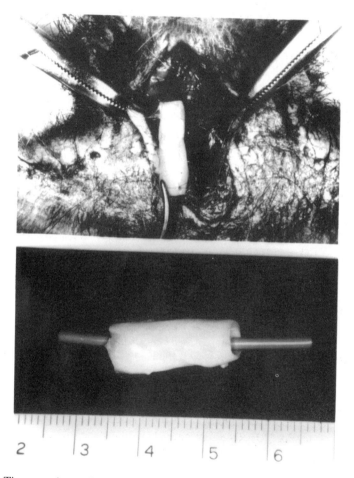

Fig. 3. Tissue engineered cartilage cylinder. The upper photograph shows a tissue engineered cartilage cylinder being sutured into a tracheal defect in a nude rat. The lower photograph gives a close up of the specimen.

7 Future directions

Future efforts in the engineering of cartilage should emphasize the development of methods to allow adequate human chondrocyte proliferation in short periods of time while preserving cellular phenotype. A key development in this effort may be the use of growth factors, which are fairly specific for the type of cartilage [55, 56]. According to the type of clinical applications, attempts should be made to find the ideal concentration of cells and polymer configuration. Biomechanical and physical properties of the constructs at different time points will give a clue concerning the feasibility of the clinical use. *In vivo* studies should be directed to the creation of autologous custom-designed replacements that mimic the ideal of tissue engineering. Studies related to the ability of tissue engineered cartilage generated in specific applications, as well as its potential to provide long term functional improvement need to be carried out. It is also belief that tissue engineering of not only cartilage, but many other tissues will be benefitial to mankind and make significant contributions to the advancement of medical sciences.

References

1. Campbell, C. J., The healing of cartilage defects. *Clinical Orthopaedic and Related Research*, 1969, **64**, 45–63.
2. Fuller, J. A. and Ghadially, F. N., Ultrastructural observations on surgically produced partial thickness defects in articular cartilage. *Clinical Orthopaedics, and Related Research*, 1972, **86**, 193–205.
3. Ghadially, F. N., Thomas, I., Oryschak, A. F. and Lalonde, J. M., Long term results of superficial defects in articular cartilage: A scanning electron-microscope study. *The Journal of Pathology*, 1977, **121**, 213–217.
4. Mankin, H. J., Current concepts Review. The response of articular cartilage to mechanical injury. *Journal of Bone and Joint Surgery*, 1982, **64**A, 460–466.
5. Meachim, G., The effect of scarifcation on articular cartilage in the rabbit. *Journal of Bone and Joint Surgery*, 1963, **45**B, 150–161.
6. Zalzal, G. H., Cotton, R. T. and McAdams, A. J., Cartilage grafts-present status. *Head and Neck Surgery*, 1986, **8**, 363–374.
7. Donald, P. J., Cartilage grafting in facial reconstruction with special consideration of irradiated grafts. *Laryngoscope*, 1986, **96**, 786–807.
8. Brittberg, M., Lindahl, A., Nilsson, A., Ohlsson, C., Isaksson, O. and Peterson, L., Treatment of deep cartilage defects in the knee with autologous chondrocyte transplantation. *The New England Journal of Medicine*, 1994, **331**(14), 889–895.
9. Langer, R. and Vacanti, J. P., Tissue Engineering. *Science*, 1993, **260**, 920–926.
10. Vacanti, C.A., Kim, W., Schloo, B., Upton, J. and Vacanti, J. P., Joint resurfacing with cartilage grown in situ from cell-polymer structures. *American Journal of Sports Medicine*, 1994, **22**, 485–488.

11. Vacanti, C. A., Cima, L. G., Ratkowski, D., Tissue engineered growth of new cartilage in the shape of a human ear using synthetic polymers seeded with chondrocytes. *Materials Research Society,* 1992, **252**, 367–373.

12. Puelacher, W. C., Mooney, D., Langer, R., Upton, J., Vacanti, J. P. and Vacanti, C. A., Design of nasoseptal cartilage replacements synthesized from biodegradable polymers and chondrocytes. *Biomaterials,* 1994, **15**, 774–778.

13. Puelacher, W. C., Wisser, J., Vacanti, C. A., Ferraro, N. F., Jaramillo, D. and Vacanti, J. P., Temporomandibular joint disc replacement made by tissue-engineered growth of cartilage. *Journal of Maxillofacial Surgery,* 1994, **52**, 1172–1177.

14. Vacanti, C. A., Paige, K. T., Kim, W., Sakata, J., and Upton, J., Experimental tracheal replacement using tissue engineered cartilage. *Journal of Pediatric Surgery,* 1994, **29**, 201–205.

15. Vacanti, C. A. and Upton, J., Tissue-engineered morphogenesis of cartilage and bone by means of cell transplantation using synthetic biodegradable polymer matrices. *Bone Repair and Regeneration,* 1994, **21**(3), 445–462.

16. Matsusue, Y., Yamamuro, T. and Hama, H., Arthroscopic multiple osteochondral transplantation to the chondral defect in the knee associated with anterior cruciate ligament disruption. *Arthroscopy,* 1993, **9**, 318–321.

17. McKee, G. K., Total hip replacement-past, present and future. *Biomaterials,* 1982, **3**(3), 130–135.

18. Peterson, C. D., Hillberry, B. M. and Heck, D. A., Component wear of total knee prostheses using Ti-6A1-4V, titanium nitride coated Ti-6Al-4V, and cobalt-chromium-molybdenum femoral components. *Journal of Biomedical Material Research,* 1988, **22**(10), 887–993.

19. Kim, W. S., Vacanti, J. P. Cima, L. G., Mooney, D., Upton, J. and Puelacher, W. C., Cartilage engineered in predetermined shapes employing cell transplantation on synthetic biodegradable polymers. *Plastic and Reconstructive Surgery,* 1994, **94**(2), 233–237.

20. Langer, R., Vacanti, J. P., Vacanti, C. A., Atala, A., Freed, L. E. and Vunjak-Novakovic, G., Tissue engineering: Biomedical applications. *Tissue Engineering,* 1995, **1**(2), 151–161.

21. Vacanti, C. A., Langer, R., Schloo, B. and Vacanti, J. P., Synthetic polymers seeded with chondrocytes provide a template for new cartilage formation. *Plastic and Reconstructive Surgery,* 1991, **88**(5), 753–759.

22. Vacanti, J. P., Beyond transplantation. *Archives of Surgery,* 1988, **123**, 545–549.

23. Vacanti, J. P., Morse, M. A., and Saltzman, W. M., Selective cell transplantion using bioabsorbable artificial polymers as matrices. *Journal of Pediatric Surgery,* 1988, **23**(1 pt 2), 3–9.

24. Bentley, G. and Greer, R. G. III., Homotransplantation of isolated epiphyseal and articular cartilage chondrocytes into joint surfaces of rabbits. *Nature,* 1971, **230**(5293), 385–388.

25. Grande, D. A., Pitman, M. L. Peterson, L., Menche, D. and Klein, M., The repair of experimentally produced defects in rabbit articular cartilage by autologous chondrocyte transplantation. *Journal of Orthopaedic Research,* 1989, **7**(2), 208–218.

26. Green, W. T. Jr., Articular cartilage repair: Behavior of rabbit chondrocytes

during tissue culture and subsequent allografting. *Clinical Orthopaedic*, 1977, (124), 237–250.

27. Lipman, J. M., McDevitt, C. A. and Sokoloff, L., Xenografts of articular chondrocytes in the nude mouse. *Calcified Tissue International*, 1983, **35**(6), 767–772.

28. Wakitani, S., Kimura, T., Hirooka, A., Ochi, T., Yuneda, M., Owaki, H., Ono, K. and Yasui N., Repair of rabbit articular surfaces with allograft chondrocytes embedded in collagen gel. *Journal of the Japanese Orthopaedic Association*, 1989, **63**(5), 529–538.

29. Bloom, W. and Fawcett, D. W., Cartilage. In *Bloom and Fawcett A Textbook of Histology*, 10th edn, ed Dreibelbis, D., W.B. Saunders Company, Philadelphia, 1986, pp. 188–198.

30. Moskalewski, S. and Rybicka, E., The influence of the degree of maturation of donor tissue on the reconstruction of elastic cartilage by isolated chondrocytes. *Acta Anatomica*, 1977, **97**, 231–240.

31. Kawiak, J., Moskalewski, S. and Hinek, A., Reconstruction of the elastic cartilage by isolated chondrocytes in autogenic transplants. *Acta Anatomica*, 1970, **76**, 530–544.

32. Moskalewski, S. and Kawiak, J., Cartilage formation after homotransplantation of isolated chondrocytes. *Transplantation*, 1965, **3**, 737–747.

33. Wakitani, S., Goto, T., Pineda, S. J., Young, R. G., Mansaur, J. M., Caplan, A. F. and Goldberg, V. M., Mesenchymal cell-based repair of large, full-thickness defects of articular cartilage. *Journal of Bone and Joint Surgery*, 1994, **76A**, 579–592.

34. Puelacher, W. C., Kim, W., Vacanti, J. P., Schloo, B., Mooney, D. and Vacanti, C. A., Tissue engineered growth of cartilage: The effect of varying the concentration of chondrocytes seeded onto synthetic polymer matrices. *Oral and Maxillofacial Surgery*, 1994, **23**, 49–53.

35. Freed, L. and Vunjak-Novakovic, G., Tissue Engineering of cartilage. In *The Biomedical Engineering Handbook*, ed CRC press. Trinity Collegue, Hartford, 1995, pp. 1778–1806.

36. Bujia, J., Sittinger M., Minuth, W. W., Hammer, C., Burmester, G. and Kastenbaur, E., Engineering of cartilage tissue using bioresorbable polymer fleeces and perfusion culture. *Acta Otolaryngologica (Stockh)*, 1995, **115**, 307–310.

37. Horton, W. A., Machado, M. A., Ellard, J., Campbell, D., Puttam, E. A., Aulthause, A. L., Sun, X. and Sandell, L. J., An experimental model of human chondrocyte differentiation. *Progress in Clinical and Biological Research*, 1993, **383B**, 533–540.

38. Ramdi, H., Legay, C. and Lievremont, M., Influence of matricidal molecules on growth and differentiation of entrapped chondrocytes. *Experimental Cell Research*, 1993, **207**, 449–454.

39. Kato, Y. and Gospodarowicz, D. Sulfated proteoglycan synthesis by confluent cultures of rabbit costal chondrocytes grown in the presence of fibroblast growth factor. *Journal of Cell Biology*, 1985, **100**, 477–485.

40. Aulthouse, A. L., Beck, M., Griffey, E., Sandford, J., Arden, K., Machado, M. A. and Horton, W. A., Expression of the human chondrocyte phenotype in vitro. *In Vitro Cell Development and Biology*, 1989, **25**, 659–668.

41. Kolettas, E., Buluwela, L., Bayliss, M. T. and Muir, H. I., Expression of cartilage-specific molecules is retained on long term culture of human articular chondrocytes. *Journal of Cell Science*, 1995, **108**, 1991–1999.

42. Pap, S. A., Rodriguez, A. M., Ibarra, C., Cao, Y. L., Peetz, C. and Vacanti, C. A., Collagen expression as a means of distinguishing de-differentiated chondrocytes from fibroblasts in monolayer cultures. Paper presented at 1st Tissue Engineering Society meeting, Orlando, FL, 12 December 1996.

43. Kim, W., Vacanti, J.P., Upton, J., Ibarra, C. and Vacanti, C. A., Potential of cold-preserved condrocytes for cartilage reconstruction. *Tissue Engineering*, 1996, **2**, 75–81.

44. Moskalewski, S., Transplantation of isolated chondrocytes. *Clinics of Orthopedics*, 1991, **272**, 16–20.

45. Homminga, G. N., Buma, P., Koot, H. W. J., van der Kraan, P. M. and van den Berg, W. B., Chondrocyte behavior in fibrin glue. *Acta Orthopaedica Scandinavica*, 1993, **64**, 441–445.

46. Upton, J., Sohn, S. A. and Glowacki, J., New cartilage derived from transplanted perichondrium: What is it?. *Plastic and Reconstructive Surgery*, 1981, **68**, 166–174.

47. Bujia, J., Sittinger, M., Pitzke, P. et al. Synthesis of human cartilage using organotypic cell culture. *ORL*, 1992, **54**, 80–84.

48. Mooney, D. J. and Langer, R., Engineering Biomaterials for Tissue Engineering: The 10-100 Micron size scale. In *The Biomedical Engineering Handbook*, ed CRC Press. Trinity College, Hartford, 1995, pp. 1609–1616.

49. Paige, K. T., Cima, L. G. and Yaremchuk, M. J., Injectable cartilage. *Plastic and Reconstructive Surgery*, 1995, **96**(6), 1390–1398.

50. Paige, K. T., Cima, L. G., Yaremchuk, M. J. et al., De novo cartilage generation using calcium alginate-chondrocyte constructs. *Plastic and Reconstructive Surgery*, 1996, **97**(1), 168–178.

51. Rodriguez, A.M., Ibarra, C., Cao, Y., Eavey, R. and Vacanti, C. A., Analysis of Human auricular Chondrocytes for the engineering of elastic cartilage. Paper presented at 1st Tissue Engineering Society meeting, Orlando, FL, 12 December 1996.

52. Cao, Y. and Vacanti, C. A., unpublished data.

53. Kim, W. S., Vacanti, J. P., Upton, J., Vacanti, M. P. and Vacanti, C. A., The repair of cranial defects in rats with cartilage grown from chondrocytes on synthetic polymer templates. *Transactions of the Orthopaedic Research Society*, 1993, **18**, 487.

54. Kim, W. S., Vacanti, C. A., Upton, J. and Vacanti, J. P., Bone defect repair with tissue engineered cartilage. *Plastic and Reconstructive Surgery*, 1994, **94**, 580–584.

55. Guerne, P., Sublet, A. and Lotz, M., Growth factor responsiveness of human articular chondrocytes: Distinct profiles in primary chondrocytes, subcultured chondrocytes, and fibroblasts. *Journal of Cellular Physiology*, 1994, **158**, 476–484.

56. Quatela, V. C., Sherris, D. A. and Rosier, R. N., The human auricular chondrocyte, *Archives of Otolaryngology Head and Neck Surgery*, 1993, **119**, 32–37.

CHAPTER III.4

Tissue Engineered Tendon

L. LOUIE
Department of Materials Science and Engineering,
Massachusetts Institute of Technology,
Cambridge, MA 02139, USA

I.V. YANNAS
Department of Mechanical Engineering,
Massachusetts Institute of Technology,
Cambridge, MA 02139, USA

AND

M. SPECTOR
Department of Orthopedic Surgery,
Brigham and Women's Hospital, Harvard Medical School,
Boston, MA 02115, USA and
Rehabilitation Engineering Research and Development,
Brockton / West Roxbury VA Medical Center,
West Roxbury, MA 02115, USA

Tendon is a specialized dense connective tissue that links muscle to bone and allows for the transmission of muscle contraction forces to the bone for skeletal locomotion; for example, the Achilles tendon, one of the largest in the body, links the triceps surae muscle, the grouping of the gastrocnemius, the soleus, and the plantaris muscles to the calcaneous bone. As organs, the tendons of the body vary greatly in their anatomy and structural mechanical properties. An approach that might result in the regeneration of one tendon may require modification to facilitate the regeneration of another. Tendons consist of three parts: the muscle attachment region, the substance of the tendon itself, and the bone attachment region. These three parts of tendon vary in their cellular make-up, histology, and function, and may require different tissue engineering approaches. This chapter will focus on the midsubstance of tendon.

1 Introduction

The extracellular matrix of tendon is composed predominantly of collagen fibers highly aligned along the long axis of the tendon. Spindle-shaped cells — fibroblasts — aligned in columns along the direction of the collagen fibers, are sparsely dispersed through the tissue (Fig. 1). When viewed transversely, the cells appear as star-shaped figures among bundles of collagen. The mid-substance of the tendon is generally avascular and absent innervation.

That tendon is arguably the least complex of the connective tissues with respect to composition and architecture might reasonably lead to the expectation that it would be more amenable to tissue engineering approaches than other biological structures. However, decades of experience have shown how refractory tendon is to treatment. This limited capability of tendon injuries to regenerate is a better indication of the challenge it poses to being engineered, and evidences the importance of developing a procedure to do so.

An understanding of the cellular and extracellular components of tendon is essential for assessing the degree to which methods to engineer the tissue are successful, and can guide the development of analogs of extracellular matrix as implants to facilitate regeneration. Moreover, review of the process of spontaneous healing of tendon injuries and the results of selected treatment modalities can provide a guide to strategies for tissue

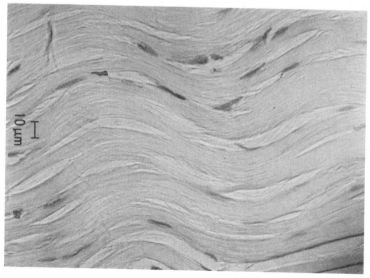

Fig. 1. Light micrograph of tendon showing the spindle-shaped fibroblasts aligned along the long axis of the tissue. Hematoxylin and eosin stain.

engineering. Finally, examination of current studies developing scaffolds as implants to facilitate tendon regeneration can serve as the foundation for future work.

2 Tendon anatomy and composition

2.1 Extracellular matrix

The major extracellular matrix constituent of tendon is type I collagen (approximately 86–90% dry weight), with the balance including type III collagen, elastin, proteoglycans and glycosaminoglycans. The collagenous component of tendon is arranged in the hierarchical microstructure typical of the type I collagen fibrils in all of the connective tissues: the primary structure of amino acids that make up the collagen chain; the secondary structure of the left-hand helical arrangement of each primary chain; and the tertiary structure, the right-handed triple helical configuration formed from hydrogen and covalent bonding of three collagen chains into a 'tropocollagen' molecule. For type I collagen, two of the three chains are identical, called 1(I), and one [2(I)] is slightly different in primary structure. The quaternary structure encompasses several sublevels. Adjacent collagen molecules (tropocollagens) are arranged in a quarter stagger such that oppositely charged segments are aligned. Five collagen molecules in the staggered configuration form a microfibril (Fig. 2). Collagen bundles are arranged in closely packed parallel bundles, oriented in a distinct longitudinal pattern to form ordered units of subfibrils, fibrils, and fascicles (Fig. 2). At the fibril unit, a characteristic 64 nm periodic banding is seen by transmission electron microscopy.

Adult tendons in many species have a bimodal distribution of collagen fibril diameters [1–4]. Collagen fibrils in adult human tendon have mean diameters of 60 and 175 nm [1]. Ultrastructural comparison [2] of collagen fibril diameters in various ages of rabbit tendon found that with aging, collagen fibril diameters increased. Collagen fibrils in newborn rabbit tendons were uniform in size and ranged between 18 and 55 nm in diameter with an average diameter of 37 nm; while the collagen fibrils of mature rabbit tendon ranged between 18 and 166 nm with the two most frequently encountered populations of fibril diameters measuring 55 and 92 nm.

A collection of collagen fibrils bundled together form a fascicle (Fig. 2). Fascicles are bound together by loose connective tissue that allows the fascicles to slide past one another. This loose connective tissue region, called the interfascicular membrane or endotenon, also supports the nervous tissue, blood vessels and lymphatics.

Tendons have a crimped, waveform appearance when seen under polarized light microscope (Fig. 3a). The periodicity of the alternating light and

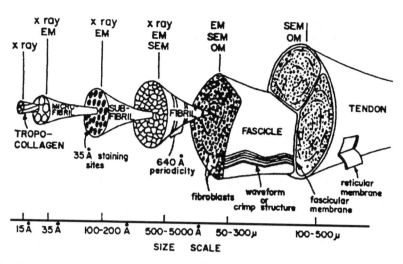

Fig. 2. Schematic representation of the hierarchical structure of tendon. Adapted from Kastekic *et al.* (Connect. Tiss. Res., 1978, 6, 11–23).

dark bands comprising the zigzag pattern (Fig. 3b) of crimped fibers can be used to determine the crimp angle (28 degrees), crimp length (65 μm), and crimp wavelength (120 μm) [5, 6]. The crimp pattern of tendon has been shown to play an important role in its mechanical properties [7,8]; Diamant *et al.* [5–7] demonstrated that the crimp pattern unfolds during initial loading of tendon.

Tendons that generally move uniaxially, such as Achilles tendon, have a loose areolar connective tissue, the paratenon, which is continuous with the tendon. The paratenon stretches several millimeters and recoils without tearing or disrupting tendon blood supply. An interlacing mesh work of thin collagen fibrils, elastic fibers and glycosaminoglycans gives the paratenon this elasticity and extensibility. On the other hand, tendons which bend sharply, such as the flexor tendons of the hand, are enclosed by a synovial sheath. The sheath helps to direct the path of tendon movement by acting like a pulley and allows low friction movement between tendon and the adjacent bones and joints. The sliding of these tendons through the sheath is assisted by the presence of synovial fluid between the outer wall of the tendon and the inner wall of the tendon sheath.

The organization of the extracellular matrix molecules of tendon at the nanometer and micrometer levels (Fig. 2) is the principal determinant of the physiological function of this tissue, its mechanical behavior, as will be discussed below. The degree to which tissue engineering approaches are successful will be reflected in the degree to which the normal composition and architecture of the extracellular matrix has been restored.

(a)

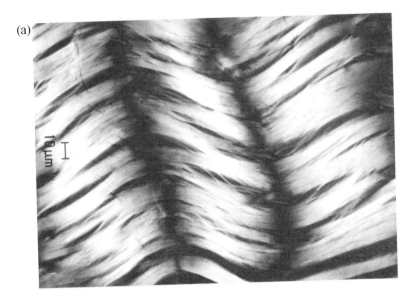

(b)

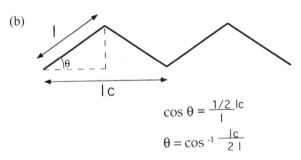

$$\cos \theta = \frac{1/2 \ \mathrm{lc}}{\mathrm{l}}$$

$$\theta = \cos^{-1} \frac{\mathrm{lc}}{2 \ \mathrm{l}}$$

Fig. 3. (a) Polarized light micrograph of tendon demonstarating the crimped structure. (b) Schematic diagram illustrating the calculation of the crimp parameters. Crimp period (lc) and crimp length (l) are determined from the polarized light image, as seen in (a). Crimp angle (θ) is calculated from the crimp wavelength and crimp length. Nomenclature adapted from Kato *et al.* [6].

2.2 Cells

Fibroblasts (also called tenocytes) are the predominant cell type in tendon. Endothelial cells and nerve processes form only a small part of the cell population. Ippolito *et al.* [9] has shown that there is a subpopulation of myofibroblast-like, contractile cells present in normal tendon and has noted their presence in healing tissue [9–13]. This subpopulation of cells has been speculated to be involved in the tensioning of the tendon and in the modulation of the contraction–relaxation of the muscle–tendon complex [9].

As the parenchymal cell of tendon, the tenocyte has the role of maintaining matrix structure through the degradative and formative processes comprising remodeling, and to some extent can contribute to healing. However, that tendon has a relatively low density of cells that display a low mitotic activity, explains the low turnover rate of this tissue and questions the degree to which these parenchymal cells can promote intrinsic healing. Moreover, this fact prepares for the likelihood that exogenous cells may be an important component of tissue engineering strategies for tendon.

2.3 Blood supply and innervation

Tendons receive their blood supply from vessels in the perimysium, the periosteal insertion, and from the surrounding tissue via vessels in the paratenon or mesotenon. In the Achilles tendon and other tendons which are surrounded by a paratenon, vessels enter from many points on the periphery and anastomose with a longitudinal system of capillaries. Blood flow is low, averaging between 0.5 and 1.5 ml per 100 grams of tissue per minute [14, 15] and concentrated mainly on the circumferential surface of the tendon with decreasing blood supply in the interior of the tendon unit. For these reasons, tendon is considered to be a tissue of low vascularity. In this regard blood-borne cells and regulators may not have access to the lesion in tendon, and may have to be replaced in their function by exogenous molecule delivered to the defect by an implant.

Tendon has a complex set of sensory nerve receptors on its surface and throughout the tendon proper. Nerve endings function as pain receptors, vasomotor efferents, and mechanoreceptors. Mechanoreceptors sense joint position, muscle tension, and loads applied to the tendon. The stretch or deformation of a tendon can trigger a reflex muscle contraction or an adjustment in muscle tension. This reflex muscle contraction may play a role in stabilizing and protecting joints from potentially injurious movements.

3 Function: mechanical properties

The main function of a tendon is to transmit muscle forces to bone. The mechanical properties of tendon, as with other tissues, is directly related to the molecular structure of the material. Because its collagen fiber structure is densely packed and highly aligned, tendon has one of the highest tensile strengths of any connective tissues in the body. At low stress the stress–strain curve is characterized by a non-linear, concave-up 'toe region' in which the tendon deforms easily without much tensile force. This is due to the 'uncrimping' of the structure as the collagen fibers align in the direction of applied stress. With increasing applied stress, the toe region is followed by linearly elastic behavior, the slope of which is generally considered the elastic stiffness, or Young's modulus of elasticity of the tendon. At high stress, plastic deformation of the tendon occurs as fiber bundles fracture, leading to eventual rupture of the tendon. The ultimate tensile strength, the maximum stress sustained, of tendons varies greatly from 5–7 MPa for mouse rat tail tendon [16, 17] to over 100 MPa for human tendon [18]. The ultimate strain, the stain at which failure occurs, ranges from 9–35% for Achilles tendon [18]. Achilles tendon undergoes strains between 10 and 12% during normal activities such as walking [17, 18]. Mechanical properties measurements vary greatly with the age, location, species, and testing procedure.

The mechanical properties of tendon have been modeled as a function of the morphology and hierarchical structure of the collagen fibers [3–5, 8, 19, 20]. Few studies, however, have investigated the relationship between morphology and the restoration of mechanical properties in healing tendon [19–21]. Frank, et al. [19] investigated the correlation of certain mechanical properties with collagen fibril diameter in healing medial collateral ligaments in the rabbit. These considerations emphasize the need for tissue engineering approaches to regenerate the molecular composition and hierarchical structure of tendon in order to be assured that the material will be able to function as normal.

4 Tendon healing

The extent to which injuries to tendon heal depends on many factors such as the anatomical location, vascularity, skeletal maturity as well as the amount of tissue loss. Spontaneous healing of tendon has been studied extensively in both the Achilles tendon and the flexor tendons of the hand. Tendon healing normally results in the formation of scar, which is different morphologically, biochemically, and biomechanically from tendon. With time, the scar tissue may assume some of the characteristics of tendon; however, complete regeneration does not appear to occur. Although both Achilles and flexor tendons respond to injury by forming scar tissue, scarring in the flexor tendon appears to have a more detrimental effect on the function of the tissue. Flexor tendons need the ability to glide within their synovial sheath to function properly and formation of adhesions to the sheath during healing interferes with this gliding function.

Following tendon injury by full transection, there is a spontaneous retraction of the cut tendon ends. This retraction has also been reported to occur in a full-transection rabbit animal model for the medial collateral ligament. In the medial collateral ligament, the retraction of the cut ligament ends produced a gap as large as 2–4 mm [22]. In the Achilles tendon of both a rat and rabbit animal model, the gap formed by the retraction of ends, with the joints held in neutral position, was observed to be approximately 9–12 mm [23]. Additional retraction of the ends can occur with movement of calcaneous and knee joints.

The response of Achilles tendon to injury, reviewed in several articles [15, 24, 25], follows a sequence similar to that found in other connective tissues such as ligament and skin. This sequence is generally considered to consist of three overlapping phases: inflammation, repair, and remodeling. Following is a summary of the response of tendon to a lesion produced by a full transection of tendon as reported by several authors [15, 17, 23, 24, 26, 27]. Within the first few hours of injury, the collagenous matrix is disrupted and tendon and blood cells die [27–29]. A hemorrhagic exudate fills the lesion site, and within minutes a fibrin clot forms and seals the wound [23]. The clot contains inflammatory products (fibrin, platelets, red cells, and nuclear and matrix debris) [30]. This clot has little tensile strength.

The inflammatory stage generally starts within hours of injury and can take from 3 to 10 days to complete. This stage is associated with 'clean-up' of the lesion site. Polymorphonuclear neutrophils (PMNs) and lymphocytes, and other acute inflammatory cells, invade and populate the wound site within hours of injury. Macrophages appear soon after to continue the phagocytosis of cell and matrix debris begun by the PMNs.

The period of dramatic fibroblast migration and proliferation and matrix synthesis, the repair phase, starts as early as 10 days post-injury and can take up 2–5 weeks to complete [23, 31]. Undifferentiated and disorganized fibroblasts containing well-developed endoplasmic reticulum, infiltrating from the wound edge and paratenon, begin to proliferate in wound site within the fibrin mesh of the clot. Simultaneously, endothelial cells of surrounding vessels enlarge and proliferate forming capillary buds that follow the migrating fibroblasts. Together the fibroblasts, macrophages and capillaries form the granulation tissue in the wound site [27]. This early stage of the repair phase is characterized by increased cellularity. The fibrin clot is gradually replaced by a collagen bridge, initially comprised predominantly of collagen type III collagen. The type III collagen fibers, which are smaller in diameter than the collagen type I to be deposited in the next stage of healing, are referred to as reticular fibers because of the network-like pattern of their deposition; the collagen type III fibers do not aggregate in a preferential direction. Both collagen production and fibroblast proliferation peak during this phase (characterized as loosely organized fibrous tissue), and subsequently decrease over the next several months [29].

The remodeling phase, which can begin as early as 3 weeks and last for over 1–2 years, is marked by a reduction in the production of type III collagen and reorganization of the type I collagen fibers [29]. During the remodeling stage, the matrix fibers reorient themselves along the long direction of the tendon. This direction coincides with the direction of tensile stress in the tendon [29, 30, 32]. The remodeling stage is also marked by a decrease in the number of fibroblasts present in the tissue and a decrease in the overall volume of the scar tissue. The tensile strength of the tendon increases through this period of remodeling even though the total volume is decreasing. This increase has been explained by the reorganization of the collagen fibers, which has been observed to occur during the same period. In the analogous case of ligament healing, however, it has shown that the tensile strength of the reparative tissue did not return to normal ligament levels even after a year [33]. These data can probably be extended to the case of the tendon.

In summary, the response of mature Achilles tendon to an injury involving a full transection of the tendon results in reparative fibrous tissue that lacks the structure of normal tendon. During the process of remodeling, collagen fibers initially arranged in random directions become reoriented in the longitudinal direction of the tendon. However, there is never a return to normal composition and architecture [23, 30], thus demonstrating the need for tissue engineering approaches.

4.1 Source of reparative cells

There is controversy surrounding the identity and location of the cells responsible for collagen synthesis during tendon repair. On one side of the controversy is the concept that tendon has the necessary cells to produce collagenous tissue (the intrinsic mechanism) [34–36], while on the other side, there is the belief that the source of collagen-producing cells is outside of the tendon (i.e. an extrinsic source such as the surrounding tissues or from the tendon sheath) [37, 38]. Some believe that both intrinsic and extrinsic sources of collagen-producing cells contribute to the healing process [39, 40].

In vitro studies [34, 35, 41] have shown that in response to tendon injury, cells within the tendon had the ability to migrate to and proliferate in the wound site. In these studies, by six weeks, the injury site appeared to be filled with collagen. *In vivo* data appear to parallel these *in vitro* studies. Matthews and Richards [42] showed that cells within the rabbit flexor tendon participated in the wound healing process when the synovial sheath was not violated and the tendon was mobilized with early controlled passive motion. However, if the tendon sheath was compromised and the tendon was immobilized, cells from external sources (e.g. tendon sheath, blood vessels, and other neighboring tissue) migrated and proliferated into the wound site. Tendon repair, in that case, involved the participation of all surrounding tissue in the healing of the entire wound.

Questions remain about whether there is a sufficient pool of parenchymal fibroblasts in tendon to adequately populate a defect. This uncertainty is rationale for the use of exogenous cultured cells in tissue engineering modalities.

4.2 Mechanical loading during tendon healing

There have been many studies investigating the effect of post-operative mobilization on the healing of tendon injuries [22, 23, 43–48]. The majority of studies investigated the mobilization of healing flexor tendon. There is added concern of tendon scar tissue attaching surrounding tissue during healing which limits the gliding ability of the tendon. Limitation in gliding of Achilles tendon is generally not as great a concern. The need for mobilization of reparative tissue in tendon to physically regulate cell behavior and the organization of extracellular matrix complicates methodologies to be implemented to regenerate tendon. Future studies need address the timing and nature of the mobilization.

4.3 Techniques for evaluation of tendon healing

Studies evaluating tendon healing have utilized structural and functional techniques. Structural methodology has included morphological assessment at the microstructural (micrometer) and ultrastructural (nanometer) levels and biochemical analysis. Historically, the extent of tendon healing was evaluated qualitatively by comparing histological sections of repair tissue to normal tendon [6, 23, 30, 46, 49–51]. Functional techniques have included biomechanical testing [52, 53] and gait analysis, with mechanical testing the most widely used method to determine the degree to which restoration of function has been achieved.

Because functional techniques developed for clinical evaluation generally involve either the active participation of the patient or verbal feedback by the patient, different 'functional' [9] techniques have been utilized for evaluation in animal models. A walking test originally developed to evaluate the functional state of sciatic nerve damage in rats [54] was adopted by investigators to evaluate the extent of Achilles tendon healing [55, 56]. Measurements of paw prints recorded from a rat walking down a narrow track included: the plantar length, toe spreading, and distance between intermediary toes. Murrell and associates [47] claimed that these measurements 'provided sensitive and reliable indicators of Achilles tendon function'. The major advantage of this method is that, unlike biomechanical evaluation, the animal does not need to be sacrificed in order to provide the measurements. This allows for temporal studies utilizing fewer animals.

4.4 Normal tendon versus 'scar'

Regeneration of tissue results in a tissue that is indistinguishable from the original tissue, i.e. the newly formed tissue is morphologically, ultrastructurally, biochemically, biomechanically, and functionally indistinguishable from the original tissue. Repair, in the classical use of the term, results in a fibrocollagenous tissue that is distinguishable from the original tissue, and is generally referred to as 'scar'. Many studies have claimed regeneration of tendon [6, 30], but close examination of the studies shows that the tissue in question may appear to fulfill the criteria of regeneration in one area but not in another. In the field of tendon healing, regeneration and repair have, in general, not been clearly distinguished in the literature. There is accordingly, a lack of consensus on the degree of functional recovery which can be considered acceptable to restore function.

There are many similarities between normal adult Achilles tendon and tissue formed in a tendon wound site. Both tissues tend to have highly aligned matrix fibers and a relatively low density of fibroblast cells present in the tissue. Morphologically, differences between these two tissue appear to be the crimp pattern and the average fibril diameter and distribution of the tissue. For example, 1 year post-injury, Kato *et al.* [6] found that the crimp length of the healing tendon in their animal model was smaller than that of normal tendon. Collagen fibril diameters were also significantly smaller than those of normal tendon [2, 30]. Biomechanically, mechanical properties of tissue formed in a tendon wound site appear to be 40–60% of normal tendon levels [23, 55].

These considerations can serve as the basis for criteria to assess the success of certain tissue engineering approaches in regenerating tendon.

5 Techniques for treatment of tendon injuries

A wide variety of treatments for tendon injuries have been studied, from suture techniques [57, 58] to effects of mechanical loading [22, 23, 43–48]. In the cases in which tendon is missing or the wound site too large to allow for reapposition of the ends, a tendon replacement is necessary. Work has focused on the use of biological replacements [59–61], permanent prostheses [62–68], or bioresorbable devices [6, 64, 69–71]. Recognition of the myriad treatment methods previously investigated can serve as the rationale for selected tissue engineering approaches.

5.1 Surgical apposition of tendon ends

Considerable attention has been directed toward methods of suturing tendon lacerations and the effects of continuous passive motion on healing of this tissue. Several techniques of suturing have been proposed to increase the immediate strength of the repair, and to facilitate subsequent healing [57, 58]. Most modern techniques of suturing are variations of a technique devised by Kessler [72]. The challenge in obtaining a suture repair of a tendon injury increases significantly the greater the segmental loss of tissue. Suturing techniques alone are not adequate for the treatment of tendon injuries with a large defect area resulting from segmental loss of tendon tissue.

5.2 Tendon grafts

The performance of soft tissue autografts, allografts and xenografts used in treating tendon injuries have been evaluated in both humans [73–76] and animals [59–61]. Questions relate to the source of the donor graft, antigenicity of grafts, methods of harvesting and preserving grafts for surgery, methods of attachment to the residual tendon fragments, and proper tensioning. These grafts are generally used in situations where there is a desire to have immediate post-operative mobilization. Kato *et al.* [6] found that although an autogenous tendon graft 'had been completely filled with neotendon that was characterized by crimped, aligned collagen fibers', the neotendon was not identical to normal Achilles tendon. The mechanical properties of the neotendon were approximately 60% of normal at 1 year and the histological morphology of the neotendon was not identical to normal; the crimp length was only 20% of normal at 1 year.

5.3 Permanent tendon prostheses

The search for a prosthesis to replace tendons and ligaments has been prompted by the desire to obtain immediate load bearing capability. Most of the efforts have focused on the replacement of ligaments, due to the greater number of overall ligament injuries that occur annually. Of the 33,000 tendon and ligament injuries that occur annually, two thirds were ligament injuries [77]. Prostheses fabricated from polytetrafluorethylene, polypropylene, and carbon fibers have been investigated in animal and human trials [62–68]. Problems generally relate to the insertion of the device into bone, proper tensioning, and issues related to the abrasion of the prosthesis against bone at sites at which the prosthesis exits a tunnel through bone. While encouraging results have been reported in the short term [62–65], questions about the long-term performance of these devices have limited their use [78].

6 Tendon tissue engineering

Tissue engineering approaches applied to tendon, as with other tissues, are founded on the use of matrices, cells, and soluble regulators, alone or in combination. Also, as with other applications, the goal might be to engineer tendon *in vitro* for subsequent implantation or to develop an implant to facilitate regeneration *in vivo*. A problematic issue associated with the former approach is the need to have the tendon tissue, engineered *in vitro*, incorporated in the host tissue when it is implanted. Work to date has focused on the implementation of absorbable synthetic and natural matrices, alone, as implants for the engineering of the tissue by the host. While the results of this approach have been promising, there are indications that exogenous cells will be necessary for more complete regeneration. This has led to a preliminary studies of methods for seeding certain matrices with tendon and mesenchymal stem cells *in vitro*.

While there have been reports of the use of soluble regulators alone, injected into healing ligaments to attempt to facilitate regeneration [79], no such approach has yet been reported as a modality to promote the engineering of tendon.

6.1 Animal models

A number of animals models have been used to study Achilles and flexor tendon healing (Table 1), with a few attempting to differentiate between intrinsic and extrinsic healing of the flexor tendon. In these animals models, lesions were produced either by a simple transverse incision [80] or an excision [6] of the tendon tissue. Species have included the rat, sheep, rabbit, dog and chicken (Table 1). In models in which immobilization was employed during healing it was achieved using a plaster cast on the operated leg [44, 61] or transection of the sciatic nerve [23].

Animal investigations have not yet been specifically directed toward determining the largest tendon gap across which regeneration can occur (i.e. the 'critical size defect'). In part this has been due to the inability of most animal models to produce and initially maintain a specific gap width, and to isolate the defect from extrinsic influences that may negatively impact the regeneration processes. Moreover, there is an indication from many clinical observations and animal studies that a gap of any size, precluding direct apposition and suturing of the tendon stumps, does not heal by regeneration with restoration of normal functional or mechanical characteristics of tendon.

Until recently, none of the tendon animal models, including those used to study intrinsic and extrinsic healing, appeared to be capable of studying tendon healing in isolation from effects due to the presence of other tissues adjacent to the injury (e.g. overlying dermal tissue, tendon sheath, neighboring blood vessels). This was, in fact, desirable in studies investigating certain clinical conditions extant during spontaneous healing. However, animal models to be employed for the evaluation of tissue engineering

Table 1. Animal models employed for the study of tendon healing

Animal	Lesion (size of lesion)	Immobilization method (time)	Reference
Rabbit	Transection	None ?*(45 days)	Postacchinni (1978) [30] Davis (1991) [69]
	Excision of proximal third	None	Aragona (1981) [102]; Shieh (1990) [64]
	Excision of 1.5 cm	None	Tauro (1991) [60]
	Excision of 2.0 cm midsection	Long plaster cast (6 weeks)	McMaster (1976) [61]
	Excision of 3 cm	None	Kato (1991) [6]
Sheep	Excision of (2.5–4 cm)	Plaster (6 weeks) None	Howard (1985) [44] Rueger (1986) [62]
Rat	Transection	? (7 days) or transection of sciatic nerve or removal of tibia and fibula external fixation knee and ankle (12 days)	Buck (1953) [23] Murrell (1994) [47]
Dog	Transection (sheath and tendon)	Shoulder -spica cast or controlled passive motion	Gelberman (1983) [46]
Chicken	Transection (sheath and tendon)	None	Garner (1989) [36]
	50% of tissue removed	? (throughout post-op)	Joyce (1992) [103]

*"?' means that the author indicated that immobilization was used, but did not describe the technique

modalities may have to isolate the defect from the influence of extrinsic factors, as they could impede processes of tendon regeneration and confound interpretation of findings.

Recently, an animal model was developed to isolate a 1 cm gap in the rabbit Achilles tendon by 'entubulating' the tendon stumps in a silicone tube [81] as was done in an early study of flexor tendon healing [31] and prior studies of bone [82] and peripheral nerve regeneration [83]. In this model, the Achilles tendon was transected at its mid-point, after which the tendon stumps retracted about 8–10 mm. The stumps were subsequently inserted into a silicone tube. A 10 mm gap in the tendon was maintained by stitches through each stump and the tube, that allowed any mechanical stresses imparted to the system to be taken up by the sutures. In this model, the tendons of the peroneus long, brevis and tertius were cut and a portion of each tendon dissected away to immobilize the tendon gap site. Moreover, the plantaris tendon was cut with a 'Z' plasty, the ends separated, and then sutured under minimum tension. The knee joint was immobilized by external fixation to reduce loading to the tendon. The spontaneous healing in the entubulated 1 cm gap after 6 weeks comprised a thin continuous cable of the fibrous tissue (Fig. 4), less than 1 mm in diameter. There was no significant increase in the diameter of the reparative tissue after 12 weeks. The tissue comprised dense aggregates of crimped collagen fibers (Figs. 5 a and b), with a wavelength (12 μm) and fiber bundle thickness that were both significantly shorter than those in normal tendon. Future studies need be directed toward answering the question: Is this crimped tissue end-stage scar?

Findings with the entubulated tendon gap model in the rabbit demonstrate that a tendon wound site can be isolated from the effects of extrinsic factors during healing, and that there is only limited capability for spontaneous intrinsic healing, with no indication of regeneration of tendon. An *in vivo* model of this type could thus be valuable for the evaluation of constructs developed for engineering tendon. Future studies need consider how certain characteristics of the tube affect the healing process. One potentially important feature is permeability. Because of the limited blood

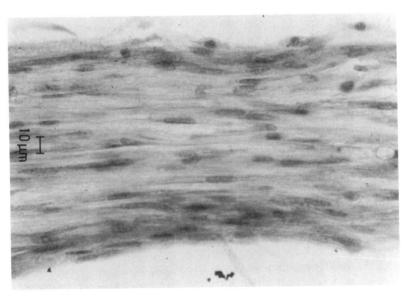

Fig. 4. Histology of tissue in an empty silicone tube bridging a 1 cm gap in a rabbit Achilles tendon, 6 weeks post-operatively. The predominantly loose fibrous tissue was aligned along the long axis of the tendon. H & E stained paraffin sections.

(a)

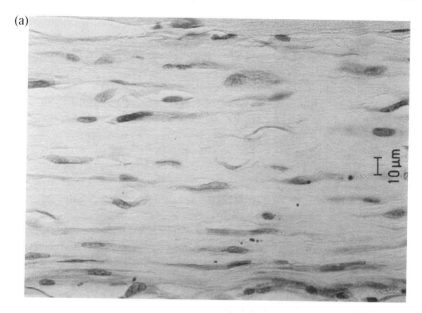

(b)

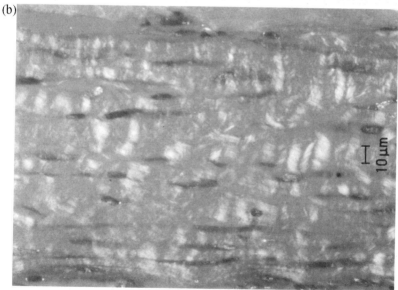

Fig. 5. (a) Micrograph of tissue in the middle of a 1 cm defect contained within a silicone tube, 12 weeks post-operatively, in the rabbit. (b) Polarized light micrograph of the area in (a). The collagen fibers and spindle-shaped fibroblasts are aligned along the long axis of the tendon (a), and the collagen fibers display a crimp pattern (b) with a wavelength approximately one-tenth that of normal tendon (see Fig. 3a for comparison). Paraffin sections stained with H & E.

supply to the mid-substance of many tendons, regenerating tissue in the tube needs to derive nutrients from the surrounding milieu. While the permeability of the tube may facilitate this nutrition of the tissue forming in the entubulated gap, it might allow for the loss of intrinsic regulators of growth and remodeling from the lesion. These considerations have been addressed in entubulated peripheral nerve regeneration [84–87], and may be as relevant for tendon healing.

6.2 Bioresorbable fibrous scaffolds and sponge-like matrices

6.2.1 Fibrous implants

Bioresorbable prostheses currently being investigated to facilitate tendon healing include: collagen fibers tows [6, 70, 71, 88], resorbable fibers tows of dimethyltrimethylene carbonate trimethylene carbonate copolymer [64], and a composite artificial tendon of poly(2-hydroxyethylmethacrylate)/poly(caprolactone) blend hydrogel matrix and poly (lactic acid) fibers [69]. Kato *et al.* [6, 70, 71] reported the use of a carbodiimide-cross-linked and a glutaraldehyde-cross-linked collagen-fiber prosthesis for the Achilles tendon of rabbits. They found that the healing in gaps in the tendon, bridged by the devices, was affected by the rate of implant degradation. The carbodiimide-cross-linked implant was resorbed by 10 weeks and was replaced with 'neotendon'. This reparative tissue was characterized by 'aligned, crimped collagen fiber bundles' as early as at 20 weeks. The slower degrading glutaraldehyde-cross-linked implant was surrounded by a 'capsule of collagenous connective tissue' at twenty weeks and both capsule and implant were still present at one year. Repair tissue infiltrated into the glutaraldehyde-cross-linked implant but the tissue was 'not as developed' (was not as aligned and was not crimped) as the carbodiimide-cross-linked implant. While these results are promising, it should be noted that Kato *et al.* stated that the neotendon was 'similar, but not identical, to normal tendon', 1 year after implantation of the prosthesis. The tissue in Kato's tendon lesion site was described to have a crimp wavelength of 10 μm. In Kato's study, it was also observed that this crimp pattern was present from 3 weeks to 52 weeks with minimal change in the crimp characteristics. That fibrous tissue with this crimp pattern did not appear to remodel significantly by 52 weeks may suggest that it is a terminal 'scar.' This raises the question of whether complete regeneration is necessary for the reparative tissue to be of functional value.

6.2.2 Collagen-glycosaminoglycan (GAG) analogs of extracellular matrix

The rationale for use of a porous collagen-GAG analog of extracellular matrix to facilitate regeneration of tendon was based, in part, on the success experienced in developing such materials for the regeneration of dermis [89–92] in animals and human subjects and the elongation of axons in ruptured peripheral nerves in rats [91, 93, 94]. Moreover, there was an indication that such an approach was successful in the regeneration of meniscus [95]. Left untreated, defects in these tissues would not heal by regeneration. For each application for which the porous collagen-GAG analog was developed, the pore diameter and degradation rate were found to be tissue specific.

In recent studies, a collagen-GAG analog was prepared for investigation as a matrix for engineering tendon [96]. The matrix consisted of type I bovine hide collagen precipitated from acid dispersion with chondroitin-6-sulfate [97]. The suspended coprecipitate suspension was injected into a silicone tube (3.8 mm inside diameter) and immersed into a coolant bath and freeze-dried, to produce a porous architecture. The matrices were then exposed to a dehydrothermal treatment for cross-linking and sterilization. The matrix was nominally 95% porous with an average pore channel diameter of from 30 to 120 μm (Figs. 6 a and b).

The effect of the collagen–GAG analog on the early stages of tendon healing was evaluated in the rabbit Achilles tendon entubulated gap model described above (Fig. 7). The presence of the collagen-GAG-matrix altered the kinetics of tendon healing. Silicone tubes filled with the collagen–GAG matrix contained a significantly greater volume of tissue at 3, 6, and 12 weeks, when compared to the empty silicone tube control. While the presence of the non-degraded collagen–GAG matrix contributed to this volume at 3 and 6 weeks, no residual collagen–GAG matrix was observed at 12 weeks. Specimens at 3 and 6 weeks were infiltrated with elements of blood clot (Fig. 8) and granulation tissue (Fig. 9), respectively. Although, collagen–GAG-filled tubes contained dense fibrous tissue by 12 weeks, the tissue had no crimp (Figs. 10 a and b). The collagen–GAG matrix may have prolonged the synthesis of granulation tissue and have affected the ability of the matrix fibers in the tissue to align.

These findings indicate that the collagen–GAG matrix altered the process of tendon healing such the formation of scar was delayed or prevented. If this is the case, the results would be paralleling the findings of studies in which a collagen–GAG matrix delayed the onset of, and reduced the severity of, scarring in full-thickness dermal wounds, and thereby facilitated regeneration [92].

It is speculated that the addition of exogenous tendon cells to the collagen–GAG matrix prior to implantation would result in a larger volume of tissue filling the lesion site. Studies by Orgill [98], in which fibroblast-seeded collagen–GAG matrices were implanted in a dermal wound healing model, demonstrated that the kinetics of dermal wound healing (viz., contraction) were different than those of sites implanted with collagen–GAG matrix only; wound contraction was arrested and the degree of final wound contraction was significantly lower in the wound sites implanted with cell-seeded matrices. These cell-seeded matrices ap-

(a)

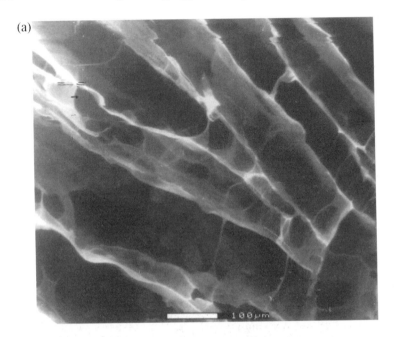

(b)

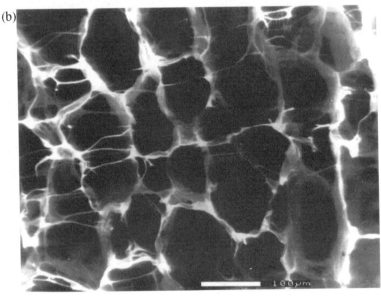

Fig. 6. Scanning electron micrographs of a porous collagen-GAG matrix employed in an investigation of tendon regeneration. Cylindrical structures comprising the material were prepared with pore channels aligned along the long axis of the cylinder. (a) Longitudinal and (b) transverse sections.

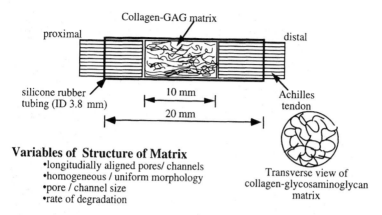

Variables of Structure of Matrix
- longitudially aligned pores/ channels
- homogeneous / uniform morphology
- pore / channel size
- rate of degradation

Transverse view of
collagen-glycosaminoglycan
matrix

Fig. 7. Schematic showing a silicone tube filled with a collagen-GAG matrix to be implanted into a defect in a rabbit Achilles tendon in order to determine the effect of the analog of extracellular matrix on tendon healing.

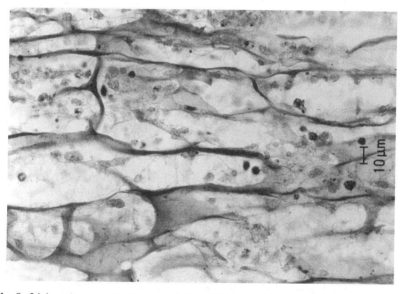

Fig. 8. Light micrograph of tissue in the middle of defects treated with a silicone tube filled with a collagen-GAG matrix, three weeks post-operatively. Neovascular granulation tissue was found in treatment site, contained within the collagen-GAG matrix. H & E stained paraffin sections.

peared to have accelerated the morphogenesis of skin. Studies of this type have served as the foundation for the isolation of tendon cells for seeding matrices *in vitro* prior to implantation.

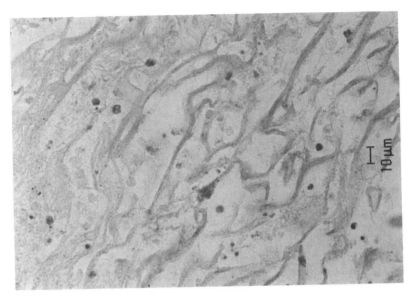

Fig. 9. Histology of 6-week tissue in a silicone tube filled with the collagen-GAG matrix. Granulation tissue persisted in the tubes filled with the matrix, and some residual matrix could be found in the defect. For comparison with an empty silicone tube see Fig. 4. H & E stained paraffin sections.

6.3 Cell-seeded matrices for engineering tendon

That the mid-substance of tendon is poorly vascular with a low density of parenchymal cells, suggests that matrices seeded with exogenous cells may be necessary to facilitate regeneration. Recent preliminary investigations have seeded fibers of a synthetic polymer [99] and a collagen–GAG sponge [100] with tenocytes *in vitro* in order to ultimately develop cell-seeded implants for tendon engineering. Another approach recently taken was to seed mesenchymal stem cells into a collagen gel for encapsulation of collagen sutures to bridge tendon gaps [101].

6.3.1 Tenocyte-seeded polyglycolic acid fiber mesh

In a recent investigation [99], samples of non-woven meshes of polyglycolic acid fibers, with interstitial spaces from 75–100 μm in diameter, were seeded with tendon cells isolated from newborn calves by collagenase treatment. After 1 week in culture, the cell-seeded specimens were implanted subcutaneously in nude mice for up to 10 weeks. Histological evaluation of the 10 week samples showed 'parallel linear organization of collagen bundles throughout the specimens.' Mechanical testing revealed that the tissue engineered neo-tendon structures had approximately one-

(a)

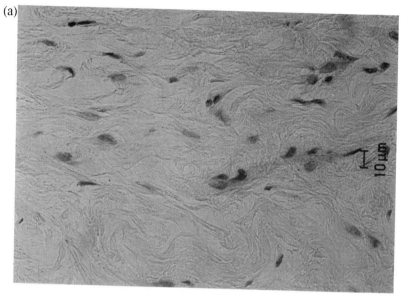

(b)

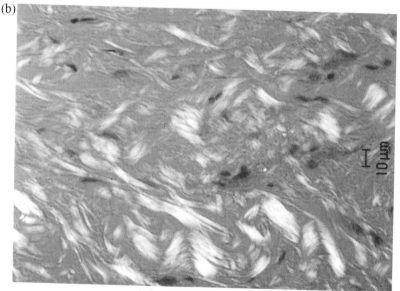

Fig. 10. (a) Micrograph of tissue in the middle of defects treated with a silicone tube filled with a collagen-GAG matrix, 12 weeks post-operatively. (b) Polarized light micrograph of the area in (a). The collagen fibers do not display the crimp pattern seen in the defect treated with the empty tube (Fig. 5). Paraffin sections stained with H & E.

third the tensile strength of normal tendon (11 versus 32 MPa), 8 weeks post-implantation. These promising findings are serving as the basis for efforts to further improve the engineered tendon. Additional studies will be required to determine if comparable results can be achieved with adult cells.

6.3.2 Seeding of cultured tenocytes into porous collagen–GAG matrices

Tenocytes, recovered by collagenase digestion of the Achilles and plantaris tendon of adult New Zealand white rabbits, were grown to confluence for 2–3 days and subsequently passaged to increase cell number. In a preliminary experiment [100] , collagen–GAG matrices (described in a previous section) were seeded with 0.75–1.5 million cells.

The number of cells and their distribution in collagen–GAG matrices varied from few cells uniformly distributed throughout the matrix (Fig. 11) to cells concentrated on the surface of the matrix. Cells were either spread out along the surface of the collagen–GAG material (Fig. 11) or aggregated. A cell count of the medium and trypsinized contents of the wells, in which the samples were seeded, revealed that from 2% to 50% of the seeded tendon cells were not incorporated into the collagen–GAG matrices after 1 day. In general, the tenocytes appeared to infiltrate to a depth of 0.35 mm into the 1 mm thick samples. In many of the matrices, the interior of the matrix was devoid of cells. The majority of cells were near or on the surface of the cells. The degree of infiltration appeared to be dependent on the pore diameter of the collagen–GAG matrices. Matrices with a larger pore diameter (120 μm) allowed for the infiltration of cells into the interior of the matrices and produced, in general, matrices with evenly distributed cells throughout the devise. While matrices with larger pore diameters would be desirable as cell transplantation devices because of the deeper cell infiltration and the uniformity of cell distribution, there might be drawbacks to using too large of a pore diameter. For the matrices with pore sizes averaging 120 μm, the cells attached and spread out along the surface of collagen–GAG fibers. The surface area of collagen–GAG sponges is inversely proportional to the matrix pore diameter; therefore, with the larger pore diameters, there is less surface area to which the cells can attach. Future studies need systematically investigate the matrix characteristics that yield optimal cell seeding.

6.3.3 Mesenchymal Stem Cell-Seeded Gel

In a recently reported study [101], autologous marrow-derived mesenchymal stem cells seeded in a collagen gel were implanted into 1 cm long defects in the lateral gastrocnemius tendons of rabbits for 3 months. The mesenchymal stem cells (MSCs) were mixed with the collagen solution and incubated for 36–40 hours in the presence of biodegradable sutures, such that the gel contracted around the suture to form an 'integrated implant.'

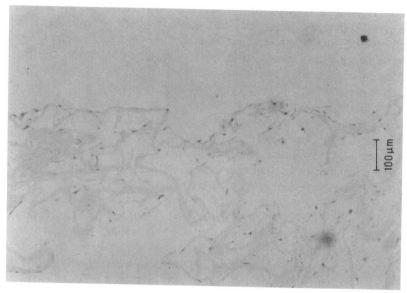

Fig. 11. Light micrograph of tendon cells seeded in a collagen-GAG matrix. H & E stain of parrafin section.

During the culture period tensile loading was applied to the sutures in order to align the cells seeded in the incorporated gel. The implant was then sutured into the gap in the rabbit tendon.

The reparative tissue after 3 months comprised 'dense bands of matrix organized in the axial direction of the tendon.' Controls with the sutures alone (but absent the unseeded gel) 'demonstrated similar attributes but with less volume and less organization than those seen in MSC-treated repair tissue.' While the mechanical properties of the cell-treated defects were improved relative to the suture-alone group, they were still significantly below normal levels. Further investigation of this novel approach will be required to determine how the system can be modified to achieve a more physiological neotendon.

7 Summary

There is a compelling need for engineered tendon. This is due to the fact that any defect large enough to preclude a direct suture repair of apposed tendon ends results in a reparative tissue with inadequate mechanical properties. Promising results have been obtained in preliminary studies of tenocyte and mesenchymal stem cell-seeded matrices. Moreover, there appear to be animal models in which the performance of implants to facilitate tendon regeneration can be critically evaluated. However, many questions have yet to be answered. Which exogenous cells and from what source, and what type of matrix, will be the most suitable for the engineering of tendon *in vitro* or for the production of cell-seeded matrices to facilitate regeneration *in vivo*? Is there a role for growth factors in accelerating certain phases of the tendon regeneration process? To what extent does the histology of normal tendon need be regenerated by tissue engineering strategies in order for the reparative tissue to offer a functional — mechanical — benefit? Answers to these questions may come not only from continuing studies of tendon healing but also from investigations of tissue engineering approaches implemented for treatment of injuries to other musculoskeletal tissues. Insights may also come from our deepening understanding of the embryological development of tendon.

Several strategies for engineering tendon, based on promising findings *in vitro* and in animal models are currently being pursued. However, that no treatment modality in human subjects or animal models has yet yielded a reparative tissue with the histological characteristics specific to tendon, indicates the magnitude of the challenge to be met.

References

1. Dyer, R. F. and Enna, C. D., Ultrastuctural features of adult human tendon. *Cell Tissue Research,* 1976, **168**, 247–259.
2. Ippolito, E., Natali, P.G., Postacchini, F., Accinni, L. and De Martino, C., Morphological, immunochemical, and biochemical study of rabbit achilles tendon at various ages. *Journal of Bone and Joint Surgery,* 1980, **62**-A, 583–598.
3. Parry, D. A. D., Barnes, G. R. G. and Craig, A. S., A comparison of the size distribution of collagen fibrils in connective tissues as a function of age and a possible relation between fibril size distribution and mechanical properties. *Proceedings of the Royal Society of London B,* 1978, **203**, 305–321.
4. Parry, D. A. D., Craig, A. S. and Barnes, G. R. G., Tendon and ligament from the horse: an ultrastructural study of collagen fibrils and elastic fibres as a function of age. *Proceedings of the Royal Society of London B,* 1978, **203**, 293–303.
5. Diamant, J., Keller, A., Baer, E., Litt, M. and Arridge, R. G. C., Collagen: ultrastructure and its relation to mechanical properties as a function of age. *Proceedings of the Royal Society B,* 1972, **180**, 293–315.
6. Kato, Y. P., Dunn, M. G., Zawadsky, J. P., Tria, A. J. and Silver, F. H., Regeneration of Achilles tendon with a collagen tendon prosthesis: results of a one-year implantation study. *Journal of Bone and Joint Surgery,* 1991, **73**-A, 561–574.
7. Dale, W. C. and Baer, E., Fibre-buckling in composite systems: a model for the ultrastructure of uncalcified collagen tissues. *Journal of Materials Science,* 1974, **9**, 369–382.

8. Comninou, M. and Yannas, I. V., Dependence of stress-strain nonlinearity of connective tissues on the geometry of collagen fibers. *Journal of Biomechanics*, 1976, **9**, 427–433.

9. Ippolito, E., Natali, P.G., Postacchini, F., Accinni, L. and De Martino, C., Ultrastructural and immunochemical evidence of actin in the tendon cells. *Clinical Orthopaedics*, 1977, **126**, 282–284.

10. Skalli, O., Schürch, W., Seemayer, T., Lagace, R., Montandon, D., Pittet, B. and Gabbiani, G., Myofibroblasts from diverse pathologic settings are heterogeneous in their content of actin isoforms and intermediate filament proteins. *Laboratory Investigation*, 1989, **60**, 275–285.

11. Gown, A. M., The mysteries of the myofibroblast (partially) unmasked. *Laboratory Investigation*, 1990, **63**, 1–3.

12. Frey, J., Chamson, A., Raby, N. and Rattner, A., Collagen bioassay by the contraction of fibroblast-populated collagen lattices. *Biomaterials, 1995,* **16**, 139–143.

13. Baur, P. S., Barratt, G., Linares, H. A., Dobrkovsky, M., De La Houssaye, A. I. and Larson, D. L., Wound contractions, scar contractures and myofibroblasts: a classical case study. *Journal of Trauma*, 1978, **18**, 8–22.

14. Hooper, G., Davies, R. and Tothill, P., Blood flow and clearance in tendons: Studies with dogs. *Journal of Bone and Joint Surgery*, 1984, **66**-B, 441–443.

15. Buckwalter, J. A., Maynard, J. A. and Vailas, A. C., Skeletal fibrous tissue: Tendon, joint capsule, and ligament. In *The Scientific Basis of Orthopedics*, eds. J. A. Albright and R. A. Brand. Appleton and Lange, Norwalk, Conn, 1987, pp. 387–405.

16. Woo, S. L.-Y., Gomez, M. A., Woo, Y.-K. and Akeson, W. H., Mechanical properties of tendon and ligaments: II. The relationships of immobilization and exercise on tissue remodeling. *Biorheology*, 1982, **19**, 397–408.

17. Woo, S. L.-Y., An, K.-N., Arnoczky, S. P., Wayne, J. S., Fithian, D. C. and Myers, D. S., Anatomy, biology, and biomechanics of tendon, ligament, and meniscus. In *Orthopaedic Basic Science*, ed. S. R. Simon. American Academy of Orthopaedic Surgeons, Park Ridge, IL, 1994, pp. 45–87.

18. Butler, D. L., Grood, B. S., Noyes, F. R. and Zernicke, R. F., Biomechanics of ligaments and tendons. In *Exercise and Sports Science*, ed. R. S. Hutton. Franklin Institute Press, Washington, 1978, pp. 125–181.

19. Frank, C., McDonald, D., Bray, D., Bray, R., Rangayyan, R., Chimich, D. and Shrive, N., Collagen fibril diameters in the healing adult rabbit medial collateral ligament. *Connective Tissue Research*, 1992, **27**, 251–263.

20. Frank, C. B., Shrive, N. G. and McDonald, D. B., Collagen fibril diameters in ligament scars: a long term assessment. In *42nd Annual Meeting, Orthopaedic Research Society*, Atlanta, Georgia, 1996.

21. Derwin, K. A. and Soslowsky, L. J., Tendon fascicle mechanical properties: A study toward elucidating structure-function relationships in a transgenic model. In *42nd Annual Meeting, Orthopedic Research Society*, Atlanta, Georgia, 1996.

22. Frank, C., MacFarlane, B., Edwards, P., Rangayyan, R., Liu, Z.-Q., Walsh, S. and Bray, R., A quantitative analysis of matrix alignment in ligament scars: a comparison of movement versus immobilization in an immature rabbit model. *Journal of Orthopaedic Research*, 1991, **9**, 219–227.

23. Buck, R. C., Regeneration of tendon. *Journal of Pathology and Bacteriology*, 1953, **66**, 1–18.

24. Amadio, P. C., Tendon and ligament. In *Wound Healing. Biochemical and Clinical Aspects*, eds. I. K. Cohen, R. F. Diegelmann and W. J. Lindbad. W.B. Saunders Co., Philadelphia, PA, 1992, pp. 384–395.

25. Lui, S. H., Yang, R.-S., Al-Shaikh, R. and Lane, I. M., Collagen in tendon, ligament, and bone healing: A current review. *Clinical Orthopaedics*, 1995, **318**, 265–278.

26. Enwemeka, C. S., Membrane-bound intracellular collagen fibrils in fibroblasts and myofibroblasts of regenerating rabbit calcaneal tendons. *Tissue Cell*, 1991, **23**, 173–190.

27. Enwemeka, C. S., Inflammation, cellularity, and fibrillogenesis in regeneration tendon: implications for tendon rehabilitation. *Physical Therapy*, 1989, **69**, 816–825.

28. Fernando, N. V. P. and Movat, H. Z., Fibrillogenesis in regenerating tendon. *Laboratory Investigation*, 1963, **12**, 214–229.

29. Flynn, J. B. and Graham, J. H., Healing of tendon wounds. *American Journal of Surgery*, 1965, **109**, 315–324.

30. Postacchini, F., Accinni, L., Natali, P. G., Ippolito, E. and DeMartino, C., Regeneration of rabbit calcaneal tendon: A morphological and immunochemical study. *Cell Tissue Research*, 1978, **195**, 81–97.

31. Gonzalez, R. I., Experimental tendon repair within the flexor tunnel: Use of polyethylene tubes for improvement of functional results in the dog. *Surgery*, 1949, **26**, 181–198.

32. McGaw, T., The effect of tension on collagen remodelling by fibroblasts: A stereological ultrastructural study. *Connective Tissue Research*, 1986, **14**, 229–235.

33. Frank, C., Woo, S.-Y., Amiel, D., Harwood, F., Gorney, M. and Akeson, W., Medial collateral ligament healing: a multidisciplinary assessment in rabbits. *American Journal of Sports Medicine*, 1983, **11A**, 379–389.

34. Mass, D. P. and Tuel, R. J., Participation of human superficialis flexor tendon segments in repair in vitro. *Journal of Orthopaedic Research*, 1990, **8**, 21–34.

35. Manske, P.R. and Lesker, P. A., Histological evidence of intrinsic flexor tendon repair in various experimental animals. *Clinical Orthopaedics*, 1984, **182**, 297–304.

36. Garner, W. L. and McDonald, I. A., Identification of the collagen-producing cells in healing flexor tendon. *Plastic and Reconstructive Surgery*, 1989, **83**, 875–879.

37. Potenza, A. D., Tendon healing within the flexor digital sheath of the dog. *Journal of Bone and Joint Surgery*, 1962, **44A**, 49–64.

38. Potenza, A. D., Concepts of tendon healing and repair. In *Symposium on Tendon Surgery in the Hand (American Academy of Orthopedic Surgeons)*. C.V. Mosby Co., St. Louis, 1975, pp. 18–47.

39. Russell, I.E. and Manske, P.R., Collagen synthesis during primate flexor tendon repair in vitro. *Journal of Orthopaedic Research*, 1990, **8**, 11–20.

40. Mason, M. L. and Allen, H. S., The rate of healing of tendons. An experimental study of tensile strength. *Annals of Surgery*, 1941, **113**, 424–459.

41. Manske, P.R., Intrinsic flexor-tendon repair. *Journal of Bone and Joint Surgery*, 1984, **66A**, 385–396.

42. Matthews, P. and Richards, H., The repair potential of digital flexor tendon after repair. *Journal of Bone and Joint Surgery*, 1974, **58**-B, 230–236.

43. Meislin, R. I., Wiseman, D. M., Alexander, H., Cunningham, T., Linsky, C., Carlstedt, C., Pitman, M. and Casar, R., A biomechanical study of tendon adhesion reduction using a biodegradable barrier in a rabbit model. *Journal of Applied Biomaterial*, 1990, **1**, 13–19.

44. Howard, C. B., McKibbin, B. and Ralis, Z. A., The use of Dexon as a replacement for the calcaneal tendon in sheep. *Journal of Bone and Joint Surgery*, 1985, **67**-B, 313–316.

45. Takai, S., Woo, S. L.-Y., Horibe, S., Tung, D. K.-L. and Gelberman, R. H., The effects of frequency and duration of controlled passive mobilization on tendon healing. *Journal of Orthopaedic Research*, 1991, **9**, 705–713.

46. Gelberman, R. H., Berg, J. S. V., Lundborg, G. N. and Akeson, W. H., Flexor tendon healing and restoration of the gliding surface: An ultrastructural study. *Journal of Bone and Joint Surgery*, 1983, **65**A, 70–80.

47. Murrell, G. A. C., Lilly, E. G., Goldner, R. D., Seaber, A. V. and Best, T. M., Effects of immobilization on Achilles tendon healing in a rat model. *Journal of Orthopaedic Research*, 1994, **12**, 582–591.

48. Enwemeka, C. S., Spielholz, N. I. and Nelson, A. I., The effect of early functional activities on experimentally tenotomized Achilles tendons in rat. *American Journal of Physiology and Medical Rehabilitation*, 1988, **67**, 264–269.

49. Stein, S. R. and Luekens, C. A., Methods and rationale for closed treatment of Achilles tendon rupture. *American Journal of Sports Medicine*, 1976, **4**, 162–169.

50. Ketchum, L. D., Tendon healing. In *Fundamentals of Wound Management in Surgery: Selected Tissues.* Chirugecom, New Jersey, 1977, pp. 122–153.

51. Gelberman, R., Goldberg, V., An, K. -N. and Banes, A., Tendon. In *Injury and Repair of the Musculoskeletal Soft Tissue*, eds. S. L.-Y. Woo and J. A. Buckwalter. American Academy of Orthopaedic Surgery, Park Ridge, IL, 1988, pp. 5–40.

52. Best, T. M., Collins, A., Lilly, E. G., Seaber, A. V., Goldner, R. and Murrell, G. A. C., Achilles tendon healing: A correlation between functional and mechanical performance in the rat. *Journal of Orthopaedic Research*, 1993, **11**, 897–906.

53. Goldin, B., Block, W. D. and Pearson, I. R., Wound healing of tendon: mechanical and metabolic changes. *Journal of Biomechanics*, 1980, **13**, 241–256.

54. de Medinaceli, L., Freed, W. J. and Wyatt, R. J., An index on the functional condition on rat sciatic nerve based on measurements made from walking tracks. *Experimental Neurology*, 1982, **77**, 634–643.

55. Murrell, G. A. C., Lilly, E. G., Davies, H., Best, T. M., Goldner, R. D. and Seaber, A. V., The Achilles functional index. *Journal of Orthopaedic Research*, 1992, **10**, 398–404.

56. Best, T. M., Collins, A., Lilly, E. G., Seaber, A. V. and Murrell, G. A. C., A functional, morphological and biomechanical study of the repair process in rat Achilles tendon. *Medical Science and Sports Exercise*, 1991, **23**, S139.

57. Urbaniak, J. R., Cahill, I. D. and Mortenson, R. A., Tendon suturing methods: analysis of tensile strengths. In *American Academy of Orthopaedic Surgeons Symposium on Tendon Surgery in the Hand*, St. Louis, 1975.

58. Kleinert, H. E., Schepel, S. and Gill, T., Flexor tendon injuries. *Surgical Clinics of North America*, 1981, **61**, 267–286.
59. Andreeff, I., Dimoff, G. and Metschkarski, S., A comparative experimental study on transplantation of autogenous and homogenous tendon tissue. *Acta Orthopaedica Scandinavica*, 1967, **38**, 35–44.
60. Tauro, J. C., Parsons, J. R., Ricci, J. and Alexander, H., Comparison of bovine collagen xenografts to autografts in the rabbit. *Clinical Orthopaedics*, 1991, **266**, 271–284.
61. McMaster, W. C., Kouzelos, J., Liddle, S. and Waugh, T. R., Tendon grafting with glutaraldehyde fixed material. *Journal of Biomedical Materials Research*, 1976, **10**, 259–271.
62. Rueger, I. M., Siebert, H. R., Wagner, K. and Pannike, A., Longterm implantation of a polyethelene ligament/tendon allograft in sheep. Mechanical and histological studies. In *Biological and Biomechical Performance of Biomaterials*, eds. P. Christel, A. Meunier and A. J. C. Lee. Elsevier Science Publishers B.V., Amsterdam, 1986, pp. 135–140.
63. Gleason, T. F., Barmada, R. and Ghosh, L., Can carbon fiber implants substitute for collateral ligament? *Clinical Orthopaedics*, 1984, **191**, 274–280.
64. Shieh, S., Zimmerman, M. C. and Parsons, I. R., Preliminary characterization of bioresorbable and nonresorbable synthetic fibers for the repair of soft tissue injuries. *Journal of Biomedical Materials Research*, 1990, **24**, 789–808.
65. Park, J. P., Grana, W. A. and Chitwood, J. S., A high-strength Dacron augmentation for cruciate ligament reconstruction. A two-year canine study. *Clinical Orthopaedics*, 1985, **196**, 175–185.
66. Mendes, D. G., Lusim, M., Angel, D., Rotem, A., Roffman, M., Grishkan, A., Mordohohovich, D. and Boss, J., Histologic pattern of biomechanic properties of the carbon fiber-augmented ligament tendon. A laboratory and clinical study. *Clinical Orthopaedics*, 1985, 51–60.
67. Howard, C. B., Winston, I., Bell, W., Mackie, I. and Jenkins, D. H. R., Late repair of the calcaneal tendon with carbon fibre. *Journal of Bone and Joint Surgery*, 1984, **66**-B, 206–208.
68. Amis, A. A., Campbell, J. R., Kempson, S. A. and Miller, J. H., Comparison of the structure of neotendons induced by implantation of carbon or polyester fibres. *Journal of Bone and Joint Surgery*, 1984, **66**-B, 131–139.
69. Davis, P. A., Huang, S. J., Ambrosio, L., Ronca, D. and Nicolais, L., A biodegradable composite artifical tendon. *Journal of Materials Science: Materials in Medicine*, 1991, **3**, 359–364.
70. Kato, Y. P. and Silver, F. H., Formation of continuous collagen fibres: evaluation of biocompatibility and mechanical properties. *Biomaterials*, 1990, **11**, 169–175.
71. Goldstein, J. D., Tria, A. J., Zawadsky, J. P., Kato, Y. P., Christiansen, D. and Silver, F. H., Development of a reconstituted collagen tendon prosthesis. *Journal of Bone and Joint Surgery*, 1989, **71**-A, 1183–1191.
72. Kessler, I., The 'grasping' technique for tendon repair. *Hand*, 1973, **5**, 253–255.
73. Schuberth, J. M., Dockery, G. L. and McBride, R. E., Recurrent rupture of the tendo Achillis. *Journal of American Podiatry Association*, 1984, **74**, 157–162.
74. Leitner, A., Voigt, C. and Rahmanzadeh, R., Treatment of extensive aseptic defects in old Achilles tendon ruptures: Methods and case reports. *Foot and Ankle*, 1992, **13**, 176–180.

75. Bugg, E. I. and Boyd, B. M., Repair of neglected rupture or laceration of the Achilles tendon. *Clinical Orthopaedics*, 1968, **56**, 73–75.

76. Teuffer, A. P., Traumatic rupture of the Achilles tendon. *Orthopaedic Clinics of North America*, 1974, **5**, 89–93.

77. Evaluation of the musculoskeletal diseases program, Public Health Service, Washington DC, 1984.

78. Litsky, A. S. and Spector, M., Biomaterials. In *Orthopaedic Basic Science*, ed. S. R. Simon. American Academy of Githopaedic Surgeons, Park Ridge, IL, 1994, pp. 447–486.

79. Letson, A. K. and Dahners, L. E., The effect of combinations of growth factors on ligament healing. *Clinical Orthopaedics*, 1994, **308**, 207–212.

80. Postacchini, F., Natali, P. G., Accinni, L., Ippolito, E. and De Martino, C., Contractile filaments in cells of regenerating tendon. *Experimentia*, 1977, **33**, 957–959.

81. Louie, L. K., Yannas, I. V., Hsu, H.-P. and Spector, M., Healing of tendon defects implanted with a porous collagen-GAG matrix: Histological evaluation. *Tissue Engineering*, 1997, in press.

82. Narang, R. and Wells, H., Osteogenesis within polyethylene implants at fracture gaps. *Oral Surgery, Oral Medicine and Oral Pathology*, 1975, **39**, 203–209.

83. Lundborg, G., Gelberman, R. H., Longo, F.M., Powell, H. C. and Varon, S., *In vivo* regeneration of cut nerves encased in silicone tubes. *Journal of Neuropathology and Experimental Neurology*, 1982, **41**, 412–422.

84. den Dunnen, W. F. A., Stokroos, I., Blaauw, E. H., Holwerda, A., Pennings, A. J., Robinson, P. H. and Schakenraad, J. M., Light-microscopic and electron-microscopic evaluation of short-term nerve regeneration using a biodegradable poly(DL-lactide-ε-caprolacton) nerve guide. *Journal of Biomedical Materials Research*, 1996, **31**, 105–115.

85. Li, S.-T., Archibald, S. J., Krarup, C. and Madison, R., Semipermeable collagen nerve conduits for peripheral nerve regeneration. *Polymer Materials Science Engineering*, 1990, **62**, 575–582.

86. Danielsen, N. L., Dahlin, L. B., Lee, Y. F. and Lundborg, G., Axonal growth in mesothelial chambers: the role of the distal nerve segment. *Scandinavian Plastic and Reconstructive Surgery*, 1983, **17**, 119–125.

87. Lundborg, G., Dahlin, L. B., Danielsen, N., Gelberman, R. H., Longo, F.M., Powell, H. C. and Varon, S., Nerve regeneration in silicone chambers: Influence of gap length and of distal stump components. *Experimental Neurology*, 1982, **76**, 361–375.

88. Wasserman, A. J., Kato, Y. P., Christiansen, D., Dunn, M. G. and Silver, F. H., Achilles tendon replacement by a collagen fiber prosthesis: morphological evaluation of neotendon formation. *Scanning Microscopy*, 1989, **3**, 1183–1200.

89. Burke, J. F., Yannas, I. V., Quinby, W. C., Bondoc, C. C. and Jung, W. K., Successful use of a physiologically acceptable artificial skin in the treatment of extensive burn injury. *Annals of Surgery*, 1981, **194**, 413–428.

90. Yannas, I. V., Certain biological implications of mammalian skin regeneration by a model extracellular matrix. *Cutaneous Development Aging and Repair*, 1989, **18**, 131–139.

91. Yannas, I. V., Regeneration of skin and nerves by use of collagen templates. In *Collagen Vol. III Biotechnology,* ed. M. Nimni. CRC Press, Boca Raton, FL, 1989, pp. 87–115.

92. Yannas, I. V., Burke, J. F., Orgill, D. P. and Skerabut, E. M., Wound tissue can utilize a polymeric template to synthesize a functional extension of skin. *Science,* 1982, **215**, 174–176.

93. Chang, A. S., Yannas, I. V., Perutz, S., Loree, H., Sethi, R. R., Krarup, C., Norregaard, T. V., Zervas, N. T. and Silver, J., Electrophysiological study of recovery of peripheral nerves regenerated by a collagen-glycosaminoglycan copolymer matrix. In *Progress in Biomedical Polymers,* ed. C. G. Gebelein. Plenum Press, NY, 1990, pp. 107–120.

94. Yannas, I. V., Orgill, D. P., Silver, J., Norregaard, T. V., Zervas, N. T. and Schoene, W. C., Regeneration of sciatic nerve across 15 mm gap by use of a polymeric template. In *Advances in Biomedical Polymers,* ed. C. G. Gebelein. Plenum Publishing Corporation, 1987, pp. 1–9.

95. Stone, K. R., Rodkey, W. R., Webber, R. J., McKinney, L. and Steadman, J. R., Future directions: collagen-based prostheses for meniscal regeneration. *Clinical Orthopaedics,* 1990, **252**, 129–135.

96. Louie, L. K., Yannas, I. V. and Spector, M., Development of a collagen-GAG copolymer implant for the study of tendon regeneration. In *Biomaterials for Drug and Cell Delivery,* eds. A. G. Mikos, R. M. Murphy, H. Bernstein and N. A. Peppas. MRS, Pittsburgh, PA, 1994, pp. 19–24.

97. Yannas, I. V., Lee, E., Orgill, D. P., Skrabut, E. M. and Murphy, G. F., Synthesis and characterization of a model extracellular matrix which induces partial regeneration of adult mammalian skin. *Proceedings of the National Academy of Science USA,* 1989, **86**, 933–937.

98. Orgill, D. P., The effects of an artficial skin on scarring and contraction in open wounds. Doctoral, MIT, Boston, MA, 1983.

99. Cao, Y., Vacanti, J. P., Ma, X., Paige, K. T., Upton, J., Chowanski, Z., Schloo, B., Langer, R. and Vacanti, C. A., Generation of neo-tendon using synthetic polymers seeded with tenocytes. *Transplantation Proceedings,* 1994, **26**, 3390–3391.

100. Louie, L. K., Schulz-Torres, D., Sullivan, L., Yannas, I. V. and Spector, M., Behavior of fibroblasts cultured in porous collagen-GAG copolymer matrices. In *Transactions of the Society for Biomaterials, ,* New Orleans, LA, 1997.

101. Young, R. G., Butler, D. L., Weber, W., Gordon, S. L. and Fink, D. J., Mesenchymal stem cell-based repair of rabbit Achilles tendon. In *Transactions of the Orthopaedic Research Society,* San Francisco, CA, 1997.

102. Aragona, J., Parsons, J. R., Alexander, H. and Weiss, A. B., Soft tissue attachment of a filamentous carbon absorbable polymer tendon and ligament replacement. *Clinical Orthopaedics,* 1981, **160**, 268–278.

103. Joyce, M. E., Pduitt, D. L. and Manske, P. R., Effect of growth factors in tendon sealing. On Transactions of the Orthopaedic Research Society, Washington, DC, 1992.

CHAPTER III.5

Biodegradable Polymer Matrices in Dental Tissue Engineering

LONNIE D. SHEA
ISAAC C. YUE
Department of Chemical Engineering,
The University of Michigan,
Ann Arbor, MI 48109, USA

DAVID J. MOONEY
Department of Biologic and Materials Science,
Department of Chemical Engineering,
The University of Michigan,
Ann Arbor, MI 48109, USA

1 Introduction

Oral diseases are extremely widespread, and virtually every individual suffers from some type of oral tissue damage or loss during his/her lifetime. The affected tissues can include both soft (e.g. periodontal ligaments) and hard tissue (e.g. dentin). The major oral diseases include caries, periodontitis, oral cancer and congenital defects. Traditional therapies to these problems typically rely on permanent (non-degradable) synthetic materials (e.g. dental amalgam) to replace lost tissue structure and function, but these therapies suffer from a variety of limitations. The use of biodegradable polymer matrices to regenerate or engineer oral tissues is a newer approach, and this approach is the topic of the current chapter.

Dental caries is one of the most common oral diseases, and is generally reported as the average number of teeth in decay, missing, or previously repaired as a result of bacterial infection. At the age of 17, 84% of Americans have at least one decayed, missing or repaired tooth, and by this age the median number of tooth surfaces which fall into this category is 8 [1]. It is important to realize that caries, and the other dental diseases discussed below, do not effect all population groups equally, and this may alter the treatment options. Low income and minority group are often disproportionately affected by these diseases. For example, approximately 25% of children have 75% of the carious lesions in the USA [1]. The therapy most often utilized to treat dental caries is replacement of the effected tooth structure with a synthetic material, typically dental amalgam for load-bearing regions and composites for non-load bearing anterior teeth [2]. Approximately 100 million amalgam restorations are placed each year in the United States [2]. This therapy is truly one of the success stories of modern dentistry, but these restorations only have an anticipated life of 6–12 years. These synthetic restorations will need to be replaced multiple times in an individuals lifetime, and each replacement will involve the removal of increasingly more healthy tooth structure and subsequent loss of tooth viability.

Periodontal disease is one of the most common diseases affecting human dentition. Gingivitis, a mild form of periodontal disease, refers to inflammation of the gingiva only, while periodontitis affects deeper structures, including alveolar bone. Loss of clinical periodontal attachment is observed in over 90% of people aged 13 or older, but only 15% exhibited advanced destruction [3]. Periodontal destruction increases with age and more than 40% of individuals over 40 exhibit advanced loss of periodontal attachment [3]. Bacterial accumulation on teeth has long been appreciated as important in periodontal disease, and the bacterial likely to be the primary etiologic agents have been identified [4]. Periodontal disease has been previously addressed through both preventive approaches (e.g. tooth brushing), and removal of supra- and subgingival plaque and calculus, but neither of these approaches addresses the concerns of a patient who has lost significant periodontal attachment.

Tooth loss (edentulism) can occur as a result of multiple factors, including dental caries, advanced periodontal disease, trauma, congenital absence. Only 2% of American seniors (age 65 +) have all 28 teeth, and this population averages 8–12 lost teeth [1]. A variety of treatments involving synthetic materials, including bridges, partial or complete dentures and implants have been utilized to replace the function of lost teeth [5]. However, these materials have a limited lifetime and only partially replace the mechanical function of teeth.

Oral cancer and congenital defects often result in the loss or absence of significant oral tissue mass, and this causes impaired function, pain, and disfigurement. Approximately 30,000 cases of oral cancer are diagnosed each year in the U.S., and orofacial clefts occur in 1 out of 700 total births [1]. Replacement of lost tissue structure often involves the use of auto-/allografts of bone to replace lost hard tissue, and synthetic materials to replace soft tissues. Autogenous bone grafts are limited due to donor site morbidity and availability. Allogenous bone grafts pose a risk of disease transmission and are subjected to immunological responses [6]. Synthetic prosthetics are expensive, have a limited lifetime, and typically do not replicate lost tissue structure in a realistic manner [5].

Dental research has traditionally focused on the development of effective clinical approaches to promote the regeneration of oral tissues following various insults or diseases. Synthetic materials (polymer, ceramics, and metals) have been developed as replacements for lost oral tissue structure and function. This approach has given hundreds of millions of patients a vastly improved quality of life. However, these materials do not replace the normal structure and function of the lost tissue and are incapable of remodeling/repairing in the face of ongoing insult or stimulation. Engineering new tissues from biodegradable polymer matrices and cells represents a new approach to treat patients suffering from the loss or malfunction of tissues [7], and may provide a new approach to replace lost oral tissue structure and function. Synthetic, biodegradable polymer matrices are currently being investigated as barriers, drug delivery, and cell transplantation devices. Once the desired tissue structure is regained by the patient, the matrix can be designed to degrade. This leaves no permanent synthetic element in the patient.

2 Physiology

The structure and function of oral and maxillofacial tissues vary widely due to the numerous functions that are performed in the oral environment; however, since the most common diseases involve teeth or the surrounding periodontal tissue, our focus is on the physiology of these two tissues. A schematic of a tooth and the surrounding tissue is given in Fig. 1. Each tooth is similar in structure, consisting of a core of dentin surrounding neurovascular connective tissue called pulp. Dentin is overlaid by enamel in the crown and by cementum in the root. The tooth is attached at the cementum to the alveolar bone by the periodontal ligament. The alveolar bone and periodontal ligament are covered by the gingiva. These elements are described briefly below. A more detailed description can be found in [8, 9].

2.1 Tooth

A tooth is a bonelike structure that functions to cut and grind food during mastication.

2.1.1 Enamel

Enamel, the hardest biologic tissue in the human body, covers the tooth surface exposed to the oral environment and protects the underlying tissue

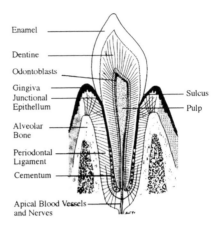

Fig. 1. Schematic of tooth and surrounding periodontal tissue structure. Major components of these tissues are noted. This figure was adapted from [9].

from dental caries. Enamel is highly mineralized, approximately 96% by weight hydroxyapatite, 3% water and 1% organic matter (enamelin). By itself, enamel is brittle. However, the elasticity of the underlying dentin provides support during mastication. Enamel is comprised of interlocking rods, formed by ameloblasts during tooth development, that run from the dentinoenamel junction to the outer surface. Ameloblasts can take different paths during enamel deposition which leads to a bending of the rods. The sites at which the rods bend may have minute spaces or gaps which may allow small particles to penetrate the enamel which can ultimately lead to tooth destruction. It is believed that following tooth maturation, no further enamel can be deposited due to apoptosis of the ameloblasts.

2.1.2 Dentin

Dentin provides support to the brittle overlying enamel during mastication. Dentin, which constitutes the bulk of the tooth, is composed of approximately 70% hydroxyapatite crystals, 20% organic collagen fibers along with small amounts of other proteins, and 10% water by weight. Dentin is classified into three major categories (primary, secondary, and tertiary or reparative) based on the time of its development and the histological appearance of the tissue. Primary dentin, which constitutes most of the tooth, is formed until the tooth reaches occlusion and is functioning. Secondary dentin is continuously formed internal to the primary dentin, but at a rate that is much slower than for primary dentin. Tertiary or reparative dentin forms in response to trauma and is formed only in stimulated areas.

Dentin is formed by odontoblasts which are present at the interface between the pulp and the dentin. Dentin is laid down throughout life and the rate of production of dentin can be altered by dental caries, attrition, and abrasion. Branching processes from the odontoblast extend toward, and possibly into, the dentinoenamel junction. These processes are approximately 3–4 μm in diameter at the dentin-pulp interface and taper to approximately 1 μm in diameter at the dentinoenamel junction. The odontoblastic processes contain microtubules, small filaments, and occasional mitochondria which is indicative of the protein-secreting nature of the odontoblast. Enclosure of these processes by calcified dentin matrix during its formation imparts a tubular structure to the dentin. The resilient properties of the dentin are due in part to the presence of the tubules throughout the matrix.

2.1.3 Pulp

Dental pulp is a neurovascular connective tissue that provides nutritive, sensory, formative, protective and reparative functions for the tooth. The pulp provides nutrients to the tooth by way of blood vessels. The nerves in the pulp act to protect the tooth by responding to stimuli such as heat, cold, and pressure. Pulp is both reparative and formative because of the presence of the odontoblastic cells which form the surrounding dentin.

Pulp has a central zone and a peripheral zone. The central zone contains large arteries, veins, and nerve trunks surrounded by fibroblasts and an intercellular substance of glycoaminoglycans and collagen fibers. The peripheral zone contains the formative cells of dentin and odontoblasts, and is characterized by an odontogenic region composed of odontoblasts, a cell-free and cell-rich regions. A parietal layer of nerves is adjacent to the cell-rich region.

2.2 Periodontium

The function of the components of the periodontium (cementum, periodontal ligament, alveolar bone, and gingiva) is to support the tooth during mastication, speech and swallowing. The periodontium maintains the spatial position of each tooth with respect to its surroundings and the other teeth in the arch.

2.2.1 Cementum

Cementum is a mineralized connective tissue which anchors the collagen fibers of the periodontal ligament to the tooth and seals the tubules of the dentin in the root of the tooth. Cementum is approximately 50% hydroxyapatite by weight with an organic matrix comprised of type I collagen and some proteoglycans. Two types of cementum are observed on the root surface: intermediate cementum and cellular cementum. Intermediate cementum is a thin, non-cellular, amorphous layer that is deposited by the inner layer of epithelial cells of the root sheath. This layer is the first layer deposited and is responsible for sealing the tubules of dentin. Cellular cementum, which is deposited directly on intermediate cementum, contains cementocytes. Cellular cementum resembles bone in that it is a hard, dense tissue with cells (cementocytes) contained within lacunae and having canaliculi; however, unlike bone, cellular cementum does not contain blood vessels or nerves. The lack of vascularity in cementum may be part of the reason why cementum resists resorption more than bone.

2.2.2 Alveolar bone

The alveolar process extends from the maxilla or mandible, and has the primary function of supporting the roots of teeth. It is composed of alveolar bone proper, and supporting bone. Alveolar bone proper is the bone lining the tooth socket, and provides the site of attachment for the periodontal ligament. Supporting bone is comprised of the compact cortical plate on the outer surface of the alveolar process, and spongy bone between the cortical plates. Alveolar bone matures as teeth gain functional occlusion and typically remains functional throughout life. However, tooth loss can result in the disappearance of alveolar bone.

2.2.3 Gingiva

The gingiva protects the other periodontal tissues from the oral environment by preventing the entry of microorganisms and toxic substances from the oral cavity. The gingiva surrounds the necks of teeth and extends apically to the mucogingival junction. The gingiva is divided into three zones: the free or marginal zone, the attached gingiva, and the interdental zone. The free or marginal zone encircles the tooth and defines the gingival sulcus, which is the trench around a tooth that is deepened in periodontal disease. The attached gingiva is that portion of the epithelium that is attached to the neck of the tooth, and the interdental zone lies between two adjacent teeth.

2.2.4 Periodontal ligament

The periodontal ligament, which is a fibrous connective tissue, attaches the alveolar bone to the tooth. It absorbs mechanical loads, prevents gross movement of the teeth, and protects the vessels and nerves of the periodontium. Additionally, receptors and nerves within the ligament sense pressure and movement associated with mastication. The ligament covers the root of the tooth and connects with the gingiva. The ligament is composed of two groups of fibers: the gingival group, located around the necks of teeth, and the dentoalveolar group, which surrounds the roots of teeth. The interstitial spaces between fibers provide a space for a variety of cells and networks of blood vessels and nerves. A rich vascular supply to the ligament is essential to its health and function. Fibroblasts are the most numerous cell type in this tissue. Osteoblasts are located along the surface of the alveolar bone, and cementoblasts appear near the cementum.

3 Engineering oral tissues

Tissue engineering has been defined as the interdisciplinary field that applies the principles of engineering and the life sciences toward the development of biological substitutes that restore, maintain, and improve the function of damaged tissues and organs [10]. Tissue engineering, in contrast to traditional therapies, opens up the possibility of creating a new, natural tissue which performs all functions of the original tissue and is capable of adjusting its function to the changing requirements of the body. The use of biodegradable polymer matrices leads to eventual degradation of the synthetic element of the new tissue, and thus no long-term synthetic component to the tissue. This may avoid many of the difficulties (e.g. chronic inflammation, bacterial infections) often observed when a permanent material is implanted in the body. A variety of naturally occurring materials (e.g. type I collagen) have been used as tissue engineering matrices [11], but naturally derived materials must be isolated from animals, humans, or plants. These materials often suffer from large batch-to-batch variability, high expense, and a limited range of physical properties. Synthetic polymers, in contrast, can be synthesized reproducibly and processed with a variety of techniques. Simple changes in synthetic chemistry or processing can be exploited to obtain a wide range of physical properties [12]. This chapter focuses on biodegradable synthetic polymer matrices which function as barriers, drug delivery devices, and cell transplantation vehicles in a variety of dental applications.

3.1 Biodegradable polymer barriers

A significant challenge in tissue engineering is the localization of specific cell types in an anatomic location to promote tissue regeneration while excluding non-desirable cell types from this site. The potential barrier function of polymer matrices has been exploited in the design of biodegradable polymer matrices for guided tissue regeneration (GTR). This approach is based on the premise that progenitor cells responsible for regeneration reside in the underlying healthy tissue and can be induced to migrate into a tissue defect and regenerate the lost tissue. The migration of other cell types is prevented by the polymer matrix. The two primary GTR applications have been the replacement of connective tissue attachment lost as a consequence of periodontal disease and the regeneration of bone (e.g. to provide sufficient mass for implant placement). In the formation of new connective tissue attachment, a mechanical barrier (GTR membrane) is placed between the root surface and the muco-gingival flap [13]. Bone cells, gingival connective tissue cells, and gingival epithelial cells are prevented by the barrier from invading the healing area. In contrast, cells of the periodontal ligament can repopulate the root surface thus producing new attachment. In bone regeneration, the membrane acts in a similar manner by preventing the ingrowth of non-osteogenic soft tissue derived cells into the membrane-protected space [13]. Angiogenic and osteogenic cells are able to populate and regenerate the bone defect.

The critical feature of matrices for GTR is their ability to block cell transport across the matrix. The porosity, pore size distribution and pore continuity dictate the ability of cells to migrate through a matrix [14]. Cells can invade a matrix if the pores are larger than approximately 10 μm, and the rate of invasion will increase with the pore size and total porosity of a device. It should be noted that the capillary network formed in the developing tissue will also only be able to invade from certain directions. The degradation of the GTR matrix should be matched to the time required for regeneration of the tissue.

The most commonly used membrane for GTR is made from expanded polytetrafluoroethylene (ePTFE) (Gore-Tex, W.L. Gore and Associates, Flagstaff, AZ, USA). ePTFE is a stable polymer that is chemically and biologically inert and able to resist enzymatic and microbiological attack. ePTFE has been used with positive results for both the regeneration of periodontal attachment [15, 16] and the regeneration of bone [17]. However, ePTFE is non-biodegradable and must be removed by a second surgical technique. This second surgical technique may jeopardize regenerative healing, increases problems relating to infection control post-surgery, and is traumatic for the patient. These problems can be avoided by using a biodegradable membrane.

Biodegradable membranes composed of poly(lactic acid) (PLA) have been used to generate new periodontal attachment. Studies of these membranes in baboons have shown the membrane was histologically acceptable and was compatible for a wound-healing barrier [16]. A membrane composed of poly(lactic-co-glycolic acid) (PLGA) was used to form new attachment for large defects in dogs [15]. A clinical study of 29 patients has also been performed using a PLA membrane dissolved in N-methyl-2-pyrrolidone [18]. In the extra-oral environment, this barrier can be trimmed to the appropriate size of the defect and is flexible enough that it can be adapted to the defect. Once placed in situ, the membrane hardens. The overall results with this membrane indicated favorable clinical outcomes.

Membranes comprised of PLA have also been used for GTR of bone defects in the adult beagle dog [19]. Epithelial attachment, alveolar bone regrowth, connective tissue attachment, and new cementum were observed, and these results suggest that the PLA membranes could be one alternative for GTR in advanced periodontal lesions in humans. A number of other biodegradable membranes have also been examined for application to bone defects. Collagen barriers have been used for GTR with variable results [20–25]. These membranes are hemostatic, chemotactic for fibroblasts, and form a scaffold for migrating cells. However, these materials were weak immunogens and only a limited amount of tissue regeneration was observed before degradation of the membrane. Membranes of PLGA [26, 27] and polyurethane [28] have also been tested and yielded mixed results in periodontal regeneration.

Important engineering considerations in the design of GTR membranes include the appropriate mechanical properties and degradation rates. Membranes must be elastic to allow adaptation to the defect. However, the membranes must not collapse during healing as this would reduce the volume of the membrane-protected defect. The application of GTR to large defects is uncertain because the appropriate cells may not be able to repopulate the entire defect; hence inductive or conductive proteins may need to be added.

3.2 Drug delivery devices

A critical element of tissue engineering is providing a suitable environment for both the resolution of disease progression and promotion of appropriate tissue regeneration. Controlled drug delivery technology offers the potential of precisely controlling both the type and local concentration of various soluble signals present in a tissue. Controlled drug delivery infers a pre-determined release rate which typically continues for time periods ranging from hours to years. There are three main driving forces behind developing drug delivery matrices for dental applications in general, and periodontal applications in specific. These are a desire to place high concentrations of drugs locally in periodontal pockets in order to increase the efficacy of the active agent, to minimize the side-effects which frequently accompany systemic delivery of drugs, and to protect the biological factor from degradation over extended release periods.

A critical issue in controlled drug delivery is the design of the polymer matrix, as it will dictate the kinetics of drug release in the tissue. In general, drug release can be designed to occur primarily via polymer degradation or scission of a covalent bond between the drug and polymer matrix (reaction controlled release), or via diffusion of the drug through the polymer matrix (diffusion controlled release) [29]. Drug release is usually regulated by polymer degradation in dental applications, and this requires that a polymer with the appropriate degradation rate and mechanism be chosen to fabricate the matrix. Bulk degrading polymers will exhibit relatively even polymer chain scission throughout their bulk after being placed *in vivo*, and drug release will typically occur at a low rate until the point at which the polymer is significantly degraded and the matrix rapidly releases both polymer degradation products and the drug. In contrast, surface eroding polymers will exhibit chain scission mainly at the surface of the polymer matrix, and drug release will occur continuously as the matrix degrades and loses mass from the exterior to the interior [29]. The two major applications of controlled drug delivery in dentistry is delivering antibiotics to slow or block periodontal disease progression, and delivery of protein growth factors to promote tissue regeneration.

3.2.1 Antibiotic delivery

The delineation of specific bacteria species responsible for periodontal disease has created an opportunity to block disease progression by controlled delivery of antibiotics to the periodontal pocket [4]. There are several controlled antibiotic delivery products currently available for clinical use, and a variety of matrices have been utilized in these products [30]. Non-degradable ethylene vinyl acetate co-polymer fibers have been utilized to release tetracycline, and have shown clinical efficacy similar to scaling and root planing [31]. Acrylic and lipid-based gels have also been utilized to deliver antibiotics, and shown to have a clinically significant effect [30]. Degradable systems are under development, and these systems include polymer microspheres of lactide-glycolide copolymers in which drugs are dispersed [30]. A new biodegradable polymer based on poly(ortho esters) has recently been described that may provide an injectable matrix with high periodontal pocket retention and a desirable degradation rate and drug release profile [32].

3.2.2 Growth factor delivery

Advances in molecular biology have led to an understanding of the role of specific molecules in tissue (e.g. bone) regeneration, and create the opportunity to manipulate these processes to engineer new tissues in desired anatomic locations via controlled delivery of specific molecules. A number of factors have been investigated to achieve tissue regeneration, but one of the most widely studied family of proteins is the bone morphogenetic proteins (BMPs). Demineralized bone powder was first recognized to possess bone inductive capacity, and this activity was subsequently localized to BMPs present in this preparation [33]. Recombinant forms of these molecules are now available in large quantities. These factors may allow regeneration of alveolar bone lost to periodontal disease, and prevent or delay tooth loss or removal. These molecules may also be utilized to build up bone tissue in cranial defects and mandibles, aid in the placement of dental implants, and promote reparative dentin formation [34–39]. A

Fig. 2. Polymer microsphere fabricated from a copolymer of lactic and glycolic acids. These microspheres have been utilized for controlled delivery of growth factors *in vivo* [51].

variety of other factors (e.g. PDGF) have also been delivered to promote tissue regeneration [40].

A variety of matrices have been utilized to deliver protein growth factors *in vivo*, including collagen-based matrices and biodegradable polymers such as PLA, PLGA, and polyorthoesters. Growth factors incorporated into microspheres fabricated from biodegradable polymers, such as those shown in Fig. 2, will be released as the polymer degrades. The degradation rate of copolymers comprised of lactide and glycolide can be varied from weeks to years [12]. The release profile will also be controlled by the physical form of the matrix (e.g. microsphere size). These features may prove to be critically important, as the time period during the regenerative process in which specific factors are required may vary considerably. Controlled growth factor delivery is likely to be combined with GTR matrices in clinical practice in order to promote population of the potential tissue space with the desired cell population that will be acted on by the growth factor, while excluding undesirable cells from this space.

3.3 Cell transplantation vehicles

In many disease situations an inductive approach to promote tissue regeneration may not be sufficient, and transplantation of a desired cell population into the potential tissue space may be required. Specific molecules required for induction may not be identified in these situations, the inducible cell population may not exist, or a more rapid regeneration may be required than is possible in induction alone. One of the largest challenges in cell transplantation is transport of nutrients. Transplanted cells must survive by diffusion until a vascular supply is established in the engineered tissue, and this often limits cell viability [41]. A variety of approaches are being investigated to enhance transport in these situations, including pre-vascularization of the matrix prior to cell transplantation [42] and co-transplantation of endothelial cells capable of forming new vascular structures [43].

One potential application of cell transplantation is the engineering of dental pulp tissue. It is possible to regenerate dentin using BMP delivery as described earlier in this review, but this process is dependent on the presence of healthy dental pulp. Pulp infection often necessitates a root canal (approximately 15 million/year in US), and loss of the pulp. Dental pulp-like tissue has recently been engineered from cells derived from adult human third molars utilizing a matrix of poly(glycolic acid) fibers [44]. The cells adhere to the polymer fibers (Fig. 3), and proliferate while the synthetic matrix degrades to form a new tissue which histologically resembles normal adult pulp in overall appearance and cellularity. These results open up the possibility of engineering both the pulp and dentin components of teeth.

New bone tissue has also been engineered utilizing cell transplantation on biodegradable polymer matrices. PLA and PLGA foams have been utilized to engineer new bone tissue *in vitro* [45], and poly(glycolic acid) fiber-based matrices have been utilized to engineer new bone in a variety of animal models [46, 47]. Cells derived from periosteal tissue and multiplied *in vitro* were utilized for these studies. Cells derived from marrow have also been transplanted on ceramic matrices to engineer new bone tissue. Cartilage can also be readily engineered *in vivo* using chondrocyte transplantation on biodegradable polymer matrices. Cartilage histologically and biochemically similar to native cartilage can be engineered [48], and the size and shape of the new tissue can be precisely controlled with the polymer matrix [49] to yield new cartilaginous structures with a desired structure (e.g. nasoseptal cartilage) [50].

4 Summary

The large number of patients suffering from oral diseases, and the limitations of current therapies provide a large driving force for the development of tissue engineering approaches to replace lost tissue structure and

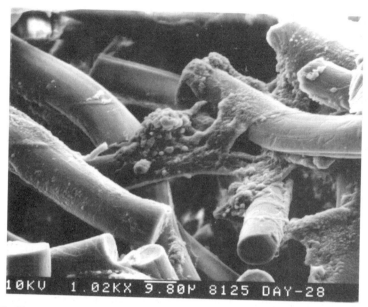

Fig. 3. Photomicrograph of fibroblasts derived from adult human third molars adherent to fibers of poly(glycolic acid) in vitro (DJ Mooney, unpublished photomicrograph).

function. Biodegradable polymer matrices are under development for applications as GTR membranes, drug delivery devices, and cell transplantation vehicles. It is likely that these different applications will not be used in isolation of each other, but instead used in combination. For example, controlled drug delivery will likely be combined clinically with both GTR membranes to enhance tissue regeneration, and cell transplantation to regulate the engraftment and function of transplanted cells [51]. The limited clinical trials performed to date with biodegradable matrices support the potential for these matrices, and suggest that tissue engineering may become an important tool in clinical dentistry.

References

1. Caplan, D. J. and Weintraub, J. A., The oral health burden in the United States: A summary of recent epidemiologic studies. *Journal of Dental Education*, 1993, **57**(12), 853–862.
2. Dental amalgam: A scientific review and recommended public health service strategy for research, education and regulation. Dept. Of Health and Human Services (Public Health Service), January, 1993.

3. Brown, L. J., Brunelle, J. A. and Kingman, A., Periodontal status in the United States, 1988–91: Prevalence, extent and demographic variation. *Journal of Dental Research*, 1996, **75**(Suppl.), 672–683.

4. Williams, R. C., Beck, J. D. and Offenbacher, S. N., The impact of new technologies to diagnose and treat periodontal disease: A look to the future. *Journal of Clinical Periodontology*, 1996, **23**, 299–305.

5. Craig, R. G., *Restorative dental materials*, Mosby Co, 8th edition, 1989.

6. Friedlander, G. E., Bone allografts: the biological consequences of immunological events. *Journal of Bone Joint Surgery*, 1991, **73**-A, 1119–1122.

7. Langer, R. and Vacanti, J. P., Tissue engineering, *Science*, 1993, **260**, 920–926.

8. Atkinson, M. E. and White, F. H., *Principles of anatomy and oral anatomy for dental students.*, Churchill Livingstone, 1st edition, 1992.

9. Avery, J. K., *Essentials of Oral Histology and Embryology*, Mosby Year Book, 1992.

10. Nerem, R. M. and Sambanis, A., Tissue engineering: From biology to biological substitutes. *Tissue Engineering*, **1**, 3–13.

11. Cavallaro, J. F., Kemp, P. D. and Kraus, K. H., Collagen fabrics as biomaterials. *Biotechnology and Bioengineering*, 1994, **43**, 781–791.

12. Wang, W. H. and Mooney, D. J., Synthesis and properties of biodegradable polymers used in tissue engineering. In *Tissue Engineering*, eds. A. Atala and D. J. Mooney. Birkhauser Press, 1997, 51–82.

13. Aaboe, M., Pinholt, M. and Hjorting-Hansen, E., Healing of experimentally created defects: a review. *British Journal of Oral and Maxillofacial Surgery*, 1995, **33**, 312–318.

14. Mooney, D. J. and Langer R., Engineering biomaterials for tissue engineering: The 10–100 micron scale. In *Biomedical Engineering Handbook*, ed. J. Bronzino. CRC Press, 1995, 1609–1618.

15. Lindhe, J., Pontoriero, R., Berglundh, T. and Araujo, M., The effect of flap management and bioresorbable occlusive devices in GTR treatment of degree III furcation defects: An experimental study in dogs. *Journal of Clinical Periodontology*, 1995, **22**, 276–283.

16. Vernino, A. R., Jones, F. L., Holt, R. A., Nordquist, R. E. and Brand, J. W., Evaluation of the potential of a polylactic acid barrier for correction of periodontal defects in baboons: A clinical and histologic study. *International Journal of Periodontology and Restorative Dentistry*, 1995, **15**, 85–101.

17. Buser, D., Dula, K., Hirt, H. P. and Schenk, R. K., Lateral ridge augmentation using autografts and barrier membranes: A clinical study with 40 partially edentulous patients. *Journal of Oral Maxillofacial Surgery*, 1996, **54**, 420–432.

18. Polson, A. M., Garrett, S., Stoller, N. H., Greenstein, G., Polson, A. P., Harrold, C. Q. and Laster, L., Guided tissue regeneration in human furcation defects after using a biodegradable barrier: A multi-center feasibility study. *Journal of Periodontology*, 1995, **66**, 377–385.

19. Robert, P. M. and Frank, R. M., Periodontal guided tissue regeneration with a new resorbable polylactic acid membrane. *Journal of Periodontology*, 1994, **65**(5), 414–422.

20. Minabe, M., A critical review of the biologic rationale for guided tissue regeneration. *Journal of Periodontology*, 1991, **62**, 171–179.

21. Pitaru, S., Tal, H., Soldinger, M. and Noff, M., Collagen membranes prevent apical migration and support new connective tissue attachment during periodontal wound healing in dogs. *Journal of Periodontology Research*, 1989, **24**, 247–253.

22. Blumenthal, N. and Steinberg, J., The use of collagen barriers in conjunction with combined demineralized collagen gel implants in human infra-bony defects. *Journal Periodontology*, 1990, **61**, 319–327.

23. Galgut, P. N., Guided tissue regeneration: Observations from five treated cases. *Quintessence International*, 1990, **21**, 713–721.

24. Chung, K. M., Salking, L. M., Stein, M. D. and Freedman, A. L., Clinical evaluation of biodegradable membrane in guided tissue regeneration. *Journal of Periodontology*, 1990, **61**, 732–736.

25. Quteisch, D. and Dolby, A. E., The use of irradiated-cross linked human collagen membrane in guided tissue regeneration. *Journal of Clinical Periodontology*, 1992, **19**, 476–484.

26. Gager, A. H. and Schultz, A. J., Treatment of periodontal defects with an absorbable (polyglactin 910) with and without osseous grafting: case reports. *Journal of Periodontology*, 1991, **62**, 276–283.

27. Vuddkakanok, S., Solt, S. W., Mitchell, J. C., Forman, D. W. and Alger, F. A., Histologic evaluation of periodontal attachment apparatus following the insertion of a biodegradable copolymer barrier in humans. *Journal of Periodontology*, 1993, **64**, 202–210.

28. Warrer, K., Karring, T., Nyman, S. and Gogolewski, S., Guided tissue regeneration using biodegradable membranes of polylactic acid or polyurethane. *Journal of Clinical Periodontology*, 1992, **19**, 633–640.

29. Langer, R., New methods of drug delivery, *Science*, **249**, 1527–1533.

30. Needleman, I. G., Pandya, N. V., Smith, S. R. and Foyle, D. M., The role of antibiotics in the treatment of periodontitis (part 2 — controlled drug delivery). *European Journal of Prosthodontology and Restorative Dentistry*, 1995, 3(3), 111–117.

31. Goodson, J. M., Cugini, M. A., Kent, R. L. et al., Multicenter evaluation of tetracycline fiber therapy: II. Clinical response. *Journal of Periodontology Research*, 1991, **26**, 371–379.

32. Roskos, K. V., Fritzinger, B. K., Rao, S. S., Armitage, G. C. and Heller, J., Development of a drug delivery system for the treatment of periodontal disease based on bioerodible poly(ortho esters). *Biomaterials*, 1995, **16**(4), 313–317.

33. Wozney, J. M., The potential role of bone morphogenetic proteins in periodontal reconstruction. *Journal of Periodontology*, 1995, **66**, 506–510.

34. Boyne, P. J., Advances in preprosthetic surgery and implantation. *Current Opinion in Dentistry* 1991, **1**, 277–281.

35. Marden, L. J., Hollinger, J. O., Chaudhari, A., Turek, T., Schaub, R. G. and Ron, E., Recombinant human bone morphogenetic protein-2 is superior to demineralized bone matrix in repairing craniotomy defects in rats. *Journal of Biomedical Materials Research*, 1994, **28**, 1127–1138.

36. Kenley, R., Marden, L., Turek, T., Jin, L., Ron, E. and Hollinger, J. O., Osseous regeneration in the rat calvarium using novel deliver systems for

recombinant human bone morphogenetic protein-2 (rhBMP-2). *Journal of Biomedical Materials Research*, 1994, **28**, 1139–1147.

37. Mayer, M., Hollinger, J., Ron, E. and Wozney, J., Maxillary alveolar cleft repair in dogs using recombinant human bone morphogenetic protein-2 and a polymer carrier. *Plastic and Reconstructive Surgery*, 1996, **98**(2), 247–257.

38. Sugurdsson, T. J., Lee, M. B., Kubota, K., Turek, T. J., Wozney, J. M. and Wikesjo, U. M. E., Periodontal repair in dogs: Recombinant human bone morphogenetic protein-2 significantly enhances periodontal regeneration. *Journal of Periodontology*, 1995, **66**, 131–138.

39. Rutherford, R. B., Spangberg, L., Tucker, M., Rueger, D. and Charette, M., The time-course of the induction of reparative dentine formation in monkeys by recombinant human osteogenic protein-1. *Archives of Oral Biology*, 1994, **39**(10), 833–838.

40. Rutherford, R. B., Ryan, M. E., Tucker, M. M. and Charette, M. F., Platelet-derived growth factor and dexamethasone combined with a collagen matrix induce regeneration of the periodontium in monkeys. *Journal of Clinical Periodontology*, 1993, **20**, 537–544.

41. Mooney, D. J., Park, S., Kaufmann, P. M., Sano, K., McNamara, K., Vacanti, J. P. and Langer, R., Biodegradable sponges for hepatocyte transplantation. *Journal of Biomedical Materials Research*, 1995, **29**, 959–966.

42. Uyama, S., Kaufmann, P. M., Takeda, T. and Vacanti, J. P., Deliver of whole liver equivalent hepatic mass using polymer devices and hepatotrophic stimulation. *Transplantation*, 1993, **55**, 932–935.

43. Holder, W. D., Gruber, H. E., Roland, W. E., Moore, A. L., Culberson, C. R., Loeback, A. B., Burg, K. J. and Mooney, D. J., Unique vascular structures occurring in polyglycolide matrices containing aortic endothelial cells implanted in the rat, *Tissue Engineering*, 1997, **3**, 149–160.

44. Mooney, D. J., Powell, C., Piana, J. and Rutherford, B., Engineering dental pulp-like tissue in vitro. *Biotechnology Progress*, 1996, **12**(6), 865–868.

45. Ishaug, S. H., Crane, G. M., Miller, M. J., Yasko, A. W., Yaszemski, M. J. and Mikos, A. G., Bone formation by three-dimensional stromal osteoblast culture in biodegradable polymer scaffolds. *Journal of Biomedical Materials Research*, 1997, **36**, 17–28.

46. Vacanti, C. A., Kim, W., Vacanti, M. P., Mooney, D., Schloo, B. and Vacanti, J. P., Comparison of tissue engineered bone and cartilage using transplanted cells on synthetic, biodegradable polymer scaffolds. *Transplantation Proceedings*, 1993, **25**, 1019–1021.

47. Vacanti, C. A., Kim, W., Upton, J., Mooney, D. and Vacanti, J. P., Efficacy of periosteal cells compared to chondrocytes in the tissue engineered repair of bone defects. *Tissue Engineering*, 1995, **1**, 301–308.

48. Puelacher, W. C., Mooney, D., Langer, R., Vacanti, J. P. and Vacanti, C. A., Tissue engineered growth of cartilage: the effect of varying the concentration of chondrocytes seeded onto synthetic matrices. *International Journal of Oral Maxillofacial Surgery*, 1994, **23**, 49–53.

49. Kim, W. S., Vacanti, C. A., Puelacher, W., Cima, L. G., Upton, J., Mooney, D. and Vacanti, J. P., Cartilage configuration in pre-determined shapes employing cell transplantation on prosthetic biodegradable synthetic polymers. *Plasic and Reconstructive Surgery*, 1994, **94**, 233–237.

50. Puelacher, W. C., Mooney, D., Langer, R., Upton, J., Vacanti, J. P. and Vacanti, C. A., Design of nasoseptal cartilage replacements synthesized from biodegradable polymers and chondrocytes. *Biomaterials*, 1994, **15**, 774–778.

51. Mooney, D. J., Kaufmann, P. M., Sano, K., McNamara, K., Schwendeman, S., Vacanti, J. P. and Langer, R., Localized delivery of epidermal growth factor improves the survival of transplanted hepatocytes. *Biotechnology and Bioengineering*, 1996, **50**(4), 422–429.

Hematopoietic Cells

BERNHARD PALSSON

Department of Bioengineering,
University of California–San Diego,
La Jolla, CA 92093-0412, USA

1 Introduction

One of the primary motivations for developing tissue engineering is for the implementation of cellular therapies. Cellular therapies use transplanted or grafted human cells as therapeutic agents to affect a pathological condition. Cellular therapies are not new. Blood transfusion, basically of red blood cells, into anemic patients to restore adequate oxygen transport, has been practiced for decades with much therapeutic success and benefit. Similarly, platelets have been successfully transfused into patients that have blood clotting problems. Bone marrow transplantation (BMT) has been practiced for about two decades, and currently tens of thousands of cancer patients annually undergo high dose chemo- and radio-therapies with BMT. More recently transplantation of mobilized peripheral blood stem cells has grown rapidly, and is expanding the clinical use of hematopoietic stem cell transplants. These are all applications of cell therapies associated with the blood cells and blood cell generation (hematopoiesis; Greek for; hemato = blood, poiesis = generation of). Hematopoietic cell biology has been at the forefront of developments in cellular biology in terms of unraveling roles of growth factors, their receptors, and signal transduction pathways. Advances in cell separation and cell culture technology have led to a variety of new experimental cellular therapies associated with hematopoiesis, many of which require *ex vivo* cell manipulation and cell cultivation.

The field of tissue engineering has grown significantly since 1988 when the first symposium was held on this subject [1]. The ability to engineer tissue function and produce cells in clinically meaningful numbers has a wide spectrum of applications [2–8]. Not surprisingly, the history of clinical uses of hematopoietic cellular therapies has lead to early developments in tissue engineering and their reduction to practice [9–13]. However, as evidenced by many chapters in this book, tissue engineering of may other tissue types is leading to new cellular therapies, such as newly FDA approved skin products.

In this chapter we will address several important issues associated with the delivery of tissue engineered products in general, and the experience with hematopoietic products in particular. We will describe the multilayered series of issues that must be addressed for the clinical implementation of a tissue engineered product. Then we will describe a generic model that describes tissue dynamics which has its roots in hematopoietic research. The cell biology of hematopoiesis will then be described highlighting the central role that the hematopoietic stem cell (HSC) plays. We will then describe current issues associated with understanding HSC behavior and how the tissue microenvironment influences the HSC.

2 Delivering a tissue engineered product

What is involved in developing and clinically implementing a tissue engineered product? Experience shows that the answer to this question lies at four levels (Fig. 1). First, the cell and tissue biology of the tissue of interest needs to be understood. For complex tissues such as bone marrow, that have many functional cellular elements, this is a difficult problem since the details of how cells interact is not known [14–17]. Engineering tissue behavior at this level must be approached phenomenologically. Using the dynamics of the *in vivo* situation as a guide has proven to be a rewarding approach [18]. Simpler applications involve the growth of pure cell cultures, such as fibroblast for skin applications and condrocytes for cartilage repair. In some cases, use of structural biomaterials is important.

The second challenge, is to design and implement a cell culture device. In the simplest case, this device is simply a cell culture bag. Such bags have been used for the growth of T cells. Conversely, specific and complex bioreactor designs have been proposed for other hematopoietic applications [19–21]. Once a suitable bioreactors has been designed and evaluated, an automated platform on which it operates must be provided. Such platforms need to meet certain FDA requirements, such as detailed documentation for sample tracking, recording operating variables, and so on. For T-cells grown in bags, these needs can be provided for by a regular cell culture incubator, whereas specific and complex bioreactors require a tailor-made and more specific platform design [22]. Finally, the process must be implemented at a clinical site. For hematopoietic applications, cell separation and growth procedures are typically performed in a hospital's Blood Bank or a specialized facility therein.

The following will focus on the first level in Fig. 1, since many of the frontiers on hematopoietic tissue engineering are associated with tissue dynamics and how hematopoietic stem cells function.

The Four Principal Size Scales in
Tissue Engineering & Cellular Therapies

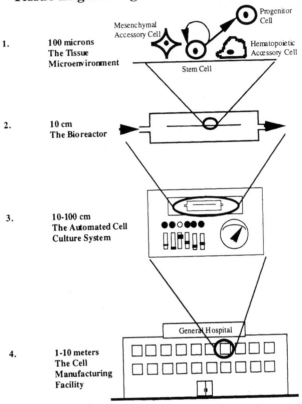

Fig. 1. The four levels of challenges that need to be dealt with in developing and implementing a tissue engineered product.

3 Hematopoietic tissue dynamics

Tissues have their characteristic turnover rates. The bone marrow is the most proliferative tissue in the body, followed by the lining of the small intestine, followed by the epidermis. The turnover rate of these tissues is on the order of a 2–3 days, 5–6 days, and 1–2 weeks, respectively. Other tissues are much slower in their cell turnover times. The cellularity in the liver turns over in about 1 year.

The bone marrow produces about 400 billion myeloid cells daily in a homeostatic state [9]. Over a 70 year life time, the cell production from marrow is a staggering 10^{16} cells. This cell number is several hundred times greater than the total number of cells that are in the body at a given time. This highly proliferative nature of bone marrow renders it vulnerable to high-dose chemo- and radio- therapies. Autologous bone marrow transplantation was developed as a way to enable the use of myoablative regiments, by simply removing the marrow during this ablative regiment and then reintroducing it into the patient. Hematopoetic tissue is thus rebuilt *in vivo* from transplanted HSCs.

The cellular fate processes that underlie the formation of tissue the homeostatic function of tissue, and the repair of tissue are:

1. Cell replication, an increase in cell number;
2. Cell differentiation, changes in gene expression, and the acquisition of a particular phenotypic function;
3. Cell motility, the motion of a cell into a particular niche, or location;
4. Cell apoptosis; the controlled death of a cell, distinguished from necrotic death; and
5. Cell adhesion; the physical binding of a cell to its immediate environment, that may be a neighboring cell, extra-cellular matrix, or an artificial surface.

A general schema for the interaction of the three fate processes, of differentiation, replication, and apoptosis in the production of mature hematopoietic cells, is shown in Fig. 2. This model, derived from decades of hematopoietic research seems to apply to other rapidly proliferating tissues (skin and intestinal epithelium), and may apply to slowly replicating tissues, such as the liver [23].

HSCs form pre-progenitor cells that replicate slowly. Further differentiation and divisions make them progenitor cells (with colony forming potential in semi-solid media). Progenitor cells divide at a maximal cycling rate, with cycling times as short as 12 hours. Apoptosis is very active at this stage, and in fact, most of the hematopoietic growth factors that act at this stage are survival factors (i.e. anti-apoptotic) and permit the progenitors to pass this stage of differentiation to become precursor cells. The precursors replicate slowly and mature into fully morphologically developed cells. These cells then take on their mature cell function. They leave the bone marrow and enter circulation where they perform their function. The mature cells eventually die and have to be replaced. The rate of death of mature cells sets the need for the cell production rate in the tissue, and ultimately the number of stem cell commitments that are needed.

4 Hematopoietic stem cells and lineage development

Hematopoiesis, the regulated production of mature blood cells, is a complex scheme of multilineage differentiation, which occurs mainly in the

Model for Cell Production in Prolific Tissues

Cellular Fate Processes & Bone Marrow Differentiation

	Stem Cells →	Early Progenitors →	Late Progenitors →	Precursor Cells →	Mature Cells →
Cell Number	Potential 2^{30}-2^{50} per cell				Need about 10^{16} total over a lifetime
Cell Cycling	Very slow 1/6 weeks	Slow t_d~60-100 hrs	Very rapid t_d~12 hrs	Moderate	Zero (could be activated in specific cases)
Apoptosis	Inactive	Inactive	Very active 1:5000 survives	Slow	Inactive (can be induced)
Regulation	Cell-Cell contact	Cell-cell contact	Soluble growth factors	Soluble growth factors	Soluble growth factors

Fig. 2. A summary of cellular fate processes and their activities at different stages of hematopoietic differentiation.

bone marrow of adult mammals (Fig. 3). The eight types of mature blood cells which exist in the circulation are derived from a common small population of pluripotent HSCs which form during one short interval in early embryonic life and maintain hematopoiesis thereafter through an extensive capacity for self-renewal. During embryonic development, these cells first arise in the aortic region, are later found in the fetal liver, and at the time of delivery are found in high concentrations in umbilical cord blood [24]. In adults, these pluripotent HSCs reside predominantly in the bone marrow, but may also be found at low frequencies in peripheral blood.

The eight major types are divided into two major groups: the myeloid and lymphoid. The myeloid lineage includes erythrocytes (red blood cells), monocytes, the granulocytes (neutrophils, eosinophils, and basophils), and platelets (derived from non-circulating megakaryocytes). Thymus-derived (T) lymphocytes and bone marrow-derived (B) lymphocytes constitute the lymphoid lineage. All of these blood cell types potentially are of therapeutic value and thus target for the development of cellular therapies.

A remarkable diversity of function is demonstrated in the hematopoietic system. Mature blood cells are highly specialized and perform a variety of

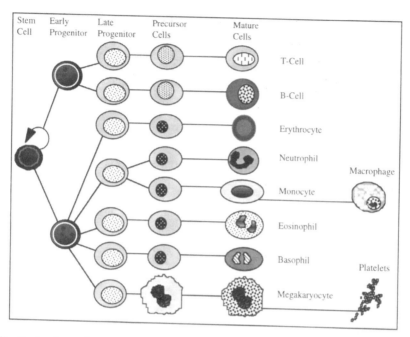

Fig. 3. A schematic representation of hematopoietic stem cells and the multi-lineage differentiation that they are capable of undergoing.

functions including oxygen transport, immune response, and blood clotting. Most mature blood cells exhibit a limited lifespan *in vivo*. Although some lymphocytes are thought to survive for many years, it has been shown that erythrocytes and neutrophils have half lives of 120 days and 8 hours, respectively. Because the vast majority of mature blood cells are destined to remain functionally active for a relatively short time, a continuous process of regeneration is required. In fact, an average human must produce nearly 400 billion mature blood cells per day to replace hematopoietic cells lost due to natural turnover.

This staggering number of mature cells which must be continuously produced under normal, unstressed conditions are derived from progenitor cells. Progenitor cells are unipotent, bipotent, or multipotent, but are not pluripotent (capable of generating both myeloid and lymphoid progeny). They are therefore capable of undergoing proliferation, differentiation, and development into only one or several of the mature cell types. These progenitor cells are designated by the term colony-forming unit, CFU (or colony-forming cell, CFC), because of their ability to form colonies of mature cells in semisolid agar or methylcellulose cultures.

To specify the type of progenitor, a suffix is simply added to the CFU designation. For example, granulocyte/macrophage colony-forming units

(CFU-GM) proliferate and develop into mature granulocytes and macrophages. Erythrocyte colony-forming units (CFU-GM) undergo growth and hemoglobinization to form mature erythrocytes. Erythrocyte burst-forming units (BFU-E) are the more commonly measured precursor of the CFU-E. A more primitive cell that can be detected in colony assay is the granulocyte/erythrocyte/macrophage/megakaryocyte colony- forming unit (CFU-GEMM or CFU-Mix).

In adult humans, these myeloid progenitor cells are located mainly in the bone marrow with small populations in the spleen and the circulation. It therefore follows that the bone marrow is the major site of myeloid blood cell production in adults. Like the mature cells they produce, most progenitor cells are short lived because as they proliferate, they concomitantly undergo differentiation and lose proliferative potential.

5 Hematopoietic stem cells

Under normal conditions, the vast majority of HSCs are not engaged in active proliferation and reside in the G_0 state (outside of the proliferative cell cycle), during which time they may repair DNA damage and maintain the genetic integrity of the stem cell population. Evidence for this concept was first provided by studies in which the administration of 5-fluorouracil (5-FU, which preferentially kills cycling cells) to mice resulted in bone marrow which was enriched in primitive cells. Subsequent experiments have shown that greater than 90% of primitive human cells are also in a non-cycling state.

Experimental evidence for the existence of hematopoietic stem cells was first provided by Till and McCulloch. In their experiments, lethally irradiated mice were injected with bone marrow cells from healthy syngeneic donor mice. The hematopoietic system of mice receiving donor cells was reconstituted, whereas control mice died within a week. Some of the injected cells seeded in the spleen and gave rise to macroscopically discernible hematopoietic colonies containing cells of the myeloid lineage.

Subsequent experiments in mice utilizing genetic marking have demonstrated that long-term engraftment of both the lymphoid and myeloid lineages can be achieved by the progeny of a single cell. These results confirm the existence of true pluripotent hematopoietic stem cells. In addition, the ability to further repopulate multiple secondary recipients with the progeny of a single clone has demonstrated that these very primitive murine cells are capable of *in vivo* expansion. More importantly, studies have shown that retrovirally marked cells grown in culture are also able to repopulate multiple secondary recipients from the progeny of a single clone. This result indicates that some individual stem cells undergo self-renewal *in vitro*. However, in these experiments, there was actually an

85% net decrease in the total number of stem cells by week 4 of culture. Therefore, while some murine stem cells undergo self-renewal *in vitro*, improvements in culture methods are required to obtain net expansion of their numbers.

Although the existence of a pluri-potent HSC has been established, many of their properties are unknown. The factors that control their replication and differentiation decisions are being hotly pursued in a variety of laboratories. The current state of knowledge of HSC biology has been reviewed recently [25]. Many of the lessons learned from HSC research apply to a growing number of identified tissue specific stem cells [26].

6 The importance of stem cells in tissue engineering

Being able to grow stem cells *ex vivo*, that is, have them divide without differentiation, would theoretically provide an inexhaustible source of stem cells. From such a stem cell supply would eliminate many of the problems with tissue procurement for tissue engineering and implementation of cellular therapies. However, as discussed below stem cells appear to age and this capability may never be fully realized. The most likely way to success is to tap the potential that early stem cells provide, i.e. in the case of hematopoiesis from fetal liver and cord blood. This capability has however not yet been established.

When linear DNA is replicated, the lagging strand is synthesized discontinuously, through the formation of the so-called Okazaki fragments. The last one cannot be initiated, and therefore the lagging strand will be shorter than the leading strand. Linear chromosomes have non-coding repeating sequences on their ends that are called telomeres. These telomeres can be rebuilt using an enzyme called telomerase. This enzyme is active in microorganisms, such as yeast, and has recently been discovered to be active in many cancer cells. Telomerase is a ribonucleoprotein DNA polymerase that elongates teleomeres in eukaryotes.

Normal human cells lack teleomerase activity and their telomeres are shortened by about 50–200 bp per replication. This shortening gives rise to the concept of the so-called mitotic clock. The total length of telomeres in an immature cell is about 11 kbp. When the teleomere length reaches about 5–7 kbp the chromosomes become unstable and replication ceases. This mechanism is believed to underlie the Hayflick limit.

Telomerase activity is found in somatic hematopoietic cells, however at a low activity level. There is evidence that telomeres in immature hematopoietic cells do shorten with ontogeny and with increased cell doublings *in vitro*. The rate of telomere shortening in stem cells is finite, but may be slower than in other somatic cells.

7 The solitary versus the social stem cell

The developmental program requires the systematic and regulated unfolding of the information on the DNA, through coordinated execution of 'genetic subroutines and programs.' Participating cells require detailed information about the activities of their neighbors. Upon completion of organogenesis, the function of fully formed organs is strongly dependent on the coordinated function of multiple cell types. The microenvironment has a major effect on the function of an individual cell; including HSCs. The accessory cell population in all tissues consists of mesenchymal cells (i.e. fibroblasts), monocytes, endothelial cells, a transient lymphocyte population. These accessory cells may be about 30% of the cellularity of tissue, while the parenchymal cells make up the balance.

The complex mechanisms that govern HSC behavior are poorly understood [27]. Although a number of cytokines have been identified that influence HSC behavior, the expansion of purified HSCs *ex vivo* using these cytokines lead to the loss of the ability to reconstitute hematopoietic function [28, 29]. On the other hand it has been shown that cultivating HSCs with their natural accessory cell population does lead to engraftment both in animals and humans [13, 28]. Preformed irradiated stroma only restores partially the function that is lost during HSC selection procedures [30]. Furthermore, it has been shown that the ratio of hematopoietic to accessory cell found in freshly aspirated bone marrow cell populations gives the best performance in culture [31] and their presence reduces the troublesome donor-to-donor variations in culture performance [32].

It thus appears, that to be able to control HSC fate decisions, one has to provide an adequately reconstituted bone marrow microenvironment. This conclusion is an important one for tissue engineering; namely to be able to advance the development and implementation of new hematopoietic cell therapies one needs to engineer the function of the whole tissue rather than to try to manipulate the behavior of HSCs in isolation using defined growth factor combinations.

8 Summary

A number of experimental and established hematopoietic cell therapies exist. The newer hematopoietic cell therapies require cell selection, manipulation, and growth. Several important engineering challenges exist for the implementation of these cellular therapies. Stem cells play a key role in all hematopoietic cell therapies and tissue engineering. As for tissue engineering in general, hematopoietic cell therapy applications are likely to focus on stem cell biology. Procuring stem cells at an early ontological age and learning how to control their replication and differentiation programs is likely to represent the key challenges in hematopoietic tissue engineering, as well as for tissue engineering in general.

References

1. Skalak, R., C. and Fox, F. eds. *Tissue Engineering*. Liss, New York, 1988.
2. Langer, R. and Vacanti, J.P., Tissue engineering. *Science*, 1993, **260**, 920–926.
3. Bell, E., Tissue engineering: A perspective. *Journal of Cellular Biochemistry*, 1991, **45**, Entire Volume.
4. Heineken, F. G. and Skalak, R., Tissue engineering: A brief overview. *Journal of Biomechanical Engineering*, 1991, **113**, 111.
5. Hubbell, J. A., Palsson, B. O. and Papoutsakis, E. T., Special issues on tissue engineering and cell therapies. *Biotechnology and Bioengineering*, 1994, **43**(7 and 8).
6. Hubbell, J. A. and Langer, R., Tissue engineering. *Chemical and Engineering News*, 1995 (March 13), p. 42.
7. Miller, W. M. and Peshwa, M. V., Special issues on tissue engineering, bioartificial organs and cell therapies. *Biotechnology and Bioengineering*, 1996, **50**(4 and 5).
8. Nerem, R. M., Tissue engineering in the USA. *Medical and Biological Engineering and Computing*, 1992, **30**, CE8–12.
9. Koller, M. R. and Palsson, B. O., Tissue engineering: Reconstitution of human hematopoiesis ex vivo. *Biotechnology and Bioengineering*, 1993, **42**, 909–930.
10. Emerson, S. G., Ex vivo expansion of hematopoietic precursors, progenitors and stem cells: the next generation of cellular therapeutics. *Blood*, 1996, **87**(8), 3082–3088.
11. McAdams, T. A., Miller, W. M. and Papoutsakis, E. T., Hematopoietic Cell culture therapies (Part I): Cell culture considerations. *Trends in Biotechnology*, 1996, **14**(September), 341–349.
12. McAdams, T. A. et al., Hematopoietic cell culture therapies (Part II): Clinical aspects and applications. *Trends in Biotechnology*, 1996, **14**(October), 388–396.
13. Stiff, P. J., O. D., Hsi, E., Chen, B., Douville, J. W., Burhop, S., Bayer, R., Peace, D., Malhotra, D., Kerger, C., Armstrong, R. D. and Muller, T. E., Successful hematopoietic engraftment following high dose chemotherapy using only ex-vivo expanded bone marrow grown in Aastrom (stromal-based) bioreactors. *Proceedings of the American Society of Clinical Oncology*, 1997, **16**(88a).
14. Long, M. W., Blood cell cytoadhesion molecules. *Experimental Hematology*, 1992, **20**, 288–301.
15. Wineman, J., Moore, K., Lemischka, I. and Muller-Sieburg, C., Functional heterogeneity of the hematopoietic microenvironment: Rare stromal elements maintain long-term repopulating stem cells. *Blood*, 1996, **87**(May 15), 4082–4090.
16. Muller-Sieburg, C. E. and E. Deryugina, The stromal cells' guide to the stem cell universe. *Stem Cells*, 1995, **13**, 477–486.

17. Deryugina, E. I. and Muller-Sieburg, C. E., Stromal cells in long-term cultures: Keys to the elucidation of hematopoietic development? *Critical Reviews in Immunology*, 1993, 13, 115–150.

18. Schwartz, R. M., Palsson, B. O. and Emerson, S. G., Rapid medium perfusion rate significantly increases the productivity and longevity of human bone marrow cultures. *Proceedings of the National Academy of Science USA*, 1991, **88**, 6760–6764.

19. Peng, C. A. and Palsson, B. O., The importance of non-homogeneous concentration distributions near walls in tissue engineering bioreactors. *Industrial and Engineering Chemistry Research*, 1995, **34**(Transport Phenomena), 3239–3245.

20. Peng, C. and Palsson, B. O., Cell growth and differentiation on feeder layers is predicted to be influenced by bioreactor geometry. *Biotechnology and Bioengineering*, 1996, **50**, 479–492.

21. Sandstrom, C. E. and Miller, W. M., Development of novel perfusion chamber to retain nonadherent cells and its use for comparison of human 'mobilized peripheral blood mononuclear cell cultures with and without irradiated bone marrow stroma. *Biotechnology and Bioengineering*, 1996, **50**, 493–504.

22. Armstrong, R. D., Ogier, W. C. and Maluta, J., Clinical systems for the production of human cells and tissues. *Bio / technology*, 1995, **13**, 449–451.

23. Sigal, S. and Brill, S., Invited review: The liver as a stem cell and lineage system. *American Journal of Physiology*, 1992, **263**, 139–148.

24. Zon, L. I., Developmental biology of hematopoiesis. *Blood*, 1995, **86**, 2876–2891.

25. Morrison, S. J. and Shah, N. M., Regulatory mechanisms in stem cell biology. *Cell*, 1997, **88**(February), 287–298.

26. Potten, C. S., *Stem cells*, Academic Press Limited, New York, 1997.

27. Ogawa, M., Differentiation and proliferation of hematopoietic stem cells. *Blood*, 1993, **81**, 2844–2853.

28. Knobel, K. M. and McNally, M. A., Long-term reconstitution of mice after ex vivo expansion of bone marrow cells: Differential activity of cultured bone marrow and enriched stem cell populations. *Experimental Hematology*, 1994, **22**, 1227–1235.

29. Peters, S. O. and Kittler, E. L., Murine marrow cells expanded in culture with IL-3, IL-6, IL-11, and SCF acquire an engraftment defect in normal hosts. *Experimental Hematology*, 1995, **23**, 461–469.

30. Koller, M. R., Palsson, M. A., Manchel, I. and Palsson, B. O., Long-term culture-initiating cell expansion is dependent on frequent medium exchange combined with stromal and other accessory cell effects. *Blood*, 1995, **86**, 1784–1793.

31. Koller, M. R., Manchel, I. and Palsson, B. O., *Importance of parenchymal:stromal ratio for ex vivo reconstitution of Human Hematopoiesis*. submitted, 1996.

32. Koller, M. R., Manchel, I., Brott, D. A. and Palsson, B. O., Donor-to-donor variability in the expansion potential of human bone marrow cells is reduced by accessory cells but not by soluble growth factors. *Experimental Hematology*, 1996, **24**, 1484–1493.

CHAPTER III.7

The Regeneration of Skeletal Tissues With Mesenchymal Stem Cells

ARNOLD I. CAPLAN
Skeletal Research Center,
Biology Department,
Case Western Reserve University,
2080 Adelbert Road,
Cleveland, OH 44106, USA

DAVID J. FINK,
SCOTT P. BRUDER,
RANDELL G. YOUNG
Osiris Therapeutics, Inc.,
2001 Aliceanna Street,
Baltimore, MD 21231, USA

1 Introduction

1.1 Development of mesenchymal tissues

The adult mesenchymal tissues (bone, cartilage, tendon, ligament, muscle, etc.) have distinct molecular compositions and architectures that are optimized for their unique mechanical functions. Yet all of these tissues are derived from relatively common mesenchymal progenitor cells in the mesodermal layer of the developing embryo. The genesis of these mesenchymal tissues involves several basic principles:

1. Differentiated tissues are derived from a highly cellular progenitor tissue, generally termed the *anlage*, composed of apparently uniform, undifferentiated mesenchymal cells that are housed within a relatively simple extracellular matrix (ECM), which has a high water content [1].

2. Differentiation of a particular mesenchymal tissue involves the establishment of distinct boundaries and the subsequent differentiation of that tissue within those boundaries [2]. Although these

boundaries are definitive and physical, they also involve a specialized series of signaling genes exemplified by the homeotic genes of *Drosophila* and mouse [3].

3. Continued growth and development of each of these tissues involve the continual proliferation of cells and the expansion of the ECM, which becomes complex and specialized in the process of maturation [1].

4. The matrix-to-cell ratio continually increases during maturation of each skeletal tissue, which leads to specialized tissues with very low cellularity compared to organs like the liver, pancreas or kidney.

5. Each embryonic tissue gradually changes its molecular composition and architecture during maturation, which leads to tissue-specific adult isoforms [4].

6. During the maturation and modulation process, these tissues become functionally integrated with their connecting tissues through neural and mechanical elements, which themselves continually change until the adult configuration is established.

These observations are common to all of the mesenchymal tissues. It is the hypothesis of our laboratories that these principles can serve as important guides for the design and engineering of *regenerated* mature mesenchymal tissues by the recapitulation of embryonic events.

1.2 Intrinsic repair of mature mesenchymal tissues

The sequence of events involved in adult tissue repair following injury has evolved to bring about a focused and rapid process to attain at least partial function of the tissue. A number of common features are involved in repair of traumatized mesenchymal tissues [5]:

1. Injuries to mesenchymal tissues elicit an intrinsic *stasis* response that floods the injury site with fluids and cells to protect against infection and long-term damage. This stasis response is followed by resorptive cells that cleanse the injury site in anticipation of the repair events. These acute events are generally characterized as the *inflammatory* response of the tissue.

2. Each tissue also presents an intrinsic *segregation* response that focuses the tissue damage and limits its spread, generally based on

vascular limits. This response also sets limits and boundaries for tissue repair.

3. Injured mesenchymal tissues then undergo an intrinsic *repair* or *wound closure* response that involves the filling of the site with matrix-forming cells, which secrete an ECM that rapidly spans the defect and binds the edges of the tissue together. The formation of a repair blastema or scar track produces an interim, highly cellular tissue and results in partial function of the tissue.

4. Finally, every mesenchymal tissue conducts an intrinsic *remodeling* response that leads to the resculpting of the over-abundant blastema or scar tissue to a more normal tissue composition and architecture. This remodeling capacity is a reflection of the normal turnover dynamics of each tissue, and it is sensitive to the normal homeostatic cues that re-establish the normal tissue geometry, cellularity and limits. The remodeling process is slow, it is physical activity- and age-related, and it is generally influenced by the physiology of the animal.

1.3 Approach to tissue reconstruction by a regenerative process

Our approach for effecting successful reconstruction of injured or pathologic lesions in mesenchymal tissues is to foster developmental or regenerative processes and, at the same time, to inhibit or to repress the intrinsic repair process. Principles involved in this approach include:

1. Simulate the first stage of embryonic development by introducing a *synthetic anlage* to the injured site. This implant contains a high concentration of mesenchymal progenitor cells in a loose 'embryonic-like' matrix. This structure also serves to fill the lesion and thereby inhibit the progression of the intrinsic repair process and minimize production of disorganized scar tissue. The progenitor cells are derived from the animal's own marrow and have been mitotically expanded in culture. These cells are referred to as *mesenchymal stem cells* (MSCs) because they have the capacity to differentiate into a number of different mesenchymal tissues [6].

2. Allow the implanted progenitor cells in the *synthetic anlage* to recapitulate the sequence of tissue morphogenesis and form a mature tissue *in situ* as this sequence proceeds. This development is influenced by a variety of factors, including:
   - the mechanical environment of the tissue
   - the age of the animal, which affects its regenerative capacity

- local cytokines and growth factors
- systemic factors

3. Segregate the developing tissue from surrounding tissues by establishing definitive boundaries within which tissue development occurs. It is likely that physical and/or biological factors, such as tissue mechanical integrity, vascular supply or cell nutrition, will play prominent roles in establishing such boundaries for tissue repair.

1.4 Isolation of mesenchymal stem cells (MSCs)

We have reported methods for the isolation, *in vitro* expansion and characterization of mesenchymal stem cells from tissues rich in progenitor cells, such as periosteum or bone marrow, and from many species, including chick, mouse, rat, rabbit, dog, sheep and man [6–10]. MSCs isolated by these methods appear to be progenitor cells capable of differentiating into a variety of mesenchymal tissue-producing cells including osteoblasts, chondrocytes, tendon fibroblasts, myoblasts, stromal fibroblasts and adipocytes. These cells proliferate in culture without loss of their ability to form a mature differentiated phenotype [10]. When harvested from bone marrow, they are therefore considered to offer certain advantages for tissue regeneration. These cells can serve as a practical source of autologous cells with multi-lineage potential and consequently eliminate the need for isolating committed cells from a sacrificial tissue, which minimizes harvest site morbidity. As the cells produce new tissue, they have the potential to recapitulate the embryonic lineage transitions involved in the formation of each of the skeletal tissues as described in the sections above.

In the following sections, our initial efforts to engineer skeletal tissue regeneration will be illustrated by examples from three diverse tissues: bone, cartilage and tendon.

2 Bone regeneration

We have investigated the use of marrow-derived MSCs for the repair of large bone defects in the spine and the femur of experimental animals. A variety of cell delivery vehicles have been evaluated, and selection of the ideal carrier for repair of such local defects is based on several criteria. First, the material should allow uniform loading and retention of MSCs. Second, the carrier should support rapid vascular ingrowth. Third, the matrix should be composed of materials which are incorporated into the normal remodeling cycle of bone turnover. Fourth, the material should encourage or enhance osteoconductive bridging of host bone with the developing donor tissue. And fifth, the cell–matrix combination should be easy for the physician to handle in a clinical setting.

Table 1. Average bone fill (standard deviation) in HA/TCP implants as a percentage of available space

	Carrier alone	Carrier + marrow	Carrier + MSCs	Carrier + BMP
4 Weeks	2.3 (1.5)	2.9 (1.7)	19.3* (3.7)	21
8 Weeks	10.4 (2.4)	17.2 (6.0)	43.3* (7.7)	22

Histomorphometric measurements were obtained on the bone formed within the confines of a segmental resection in rat femurs, excluding the implant material itself and the medullary canal; $N = 3$, except for the data derived from [13]. * = significantly greater ($p < 0.01$) than other groups at the corresponding time points according to *post hoc* Student-Newman-Keuls tests.

One material which fulfills these criteria is porous hydroxyapatite/tricalcium phosphate (HA/TCP) ceramic. We have recently shown that syngeneic rat MSCs loaded into an HA/TCP cylinder are able to regenerate bone in a critical-sized segmental femoral defect [11, 12]. Bone rapidly forms throughout the biomatrix as a result of the osteoblastic differentiation of the implanted cells. As noted in Table 1, quantitative histomorphometry documents that implanted MSCs produce significantly more bone than either whole marrow or purified bone morphogenetic protein loaded into the same matrix [13]. Additionally, this biomatrix supports the bridging and infiltration of host-derived osteogenic elements at the bone–ceramic interface. Using the same critical-sized defect in an immunocompromised rat model, we have demonstrated that human MSCs possess the same osteo-regenerative potential, and that they fabricate biomechanically stable bone within 12 weeks [14].

As an extension of these studies, we are in the process of developing biomechanically stringent, large animal models for the evaluation of MSC-based bone regeneration methods. Thus far, we have established techniques for the isolation and purification of canine marrow-derived MSCs, and identified several cell–matrix formulations which support osteoblastic differentiation when implanted into autologous hosts [15]. Currently, we are investigating the use of autologous canine MSCs for the augmentation of bone formation in spinal fusion and in the repair of large segmental defects.

Together, these experiments illustrate that the process of tissue regeneration with MSCs simulates the events of bone morphogenesis and bypasses the early events of natural bone repair. The implantation of MSCs serves to immediately provide biosynthetic cellular machinery, which normally takes several weeks to develop in a traditional repair blastema. This bone results from the direct conversion of the undifferentiated precursors into osteoblasts, which secrete and mineralize their extracellular matrix. The pores within the HA/TCP carrier become filled with bone, whose synthesis is oriented, in part, by the invading vasculature. The expectation is that the matrix will eventually be resorbed by osteoclastic activity and replaced by host-derived bone. In contrast, HA/TCP implants that did not contain MSCs were incapable of healing the bone defect. Radiographs and histologic sections not only show a lack of new bone formation, but reveal fracture and crumbling of the ceramic material itself [12, 14].

3 Cartilage regeneration

The use of MSCs isolated from periosteum or bone marrow has also been investigated for the repair of full-thickness osteochondral defects in the femoral condyles of rabbits [16]. Because collagen gels have been used successfully as delivery vehicles in the transplantation of cells and are of low antigenicity, MSCs were embedded into bovine skin type I collagen gels. These cellular grafts were then implanted into large (3 mm × 6 mm × 3 mm-deep), full-thickness defects in the weight-bearing articular surfaces of rabbit knees. The sequential repair of these full-thickness defects was studied by histological and biomechanical evaluations from 2 to 24 weeks after implantation of the MSCs.

This analysis demonstrated, in many of the defects, the uniform differentiation of the implanted cells into chondrocytes as early as 2 weeks post-implantation (Fig. 2A). Subsequently, the subchondral plate reformed by means of endochondral replacement of the basal cartilage in a proximal-to-distal direction, which led to nearly complete replacement of subchondral bone at 4 weeks (Fig. 2B). Twenty-four weeks after MSC implantation, the reparative tissue was stiffer and less compliant than the tissue

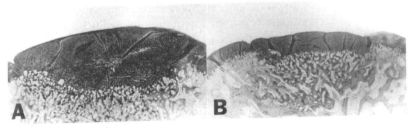

Fig. 1. Photomicrographs of sagittal sections of osteochondral defects treated with bone marrow-derived MSCs, evaluated at 2 weeks (A) and 4 weeks (B) post-implantation. Note early production of *synthetic anlage* (A), followed by gradual remodeling of subchondral bone from below (B). Photos used with permission.

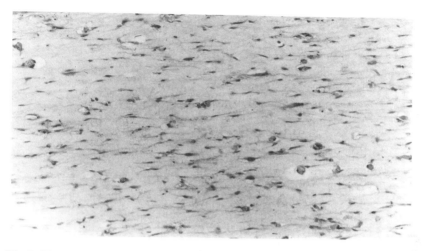

Fig. 2. Photomicrograph of the *in vitro* tendon construct to illustrate orientation of cells and ECM along the direction of the tensile field. Original magnification = 10X.

derived from unfilled defects, but less stiff and more compliant than normal cartilage [16].

4 Tendon regeneration

As noted above, a number of variables, including mechanical fields, may modulate a progenitor cell's potential to recapitulate the sequence of events in tissue morphogenesis. We have tested the hypothesis that implanted MSCs would respond to the mechanical forces exerted upon them in tendon defects, in combination with cues from local tissues, to produce and organize a matrix into a structure consistent with tendon developmental architecture [17].

MSCs maintain a fibroblastic morphology and adhere to plastic in monolayer culture. These same cells, when released via trypsin and suspended in a type I collagen matrix, have been observed to contract that matrix similar to contraction reported for committed human fibroblasts derived from tissues as diverse as skin, ligament, lung, synovium and heart [18, 19].

Autologous, culture-expanded, bone marrow-derived MSCs were suspended in a bovine skin-derived type I collagen matrix and immediately implanted into a gap in rabbit Achilles tendon in which a 1 cm tenotomy had been performed. At 3, 6 and 12 weeks post-implantation, histology was evaluated for cell population density, orientation of the cells and ECM and the structure of the collagen fibers in the ECM. After 3 weeks, nearly all of the implanted cells retained a rounded morphology, which progressed to the typical, elongated tendon fibroblastic shape over the course of 12 weeks. Cell density progressively decreased as the neo-tendon was remodeled. At the earliest time points, MSC-loaded implants contained collagen fibers oriented parallel to the load-bearing axis of the tendon. Those fibers were structured with short crimp periods characteristic of immature tendon fibers. This organization appeared to be advanced compared to the non-MSC containing control implants.

Recent experiments in our laboratories have investigated a progenitor-rich construct produced by MSC-induced contraction of a collagen gel around a biodegradable suture, which is then used to surgically implant the device [20]. The contraction occurred under controlled physical constraints, which caused the gel to be under tension in the load-bearing axis of the implant, parallel to the suture, in a manner similar to the 'ligament equivalent' devices described by Huang et al. [19]. During 24–48 hour culture prior to implantation, the matrix and cells were observed to become oriented in the direction of the tensile load (Fig. 2). These tendon equivalents were then implanted into a rabbit Achilles tendon gap defect and evaluated at 4, 8 and 12 weeks post-implantation [20].

Biomechanical testing performed on these implants demonstrated that repair tissues regenerated from MSC-contracted implants have significantly improved values for all structural and material properties measured relative to contralateral, non-MSC controls. These biomechanical data were supported by histologic observations that MSC-loaded implants have a greater degree of organization of the cells and ECM compared to controls. ECM organization was associated with crimp patterns characteristic of immature 'neo-tendon'.

Continuing refinements of this implant technology will broaden the scope of possible clinical applications and hold promise for improved repair modalities for a variety of soft tissue structures including flexor tendons, cruciate and collateral ligaments of the knee, meniscus and fibrocartilaginous discs.

5 Summary

In considering the options for reconstruction of full or partial lesions in structural tissues, three distinct processes can be identified. *Development* is the embryonic process by which tissues are formed and mature in response to *in situ* chemical and mechanical fields. *Repair* is the intrinsic process by which tissues attempt to minimize the consequences of injury and provide maximum tissue function by the most rapid method. The authors define *regeneration* in this context as the process of reproducing an injured tissue's molecular composition and architecture as completely as possible, so as to return the tissue to normal function. The authors believe that the recapitulation of embryonic developmental processes provides the framework for successful tissue regeneration and restoration of full tissue function.

This chapter outlines a general approach to structural tissue reconstruction that attempts to follow many of the steps observed in tissue development. Generally, this approach involves the implantation of a *synthetic anlage* produced from high densities of mesenchymal stem cells (MSCs) derived from pools of the patient's progenitor cell population in combination with appropriate carrier vehicles that will transfer the cells to the site of implantation. This approach has been illustrated by examples of implants in three diverse tissues: bone, cartilage and tendon.

References

1. Caplan, A. I., The extracellular matrix is instructive. *BioEssay*, 1986, **5**, 129–132.
2. Caplan, A. I., The vasculature and limb development. *Cell Differentiation*, 1985, **16**, 1–11.
3. Dollé, P., Dierich, A., LeMeur, M., Schimmang, T., Schuhbaur, B., Chambon, P. and Duboule, D. Disruption of the Hoxd-13 gene induces localized heterochrony leading to mice with neotenic limbs. *Cell*, 1993, **75**, 431–441.
4. Caplan, A. I., Fiszman, M. Y. and Eppenberger, H. M., Molecular and cell isoforms during development. *Science*, 1983, **221**, 921–927.
5. Caplan, A. I., Cartilage begets bone *versus* endochondral myelopoiesis. *Clinical Orthopaedics and Related Research*, 1990, **261**, 257–267.
6. Caplan, A. I., The mesengenic process. *Clinical Plastic Surgery*, 1994, **21**, 429–435.
7. Haynesworth, S. E., Baber, M. A. and Caplan, A. I., Cell surface antigens on human marrow-derived mesenchymal cells are detected by monoclonal antibodies. *Bone*, 1992, **13**, 81–88.
8. Nakahara, H., Goldberg, V. M. and Caplan, A. I., Culture-expanded periosteal-derived cells exhibit osteochondrogenic potential in porous calcium phosphate ceramic blocks. *Clinical Orthopaedics and Related Research*, 1992, **276**, 291–298.
9. Dennis, J. E. and Caplan, A. I., Differentiation potential of conditionally immortalized mesenchymal progenitor cells from adult marrow of a h-2kb-tsa58 transgenic mouse. *Journal of Cellular Physiology*, 1996, **167**, 523–538.
10. Bruder, S. P., Jaiswal, N. and Haynesworth, S. E., Growth kinetics, self-renewal and the osteogenic potential of purified human mesenchymal stem cells during extensive subcultivation and following cryopreservation. *Journal of Cellular Biochemistry*, 1997, **64**(2), 278–294.
11. Liebergall, M., Young, R. G., Ozawa, N., Reese, J. H., Davy, D. T., Goldberg, V. G. and Caplan, A. I., The effects of cellular manipulation and TGF-β in a composite graft, In: *Bone Formation and Repair*, ed. C. T. Brighton, G. E. Friedlaender and J. M. Lane. American Academy of Orthopaedic Surgeons

Symposium, November 13–16, 1993, Tampa, Florida, Section 4, Chapter 26, pp. 367–378.

12. Kadiyala, S., Jaiswal, N. and Bruder, S. P., Culture-expanded, bone marrow-derived mesenchymal stem cells can regenerate a critical-sized segmental bone defect. *Tissue Engineering*, 1997, in press.

13. Stevenson, S., Cunningham, N., Toth, J., Davy, D. and Reddi, A. H., The effect of osteogenin (bone morphogenetic protein) on the formation of bone in orthotopic segmental defects in rats. *Journal of Bone and Joint Surgery*, 1994, **76**(11), 1676–1687.

14. Bruder, S. P., Kurth, A. A., Shea, M., Hayes, W. C. and Kadiyala, S., Quantitative parameters of human mesenchymal stem cell-mediated bone regeneration in an orthotopic site. *Transactions of the Orthopaedic Research Society*, 1997, **22**, 250.

15. Kadiyala, S., Young, R. G., Thiede, M. A. and Bruder, S.P., Culture-expanded canine mesenchymal stem cells possess osteochondrogenic potential *in vivo* and *in vitro*. *Cell Transplantation*, 1997, in press.

16. Wakitani, S. Goto, T., Pineda, S., Young, R. G., Mansour, J. M., Goldberg, V. M. and Caplan, A. I., Mesenchymal cell-based repair of large full-thickness defects of articular cartilage and underlying bone. *Journal of Bone and Joint Surgery*, 1994, **76**, 579–592.

17. Caplan, A. I., Fink, D. J., Goto, T., Linton, A. E., Young, R. G., Wakitani, S., Goldberg, V. M. and Haynesworth, S. E., Mesenchymal stem cells and tissue repair. In: *The Anterior Cruciate Ligament: Current and Future Concepts*, eds. D. Jackson, S. Arnoczky, S. Woo and C. Frank. Raven Press, New York, 1993, pp. 405–417.

18. Bell, E., Ivarsson, B. and Merrill, C., Production of a tissue-like structure by contraction of collagen lattices by human fibroblasts of different proliferative potential *in vitro*. *Proceedings of the National Academy of Science USA*, 1979, **76**(3), 1274–1278.

19. Huang, D., Chang, T. R., Aggarwal, A., Lee, R. C. and Ehrlich, H.P., Mechanisms and dynamics of mechanical strengthenng in ligament-equivalent fibroblast-populated collagen matrices. *Annals of Biomedical Engineering*, 1993, **21**, 289–305.

20. Young, R. G., Butler, D. L., Weber, W., Gordon, S. L. and Fink, D. J., Mesenchymal stem cell-based repair of rabbit achilles tendon. *Transactions of the Orthopaedic Research Society*, 1997, **22**, 249.

CHAPTER III.8

Hemoglobin-Based Blood Substitutes

ALAN S. RUDOLPH

Center for Biomolecular Science and Engineering,
Naval Research Laboratory,
Washington DC 20375, USA

1 Introduction

Transfusion medicine has intrigued man throughout history. In the last century, as the many physiologic roles of blood have been defined, and fractionation methods to separate blood into its cellular and molecular components have progressed, many blood products which dramatically improve human health have been developed. With increasing transfusion practices in the latter part of this century, the increased incidences of disease transmission and the shortages of blood in urban centers and certain areas of the world, have heightened the search for substitutes which mimic various functions of blood.

Biomimetic research and development of red blood cell function toward development of a resuscitative artificial oxygen carrying fluid has been one active area of blood substitute research. There is clearly a need for such a product as current estimates of red blood cell usage in the United States are between 12 and 13 million units per year. The efforts to develop such a fluid have focused on using the allosteric oxygen binding properties of the protein hemoglobin or the solublization of oxygen by fluorocarbons. Both of these strategies have been fertile areas of research and have resulted in the development of products which have been, and are now being tested in the clinic. The study of the oxygen delivery properties of these solutions have raised new questions over the appropriate physiologic need and delivery of oxygen both systemically and in the microcirculation. In addition, the development of these products have raised new regulatory challenges as these agencies have struggled with the determination of efficacy of these products in spite of the assumption that a plasma expander that carried oxygen was better than one that did not. Because of the difficulty in clearly demonstrating the efficacy of red blood cell transfusion in the clinic, and the differing clinical practices of red blood cell use, indirect demonstrations of efficacy (surrogate end points) through the reduction in the use of allogeneic blood in the clinic have been allowed [1].

There have been numerous proposed uses of a blood substitute. These include elective surgery, emergency surgery and trauma, trauma resuscitation, treatment of anemias, oxygenation of ischemic tissues in stroke, post-angioplasty treatment, peripheral vascular disease, extracorporeal oxygenation in cardiopulmonary bypass, organ preservation solutions, and enhanced oxygen delivery to tumors for irradiation therapies. The ultimate use of any blood substitute will depend on the clinical indication proposed and the physiologic properties of the solution used.

The purpose of this paper is to review the recent research and development in the field of hemoglobin-based blood substitutes. The pursuit of an artificial oxygen carrying fluid has been at times full of promise and disappointment, with a history of successes and failures in the clinic. The realities and perceptions of the safety of the blood supply have at times driven research support and clinical development in the field. In spite of this, with new products now entering late phase clinical trials, there is once again hope that artificial oxygen carrying fluids will soon be available.

2 Design considerations in the fabrication of a hemoglobin-based blood substitute

There are a number of important design considerations in the fabrication of a safe and effective red cell substitute. It is important to recognize that the clinical utility of a red blood cell substitute is likely to require administration of a very large therapeutic dose, with grams of protein or fluorocarbon administrated over a short time period. The final design of the system will of course depend on the clinical use of the substitute, but stringent quality control as well as physiologic effects in these dose ranges must be understood. Design considerations include engineering a substitute that is metabolized by the body but has vascular persistence commensurate with desired oxygen delivery characteristics, does not contain histocompatibility antigens and is therefore blood type free resulting in wide patient acceptability, is free of transmittable antigens, can be scaled-up, and has shelf-life properties consistent with the logistics of standard hospital or military use of blood. For hemoglobin based artificial oxygen carrying fluids, there has been considerable progress in achieving these figures of merit. More recently, with the increased concern about the rising costs of health care administration in the US, there has been additional pressures to design a red cell subsitute that is cost competitive and competes favorably with the fully loaded cost of a tested unit of red cells. These design considerations will be used as a basis to evaluate the recent progress of hemoglobin based blood substitutes.

3 Research issues in the development of hemoglobin-based blood substitutes

Hemoglobin is a tetrameric allosteric protein made up of two α and two β subunits. The two $\alpha\beta$ dimeric subunits are loosely bound and dynamically participate in the reversible binding of oxygen by forming the tense (T) and relaxed (R) structures. The cooperativity of oxygen binding and the large oxygen binding capacity of hemoglobin in blood (the normal hemoglobin content of blood in mammals in 15 g/dl) made the choice of using hemoglobin in the design of blood substitutes natural. Attempts as early as 1920 to use hemolysed blood in transfusion demonstrated toxicities and initiated the search for the mechanism of hemoglobin based toxicities. This search was obscured by the difficult task of purifying the molecule. Many of the toxicities observed have since been attributed to trace stromal or other contaminants [2, 3]. Methods to remove stromal contaminants, viral inactivation, and sterile filtration of hemoglobin improved outcomes following its administration. These efforts allowed more definitive studies on the effects of hemoglobin with new important information relevant to the pursuit of a blood substitute still being generated. The three main strategies of hemoglobin-based blood substitutes are seen in Fig. 1.

3.1 Metabolism, biodistribution, and vascular persistence of native and modified hemoglobins

The loose binding of the dimeric subunits of native tetrameric hemoglobin

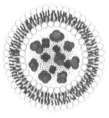

Liposome Encapsulated
Hemoglobin

Polymerized
Hemoglobin

Chemically Modified
Tetrameric Hemoglobin

Fig. 1. Three most prominent strategies for hemoglobin-based blood substitutes.

has been an important characteristic in the biodistribution and metabolism of hemoglobin-based red cell substitutes. Early attempts to transfuse hemoglobin showed a rapid dissociation of the tetramer and clearance of the dimeric subunits in the kidney, with precipitation in the proximal tubulae resulting in significant nephrotoxicity observed in animal studies [6–8]. This observation was dose dependent as the mechanism of hemoglobin metabolism involves opsonization with serum proteins (haptoglobin, hemopexin, albumin), uptake by the liver and transport to microsomes where heme oxygenase initiates its conversion to bilirubin. With high dose administration, this system is saturated resulting in filtration by the kidney. Early quantitative biodistribution studies with native hemoglobin show the majority in the renal cortex and renal medulla, with significant distribution to the liver, spleen, marrow, and lymph, indicating a considerable extravasation of the protein [9]. Much debate on the effects of hemoglobin has focused on the purification and preparation of the protein as the large dose administrations amplified small impurities present in the solutions. Purification methods introduced by Rabiner et al., 1967 demonstrated that infusion of purified hemoglobin into cynomologous monkeys was not nephrotoxic, while stroma from the same preparation caused renal failure [10, 11]. In spite of this, the clinical trial with purified native tetrameric hemoglobin of Savitsky in 1978 was a major set back for the field as human volunteers showed a fall in urinary output and creatinine clearance following the infusion of small doses of purified hemoglobin [12].

The results of the Savitsky clinical trial led investigators in the early 1980s to explore protein engineering as a means to modify hemoglobin using chemical and recombinant techniques with the purpose of stabilizing the tetramer and preventing dissociation into dimers with subsequent clearance by the kidneys. Chemical modification to intramolecularly crosslink the dimer subunits of hemoglobin or attach modifiers of oxygen carrying capacity, resulting in a stabilized 64 kDa protein, have been demonstrated by many groups with a variety of cross-linking agents [13–15]. These reactions take advantage of various reactive sites on the surface, internal oxygen binding cavity, and amino terminus of hemoglobin. These agents include aldehydes, pyridoxyl derivatives, trimesic acid, polyethylene glycol, and dextrans. The use of the non-specific aldehydes also results in the formation of intermolecular cross-links resulting in a heterogenous distribution of higher molecular weight polymerized hemoglobin species [16]. Recombinant techniques have also been used to express intramolecularly cross-linked hemoglobin [17].

These modifications result in an increased vascular persistence of hemoglobin and reduced filtration by the kidney, alleviating one of the serious toxicities associated with native hemoglobin infusion [18]. The circulation half-life of the modified hemoglobins is species and dose dependent, but has been reported between 8 and 12 hours, while polymerization extends the half-life to 12–16 hours [19]. The biodistribution of chemically stabilized tetrameric hemoglobin may still result in a significant level of extravasated protein which may play a major role in the physiologic effects of hemoglobin. An alternative strategy to shifting the biodistribution and metabolism of native or chemically modified hemoglobins which has recieved more limited attention is the encapsulation of hemoglobin in biodegradable delivery vehicles [20]. The encapsulation of hemoglobin in phospholipid vesicles has been shown to shift the metabolism of hemoglobin to the reticuloendothelial system, specifically sites of phagocytic removal, the liver and spleen [21]. These studies also indicated that there is minimal (< 3%) extravasation of free hemoglobin when encapsulated in 0.2 μm or larger vesicles.

3.2 Interactions of hemoglobin with cellular and molecular species

The interactions of hemoglobin with cellular and molecular species in the vascular compartment also plays a dominant role in the safety of efficacy of hemoglobin-based blood substitutes. Over the last decade, as chemical modifications demonstrated reduced renal handling of hemoglobin, one focus of much work has been on direct effects of hemoglobin in the vascular compartment, with observations of hemoglobin induced vasoactivity manifest as systemic and pulmonary hypertension following the administration of native or chemically modified hemoglobins in animals and humans [22–25], and direct effects on endothelial cells due to the free radical damage from the uptake of heme and iron [26, 27].

The discovery of endothelium-derived relaxation factor as nitric oxide (NO) suggested a possible mechanism for the observed vasoconstrictor response of hemoglobin as hemoglobin is a potent ligand of NO [28]. While there is some debate on changes in flow in vessels and the clinical significance of the hypertension caused by hemoglobin-NO interactions, the extravasation of the chemically modified 64 kDa forms of hemoglobin is thought to result in the observed lack of vasorelaxation and resultant hypertension. The effects of hemoglobin on vascular flow in the microcirculation has been documented as a decrease in the functional capillary density in mesenteric hamster skin fold model [29] while other groups have shown that chemically modified hemoglobin does not alter the flow to various organs in the rat [30]. This loss of functional vasculature as a result of hemoglobin-induced vasoconstriction has also been observed in a high flow tissues such as the vascular plexus of the rabbit eye [31]. Larger molecular forms of hemoglobin, either through polymerization or encapsulation may not extravasate to the same extent. Encapsulation of chemically modified hemoglobin was recently shown to result in significantly less vasoconstrictor activity (60–100 fold) than cell free hemoglobin (chemically modified) in an isolated vessel segment model (Fig. 2) [32]. The encapsulation of hemoglobin also has been shown to reduce hemoglobin induced impaired flow in the choroidal plexus of the rabbit eye compared to equivalent concentrations of chemically cross-linked 64 kDa hemoglobin [31]. Preliminary results in our laboratory suggest that the lipid bilayer in liposome encapsulated hemoglobin is also a barrier to heme transfer and may reduce heme oxygenase activity in endothelial cells exposed to encapsulated hemoglobin.

The binding of NO by hemoglobin may have consequences for other cell and subcellular elements as well which may have import in surgical settings. Recent reports have examined the effect of hemoglobin on platelet deposition as reduced NO availability increases platelet deposition. In a rat carotid endarterectomy model, infusion of hemoglobin (0.9 g/kg) increased plaletet deposition by 71% [34]. In some surgical procedures, enhanced platelet deposition may be advantageous but this effect warrants further investigation. Hemoglobin binding of NO produced by macrophages as a result of upregulated iNOS activity in bacteremia or endotoxemia models has been another area of active investigation [35, 36]. This activity has resulted in the exploration of chemically modified hemoglobin as an adjunct therapy for sepsis [37]. Finally, there has been recent important evidence that hemoglobin may play a regulatory role in shuttling NO from the lungs to the tissue based on the reactivity of hemoglobin cystein groups and the formation of S-nitrosylthiols [38]. How this proposed activity of hemoglobin may play a role in the development of a hemoglobin based blood substitute remains to be determined.

Hemoglobin has been proposed to be a binding protein for endotoxin [39]. Enhanced endotoxin activity in the limulus ameoba lysate assay has been demonstrated (same). The *in vivo* consequences of such binding has been the subject of some debate. Early reports of hemoglobin exacerbating endotoxemias in animal models were performed with solutions that were suspect for stromal impurities [40, 41]. More recent studies have also demonstrated exacerbated septic consequences in mice that have been infused with pyridoxylated polymerized hemoglobin [42]. Additional considerations in this effect is the nutritional supplement provided by the large dose infusion of the iron containing protein. While encapsulation may limit the exposure of hemoglobin to endotoxin, the liposome surface of liposomal encapsulated hemoglobin has been shown to interact and bind endotoxin as a result of surface adherent hemoglobin [43, 44]. The phagocytosis of liposome encapsulated hemoglobin has also been shown to reduce the inflammatory cytokine response from macrophages exposed to endotoxin [45, 46]. The clinical significance of this effect has not been addressed.

3.3 Oxygen delivery

The standard transfusion practice of giving patients packed red cells if their hemoglobin drops below 10 g/dl is based on the assumption that an O_2 reserve is needed and can only be maintained by increasing hemoglobin content. With the blood safety issues of the 1980s these practices have been re-examined in the context of the 'transfusion trigger'. The development of hemoglobin based blood substitutes has also presented the clinician with a new set of issues. The point distribution of the oxygen carrying unit is no longer a 6–7 μm package, but a 64 kDa protein. The transit time of this protein is relatively short (on the order of hours) compared to days for transfused red cells. The optimal oxygen carrying capacity of the protein in the microcirculation is also unknown. In spite of these issues, the most important characteristic of a hemoglobin based blood substitute is the ability of the solution to participate in the cycling of oxygen and carbon dioxide. Removal of hemoglobin from the red blood cell results in a significant decrease in the oxygen carrying capacity (P_{50}) of the hemoglobin due to the loss of DPG. In spite of this, there were numerous early studies demonstrating that native bovine hemoglobin when infused into animals delivers oxygen [47,48]. The derivatization of hemoglobin with pyridoxyl phosphates (and other organic phosphates) results in an increase in P_{50}, yet there is much debate on what is the optimal oxygen carrying capacity, which may depend on the clinical indication desired. Early demonstrations of efficacy (as defined by oxygen delivery) with purified native hemoglobin preparations also showed that animals could be kept alive after total exchange of red cells with hemoglobin solutions [49, 50]. A few studies with chemically modified or encapsulated hemoglobin have measured direct oxygenation of tissues with oxygen electrodes, or by oxygen-dependent quenching of phosphorescence emitted by metalloporphyrins bound to albumin, and demonstrated oxygen delivery in hemodilution and exchange transfusion models [51, 52]. The encapsulation of hemoglobin does not significantly alter the oxygen binding characteristics of hemoglobin and offers the advantage of cocompartmentalizing modifiers of hemoglobin in order to engineer the desired oxygen carrying capacity. The distribution of oxygen in the microvascular bed has also been a recent focus of attention as the vasoconstrictor properties of hemoglobin have been proposed to limit the oxygen delivery capabilities of these solutions [53]. The alteration in blood viscosity upon red blood cell loss is also thought to be an important determinant in the hemodynamics of oxygen transport in the microvasculature [54].

3.4 Hemoglobin products and clinical trials

Over the last 5 years, the clinical development of hemoglobin-based blood substitute products has progressed considerably. There are at least 6 companies in phase I–III clinical trials (see Table 1). As expected in this stage of development, there are few new reports in the refereed literature on the effects of these solutions on humans. The few reports published appear to suggest that purified, modified hemoglobin is safe at low to moderate doses. Certain side effects of stabilized tetrameric hemoglobin do persist and there is debate on the whether these side effects will limit the utility of these products.

Table 1. Hemoglobin based blood products — current status

Hemoglobin source / treatment	Product	Company	Status
Human hemoglobin purified from outdated, donated blood, cross-linked 64 kDa protein	HemAssist	Baxter International	In pivotal phase II–III in Europe and US in cardiac surgery, stroke and acute blood loss
Bovine hemoglobin heterogenous molecular weight	HemoPure	BioPure	In Phase II. Phase III studies in elective surgery expected to start in 1997
Human hemoglobin purified from otudated, donated blood; cross-linked and polymerized chemically, heterogenous molecular weight	Hemolink	Hemososl / Fresenius	Phase I–II. Phase II trials expected to start soon. Indication is orthopedic surgery in patients who have pre-donated blood (much like hemodilution in intent).
Human hemoglobin purified from outdated, donated blood; chemically polymerized molecular weight > 64 kDa	PolyHeme	Northfield	In late Phase II (randomized and controlled, but not blinded) in acute blood loss (trauma and urgent surgery). Phase III expected soon.
Genetically engineered human hemoglobin produced in E. coli cross-linked 64 kDa hemodilution	Optro	Somatogen	Concluded early Phase II studies, in hemodilution, surgery, and surgical patients who have pre-donated blood, indications include and acute blood loss

In a phase I study of diaspirin cross-linked 64 kDa hemoglobin, a half-life of 3.3 hours at a dose of 100 mg/kg and a dose related increase in lactate dehydrogenase-5-isozyme and serum creatinine kinase was observed in a randomized double blind, cross-over study in 42 healthy adult volunteers [55]. All doses tested were well tolerated but a dose related increase in blood pressure was observed with no evidence of decreased peripheral perfusion at all doses administered. Phase III clinical trials have been initiated with this product. There are other tetrameric hemoglobins stabilized by chemical or recombinant techniques that are also making progress in the clinic with similar side effect profiles. The likely clinical indications for these products will be as adjunct therapies at low to moderate dose application with surrogate efficacy demonstrations that show reduced use of allogeneic blood. Higher molecular weight chemically polymerized human Hb are also making progress in clinical trials specifically aimed at demonstrating efficacy in trauma. Phase I–II clinical trials have been performed and no signficant side effects have been reported. Other products are heterogenous mixtures of cross-linked hemoglobins of varying molecular weight. The larger molecular weight form of hemoglobin is proposed to have much reduced extravasation compared to the 64 kDa form [56]. However, recent data in a random clinical trial on polymerized bovine hemoglobin in patients undergoing pre-operative hemodilution for elective abdominal aortic surgery indicated signficantly increased mean arterial pressure, vascular resistance, and reduced cardiac output, which led the authors to conclude that this preparation impaired oxygen delivery at the dose given (3 ml/kg) [57]. While continuing progress is reported in the popular press, peer-reviewed reports of the efficacy of these products remain to be evaluated. A summary of the status of clinical trials in hemoglobin-based red cell substitutes is seen in Table 1.

4 Future directions in the development of a hemoglobin-based blood substitute

This is an exciting time for the field of blood substitutes as the long awaited promise of a substitute oxygen carrying fluid may soon be realized. These products could dramatically change transfusion practices and offer the clinician new strategies in treating patients that were good candidates for traditional blood therapies. In spite of this progress, the large dose application of hemoglobin-based blood substitutes may not be realized with the first generation products which could limit the clinical utility in unplanned, emergency medicine, especially in military settings where modern hospital practices may not be available to monitor and treat the potential adverse side effects. Research is ongoing into modifications to hemoglobin to widen its clinical utility, reduce observed adverse responses and extend its circulation persistence. The encapsulation of hemoglobin has been actively pursued by a number of investigators as this is clearly a more biomimetic strategy for the delivery of hemoglobin. There is wide agreement that encapsulation of hemoglobin offers considerable potential for extending the utility of hemoglobin, increasing the circulation persistence, and shielding hemoglobin from cellular and molecular interactions. Coencapsulation and surface modification also offer the potential to further engineer an oxygen carrying fluid. To date, however, the advantages of the encapsulation of hemoglobin have not outweighed the increased cost of development, which thus have left very few groups actively developing an encapsulated hemoglobin product. One group is the US Navy which actively has pursued the encapsulation of hemoglobin (and the long-term storage of freeze-dried encapsulated hemoglobin) over the last 15 years. Additional strategies for the development of a second generation hemoglobin-based blood substitute include further efforts to protein engineering of the hemoglobin molecule to alter oxygen binding characteristics and decrease reactivity with nitric oxide.

References

1. Fratantoni, J. C., Demonstration of the efficacy of a therapeutic agent. In *Blood Substitutes: Physiological Basis of Efficacy*, ed. R. M. Winslow, K. D. Vandegriff and M. Intaglietta. Birkhauser, Boston, 1995, pp. 20–24.

2. Rabiner, S. F., Helbert, J. R., Lopas, H. and Friedman, L. H., Evaluation of stroma-free haemoglobin for use as a plasma expander. *Journal of Experimental Medicine*, 1967, **126**, 1127–1142.

3. Feola, M., Simoni, J., Tran, R. and Canizaro, P. C., Mechanisms of toxicity of hemoglobin solutions. *Biomaterials, Artificial Cells and Artificial Organs*, 1988, **16**, 217–226.

4. Baker, S. L. and Dodds, E. C., Obstruction of the renal tubules during the excretion of haemoglobin. *British Journal of Experimental Pathology*, 1925, **6**, 247–260.

5. Ayer, G. D. and Gauld, A. G., Uremia following blood transfusion. *Archives of Pathology*, 1942, **33**, 513–532.

6. Oken, D. E., Arce, M. L. and Wilson, D. R., Glycerol-induced hemoglobinuric acute renal failure in the rat. I. Micropuncture study of the development of oliguria. *Journal of Clinical Investigations*, 1966, **45**, 724–735.

7. Anderson, M. N., Mouritzen, C. V. and Gabrielli, E. R., Mechanisms of plasma hemoglobin clearance after acute hemolysis in dogs: Serum haptoglobin levels and selective deposition in liver and kidney. *Annals of Surgery*, 1966, **164**, 905–912.

8. Birndorf, N. I., Lopas, H. and Robboy, S. J., Disseminated intravascular coagulation and renal failure. *Laboratory Investigations*, 1971, **25**, 314–319.

9. Relihan, M. and Litwin, M. S., Effects of stroma-free haemoglobin solutions on clearance rate and renal function. *Surgery*, 1972, **71**, 395–399.

10. Savitsky, J. P., Doczi, J., Black, J. and Arnold, J. D., A clinical safety trial of stroma-free hemoglobin. *Clinical Pharmacology Therapy*, 1978, **23**, 73–80.

11. Benesch, R. E. and Kwong, S., The stability of the heme-globin linkage in some normal, mutant, and chemically modified hemoglobins. *Journal of Biological Chemistry*, 1990, **265**, 14881–14885.

12. Bucci, E., Razynska, A., Urbaitis, B. and Fronticelli, C., Pseudo-cross link of human hemoglobin with mono-(3,5-dibromosalicyl)-fumarate. *Journal of Biological Chemistry*, 1989, **111**, 777–784.

13. Walder, J. A., Zaugg, R. H., Walder, R. Y., Steele, J. M. and Klotz, I. M., Diaspirins that cross-link α chains of hemoglobin: bis(3,5-dibromosalicyl) succinate and bis(3,5-dibromosalicyl)fumarate. *Biochemistry*, 1979, **18**, 4265–4270.

14. Sehgal, L. R., Rosen, A. L., Gould, S. A., Seghal, H. L. and Moss, G. S., Preparation and in vitro characteristics of polymerized pyridoxylated hemoglobin. *Transfusion*, 1983, **23**, 158–162.

15. Looker, D. D., Abbott-Brown, D., Cozart, P., Durfee, S., Hoffman, S., Mathews, A. J., Miller-Roehrich, J., Shoemakder, S., Trimble, S., Fermi, G., Komiyama, N. H., Nagai, K. and Stetler, G. H., A human recombinant haemoglobin designed for use as a blood substitute. *Nature (Lond)*, 1992, **356**, 258–260.

16. Blantz, R. C., Evan A. P. and Gabbai, F. B., Red cell substitutes in the kidney. In *Blood Substitutes: Physiological Basis of Efficacy*, ed. R. M. Winslow, K. D. Vandegriff and M. Intaglietta. Birkhauser, Boston, 1995, pp. 132–42.

17. Keipert, P. E. and Chang, T. M. S., Effects of partial and total isovolemic exchange transfusion in fully conscious rats using pyridoxylated polyhemoglobin solution as a colloidal oxygen-delivering blood replacement fluid. *Vox Sang*, 1987, **53**, 7–14.

18. Rudolph, A. S., Encapsulation of hemoglobin in liposomes. In *Blood Substitutes: Physiological Basis of Efficacy*, ed. R. M. Winslow, K. D. Vandegriff and M. Intaglietta. Birkhauser, Boston, 1995, **7**, 90–104.

19. Rudolph, A., Spatially controlled adhesion, spreading, and differentiation of endothelial cells on self-assembled molecular monolayers. *Proceedings of the National Academy of Science*, 1991, **91**, 11070–11074.

20. Savitsky, J. P., Doczi, J., Black, J. and Arnold, J. D., A clinical safety trial of stroma-free hemoglobin. *Clinical Pharmacology Therapy*, 1978, 23, 73–80.

21. Gould, S. A., Sehgal, L. R., Rosen, A. L., Sehgal, H. L. and Moss, G. S., The development of polymerized pyridoxylated hemoglobin solution as a red cell substitute. *Annals of Emergency Medicine*, 1986, **15**, 1416–1419.

22. White, C. T., Murray, A. J., Greene, J. R., Smith, D. J., Medina, F., Makovec, G. and Bolin, R. B., Toxicity of human hemoglobin solutions infused into rabbits. *Journal of Laboratory Clinical Medicine*, 1986, **108**, 121–131.

23. Mohanty, M., Kumari, T. V., Rathinam, K. and Vijayakumari, Pulmonary vasospasm in rabbits infused with stroma free hemoglobin solution. *Indian Journal of Experimental Biology*, 1989, **27**, 265–268.

24. Balla, J., Nath, K. A., Balla, G., Juckett, M. B., Jacob, H. S. and Vercellotti, G. M., Endothelial cell heme oxygenase and ferritin induction in rat lung by

hemoglobin in vivo. *American Journal of Physiology (Lung Cell Mol Physiol 12)*, 1995, **268**, L321–L327.

25. Motterlini, R., Interaction of hemoglobin with nitric oxide and carbon monoxide: physiological implications. In *Blood Substitutes: New Challenges*, ed. R. M. Winslow, K. D. Vandegriff and M. Intaglietta. Birkhauser, Boston, 1996, pp. 74–98.

26. Olson, S. B., Tang, D. B., Jackson, M. R., Gomez, E. R., Ayala, B. and Alving, B. M., Enhancement of platelet deposition by cross-linked hemoglobin in a rat carotid endarterectomy model. *Circulation*, 1996, **93**, 327–332.

27. Tsai, A. G., Kerger, H. and Intaglietta, M., Microvascular oxygen distribution: effects due to free hemoglobin in plasma. In *Blood Substitutes: New Challenges Challenges*, ed. R. M. Winslow, K. D. Vandegriff and M. Intaglietta. Birkhauser, Boston, 1996, pp. 124–131.

28. Sharma, A. and Gulati, A., Effect of diaspirin cross-linked hemoglobin and norepinephrine on systemic hemodynamics and regional circulation in rats. *Journal of Laboratory Clinical Medicine* 1994, **123**, 299–308.

29. Rudolph, A. and Flower, R., Effects of hemoglobin-based blood substitutes on vascular plexus hemodynamics. *Artificial Cells, Blood Substitutes and Immobilization Biotechnology*, 1996, **24**, 337.

30. Szebeni, J., Wassef, N. M., Hartman, K. R., Rudolph, A. S. and Alving, C.R., Complement activation in vitro by the red cell substitute, liposome-encapsulated hemoglobin-mechanism of activation and inhibition by soluble complement receptor type 1. *Transfusion*, 1997, **37**(2), 150–159.

31. Flower, R. W. and Rudolph, A. S., Effects of hemoglobin-based blood substitutes on vascular plexus hemodynamics. *Artificial Cells, Blood Substitutes, and Immobilization Biotechnology*, 1996, 337.

32. Olsen, S. B., Tang, D. B., Jackson, M. R., Gomez, E. R., Ayala, B. and Alving, B. M., Enhancement of platelet deposition by cross-linked hemoglobin in a rat carotid endarterectomy model. *Circulation*, 1996, **93**, 327–332.

33. Heneka, M. T., Loschmann, P. A. and Osswald, H., Polymerized hemoglobin restored cardiovascular and kidney function in endotoxin induced shock in the rat. *Journal of Clinical Investigation*, 1997, **99**, 47–54.

34. Otterbein, L., Sylvester, S. L. and Choi, A. M., Hemoglobin provides protection against lethal endotoxiema in rats: the role of heme oxygenase-1. *American Journal of Respiratory Cell and Molecular Biology*, 1995, **13**, 595–601.

35. Mourelatos, M. G., Enzer, N., Ferguson, J. L., Ripins, E. B., Burhop, K. E. and Law, W. R., The effects of disaspirin cross-linked hemoglobin in sepsis. *Shock*, 1996, **5**, 141–148.

36. Jia, L., Bonaventura, J. and Stamler, J. S., S-nitrosohaemoglobin: a dynamic activity of blood involved in vascular control. *Nature*, 1996, **380**(6571), 221–226.

37. Kaca, W. and Roth, R. I., Activation of complement by human hemoglobin and by mixtures of hemoglobin and bacterial endotoxin. *Biochemica Biophysica Acta*, 1995, **1245**, 49–56.

38. White, C. T., Murray, A. J., Smith, D. J., Greene, J. R. and Bolin, R. B., Synergistic toxicity of endotoxin and hemoglobin. *Journal of Laboratory Clinical Medicine*, 1996, **108**, 132–137.

39. Hoyt, D. B., Greenburg, A. G., Peskin, G. W., Forbes, S. and Reese, H., Hemorrhagic shock and resuscitation: improved survival with pyridoxylated

stroma-free hemoglobin. *Surgical Forum*, 1980, 31, 15–17.

40. Griffiths, E., Cortes, A., Gilbert, N., Stevenson, P., MacDonald, S. and Pepper, D., Haemoglobin-based blood substitutes and sepsis. *Lancet*, 1995, 345(8943), 158–160.

41. Cliff, R. O., Kwasiborski, V. and Rudolph, A. S., A comparative study of the accurate measurement of endotoxin in liposome-encapsulated hemoglobin. *Artificial Cells, Blood Substitutes, and Immobilization Technology*, 1995, 23(3), 331–336.

42. Leslie, S. B., Puvvada, S., Ratna, B. R. and Rudolph, A. S., Encapsulation of hemoglobin in a bicontinuous cubic phase lipid. *Biochemica Biophysica Acta*, 1996, 1285, 246–254.

43. Langdale, L. A., Maier, R. V., Wilson, L., Pohlman, T. H., Williams, J. G. and Rice, C. L., Liposome encapsulated hemoglobin inhibits tumor necrosis factror release from rabbit alveolar macrophages by a posttranscriptional mechanism. *Journal of Leukocyte Biology*, 1992, 52, 679–686.

44. Rudolph, A. S., Cliff, R., Kwasiborski, V., Neville, L., Abdullah, F. and Rabinovici, R., Liposome encapsulated hemoglobin modulates lipopolysaccharide-induced tnf-α production in mice. *Critical Care Medicine*, 1997, (in press).

45. Buttle, G. A. H., Kekwick, A. and Schweitzer, A., Blood substitutes in treatment of acute hemorrhage. *Lancet*, 1940, 2, 507–510.

46. Amberson, W. R., Jennings, J. J. and Rhode, C. M., Clinical experience with hemoglobin-saline solutions. *Journal of Applied Physiology*, 1949, 1, 469–489.

47. Palani, C. K., DeWoskin, R., Michuda, M., Rosen, A. L. and Moss, G. S., In vivo oxygen transport with a stroma-free hemoglobin solution in primates. *Surgical Forum*, 1974, 25, 199–201.

48. DeVenuto, F., Friedman, H. I. and Mellick, P. W., Massive exchange transfusions with crystalline hemoglobin solution and blood volume. *Surgery*, 1980, 151, 361–365.

49. Rabinovici, R., Rudolph, A. S., Vernick, J. and Feuerstein, G., A salutary resuscitative fluid: liposome encapsulated hemoglobin/hypertonic saline solution. *Journal of Trauma*, 1993, 35(1), 121–127.

50. Tsai, A. G., Kerger, H. and Intaglietta, M., Microcirculatory consequences of blood substitution with $\alpha\alpha$-hemoglobin. *Blood Substitutes: Physiological Basis of Efficacy*, ed. R. M. Winslow, K. D. Vandegriff and M. Intaglietta. Birkhauser, Boston, 1995, 11, 155–174.

51. Tsai, A. G., Kerger, H. and Intaglietta, M., Microvascular oxygen distribution: effects due to free hemoglobin in plasma. *Blood Substitutes: New Challenges* ed. R. M. Winslow, K. D. Vandegriff and M. Intaglietta. Birkhauser, Boston, 1996, 8, 124–131.

52. Johnson, P. C., Richmond, K., Shonat, R. D., Toth, A., Pal, M., Tischler, M., Tischler, E. and Lynch, R. M., Oxygen delivery regulation: implications for blood substitutes. *Blood Substitutes: Physiological Basis of Efficacy*, ed. R. M. Winslow, K. D. Vandegriff and M. Intaglietta. Birkhauser, Boston, 1995, 12, 175–186.

53. Pryzebelski, R. J., Daily, E. K., Kisicki, J. C. et al., Phase I study of the safety and pharamacologic effects of diaspirin cross-linked hemoglobin solution. *Critical Care Medicine*, 1996, 24, 1993–2000.

54. Gould, S. A. and Moss, G. S., Clinical development of human polymerized hemoglobin as a blood substitute. *World Journal of Surgery,* 1996, **20**(9), 1200–1207.

55. Kasper, S. M., Walter, M., Grune, F., Bischoff, A., Erasmi, H. and Buzello, W., Effects of a hemoglobin-based oxygen carrier (hboc-201) on hemodynamics and oxygen transport in patients undergoing preoperative hemodilution for elective abdominal aortic surgery. *Anesthesia and Analgesia,* 1996, **83**(5), 921–927.

CHAPTER III.9

Muscle Tissue Engineering

PATRICIA PETROSKO[1],
TAHSIN OGUZ ACARTURK[1],
PAUL A. DIMILLA[2],
PETER C. JOHNSON[1,3]

[1]*Division of Plastic and Maxillofacial
Reconstructive Surgery, University of
Pittsburgh Medical Center,
676 Scaife Hall,
3550 Terrace Street,
Pittsburgh, PA 15261, USA*

[2]*Department of Chemical Engineering,
Carnegie Mellon University,
3101A Doherty Hall,
Pittsburgh, PA 15213-3890, USA*

[3]*Pittsburgh Tissue Engineering Initiative, Inc.,
Center for Biotechnology and Bioengineering,
300 Technology Drive,
Pittsburgh, PA 15219, USA*

1 Introduction

The goals of muscle tissue engineering are to treat disorders for which present medical and surgical therapies are either ineffective or impractical. Some of these disorders include cardiac myopathies, intrinsic skeletal muscle diseases, intestinal and urinary tract smooth muscle disorders and the loss of skeletal or cardiac muscle following trauma or ischemia. The principal methods are designed to introduce foreign genes into native muscle, generate and transfer muscle tissue to augment the function of existing muscle and control muscle function at the cellular level via autocrine and paracrine pathways. Each type of muscle presents unique challenges for tissue engineering, and while there is some overlap of techniques used for treatment of muscle disorders, a review of muscle tissue engineering is best understood by describing new advances in each tissue type separately. As such, the format for this chapter will include a separate section for each of the three types of muscle. The sections will include a brief description of form and function that is characteristic of each tissue, followed by the application of muscle tissue engineering methods to treat specific muscle diseases. This chapter is not intended to be an exhaustive review, but rather a sampling of the scope of muscle tissue engineering and its potential to resolve many of the diseases that plague the cardiac, skeletal and smooth muscle systems of the human body.

2 Skeletal muscle

Mature skeletal muscle fibers cannot regenerate if damaged, but contain a regeneration-capable population of adjacent satellite cells, or myoblasts, that comprise 1–5% of the total nuclei of a mature muscle [1]. Satellite cells are mononuclear cells that inhabit shallow depressions between mature muscle fibers and their associated basal lamina sheaths. Unstimu-lated satellite cells are mitotically quiescent in normal adult muscle, but are induced to proliferate and fuse to form new myofibers following muscle injury. During regeneration, satellite cells also are capable of migrating through the basal lamina sheaths that encase mature muscle fibers, where they fuse to form myotubes [2].

Skeletal muscle tissue engineering, including myoblast therapy, gene therapy and *in vitro* muscle formation, have developed directly from the unique regenerative properties of the satellite cells of mature muscle. Unlike cardiac muscle cells, human primary skeletal muscle cells can be harvested and successfully grown *in vitro* from either fetal muscle or from satellite cells of adult muscle [3–5]. In addition, a number of myoblast cell lines have been developed to study skeletal muscle development, such as the rat L6 and mouse C2, C2C12 and MM14 cell lines. Skeletal muscle development, both *in vivo* and *in vitro*, has been well characterized and several excellent reviews are present in the literature [6, 7].

Myoblast transfer therapy (MTT) was first performed twenty years ago [8] and was developed as a method to augment dysfunctional muscle or muscle of insufficient mass. The hypothesis for MTT is that the addition of normal myoblasts to diseased muscle will result in fusion, thereby introduc-ing normal DNA to deficient tissue. The resulting mosaic fibers will theoretically deliver the normal genome into myopathic muscle and stimu-late regeneration of atrophied muscle. Primary neonatal myoblasts can be expanded *in vitro*, then transplanted into recipient muscle where they fuse with host myofibers [9]. MTT has been used primarily as a method for treatment of Duchenne's muscular dystrophy (DMD) patients [10–12]. MTT has progressed to the human clinical trial stage [13–15] due in part to the availability of a relatively suitable animal model for study, the *mdx* mouse. This mouse model serves as a genetic and biochemical homolog for human DMD, if *mdx* muscle tissue is first irradiated to destroy regenerating satellite cells [16, 17]. Human dystrophin gene cDNA can be inserted into plasmid expression vectors [18], into myoblast cell lines [19] or, more recently, into autologous myoblasts [20] for injection into muscle. Research in this area has been hampered by host immune system attack upon injected myoblasts [21]. Control of the immune response will translate into more effective gene therapy for the therapeutic intervention in this disease.

Although myoblast transfer therapy has potential benefits for muscle tissue replacement or augmentation, there are certain theoretical limitations to its success. Transplanted myoblasts remain localized to the implantation area [12, 22, 23], thereby reducing the effectiveness of the therapy in disorders where large, multiple muscle groups are involved. In one DMD clinical trial, transplanted myoblasts failed to improve strength in patients who received a series of myoblasts injections over a six month period [14]. Injected myoblasts are often distributed unequally among host fibers, so that gene transfer is unpredictable [12]. Most of the injected myoblasts are lost within the first 48 hours and the remainder fuse with host myofibers in a limited area surrounding the injection site [24]. Significant loss occurs even if autologous myoblasts are injected, presumably due to host rejection of the foreign DNA that they contain. Rejection of allogeneic myoblasts, even of those that have fused with host fibers, is well documented [22, 25]. MTT coupled with immunosuppressant therapy may reduce the host immune response, but leads to undesirable side effects. The use of autologous myoblasts for MTT may reduce the need for immunosuppressant therapy, once the foreign DNA within the myoblasts can be rendered invisible to the immune system.

Skeletal muscle gene therapy has developed rapidly within the past decade; methods include direct DNA injection, as well as the injection of viral vectors and recombinant myoblasts into diseased skeletal muscle. Skeletal muscle may be suited to gene therapy due to the transverse tubule system, presence of caveolae and ability of skeletal muscle cells to fuse and form syncytia. Direct gene transfer via plasmid vectors was first described by Wolff [26]. Direct DNA injection is an efficient method for the study of myogenesis, but is insufficient for therapeutic intervention of muscle disease. Adenovirus vectors have been used for skeletal muscle gene therapy, due to their ability to theoretically infect non-mitotic cells without becoming integrated into the host genome. Recombinant adenovirus vectors have not infected mature muscle fibers well, perhaps because the virions are taken up via receptor-mediated endocytosis and must escape from endosomes to be effective [27]. Retroviral and herpesvirus vectors are expressed well in the myotubes of fetal and neonatal skeletal muscle, but not in mature muscle, possibly due to the inability of these viruses to pass through the extensive connective tissue network that is present in adult skeletal muscle [20].

Myoblast-mediated gene therapy is an alternate approach for treatment of myopathies or to augment the production of secreted proteins. Myoblasts are infected with a recombinant vector *in vitro*, then injected into developing or mature muscle. Recombinant human growth hormone (hGH) was introduced into muscle using C2C12 myoblasts infected with a retroviral vector and was secreted into the circulation for three weeks following implantation in one study [28] and for 12 weeks in another study [29]. Myoblast-mediated gene complementation may be used to restore coagulation factors to the serum [30] or to potentially deliver antioncogenic agents to the circulatory system [29].

The *in vitro* development of skeletal muscle as an alternative to muscle transplantation is in the initial stages of study. Specific areas of *in vitro* muscle culture that are presently under investigation are the effects of mechanical and electrical stimulation upon cultured muscle cells, the role of the extracellular matrix (natural or synthetic) in the migration, attachment and differentiation of myoblasts and the parallel alignment of myoblasts on a biomaterial surface. A number of researchers, including those in our own laboratory, are working on one or more of these areas of study to achieve the goal of *in vitro* muscle formation.

During myogenesis, both passive and active mechanical forces play an important role in the transition of skeletal muscle from the embryonic to the mature state [31]. Passive forces are those forces that are applied to muscle from the growing skeleton, while active forces include muscular contractions of the fetus (movement in utero). *In vivo-* directed mechanical tension helps to organize myoblasts into functional, aligned myotubes and provides a stimulus for the expression of mature isoforms of myofibrillar proteins. None of these forces are present in a static culture system. Vandenburgh has developed an *in vitro* cell culture system that mimics mechanical forces applied *in vivo* [32]. This model has been used to characterize the cellular effects of mechanical stimulation, particularly those associated with cytoskeletal rearrangements. Mechanical forces have been shown to directly alter the transcription of myosin heavy chain genes [33] and can alter the flux rates of prostaglandins E_2 and F_2 through the activation of phospholipases and the arachidonic acid pathway [34, 35]. These prostaglandins act as second messengers to regulate muscle protein turnover rates; PGE_2 causes protein degradation, while PGF_2 stimulates protein synthesis [36, 37]. Each of these second messengers are further controlled by the levels of calcium and the glucocorticoid dexamethasone [35, 38]. In addition, mechanical stretch has also been shown to activate the plasma membrane Na^+ pump [39, 40] and to increase growth factor receptor sensitivity [41]. It is clear that the cellular effects of mechanical stimulation upon myogenesis are quite complex and involve multiple signalling pathways. Applied mechanical force may be essential to the *in vitro* development of aligned, functional skeletal muscle tissue.

Muscle is a highly ordered tissue, wherein, each fiber is arranged in a linear fashion and adjacent fibers constitute a single plane. This structure is essential for effective muscle contraction; individual fibers must act as a unit during contraction and transmit their collective generated force to the object in motion. Cultured myoblasts grown as a monolayer lack an ordered structure. As myoblasts differentiate and fuse to form myotubes in culture, their arrangement is not ordered so that directional force cannot be generated. In order to achieve a functional vector of force, myoblasts should be aligned within the same spatial plane [32]. When unidirectional mechanical stretch is applied to adult cardiac myocytes in culture, the cells are induced to orient themselves parallel to the direction of force. Alignment only occurs if stretch is applied before cells attach [42]. In contrast, on a substratum undergoing continuous stretch-relax cycling, myotubes orient perpendicular to the direction of movement [43].

In addition to the application of mechanical force, the extracellular matrix (ECM) plays an essential role in the attachment, alignment and differentiation of myoblasts [44–46]. Various biomaterial substrates have been used to replace the ECM *in vitro*, either to enhance the attachment of myoblasts or to alter their growth. One of the first biomaterials used were DEAE-cellulose supports [47]. More recently, dermal sheep collagen discs were seeded with C2C12 cells for reconstructive surgery of the abdominal wall. Ultrastructural examination revealed up to 45% of surviving myoblasts after seven days, with extracellular matrix production and occasional myotube formation [48]. In our laboratory, C2C12 myoblasts are grown on patterned surfaces, enabling cells to align parallel to each other. Patterned surfaces are created by alternating areas (each being approximately the diameter of a mature myofiber) of adhesiveness with areas of non-adhesiveness. The ability of cells to attach to a substrate is dependent upon the hydrophilic nature of that surface. Generally, C2C12 myoblasts adhere well to hydrophilic substrata, and significantly less to hydrophobic substrata. We have used organosilane-coated glass slides to control C2C12 myoblast morphology, attachment and proliferation. The hydrophilic aminosilane EDA promotes these properties, whereas, the hydrophobic fluorosilane 13F resists myoblast attachment and proliferation [49]. These two organosilanes can be combined on the same surface in the form of a pattern using laser technology [50, 51]. The pattern thus formed will be used to align fusing mononuclear myotubes in the same direction. Figure 1 illustrates selective attachment of C2C12 myoblasts to the adhesive regions of a biomaterial pattern created in our laboratory. Once myoblasts are aligned and their growth controlled, research will proceed with vascularization and innervation of this neotissue. Our ultimate goal is to create an *in vitro* muscle for later transplantation (using autologous satellite cells) to

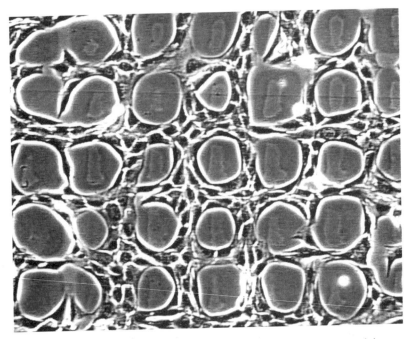

Fig. 1. Selective attachment of C2C12 myoblasts on patterned biomaterials.

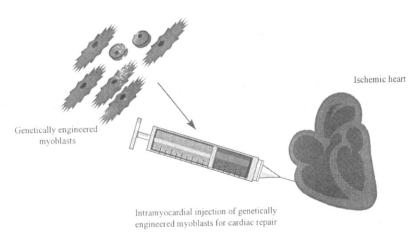

Ischemic heart

Genetically engineered
myoblasts

Intramyocardial injection of genetically
engineered myoblasts for cardiac repair

Fig. 2. A method of cardiac tissue engineering.

significantly enhance the treatment of those conditions presently requiring free muscle transfer for reconstruction.

3 Smooth muscle

Of the three types of muscle tissue, smooth muscle has the greatest regenerative capacity. Normal, adult smooth muscle cells (SMCs) are quiescent and possess a contractile phenotype. Upon injury, SMCs enter into a synthetic state, proliferate and produce new matrix proteins. SMCs in culture behave similarly. When SMCs are initially cultured, they lose their contractile phenotype and undergo logarithmic cell division. New endoplasmic reticulum appears and extracellular matrix production increases. As cells reach confluence, their total protein synthesis decreases, and the quiescent, contractile phenotype eventually replaces the synthetic phenotype [52]. Cultured embryonic smooth muscle cells display a synthetic phenotype. The amount of SMC proliferation is measured by the replicative index, defined as the number of cells undergoing mitosis in a population of SMCs per unit of time. A typical index of embryonic SMCs is 70–80%, while for quiescent adult SMCs, the replicative index is less than 0.06% per day. Embryonic SMCs are capable of proliferation in serum-free conditions [53].

The signal transduction pathways that control SMC proliferation are not known in detail. Mitogenic stimuli result in induction of the protooncogenes *c-myc*, *c-fos* and *c-myb* mRNA in less than 12 hours [54]. *C-myb* encodes a DNA-binding protein that functions as an intracellular growth mediator. *C-myb* may regulate calcium influx during the G_1/S interface of the cell cycle [55]. Mitogens for SMCs include numerous growth factors, mechanical stimuli, neighboring cells and other exogenous factors, depending on the anatomical location of the smooth muscle tissue.

Many smooth muscle disease syndromes are associated with overproliferation of SMCs and overproduction of matrix proteins. These include, but are not limited to, intimal hyperplasia resulting from arterial wall injury or from vascular reconstructive surgery and SMC hypertrophy occurring in inflammatory bowel diseases. Smooth muscle tissue engineering has arisen from the need to understand the mechanisms that underlie proliferative smooth muscle diseases, and of the need to develop vascular grafts to replace occluded vessels (where occlusion may be a sequela of intimal hyperplasia). There appear to be two major goals in the field of smooth muscle tissue engineering. The first goal is to elucidate the mechanisms that cause hyperplasia and concomitant matrix production in vascular and intestinal smooth muscle disease. By understanding the molecular mechanisms that underlie hyperplasia and hypertrophy of SMCs, growth-suppressive (antisense oligonucleotide) therapies can be developed to prevent SMC proliferation following injury or inflammation. A second aim of smooth muscle tissue engineering is to manipulate SMC proliferative and matrix production capabilities to develop new tissues for bladder and vessel reconstruction.

Abnormal proliferation of vascular SMCs is an important component of many cardiovascular diseases. In hypertension, elevated blood pressure causes an increase in mass of the smooth muscle layer of the arterial wall due to local migration, hyperplasia and hypertrophy of affected SMCs. An initial stimulus for abnormal SMC proliferation in hypertension may be the vasoactive hormone, angiotensin [56]. Abnormal SMC proliferation aids in restenosis of vascular grafts and following angioplasties. Angioplasty injury strips endothelial cells and a portion of SMCs from a treated artery. Within 24 hours, SMCs at the site of injury replicate. Several days after injury, SMCs begin to migrate from the media to the intima where they continue to proliferate. Proliferation and matrix production continues for approximately 12 weeks. At this time, 80% of the mass of the neointima is comprised of matrix. The stimulus for initial SMC migration may be a growth factor released by damaged SMCs, such as β-FGF. *In vitro*, β-FGF is a potent stimulus for SMC proliferation [57]. Alterations in hemodynamic forces on the vessel wall also affect SMC proliferation and subsequent intimal thickening [58]. Extracellular matrix (ECM) components, such as collagen type I, and ECM cell surface receptors, such as integrin β-1 [59], may promote the initial attachment, phenotypic modulation and migration of vascular SMCs. It has recently been postulated that the abnormal increase in collagen synthesis of injured SMCs may be due to a decrease in or lack of production of collagen-degrading enzymes, the metalloproteases [52]. SMCs of mechanically injured arteries secrete and respond to a number of mitogens, including PDGF and β-FGF [55]. Concentrations of heparan sulfate in excess of those of FGF-1 may suppress SMC proliferation [60]. In summary, intimal hyperplasia depends on SMC proliferation, migration and ECM production and is regulated by a complex of positive and negative factors from blood and cellular components of the vascular wall, as well as by hemodynamic forces on the SMCs of the arterial wall.

Antisense oligodeoxynucleotide (ODN) therapy has recently been developed to manage SMC proliferation in vascular disease. Retroviral or liposome-mediated gene transfer techniques can deliver DNA directly to replicating SMCs of the arterial wall, or endothelial cells can be infected *in vitro*, then injected into the diseased site [61]. *C-myc* expression has been targeted as an essential component of the SMC proliferative process. In a recent study, antisense ODNs directed to *c-myc* mRNA inhibited SMC proliferation by 50% and inhibited SMC migration by 90% in a dose-dependent manner [62]. A similar study has been carried out on SMCs using antisense ODNs to the proliferating cell nuclear antigen [63]. An *in vitro* study of aortic and graft SMCs incubated with antisense *c-myb* reduced proliferation by approximately 50% in both tissues [55]. SMC proliferation may be prevented by using antisense molecules to block cell division. An antisense ODN to cell division cycle 2 kinase (cdc2 kinase) has been shown to prevent intimal hyperplasia in response to angioplasty [64]. For clinical applications, antisense molecules may have to be delivered in slow-release formulations, because the length of time that antisense molecules exist in living cells before they are degraded is not precisely known.

Intestinal SMCs are important components of inflammatory bowel diseases. Few studies have characterized the molecular mechanisms responsible for SMC proliferation and hypertrophy in these diseases. In Crohn's disease, SMCs respond to chronic inflammation by proliferating and depositing collagen in the submucosal layer of the small intestine. Interleukin 1-β is present in increased concentrations in inflamed intestine and is known to be mitogenic for SMCs [65]. Other studies have concentrated on creating organotypic cultures for the small intestine as a model for the study of gut innervation [66]. Organotypic culture can potentially be used to develop engineered intestinal tissue for treatment of diseases presently requiring removal of the affected area of the digestive tract. Most tissue engineering studies of the small intestine have focused on the excision of the submucosa as a scaffold for cell seeding, or upon the seeding of intestinal enterocytes onto an artificial polymeric scaffold. The smooth muscle layer of the intestine is not usually included in studies designed to replace damaged intestinal tissue [67].

Smooth muscle and mucosal components of the intestine have been used in urinary reconstruction experiments. The smooth muscle layer of the ileum was used to create a continent urinary conduit in male humans [68]. The small intestinal submucosal layer (SIS) is essentially an acellular layer of matrix proteins harvested from the jejunum that is used as a scaffold upon which cells can be seeded. SIS is not immunogenic, and promotes angiogenesis in various *in vivo* model systems [69,70]. SIS has been used as a bladder patch in rats, and was found to promote smooth muscle and uroepithelial healing [71]. More recently, human bladder muscle and uroepithelial cells have been seeded onto biocompatible polymeric matrices and implanted onto the backs of athymic mice. These tissue engineered constructs may potentially be useful for genitourinary reconstruction [72].

4 Cardiac muscle

Cardiac muscle lacks stem cells. The inability of the myocardium to regenerate following injury is the primary reason that tissue engineering of cardiomyocytes may be beneficial. There have been considerable advancements made in the fields of heart transplantation and artificial heart development. As these therapies do not include alteration of cardiomyocytes themselves, the focus of this section will be on tissue engineering of cardiomyocytes including exogenous myoblast implantation to replace damaged myocardium and cardiomyocyte gene therapy.

Cardiac muscle develops from the splanchnopleuronic mesoderm. The development of the myocardium is under complex genetic regulatory control and may include many of the same factors as those involved in skeletal myogenesis. Shortly after birth, cardiomyocytes withdraw from the cell cycle and terminally differentiate. Further response to growth, increased workload and injury is limited to hypertrophy of cardiomyocytes, and not to hyperplasia. During normal development, hypertrophy can result in a 30–40-fold increase in overall mass of the myocardium [73].

Terminal differentiation of cardiomyocytes is proposed to be under the control of proteins that mediate exit and entry into the cell cycle, of the myogenic basic helix-loop-helix-motif (bHLH) proteins and of other transcription factors. A number of *cis*-acting consensus sequences may play a role in cardiac development, including the CArG box, A/T-rich sequences, the muscle-CAT element (transcriptional enhancing factor-1, or TEF-1), the GATA box, and steroid response elements [74, 75]. During terminal differentiation, a large number of genes are transcribed, such as those that encode the sarcomeric proteins (more than 40), channels and receptors, and metabolic enzymes. Cardiac function and gene expression also are subject to trophic stimuli during development. In the adult organism, the heart can modulate gene transcription in response to mechanical forces such as hemodynamic load [76]. An *in vitro* model of myocardial stretch has been developed using cells grown on a deformable silicon sheet that can mimic the heart's responses to pressure overload [76, 77]. Simpson *et al.* [78] have discovered that cultured cardiomyocytes undergo cardiac hypertrophy in response to alpha adrenergic stimulation. Essentially, cardiac development is controlled by a complex regulatory system, with no one group of transcription factors being solely responsible for cardiogenesis, as there is for skeletal myogenesis. Different transcription factors bind to multiple sites on promoters, thus, modulating each promoter's activity in response to external and internal stimuli [75].

Atrial and ventricular cardiomyocytes display different phenotypes early in development of the myocardium [79]. Recent studies of transgenic mice bearing lacZ reporter genes under control of the human desmin gene promoter have shown that the expression of lac Z is specific to a subset of ventricular myocytes [80]. In addition, lac Z expression of the fast skeletal muscle MLC-3F promoter when combined with a muscle-specific enhancer was observed in the atria and left ventricle, but not in the right ventricle [81]. These observations illustrate the heterogeneity of gene expression in the myocardium.

No permanent cardiogenic cell lines are available for the study of cardiac gene expression or for gene therapy. Primary cardiomyocyte cultures can be used, but must be harvested from fetal tissue that may not be readily available. Embryonal carcinoma (P19) cells are multipotent stem cells which can be induced to display a cardiac phenotype when exposed to DMSO [82]. A second cardiomyocyte cell line, AT-1, is derived from transgenic mice expressing the SV40 T-antigen oncogene in the atrium. AT-1 cells retain their oncongenic potential, so they may not be suitable for gene therapy [83]. Skeletal satellite cells or myoblast cell lines may be used for implantation into damaged myocardium, and also serve as vehicles for recombinant gene transfer [84].

Implantations of primary cardiomyocyte and cardiomyocyte cell lines into the myocardium have produced variable results. Recently, Soonpa *et al.* [85] have reported the implantation of fetal cardiac myoblasts into the murine myocardium. Fetal myoblasts differentiated into cardiac myocytes, forming gap junctions with native myocardium. Delcarpio and Claycomb [86] were able to demonstrate that AT-1 cells, when injected into both syngeneic and immunosuppressed mice, were capable of forming nascent intercalated discs and morphologically rudimentary junctions with adjacent host myocytes. However, immunologic rejection occurred in the cell line, and there was concern about increased oncogenic potential.

Myoblast therapy has a potential for resolving heart disease that involves ischemia. It is known that satellite cells can survive up to 24 hours in an ischemic environment [2]. Chiu *et al.* [87] implanted skeletal satellite cells into cryoinjured myocardium to assess their effect on regeneration of the injured heart tissue. Myoblasts were labelled with either Tritium or the LacZ reporter gene with a CMV promoter. β-gal was produced at the site of implantation, indicating that the cells survived the transfer. New striated muscles containing intercalated discs and centrally placed nuclei were tritium labelled. In a similar study, skeletal muscle satellite cells were injected into cryoinjured myocardium, but failed to repair the injured heart tissue, possibly due to the inability of the myotubes to survive within the dense scar tissue that formed as a result of the injury (inadequate microcirculation) [88].

Intramyocardial injection or grafting assures delivery of exogenous cells, but is an invasive technique. Arterial delivery of myoblasts is a recently developed method of cell transfer which has shown promising results. Robinson *et al.* [89] have delivered C2C12 myoblasts to the myocardium in a mouse model via transventricular injection. Initially, the cells were entrapped in myocardial capillaries, but then translocated to the interstitium, where they developed a phenotype that was more typical of a cardiomyocyte. Neumeyer *et al.* [90] have reported the successful arterial delivery of L6 skeletal myoblasts to rat skeletal muscle.

The use of myoblast implantation to repair damaged myocardium is in the initial stages of research. For repair to be effective, myoblasts must differentiate and be electrically coupled to the host cardiomyocytes. Integration into the host's electrical conduction system may be verified by the presence of the major gap junction protein, connexin43 [91], and, possibly by cell adhesion molecules [84]. Myotubes must be oriented so that their contractile ability enhances the function of the myocardium. It has been demonstrated that cardiomyocytes cultured under conditions of oscillating tension will spontaneously align themselves parallel to the direction of force [42]. The preexisting alignment of cardiomyocytes may play a role in the organization of implanted myotubes. Myoblast injection or grafting must not cause arrhythmias in the host myocardium (at the border of the graft or injection site and healthy myocardium). Finally, implanted cells must not be tumorigenic or elicit an immune response in the host. Studies of myoblast implantation using C2C12 cells have not led to tumorigenesis in host animals [29]. C2 myoblast implantation, however, can lead to tumorigenesis in mice [92].

Implanted myoblasts or other cells may simply serve as vehicles for delivery of recombinant proteins to the myocardium. In this way, angiogenic factors or neurotrophic agents may be delivered to damaged myocardium. In one study, transplanted AT-1 cells caused an increase in circulation in the grafted area, possibly by secreting as yet unidentified angiogenic factors [86]. Koh *et al.* [93] have used intramyocardial grafts to deliver a secreted protein directly into the myocardium. Myoblasts are first infected with a recombinant viral vector containing the desired recombinant protein sequence and an appropriate strong promoter. Alternately, retroviral vectors may be useful to directly transform the non-myocyte population of the myocardium, causing these cells to secrete factors needed for repair of damaged cardiomyocytes [94]. In a similar study, MyoD was transferred into an existing myocardial infarct and resulted in the conversion of non-myocytes to the skeletal muscle phenotype [95]. Using this methodology, a region of skeletal muscle potentially may be created to strengthen the damaged heart.

A final potential therapy for damaged myocardium is to induce hyperplastic growth in existing cardiomyocytes by the adenoviral delivery of recombinant proteins involved in the reentry of the cell cycle [96]. A recent study by Katz *et al.* [97] has shown promising results in this area of myocardial tissue engineering: they have demonstrated that expression of the large T-antigen of SV40 under control of the strong cardiac-specific α-myosin heavy chain promoter has resulted in proliferation of cardiomyocytes in the adult mouse myocardium. Very little is known about regulation of terminal differentiation in cardiomyocytes, so application of this technique must necessarily await a further understanding of cardiogenesis.

References

1. Alameddine, H. S., Dehaupas, M. and Fardeau, M., Regeneration of skeletal muscle fibers from autologous satellite cells multiplied in vitro. An experi-

mental model for testing cultured cell myogenicity. *Muscle and Nerve*, 1989, (12), 544–555.

2. Schultz, E., Albright, D. J., Jaryszak, D. L. and David, T. L., Survival of satellite cells in whole muscle transplants. *Anatomical Record*, 1988, (222), 12–17.

3. Webster, C., Pavlath, G. K., Park, D. R., Walsh, F. S. and Blau, H. M., Isolation of human myoblasts with the fluorescence-activated cell sorter. *Experimental Cell Research*, 1988, (174), 252–265.

4. Yaffe, D. and Saxel, O., Serial passaging and differentiation of myogenic cells isolated from dystrophic mouse muscle. *Nature*, 1977, (270), 725–727.

5. Blau, H. M. and Webster, C., Isolation and characterization of human muscle cells. *Proceedings of the National Academy of Sciences, USA*, 1981, (78), 5623–5627.

6. Weintraub, H., The MyoD family and myogenesis: Redundancy, networks and thresholds. *Cell*, 1993, (75), 1241–1244.

7. Emerson, C., Jr., Embryonic signals for skeletal myogenesis: Arriving at the beginning. *Current Opinion, in Cell Biology*, 1993, (5), 1057–1064.

8. Partridge, T. A., Grounds, M. and Sloper, J. C., Evidence of fusion between host and donor myoblasts in skeletal muscle grafts. *Nature*, 1978, (273), 306–308.

9. Watt, D. J., Lambert, K., Morgan, J. E., Partridge, T. A. and Sloper, J.C., Incorporation of donor muscle precursor cells into an area of muscle regeneration in the host mouse. *Journal of Neurological Science*, 1982, (57), 319–331.

10. Law, P. K., Goodwin, T. S. and Wang, M. G., Normal myoblast injections provide genetic treatment for murine dystrophy. *Muscle and Nerve*, 1988, (11), 525–533.

11. Karpati, G., Pouliot, Y., Zubrzycka-Gaarn, E. E., Carpenter, S., Ray, P. N., Worton, R. G. and Holland, P., Dystrophin is expressed in *mdx* skeletal muscle fibers after normal myoblasts implantation. *American Journal of Pathology*, 1989, (135), 27–32.

12. Rando., T. A., Pavlath, G. K. and Blau, H. M., The fate of myoblasts following transplantation into mature muscle. *Experimental Cell Research*, 1995, (220), 383–389.

13. Law, P. K., Goodwin, T. G., Fang, Q., Deering, M. B., Duggirala, V., Larkin, C., Florendo, J. A., Kirby, D. S., Li, H. J., Chen, M., Cornett, J., Li, L. M., Shirzad, A., Quinley, T., Yoo, T. J. and Holcomb, R., Cell transplantation as an experimental treatment for Duchenne Muscular Dystrophy. *Cell Transplantation*, 1993, (2), 485–505.

14. Mendell, J. R., Kissel, J. T., Amato, A. A., King, W., Signore, L., Prior, T. W., Sahenk, Z., Benson, S., McAndrew, P. E., Rice, R., Nagaraja, H., Stephens, R., Lantry, L., Morris, G. E. and Burghes, A. H. M., Myoblast transfer in the treatment of Duchenne's Muscular Dystrophy. *New England Journal of Medicine*, 1995, (333), 832–838.

15. Tremblay, J. P., Malouin, F., Roy, R., Huard, J., Bouchard, J. P., Satoh, A. and Richards, C. L., Results of a triple blind clinical study of myoblast transplantations without immunosuppressive treatment in young boys with Duchenne Muscular Dystrophy. *Cell Transplantation*, 1993, (2), 99–112.

16. Morgan, J. E., Hoffman, E. P. and Partridge, T. A., Normal myogenic cells from newborn mice restore normal histology to degenerating muscles of the *mdx* mouse. *Journal of Cell Biology*, 1990, (111), 2437–2449.

17. Wakeford, S., Watt, D. J. and Partridge, T. A., X-irradiation improves *mdx* mouse muscle as a model of DMD. *Muscle and Nerve*, 1991, (14), 42–50.

18. Acsadi, G., Dickson, G., Love, D. R., Jani, A., Walsh, F. S., Gurusinghe, A., Wolff, J. A. and Davies, K. E., Human dystrophin expression in *mdx* mice after intramuscular injection of DNA constructs. *Nature*, 1991, (352), 815–818.

19. Sopper, M. M., Hauschka, S. D., Hoffman, E. and Ontell, M., Gene complementation using myoblast transfer into fetal muscle. *Gene Therapy*, 1994, (1), 108–113.

20. Floyd, S., Mudigonda, S., Clemens, P., Yang, J., Ontell, M., Kochanek, S., Moreland, M. and Huard, J. [Abstract], Myoblast-mediated gene transfer to muscle. First Tissue Engineering Society Meeting, Orlando, FL; Dec. 12–16, 1996.

21. Huard, J., Bouchard, J. P., Roy, R., Malouin, F., Dansereau, G., Labrecque, C., Albert, N., Richards, C. L., Lemieux, B. and Tremblay, J. P., Human myoblast transplantation: Preliminary results of 4 cases. *Muscle and Nerve*, 1992, (15), 550–560.

22. Rando, T. A. and Blau, H. M., Primary myoblast purification, characterization and transplantation for cell-mediated gene therapy. *Journal of Cell Biology*, 1994, (125), 1275–1287.

23. Satoh, A., Huard, J., Labrecque, C. and Tremblay, J. P., Use of fluorescent latex microspheres (FLMs) to follow the fate of transplanted myoblasts. *Journal of Histochemistry and Cytochemistry*, 1993, (41), 1579–1582.

24. Huard, J., Acsadi, G., Jani, A., Massie, B. and Karpati, G., Gene transfer into skeletal muscles by isogenic myoblasts. *Human Gene Therapy*, 1994, (5), 949–958.

25. Pavlath, G. K., Rando, T. A. and Blau, H. M., Transient immunosuppressive treatment leads to long term retention of allogenic myoblasts in hybrid myofibers. *Journal of Cell Biology*, 1994, **127**(6 Pt 2), 1923–1932.

26. Wolff, J. A., Malone, R. W., Williams, P., Chong, W., Acsadi, G., Jani, A. and Felgner, P. L., Direct gene transfer into mouse muscle in vivo. *Science*, 1990, (247), 1465–1468.

27. Kass-Eisler, A., Li, K. and Leinwand, L. A., Prospects for gene therapy with direct injection of polynucleotides. *Annals of the New York Academy of Sciences*, 1995, (245), 232–240.

28. Barr, O. and Leiden, J. M., Systemic delivery of recombinant proteins by genetically modified myoblasts. *Science*, 1991, (254), 1507–1509.

29. Dhawan, J., Pan, L. C., Pavlath, G. K. and Travis, M. A., Lanctot, A.M., Blau, H. M., Systemic delivery of human growth hormone by injection of genetically engineered myoblasts. *Science*, 1991, (254), 1509–1512.

30. Yao, S. N. and Kurachi, K., Expression of human factor IX in mice after injection of genetically modified myoblasts. *Proceedings of the National Academy of Sciences*, 1992, (89), 3357–3361.

31. Vandenburgh, H. H., Karlisch, P. and Farr, L., Maintenance of highly contractile tissue in cultured avian skeletal myotubes in a collagen gel., *In vitro Cellular Developmental Biology* 1988, **24**(3), 166–174.

32. Vandenburgh, H. H. and Karlisch, P., Longitudinal growth of skeletal myotubes in vitro in a new horizontal mechanical cell stimulator. *In Vitro Cellular Developmental Biology*, 1989, (25), 607–616.

33. Periasamy, M., Gregory, P., Martin, B. J. and Stirewalt, W. S., Regulation of myosin heavy chain gene expression during skeletal muscle hypertrophy. *Biochemical Journal*, 1989, (257), 691–698.

34. Vandenburgh, H. H., Hatfaludy, S., Sohar, I. and Shansky, J., Stretch-induced prostaglandins and protein turnover in cultured skeletal muscle. *American Journal of Physiology*, 1990, (256), C674–C682.

35. Vandenburgh, H. H., Shansky, J., Karlisch, P. and Solerssi, R. L., Mechanical stimulation of skeletal muscle generates lipid-related second messengers by phospholipase activation. *Journal of Cellular Physiology*, 1993, (155), 63–71.

36. Goldberg, A. L., Baracos, V., Rodemann, P., Waxman, L. and Dinarello, C., Control of protein degradation in muscle by prostaglandins, calcium and leukocytic pyrogen (interleukin 1). *FASEB Proceedings*, 1984, (43), 1301–1306.

37. Vandenburgh, H. H., Shansky, J., Solerssi, R. and Chromiak, J. A., Mechanical stimulation of skeletal muscle increases PGF_{2a} production, cyclooxygenase activity and cell growth by a pertussis toxin-sensitive mechanism. *Journal of Cellular Physiology*, 1995, **163**(2), 285–294.

38. Chromiak, J. A. and Vandenburgh, H. H., Mechanical stimulation of skeletal muscle mitigates glucocorticoid-induced decreases in prostaglandin production and prostaglandin synthase activity. *Journal of Cellular Physiology*, 1994, **159**(3), 407–414.

39. Vandenburgh, H. H., In vitro skeletal muscle hypertrophy and sodium pump activity. In *Plasticity of Muscle*, ed. D. Patte. Walter de Gruyter, Berlin, 1980, pp. 493–506.

40. Vandenburgh, H. H., Stretch induced growth of skeletal myotubes correlates with activation of the Na pump. *Journal of Cellular Physiology*, 1981, (109), 205–214.

41. Vandenburgh, H. H., Cell shape and growth regulation in skeletal muscle: exogenous versus endogenous factors. *Journal of Cellular Physiology*, 1983, (116), 363–371.

42. Samuel, J.-L. and Vandenburgh, H.H., Mechanically induced orientation of adult rat cardiac myocytes in vitro. *In Vitro Cellular and Developmental Biology*, 1990, (26), 905–914.

43. Vandenburgh, H. H., A computerized mechanical cell stimulator for tissue culture: effects on skeletal muscle organogenesis. *In Vitro Cellular and Developmental Biology*, 1988, **24**(7), 609–619.

44. Adams, J. C. and Watt, F. M., Regulation of development and differentiation by the extracellular matrix. *Development*, 1993, (117), 1183–1198.

45. Maley, M. A., Davies, M. J. and Grounds, M. D., Extracellular matrix, growth factors, genetics: Their influence on cell proliferation and myotube formation in primary cultures of adult mouse skeletal muscle. *Experimental Cell Research*, 1995, (219), 169–179.

46. Lin, C. Q. and Bissel, M. J., Multi-faceted regulation of cell differentiation by the extracellular matrix. *FASEB*, 1993, (7), 734–743.

47. Shainberg, A., Isac, A., Reuveny, S., Mizrahi, A. and Shahar, A., Myogenesis on microcarriers. *Cell Biology International Reports*, 1983, (7), 727–734.

48. van Wachem, P. B., van Luyn, M. J. A. and Ponte de Costa, M. L., Myoblast seeding on a collagen matrix evaluated in vitro. *Journal of Biomedical Materials Research*, 1996, (30), 353–360.

49. Acarturk, O., Petrosko, P., Johnson, P. C. and DiMilla, P. [Abstract], Biomaterial driven control of myoblast morphology, growth and attachment on silanized substrates. First Tissue Engineering Society Meeting, Orlando, FL, Dec. 12–16, 1996.

50. Dulcey, C. S., Georger, J. H., Krauthamer, V., Stenger, D. A., Fare, T.L. and Calvert, J. M., Deep UV photochemistry of chemisorbed monolayers: Patterned coplanar molecular assemblies. *Science*, 1991, (252), 551–554.

51. Stenger, D. A., Georger, J. H., Dulcey, C. S., Hickman, J. J., Rudolph, A. S., Nielsen, T. B., McCort, S. M. and Calvert. J. M., Coplanar molecular assemblies of amino- and perfluorinated alkylsilanes: Characterization and geometric definition of mammalian cell adhesion and growth. *Journal of the American Chemical Society*, 1992, (114), 8435–8442.

52. Mesh, C. L., Majors, A., Mistele, D., Graham, L. M. and Ehrhart, L. A., Graft smooth muscle cells specifically synthesize increased collagen. *Journal of Vascular Surgery*, 1995, (22), 142–149.

53. Cook, C. L., Weiser, M. C. M., Schwartz, P. E., Jones, C. L. and Majack, R.A., Developmentally timed expression of an embryonic growth phenotype in vascular smooth muscle cells. *Circulation Research*, 1993, (74), 189–196.

54. Kindy, M. S. and Sonenshein, G. E., Regulation of oncogene expression in cultured aortic smooth muscle cells: post-transcriptional control of *c-myc* mRNA. *Journal of Biological Chemistry*, 1986, (261), 12865–12868.

55. Pitsch, R. J., Goodman, G. R., Minion, D. J., Madura, J. A., Fox, P.L. and Graham, L. M., Inhibition of smooth muscle cell proliferation and migration in vitro by antisense oligonucleotides to *c-myb*. *Journal of Vascular Surgery*, 1996, (23), 783–791.

56. Chen, Y.-Q., Gilliam, D. M., Rydzewski, B. and Naftilan, A. J., Multiple enhancer elements mediate induction of *c-fos* in vascular smooth muscle cells. *Hypertension*, 1996, (27), 1224–1233.

57. Clowes, A. W. and Reidy, M. A., Prevention of stenosis after vascular reconstruction: Pharmacologic control of intimal hyperplasia-A review. *Journal of Vascular Surgery*, 1991, (13), 885–891.

58. Clowes, A. W. and Clowes, M. M., Influence of chronic hypertension on injured and uninjured arteries in spontaneously hypertensive rats. *Laboratory Investigation*, 1980, (6), 535–541.

59. Yamamoto, K. and Yamamoto, M., Cell adhesion receptors for native and denatured type I collagens and fibronectin in rabbit arterial smooth muscle cells in culture. *Experimental Cell Research*, 1994, (214), 258–263.

60. Greisler, H. P., Gosselin, C., Ren, D., Kang, S. S. and Kim, D. U., Biointeractive polymers and tissue engineered blood vessels. *Biomaterials*, 1996, (1), 329–336.

61. Nabel, E. G., Plautz, G. and Nabel, G. J., Site-specific gene expression in vivo by direct gene transfer into the arterial wall. *Science*, 1990, (249), 1285–1288.

62. Biro, S., Fu, Y. M., Yu, Z.-X. and Epstein, S. E., Inhibitory effects of antisense oligodeoxynucleotides targeting *c-myc* mRNA on smooth muscle

cell proliferation and migration. *Proceedings of the National Academy of Sciences*, 1993, (90), 654–658.

63. Speir, E and Epstein, S. E., Inhibition of smooth muscle cell proliferation by an antisense oligodeoxynucleotide targeting the messenger RNA encoding proliferating cell nuclear antigen. *Circulation*, 1992, (86), 538–547.

64. Morishita, R., Gibbons, G. H., Ellison, K. E., Nakajima, M., Zhang, L., Kneda, Y., Ogihara, T. and Dzau, V. J., Single intraluminal delivery of antisense cdc2 kinase and proliferating-cell nuclear antigen oligonucleotides results in chronic inhibition of neointimal hyperplasia. *Proceedings of the National Academy of Sciences*, 1993, (90), 8474–8478.

65. Graham, M. F., Willey, A., Adams, J., Yager, D. and Diegelmann, R. F., Interleukin 1β down-regulates collagen and augments collagenase expression in human intestinal smooth muscle cells. *Gastroenterology*, 1986, (110), 344–350.

66. Song, Z.-M., Brookes, S. J. H., Llewellyn-Smith, I. J. and Costa, M., Ultra-structural studies of the myenteric plexus and smooth muscle in organotypic cultures of the guinea pig small intestine. *Cell and Tissue Research*, 1995, (280), 627–637.

67. Organ, M., Mooney, D. J., Hansen, L. K., Schloo, B. and Vacanti, J. P., Transplantation of enterocytes utilizing polymer-cell constructs to produce a neointestine. *Transplantation Proceedings*, 1992, (24), 3009–3011.

68. Atta, M. D. A new technique for continent urinary reservoir reconstruction. *Journal of Urology*, 1991, (145), 960–962.

69. Lantz, G. C., Badylak, S. F., Hiles, M. C., Coffey, A. C., Geddes, L.A., Kokoni, K., Sandusky, G. E. and Morff, R. J., Small intestinal submucosa as a vascular graft: A review. *Journal of Investigative Surgery*, 1993, (6), 297–310.

70. Sandusky, G. E., Lantz, G. C. and Badylak, S. F., Healing comparison of small intestine submucosa and ePTFE grafts in the canine carotid artery. *Journal of Surgical Research*, 1995, (58), 415–420.

71. Kropp, B. P., Eppley, B. L., Prevel, C. D., Rippy, M. K., Harruff, R.C., Badylak, S. F., Adams, M. C., Rink, R. C. and Keating, M. A., Experimental assessment of small intestine submucosa as a bladder wall substitute. *Urology*, 1995, (46), 396.

72. Atala, A., Freeman, M. R., Vacanti, J. P., Shepard, J. and Retik, A.B., Implantation in vivo and retrieval of artificial structures consisting of rabbit and human uroethelium and human bladder muscle. *Journal of Urology*, 1993, (150), 608–612.

73. Oparil, S., Pathogenesis of ventricular hypertrophy. *Journal of American College of Cardiology*, 1985, 57B–65B.

74. Farrance, I. K., Mar, J. H. and Ordahl, C. P., M-CAT binding factor is related to the SV40 enhancer binding factor, TEF-1. *Journal of Biological Chemistry*, 1992, (267), 17234–17240.

75. Robbins, J., Regulation of cardiac gene expression during development. *Cardiovascular Research*, 1996, (31), E2–E16.

76. Komuro, I., Kaida, T., Shibazaki, Y., Kurabayashi, M., Katoh, Y., Hoh, E., Takakuu, F. and Yazaki, Y., Stretching cardiac myocytes stimulates pro-tooncogene expression. *Journal of Biological Chemistry*, 1990, (265), 3595–3598.

77. Sadoshima, J., Jahn, L., Takahashi, T., Kulik, TJ. and Izumo, S., Molecular characterization of the stretch-induced adaptation of cultured cardiac cells. An in vitro model of load-induced cardiac hypertrophy. *Journal of Biological Chemistry*, 1992, (267), 10551–10560.

78. Simpson, P. C., Kariya, K., Karns, L. R., Long, C. S. and Karliner, J. S., Adrenergic hormones and control of cardiac myocyte growth. *Molecular and Cellular Biochemistry*, 1991, (104), 35–43.

79. Melnik, N., Yutzey, K. E., Gannon, M. and Bader, D. Commitment, differentiation, and diversification of avian cardiac progenitor cells. *Annals of the New York Academy of Science*, 1995, (245), 1–7.

80. Li, Z., Marchand, P., Humbert, J., Babinet, C. and Paulin, D., Desmin sequence elements regulating skeletal muscle-specific expression in transgenic mice. *Development*, 1993, (117), 947–959.

81. Kelly, R., Alonso, S., Tajbakhsh, S., Cossu, G. and Buckingham, M., Myosin light chain 3F regulatory sequences confer regionalized cardiac and skeletal muscle specific expression in transgenic mice. *Journal of Biological Chemistry*, 1995, (129), 383–396.

82. Rudnicki, M. A., Jackowski, G., Saggin, L. and McBurney, M. W., Actin and myosin expression during development of cardiac muscle from cultured embryonal carcinoma cells. *Developmental Biology*, 1990, (138), 348–358.

83. Delcarpio, J. B., Lanson, N. A. Jr., Field, L. J. and Claycomb, W. C., Morphological characterization of cardiomyocytes isolated from a transplantable cardiac tumor derived from transgenic mouse atria. (AT-1 cells). *Circulation Research*, 1991, (69), 1591–1600.

84. Koh, G. Y., Klug, M. G., Soonpa, M. H. and Field, L. J., Differentiation and long-term survival of C2C12 myoblast grafts in heart. *Journal of Clinical Investigation*, 1993, (92), 1548–1554.

85. Soonpa, M. H., Koh, G. Y., Klug, M. G. and Field, L. J., Formation of nascent intercalated disks between grafted fetal cardiomyocytes and host myocardium. *Science*, 1994, (264), 98–101.

86. Delcarpio, J. B. and Claycomb, W. C., Cardiomyocyte transfer into the mammalian heart. *Annals of the New York Academy of Science*, 1995, (245), 267–285.

87. Chiu, R. C. -J., Zibaitis, A. and Kao, R. L., Cellular cardiomyoplasty: Myocardial regeneration with satellite cell implantation. *Annals of Thoracic Surgery*, 1995, (60), 12–18.

88. Marelli, O., Ma, F. and Chiu, R. C. -J., Satellite cell implantation for neomyocardial regeneration. *Transplantation Proceedings*, 1992, **24**(6), 2995.

89. Robinson, S. W., Cho, P. W., Levitsky, H. L., Olson, J. L., Hruban, R.H., Acker, M. A. and Kessler, P. D., Arterial delivery of genetically labelled skeletal myoblasts to the murine heart: Long-term survival and phenotypic modification of implanted myoblasts. *Cell Transplantation*, 1996, **5**(1), 77–91.

90. Neumeyer, A. M., DiGregorio, D. M. and Brown, R. H., Arterial delivery of myoblasts to skeletal muscle. *Neurology*, 1992, (42), 2258–2262.

91. Beyer, E. C., Paul, D. L. and Goodenough, D. A., Connexin43: A protein from rat heart homologous to a gap junction protein from liver. *Journal of Cellular Biology*, 1987, (105), 2621–2629.

92. Morgan, J. E., Moore, S. E., Walsh, F. S. and Partridge, T. A., Formation of skeletal muscle in vivo from the mouse C2 cell line. *Journal of Cell Science,* 1992, **102**(Pt. 4), 779–787.

93. Koh, G. Y., Kim, S. -J., Klug, M. G., Park, K., Soonpa, M. and Field, L. J. Targeted expression of transforming growth factor-$\beta 1$ in intra-cardiac grafts promotes vascular endothelial cell DNA synthesis. *Journal of Clinical Investigation,* 1995, (95), 114–121.

94. McDonald, P., Hicks, M. H., Cobbe, S. M. and Prentice, H. Gene transfer for myocardial ischemia. *Annals of the New York Academy of Science,* 1995, (245), 455–459.

95. Prentice, H., Kloner, R. A., Sartorelli, V., Bellows, S. D., Alker, K. and Kedes, L., Transformation of cardiac fibroblasts into the skeletal muscle phenotype by injection of a MyoD-expressing retrovirus into ischemic heart. *Circulation,* 1993, (88), 1475.

96. Soonpa, M. H., Daud, A. I., Koh, G. Y., Klug, M. G., Kim, K. K., Wang, H. and Field, L. J., Potential approaches for myocardial regeneration. *Annals of the New York Academy of Science,* 1995, (245), 446–453.

97. Katz, E. B., Steinhelper, M. E., Delcarpio, J. B., Daud, A. I., Claycomb, W. C. and Field, L. J., Cardiomyocyte proliferation in mice expressing alpha-cardiac myosin heavy chain-SV40 T-antigen transgenes. *American Journal of Physiology,* 1992, (262), H1867–1876.

Tissue Engineering of the Peripheral Nervous System

ELIZABETH J. FURNISH,
CHRISTINE E. SCHMIDT
Department of Chemical Engineering,
The University of Texas-Austin,
Austin, TX 78712, USA

Tissue engineering in the nervous system presents a major challenge, especially when one considers that the chemical and physiological bases of many neurological disorders are not fully understood. However, advances in neurobiology, genetic techniques, and nerve cell culture, along with the development of novel biomaterials, are making tissue engineering in the central and peripheral nervous systems a reality. This chapter will focus on research efforts in the peripheral nervous system, and particular emphasis will be placed on recent approaches for the repair of damaged peripheral nerves.

Current clinical treatment of a severed peripheral nerve involves either surgical realignment of the two nerve ends, if they are directly adjacent, or use of an autologous nerve or vein graft to bridge an existing gap. However, the success of regeneration is variable and never complete. Additional drawbacks of nerve or vein autografts include the need for multiple surgical procedures and loss of function at the donor site. Thus, natural and synthetic tubular nerve guidance channels as alternatives to autografts have been the subject of intensive research. Guidance channels help direct axons sprouting off the regenerating nerve end, provide a conduit for diffusion of neurotrophic factors secreted by the damaged nerve stump, and minimize infiltrating fibrous tissue. In addition to efforts to control these passive characteristics of nerve guidance channels, researchers are currently making great progress in engineering interactive biomaterials that can specifically stimulate the regeneration process.

This chapter will review the physiology of the peripheral nervous system, the factors that are critical for peripheral nerve regeneration, and the current approaches that are being explored to aid the regeneration process. In addition to a detailed discussion of nerve guidance channels, which constitute the largest area of tissue engineering research in the peripheral nervous system, a brief review of other engineering efforts in this area will be presented.

1 Physiology of the peripheral nervous system

1.1 Classification of the nervous system

The nervous system is divided into the central nervous system (CNS), which includes the brain and spinal cord, and the peripheral nervous system (PNS), which consists of the cranial nerves arising from the brain and the spinal nerves arising from the spinal cord. Neurons in the peripheral nervous system are classified according to the direction in which the neuron conducts impulses. Sensory (afferent) nerves direct impulses from sensory receptors to the CNS, and motor (efferent) nerves conduct impulses out of the CNS to effector tissues such as muscles or glands. In the PNS, motor neurons are classified as somatic or autonomic. The cell bodies of somatic neurons are in the spinal cord; these neurons send axons directly to their skeletal muscle targets, which are usually under voluntary control. In contrast, autonomic motor control involves two neurons: the first has its cell body in the CNS and its axon synapses with a second neuron located outside the CNS in an autonomic ganglion (collection of nerve cell bodies). It is this second post-ganglionic neuron that synapses with, or innervates, involuntary effectors such as smooth muscle, cardiac muscle, and glands. The autonomic nervous system is further divided into the sympathetic, parasympathetic, and enteric systems. In general, the sympathetic and parasympathetic systems dominate during action and rest, respectively. The enteric nervous system innervates the intestines and is unique in that it can operate independently of CNS control.

1.2 Cellular components of the peripheral nervous system

The two cell types that comprise the nervous system are neurons and neuroglia. Neurons are the basic functional elements of the nervous system and are characterized by a cell body containing the nucleus, a long cell process (axon) stretching from the cell, and numerous short cell processes (dendrites). Dendrites transmit electrical signals to the cell body, whereas the axon conducts impulses away from the cell body. Neuroglia, or glial cells, are non-neural support cells that aid the function of neurons. Examples include Schwann cells in the PNS and oligodendrocytes and astrocytes in the CNS. Glial cells are considerably more abundant than neurons, and unlike neurons which cannot undergo mitosis, glial cells have some capacity for cell division.

Axons in the PNS are surrounded by a living sheath of Schwann cells. The outer surface of this layer of Schwann cells consists of a glycoprotein basement membrane, the neurilemma, which is analogous to the basement membrane in epithelial layers. In contrast, axons in the CNS have neither a sheath of Schwann nor a continuous basement membrane, characteristics which prove to be significant in terms of regeneration. The most striking difference between the CNS and PNS is the capacity for peripheral nerves to regenerate; axons in the CNS do not regenerate significantly in their natural environment. Many axons, especially those that extend great distances (up to 1 m), are surrounded by an insulating myelin sheath, which serves to increase the propagation velocity of the nerve impulse. Myelin sheaths are formed from dense layers of successive wrappings of the cell membrane of Schwann cells (PNS) or oligodendrocytes (CNS).

1.3 Anatomy of a nerve

A nerve is a bundle of motor and/or sensory axons that are collectively linked together by support tissue into an anatomically defined trunk. Endoneurium surrounds individual axons and their Schwann cell sheaths, and is composed of oriented collagen fibers and some fibroblasts. Next, the perineurium, formed from multiple layers of flattened cells and collagen, surrounds groups of axons and endoneurium to form small bundles called fascicles. Finally, epineurium is an outer sheath of loose fibrocollegenous tissue which binds individual nerve fascicles into a nerve trunk. Peripheral nerves are well vascularized by capillaries within the support tissue of the nerve trunk or by vessels which penetrate the nerve from surrounding arteries and veins. Figure 1 illustrates the arrangement of the axons and support tissue in a peripheral nerve. In an enlarged view of a nerve fascicle from the rat sciatic nerve (Fig. 2), myelinated and non-myelinated axons, basal lamina, and Schwann cells are visible using transmission electron microscopy. For a more thorough discussion of the peripheral nervous system, refer to Stevens and Lowe [1].

2 Peripheral nerve injuries

2.1 Types of PNS injuries

Nerve injury can result from mechanical, thermal, chemical, or pathological damage. Nerve malignancies usually require the removal of a segment of the nerve. Injuries to sensory nerves can result in sensory loss and phantom sensations. Hand or facial reconstruction, either after an accident or to repair an existing defect, often requires that nerves be rerouted or replaced. In the case of facial paralysis, the origin of the injury may be congenital, or may result from numerous causes, including surgery, trauma, tumor and diseases of the CNS, and viral infections of the nerve. It is estimated that about 1 in 60 or 70 people in a lifetime suffer from partial or complete facial paralysis, such as Bell's palsy [2]. Aside from the

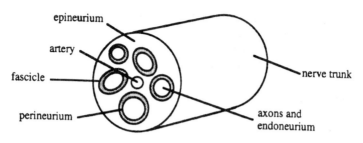

Fig. 1. Schematic of a peripheral nerve showing the arrangement of the axons and support tissue.

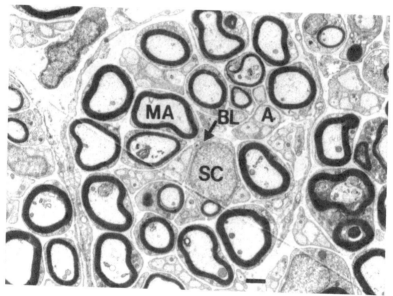

Fig. 2. Transmission electron micrograph (cross section) of part of a nerve fascicle. Myelinated axons (MA) are insulated by sense myelin (black). A, non-myelinated axon; SC, Schwann cell; BL, basal lamina (neurilemma). Bar, 1 μm.

cosmetic issues associated with facial paralysis, loss of muscle control around the eyes can lead to risk of exposure and possible corneal ulceration.

2.2 Nerve response to injury

The most severe injury is a complete nerve transection. When an axon is initially severed, the distal portion begins to degenerate as it is separated from the metabolic resources of the cell body. At the proximal end, which is still associated with the cell body, the nerve stump swells and retrograde degeneration occurs. The extent of this retrograde degeneration depends on the type and position of the injury, but is only slight after a cut with a sharp instrument. Wallerian degeneration begins one to two hours after injury and continues over the next four days after transection. During this time, the axons and myelin beyond the site of injury are phagocytosed by proliferating Schwann cells and macrophages which migrate into the nerve. The target muscle, devoid of innervation, begins to atrophy and the neuron cell body undergoes chromatolysis, with cytoplasmic swelling and lateral migration of the nucleus and Nissl substance (rough endoplasmic reticulum). During this period, Schwann cells proliferate and migrate, forming cellular bridges and processes upon which axons regenerate, and releasing tropic and trophic factors that aid in axonal outgrowth [3, 4]. Revascularization of the injured area is also critical for regeneration. Finally, undamaged endoneurial tubes formed by Schwann cell sheaths remain intact, serving as templates for regeneration of the axons to the proper target tissue.

2.3 Axonal regeneration

At the proximal end of the damaged nerve, the axons regenerate by sprouting new processes which elongate down Schwann cell guides at approximately 2–5 mm/day [5]. Eventually a nerve fiber reaches the target muscle, forms a functional synapse, and becomes remyelinated. Reinnervation is not a perfect process; some axons continue to grow past their muscle endplate and innervate nearby muscle [6]. Terminal Schwann cells overlying the neuromuscular junction have been shown to sprout elaborate processes, which in turn permit nerve fibers to escape the confines of their muscle endplates, giving rise to polyneuronally innervated muscle.

The extension of the axon is an active migratory event guided by the axon's growth cone, a cytoplasmic and actin-rich protrusion at the end of the axon. Guidance information such as soluble chemoattractants (e.g. nerve growth factor, NGF) or chemorepellents [7], as well as insoluble signals located on other cells or in the extracellular matrix (e.g. laminin), is first perceived by filopodia and lamellipodia, extensions at the growth cone's leading edge [8]. In response to these cues, the growth cone alters its path and speed by reorganizing its contractile apparatus, the actin cytoskeleton. The ability of the growth cone to move forward and to change directions easily, and therefore guide the extension of the axon, has also been attributed in part to adhesion receptors such as the integrins [9].

After the growth cone of a neuron reaches its target tissue, it innervates the tissue by forming a synapse, which is defined as a functional connection between a neuron (pre-synaptic cell) and a second cell (post-synaptic cell). This second cell is either another neuron or an effector cell in a target muscle or gland. In a functional synapse, the action potential or nerve impulse stops at the pre-synaptic axon ending, where it stimulates the release of neurotransmitters (e.g. acetylcholine and norepinephrine) that diffuse across the synaptic cleft to stimulate the post-synaptic cell. For a damaged nerve to regenerate, the axon must re-establish a functional synapse with the target tissue.

2.4 Clinical repair of damaged peripheral nerve

The current clinical treatments for nerve transection or gap injury are surgical end-to-end anastomosis and autologous nerve grafts. Surgical end-to-end repair involves the direct reconnection of individual nerve fascicles or bundles of fascicles. Because tension in the nerve cable has been shown to prohibit regeneration [10], surgical end-to-end repair is useful only if nerve stumps are directly adjacent. If the nerve ends cannot be reconnected, occluding connective tissue infiltrates the path of regenerating axons, decreasing the likelihood that sprouting axons will locate the distal stump. Thus, to repair a peripheral nerve over a gap, autologous nerve grafts are used clinically [11]. Nerve autografts can be derived from one of several cutaneous nerves such as the sural or saphenous nerve, with an available length up to about 40 cm and a cable diameter of 2–3 cm [12, 13]. For an interesting synopsis of the early history of nerve repair, refer to the review by Chiu [14].

Complete sensory and motor recovery after end-to-end nerve anastomosis and autografting never occurs, although partial recovery rates of up to about 80% can be achieved with current surgical techniques [14]. With nerve autografts, the return of motor and sensory function is dependent on many factors including the position of the injury, the delay between injury and repair, and defect length [12, 13]. The closer the injury to the nerve cell body, the greater the damage to the neuron via retrograde degeneration and the lower the probability for optimal functional recovery. For similar reasons, delay between injury and repair is detrimental in nerve grafting. Recovery of function after a delay is variable and difficult to generalize, especially since the impact of delay on regeneration is dependent on other factors such as the type and position of the injury. For nerves in the arm, sensory and motor recovery has been observed following a 2 month delay, while no recovery has been observed for a 9 month delay; in other cases, some recovery still occurs for delay periods up to 18 months and longer. Typically, little is lost by a delay of only 3–5 weeks [15]. The length of the nerve defect also affects regeneration success, but in general, regeneration occurs through grafts up to and just over 15 cm in length [12].

An interesting clinical application of nerve grafts is the surgical correction of facial paralysis, which can be almost completely rectified using an approach pioneered at Mount Vernon Hospital (UK). First, a nerve from the leg is grafted onto the facial nerve of the unparalyzed side of the face. After new nerve fibers have grown along the graft, this nerve graft is then rotated and rerouted to innervate a muscle transplant on the paralyzed side of the face [16].

Disadvantages to nerve grafts include the need for multiple surgical procedures, loss of function at the donor site, and mismatch of nerve cable dimensions between the donor graft and recipient nerve. In addition to nerve autografts, autologous vein grafts (e.g. saphenous vein) are used in the clinic to bridge the gap between damaged nerve ends, but often with slightly lower success rates [14, 17]. Similar drawbacks exist when using any autologous tissue graft, although donor site morbidity and functional loss associated with vein grafts are lower than for nerve grafts. In an effort to bypass these limitations, the use of tubular guidance channels to bridge the gap between severed nerve ends is being extensively explored [18].

3 Nerve guidance channels

3.1 Advantages of nerve guidance channels

Nerve guidance channels (NGCs) are either natural or synthetic tubular conduits that are used to bridge the gap between injured nerve stumps. NGCs serve to direct axons sprouting from the regenerating (proximal) nerve end, to reduce the infiltration of fibrous tissue, and to provide a conduit for diffusion of neurotrophic factors secreted by the damaged nerve stump [19]. Additional benefits of NGCs include reducing suture line tension, increasing the concentration of endogenous proteins, providing a barrier to permit or inhibit the diffusion of macromolecules, and controlling the microenvironment. In addition to this 'passive' or purely mechanical role, the NGC may be fashioned to stimulate the regeneration process actively, as described in more detail in the following sections.

Although NGCs do not currently play a critical role in the clinic, they are a valuable tool for experimental studies since the channel properties (permeability, length, conduit material, adhesion properties, etc.) can be explicitly specified and the lumen contents and axonal regeneration can be analyzed. Since the microenvironment can be precisely controlled by the NGC, it is possible to determine specifically the effects of various biologically active macromolecules on regeneration.

3.2 Regeneration in nerve guidance channels

Most experimental studies with NGCs have used a 10 or 15 mm defect in the sciatic nerve of the rat. However, mice, rabbits, primates, and other animal models, with different nerves and lengths of defects, have also been used. For the surgical reconnection of a 10 mm defect in the sciatic nerve of rat, the NGC should have approximate dimensions of 12 mm × 1.2 mm (length × diameter). First, a clean cut is made in the sciatic nerve of the animal's femur. A length of nerve is removed so that the nerve, when sutured to the NGC, will neither kink nor be exposed to tension. Next, approximately 1 mm of the proximal portion and 1 mm of the distal portion of the transected nerve are inserted into the two ends of the conduit, leaving the desired length (e.g. 10 mm) between the two nerve ends (Fig. 3). The tube is then fixed to each nerve stump using a single stitch of 10-0 suture through the epineurium of the nerve trunk. Finally, a syringe needle is used to fill the conduit with saline and remove any occluding air pockets.

For the general case of an inert synthetic conduit (e.g. silicone) or an arterial graft, regeneration proceeds as follows. Within hours, the tube or graft is filled with a clear fluid released by the severed blood vessels in the nerve cable. This fluid contains many proteins, including the clotting factor fibrin and many soluble factors that enhance regeneration. Within a week, fibrin forms a longitudinally oriented matrix that bridges the gap between the two nerve stumps. During the second week, fibroblasts, Schwann cells, macrophages, and endothelial cells permeate the fibrin matrix. Axons sprouting from the proximal nerve stump elongate along the matrix, and after about four weeks, some axons have reached the distal stump and have become myelinated [20, 21]. After bridging the gap, the axons now elongate down the intact endoneurial tubes until they reach their final muscle target. Success rates of about 60% have been reported for regeneration across 10 mm defects using silicone tubes [22]. However, silicone guidance channels do not support regeneration if the distal nerve stump is left out of the NGC, nor do they support regeneration across defects larger than 10 mm, regardless of the presence of the distal nerve stump.

3.3 Analysis of regeneration

Given the different regeneration models, the various times for monitoring

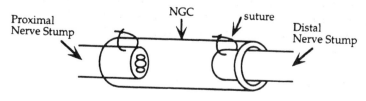

Fig. 3. Schematic of surgical placement of nerve guidance channel (NGC).

results (2–16 weeks), and the wide variety of methods for analysis, inconsistencies arise when trying to compare the success of regeneration for different studies. Most commonly, the progress of regeneration is monitored histologically using either light or electron microscopy to determine the overall diameter of the regenerated nerve cable, numbers and sizes of axons, axon morphologies, presence of Schwann cells, and degree of myelination at the midpoint of the regenerated nerve cable (see Fig. 2). Interpretation of histological samples can be somewhat subjective and is not necessarily indicative of innervation or functional recovery. In contrast, retrograde labeling allows one to discern reinnervation and the extent of axonal branching directly. With this technique, the target muscle is stained and the axonal processes leading back to the cell bodies are traced by the retrograde transport of the dye. For both histological analysis and retrograde labeling, the animal must be sacrificed, prohibiting a temporal measurement of regeneration in the same animal.

Electrophysiological responses are also commonly reported, and can be performed in the living animal while under anesthesia. In some cases, electrodes placed proximal and distal to the transection are used to measure conduction velocity through the regenerated nerve cable. In contrast, electromyography (EMG) provides information on actual reinnervation by measuring the generation of compound action potentials in the muscle after stimulation of the nerve. The primary advantage of EMG over morphological studies is that it is a functional, rather than a structural, measure of regeneration that can be monitored over the duration of the regeneration process.

Other functional tests involve assessing reflex and sensory perception within the animal; however, physical movements of animals during handling can introduce significant uncertainty into the interpretation. Many studies report only histological findings, which have the advantage of relatively quick feedback (4–8 weeks), while some studies will combine structural and functional methods to ensure a more accurate assessment of true regeneration (8–12 weeks and longer). There are several detailed descriptions [22–25] of these analyses in the literature.

3.4 Properties of nerve guidance channels

There are several key properties that all NGCs should possess based on the physiology of the nerve. First, it is necessary that the material of choice be readily formed into a tubular conduit having a desired diameter (e.g. 1.2 mm for the rat sciatic nerve). Second, it is imperative that the NGC be simple to implant surgically; the NGC must be able to be sterilized and to be sutured with microsurgical tools. Next, permanent materials are not desired; they pose a more severe risk for infection, tend to provoke connective tissue responses, and are especially risky if the material dislodges and damages the nerve. Thus, the ideal NGC should degrade, or be resorbed, as the nerve regenerates, providing support for only as long as needed (for at least the first three to four weeks of regeneration). Finally, rigid channels also create problems, including possible injury to the nerve with extreme movement; therefore, NGCs should be made of pliable materials. Still, the channel needs to maintain its shape over the period of regeneration and also needs to be firm enough not to kink during implantation or during normal activity of the animal.

Aside from its purely mechanical role, the NGC may be fashioned to stimulate regeneration actively. In particular, insights into the critical biological factors and advances in the synthesis of biocompatible polymers have created many opportunities to engineer NGCs with desirable and predictable properties. Thus, in addition to the requirements listed above, other important factors to be considered when designing an NGC include: conduit material, surface texture, porosity, electrical properties, incorporation of insoluble factors and matrices, release of soluble factors, and seeding of support cells.

3.4.1 Conduit material
Natural materials —
Conduit materials can be of either natural or synthetic origin (for a review see [18]). Various natural extracellular matrix (ECM) molecules such as laminin, fibronectin, and collagen promote axonal extension [26], and have been investigated as materials for NGCs. Prosthetic tubes derived from laminin, fibronectin, and collagen [27, 28] show improved regeneration over a 4 mm or 10 mm rat sciatic nerve gap compared to silicone controls, as assessed by an increased number of myelinated axons or recovery of ascending and descending compound action potentials. In addition, vein grafts with intraluminal collagen coating [29] and veins that have been turned inside out to expose the rich collagen layer to the lumen [24] both show enhanced regeneration across a 10 mm defect in the rat sciatic nerve compared to standard vein grafts, as demonstrated by histology, EMG, and retrograde labeling. In other cases, preformed tissue tubes, such as those formed from synovial [30] or perineurial [31] tissue, have been used to guide regeneration.

Natural materials have the advantages of being biocompatible and providing a good surface for cell adhesion, but may induce an undesirable immune response, depending on the source. In addition, the use of natural proteins is limited by the poor mechanical properties of the biological ECM molecules and potential batch-to-batch variations in large-scale isolation procedures. The existence of these limitations has driven the development of novel biodegradable synthetic materials that can be used for NGCs.

Synthetic Materials —

Several biodegradable synthetic materials such as polyesters (e.g. poly-lactic and polyglycolic acids), degradable polyurethanes, and poly(organo)phosphazenes have been shown to support nerve regenera-tion. One of the first synthetic biodegradable polymers studied as a material to guide nerve regeneration over a significant defect was poly(lactic-co-glycolic acid) (PLGA) or polygalactin [32]. Specifically, poly-galactin or Vicryl® suture mesh was used to bridge 7–9 mm defects in the tibial nerves of rabbits. Although the regenerated nerve cable had a slightly different morphology from normal nerve, these studies demon-strated that PLGA is a low irritant to the regenerating nerve and that synthetic materials can be used as NGCs. Another polyester, polylactic acid (PLA), has also been investigated for its ability to support nerve regeneration [33]. In these studies, retrograde labeling of the nerve distal to the defect was used to demonstrate a steady increase over time in the number of motor and sensory fibers regenerated through a polylactate conduit. Interestingly, motor nerve fibers always appeared to regenerate faster than sensory axons, which is consistent with results using perineurial tubes [31]. Furthermore, biodegradable conduits formed from poly (lactide-co-caprolactone), a copolymer of lactic acid and caprolactone, have recently demonstrated promise for nerve regeneration [34, 35].

In addition to polyesters, biodegradable polyurethanes and polyphos-phazenes have also shown a capacity for guiding regeneration. Degradable polyurethane-based nerve guides support regeneration across an 8 mm gap in the rat sciatic nerve [36]. For one type of polyphosphazene tube, axonal elongation is initially delayed about 10 days relative to a silastic tube, but after 45 days, the lag disappears [37]. The reason for the delay is not known, although it is hypothesized that hydrolysis of the polyphosphazene interferes with the local microenvironment and the formation of the fibrin matrix. However, this delay could potentially be overcome by taking advantage of the slow, controlled degradation of polyphosphazenes as a means to deliver drugs or various bioactive molecules that stimulate nerve regeneration [38].

Despite potential barriers of biocompatibilty and degradation, synthetic materials are attractive from an engineering standpoint because their physical properties (porosity, degradation rate, etc.) can be specifically altered to design an optimal NGC. In addition, the manufacture of synthetic matrices has several advantages over the isolation of natural materials, including less problematic scale-up and tighter control of product properties.

3.4.2 Surface texture

Conduit wall texture has been shown to have a profound impact on the formation of the fibrin matrix and axonal regeneration. In one study, non-resorbable polyacrylonitrile/polyvinyl chloride (PAN/PVC) tubes were constructed to have identical chemical composition and permeability characteristics and either smooth or rough inner surface texture [39]. Morphology of the regenerated cable over a 4 mm sciatic nerve gap in mice was much improved in smooth tubes relative to rough tubes. Smooth tubes permitted the formation of a centralized nerve cable surrounded by an epineurial-like layer and containing myelinated and unmyelinated axons grouped into fascicles, whereas rough tubes were completely filled with a loose connective tissue stroma with only a few axons and no distinct outer sheath. This trend also held for tubes with alternating smooth and rough sections, regardless of which surface texture was positioned at the distal or proximal end of the regenerating nerve. These results suggest that surface microgeometry plays a critical role in regeneration and that changes in tube surface texture influence the three-dimensional organization of the longitudinal fibrin cable bridge that precedes axonal elongation.

3.4.3 Porosity

A key physical property of the channel wall is its porosity, which affects the diffusion of soluble factors into and out of the NGC. To assess the importance of soluble factors secreted from the surrounding tissue, semipermeable and impermeable conduits were compared for their ability to support sciatic nerve regeneration with and without the distal nerve stump [40]. Semipermeable acrylic copolymer channels having a molecular weight cutoff of 50 kDa supported regeneration over a 4 mm sciatic nerve gap in mice both with and without the presence of the distal nerve stump, although the channel containing the distal stump showed improved regeneration. In addition, semipermeable conduits plugged with the same polymer at the proximal end showed improved regeneration over conduits that were left completely open. In contrast, impermeable silicone conduits required the presence of the distal nerve stump for any regeneration, indicating that outer wound-healing environment factors (e.g. secreted from macrophages, etc.) were required for regeneration in the absence of the distal stump. The use of semipermeable channels allowed the influx of nutrients and growth factors from the external environment and presumably helped concentrate factors released by the proximal nerve stump.

A further investigation used semipermeable polysulfone tubes with porosities of 100 and 1000 kDa to determine a more precise range of critical stimulatory and/or inhibitory factors from the external environment [41]. The 1000 kDa channels contained fewer myelinated axons compared to the 100 kDa channels, conceivably as a result of an influx of inhibitory molecules (possibly protease inhibitors) with molecular weights between 100 and 1000 kDa. Thus, a semipermeable NGC could be engineered with an optimum porosity to permit diffusion into the guidance channel of beneficial growth factors secreted by surrounding macrophages, to retain trophic factors released by the nerve stumps within the intraluminal space, and at the same time, to exclude inhibitory molecules found in the external environment.

3.4.4 Electrical properties

Past work has demonstrated that electrical charges play an important role in the stimulation of both the proliferation or differentiation of various cell types including neurons. For example, it has been shown that neurite extension *in vitro* and *in vivo* is enhanced on electrets such as poled polyvinylidene fluoride (PVDF) [42, 43] and polytetrafluoroethylene (PTFE) [44]. Electrets are broadly defined as materials possessing quasi-permanent surface charge because of trapped monopolar charge carriers. In piezoelectric materials (one class of electrets), such as PVDF, transient surface charges are generated as a result of minute mechanical deformations of the material, whereas PTFE displays a static surface charge.

Polypyrrole, an inherently electrically conductive polymer, can be used for localized stimulation of nerve regeneration [45, 46]. *In vitro* studies show that an electric current passed through polypyrrole films significantly enhances neurite outgrowth. Polypyrrole conduits have also been shown to guide nerve regeneration across 10 mm defects in rats.

Several theories have been suggested to explain the effect of electric stimulation on nerve regeneration, including (a) electrophoretic redistribution of cell membrane growth factor and adhesion receptors or cytoskeletal proteins such as actin, all of which are involved in growth cone migration, (b) favorable membrane or extracellular matrix protein conformational changes, (c) direct depolarization or hyperpolarization of nerves, (d) promotion of electrical communication between nerve stumps, (e) enhancement of protein synthesis, and (f) field-induced gradients of ions and molecules in the culture medium or tissue fluid.

3.4.5 Insoluble factors and intraluminal support matrices

Once a suitable conduit material is found, the lumen of the NGC can be altered to enhance regeneration further. Fibrin plays a major role in nerve regeneration; empty guidance channels naturally fill with an oriented fibrin matrix within weeks after placement across a nerve defect. To take advantage of this effect, Williams *et al.* [47] incorporated fibrin precursors from plasma directly into the lumen of NGCs and found enhanced regeneration compared to similar conduits filled with saline. Along similar lines, hyaluronic acid, a compound known to improve fibrin matrix formation, also enhances sciatic nerve regeneration in the rat [48].

In addition, as previously mentioned, various insoluble extracellular matrix molecules, including laminin, fibronectin and some forms of collagen, promote axonal extension [26, 49], and are therefore excellent candidates for incorporation as matrices into the lumen of NGCs. Laminin promotes neurite extension, stimulates Schwann cell mitosis, and may need to be matrix-bound in order to augment regeneration [50]. Collagen tubes containing laminin- and fibronectin-coated collagen fibers show increased numbers of myelinated axons and an improved electrophysiological response compared to tubes containing uncoated collagen fibers [28]. Recent studies have used magnetically aligned collagen matrices to promote regeneration *in vitro* by harnessing the contact guidance response of axonal growth cones [51]. In addition, it has been shown that silicone tubes encasing a collagen-glycosaminoglycan matrix yielded improved regeneration across a 10 mm defect in a rat sciatic nerve compared to saline-filled tubes [23]. However, in other cases, intraluminal matrices such as collagen and laminin have been shown to hinder regeneration in semipermeable and biodegradable NGCs [52]. This negative phenomenon is more severe for collagen and presumably results from acellular gel remnants blocking essential molecular diffusion, cellular migration, or axonal elongation. This barrier may be overcome by drastically reducing the concentration of the gel.

Short oligopeptide sequences (3–10 amino acids) that are recognized by specific adhesion receptors can be used in place of macromolecules such as laminin. These minimum recognition sequences are easier to manage and, because of their small size, are less likely to invoke an immune response. Two different amino acid sequences derived from laminin, YIGSR and IKVAV, were covalently bound to fluorinated ethylene propylene films and shown to promote neuronal cell attachment *in vitro* [53]. Three dimensional hydrogels chemically modified with laminin-derived oligopeptides promote neurite extension *in vitro* [54] and are excellent candidates for intraluminal matrices for nerve regeneration.

Additional benefits of matrices are the ease with which exogenous factors and non-neural support cells can be incorporated into the lumen to optimize axonal regeneration. A more detailed description of three-dimensional gels is presented in section 4.

3.4.6 Soluble factors

Soluble neurotropic and neurotrophic factors can be incorporated directly into NGCs to stimulate nerve regeneration further. Factors that are beneficial for nerve regeneration include nerve growth factor (NGF), brain-derived neurotrophic factor (BDNF), insulin-like growth factors (IGF-1 and IGF-2), transferrin, platelet-derived growth factors (PDGF-BB and PDGF-AB), basic fibroblast growth factor (bFGF), acidic fibroblast growth factor (aFGF), ciliary neurotrophic factor (CNTF), and interleukin-1 (IL-1) [55]. Some of these molecules enhance neuronal survival (e.g. CNTF, PDGF, and BDNF), while others promote axonal extension and neurite sprouting (e.g. IGF-1, IGF-2, aFGF, and bFGF). These factors are naturally produced during the regeneration process; for example, Schwann cells secrete NGF, BDNF, IGF-1, and CNTF, and macrophages release PDGF and IL-1. However, the presence of exogenous factors has also been shown to enhance nerve regeneration.

In one study, ethylene-vinyl acetate (EVA) guidance channels impregnated with bFGF were used to bridge a 15 mm gap in the rat sciatic nerve [56]. Electron microscopy revealed an improved morphology of the regenerated nerve cable compared to nerve regenerated in a plain EVA channel. Polyethylene tubes filled with a gel of collagen and aFGF demonstrated greater numbers of myelinated axons compared to collagen-filled tubes [25]. As more neurotropic and neurotrophic factors and their roles are identified, NGCs could be precisely tailored to improve nerve regeneration for various applications.

3.4.7 Support cell seeding

The addition of neuronal support cells can provide the regenerating nerve with a long-term source of various neurotropic and neurotrophic factors and a substrate upon which the axons can elongate. Schwann cells are known to secrete neurotrophic factors (e.g. CNTF, BDNF, NGF, and IGF-1) and, as mentioned earlier, appear to play a causal role in axonal regeneration [4]. Schwann cells isolated from a 10 mm segment of rat sciatic nerve were cultured on the internal lumen of poly-L-lysine-coated polyethylene tubes [57]. One week later, the seeded guides were implanted into the gap from which the Schwann cells were isolated. After two weeks of regeneration, a significantly greater number of total axons and myelinated axons were observed in guides containing pre-established Schwann cell cultures. In another study, semipermeable PAN/PVC channels containing human Schwann cells in Matrigel® promoted increased axonal regeneration over an 8 mm gap in rat sciatic nerve [58]. To enhance the effects that support cells have on axonal regeneration, cells seeded within a NGC could be genetically altered to secrete desired levels of neurotrophic factors or to express specific extracellular matrix proteins [59]. For example, fibroblasts genetically modified to produce NGF, BDNF, and bFGF have been transplanted into the adult rat spinal cord [60]. Neurites were shown to penetrate NGF- and bFGF-secreting grafts, while BDNF-secreting grafts elicited no growth response.

4 Other tissue engineering efforts in the PNS

The creation of NGCs constitutes the largest area of tissue engineering efforts in the peripheral nervous system. However, there are other areas of active research. In particular, there has been progress in the development of three-dimensional and/or micropatterned *in vitro* nerve cell cultures.

Although much of this work is performed on CNS neurons with the hope of using the final systems for replacement tissue in the brain or for neural-based circuits and biosensors, many of the techniques could also be utilized for the PNS. Three-dimensional cultures and *in vitro* patterning of nerves could aid in diagnostics and could also serve as a more representative model of the body for fundamental studies.

Previous research has demonstrated that both CNS and PNS neurons can extend neurites in three-dimensional agarose hydrogels [61]. Neurite extension is highly dependent on agarose concentration; an agarose concentration less than a threshold concentration of 1.25% wt/vol. (150 mm pore radius) is required for neuronal extension. Gels of higher concentration, i.e. lower porosity, inhibited neurite extension. Agarose gels modified with laminin-derived oligopeptides have been shown to promote neurite extension of *in vitro* cultures of dorsal root ganglia and PC12 cells [54]. In a similar manner to agarose gels, three-dimensional fibrin gels supported the extension of neurites from dorsal root ganglia in a concentration-dependent manner; neurite length was increased for cases in which proteolysis of fibrin, and hence porosity, were enhanced [62]. Additionally, it was shown that results within three-dimensional fibrin gels were opposite of those on two-dimensional fibrin films, further supporting the requirement for adequate three-dimensional culture systems to model *in vivo* neurite extension. Porous hydrophilic sponges made from 2-hydroxyethyl methacrylate have also been used as three-dimensional scaffolds to promote axonal regeneration of the optic nerve [63]. As a final example, biodegradable macroporous gelatin microcarriers have been used to support the growth of cerebral neurons in a three-dimensional, porous environment [64].

In vitro neuronal patterning is another area of active research. Results from such studies have application in the CNS, especially with respect to the generation of *in vitro* neural circuits or cell-based neural biosensors, but can also be useful for fundamental studies in the PNS. These studies could provide insights into the mechanisms for directional guidance by growth cones (neuronal pathfinding). Selective nerve fiber growth has been demonstrated on two-dimensional patterns of metal oxide [65], amine derivatives [66], and various biomolecules [67]. Photolithographic techniques are typically employed to generate the desired patterns. In addition, self-assembled monolayers with alkanethiolates on gold can be used in conjunction with photolithography to help control surface chemistry [68]. These studies have focused on patterning in two dimensions, although three-dimensional patterning would have major implications in tissue engineering. Three-dimensional neural tracts *in vitro* have been created by lamination of agarose gel layers of alternating non-permissive and permissive character [61].

5. Summary

Despite recent efforts, there is still no clinically attractive alternative to nerve or vein autografts for repair of peripheral nerve defects. Much research is currently being devoted to the creation of the optimum nerve guidance channel, which will most likely incorporate multiple forms of stimuli for the regenerating nerve. Present data suggest that the ideal NGC:

1. Should be biocompatible, biodegradable, non-immunogenic, suturable and pliable;
2. Should have a smooth inner lining of the conduit wall;
3. Should have a molecular weight cutoff between 50 and 100 kDa to permit diffusion of small growth factors but exclude larger inhibitory molecules;
4. Should have inherent electrical properties;
5. Should have a highly porous intraluminal matrix to support cell attachment and migration; and
6. Should release soluble neurotrophic factors either by direct incorporation of the molecules or by inclusion of support cells (e.g. Schwann cells).

In the near future, the synthesis of additional novel, and interactive, biomaterials will open many doors for tissue engineering efforts in the nervous system. Biotechnology will also play a larger role, either by incorporation of cells genetically engineered to secrete neurotrophic or survival factors or by direct incorporation of specifically tailored recombinant peptides or proteins (e.g. adhesive ligand sequences, growth factors, enzyme pockets, catalytic or inhibitory antibody fragments). In addition, drug delivery technology will become more critical for the design of the ideal NGC. Techniques by which a series of bioactive molecules could be released over time in sequence with stages of regeneration may ultimately enhance repair in the PNS, and potentially even in the CNS. In addition, more quantitative and uniform methods of analyzing regeneration need to be realized. Quantitative molecular analyses must be conducted in parallel with more phenomenological studies, so that ultimately, mechanistic models can be developed to make *a priori* predictions based on the biological responses of the regenerating nerve to stimulatory and inhibitory cues.

References

1. Stevens, A. and Lowe, J., *Histology*, Mosby-Year Book Europe Limited, London, 1993, pp. 206–225.

2. Wilson, J. D., Braunwald, E. et al., *Harrison's Principles of Internal Medicine*, 12th ed., McGraw-Hill, Inc., New York, NY, 1991, pp. 2078–2079.

3. Hall, S. M., The effect of inhibiting Schwann cell mitosis on the reinnervation of acellular autografts in the peripheral nervous system of the mouse. *Neuropathology and Applied Neurobiology*, 1986, **12**, 401–414.

4. Son, Y. -J. and Thompson, W. J., Schwann cell processes guide regeneration of peripheral axons. *Neuron*, 1995, **14**, 125–132.

5. Jacobson, S. and Guth, L., An electrophysiological study of the early stages of peripheral nerve regeneration. *Experimental Neurology*, 1965, **11**, 48–60.

6. Gorio, A., Carmignoto, G., Finesso, M. Polato, P. and Nunzi, M. G., Muscle reinnervation. II. Sprouting, synapse formations and repression. *Neuroscience*, 1983, **8**, 403–416.

7. Pini, A., Axon guidance. Growth cones say no. *Current Biology*, 1994, **4**, 131–133.

8. Bray, D. and Hollenbeck, P. J., Growth cone motility and guidance. *Annual Review of Cell Biology*, 1988, **4**, 43–61.

9. Schmidt, C. E., Dai, J., Lauffenburger, D. A., Sheetz, M. P. and Horwitz, A. F., Integrin-cytoskeletal interactions in neuronal growth cones. *Journal of Neuroscience*, 1995, **15**, 3400–3407.

10. Millesi, H., Meissl, G. and Berger, A., The interfascicular nerve-grafting of the median and ulnar nerves. *Journal of Bone and Joint Surgery*, 1972, **54A**, 727.

11. Millesi, H., Indications and techniques of nerve grafting. In *Operative Nerve Repair and Reconstruction*, ed. R. H. Gelberman. J.B. Lippincott, Philadelphia, 1991, pp. 525–544.

12. Seddon, H. J., *Surgical Disorders of the Peripheral Nerves*, 2nd edn, Churchill Livingstone, New York, 1975, pp. 287–302.

13. Sunderland, S. *Nerve Injuries and Their Repair*, Churchill Livingstone, London, 1991.

14. Chiu, D. T., Special article: The development of autogenous venous nerve conduit as a clinical entity. P and S Medical Review, Columbia-Presbyterian Medical Center, Vol. 3, No. 1, Dec., 1995.

15. Schwartz, S. I., Shires, G. T., Spencer, F. C., Storer, E. H., *Principles of Surgery*, 3rd edn, McGraw-Hill Book Co., New York, 1979, p. 1775.

16. Jacobs, J. M., Laing, J. H. and Harrison, D. H., Regeneration through a long nerve graft used in the correction of facial palsy. A qualitative and quantitative study. *Brain*, 1996, **119**, 271–279.

17. Chiu, D. T., Janecka, I., Krizek, T. J., Wolff, M., Lovelace, R. E., Autogenous vein grafts as a conduit for nerve regeneration. *Surgery*, 1982, **91**, 226–233.

18. Valentini, R. F., Nerve guidance channels. In *The Biomedical Engineering Handbook*, ed. J. D. Bronzino. CRC Press Inc., Boca Raton, 1995, pp. 1985–1996.

19. Williams, L. R., Powell, H. C., Lundborg, G. and Varon, S., Competence of nerve tissue as distal insert promoting nerve regeneration in a silicone chamber. *Brain Research*, 1984, **293**, 201–211.

20. Weiss, P. and Taylor, A. C., Histomechanical analysis of nerve reunion in the rat after tubular splicing. *Archives of Surgery*, 1943, **47**, 419–450.

21. Williams, L. R., Longo, F. M., Powell, H. C., Lundborg, G. and Varon, S., Spatial-temporal progress of peripheral nerve regeneration within a silicone chamber: parameters for a bioassay. *Journal of Comparative Neurology*, 1983, **218**, 460–470.

22. Fields, R. D. and Ellisman, M. H., Axons regenerated through silicone tube splices. I. Conduction properties. *Experimental Neurology*, 1986, **92**, 48–60.

23. Chang, A., Yannis, I. V., Perutz, S., Loree, H., Sethi, R. R., Krarup, C., Norregaard, T. V., Zervas, N. T. and Silver, T., Electrophysiological study of recovery of peripheral nerves regenerated by a collagen-glycosaminoglycan copolymer matrix. In *Progress in Biomedical Polymers*, ed. C. G. Gebelein and R. L. Dunn. Plenum Press, New York, 1990, pp. 107–120.

24. Wang, K. -K., Costas, P. D., Bryan, D. J., Jones, D. S. and Seckel, B. R., Inside-out vein graft promotes improved nerve regeneration in rats. *Microsurgery*, 1993, **14**, 608–618.

25. Cordeiro, P. G., Seckel, B. R., Lipton, S. A., D'Amore, P. A., Wagner, J. and Madison, R., Acidic growth factor enhances peripheral nerve regeneration in vivo. *Plastic and Reconstructive Surgery*, 1989, **83**, 1013–1019.

26. Rutishauser, U., Adhesion molecules of the nervous system. *Current Opinion in Neurobiology*, 1993, **3**, 709–715.

27. Archibald, S. J., Krarup, C., Shefner, J., Li, S. -T. and Madison, R. A., A collagen-based nerve guide conduit for peripheral nerve repair: an electrophysiological study of nerve regeneration in rodents and non-human primates. *Journal of Comparative Neurology*, 1991, **306**, 685–696.

28. Tong, X., Hirai, K., Shimada, H., Mizutani, Y., Izumi, T., Toda, N., Yu, P., Sciatic nerve regeneration navigated by laminin-fibronectin double coated biodegradable collagen grafts in rats. *Brain Research*, 1994, **663**, 155–162.

29. Wang, K. -K., Costas, P. D., Jones, D. S., Miller, R. A. and Seckel, B. R., Sleeve insertion and collagen coating improve nerve regeneration through vein conduits. *Journal of Reconstructive Microsurgery*, 1993, **9**, 39–48.

30. Lundborg, G. and Hansson, H. -A., Regeneration of peripheral nerve through preformed pseudosynovial tubes. *Hand Surgery*, 1980, **5**, 35–38.

31. Restrepo, Y., Merle, M., Michon, J., Folliguet, B. and Petry, D., Fascicular nerve graft using an empty perineurial tube: an experimental study in the rabbit. *Microsurgery*, 1983, **4**, 105–112.

32. Molander, H., Olsson, Y., Engkvist, O., Bowald, S. and Eriksson, I., Regeneration of peripheral nerve through a polyglactin tube. *Muscle and Nerve*, 1982, **5**, 54–57.

33. DaSilva, C. F., Madison, R. Dikkes, P., Chiu, T. -H. and Sidman, R. L., An in vivo model to quantify motor and sensory peripheral nerve regeneration using bioresorbable nerve guide tubes. *Brain Research*, 1985, **342**, 307–315.

34. Nicoli, A. N., Perego, G., Cella, G. D., Maltarello, M. C., Fini, M., Rocca, M. and Giardino, R., Effectiveness of a bioabsorbable conduit in the repair of peripheral nerves. *Biomaterials*, 1996, **17**, 959–962.

35. den Dunnen, W. F., Stokroos, I., Blaauw, E. H., Holwerda, A., Pennings, A. J., Robinson, P. H., Schakenraad, J. M., Light microscopic and electron-microscopic evaluation of short-term nerve regeneration using a biodegradable poly(DL-lactide-epsilon-caprolacton) nerve guide. *The Journal of Biomedical Materials Research*, 1996, **31**, 105–115.

36. Robinson, P. H., Van der Lei, B., Hoppen, H. J., Leenslag, J. W., Pennings, A. J. and Nieuwenhuis, P., Nerve regeneration through a two-ply biodegradable nerve guide in the rat. *Microsurgery*, 1991, 12, 412–419.

37. Langone, F., Lora, S., Veronese, F. M., Caliceti, P., Parnigotto, P. P., Valenti, F. and Palma, G., Peripheral nerve repair using a poly(organo)phosphazene tubular prosthesis. *Biomaterials*, 1995, 16, 347–353.

38. Laurencin, C. T., Koh, H. J., Neenan T. X., Allcock H. R. and Langer, R., Controlled release using a new bioerodible polyphosphazene matrix system. *Journal of Biomedical Materials Research*, 1987, 21, 1231–1246.

39. Aebischer, P., Gunard, V., Valentini, R. F., The morphology of regenerating peripheral nerves is modulated by the surface microgeometry of polymeric guidance channels. *Brain Research*, 1990, 531, 211–218.

40. Aebischer, P., Gunard, V., Winn, S. R., Valentini, R. F. and Galletti, P. M., Blind-ended semipermeable guidance channels support peripheral nerve regeneration in the absence of a distal nerve stump. *Brain Research*, 1988, 454, 179–187.

41. Aebischer, P., Gunard, V. and Brace, S., Peripheral nerve regeneration through blind-ended semipermeable guidance channels: effect of the molecular weight cutoff. *Journal of Neuroscience*, 1989, 9, 3590–3595.

42. Aebischer, P., Valentini, R, F., Dario, P., Domenici, C. and Galletti, P. M., Piezoelectric guidance channels enhance regeneration in the mouse sciatic nerve after axotomy. *Brain Research*, 1987, 436, 165–168.

43. Valentini, R. F., Vargo, T. G., Gardella, Jr., J. A., Aebischer, P., Electrically charged polymeric substrates enhance nerve fibre outgrowth in vitro. *Biomaterials*, 1992, 13, 183–190.

44. Valentini, R. F., Sabatini, A. M., Dario, P. and Aebischer, P., Polymer electret guidance channels enhance peripheral nerve regeneration in mice. *Brain Research*, 1989, 480, 300–304.

45. Shastri, V. R., Schmidt, C. E., Kim, T. -H., Vacanti, J. P. and Langer R., Polypyrrole - a potential candidate for stimulated nerve regeneration. *Proceedings of the Materials Research Society Conference*, 1996, 414, 113–118.

46. Schmidt, C. E., Shastri, V. R., Vacanti, J. P. and Langer, R., Stimulation of neurite outgrowth using an electrically conducting polymer. *Proceedings of the National Academy of Science*, 1997, 94, 8948–8953.

47. Williams, L. R., Danielsen, N., Muller, H. and Varon, S., Exogenous precursors promote functional nerve regeneration across a 15-mm gap within a silicone chamber in the rat. *The Journal of Comparative Neurology*, 1987, 264, 284–290.

48. Seckel, B. R., Jones, D., Hekimian, K. J., Wang, K. K., Chakalis, D. P. and Costas, P. D., Hyaluronic acid through a new injectable nerve guide delivery system enhances peripheral nerve regeneration in the rat. *Journal of Neuroscience Research*, 1995, 40, 318–324.

49. Madison, R. D., DaSilva, C., F., Dikkes, P., Entubulation repair with protein additives increases the maximum nerve gap distance successfully bridged with tubular prostheses. *Brain Research*, 1988, 447, 325–334.

50. Bryan, D. J., Miller, R. A., Costas, P. D., Wang, K. and Seckel, B. R., Immunocytochemistry of skeletal muscle basal lamina grafts in nerve regeneration. *Plastic and Reconstructive Surgery*, 1993, 92, 927–940.

51. Dubey, N., Letourneau, P. C., Tranquillo, R. T., Development and characterization of magnetically aligned collagen gel rods for guided peripheral nerve regeneration. Paper presented at the 1996 American Institute of Chemical Engineers Annual Meeting, Chicago, IL, 12 November 1996.

52. Valentini, R. F., Aebischer, P., Winn, S. R., Galletti, P. M., Collagen- and laminin-coating gels impede peripheral nerve regeneration through semipermeable nerve guidance channels. *Experimental Neurology*, 1987, **98**, 350–356.

53. Ranieri, J. P., Bellamkonda, R., Bekos, E. J., Vargo, T. G., Gardella, Jr., J. A. and Aebischer, P., Neuronal cell attachment to fluorinated ethylene propylene films with covalently immobilized laminin oligopeptides YIGSR and IKVAV. II. *The Journal of Biomedical Materials Research*, 1995, **29**, 779–785.

54. Bellamkonda, R., Ranieri, J. P. and Aebischer, P., Laminin oligopeptide derivatized agarose gels allow three-dimensional neurite extension in vitro. *Journal of Neuroscience Research*, 1995, **41**, 501–509.

55. Raivich, G. and Kreutzberg, G. W., Peripheral nerve regeneration: role of growth factors and their receptors. *The International Journal of Developmental Neuroscience*, 1993, **11**, 311–324.

56. Aebischer, P., Salessiotis, A. N. and Winn, S. R., Basic fibroblast growth factor released from synthetic guidance channels facilitates peripheral nerve regeneration across long nerve gaps. *Journal of Neuroscience Research*, 1989, **23**, 282–289.

57. Bryan, D. J., Eby, P. L., Costas, P. D., Wang, K., Chakalis, B. P. and Seckel, B. R., Living artificial nerve guides. American College of Surgeons 1992 Surgical Forum, Volume XLIII, 1992, pp. 651–652.

58. Levi, A. D. O., Guénard, V., Aebischer, P. and Bunge, R. P., The functional characteristics of Schwann cells cultured from human peripheral nerve after transplantation into a gap within the rat sciatic nerve. *Journal of Neuroscience*, 1994, **14**, 1309–1319.

59. Shenaq, S. M. and Rabinovsky, E. D., Gene therapy for plastic and reconstructive surgery. *Clinics in Plastic Surgery*, 1996, **23**, 157–171.

60. Nakahara, Y., Gage, F. H. and Tuszynksi, M. H., Grafts of fibroblasts genetically modified to secrete NGF, BDNF, NT-3, or basic FGF elicit differential responses in the adult spinal cord. *Cell Transplantation*, 1996, **5**, 191–204.

61. Bellamkonda, R., Ranieri, J. P., Bouche, N. and Aebischer, P., Hydrogel-based three-dimensional matrix for neural cells. *Journal of Biomedical Materials Research*, 1995, **29**, 663–671.

62. Herbert, C. B., Bittner, G. D. and Hubbell, J. A., Effects of fibrinolysis on neurite growth from dorsal root ganglia cultured in two- and three-dimensional fibrin gels. *Journal of Comparative Neurology*, 1996, **365**, 380–391.

63. Plant, G. W., Harvey, A. R. and Chirila, T. V., Axonal growth within poly (2-hydroxyethyl methacrylate) sponges infiltrated with Schwann cells and implanted into the lesioned rat optic tract. *Brain Research*, 1995, **671**, 119–130.

64. Shahar, A., Reuveny, S., David, Y., Budu, C. and Shainberg, A., Cerebral neurons, skeletal myoblasts, and cardiac muscle cells cultured on macroporous beads. *Biotechnology and Bioengineering*, 1994, **43**, 826–831.

65. Torimitsu, K. and Kawana, A., Selective growth of sensory fibers on metal oxide pattern in culture. *Developmental Brain Research*, 1990, **51**, 128–131.

66. Kleinfeld, D., Kahler, K. H. and Hockberger, P. E., Controlled outgrowth of dissociated neurons on patterned substrates. *Journal of Neuroscience*, 1988, **8**, 4098–4120.

67. Lom, B., Healy, K. E. and Hockberger, P. E., A versatile technique for patterning biomolecules onto glass coverslips. *Journal of Neuroscience Methods*, 1993, **50**, 385–397.

68. Chen, C. S., Mrksich, M., Whitesides, G. M., Ingber, D. E., Micropatterned adhesive surfaces identify geometric determinants of cell shape and growth. Paper presented at the 1996 American Institute of Chemical Engineers Annual Meeting, Chicago, IL, 12 November 1996.

CHAPTER III.11

Site-Specific Treatment of Central Nervous System Disorders Using Encapsulated Cells

FRANK T. GENTILE,
DWAINE F. EMERICH
CytoTherapeutics, Inc.,
2 Richmond Square,
Providence, RI 02906, USA

1 Introduction

The goal of encapsulated cell therapy research is to develop implants containing living xenogeneic cells to treat serious and disabling human conditions. The concept is straightforward: cells or small clusters of tissue are surrounded by a selective membrane barrier which admits oxygen and required metabolites, releases bioactive cell secretions but restricts the transport of the larger cytotoxic agents of the body's immune defense system. Use of a selective membrane both eliminates the need for chronic immuno suppression in the host and allows cells to be obtained from non-human sources, thus avoiding the cell-sourcing constraints which have limited the clinical application of generally successful investigative trials of *unencapsulated* cell transplantation for chronic pain [1], Parkinson's disease [2], and Type I diabetes [3, 4]. Cross-species immunoisolated cell therapy has been validated in small and large animal models of chronic pain [5, 6], Parkinson's disease [7, 8], and Type I diabetes [9–12], and is under active investigation in animal models of Huntington's disease [13–16], ALS [17] and Alzheimer's disease [18–22]. In addition, the first encapsulated cell therapy using xenografts in humans has been successfully performed in chronic pain [23, 24] and ALS [25, 26].

Recent reports have demonstrated the effectiveness of small diameter, ultrafiltration (UF) grade, hollow fibers as immunoisolation devices for the protection of allogeneic [27, 28] and xenogeneic [6–26] transplantation of cells. Although of varying composition, these devices typically consist of cells imbedded in an extracellular matrix (usually a hydrogel) and enclosed in an immunoisolation membrane. In addition to providing biocompatibility and permitting retrieval of appropriately designed implants, the role of the membrane portion of the device is to restrict the passage of large immune-system molecules and to shed antigens, while permitting relatively unhindered passage of small molecules required for cell metabolism and release of bioactive products to the host. Figure 1 depicts a hollow fiber membrane device that has been used in a US Phase I IND trial [24]. The device fiber contains bovine adrenal chromaffin (BAC) cells surrounded by Ca^{2+} crosslinked sodium alginate. Bovine chromaffin cells secrete a number of substances including catecholamines and opiod peptides. These substances have been hypothesized to be useful to ameliorate or alleviate chronic neuropathic or cancer related pain [1, 24].

Here we examine some of the factors which affect design and application of these devices. First, we will discuss the cell and extracellular matrix issues for these devices. Second, we will address membrane formulation and properties with particular emphasis on diffusive transport characterization. Third, we will present a case study of the evaluation of these devices in small and large animal models of Huntington's disease.

2 Cells and extracellular matrices used in encapsulation

2.1 Cells

Xenogeneic cells used within these membrane based devices fall into one of three basic types: (1) primary post-mitotic cells such as porcine islets of Langerhans, BAC cells or porcine hepatocytes; (2) immortalized (or divid-

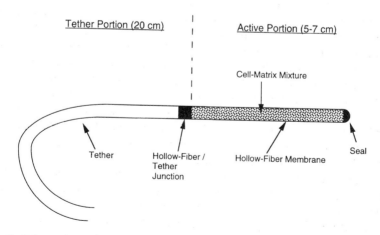

Fig. 1. Schematic representation of a hollow fiber membrane-based bioartificial organ for use in lumbar intrathecal, lateral ventricular, or parenchymal implantation.

ing) cells such as PC12 (derived from a rat pheochromacytoma) or fibroblasts such as BHK (baby hamster kidney); and (3) cell lines that have been engineered to secrete a bioactive substance such as BHK cells which secrete human nerve growth factor (hNGF), or human ciliary neurotrophic factor (hCNTF). Dividing tissue has advantages over post-mitotic tissue in that it can be banked and thus more easily tested for sterility and contaminants. It may also be possible to use a conditional immortalization technique for *in vitro* expansion of cell lines, including perhaps, allograft cell lines.

2.2 Extracellular matrices

In vivo, extracellular matrices (ECMs) provide control of cell function through the regulation of morphology, proliferation, differentiation, migration and metastasis [29–32]. Within a capsule, ECMs were originally employed simply to prevent aggregation of cells *(immobilization)* and resultant central necrosis, but have since been found to be beneficial to the viability and function of cells that require immobilization as well as a *scaffolding* for anchorage-dependent cell lines. Currently employed matrices are derived from naturally occurring polysaccharides (agarose, sodium alginate, chitin, chitosan or hyaluronic acid) or biologically-derived materials (Matrigel, vitrogen, Type I and IV collagen). Sodium alginate has been used with a wide variety of cell types (e.g. rat, canine and porcine islets and bovine adrenal chromaffin cells) [6, 9, 11, 12]. Chitosan is deacetylated chitin and has been used as a substrate for PC12 cells [8, 33]. Matrigel is a complex matrix consisting of growth factors and basement membrane, which has been used to study attachment, migration and differentiation of many cell types including neuronal, epithelial and endothelial cells. Low concentrations of laminin, fibronectin, Type I and Type IV collagen have been reported to support attachment, but not cell spreading, of hepatocytes [34, 35]. Collagen has also been reported as a matrix for baby hamster kidney cells genetically engineered to secrete human nerve growth factor (hNGF) and human ciliary neurotrophic factor (hCNTF) [13–15].

More recently, experiments have been performed to determine diffusive transport rates in matrix material used for encapsulation. Li *et al.* [36] showed that relatively low concentration hydrogels of crosslinked alginate (1.5–3%) had a significant diffusive resistance to a 150,000 MW marker when compared to water alone (10–100-fold). A similar preparation using agarose did not show such a dramatic change in diffusive transport.

3 Preparation of membranes used for immunoisolation

The majority of thermoplastic ultrafiltration (UF) and microfiltration (MF) membranes used to encapsulate cells are manufactured from homogenous polymer solutions by phase inversion. Ultrafiltration membranes have pore sizes ranging from 5 nm–0.1 μm while microfiltration (or microporous) membranes have pores ranging from 0.1–3 μm. Phase inversion is a very versatile technique allowing for the formation of membranes with a wide variety of nominal molecular weight cutoffs, permeabilities and morphologies [37–39]. The morphology and membrane properties depend on the thermodynamic and kinetic parameters of the process. The polymer is dissolved in an appropriate solvent. The solution is then cast as a flatsheet or extruded as a hollow fiber. As part of the casting or extrusion procedure, the polymer solution is precipitated by a phase transition which can be brought about by either a change in temperature or solution composition. This process involves the transfer of a single phase liquid polymer solution into a two phase system consisting of a polymer-rich phase that forms the membrane structure and a second liquid polymer-poor phase that forms the membrane pores. Any polymer that will form a homogenous solution which under certain temperatures and compositions will separate into two phases can be used. Thermodynamic and kinetic parameters such as chemical potential of the components and the free energy of mixing of the components determine the manner in which the phase separation takes place [38]. The process can be described by a polymer/solvent/non-solvent ternary phase diagram.

If membrane strength is limiting for the overall device strength, then the membrane must be manufactured with certain considerations in mind. For example, the membrane dimensions, composition and structure may have to be altered to increase the strength. Choosing a material that is inherently stronger (i.e. more ordered), or has a higher molecular weight, with which to cast the membrane should increase the overall mechanical properties. UF or MF membranes can be fabricated with macrovoids within the wall or as an open-cell foam where the microvoids are interconnected. By incorporating techniques that increase this isoreticulated structure within the membrane wall, the tensile strength can be increased generally at the same overall membrane porosity, thus maintaining the same overall diffusive transport. The strength can also be improved by increasing the cross-sectional area of the membrane by thickening the walls. Decreasing the overall membrane porosity will also serve to increase the overall membrane strength. Examples of both macrovoid containing and isoreticulated structures are shown in Fig. 2.

The outer morphology of the membranes can be altered during fabrication or by a post treatment to improve the reaction required for a successful implant. Using various phase inversion techniques, the outer surface of the membrane can range from a rejecting skin to a structure that is large enough to allow cells to enter into the wall itself (approximately 10 μm in diameter). The combination of proper membrane transport and outer morphologies may also be achieved using composite

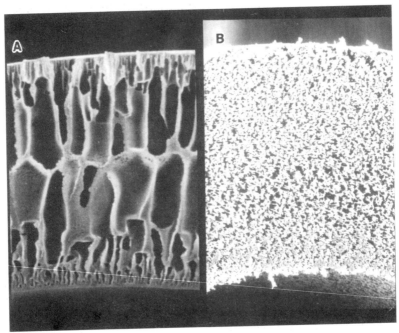

Fig. 2. (A) Scanning electron micrograph of macrovoid containing membrane. (B) Scanning electron micrograph of isoreticulated (open-cell foam) membrane structure.

membranes. Brauker [40, 41] and Boggs *et al.* [42] used such membranes for the treatment of Type I diabetes.

4 Membrane diffusive transport characterization

Transport of these membranes is typically characterized in terms of their capability to retain marker molecules in convective sieving experiments that compare which species are past or are retained by the membrane. Here the rejection coefficient, R, is defined as:

$$R = 1 - C_f/C_r \qquad (1)$$

where C_f and C_r are the concentrations of a marker in the filtrate and retentate, respectively.

Although standard convective measurements give an approximate idea of what size molecules can pass through a membrane, they are unable to predict the rate or selectivity in diffusion-based devices. Since all currently successful immunoisolation systems are primarily diffusion-based [24, 26,

42–47] it is essential to know diffusive transport properties in order to understand the intra-capsular environment of encapsulated cells, optimize the passage of essential metabolic species and delivery of product, and assess the degree of immune exposure.

4.1 General theory

In the absence of electrochemical or buoyant effects, the one-dimensional flux, N [g/cm^2 s], of a dissolved solute, i, may be represented by the following equation:

$$N_i = (-D\partial C/\partial x)_i + (1 - \sigma_i)\omega_i \qquad (2)$$

where D is the diffusion coefficient [cm^2/s], C is the concentration [g/cm^3], x is the distance [cm], σ is the Stavermen reflection coefficient [dimensionless], and ω is the mass velocity per surface area [g/cm^2 s]. The concentration and pressure dependent terms above represent the diffusive and convective components of solute transport. Equation (2) can usually only be integrated for special cases where either the interdependence of convection and diffusion is known or where the two components can be assumed to be independent and additive. However, in the case of no transmembrane convection (velocity, $\omega_i = 0$), applicable for many immunoisolation devices, the second term drops out, and Equation (2) becomes Fick's first law which can be integrated assuming a homogenous membrane and a solute concentration independent diffusion coefficient [48], yielding:

$$N_i = [k_t(C_1 - C_b)]_i = (k_t \Delta C)_i \qquad (3)$$

where ΔC is the bulk solution concentration difference between the fiber lumen (l) and the bath (b) [g/cm^3], and k_t is the overall mass transfer coefficient [cm/s]. Although non-linearities in the concentration profile caused by chemical, charge, or other interactions of the solute with the membrane material may make the internal concentration impossible to predict, they can usually be taken into account by using overall 'lumped-parameter' mass transport coefficients. Ideally, solute concentration is uniform in the liquid region on either side of the membrane and decreases linearly across the membrane. In actuality, such profiles are approximated only when the membrane resistance is large compared to the unstirred liquid resistances. Even with aggressive stirring, the transport of small, rapidly diffusing solutes can lead to gradients within the media which, if neglected, may cause large errors in calculated membrane mass transfer coefficients. The contribution of the unstirred liquid regions, often referred to as boundary layers, may be accounted for by defining the reciprocal overall mass transfer coefficient as an inverse sum of the coefficients for each region:

$$1/k_t = 1/k_b + 1/k_m + 1/k_l \qquad (4)$$

where the subscripts 'b', 'm', and 'l' refer to the outside bath, membrane and fiber lumen, respectively.

4.2 Small solutes

To minimize the effect of boundary layers, the membrane diffusivity of rapidly diffusing small solutes was measured using the flowing-type system in which solute diffuses from a recirculating bath fluid through the membrane wall and is carried away by buffer that is slowly pumped through the fiber lumen at a known flow rate. The mixing cup concentration of solute 'i', $(C_{mc})_i$, which is equal to the flux divided by the average velocity, was measured in the collected lumen fluid. Using Equation (3), assuming the buffer entering the fiber lumen is solute-free, the overall device-averaged mass transfer coefficient can be expressed in terms of the following easily measured variables:

$$(k_t)_i = (C_{mc})_i Q_f / (A \Delta C_i) \tag{5}$$

where A is the log mean membrane surface area calculated for an annulus ($2\pi(r_2 - r_1)z/(\ln(r_2/r_1))$, where r_2 and r_1 are the outer and inner radii, respectively, z is the length), and Q_f is the volumetric flowrate through the fiber lumen in $/\mu\text{lmin}$.

4.2.1 Membrane $(k_m)_i$

The transport resistance of the membrane is a complicated function of bulk diffusivity, D; membrane thickness; tortuosity; equilibrium partition coefficient (which in turn is a function of porosity, pore size distribution and membrane/solute interactions); and reduced pore diffusivity. In most practical situations, $(k_m)_i$ is impossible to calculate from fundamental principles and is instead calculated from measurements of overall mass transport. Effectively, $(k_m)_i$ can be viewed as the proportionality constant relating the measured flux, N_i, to the concentration difference of solute 'i' across the membrane if the solution concentration at the membrane/solution interface were known.

4.2.2 Bath side boundary layer $(k_b)_i$

A boundary layer develops because of the mass transport resistance, $(1/k_b)_i$, of the fluid immediately adjacent to either side of the membrane. The result of this resistance is that a concentration gradient must exist between the surface solution and the bulk solution in order to drive the species across this region. Stirring or flow can reduce, but not eliminate, the boundary layer resistance. As will be shown later, the boundary layer is numerically important only for rapidly diffusing small molecules where R_b, R_m and R_l are within an order of magnitude of each other.

The experimental system for the measurement of small solute diffusion is designed such that the bath solution flows tangential to the surface of the hollow fiber. (This provides a simpler method for estimating extra-lumenal flow than for most hollow fiber dialyzers). In this cross-flow configuration, the bath side mass transfer coefficient can be calculated using a Sherwood number (Sh) analogy for flow across a tube [49] such that:

$$Sh_{bi} = \{k_{bi}\, d_2/D\}_i = 0.911\, Re^{0.385}\, Sc^{0.333}$$

$$Re = vd_2/v \quad \text{Reynolds number}$$

$$Sc = v/D_i \quad \text{Schmidt number} \tag{6}$$

valid for $4 \le Re \le 40$; where d_2 is the outer fiber diameter [cm]; D_i is the bulk solution diffusivity of solute 'i' [cm^2/s]; v is the bath fluid velocity relative to the stationary fiber [cm/s]; and v is the kinematic viscosity of the bath fluid [cm^2/s]. The value of v used in this analysis is defined as the volumetric flowrate through the bath divided by the cross-sectional area of the fiber.

4.2.3 Fiber lumen boundary layer $(k_l)_i$

The transport resistance of a liquid flowing inside of a hollow fiber has been studied extensively in the field of dialysis. This resistance can be represented by a Sherwood number:

$$(k_l)_i = (Sh_l)_i D_i d_l \tag{7}$$

where d_l is the fiber inner diameter. Colton and Lowrie [50] present the graphical solution for $(Sh_l)_i$ for the case of axial laminar flow through a hollow tube, with no ultrafiltration, and neglecting axial diffusion. $(Sh_l)_i$ is shown to be a strong function of the dimensionless length, χ:

$$\chi = zD/vd^2 = zD\pi/4Q_f \tag{8}$$

where z is the fiber length [cm], and v is the bulk average velocity of fluid in the fiber [cm/s], and Q_f is the volumetric flowrate through the fiber lumen [cm^3/s]. $(Sh_l)_i$ is also a function of the wall Sherwood number, Sh_w, which is the sum of the membrane and bath side Sherwood numbers.

In most practical cases, for set geometry, flowrates, fiber diameter, and solute, 'i': $(k_b)_i$ is a constant, and $(Sh_l)_i$ can often be taken as a pseudo constant (numerically equal to 4.0) because of the low dependence on $(Sh_w)_i$, in which case k_l and k_b can be explicitly calculated and $(k_m)_i$ can then be calculated in a single equation:

$$(k_m)_i = 1/\{(A\,\Delta C/Q_f\,C_{mc}) - (1/k_l + 1/k_b)\}_i \tag{9}$$

The log mean concentration difference, ΔC_{lm} [g/cm^3], is usually used in Equation (9). After steady state has been achieved, k_m is given by:

$$(k_m)_i = 1/\{(A\,/Q_f\, \ln\,[C_b/(C_b - C_{mc})] - (1/k_l + 1/k_b)\}_i \tag{10}$$

4.3 Large solutes

'Large solutes' are practically defined as solutes with a molecular weight or size equal to or greater than the nMWCO of the membrane (defined here as the molecular weight at 90% rejection in a convective measurement). These solutes are sterically excluded from all but the largest pores, creating a membrane transport resistance which is at least an order of magnitude greater than the surrounding bath and lumen resistances. Since k_l and k_b were calculated to be less than 10% of the membrane resistance, this contribution to the overall resistance was assumed to be insignificant and the elaborate calculations introduced for the smaller solutes to compensate for the boundary layer contributions were not necessary. The large solutes used in the current diffusion study included both proteins (Bovine Serum Albumin, Immunoglobulin-G and Apoferritin) and fluorescein-tagged dextrans (10–2,000 kDa). Because long measurement times were necessary to attain detectable quantities of these large solutes, the membrane mass transfer coefficient, $(k_m)_i$, was measured in the static system, wherein marker solution was loaded inside the fiber and allowed to diffuse through the membrane into the surrounding bath which was initially filled with solute-free buffer. Stirring was not required to reduce boundary layers as diffusion through the membrane was the rate limiting step in transport. Equating the flux, N_i, from Equation (3) with the flux across the membrane one obtains:

$$N_i = V_b(dC_b/dt)/A = \{k_t \Delta C\}_i \tag{11}$$

A mass balance equates the amount of solute leaving the fiber to the amount entering the bath:

$$V_b dC_b/dt = -V_l \, dC_l/dt \tag{12}$$

Assuming that $k_t \approx k_m$, combining equation (12) with (11), integrating, and taking sample dilution into account, yields the following pseudo-steady-state transport expression presented by Colton *et al.* [51]:

$$\ln\left[\frac{(C_l - C_b)_{n+1}}{(C_l - C_b')_n}\right] = -(k_m)_i \, At_n\left[\frac{1}{V_l} + \frac{1}{V_b}\right] \tag{13}$$

with,

$$(C_b)'_n = (C_b)_n(1 - V_r/V_b) \tag{14}$$

$$(C_l)_{n+1} = (C_l)_n + (V_b/V_l)\{(C_b)'_n - (C_b)_{n+1}\} \tag{15}$$

where the mass transfer coefficient is in [cm/s], V_b is the bath volume [cm^3], V_l is the lumen volume [cm^3], t_n is time of sampling of the nth sample [s], $(C_l)_n$ and $(C_b)_n$ are the lumen and bath concentrations of the nth sample respectively, and V_r is the replacement volume. For the dextran diffusion experiments, it was not necessary to account for dilution of the bath since fewer samples were taken and smaller sample volumes were required due to the increased sensitivity in assaying of the fluorescently tagged dextrans.

4.4 Results

All measurements were performed on a single formulation of a hollow fiber prepared from solutions of poly(acrylonitrile-co-vinyl chloride) in an organic solvent via phase inversion. The results validating the system can be found in Dionne et al. [52]. Here, we show an example of the final diffusive characterization of a single membrane type used in transplantation studies.

Figure 3 contains the membrane mass transfer coefficient and relative membrane diffusivity (D_m/D_{H_2O}) for all markers plotted versus molecular size. The convective sieving curve using protein markers is also shown.

5 Case study: Use of encapsulated cells in the central nervous system

Much attention has been focused on cell transplantation in the Central Nervous System (CNS). The use of this approach in both the parenchymal and intrathecal spaces (lateral ventricle and lumbar intrathecal) has particularly strong advantages. First, it provides site specific delivery of various agents. Many of the therapeutic agents of interest do not cross the blood–brain barrier (e.g. dopamine or trophic factors for Parkinson's patients) or have a relatively short half-life. An example of this is ciliary neurotrophic factor (CNTF) [53, 54] for amyotrophic lateral sclerosis (ALS) or Huntington's disease (HD) patients. Second, encapsulation provides the added safety benefit of being, in some forms, retrievable. Third, it may be the only method to deliver any agent to the parenchymal site.

Table 1 shows the patient population for some of the CNS disorders for which delivery of various factors may be possible through encapsulated cells [55]. The following is a case study for the use of encapsulated cells in the treatment of HD.

Huntington's disease is an inherited neurodegenerative disease characterized by a relentlessly progressive movement disorder with psychiatric and cognitive deterioration. HD is invariably fatal with an average of 17 years of symptomatic illness. HD is found in all regions of the world, even in remote locations and occurs in an overall prevalence of between 5 and 10 per 100,000 [56]. Clinically, HD is characterized by an involuntary choreiform (dance-like) movement disorder, psychiatric and behavioral changes, and dementia [57]. The age of onset is usually between the mid thirties to late fifties, although juvenile (< 20 years of age) and late onset (> 65 years of age) HD occurs [56]. There are no effective treatments.

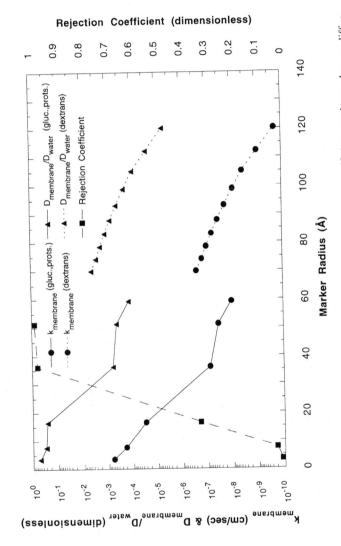

Fig. 3. Convective rejection coefficient, membrane mass transfer coefficient and membrane diffusivity relative to water vs. molecular size at 37°C for glucose and proteins and at 4°C for dextrans.

Table 1. Prevalence of central nervous system disorders which may be treated via encapsulated cells

Indication	Prevalence (US only)[a]
Dementias	4,000,000–5,000,000
(Alzheimer's)	(3,000,000–4,000,000)
Parkinson's	500,000
Amyotrophic lateral sclerosis (ALS)	25,000
Huntington's disease	25,000
Brain tumor (all types)	1,000,000
Chronic pain	600,000
(Cancer related pain)	(300,000)
Multiple sclerosis	200,000
Epilepsy	2,000,000

[a] Sources: National Parkinson Foundation, Parkinson's Disease Foundation, National Information Center for Orphan Drugs and Rare Diseases (ODPHP), National Health Information Center, and National Institute of Neurological Disorders and Statistics.

Although medications may reduce the severity of chorea or diminish behavioral symptoms, they do not increase survival or substantially improve quality of life as it relates to cognitive state, gait disorder or dysphagia [58].

Pathogenetically, HD is a disorder characterized by a programmed premature death of cells, predominantly in the caudate nucleus and the putamen. Degenerative changes can be visualized radiologically in advance of clinical symptoms [59, 60]. Initially, the disease affects GABA-ergic medium-sized spiny neurons, which receive afferent input from cortical glutamatergic and nigral dopaminergic neurons and provide efferent projections to the globus pallidus. Large aspiny interneurons and medium aspiny projection neurons are less affected and degenerate later in the disease process [61, 62]. Other subcortical and cortical brain regions are involved, but the degree of degeneration varies. This also does not correlate with the severity of the disease [62] and is dwarfed by the striatal changes.

Currently, therapy for HD is limited and does not favorably influence the progression of the disease. However, the recent increase in understanding of the genetic and pathogenetic events which cause neural degeneration suggests that there may be reason for optimism that novel experimental therapeutic strategies may be devised for the treatment of HD. The ability to devise novel therapeutic strategies for HD is tightly linked to the characterization of highly relevant animal models. Recently, it has been suggested that the neuronal death occurring in HD is related to an underlying endogenous excitotoxic process [63]. This hypothesis has given rise to models of HD in which excitotoxins are injected into the striatum to mimic the neuropathology and behavioral symptoms that occur in HD. While initial studies employed kainic acid [64], the excitotoxic model has recently been refined when it was shown that quinolinic acid more accurately modeled the pathology of HD [65, 66]. Excitotoxic lesions of the striatum in experimental animals also mimic motor and cognitive changes observed in HD [67, 68].

The delivery of neurotrophic molecules to the CNS has gained considerable attention as a potentially effective and rational strategy for the treatment of neurological disorders such as HD. Indeed, the use of trophic factors in a neural protection strategy may be particularly relevant for the treatment of HD. Unlike other neurodegenerative diseases, genetic screening can identify virtually all individuals at risk which will ultimately suffer from HD. This provides a unique opportunity to design treatment strategies which can intervene prior to the onset of striatal degeneration. Thus, instead of augmenting neuronal systems which have already undergone extensive neuronal death, trophic factor strategies can be designed to support host systems destined to die at a later time in the organism's life.

Ciliary neurotrophic factor (CNTF), is a member of the alpha-helical cytokine superfamily which has well-documented functions in the peripheral nervous system [54, 69–71]. Recently, it has become clear that CNTF also influences a wide range of CNS neurons. CNTF administration prevents the loss of cholinergic, dopaminergic and GABA-ergic neurons in different CNS lesion paradigms [72–74]. Importantly, an initial study demonstrated that infusions of CNTF into the lateral ventricle prevented the loss of Nissl-positive striatal neurons following QA administration. To confirm and expand upon these studies we conducted a series of studies examining the ability of CNTF to protect against QA lesions [14]. In these studies, animals received intraventricular implants of encapsulated human CNTF-producing (hCNTF) BHK fibroblasts followed by QA lesions of the striatum. An analysis of Nissl-stained sections demonstrated that the size of the lesion produced by QA was significantly reduced in those animals receiving BHK-hCNTF cells (1.44 ± 0.34 mm^2) compared with those animals receiving control implants (2.81 ± 0.25 mm^2). Quantitative analysis of striatal neurons further demonstrated that both choline acetyltransferase (ChAT) and glutamic acid decarboxylase (GAD)-immunoreactive neurons were protected by BHK-hCNTF implants. The loss of ChAT immunoreactive neurons in those animals receiving hCNTF implants was 12% compared to 81% in those animals receiving control cell implants. Similarly, the loss of GAD-immunoreactive neurons was attenuated in animals receiving hCNTF-producing cells (20%) compared to those animals receiving control cell implants (72%). In contrast, a similar loss of NADPH-diaphorase-positive cells was observed in the striatum of both implant groups (65%–78%). Analysis of retrieved capsules revealed numerous viable and mitotically active BHK cells that continued to secrete hCNTF.

A separate series of animals were tested on a battery of behavioral tests to determine the extent or behavioral protection produced by hCNTF. Bilateral infusions of QA produced a significant loss of body weight and mortality that was prevented by prior implantation with hCNTF-secreting cells. Moreover, QA produced a marked hyperactivity, an inability to use the forelimbs to retrieve food pellets in a staircase test, increased the latency of the rats to remove adhesive stimuli from their paws, and decreased the number of steps taken in a bracing test that assessed motor rigidity. Finally, the QA-infused animals were impaired in tests of cognitive function — the Morris water maze spatial learning task, and the delayed non-matching to position operant test of working memory. Prior implantation with hCNTF-secreting cells prevented the onset of all of the above deficits such that implanted animals were non-distinguishable from sham-lesioned controls [15]. At the conclusion of behavioral testing, nineteen days following QA, the animals were sacrificed for neurochemical determination of striatal ChAT and GAD levels. This analysis revealed that QA decreased striatal ChAT levels by 35% and striatal GAD levels by 45%. In contrast, hCNTF-treated animals did not exhibit any decrease in ChAT levels and only a 10% decrease in GAD levels.

An essential prerequisite before the initiation of clinical trials for HD is the demonstration that trophic factors can provide neuroprotection in a non-human primate model of HD. Toward this end, three cynomolgus monkeys received unilateral intrastriatal implants of polymer-encapsulated BHK-hCNTF cells [16]. The remaining three monkeys served as controls and received BHK-Control cells. One week later, all animals received injections of QA into the ipsilateral caudate and putamen. All monkeys were sacrificed 3 weeks later for histological analysis.

Within the host striatum, QA induced a characteristic lesion of intrinsic neurons together with a substantial atrophy of the striatum. In BHK-Control-implanted monkeys, Nissl-stained sections revealed extensive lesions in the caudate nucleus and putamen which were elliptical in shape. The lesion volume in the three BHK-Control animals averaged 317.72 mm^3 (± 26.01) in the caudate and 560.56 mm^3 (± 83.58) in the putamen. Many of the neurons that could be identified were shrunken and displayed a dystrophic morphology. In contrast, the size of the lesion was significantly reduced in BHK-hCNTF-implanted monkeys. The lesion volume in the three BHK-hCNTF monkeys averaged 83.72 mm^3 (± 24.07) in the caudate and 182.54 mm^3 (± 38.45) in the putamen. Numerous healthy appearing Nissl-stained neurons were observed within the striatum of BHK-hCNTF-implanted rats following the QA lesion even in regions proximal to the needle tract.

The neuroprotective effects of hCNTF were examined in further detail by quantifying the loss of specific cell types within the striatum (Table 2). Lesioned monkeys receiving BHK-Control implants displayed a significant loss (caudate = 89.0%; putamen = 94.1%) of GAD-immunoreactive neurons within the striatum ipsilateral to the transplant. Many remaining GAD-immunoreactive striatal neurons appeared atrophic relative to neurons on the contralateral side. The excitotoxic degeneration of GAD-immunoreactive neurons in both the caudate nucleus and putamen was significantly attenuated in monkeys receiving implants of polymer-encapsulated BHK-hCNTF cells as these monkeys displayed only a 64.0% and 64.1% reduction in GAD-ir neurons in the caudate and putamen, respectively, relative to the contralateral side. In addition to protecting the GABA-ergic cell bodies, hCNTF implants sustained the striatal GABA-ergic efferent pathways. Enhanced DARPP-32 immunoreactivity (a marker for GABA-ergic terminals) was seen within the globus pallidus and pars reticulata of the substantia nigra. Optical density measurements revealed a significant reduction (49.0 ± 5.2%) in DARPP-32 immunoreactivity within the globus pallidus in BHK-Control animals ipsilateral to the lesion. This reduction was significantly attenuated 12.0 ± 4.3%) in BHK-hCNTF implanted animals. Likewise, the reduction in the optical density of DARPP-32-immunoreactivity in the pars reticulata of animals receiving BHK-Control implants (17 ± 1.8%) was significantly attenuated (4.3 ± 0.4%) by hCNTF implants.

ChAT-immunoreactive and diaphorase-positive neurons were also protected by hCNTF administration (Table 2). Monkeys receiving BHK-Control implants displayed significant reductions in ChAT-positive neurons with the caudate (80.3%) and putamen (87.9%) compared to only a 41.2% (caudate) and 54.1% (putamen) in monkeys receiving BHK-hCNTF implants. Similarly, the loss of NADPH-d-positive neurons in BHK-Control-implanted monkeys (caudate = 83.9%; putamen = 86.6%) was significantly attenuated (caudate = 48.1%; putamen = 53.1%) in monkeys receiving encapsulated hCNTF implants.

These data clearly demonstrate that intrastriatal grafts of encapsulated hCNTF-producing fibroblasts protect GABAergic, cholinergic, and NADPH-d-containing neurons. However, the use of trophic factors, or any novel therapeutic strategy for HD, has been impeded by our inability to provide a rationale for how these approaches would influence critical non-striatal regions such as the cerebral cortex that also degenerate in this disorder. Accordingly, we quantified the size of neurons from layer V of monkey motor cortex from a series of Nissl-stained sections in each of the control and hCNTF-treated monkeys. QA produced a marked retrograde degeneration of cortical neurons in a region known to project to the striatum. While neuron number was unaffected, monkeys receiving BHK-Control implants displayed a significant atrophy (27%) of neurons in layer V of the motor cortex ipsilateral to the lesion. This atrophy was signifi-

Table 2. Neuronal cell counts in QA-lesioned monkey striatum

	Intact side	Caudate Lesioned/ implanted side	% loss
GAD-positive neurons			
hCNTF	18,088 (437)	6,516 (1,233)	64.0*
Control	19,298 (1,691)	2,133 (784)	89.0
ChAT-positive neurons			
hCNTF	2,259 (188)	1,328 (147)	41.2*
Control	2,239 (164)	441 (189)	80.3
NADPH-d-positive neurons			
hCNTF	3,339 (553)	1,733 (489)	48.1*
Control	3,405 (428)	548 (318)	83.9
	Intact side	Putamen Lesioned/ implanted side	% loss
GAD-positive neurons			
hCNTF	17,731 (604)	6,374 (851)	64.1*
Control	16,674 (867)	978 (545)	94.1
ChAT-positive neurons			
hCNTF	2,327 (209)	1,069 (36)	54.1*
Control	2,933 (172)	354 (253)	87.9
NADPH-d-positive neurons			
hCNTF	3,508 (824)	1,647 (389)	53.1*
Control	3,393 (106)	458 (347)	86.5

Cell counts in monkeys receiving implants of polymer-encapsulated hCNTF-secreting cells. Data are expressed as Mean($\pm$SEM) neurons on the intact side vs. the lesioned/implanted side. * $P < 0.05$, hCNTF vs. Control.

cantly attenuated (6%) in BHK-hCNTF implanted animals. Further analysis demonstrated that the atrophy of cortical neurons was not due to general volumetric changes but rather occurred preferentially in the medium sized (300–400 μm and 400–500 μm cross-sectional area) neurons of the motor cortex that project to the striatum.

These data provide the first demonstration that a therapeutic intervention can influence the degeneration of striatal neurons and disruption of basal ganglia circuitry in both rodent and primate models of HD. A number of different strategies have demonstrated protection of striatal neurons from excitotoxicity and mitochondrial dysfunction [13, 75, 76]. However, none of these have directly demonstrated protection of the GAD-immunoreactive neurons, the cell type most central in basal ganglia circuitry. Not only are GABAergic neurons viable in hCNTF-treated animals, but DARPP-32-immunoreactivity reveals that the two critical GABAergic efferent projections from striatum to globus pallidus and the pars reticulata of the substantia nigra are sustained following hCNTF treatment. Moreover, hCNTF implants exerted a robust neuroprotective effect on the cortical neurons innervating the striatum. These data indicate that a major component of the basal ganglia loop circuitry, the cortical → triatal → globus pallidus/substantia nigra outflow circuitry is sustained by cellular delivery of hCNTF. Since excitotoxicity has been implicated in a variety of pathological conditions including ischemia, and neurodegenerative diseases such as Parkinson's, Alzheimer's and Huntington's [76, 77], encapsulated cell based delivery of hCNTF may be one means of treating these and other CNS abnormalities.

References

1. Sagen, J., Chromaffin cell transplants for alleviation of chronic pain. *ASAIO Journal*, 1992, **38**, 24.
2. Lindvall, O., Rehncrona, S., Brundin, P., Gustavii, P., Astedt, B., Widner, H., Lindholm, T., Bjorklund, A., Leenders, K. L. and Rothwell, J. C., Neural transplantation in Parkinson's disease: The Swedish experience. *Progress in Brain Research*, 1990, **82**, 729.
3. Scharp, D. W., Lacy, P. E., Santiago, J. V., McCullough, C. S., Weide, L. G., Falqui, L., Marchetti, L., Gingerich, R. L., Jaffe, A. S., Cruer, P. E., Anderson, C. B. and Five, M. W., Insulin independence after islet transplantation into a Type I diabetes patient. *Diabetes*, 1990, **39**, 515.
4. Posselt, A. M., Narker, C. F., Tomaszewki, J. E., Markmann, J. F., Choti, M. A. and Naji, A., Induction of donor-specific unresponsiveness by intrathymic islet transplantation, *Science*, 1990, **249**, 1293.
5. Sagen, J., Wang, H., Tresco, P. A. and Aebischer, P., Transplants of immunologically isolated xenogeneic chromaffin cells provide a long-term source of pain-reducing neuroactive substances. *Journal of Neuroscience*, 1993, **13**(6), 2415.
6. Joseph, J. M., Goddard, M. B., Mills, J., Padrun, V., Zurn, A., Zelinski, B., Favere, J., Gardaz, J. P., Mosimann, F., Sagen, J., Christenson, L. and Aebischer, P., Transplantation of encapsulated bovine chromaffin cells in the sheep subarachnoid space: A preclinical study for the treatment of chronic pain. *Cell Transplantation*, 1994, **3**, 355.
7. Aebischer, P., Winn, S. R. and Galletti, P. M., Transplantation of neural tissue in polymer capsules. *Brain Research*, 1988, **448**, 364.

8. Aebischer, P., Goddard, M., Signore, P. and Timpson, R., Functional recovery in hemiparkinsonian primates transplanted with polymer encapsulated PC12 cells. *Experimental Neurology*, 1994, **126**, 1.

9. Chick, W. L., Perna, J. J., Lauris, V., Law, D., Galletti, P. M., Panol, G., Whittemore, A. D., Like, A. A., Colton, C. K. and Lysaght, M. J., Artificial pancreas using living beta cells: Effects of glucose homeostasis in diabetic rats. *Science*, 1977, **197**, 780.

10. Scharp, D. W., Mason, N. S. and Sparks, R. E., Islet immuno-isolation: The use of artificial organs to prevent tissue rejection. *World Journal of Surgery*, 1984, **8**, 221.

11. Lacy, P. E., Hegre, O. H., Gerasimidi-Vazeou, A., Gentile, F. T. and Dionne, K. E., Maintenance of normoglycemia in diabetic mice by subcutaneous xenografts of encapsulated islets. *Science*, 1991, **254**, 1782.

12. Sullivan, S. J., Maki, T., Borland, K. M., Mahoney, M. D., Solomon, B. A., Mueller, T. E., Monoco, A. P. and Chick, W. L., Biohybrid artificial pancreas: Long-term implantation studies in diabetic pancreatectomized dogs. *Science*, 1991, **252**, 718.

13. Emerich, D. F., Hammang, J. P., Baetge, E. E. and Winn, S. R., Implantation of polymer-encapsulated human nerve growth factor-secreting fibroblasts attenuates the behavioral and neuropathological consequences of quinolinic acid injections into rodent striatum. *Experimental Neurology*, 1994, **130**, 141.

14. Emerich, D. F., Winn, S. R., Lindner, M. D., Frydel, B. R. and Kordower, J. H., Implants of encapsulated human CNTF-producing fibroblasts prevent behavioral deficits and striatal degeneration in a rodent model of Huntington's disease. *Journal of Neuroscience*, 1996, **16**, 5168.

15. Emerich, D. F., Cain, C. K., Greco, C., Saydoff, J. A., Hu, Z. -Y., Liu, H. and Lindner, M. D., Cellular delivery of human CNTF prevents motor and cognitive dysfunction in a rodent model of Huntington's disease. *Cell Transplantation* (in press).

16. Emerich, D. F., Winn, S. R., Chen, E -Y., Chu, Y., McDermott, P. M., Baetge, E. E. and Kordower, J. H., Encapsulated CNTF-producing cells protect monkeys in a model of Huntington's disease. *Nature* (in press).

17. Sagot, Y., Tan, S. A., Baetge, E. E., Schmalbruch, H., Kato, A. C. and Aebischer, P., Polymer encapsulated cell lines genetically engineered to release ciliary neurotrophic factor can slow down progressive motor neuronopathy in the mouse. *European Journal of Neuroscience*, 1995, **7**, 1313.

18. Hoffman, D., Breakfield, X. O., Short, M. P. and Aebischer, P., Transplantation of polymer-encapsulated cell line genetically engineered to release NGF. *Experimental Neurology*, 1993, **122**, 100.

19. Kordower, J. H., Winn, S. R., Liu, Y. -T., Mufson, E J., Sladek, J. P., Hammang, J. P., Baetge, E. E. and Emerich, D. F., The aged monkey basal forebrain: Rescue and sprouting of axotomized basal forebrain neurons following grafts of encapsulated human NGF-secreting cells. *Proceedings of the National Academy of Science*, 1994, **91**, 10898.

20. Winn, S., Hammang, J., Emerich, D. F., Lee, S. R., Palmiter, D. and Baetge, E. E., Polymer-encapsulated cells genetically modified to secrete human nerve growth factor to promote the survival of axotomized septal cholinergic neurons. *Proceedings of the National Academy of Science*, 1994, **91**, 2324.

21. Emerich, D. F., Winn, S. R., Harper, J., Hammang, J., Baetge, E. E. and Kordower, J. H., Implants of polymer-encapsulated human NGF-secreting cells in the nonhuman primate: rescue and sprouting of degenerating cholinergic basal forebrain neurons. *Journal of Comparative Neurology*, 1994, **349**, 148.

22. Lindner, M. D., Kearns, C. E., Winn, S. R., Frydel, B. R. and Emerich, D. F., Effects of intraventricular encapsulated hNGF-secreting fibroblasts in aged rats. *Cell Transplantation*, 1995, **5**(2), 205.

23. Aebischer, P., Buschser, E., Joseph, J. M., Favre, J., deTribolet, N., Lysaght, M. J., Rudnick, S. A. and Goddard, M. B., Transplantation in humans of encapsulated xenogeneic cells without immunosuppression — A preliminary report. *Transplantation*, 1994, **58**(11), 1.

24. Burgess, F. W., Goddard, M. B., Savarese, D. and Wilkinson, H., Subarachnoid bovine adrenal chromaffin cell implants for cancer pain. Paper presented at American Pain Society Meeting, Washington, DC, November, 1996.

25. Pochon, N. A -M., Heyd, B., Déglon, N., Joseph, J. -M., Zurn, Z., Baetge, E. E., Hammang, J. P., Goddard, M., Lysaght, M., Kaplan, F., Kato, A. C., Schluep, M., Hirt, L., Regli, F., Prochet, F., deTribolet, N. and Aebischer, P., Gene therapy for Amyotrophic Lateral Sclerosis (ALS) using a polymer encapsulated xenogeneic cell line engineered to secrete hCNTF. *Human Gene Therapy*, 1995, **7**, 851.

26. Aebischer, P. Schluep, M., Déglon, N., Joseph, J. -M., Hirt, L., Heyd, B., Goddard, M. B., Hammang, J. P., Zurn, A. D., Kato, A. C., Baetge, E. E. and Regli, F., Intrathecal delivery of CNTF using encapsulated genetically modified xenogeneic cells in Amyotrophic Lateral Sclerosis patients. *Nature Med.*, 1996, **2**(6), 696.

27. Scharp, D. W., Swanson, C. J., Olack, B. J., Latta, P. P., Hegre, O. D., Doherty, E. J., Gentile, F. T., Flavin, K. S., Ansara, M. F. and Lacy, P. E., Protection of encapsulated human islets without immunosuppression in patients with Type I and Type II diabetes and in non-diabetic controls. *Diabetes*, 1994, **43**, 1167.

28. Soon-Shiong, P., Feldmen, E., Nelson, R., Komtebedde, J., Smidrod, O., Skjak-Braek, G., Espevik, T., Heintz, R. and Lee, M., Successful reversal of spontaneous diabetes in dogs by intraperitoneal microencapsulated islets. *Transplantation*, 1992, **54**, 769.

29. Dunn, J. C. Y., Tompkins, R. G. and Yarmush, M. L., Long-term *in vitro* function of adult hepatocytes in a collagen sandwich configuration. *Biotechnology Progress*, 1991, **7**, 237.

30. Mooney, D. J., Hansen, L., Vacanti, J., Langer, R., Farmer, S. and Ingber, D., Switching from differentiation to growth in hepatocytes: Control by extracellular matrix. *Journal of Cell Physiology*, 1992, **151**, 497.

31. Rotem, A., Toner, M., Tompkins, R. A. and Yarmush, M. C., Oxygen uptake rates in cultured rate hepatocytes. *Biotechnology Progress*, 1992, **40**, 286.

32. Uyama, S., Kaufmann, P. -M., Takeda, T. and Vacanti, J. P., Delivery of whole liver-equivalent hepatocyte mass using polymer devices and hepatotrophic stimulation. *Transplantation*, 1995, **55**(4), 932.

33. Emerich, D. F., Frydel, B. R., Flanagan, T. R., Palmatier, M., Winn, S. R. and

Christenson, L., Transplantation of polymer encapsulated PC12 cells: Use of chitosan as an immobilization matrix. *Cell Transplantation*, 1993, **2**, 241.

34. Tachibana, M., Nagamatsu, G. R. and Addonizio, J. C., Ureteral replacement using collagen sponge tube grafts. *Journal of Urology*, 1988, **133**, 866.

35. Atala, A., Vacanti, J. P., Peters, C. A., Mandell, J., Retick, A. B. and Freeman, M. R., Formation of urothelial structures *in vivo* from dissociated cells attached to biodegradable polymer scaffolds *in vitro*. *Journal of Urology*, 1992, **148**, 658.

36. Li, R. H., Altreuter, D. H. and Gentile, F. T., Transport characterization of hydrogel matrices for cell encapsulation. *Biotechnology and Bioengineering*, 1997, **50**, 365.

37. Michaels, A. S., US Patent No. 3,615,024, 1971.

38. Strathmann, H., Production of microporous media by phase inversion processes. In *Material Science of Synthetic Membranes*. American Chemical Society, 1985, p. 165.

39. Gentile, F. T., Shoichet, M. S., Rein, D. H., Doherty, E. J. and Winn, S. R., Structure and properties of polymers used in xenograft encapsulation. *Journal of Reactive Polymers*, 1995, **25**, 207.

40. Brauker, J. H., Bioarchitecture of polymers determines soft tissue response to implants. Paper presented at ACS Polymer Sci. and Engineering — Biomaterials for the 21st Century Meeting, Palm Springs, CA, 1992.

41. Brauker, J. H., Martinson, L. A., Hill, R. S., Young, S. K., Carr-Brendel, V. E. and Johnson, R. C., Neovasuclarization of immunoisolation membranes: The effect of membrane architecture and encapsulation tissue. *Transplant Proceedings*, 1992, **24**, 2924.

42. Boggs, D., Khare, A., McLarty, D., Pauley, R. and Sternberg, S. M., Membrane for immunoisolation. Paper presented at the Proceedings of the North American Membrane Society, 6th Annual Meeting, Breckenridge, CO, 1994.

43. Colton, C. K. and Augoustiniatos, E. S., Bioengineering in development of the hybrid artificial pancreas. *Journal of Biomechanical Engineering (Trans ASME)*, 1991, **113**, 152.

44. Christenson, L, Dionne, K. E. and Lysaght, M. J., Biomedical applications of immobilized cells, in *Fundamentals of Animal Cell Encapsulation and Immobilization*,' ed. M. F. A. Goosen. C.R.C. Press, Boca Raton, 1993.

45. Langer, R. and Vacanti, J. P., Tissue engineering. *Science*, 1993, **260**, 920.

46. Mullon, C., Dunleavy, K., Foley, A., O' Neil, J., Rudolph, J., Toscone, C., Otsu, I., Maki, T., Monoco, A., Gagnon, K., Naik, S., Santangini, H., Trenkler, D., Jauregui, H. and Solomon, B., Use of synthetic membranes in cell based medical devices — Artificial pancreas and liver assist device. *Polymer Materials Science Engineering*, 1994, **70**, 221.

47. Lysaght, M. J., Frydel, B., Gentile, F. T., Emerich, D. F. and Winn, S. R., Recent progress in immunoisolated cell therapy. *Journal of Cell Biochemistry*, 1994, **56**, 196.

48. Lysaght, M. J. and Baurmeister, U., Dialysis. In *Kirk-Othmer Encyclopedia of Chemical Technology*, 4th edn., 8", John Wiley, New York, 1993.

49. Keith, K. and Block, W. Z., *Basic Heat Transfer*. Harper and Row, New York, 1980.

50. Colton, C. K. and Lowrie, E. G., Hemodialysis: Physical principles and

technical considerations. In *The Kidney*, eds. B. M. Brenner and F. C. Rector. W.B. Saunders Co., Philadelphia, 1981.

51. Colton, C. K., Smith, K. A., Stroeve, P. and Merrill, E. W., Laminar flow mass transfer in a flat duct with permeable walls. *AICHE Journal*, 1971, **17**, 773.

52. Dionne, K. E., Cain, B. M., Li, R. H., Doherty, E. J., Lysaght, M. J., Rein, D. H. and Gentile, F. T., Transport characterization of membranes for immunoisolation. *Biomaterials*, 1996, **17**, 257.

53. Sendtner, M., Kreutzberg, G. W. and Thoenen, H., Ciliary neurotrophic factor prevents the degeneration of neurons after axotomy. *Nature*, 1990, **345**, 440.

54. Sendtner, M., Schmalbruch, H., Stockli, K. A., Carroll, P., Kreutzberg, G. W. and Thoenen, H., Ciliary neurotrophic factor prevents degeneration of motor neurons in mouse mutant progressive motor neuropathy. *Nature*, 1992, **358**, 502.

55. Swen, J. S., Flanagan, T. R. and Wiggans, T. G., Assessing commercial potential of central nervous system delivery approaches. *Methods in Neuroscience*, 1994, **21**, 485.

56. Conneally, P. M., Huntington's disease: genetics and epidemiology. *American Journal of Human Genetics*, 1984, **36**, 506.

57. Emerich, D. F. and Sanberg, P. R., Animal models in Huntington's disease. In *Neuromethods, Animal Models of Neurological Disease*, eds. A. A. Boulton, G. B. Baker and R. F. Butterworth, Vol. 17. Humana Press, NJ, 1992, pp. 65–134.

58. Shoulson, I., Huntington's disease: Functional capacities in patients treated with neuroleptic and antidepressant drugs. *Neurology*, 1981, **31**, 1333.

59. Grafton, S. T., Mazziotta, J. C. and Pahl, J. J., Serial changes of cerebral glucose metabolism and caudate size in persons at risk for Huntington's disease. *Archives of Neurology*, 1992, **49**(11), 1161.

60. Sedvall, G., Karlsson, P., Lundin, A., Anvret, M., Suhara, T., Haldin, C. and Farde, L., Dopamine D1 receptor number — A sensitive PET marker for early brain degeneration in Huntington's disease. *European Archives of Psychiatry and Clinical Neuroscience*, 1994, **243**(5), 249.

61. Ferrante, R. J., Beal, M. F., Kowall, N. W., Richardson, E. P. and Martin, J. B., Sparing of acetylcholinesterase-containing striatal neurons in Huntington's disease. *Brain Research*, 1987, 415.

62. Kowall, N. W., Ferrante, R. J. and Martin, J. B., Pattern of cell loss in Huntington's disease. *Trends in Neurosciences*, 1987, **10**, 24.

63. Beal, M. F., Does impairment of energy metabolism result in excitotoxic neuronal death in neurodegenerative illnesses? *Annals of Neurology*, 1992, **31**, 119.

64. Coyle, J. T. and Schwarcz, R., Lesion of striatal neurons with kinaic acid provides a model for Huntington's chorea. *Nature*, 1976, **263**, 244.

65. Beal, M. F., Kowall, N. W., Ellison, D. W., Mazurek, M. F., Swartz, K. J. and Martin, J. B., Replication of the neurochemical characteristics Huntington's disease by quinolinic acid. *Nature*, 1986, **321**, 168.

66. Beal, M. F., Ferrante, R. J., Swartz, K. J. and Kowall, N. W., Chronic quinolinic acid lesions in rats closely resemble Huntington's disease. *Journal of Neuroscience*, 1991, **11**, 1649.

67. Sanberg, P. R., Calderon, S. F., Giordano, M., Tew, J. M. and Norman, A. B.,

The quinolinic acid model of Huntington's disease: Locomotor abnormalities. *Experimental Neurology*, 1989, **105**, 45.

68. Block, F., Kunkel, M. and Schwarz, M., Quinolinic acid lesion of the striatum induces impairment in spatial learning and motor performance in rats. *Neuroscience Letters*, 1993, **149**, 126.

69. Stockli, K. A., Lottspeich, F., Sendtner, M., Masiakowski, P., Carrol, P., Gotz, R., Lindholm, D. and Thoenen, H., Molecular cloning, expression, and regional distribution of rat ciliary neurotrophic factor. *Nature*, 1989, **342**, 920.

70. Apfel, S. C., Arezzo, J. C., Moran, M. and Kessler, J. A., Effects of administration of ciliary neurotrophic factor on normal motor and sensory peripheral nerves *in vivo*. *Brain Research*, 1993, **604**, 1.

71. Masu, Y., Wolf, E., Holtmann, B, Sendtner, M., Brem, G. and Thoenen, H., Disruption of the CNTF gene results in motor neuron degeneration. *Nature*, 1993, **365**, 27.

72. Clatterbuck, R. E., Price, D. L. and Koliatsos, V. E., Ciliary neurotrophic factor prevents retrograde neuronal death in the adult central nervous system. *Proceedings of the National Academy of Science*, 1993, **90**, 2222.

73. Hagg, T. and Varon, S., Ciliary neurotrophic factor prevents degeneration of adult rat substantia nigra dopaminergic neurons *in vivo*. *Proceedings of the National Academy of Science*, 1993, **90**, 6315.

74. Hagg, T., Quon, D., Higaki, J. and Varon, S., Ciliary neurotrophic factor prevents neuronal degeneration and promotes low affinity NGF receptor expression in the adult rat CNS. *Neuron*, 1993, **8**, 145.

75. Schumacher, J. M., Short, M. P. and Hyman, B. T., Breakefield, X. O., Isacson, O., Intracerebral implantation of nerve growth factor-producing fibroblasts protects striatum against neurotoxic levels of excitatory amino acids. *Neuroscience*, 1991, **45**, 561.

76. Frim, D. M., Short, M. P., Rosenberg, W. S., Simpson, J., Breakefield, X. O. and Isacson, O., Local protective effects of nerve growth factor-secreting fibroblasts against excitotoxic lesions in the rat striatum. *Journal of Neurosurgery*, 1993, **78**, 267.

77. Choi, D. W., Glutamate neurotoxicity and diseases of the nervous system. *Neuron*, 1988, **1**, 623.

78. Olney, J. W., Excitatory amino acids and neuropsychiatric disorders. *Biological Psychiatry* 1989, **26**, 505.

SECTION IV

TISSUE ENGINEERING APPLIED TO ORGANS

CHAPTER IV.1

Tissue Engineering and the Cardiovascular System

ROBERT M. NEREM,
LINDA G. BRADDON,
DROR SELIKTAR

Parker H. Petit Institute for Bioengineering and Bioscience,
Georgia Institute of Technology,
Atlanta, GA 30332-0363, USA

1 Introduction

There has been a long-standing interest in the development of blood vessel and heart valve substitutes for use as vascular implants. In this some success has been achieved using what might be called artificial or non-biological approaches. This is certainly true of prosthetic heart valves where both mechanical valves and bioprosthetic valves are widely used.

In contrast, the success of currently available vascular grafts has been much more mixed. Synthetic materials have been successfully used for large diameter grafts, i.e. grafts with diameters of the order of 10 mm; however, for smaller diameter needs, long-term patency has not been good. Grafts of the order of 6 mm diameter fail largely due to intimal hyperplasia, while small bore grafts, i.e. 3–4 mm in diameter, fail due to thrombosis. For this reason, such common procedures as coronary bypass surgery generally use native blood vessels, either the sapphenous vein or the internal mammary artery.

Unfortunately, the use of native vessels does not take care of all the need. This is because there are patients for whom such native vessels are not satisfactory. There are others for whom the procedure may be a second or third bypass operation and where, as a result, native vessels are not available for use. Thus, there remains a clear need for an alternative to the use of native vessels, particularly for small diameter applications. It is to this challenge that the field of tissue engineering is attempting to respond.

2 Historical perspective

In the early 1950s the use of synthetic biomaterials as vascular substitutes was initiated. Over the years these efforts have included the development of plastics and other polymeric substances, and the use of non-solid, porous, woven, braided, or knitted fabrics. The first such biomaterial to be used clinically as a vascular prosthesis was made of the woven fabric Vinyon N [1, 2]. Since then vascular prostheses have been made from a variety of materials including nylon, Teflon, Orlon, Dacron, polyethylene, and polyurethane. Nylon was eliminated due to its rapid degeneration *in vivo*, and of the remaining materials, the mechanical properties of Dacron distinguished it as the most suitable for a blood vessel substitute, even though it is much less compliant than a native vessel. In the mid-1970s expanded polytetraflouroethylene (ePTFE) was introduced as another suitable material for a vascular prosthesis, and it is Dacron and ePTFE which are the most widely used today.

Using synthetic materials, a variety of approaches have been developed over the past 20 years. One is the so-called biohybrid graft in which a biological substance is applied to a synthetic material in order to alter the body's responses so as to help prevent graft failure. Knitted Dacron grafts with albumin, collagen, and gelatin are available currently for clinical use (see for example references 3–5). A different approach uses antibiotics, such as penicillins and Cephalosporins, bound to Dacron or ePTFE in order to create a resistance to bacterial infection [6]. Various substances also have been attached to grafts to decrease thrombogenicity [7], this with limited success.

In the early 1980s a novel approach to vascular grafts was introduced with the development of bioabsorbable materials. The idea behind this concept was to induce rapid and extensive tissue ingrowth, while the material gradually degraded. The cellular and extracellular matrix components being produced would over time take up the structural role [8–10]. The advantage of the bioresorbable graft is that no synthetic material is left behind.

Early attempts at tissue engineering a blood vessel substitute involved the endothelial seeding of a vascular graft made from a synthetic material. Credit for this concept goes to Herring *et al.* [11]. The purpose of the endothelial lining was to provide a 'natural' interface between the flowing blood and the underlying synthetic graft material. This in principle would both improve thromboresistance and possibly prevent intimal hyperplasia. Although some progress has been made with this approach [12–15], for small diameter vascular grafts such as needed for coronary bypass surgery the success achieved to date still leaves much to be desired.

More recently the focus of tissue engineering has moved to the development of a blood vessel substitute based on a construct composed of both a living co-culture of cells and at least in part natural biological materials. It is the recent developments in this field which are discussed in this chapter. In addressing this, it should be noted that, although the primary function of the vascular system is to transport blood, nutrients, and waste products, it is not simply a plumbing system. It is one which must function under a variety of physiological conditions, and this requires that it have the ability to adapt in a very controlled manner. Thus, the requirements of a blood vessel substitute are not simply that it be non-thrombogenic and able to handle the mechanical load imposed by the hemodynamics of the vascular system, it must also have the ability to adapt in a controlled way.

In addressing the progress achieved to date in developing tissue-engineered blood vessel substitutes, this review has been structured into three primary sections, each representing what might be called a core area of technology. These are: (i) cell technology, (ii) construct technology, and (iii) the technology required for the integration of constructs into living systems. These will each be addressed in turn.

3 Cell technology

In any 'living' blood vessel substitute, a critical issue is the selection of the cells to be employed. The 'natural' choices would be endothelial cells (EC) and smooth muscle cells (SMC); however, there are many factors which must go into this selection process. Critical areas of cell technology thus must include the following:

- Cell isolation and amplification
- Cell–substrate interactions
- Structure–function relations
- Cell function under mechanical stress
- Cell function under chemical stress
- Genetic engineering of cells

These will not all be dealt with here; rather, only a few key points will be made.

Certainly a critical issue is cell availability and that in turn is related to the issue of immune acceptance. Although the natural choice may be the use of autologous EC and SMC, this precludes having an off-the-shelf construct available for surgical implantation. What are other options? One possibility is to use allogenic EC and SMC, i.e. cells from another human. Unfortunately, these would not be accepted immunologically unless they had been modified so as not to express the MHC antigens which cause immune rejection, e.g. perhaps through genetic engineering. Whatever the likelihood of this, it should be noted that the field of immunology is one undergoing significant advances and fundamental changes. Thus, a greater range of options for immune acceptance may be available in the foreseeable future. A second possibility is to choose a non-vascular cell which normally does not express the MHC antigens. One such candidate cell is the fibroblast. Human foreskin fibroblasts have been used successfully in the development of skin substitutes, several of which will soon receive final approval from FDA. However, critical to the use of the fibroblast is whether or not the normal function of an SMC or an EC will be exhibited or can be engineered into the cell.

A second critical issue is cell function. For the cells selected, their ability to function both in their *in vitro* environment and in their *in vivo* environment will be important. Cells are the effectors of change, but how they function depends not only on their intrinsic genetic makeup, but also on their environment. Although it is tempting to draw a one-to-one relationship between genes and biological function, i.e. genes = biology; this is not the case. A more accurate statement is that genes + environment = biology. It is not simply the intrinsic genetic makeup of a cell which is important, but also the regulation of gene expression by the environment in which a cell resides.

The issue of the adaptation of a tissue-engineered blood vessel substitute has already been introduced. Tissues, and the cells within them, respond to a change in environment through physiological/biological adaptation. Cell adaptation is a response to environmental cues. Furthermore, cells in adapting modify their environment, e.g. through modification of their extracellular matrix. Equally important is the influence of environment on the phenotypic expression of a cell.

There is no better example of the above than a blood vessel where the vascular cells, i.e. EC and SMC, reside in an environment where mechanical factors, imposed by the hemodynamics of the vascular system, are an extremely important component. Taking EC as an example, this monolayer is exposed both to flow and the associated shear stress and to pressure. The latter, i.e. pressure, acts directly on the cell but also cyclically distends the vessel wall and the basement membrane to which the EC are attached. SMC, on the other hand, whether within the media or a thickened intima, are exposed to the stresses within the vessel wall, with the circumferential hoop stress being the major component of stress.

Over the last 15 years, much has been learned about the influence of flow and of cyclic stretch on vascular biology [16, 17]. For example, there is now considerable evidence indicating an important role for shear stress in regulating endothelial biology. Much of what is known results from cell culture experiments. In these studies endothelial monolayers are exposed to a well defined flow condition, e.g. the sudden onset of a known flow rate and shear stress. From such investigations, a variety of observations have been made. One of these is cell morphology (i.e. shape and orientation) where, as can be seen in Fig. 1, EC elongate in response to flow and orient their major axis with the direction of flow [18–20]. This is in contrast to EC in static culture where cell morphology is characterized by a cobblestone appearance. This difference in EC shape is also observed *in vivo*, where EC in regions of high shear are elongated while EC in regions of low shear, or even flow stasis, are polygonal or round in shape [21]. From cell culture studies it is clear that elongation is an active response, one involving a reorganization of the F-actin microfilament structure. This also is shown in Fig. 1 [22].

As dramatic as this elongation response is, there are other effects which are more important biologically. One such effect is the decrease in cell proliferation. Studies indicate that, the higher the shear stress, the lower the rate of cell growth, with this being due to an inhibition of entry into S-phase [23]. There also is an influence of flow on the synthesis and secretion of various biologically active molecules, with this effect extending to the regulation of gene expression. This influence on gene expression is a differential one. In some cases there is an upregulation, in others a downregulation, and in still others no effect at all [17].

There are several examples of the role of flow and the associated shear stress in the regulation of biologically active molecules. One is nitric oxide (NO), a molecule which is a potent vasodilator and also possibly linked to the nature of the vessel wall's oxidative environment. In cell culture studies, the sudden onset of flow results in an acute, dramatic increase in NO release [24], with the amount of NO released being regulated by the shear stress level. There also is a chronic increase which appears to be the result of an upregulation in nitric oxide synthase (NOS) which is required for the conversion of L-arginine to NO [25].

Another important molecule which is influenced by flow is the vasoconstrictor endothelin-1 (ET-1); however, here the effect of flow is more complicated. In contrast to low shear stress (5 dynes/cm^2) which increases both mRNA expression and ET-1 secretion [26], high shear stress (25 dynes/cm^2) decreases mRNA expression and secretion [27]. There also appears to be a threshold level of shear stress which induces a change in the trend of secretion [28]. The influence of flow on the regulation of vasoactive molecules thus is a very differential one, and this is true in general in terms of an effect of flow on the expression of biologically active molecules.

Cyclic stretch also influences vascular biology. For the endothelial cell [16, 17] this includes cell shape and orientation, cytoskeletal structure, cell proliferation, and the synthesis and secretion of various biologically active molecules. In some cases the influence of cyclic stretch is similar to that of flow. In a recent study using elastic tubes, it was shown that shear stress and cyclic stretch act synergistically to enhance EC elongation and alignment with the tube long axis [29]. In other cases, however, the effect of

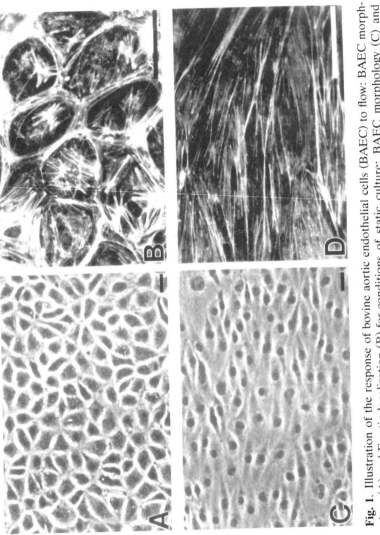

Fig. 1. Illustration of the response of bovine aortic endothelial cells (BAEC) to flow: BAEC morphology (A) and F-actin localization (B) for conditions of static culture; BAEC morphology (C) and F-actin localization (D) after exposure to a shear stress of 30 dynes/cm^2 for 24 hours.

cyclic stretch may produce a different, even opposite effect to that of flow. Thus, the regulation of endothelial structure and function by the hemody-namically-imposed mechanical environment is a very complex one.

The influence of cyclic stretch on vascular SMC also has been studied in cell culture. For SMC, as compared to EC, there are equally dramatic effects. Cyclic stretch influences cell orientation and a variety of other functional indicators. The latter includes actin reorganization [30], in-creased collagen and total protein synthesis [31, 32], and either unchanged or decreased cell proliferation, with there being some discrepancy in the results reported in the literature. Here also the effects of cyclic stretch extend to the gene expression level. For example, recent studies conducted at Georgia Tech shows that the mRNA for the IGF-1 receptor was downregulated, while mRNA for MCP-1, monocyte chemotactic protein 1, was upregulated [33]. Finally, there is an indication that cyclic stretch may modulate the phenotype of SMC in culture [34, 35]. This is an important issue which will be discussed later.

A final issue under this topic of cell technology is the possible use of genetically-engineered cells. Genetic engineering was introduced earlier as a way of perhaps altering cells to make them immune acceptable; however, for tissue engineering there are many other features of cells which could be altered through this technique. Some of these are as follows:

- Alter matrix/connective tissue synthesis
- Promote angiogenesis/vascularization
- Inhibit thrombogenesis
- Alter cell–substrate adhesion
- Promote cell proliferation
- Alter cell surface characteristics
- Promote cell migration

In addition to inhibiting the immune response of cells, there are two other examples which will be mentioned. The first is the use of genetic engineering to inhibit thrombogenesis. This might be done by engineering the cell so as to enhance its synthesis of a biologically active molecule which will inhibit platelet adherence. For example, endothelial cells have been transfected with the gene for tissue plasminogen activator which may enhance their thromboresistance [36]. A second example would be the use of genetic engineering to promote endothelialization. This could be done by enhancing a cell's migratory capabilities. As indicated above, there are other characteristics of a cell which might be altered through genetic engineering. The purpose here is not to discuss all of these, but rather to recognize that genetic engineering can be an ally of tissue engineering.

4 Construct technology

A blood vessel substitute is more than the product of cell technology. It involves the assembly of a three-dimensional, multi-cellular system that can be available as an off-the-shelf product for implantation into a patient. Although there are many factors involved, any list of critical areas for construct technology should include the following:

- Culture of multicellular systems
- Architecture for constructs
- Transport within 3-D constructs
- *In vitro* construct maintenance and growth
- *In vitro* monitoring of construct viability/function
- Construct preservation

As before, these will not all be discussed here, but rather only a few key comments will be made.

In the assembly of a blood vessel construct, there are several approaches which are being taken. Whereas some groups are employing a synthetic material as a scaffold for cell seeding, others are using collagen gels into which an appropriate cell type has been seeded. Bell *et al.* [37] were the first to create a tissue-like substance using the collagen gel approach. In this they employed rat tail collagen and fibroblasts, with the fibroblasts causing a contraction of the collagen lattice. The application of immediate interest for this work with fibroblasts in a collagen gel was the development of a skin equivalent; however, in the mid-1980s Weinberg and Bell [38,39] made a significant contribution to the work on blood vessel substitutes through the development of a construct made with vascular cells and collagen. In this construct the intimal region of the vascular equivalent was made with collagen and bovine smooth muscle cells, and a similar collagen gel, this time with fibroblasts, was molded to the outer surface of the substrate. The end result was a vascular equivalent with two distinct regions mimicking the media and the adventia. The inner surface was then seeded with endothelial cells, producing a monolayer which achieved reasonable confluency. A Dacron mesh sleeve was integrated into the structure in order to provide structural integrity. Bursting strength tests indicated that the collagen with cells alone was not strong enough to withstand physiological pressures; however, layering of multiple collagen sections and reinforcement with Dacron meshes produced a much stronger composite material capable of withstanding physiological pressures. Others also have contributed to our knowledge of how to build a blood vessel [40–42]. Significant in this has been the expansion of the work of Weinberg and Bell by Auger *et al.* using human cells and collagen to construct a triple-layer tubular vascular construct without the reinforcement of a

Dacron sleeve [43]. By culturing their media equivalent on a mandrel, they observed alignment of the smooth muscle cells in a circumferential manner with the occurrence of intercellular junctions. The degree of SMC orientation was observed to be a function of both their radial location and their time in culture. SMC which were adjacent to the casting mandrel exhibited a higher degree of circumferential alignment than those cells on the outer radial periphery. Likewise, a higher percentage of the SMC showed circumferential alignment at longer times in culture. When this tubular construct was seeded with human umbilical vein EC, the cells became confluent on the lumenal surface of the construct.

Matsuda *et al.* have used a variety of approaches in the development of a blood vessel construct [44–48]. In one approach they use a compliant, porous polyurethane graft to reinforce a hybrid medial layer comprised of SMC, type I collagen, and dermatan sulfate [44]. In a second approach, a hybrid medial layer is prepared with a mixture of SMC and type I collagen molded between a mandrel and a glass tube [45]. The hybrid vascular tissue developed by the second method resembles the native structure more precisely, but lacks the mechanical strength to withstand physiological pressures; thus, the tissue-engineered tubular constructs developed under the first approach yielded a more suitable prothesis for *in vivo* studies. In this approach a solution of collagen, dermatan sulfate, and cells is applied onto the lumenal surface of the polyurethane graft by repeated cycles of coating and thermal gelation. The hybrid tissue is then seeded with an EC monolayer on the lumenal surface. The results from such studies reveal that on the lumenal surface of the graft, an endothelial cell monolayer orients itself parallel to the direction of the blood flow. The medial layer is observed to increase in thickness and the SMC redistribute, becoming circumferencially oriented. The resulting construct thus resembles the arterial structure of a native vessel more closely than its unmodified counterpart.

More recently Matsuda's group has evaluated three different models prepared using an artificial scaffold and a sequential layering technique [46, 47]. Included were a monolayer (intima) model, a bilayer (intima/media) model, and a trilayer (intima/media/adventia) model. These models were each successfully implanted into mongrel dogs and followed for up to 52 weeks. Of particular interest was the observed phenotypic reversion of the SMC after implantation. This occurred earliest in the trilayer model, but by 52 weeks had occurred in all three models.

The approach being used at Advanced Tissue Sciences, Inc. in LaJolla, California, is one where a fibroblast/biodegradable vicryl material serves as a substrate for the adhering EC and as such is part of the local cellular environment. In this case EC function will be determined not only by the biochemical environment and the imposed hemodynamic loading, but also by the cell-seeded substrate to which the cells are anchored [48]. For the conditions of this study, it was observed that the morphology is considerably different than the cobblestone pattern normally observed in static culture. Clearly, the pattern of the vicryl substrate has an effect on EC shape and orientation. When such an EC-substrate construct is exposed to flow, there is a tendency to orient as is normally seen; however, the degree of alignment is influenced both by the direction of flow and the associated shear stress and also the pattern of the underlying biomaterial. It must be assumed that this influence of the underlying substrate is not just one on morphology, but also and even more importantly on cell function, e.g. the synthesizing and secretion of vasoactive substances and growth factors and the expression of cell adhesion molecules.

At Georgia Tech the approach being used involves a collagen gel into which SMC have been embedded and over which an EC monolayer has been plated [49]. This effort builds on the earlier work of Weinberg and Bell, with two versions of this model having been developed to date. One is a 'slab' model which can be incorporated into a parallel plate flow chamber for basic studies. The other is a 'tubular' construct which looks much like a blood vessel. In this co-culture model, porcine aortic SMC are seeded with soluble type I collagen. The collagen is allowed to polymerize in order to obtain a three-dimensional matrix within which SMC reside. The cells bind to the collagen fibers and contract the gel. The final contracted volume of the gel depends on many factors including the number of SMC seeded. When the collagen lattice has fully contracted, porcine aortic EC are plated at a very high density. These cells become confluent very rapidly. SMC grown in a collagen gel are much more elongated than cells grown on plastic and appear to have a phenotype that may be closer to contractile. SMC also are observed to grow very slowly in the gel as opposed to cells grown on a dish coated with collagen. EC seeded on top of the SMC-collagen gel construct are much more elongated than EC grown on plastic which have a cobblestone shape at confluence. Complete EC coverage and their morphology has been confirmed by scanning electron microscopy [50].

The influence of flow and the associated shear stress on such EC-SMC collagen constructs have been studied, with experiments to date focusing on morphology and on cell proliferation. In regard to the former, since even under static culture conditions EC are elongated, there is little further elongation with flow; however, the cells do align with the direction of flow. This occurs more rapidly at 30 dynes/cm^2 than at 10 dynes/cm^2.

Of particular interest to us have been the results of our studies on EC proliferation [49]. *In vivo* the endothelium may be characterized as being quiescent; however, this is not true for EC monolayers in culture. A typical EC monolayer in static culture, even when confluent, exhibits a cell turnover rate far in excess of that observed in animal studies. Such a monolayer, cultured on tissue culture plastic and exposed to the onset of a steady flow, does exhibit a decrease in cell proliferation; however, the turnover rate still is high compared to *in vivo* data. For this new co-culture model of the blood vessel wall the question then was what would be the proliferative activity of an EC monolayer seeded on top of a collagen-SMC gel? As presented in Table 1, the results indicate that, even under static culture conditions, the endothelium on a collagen–SMC gel exhibits a low cell turnover rate. As measured by [^{3}H]thymidine incorporation, the growth rate was 35 cpm/1000 cells. This was only slightly lower than that for a collagen gel without SMC; however, the measured EC growth rate for both of these cases represented a factor of five decrease compared with PAEC on plastic. This level of cell turnover corresponds to the presence of approximately 1% dividing cells and is very similar to that found in the endothelium of blood vessels [49]. As illustrated in Table 1, the effect of flow on the growth rate of co-cultured EC at best represented only a slight decrease, with the effect being much less than that noted earlier.

These initial studies demonstrate that it is possible to reconstitute a model of the vascular wall in cell culture, where the EC make a monolayer which covers the surface of the gel and within which reside SMC in a quiescent state. Most importantly, the endothelium also appears to be in a quiescent state. However, there are at least two issues critical to the success of a blood vessel substitute. One of these is to engineer such a substitute to have the appropriate mechanical properties so as to withstand the hemodynamic load. Blood vessels exhibit very complicated material property responses, being viscoelastic in nature. These properties include non-linear stress-strain characteristics, stress relaxation, creep and hysteresis [51]. The ideal vascular graft would not only be strong enough to

Table 1. Effect of culture conditions on the thymidine uptake by porcine EC

Culture condition	[^{3}H]Thymidine incorporation cpm/1000 cells (mean ± SEM)
Confluent EC on plastic 96 hours after seeding at 3.2×10^4 cells/cm^2	106 ± 37
EC exposed to 30 dyn/cm^2 for 24 hours	109 ± 23
Confluent EC on collagen gels 96 hours after seeding at 3.2×10^4 cells/cm^2	42 ± 22
Co-culture of EC and SMC	35.5 ± 10.9
Co-culture exposed to 30 dynes/cm^2 for 24 hours	25.3 ± 9.4

withstand physiological pulsatile pressures, but would also exhibit the compliance and viscoelastic characteristics of a native artery.

The proper characterization of the mechanical properties of collagen-based cellular substitutes has in fact been somewhat neglected. In preliminary studies conducted in our laboratory to determine the mechanical properties of a collagen–SMC lattice, a non-linear, plastic-like material response was exhibited [52]. The critical breaking stress of the gel was found to be largely dependent on cell seeding density. The stress–strain characteristics depended mostly on strain rate. In addition, the results of this study indicate that such gels do exhibit certain viscoelastic properties characteristic of biological tissue, such as stress relaxation. The strength of the collagen gel also was found to be a function of additives to the collagen–SMC construct. For example, the addition of the glycosaminoglycan, chitosan, inhibited gel contraction and altered the strength of the material.

A second critical issue relates to the ability of such a blood vessel substitute to adapt to a wide variety of physiological conditions. Related to this is vasoactivity, the ability to either vasoconstrict or vasodilate. Vasoactivity involves EC acting as a sensor of flow conditions within the vessel, the flow regulation of the EC synthesis of vasoactive molecules, and the presence of SMC of a contractile phenotype which, as a result, can either contract or relax depending on the signal received. A critical question thus is the phenotype of vascular SMC within a blood vessel substitute. It is well known that when SMC are put into culture, they undergo a modulation from their normal *in vivo* contractile phenotype to a synthetic phenotype. For a blood vessel substitute which incorporates SMC, it clearly would be desirable to induce a phenotypic reversion back to the contractile state. There are many factors which might contribute to such a modulation. This includes the presence of the neighboring EC [53], the three-dimensional organization of the vessel [54], and the influence of the mechanical environment [34, 35].

Also important to the mechanical properties of an SMC–collagen construct is the orientation of the SMC. As noted earlier, it appears that proper SMC orientation may be achieved through the stresses imposed during gel contraction, e.g. using a mandrel [43]. Tranquillo *et al.* [55] also have shown that it is possible to orient the SMC by applying a strong magnetic field during fibril formation. This not only increases the stiffness of the construct compared to control specimens, but also results in reduced creep in the circumferential direction.

Yet another critical question is how tissue constructs, once manufactured, can be preserved so as to be available as an off-the-shelf product? Here the solution seems to reside with the cryopreservation of such constructs, and although there is little in the literature in regard to cryopreserving tissue engineered constructs, there is a large literature in the area of the cryopreservation of native tissues. For example, Brossolet and Vito [56] have shown that the cryopreservation of blood vessels has little effect on their mechanical properties. Thus, this suggests that cryopreservation may represent an effective approach to preserving tissue constructs so as to provide off-the-shelf capability.

5 Integration into the living system

Once a blood vessel substitute has been constructed and preserved for future use, the next series of issues will arise when the construct is implanted into a patient. Critical areas of technology for the integration of constructs into living systems include the following:

- Construct/host tissue interfacing
- Immune acceptance
- Vascularization around/within implant
- Construct remodeling *in vivo*
- *In vivo* monitoring of construct viability/function

Again, these will not all be discussed here, and only a few key comments will be made.

It is only when you have implanted the construct that the immune acceptability of this blood vessel substitute becomes an issue. The reason we dealt with this issue in the earlier section on Cell Technology is because it is at the point when you select the cells to be employed that this issue needs to be first considered.

Once implanted there are issues related to the interfacing of the construct with the host tissue. One question for a blood vessel substitute is whether or not the construct will become vascularized? Even more important, particularly for a construct which relies on endothelialization to take place after implantation, will this in fact happen? Related questions include how rapidly will the host's EC migrate into the construct, how far will they move, and are there ways to accelerate the endothelization process? To date there is little information to help in answering such questions, although it is generally thought that grafts implanted into humans exhibit only limited endothelial cell ingrowth, not much more than 1–2 cm from either anastomosis. Clearly there is much here that needs to be done.

An even more critical issue related to the interfacing of the construct with the host is the extent of any adaptation or remodeling which will take place. In this case there also is very little information, particularly of a direct type; however, we can attempt to apply what we know about intimal thickening and intimal hyperplasia to the adaptation that might occur with an implanted blood vessel substitute. In doing this we start with the normal artery and the physiological adaptation that occurs in response to flow. What we know here is that both acute and chronic blood vessel responses to flow are endothelium dependent. For the chronic case, the adaptation process is one of intimal thickening. Here the lower the level of shear stress, the greater the intimal thickening. A very similar relationship has been observed for intimal hyperplasia by Ku *et al.* [57]. Again, the lower the level of shear, the greater the intimal hyperplasia. More recently it has been suggested that there may exist a similar relationship for the problem of restenosis. Although this is yet to be proven, it represents an interesting possibility.

If we apply this to the adaptation or remodeling that will take place when a tissue-engineered blood vessel substitute is implanted into a living system, then we would expect the proliferation of smooth muscle cells (or some other cell type) to be enhanced under conditions of low flow and the associated low shear stress. This would suggest that such a blood vessel substitute should be designed to maintain a high flow rate. However, Ku *et al.* [58] have shown that thrombosis becomes enhanced with increasing flow. Thus, Ku has argued that there may be an intermediate level of shear that will optimize patency.

Our final comment is in regard to the *in vivo* monitoring of the viability and function of an implanted construct. This also is a critical issue, one which has not received particular attention. However, with the diagnostic imaging modalities now available, e.g. magnetic resonance imaging, there would appear to be some real possibilities for future development.

6 Concluding discussion

This discussion has focused on the tissue engineering of blood vessel substitutes. Clearly there is still much more to be done before we will be able to have an off-the-shelf product available for clinical use.

In addition to blood vessel substitutes, there also is considerable interest in the development of tissue-engineered heart valves. As noted earlier, for prosthetic heart valves current approaches include both mechanical and bioprosthetic heart valves. More than 75,000 heart valve replacements are done annually in the U.S. alone, with two-thirds of these being mechanical valves. At first glance the bioprosthetic valve, being made of porcine tissue, would appear to be a natural material; however, these are fixed in glutalderhyde, totally decellularized, and thus represent little more than dead tissue.

For a tissue engineered heart valve to be successful, it may take even longer than a decade for at least two reasons. First, there is no environment in which a tissue engineered substitute must operate that is more dynamic than that of a heart valve. Opening and closing once a second, with both high shear stress levels and large bending stresses, the mechanical environment of a heart valve is a severe one. Secondly, whereas currently we do not have a suitable small diameter vascular graft, one with good long term patency, for heart valves we have adequate prostheses. Thus, for a tissue engineered heart valve to be successful, it must be more than adequate, it must exceed the performance of the currently available and successful mechanical and bioprosthetic valves.

In an interesting study carried out a few years ago by Breuer *et al.* [59], a tissue engineered heart valve was constructed by seeding autologous myofibroblasts onto a biodegradable polymeric scaffold to form a tissue core. Autologous endothelial cells were then formed as a monolayer on the outer leaflet surface. These tissue engineered leaflets were surgically implanted in the position of the right posterior pulmonary valve leaflet in the donor lamb. The animals were monitored for up to 3 weeks. There was no evidence of stenosis, and only minor flow regurgitation was observed. Histologically the tissue engineered leaflets resembled native valve leaflet tissue, although the tissue engineered ones were thicker and did not move as freely. This approach shows considerable promise; however, it should be remembered that the pulmonary side of the system is one at low pressure, and the mechanical environment here is much less severe than that on the arterial side. Furthermore, it may be easier to tissue engineer a leaflet than a complete valve.

There thus is still much to be done if we are to move tissue engineered cardiovascular substitutes into routine clinical use [60]. To address the challenges that must be faced will require interdisciplinary, in fact multidisciplinary teams. These will include bioengineers, cell biologists, immunologists, material scientists, and surgeons. Although one would like to be somewhat optimistic, and we believe we are, realistically the day of routine clinical availability is more than a decade away for blood vessel substitutes and perhaps even longer for tissue-engineered heart valves.

References

1. Voorhees, A. B. Jr., Jareczki, A. and Blakemore, A. H., The use of tubes constructed of Vinyon N cloth in bridging arterial defects. *Annals of Surgery*, 1952, **135**, 332.
2. Blakemore, A. H. and Voorhees, A. B. Jr., The use of Vinyon 'N' cloth in bridging arterial defects: Experimental and clinical. *Annals of Surgery*, 1954, **140**, 324.
3. Munro, M. S., Quattrone, A. J., Ellsworth, S. R., Kullcarni, P. V. and Eberhart, R. C., Alkyl-substituted polymers with enhanced albumin affinity. *Transactions of the American Society for Artificial Internal Organs*, 1981, **27**, 499.

4. Sipheia, R. and Chawla, A. S., Albuminated polymer surfaces for biomedical applications. *Biomaterials Medical Devices and Artificial Organs*, 1982, **10**, 229.

5. Kottke-Marchant, K., Anderson, J. M., Umemura, Y. and Marchant, R. E., Effect of albumin coating of the in vitro blood compatibility of Dacron arterial prostheses. *Biomaterials*, 1989, **10**, (34) 14.

6. Goldstone, J. and Moore, W. S., Infections in vascular prostheses. *American Journal of Surgery*, 1974, **128**, 325.

7. Park, K. D., Okano, T. and Jojitri, C. et al., Heparin immobilization onto segmented polyurethane urea surface: Effect of hydrophyllic spacers. *Journal of Biomedical Materials Research*, 1988, **22**, 977.

8. Bowald, S., Busch, C. and Eriksson, I., Absorbable material in vascular prostheses: A new device. *Acta Chirurgica Scandinavia*, 1980, **146**, 391.

9. Galletti, P. M., Aebischer, P., Sasken, H. F., Goddard, M. B. and Chiu, T. H., Experience with fully bioresorbably aortic grafts in the dog. *Surgery*, 1986, **103**, 231.

10. Pham, S., Durham, S., Johnson, R., Showalter, D., Endean, E. D., Vorp, D. A., Kim, D. V., Borovetz, H. S. and Greisler, H. P., Compliance changes in bioresorbably vascular prostheses following implantation. *Surgical Forum*, 1988, **39**, 440.

11. Herring, M. B., Gardner, A. L. and Glover, J., A single staged technique for seeding vascular grafts with autogenous endothelium. *Surgery*, 1978, **84**, 498.

12. Zilla, P. P., Fasol, R. D. and Deutsch, M., eds.. *Endothelialization of Vascular Grafts*, S. Karger, Basel, Switzerland, 1986.

13. Williams, S. K., Jarrell, B. E. and Kleinert, L. B., Endothelial cell transplantation onto porcine arteriovenous grafts evaluated using a canine model. *Journal of Investigative Surgery*, 1994, **7**, 503–517.

14. Zilla, P., Deutsch, M., Meinhart, J., Puschmann, R. et al., Clinical in vitro endothelialization of femoropopliteal bypass grafts: An actuarial follow-up. *Journal of Vascular Surgery* , 1994, **19**(3), 540.

15. Pasic, M., Muller-Glauser, W., von Segesser, L., Lachac, M., Mihaljevic, T. and Turina, M., Superior late patency of small-diameter Dacron grafts seeded with omental microvascular cells: An experimental study. *Annals of Thoracic Surgery*, 1994, **58**, 677.

16. Nerem, R. M. and Girard, P. R., Hemodynamic influences on vascular endothelial biology. *Toxicologic Pathology*, 1990, **18**(4/1), 572–582.

17. Nerem, R. M., Hemodynamics and the vascular endothelium. *ASME Journal of Biomechanical Engineering*, 1993, **115**, 510–514.

18. Dewey, C. F., Bussolari, S. R., Gimbrone, M. A. Jr. and Davies, P. F., The dynamic response of vascular endothelial cells to fluid shear stress. *ASME Journal of Biomechanical Engineering*, 1981, **103**, 177–181.

19. Eskin, S. G., Ives, C. L., McIntire, L. V. and Navarro, L. T., Response of cultured endothelial cells to steady flow. *Microvascular Research*, 1984, **28**, 87–94.

20. Levesque, M. J. and Nerem, R. M., The elongation and orientation of cultured endothelial cells in response to shear stress. *ASME Journal of Biomechanical Engineering*, 1985, **176**, 341–347.

21. Levesque, M. J., Liepsch, D., Moravec, S. and Nerem, R. M., Correlation of

endothelial cell shape and wall shear stress in a stenosed dog aorta. *Arteriosclerosis*, 1986, **6**, 220–229.

22. Girard, P. R. and Nerem, R. M., Shear stress modulates endothelial cell morphology and F-actin organization through the regulation of focal adhesion-associated proteins. *Journal of Cell Physiology*, 1995, **163**, 179–193.
23. Levesque, M. J., Nerem, R. M. and Sprague, E. A., Vascular endothelial cell proliferation in culture and the influence of flow. *Biomaterials*, 1990, **11**(9), 702–707.
24. Taylor, W. R., Harrison, D. G., Nerem, R. M., Peterson, T. E. and Alexander, R. W., Characterization of the release of endothelium-derived nitrogen oxides by shear stress. *FASEB Journal*, 1991, **56**(6), A1727 (Abstract).
25. Uematsu, M., Ohara, Y., Navas, J. P., Nishida, K., Murphy, T. J., Alexander, R. W., Nerem, R. M. and Harrison, D. G., Regulation of endothelial cell nitric oxide synthase mRNA expression by shear stress. *Americal Journal of Physiology: Cell Physiology*, 1996, **269**, C1371–C1378.
26. Morita, T., Kurihara, H., Maemura, K., Yoshizumi, M. and Yazaki, Y., Disruption of cytoskeletal structures mediates shear stress-induced endothelin-1 gene expression in cultured aortic endothelial cells. *Journal Clinical Investigation*, 1993, **92**, 1706–1712.
27. Sharefkin, J. B., Diamond, S. L., Eskin, S. G., Dieffenbach, C. and McIntire, L. V., Fluid flow decreases endothelin mRNA levels and suppresses endothelin peptide release in human endothelial cells. *Journal of Vascular Surgury*, 1991, **14**, 1.
28. Kuchan, M. J. and Frangos, J. A., Shear stress regulates endothelin-1 release via protein kinase C and cGMP in cultured endothelial cells. *American Journal of Physiology*, 1993, **264** (33), H150–H156.
29. Zhao, S., Suciu, A., Ziegler, T., Moore, J. E. Jr., Burki, E., Meister, J. J. and Brunner, H. R., Synergistic effects of fluid shear stress and cyclic circumferential stretch on vascular endothelial cell morphology and cytoskeleton. *Arteriosclerosis, Thrombosis and Vascular Biology*, 1995, **15**(10), 1781–1786.
30. Dartsch, P. C. and Hammerlee, H., Orientation of cultured arterial smooth muscle cells growing on cyclically stretched substrates. *Acta Anatomica*, 1986, **125**, 108–113.
31. Leung, D. Y. M., Glagov, S. and Mathews, M. B., Cyclic stretching stimulates synthesis of matrix components by arterial smooth muscle cells in vitro. *Science*, 1975, **191**(4226), 475–477.
32. Sumpio, B. E., Banes, A. J., Link, W. G. and Johnson, G., Enhanced collagen production by smooth muscle cells during repetitive mechanical stretching. *Archives of Surgury*, 1988, **123**, 1233–1236.
33. Schnetzer, K. J., Delafontaine, P. and Nerem, R. M., Uniaxial cyclic stretch of rat aortic smooth muscle cells. *Annals of Biomedical Engineering*, 1995, **23**(1)(Suppl.), S-42.
34. Birukov, K. G., Shirinsky, V. P., Stepanova, O. V., Tkachuk, V. A., Hahn, A. W., Resink, T. J. and Smirnov, V. N., Stretch affects phenotype and proliferation of vascular smooth muscle cells. *Molecular and Cellular Biochemistry*, 1995, **144**(2), 131–139.
35. Shirinsky, V. P., Birukov, K. G., Stepanova, O. V., Tkachuk, V. A., Hahn, A. W. A. and Resink, T. J., Mechanical stimulation affects phenotype features of

vascular smooth muscle cells. In *Proceedings of the 10th International Symposium on Atherosclerosis*, Montreal, 1995.

36. Dichek, D. A., Neville, R. F., Zwiebel, J. A., Freeman, S. M., Leon, M. B. and Anderson W. F., Enhancement of the fibrinolytic activity of sheep endothelial cells by retroviral-mediated gene transfer. *Blood*, 1991, **77**, 533.

37. Bell, E., Ivarsson, B. and Merrill, C., Production of a tissue-like structure by contraction of collagen lattices by human fibroblasts of different proliferative potential *in vitro*. *Proceedings of the National Academy of Science*, 1979, **76**, 1274–1278.

38. Weinberg, C. B. and Bell, E., Regulation of proliferation of bovine aortic endothelial cells, smooth muscle cells and dermal fibroblasts in collagen lattices. *Journal of Cell Physiology*, 1985, **122**, 410–414.

39. Weinberg, C. B. and Bell, E., A blood vessel model constructed from collagen and cultured vascular cells. *Science*, 1986, **231**, 397–399.

40. Jones, P. A., Construction of an artificial blood vessel wall from cultured endothelial and smooth muscle cells. *Cell Biology*, 1979, **76**, 1882–1886.

41. Lei, B. V. D., Wildevuur, C. R. H. and Nieuwenhuis, P., Compliance and biodegradation of vascular grafts stimulate the regeneration of elastic laminae in neoarterial tissue: An experimental study in rats. *Surgery*, 1986, **99**, 45–52.

42. Van Buul-Wortelboer, M. F., Brinkman, H. J. M., Dingemans, K. P., DeGroot, P. G., van Aken, W. G. and van Mourik, J. A., Reconstruction of the vascular wall in vitro: A novel model to study interactions between endothelial and smooth muscle cells. *Experimental Cell Research*, 1986, **162**, 151–158.

43. L'Heureux, N., Germain, L., Labbe, R. and Auger, F. A., In vitro construction of a human blood vessel from cultured vascular cells: A morphological study. *Journal of Vascular Surgery*, 1993, **17**(3), 499–509.

44. Miwa, H., Matsuda, T. and Iida, F., Development of a hierarchically structured hybrid vascular graft biomimicking natural arteries. *ASAIO Journal*, 1993, **39**, M273–M277.

45. Hirai, J., Kanda, K., Oka, T. and Matsuda, T., Highly oriented, tubuler hybrid vascular tissue for a low pressure circulatory system. *ASAIO Journal*, 1994, **40**, M383–M388.

46. Ishibashi, K. and Matsuda, T., Reconstruction of a hybrid vascular graft hierarchically layered with three cell types. *ASAIO Journal*, 1994, **40**(3), M284–M290.

47. Matsuda, T. and Miwa, H., A hybrid vascular model biomimicking the hierarchic structure of arterial wall: Neointimal stability and neoarterial regeneration process under arterial circulation. *Journal of Thoracic and Cardiovascular Surgery*, 1995, **110**(4/1), 988–997.

48. Braddon, L. G., Karolyi, D. R. and Nerem, R. M., F-Actin organization of endothelial cells on a fibroblast/biodegradable polymer construct. *Proceedings of the Workshop on Biomaterials and Tissue Engineering*, Hilton Head, SC, 19–23 February 1997.

49. Ziegler, T., Alexander, R. W. and Nerem, R. M., An endothelial cell-smooth muscle cell co-culture model for use in the investigation of flow effects on vascular biology. *Annals of Biomedical Engineering*, 1995, **23**, 216–225.

50. Ziegler, T., Robinson, K. A., Alexander, R. W. and Nerem, R. M., Co-culture of endothelial cells and smooth muscle cells in a flow environment: An

improved culture model of the vascular wall? *Cells and Materials*, 1995, **5**, 115–124.

51. Fung, Y. C., *Biomechanics: Mechanical Properties of Living Tissues*. Springer-Verlag, New York, 1981.

52. Greer, L. S., Vito, R. P. and Nerem, R. M., Material property testing of a collagen-smooth muscle cell lattice for the vonstruction of a bioartificial vascular graft. In *ASME Advances in Bioengineering*, BED-Vol. 28 , ed. M. J. Askew. 1994, 69–70.

53. Campbell, J. H. and Campbell, G. R., Endothelial cell influences on vascular smooth muscle phenotype. *Annual Review of Physiology*, 1986, **48**, 295–306.

54. Kanda, K., Matsuda, T., Miwa, H. and Oka, T., Phenotypic modulation of smooth muscle cells in intima-media incorporated hybrid vascular prostheses. *ASAIO Journal*, 1993, **39**, M278–M282.

55. Tranquillo, R. T., Girton, T. S., Bromberek, B. A., Triebes, T. G. and Mooradian, D. L., Magnetically-oriented tissue-equivalent tubes: Application to a circumferentially-oriented media-equivalent. *Biomaterials* (in press).

56. Brossolet, L. J. and Vito, R. P., The effects of cryopreservation on the biaxial mechanical properties of canine saphenous veins. *ASME Journal of Biomechanical Engineering*, 1997, **119**(1), 1–5.

57. Salam, T. A., Lumsden, A. B., Suggs, W. D. and Ku, D. N., Low shear stress promotes intimal hyperplasia thickening. *Journal of Vascular Investigation*, 1996 (in press).

58. Siegel, J. M., Markou, C. P., Ku, D. N. and Hanson, S. R., A scaling law for wall shear stress through an arterial stenosis. *ASME Journal of Biomechanical Engineering*, 1994, **116**, 446–451.

59. Breuer, C. K., Shinoka, T., Tanel, R. E., Zund, G., Mooney, D. J., Ma, P. X., Miura, T., Colan, S., Langer, R., Mayer, J. E. and Vacanti, J. P., Tissue engineering lamb heart valve leaflets. *Biotechnology and Bioengineering*, 1996, **50**, 562–567.

60. Anderson, J. M. et al., Tissue engineering in cardiovascular disease: A report. *Journal of Biomedical Materials Research*, 1995, **29**, 1473–1475.

CHAPTER IV.2

Tissue Engineering Applied to the Heart

CHRISTINE L. TOCK,
TIMOTHY SCOTT-BURDEN
Vascular Cell Biology Laboratory,
Texas Heart Institute,
Houston, Texas 77030, USA

1 Physiology of the heart

1.1 Normal heart physiology

The mammalian heart consists of two separate pulsatile pumps which exhibit levels of efficiency and reliability that surpass those achieved by any mechanical pump built to date [1]. The heart is comprised physically of a left and right functional unit each of which consists of two individual chambers, namely the atrium and the ventricle (see Fig. 1). The right and left hearts work through synchronous contractions to pump blood into the vascular beds of the pulmonary and systemic end organs, respectively. The periodicity of the cardiac cycle is typically divided into two phases. Diastole is the phase of ventricular filling, and systole consists of the ventricular ejection phase. During systole, the left ventricle has a stroke volume of 45 ± 13 ml at a heart rate of 60 beats per minute in a normal adult male [2]. This correlates to a cardiac output index of $2.5–4.2$ $1/min/m^2$.

Cardiac stroke volume and heart rate are controlled both through mechanisms intrinsic to the heart and through autonomic regulation via the central nervous system. Stroke volume is determined primarily by the Frank–Starling law [3]. Thus, with increasing diastolic venous blood return, the ventricular myocardium contracts more forcefully, and ejects a larger blood volume during systole. In this manner, stroke volume can double during periods of increased peripheral oxygen demands. Although the myocardium possesses an intrinsic, rhythmic contractility, heart rate is principally controlled by the autonomic nervous system. The autonomic nervous system can excite or depress the reactivity of specialized impulse conduction pathways within the heart which trigger cardiac contractions [3].

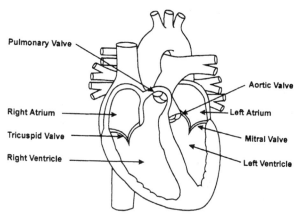

Fig. 1. Anatomical cross section of the mammalian heart.

Essentially, the human heart can be thought of as two pulsatile pumps acting in series with the main objective of this system being to physically drive the delivery of oxygenated blood into the body's organs via the systemic vasculature. Within this system, the left ventricle is the primary driving force for systemic perfusion. Therefore, impaired left ventricular function has serious implications. In summary, in spite of the succinct words of Dr. Christian Barnhard at a press conference shortly after the first successful heart transplant in 1967, 'The heart are (sic) just a pump.', the function and regulation of the heart exhibits several layers of complexity which would be hard to mimic in a mechanical system.

1.2 Congestive heart failure

Congestive heart failure (CHF) is one of the leading causes of morbidity and mortality in the United States. Epidemiological studies demonstrate that 400,000 new cases of heart failure occur annually [4]. Further, despite improved medical therapy, heart failure is the primary cause of death in approximately 40,000 people each year and is implicated as a contributing cause of death in up to an additional 250,000 cases annually.

CHF stems from many etiologies, and is a potential endpoint for all types of heart disease [1]. Braunwald defines congestive heart failure as '...the pathophysiological state in which an abnormality of cardiac function is responsible for failure of the heart to pump blood at a rate commensurate with requirements of metabolizing tissues...' [5]. The primary etiologies of this disease most often include chronic hypertension, cardiac ischemic disease, valvular disease, and cardiomyopathy [1]. Standard therapy for CHF includes treatment of the underlying cause of the CHF in addition to antihypertensive medications and diuretics to lower the systemic stress on the failing heart.

Although current medical therapy has improved, the morbidity and mortality of CHF, it continues to be a fatal disease with only a 25% 5-year survival rate for men and a 38% rate for women [6]. Currently, the only treatment available for end stage CHF is homologous cardiac transplantation. Donor availability limits transplants to 2500/year, and this is unlikely to improve significantly in spite of major efforts aimed at public awareness of the problem including recruitment into organ donor programs [4]. Encouragingly, patient's receiving a donor heart have an average 1-year survival rate of greater than 80% [7].

In an effort to prolong the window of opportunity for cardiac transplantation, heart assist devices have been developed to provide a supportive bridge between complete heart failure and cardiac transplantation. These devices include total artificial hearts (TAH) and left ventricular assist devices (LVAD). Current clinically available devices include the Novacor LVAS (Baxter Healthcare Corp., Oakland, CA), the Heartmate LVAD (ThermoCardiosystems Inc., Woburn, MA), Thoratec's LVAD and biventricular assist device (Thoratec Lab Co., Berkeley, CA) and the CardioWest TAH (CardioWest Tech. Inc., Tuscon, AZ) [8].

Recently, LVAD support has been compared to traditional medical management of end stage CHF. The results indicate that not only does LVAD support have the same efficacy as traditional management, but an increased number of LVAD recipients survived to receive subsequent heart transplants as compared to patients receiving traditional therapy alone [9]. The transplantation rate for medically managed patients was reported as 63% with an 88% discharge rate from the hospital. In comparison, patients receiving LVADs prior to transplantation had a transplantation rate of 88% and a 100% discharge rate from the hospital. Moreover, LVAD implantation in patients awaiting a donor heart has been correlated with a more favorable transplant outcome [7]. This improvement in outcome is attributed to a general increase in patient well being due to augmentation of end organ perfusion by the LVAD.

Complications associated with LVAD implantation include post-operative bleeding, device related infection, and thromboembolism [10, 11]. Approximately 39% of LVAD recipients have significant post-operative bleeding requiring further surgical intervention. Post-operative bleeding may be related to the massive initial deposition of platelets on the blood contacting surfaces of LVADs leading to a significant reduction in platelet concentration in the blood. Device related infection occurs in 25% of LVAD implantations. Interestingly, in vascular graft studies with similar infection rates, the infection rate is reduced if the graft material is seeded with endothelial cells prior to implantation [12]. Finally, thromboembolic events resulting in significant central nervous system impairment are prevalent with implantation of LVADs. With respect to all clinically available devices, thromboembolic events occur in 33–47% of LVAD implantations. The TCI Heartmate LVAD has a very low thromboembolic rate (< 1%) during the initial 100 days of implantation.

Although orthotopic cardiac transplantation is the most effective therapy for end stage CHF, it does not significantly reduce the overall mortality of this common disease due to the limitation of donor organ availability [7, 13]. In light of the clinical successes of LVAD's and TAH's, these devices are now being considered as possible long term alternatives to transplantation for individuals who do not receive a donor heart [13, 14]. Therefore, in the context of clinical utilization of the LVAD as a permanent support for chronic CHF, the physiologic complications such as thromboembolism, post-operative bleeding, and infection associated with implantation of such devices must be minimized.

1.3 Valvular disease

The tricuspid and mitral valves separate the atrium from the ventricle in the right and left heart, respectively (see Fig. 1). These atrial–ventricular (A–V) valves are one-way valves preventing backflow into the atrial chambers during ventricular contraction. Further these valves both open and close passively, driven only by the pressure differential between atrium and ventricle. The pulmonic and aortic valves are located in the outflow tracts of the right and left ventricle, respectively. These valves are anatomically distinct from the A–V valves. Although they undergo passive opening with ventricular ejection, valve closure requires some back-pressure which causes the valves to snap shut. Valvular disease results in the loss of these boundaries and disruption of the flow/pressure dynamics of the entire cardiac cycle. Ultimately, valvular disease can lead to secondary cardiac failure and consequential decline in perfusion of the body's vascular system and end organs.

The pathogenesis of valvular disease has been reviewed extensively, and therefore will be only summarized briefly [1]. The etiology of valvular disease essentially falls into the categories of congenital deformities and deformities secondary to inflammatory responses such as rheumatic heart disease. In both cases, the cardiac valves, either individually or collectively, lose their ability to perform as gatekeepers within the cardiac system. Left heart valves are more commonly affected. Valve deformities can present as either stenosis or regurgitation. Valve stenosis consists of narrowing of the valve orifice, and the subsequent reduction in blood volume delivered across the orifice during a given cardiac cycle. Valve regurgitation involves impaired valve leaflet function due to tissue damage, and this leads to incompetent closure at the end of systole. Therefore, the ejected stroke volume is effectively reduced due to backflow.

Ultimately, valvular disease may lead to significant heart failure and consequent death. Severe valvular disease is treated currently by surgical implantation of mechanical or bioprosthetic, preserved heart valves. The rate of complications for individuals with artificial heart valves is 6–9% per year [1]. The most common complication is thromboembolism (1–4% per year), and mechanical heart valve recipients must receive anticoagulant therapy. Furthermore, 20–30% of the preserved, bioprosthetic porcine valves require replacement after 8–10 years of implantation due to tissue degeneration and calcification. However, these valves have a lower incidence of thromboembolism as compared to mechanical heart valves [1, 15].

2 Tissue engineering considerations that are unique to the heart

2.1 Total heart replacement

To date, production of a tissue engineered heart has not been achieved, but it remains a much sought after goal. The primary considerations in cardiac tissue engineering include achieving a mechanical performance that equals or exceeds the *in vivo* capabilities of the native heart. The engineered organ would need to produce adequate pressures to drive tissue perfusion while maintaining a contractile environment that would not cause hemolysis. Moreover, a tissue engineered heart should be capable of responding to dynamic changes in peripheral oxygen demand with responsive changes in cardiac output. As a component of the cardiovascular system, the organ would be required to be non-thrombogenic. In addition, the tissue engineered heart should not elicit any immunologic response leading to rejection of the organ after implantation. The development of a biological, pulsatile pump would most likely involve placing the organ within an environment similar to the cardiovascular system. Physiologically appropriate load bearing would promote dynamic conditioning of the tissue engineered organ. This dynamic conditioning, unlike tranditional *in vitro* cell culture, would prepare the tissue engineered heart for its functional purpose *in vivo*. In addition to adequate cardiac contractility, heart valve competance would be critical. Functional, tissue engineered heart valves could be used both as a singular mode of therapy in cases of valvular disease as well as an important component in the development of a complete, tissue engineered heart.

2.2 Cardiac muscle replacement

All types of muscle tissue within the human body undergo remodeling secondary to applied stresses [16]. Skeletal and smooth muscle cells are capable of both hypertrophy (increase in cell size) of single cells as well as hyperplasia (increase in cell number) as means to compensate for increased work load. Adult cardiac muscle is only able to respond to increased work loads such as systemic hypertension by hypertrophy of existing tissue. Adult cardiac muscle does not undergo a hyperplastic response to such stimuli since myocardial cells have undergone terminal differentiation. Terminal differentiation occurs early in neonatal life when embryonic cardiomyocytes lose the ability to divide [17]. This accounts for the lack of cardiomyocyte regeneration following injury of adult tissue as in myocardial infarction. When muscle tissue is not exposed to a work load for an extended period of time, muscle cells undergo atrophy (decrease in cell size), and a subsequent reduction in muscle strength. Atrophy can occur with any type of muscle tissue. Significantly, muscle cells grown in traditional *in vitro* cell culture are not subjected to force bearing loads, and therefore lose the ability to contract effectively.

The use of skeletal myoblasts (undifferentiated skeletal muscle cells) for implantation into injured myocardium may provide a method to generate a viable, contractile cardiac wall. This method would result in the formation of autologous skeletal muscle fibers which may function as a replacement for infarcted myocardial wall. Alternatively, Field *et al.*, using a mouse model, have shown that cardiomyocytes can be selected from differentiating embryonic stem cells. Once implanted into adult mouse hearts, these cells remain as viable, intracardiac grafts for up to 7 weeks [18]. In both cases, the recipient heart acts as an *in vivo* scaffold for the implanted myocytes and thus facilitates their dynamic conditioning.

2.3 Valve replacement

As already stated, cardiac valves are essentially passive elements within the architecture of the heart. The stress/strain and stress/relaxation curves for an engineered valve should be comparable to those of a natural valve [19]. For instance, the aortic valve experiences both circumferential and radial stresses throughout the cardiac cycle. Moreover, the valve also undergoes deformation (strain) during the cardiac cycle [20]. The modulus of elasticity is the slope of a curve relating stress to strain in a given material. This value is a measure of the stiffness of the material (in this case the aortic valve) [21]. Thus, an engineered valve should not be too stiff as this would prevent adequate flow across the valve orifice, and could lead to detrimental effects such as valve stenosis. Conversely, a valve that was not as rigid as a natural valve would allow flow regurgitation. Thus, tissue engineered valves would need to undergo both mechanical evaluation as well as *in vitro* flow performance studies to evaluate valve integrity and reliability within simulated flow environments.

2.4 Permanent heart assist devices

As noted previously, left ventricular assist devices are currently used clinically as a bridge to cardiac transplantation (see Fig. 2). To optimize such devices for permanent implantation, the thrombogenicity of the blood contacting surfaces should be minimized to reduce the risk of thromboembolism. One potential method to improve the biocompatibility of the biomaterials currently used in the LVAD is the lining of the blood contacting surfaces of the device with a non-thrombogenic layer of cells. Ideally, this intimal cellular lining would be derived from autologous tissue prior to implantation of the LVAD in the patient. This would likely decrease the immunogenicity as well as the thrombogenicity of the blood contacting surface. In contrast to a tissue engineered heart or valve, an intimal cellular lining on the blood contacting surface should possess a strong adhesive capacity for the underlying device material. This is especially important since such a cellular lining would be exposed to dynamic, non-biologic flow regimes within the context of the intimal surface of the biolining. Previous *in vitro* flow studies have shown that shearing forces can approach 350 dynes/cm^2 within the cavity of clinically available LVADs. Further, the inlet and outlet regions of the LVAD can produce shearing forces up to 2700 dynes/cm^2 during the flow cycle [22, 23]. These shearing forces are in excess of the forces produced in the healthy human vasculature which can approach 50–100 dynes/cm^2 within the arterial system. Shearing stresses in excess of 4000 dynes/cm^2 have been shown to cause hemolysis, and it is noteworthy that shearing forces as low as 100 dynes/cm^2 can cause hemolysis in the presence of artificial surfaces [22]. Finally, when considering *in vivo* viability of a cellular lining, the diffusional limits of nutrients become important due to the avascular nature of a cellular lining. Although diffusional limitations may vary with cell type, a conservative estimate suggests that the thickness of the cell layer should be less than 150 μm thick [24].

3 Examples of tissue engineering applied to the heart

3.1 Heart replacement

Although there are published examples of tissue engineering of heart components, it is not surprising that a complete tissue engineered heart remains to be accomplished. During the 1970s, Nosé presented an extended investigation of ways to optimize formaldehyde preservation of bovine pericardium and valves so that they could be utilized as a 'biolized' ventricular assist device [25, 26]. More recently, with the emergence of genetic engineering technology, tissue engineering of hearts has taken a

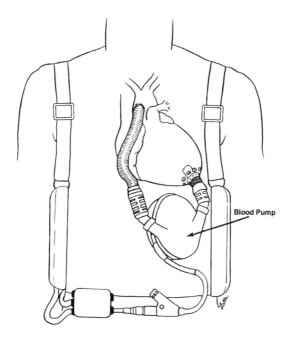

Fig. 2. The Thermocardiosystems Heartmate Left Ventricular Assist Device. The TCI Heartmate LVAD is attached to the apex of the left ventricle, and acts in series with the left ventricle to pump blood into the aorta. Percutaneous leads allow for connection of portable battery packs and venting of the pump housing.

molecular twist in the study of xenograft transplantation. A xenograft is an organ transplantation between species. Currently, the porcine model is favored for xenograft transplantations because of the comparable size and physiology of this animal to humans. Furthermore, the pig has good breeding characteristics for development of large, inbred population of genetically homogeneous donors [27]. The preliminary hurdle to be overcome in xenograft rejection is hypothesized to be endothelial injury from the host immune system by complement activation [28]. With increased molecular and cellular understanding of the spectrum of xenograft rejection, transgenic pigs have been developed which express human complement regulatory proteins. Thus, ideally the xenograft will regulate complement mediated rejection when transplanted into a human recipient. However, the xenograft recipients still retain their complement-mediated immunologic defense responses outside of the transplanted heart [28].

3.2 Valve replacement

Mayer *et al.* have presented preliminary findings for a tissue engineered pulmonic valve leaflet in a lamb model [29]. Mixed cell populations of endothelial and vascular smooth muscle cells derived from autologous and allogenic arteries were seeded onto leaflet shaped polyglycolic acid fiber matrices in static cell culture systems. The right posterior pulmonic valve leaflet was then replaced with either autologous or allogenic tissue engineered leaflets in lambs. Results at 4 weeks after implantation showed no valvular stenosis and only trivial regurgitation with these autologously derived valves. The allogenic leaflet demonstrated an acute inflammatory response. The authors concluded that autologous cell populations were preferred, and that further studies involving complete valve replacement and left heart valve replacement were needed. Furthermore, the authors noted that cell origin and cell seeding protocols would likely improve the dynamic response of the valves. A significant observation made in these studies was related to the organization of the mixed cell populations used. The authors observed that endothelial cells were actually situated on the surface of the cell impregnated polyglycolic acid matrix which contained mainly vascular smooth muscle cells. Thus, this mixed population of vascular cells formed a physiologically stratified structure with a superficial endothelial cell lining.

Advanced Tissue Sciences (LaJolla, CA) has developed a tissue engineered heart valve using human fibroblasts seeded under physiologic flow conditions onto a decellularized whole porcine valve [30, 31]. The Bio-XenoGraft™ heart valve is cultured in a bioreactor which produces variable pressure changes across the valve. This example of dynamic conditioning appears to promote extracellular matrix synthesis by the human fibroblasts. As with Mayer *et al.*, protocols for cell seeding densities are currently being optimized. Although no *in vivo* data has been reported, cell viability has been observed in the *in vitro* bioreactor at time points up to 4 weeks. Future investigations are aimed at optimizing colonization of the valve scaffold as well as determining the effects of shearing forces on cell viability.

3.3 Intimal cellular linings for ventricular assist devices

The concept of biologic intimal linings for cardiac assist devices dates back to the initial considerations in the design of ventricular assist devices. In 1966, DeBakey *et al.* proposed that nylon and dacron 'velour' fabrics provided a permanent architecture for deposition of a coagulum that would act as a autologous lining within ventricular assist pumps [32]. Further experiments with these materials in a dog model showed the development of islands of proliferating spindle cells after 10 weeks of implantation [33]. In 1969, Bernhard *et al.* presented results from studies using textured grafts and ventricular assist devices cultured with bovine fetal fibroblasts. LVADs were seeded with 3×10^7 fibroblasts for 60–120 minutes prior to implantation in a bovine model. Preliminary results showed no immunological response to the allograft cells at 50 days, and minimal cell loss from the LVAD surface [34]. Follow-up studies indicated a decreased calcification of the luminal surface of implanted LVADs that were initially seeded with fetal fibroblasts as compared to unseeded LVADs. Furthermore, after 150–160 days of implantation, cell lined LVADs displayed an adherent, thin collagenous lining containing viable cells, in contrast to unlined LVADs which displayed thrombus, white blood cell adhesion, giant cell incorporation and granular calcified material. Finally, the adhesive strength of the cellular lining was significantly greater than that of the fibrinous lining present on the control LVADs [35].

More recently, investigations have been undertaken by at least two laboratories to develop intimal cellular linings for LVADs. Lelkes *et al.* have investigated the feasibility of intimal endothelialization of an LVAD as a means to produce a biocompatible, non-thrombogenic surface for long term implantation [36]. In healthy vasculature, endothelial cells line the luminal surfaces of the blood vessels and maintain an antithrombotic environment. Utilizing endothelial cell linings in vascular prosthesis has been investigated in the past in artificial grafts with limited success [37–39]. In the case of vascular grafts, intimal endothelial cell linings have been shown to reduce thrombosis, and infection. However, the level of endothelial cell loss from these surfaces is unacceptably high both initially and long term. By extrapolation, if similar levels of cells were to be lost from the LVAD surface catastrophic downstream effects could be initiated such as cerebral vascular accidents (strokes). In order to reduce endothelial cell loss from LVAD surfaces, Lelkes *et al.* have investigated methods to promote cell monolayer stability in the dynamic flow environment of the LVAD [40]. For instance, Lelkes found that allowing endothelial cells to remain in static culture for 6 days or more allowed the elaboration of a complete extracellular matrix and decreased the level of cell loss upon exposure to flow as compared to cells seeded and exposed to flow after 24 hours. Furthermore, the importance of substrate surface characteristics became apparent. In this instance, both increased surface roughness of the underlying LVAD biomaterial and the precoating of the biomaterial surface with extracellular matrix products dramatically improved the endothelial cell adhesiveness during 6 and 24-hour cell studies within an *in vitro* LVAD flow loop.

Intuitively, endothelial cells would appear to be the ideal choice for the formation of intimal cellular linings for LVADs and other cardiovascular prostheses. However, there are several drawbacks to their use *in vitro*, including their tendency to undergo rapid senescence in culture. This has lead to the investigation of other cell types for use as intimal cellular linings in cardiovascular devices [34]. Scott-Burden *et al.* have investigated the use of autologous vascular smooth muscle cells for this purpose. The ability of smooth muscle cells to form strong attachments to biomaterials coupled to the ease with which these cells can be maintained in culture, are both significant advantages for the use of smooth muscle cells as compared to endothelial cells. However, the disadvantages associated with vascular smooth muscle cells include their propensity towards over-proliferation, and a significant thrombogenic potential. These smooth muscle cell traits would appear to be an insurmountable obstacle with respect to their use in a cardiovascular environment. Fortunately, the advent of DNA technology has provided a method to overcome these apparent drawbacks to smooth muscle use. Introduction of genes that endow smooth muscle cells with the ability to produce antithrombotic products has opened the door for their use as intimal cellular linings in the LVAD.

An example of designing an antithrombotic smooth muscle cell via DNA technology is the introduction of the genes for Nitric Oxide Synthase (NOS) and GTP cyclohydrolase into the smooth muscle cells using an episomal replication vector as a vehicle for gene expression within the cell. Expression of the NOS protein in conjunction with a required cofactor (tetrahydrobiopterin) supplied by the activity of the enzyme GTP cyclohydrolase leads to nitric oxide (NO) production by the transduced cells. Elaboration of NO promotes an antithrombotic environment surrounding the smooth muscle cells. Platelet adherence to smooth muscle cells, an obligatory step in the initiation of thrombus formation, is inhibited by NO. Moreover, NO is a demonstrative inhibitor of smooth muscle cell growth and thus transduction of NO production in such cells leads to the control of their proliferation. This added benefit of NO production is critical for maintaining a cellular lining within the LVAD both from the perspectives of nutrient diffusional limitations as well as preventing cellular overgrowth and possible occlusion of the LVAD lumen.

Longterm expression of NO synthase (coupled to GTP cyclohydrolase transduction) and consequent production of NO has been achieved in smooth muscle cells [41]. Moreover, these cells are able to adhere to biomaterials under high shear stress in a manner superior to endothelial cells (Fig. 3). The importance of extracellular matrix elaboration for cell adhesion has been pointed out by Lelkes [40]. All cultured cells elaborate an extracellular matrix to some extent, at least in the initial stages of cell culture. Smooth muscle cells have been shown to retain this ability for many population doublings and to possess a high level of matrix synthetic ability [42]. This may account for their strong adherence to a variety of biomaterials. When layers of genetically engineered smooth muscle cells producing NO were exposed to whole blood at different levels of shear stress, the numbers of platelets that adhered to their surfaces were similar or lower than observed with endothelial cell layers under the same experimental conditions (Fig. 4). Furthermore, smooth muscle cell layers that had undergone genetic engineering procedures but were devoid of the

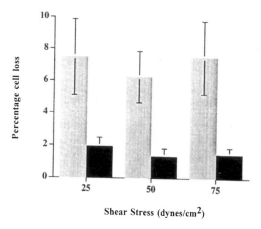

Fig. 3. Percentage Cell Loss from Sintered Titanium Microsphere Surface of LVAD. Endothelial cells (gray bars) and vascular smooth muscle cells transfected with a vector containing NOS gene (black bars) were seeded separately onto sintered titanium microsphere surfaces, and grown to confluence. Seeded samples of titanium were exposed to elevated constant shear stress for 1 hour in a short term shear stress model utilizing a centrifuge [51]. Seeded samples were then trypsinized and cell counts were compared to static culture cell counts to determine the percentage of cell loss.

ability to produce NO had large numbers of platelets associated with their surfaces. This observation is consistent with their acknowledged thrombogenic potential. These studies suggest that the natural ability of smooth muscle cells to attach to biomaterials can be exploited to form cellular intimal surfaces in cardiovascular devices as long as the thrombogenic tendency of the myocytes has been down regulated by their production of NO.

An endothelial lining to cardiovascular devices may still be desirable because of the inherent ability of endothelial cells to produce antithrombotic products in an appropriately regulated manner. Thus, current studies in our laboratory are designed to investigate the attraction of host-derived endothelial cells to the smooth muscle cell intimal lining. It is hypothesized that the endothelial cells will grow to confluence on the luminal surface of the vascular smooth muscle cell layer. Previous studies have demonstrated the strong attachment of endothelial cells to the surface of confluent layers of smooth muscle cells in culture [43]. LVADs explanted from patients after several months have been found to have areas on their luminal surfaces covered with a natural cellular lining [44]. In some instances, small islands of endothelial cells have been found associated with layers of smooth muscle-like cells. This strongly suggests that host-

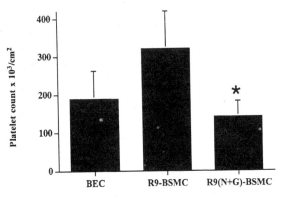

Fig. 4. Whole Blood Exposure under Elevated, Steady Shear Stress. Bovine endothelial cells (BEC), bovine smooth muscle cells transfected with replication vector only (R9-BSMC), and bovine smooth muscle cells co-transfected with replication vectors containing either the Nitric Oxide Synthase gene or the GTP cyclohydrolase gene (R9(N + G)-BSMC) were seeded separately onto tissue culture plastic microscope slides and grown to confluence in the presence of ascorbic acid. Confluent cell layers were placed in a parallel plate flow module, and exposed to freshly drawn, bovine whole blood for 1 minute (shear stress = 40 dynes/cm²). Samples were then prepared for scanning electron microscopy of the cell surface and adherent platelets. Using predetermined coordinates on the scanning electron microscope stage, 12 images of the cell surface were collected from the central flow field. Adherent platelets were quantified and expressed per unit area. Using platelet adhesion as an indicator of the relative thrombogenecity of the cell surfaces studied, the transfected cells producing NO, (R9(N + G)-BSMC), exhibited significantly less initial platelet adhesion ($p < 0.001$).

derived endothelial cells can associate with established layers of smooth muscle cells. These data have prompted an investigation using smooth muscle cells genetically engineered to produce an endothelial cell-specific cytokine, vascular endothelial growth factor (VEGF) as a 'conditioning surface' for LVAD endothelialization.

The technique envisaged for production of an endothelialized cellular lining would include smooth muscle cells engineered to produce NO mixed with cells from the same animal source that have been endowed with the capacity to produce physiologically relevant levels of VEGF. Cell blends will be seeded into LVADs to condition the surface for host-derived endothelial cell populations. In the initial stages, NO production by the engineered smooth muscle cells will provide a non-thrombogenic surface and with time, host endothelial cells will be established and eventually take over the production of NO and maintenance of an antithrombotic blood contacting surface. Such a system would eliminate the requirement for long term (months) NO synthase gene expression, but will provide an improved surface for blood compatibility during the early stages of LVAD implantation. This technique, once perfected, will be applied to human LVAD recipients. Vascular smooth muscle cells will be harvested from a peripheral vein such as the saphenous vein approximately 1 month before LVAD implantation. After isolation and transfection with the aforementioned genes, cells will be expanded in *in vitro* cell culture, and seeded into a LVAD. This seeded device will be placed in a conditioning *in vitro* flow loop prior to implantation to minimize cell loss once the seeded LVAD is implanted. In this manner, an autologous, shear conditioned cellular lining will provide an antithrombotic intimal surface within the LVAD. This intimal cellular lining would hopefully reduce the incidence of long term complications currently observed with implanted LVADs.

4 Conclusion

In the context of the heart, tissue engineering techniques are in the earliest stages of development. Frankly, the production of a viable cardiac substitute is almost inconceivable at present. However, investigations involving production of cardiac components such as heart valves have been encouraging. Continuing progress in molecular, cellular, and developmental biology as well as tissue engineering may provide unforeseen methods to produce a cardiac organ that will meet the requirements of the human body. In the interim, development of the xenograft model is very appealing. A transgenic pig model that expresses specific inhibitors of the host immunological response would allow transplantation not only of cardiac tissue, but perhaps other organs as well. Beyond this model, the understanding of early heart development, which is evolving rapidly, may lead to *in vitro* methods of cell culture of functional cardiomyocytes and subsequent generation of cardiac tissue and perhaps cardiac valves.

Dynamic conditioning of cardiac tissue will likely be critical to producing a functional, tissue engineered heart. Current examples of dynamic conditioning are evident in ventricular assist studies, skeletal muscle myoplasty, and intracardiac grafting. In the case of the LVAD, the myocardium is sufficiently unloaded to induce a remodeling response within the native ventricle. Long term implantation of LVADs has been associated with reduction of ventricular dilatation associated with endstage congestive heart failure. Moreover, reduction in myocardial injury has been demonstrated histologically with long term chronic ventricular support [45–47]. Skeletal muscle myoplasty involves conditioning the latisumis dorsi muscle with a low frequency neurostimulator. After 4–8 weeks of neurostimulation, the skeletal muscle produces increased contractile force and is more resistant to fatigue. The trained muscle may be wrapped around the failing heart and stimulated to assist ventricular contraction [48–50]. Intracardiac cellular grafting takes advantage of the *in situ* heart as a dynamic scaffold [18]. Overall, these examples provide insight into the adaptability of muscle tissue, especially cardiac muscle tissue, to the dynamic environment. This adaptability can be used in conjunction with tissue engineering techniques to produce muscle tissues that promote the specific functional needs of a failing heart. In a similar situation, clinically available devices such as the LVAD can be optimized for long term implantation by adapting host tissues at the cellular and molecular levels for the specific objectives of an intimal cellular lining.

References

1. Cotran, R. S., Kumar, V. and Robbins, S. L., *Robbins Pathologic Basis of Disease*, 4th edn. W.B. Saunders Company, Philadelphia, 1989, pp. 597–656.
2. Braunwald, E., Assessment of cardiac function. In *Heart Disease: A Textbook of Cardiovascular Medicine*, Vol. 1, 4th edn, ed. E. Braunwald. W.B. Saunders Co., Philadelphia, 1992, pp. 419–443.
3. Guyton, A. C., *Textbook of Medical Physiology*, 8th edn. W.B. Saunders Co., Philadelphia, 1991, pp. 98–106.
4. Costanzo, M. R., Augustine, S., Bourge, R., Bristow, M., O'Connell, J. B., Driscoll, D. and Rose, E., Selection and treatment of candidates for heart transplantation: A statement of health professionals from the Committtee on Heart Failure and Cardiac Transplantation of the Council on Clinical Cardiology, American Heart Association. *Circulation*, 1995, **92**(12), 3593–3612.
5. Braunwald, E., Pathophysiology of heart failure. In *Heart Disease: A Textbook of Cardiovascular Medicine*, Vol. 1, 4th edn, ed. E. Braunwald. W.B. Saunders Co., Philadelphia, 1992, pp. 393–418.
6. Ho, K. K. L., Pinsky, J. L., Kannel, W. B. and Levy, D., The epidemiology of heart failure: the Framingham study. *Journal of the American College of Cardiology*, 1993, **22**(4(Supplement A)), 6A–13A.
7. Evans, R. W., Manninen, D. L., Dong, F. B., Frist, W. H. and Kirklin, J. K., The medical and surgical determinants of heart transplantation outcomes: the results of a consensus survey in the United States. *The Journal of Heart and Lung Transplantation*, 1993, **12**(1(Part 1)), 42–45.
8. Arabia, F. A., Smith, R. G., Rose, D. S., Arzouman, D. A., Sethi, G. K. and Copeland, J. G., Success rates of long-term circulatory assist devices used currently for bridge to heart transplantation. *ASAIO Journal*, 1996, **42**(5), M542–M546.
9. Mehta, S. M., Boehmer, J. P., Pae, W. E., Aufiero, T. X., Davis, D. and Pierce, W. S., Equal extended survival for patients undergoing LVAD support when compared with long-term medical management. *ASAIO Journal*, 1996, **42**(5), M406–M410.
10. Frazier, O. H., Rose, E. A., Macmanus, Q., Burton, N. A., Lefrak, E. A., Poirier, V. L. and Dasse, K. A., Multicenter clinical evaluation of the HeartMate 1000 IP left ventricular assist device. *Annals of Thoracic Surgery*, 1992, **53**(6), 1080–1090.

11. Griffith, B. P., Kormos, R. L., Nastala, C. J., Winowich, S. and Pristas, J. M., Results of extended bridge to transplantation: window into the future of permanent ventricular assist devices. *Annals of Thoracic Surgery*, 1996, **61**(1), 396–398.

12. Birinyi, L. K., Douville, E. C., Lewis, S. A., Bjornson, H. S. and Kempczinski, R. F., Increased resistance to bacteremic graft infection after endothelial cell seeding. *Journal of Vascular Surgery*, 1987, **5**(1), 193–197.

13. Rose, E. A. and Goldstein, D. J., Wearable long-term mechanical support for patients with end-stage heart disease: A tenable goal. *Annals of Thoracic Surgery*, 1996, **61**, 399–402.

14. Collins, E. G., Pfeifer, P. B. and Mozdzierz, G., Decisions not to transplant: Futility or rationing. *Journal of Cardiovascular Nursing*, 1995, **9**(3), 23–29.

15. Akins, C. W., Results with mechanical cardiac valvular prostheses. *Annals of Thoracic Surgery*, 1995, **60**, 1836–1844.

16. Fung, Y. C., What are the residual stresses doing in our blood vessels? *Annals of Biomedical Engineering*, 1991, **19**, 237–249.

17. Soonpaa, M. H., Daud, A. I., Koh, G. Y., Klug, M. G., Kim, K. K., Wang, H. and Field, L. J., Potential approches for myocardial regeneration. *Annals of the New York Academy of Sciences*, 1995, **752**, 446–454.

18. Klug, M. G., Soonpaa, M. H., Koh, G. Y. and Field, L. J., Genetically selected cardiomyocytes from differentiating embryonic stem cells form stable intra-cardiac grafts. *Journal of Clinical Investigation*, 1996, **98**(1), 216–224.

19. Vesely, I., Gonzalez-Lavin, L., Graf, D. and Boughner, D., Mechanical testing of cryopreserved aortic allografts: Comparison with xenografts and fresh tissue. *Journal of Thoracic and Cardiovascular Surgery*, 1990, **99**(1), 119–123.

20. Silver, F. H., *Biomaterials, Medical Devices and Tissue Engineering: An Integrated Approach.* Chapman and Hall, New York, 1994, pp. 153–193.

21. Askeland, D. R., *The Science and Engineering of Materials*, 2nd edn. PWS-Kent Publishing Company, Boston, 1989, pp. 152–153.

22. Samet, M. M. and Lelkes, P. I., Flow patterns and endothelial cell morphology in a simplified model of an artificial ventricle. *Cell Biophysics*, 1993, **23**, 139–163.

23. Baldwin, J. T., Deutsch, S., Geselowitz, D. B. and Tarbell, J. M., LDA measurements of mean velocity and reynolds stress fields within an artificial heart ventricle. *Transactions of the ASME: Journal of Biomechanical Engineering*, 1994, **116**, 190–200.

24. Ezzell, C., Research in focus: Tissue engineering and the human body shop: Designing 'bioartificial' organs. *The Journal of NIH Research*, 1995, **7**, 49–53.

25. Nosé, Y., Tajima, K., Imai, Y., Klain, M., Mrava, G., Schriber, K., Urbanek, K. and Ogawa, H., Artificial heart constructed with biological material. *Transactions of the ASAIO*, 1971, **17**, 482–487.

26. Hayashi, K., Snow, J., Washizu, T., Jacobs, G., Kiraly, R. and Nosé, Y., Biolized intrathoracic left ventricular assist device (LVAD). *Medical Instrumentation*, 1977, **11**(4), 202–207.

27. Sykes, M., Immunobiology of transplantation. *The FASEB Journal*, 1996, **10**, 721–730.

28. Platt, J. L., A perspective on xenograft rejection and accommodation. *Immunological Reviews*, 1994, **141**, 127–149.

29. Shinoka, T., Breuer, C. K., Tanel, R. E., Zund, G., Miura, T., Ma, P. X., Langer, R., Vacanti, J. P. and Mayer, J. E., Tissue engineering heart valves: valve leaflet replacement study in a lamb model. *Annals of Thoracic Surgery*, 1995, **60**, S513–516.

30. Alexander, H. G., Kidd, I. D., Ilten-Kirby, B. M., Landeen, L. K., Peterson, A. and Zeltinger, J., Development of tissue engineered heart valves. In *2nd International Conference on Cellular Engineering*, International Federation for Medical and Biological Engineering, La Jolla, California, 1995, p. 2.

31. Naughton, G. K., Tolbert, W. R. and Grillot, T., Emerging developments in tissue engineering and cell technology. *Tissue Engineering*, 1995, **1**(2), 211–219.

32. Liotta, D., Hall, C. W., Akers, W. W., Villanueva, A., O'Neal, R. M. and DeBakey, M. E., A pseudoendocardium for implantable blood pumps. *Transactions of the ASAIO*, 1966, **12**, 129–138.

33. Ghidoni, J. J., Liotta, D., Adams, J. G., Hall, C. W., and O'Neal, R. M., Culture of autologous tissue fragments in paracorporeal left ventricular bypass pumps. *Archives of Pathology*, 1968, **86**, 308–311.

34. Bernhard, W. F., Husain, M., George, J. B. and Curtis, G. W., Fetal fibroblasts as a substratum for pseudoendothelial development on prosthetic surfaces. *Surgery*, 1969, **66**(1), 284–290.

35. Bernhard, W. F., Colo, N. A., Wesolowski, J. S., Szycher, M., Fishbein, M. C., Parkman, R., Franzblau, C. C. and Haudenschild, C. C., Development of collagenous linings on impermeable prosthetic surfaces. *Journal of Thoracic and Cardiovascular Surgery*, 1980, **79**, 552–564.

36. Lelkes, P. I. and Samet, M. M., Endothelialization of the luminal sac in artificial cardiac prostheses: A challenge for both biologists and engineers. *Transactions of the ASME: Journal of Biomechanical Engineering*, 1991, **113**, 132–142.

37. Rosenman, J. E., Kempczinski, R. F., Pearce, W. H. and Silberstein, E. B., Kinetics of endothelial cell seeding. *Journal of Vascular Surgery*, 1985, **2**(6), 778–784.

38. Rupnick, M. A., Hubbard, F. A., Pratt, K., Jarrell, B. E. and Williams, S. K., Endothelialization of vascular prosthetic surfaces after seeding or sodding with human microvascular endothelial cells. *Journal of Vascular Surgery*, 1989, **9**(6), 788–795.

39. Schneider, P. A., Hanson, S. R., Price, T. M. and Harker, L. A., Durability of confluent endothelial cells monolayers on small-caliber vascular prostheses in vitro. *Surgery*, 1988, **103**(4), 456–462.

40. Nikolaychik, V. V., Wankowski, D. M., Samet, M. M. and Lelkes, P. I., In vitro testing of endothelial cell monolayers under dynamic conditions inside a beating ventricular prosthesis. *ASAIO Journal*, 1996, **42**(5), M487–M494.

41. Scott-Burden, T., Regulation of nitric oxide production by tetrahydrobiopterin. *Circulation*, 1995, **95**, 248–250.

42. Scott-Burden, T., Resink, T. J., Burgen, M. and Buhler, F. R., Extracellular matrix: differential influence on growth and biosynthesis patterns of vascular smooth muscle cells from SHR and WKY rats. *Journal of Cellular Physiology*, 1989, **141**, 267–274.

43. Jones, P. A., Construction of an artificial blood vessel wall from cultured

endothelial and smooth muscle cells. *Proceedings of the National Academy of Sciences of the United States of America*, 1979, **76**, 882–886.

44. Scott-Burden, T. and Frazier, O. H., Cellular linings of ventricular assist devices. *Annals of Thoracic Surgery*, 1995, **60**, 1561–1562.

45. Bonn, D., Artificial hearts: bridges to transplantation or recovery? *The Lancet*, 1996, **347**, 960.

46. Nakatani, S., McCarthy, P. M., Kottke-Marchant, K., Harasaki, H., James, K. B., Savage, R. M. and Thomas, J. D., Left ventricular echocardiographic and histologic changes: impact of chronic unloading by an implantable ventricular assist device. *Journal of the American College of Cardiology*, 1996, **27**(4), 894–901.

47. Levin, H. R., Oz, M. C., Chen, J. M., Packer, M., Rose, E. A. and Burkoff, D., Reversal of chronic ventricular dilation in patients with end-stage cardiomyopathy by prolonged mechanical unloading. *Circulation*, 1995, **91**(11), 2717–2720.

48. Gustafson, K. J., Sweeney, J. D., Gibney, J. and Brandon, T. A., Progressive pressure expansion in skeletal muscle ventricle conditioning. *ASAIO Journal*, 1996, **42**(5), M360–M364.

49. Chekanov, V. S., Tchkanov, G. V., Rieder, M. A., Cheng, Q., Smith, L. M., Zander, G. L., Christensen, C. W., McConchie, S., Jacobs, G. and Schmidt, D. H., Skeletal muscle of a growing organism has a greater transformation after electrical stimulation than adult skeletal muscle. *ASAIO Journal*, 1996, **42**(5), M630–M636.

50. Thomas, G. A., Lelkes, P. I., Isoda, S., Chick, D., Lu, H., Hammond, R. L., Nakajima, H., Nakajima, H. and Walters, H. L., III., Endothelial cell-lined skeletal muscle ventricles in circulation. *The Journal of Thoracic and Cardiovascular Surgery*, 1995, **109**(1), 66–73.

51. McClay, D. R., Wessel, G. M. and Marchase, R. B., Intercellular recognition: quantitation of initial binding events. *Proceedings of the National Academy of Sciences of the United States of America*, 1981, **78**(8), 4975–4979.

CHAPTER IV.3

Esophagus

YOSHITO IKADA

Research Center for Biomedical Engineering,
Kyoto University,
53 Kawahara-cho, Shogoin, Sakyo-ku,
Kyoto 606, Japan

1 Introduction

The esophagus is an organ that passes through fields of the neck, thorax, and abdomen, and has the simple function of transporting food and water from mouth to stomach. However, esophageal cancer is one of the most difficult gastrointestinal (GI) malignancies to treat. After esophagectomy, reconstruction using stomach, jejunum, or colon is necessary, but these procedures are stressful for patients and often lead to post-operative complications. Some patients may not have sufficient length in the reconstructed organ, especially those who have had gastrectomy or colectomy. Loss of length and function of the GI tract after reconstruction may prevent patients from receiving proper nutrition. To decrease surgical stress, to avoid sacrificing a healthy GI tract, and to give patients another material for reconstruction, a variety of artificial esophagi have been designed [1–4]. They can be classified into three categories in terms of the materials used: natural, synthetic, and composite. However, it is impossible to prevent complications, even if an artificial esophagus is made of materials with the highest tissue compatibility, because it permanently remains as a foreign body for the host. There are two different mechanisms responsible for the occurrence of complications. One is leakage and infection in the early stage, and the other is stenosis after displacement of the prosthesis in the late stage. The former is attributable to the materials used to prepare the artificial esophagus, in other words, biocompatibility problems between the prosthesis and the host tissue. The latter is caused by the regenerated tissue itself and is, therefore, the problem of the immaturity of the regenerated tissue, especially of the submucosal tissue. In addition, although prostheses should be made from totally bioinactive materials so as to be capable of remaining in the body for a long time, implantation of artificial esophagi made of a bioinactive material such as

silicone is inevitably followed by local complications, infection, or dislocation at the site of anastomosis between the artificial esophagus and the
host tissue. Thus, it seems probable that tissue engineering is the best way
for replacement of the lost esophageal tissue.

2 Esophagus replacement by tissue engineering

It is not surprising that very few attempts have been made to regenerate
the esophagus tissue. The major reason is probably a very small need from
the surgery for the artificial esophagus or esophagus replacement. It
should be also pointed out that the esophagus tissue has been thought to
be very difficult to regenerate. Two approaches attempted by two different
research groups for the tissue engineering of the esophagus will be
represented below.

3 Tissue engineering using esophageal epithelial cells

Sato *et al.* proposed to use esophageal epithelial cells for regeneration of
the esophagus [5, 6]. Their final goal was to harvest a small amount of
normal esophageal tissue from a patient with esophageal cancer before
surgery, culture the esophageal cells, and make an artificial esophagus *in
vitro*, which would be then grafted in the same patient to replace the
cancerous esophagus. For this purpose they made a tubular artificial
esophagus from a poly(glycolic acid) (PGA) mesh-collagen complex tube
whose inner side was covered *in vitro* by cultured human esophageal
epithelial cells. After covering, the tube was grafted in latissimus dorsi
muscle flaps of athymic rats.

3.1 Preparation of a mucosal tube and grafting

The PGA mesh–collagen complex was prepared as follows. An acid-solu-
ble type I collagen solution, made from pig tendon, was mixed with Ham's
F12 medium and HEPES supplemented with $NaHCO_3$ and NaOH. Three
milliliters of this mixture was prepared in a 60 mm culture dish and a PGA
mesh (1.5 cm $\times$ 4 cm) was embedded in this solution. After warming the
collagen solution to 37°C , it became a gel, completing the PGA mesh-collagen complex. To reconstruct a mucosal tube, samples of normal mucosa
(10 cm^2) were resected immediately after surgical removal from specimens
of esophageal cancer patients. They were treated with dispase and trypsin
and disaggregated epithelial cells were pelleted by centrifugation. Approximately 2×10^6 cells were inoculated on the surface of each PGA
mesh–collagen complex and cultured in keratinocyte growth medium.

After the cultures reached confluence in 5–7 days, the medium was changed to keratinocyte growth medium supplemented with 20% fetal calf serum and the cultures were maintained for more 3–5 days. At this time, the edges of the PGA mesh were sutured to create a tube, 0.5 cm in diameter and 4 cm in length. The backs of 10 week old athymic rats were incised and the latissimus dorsi muscles were exposed and dissected to make muscle flaps with pedicles. The restructured mucosal tubes were wrapped in the flaps using a fibrin glue. The animals were killed at 4, 8, 20, and 28 days after grafting and the muscle tubes were excised for macroscopic and microscopic examination.

3.2 Results of the grafting of restructured mucosal tubes

The restructured mucosal tube for grafting was 4 cm long with a diameter of 0.5 cm [6]. The inner side wall was covered with cultured human esophageal epithelial cells. The grafted tubes were excised at 4, 8, 20, and 28 days after grafting. All of them maintained a tubular shape and the inside was covered with a white translucent membrane-like epithelium. The PGA mesh–collagen complex tubes without an epithelial cell coating used as controls collapsed and did not maintain a tubular shape. Microscopic appearance of the tube's inner surface showed growth of grafted epithelia. At 8 days after grafting the epithelium was about ten layers thick. Connective tissue infiltrated the collagen layer from muscle in a time-dependent fashion and neovascularization appeared. At 20 days after grafting, there were more cell layers and the epithelium was thicker than at 8 days. There was neither stenosis of the lumen caused by overgranulation nor contraction by means of scar. The wall structure, from the outside in, consisted of a complex, and multilayered human esophageal epithelial cells. Ultimately it was similar to a normal esophageal wall.

At 8 days after grafting, the junction between the epithelium and the collagen layer stained discontinuously and vaguely, but at 20 days after grafting this zone became continuous and thicker and it appeared to be a basement membrane. Encouraged by this successful creation of tubular artificial organs with the structure similar to normal esophagi using cultured human esophageal epithelial cells and a PGA mesh–collagen complex, Sato et al. successfully created the tubular artificial organs in athymic rat latissimus dorsi muscles.

Human esophageal epithelial cells from a small amount of mucosa could be cultured to sufficient quantities to cover the entire inside of an artificial esophagus. During a 2 week culture period a 250 cm^2 sheet of cultured epithelium could be produced from about 2.5 cm^2 of mucosa, and the cultured cells could be cryopreserved.

This development of an artificial tubular organ from a latissimus dorsi muscle, PGA mesh–collagen complex, and cultured esophageal epithelia, suggests that this hybrid artificial esophagus may have clinical applications in the future.

4 Esophageal tissue regeneration using a collagen scaffold

Natsume *et al.* developed a new type of artificial esophagus with bilayered structure made of porous collagen sponge and silicone [7–10], as shown in Fig. 1.

4.1 Preparation of a bilayered tube

The bilayered structure of the tube is depicted in Fig. 2. The outer porous collagen sponge layer serves as a template for the construction of the 'neoesophagus' and the inner silicone tube layer imparts rigidity to the artificial esophagus temporily and protects the anastomotic sites from infection, leakage, collapse, and dislocation until neoesophagus formation is complete. The silicone tube will dislocate into the stomach as a result of peristalsis induced by food consumption, no synthetic prostheses remaining in place.

The tube was prepared as follows. A silicone tube, 5 cm long with an internal diameter of 2.5 cm, was made of a medical grade 1 mm thick silicone sheet, which was reinforced by a nylon mesh embedded with silicone glue. The outer surface of the tube was exposed to corona

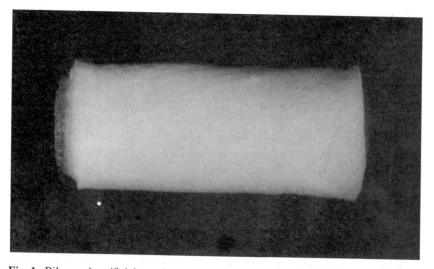

Fig. 1. Bilayered artificial esophagus consisiting of collagen sponge and silicone.

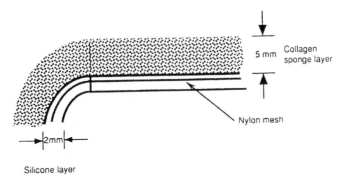

Fig. 2. Schematic representation of the artificial esophagus.

discharge to make the surface hydrophilic and then placed in a Teflon tube with an internal diameter of 3 cm. The collagen solution, bubbled under stirring in a refrigerated homogenizer, was poured between the Teflon tube and the silicone tube and then freeze-dried to form a porous sponge (5 mm thick). The collagen predominantly consisted of type I collagen (70–80%) and the rest type III. The sponge was heated at 105°C under vacuum to introduce crosslinking in collagen molecules. This process was necessary to maintain the highly porous structure during suturing and handling. The pore sizes of the sponge were always controlled to be in the range of 100–500 μm.

4.2 Implantation of the bilayered tube

The bilayered tubes were implanted into adult mongrel dogs. The trachea and cervical esophagus were carefully isolated, taking care to preserve the vagus nerves. A 5 cm length of the cervical esophagus was resected and replaced with an artificial esophagus. The edges (2–3 mm) of the prosthesis were invaginated into the inside of the esophagus so that the resected edge of the esophagus was only in contact with the outer collagen sponge layer of the prosthesis, as shown in Fig. 3, and anastomosed with interrupted sutures, using an absorbable suture (3-0 Vicryl). After esophageal replacement, an intravenous hyperalimentation (IVH) tube was inserted into the left femoral vein. Dogs received no food orally but were fed by IVH alone using an infusion pump for at least 2 weeks after implantation. Penicillin and streptomycin were given intravenously during this period. After termination of IVH, animals were given standard dry dog food.

4.3 post-operative findings

All dogs survived for more than 7 days as summarized in Table 1 and esophageal tissue regeneration was evident in all of them [7]. Four died of

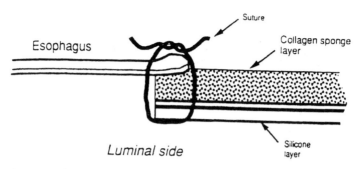

Fig. 3. Anastomotic technique to secure the implant.

unknown causes, but neither breakdown of prostheses nor leakage in anastomotic sites were observed in those animals. Microscopic studies confirmed that there was no local infection associated with the implant. The esophageal regeneration was comparable to that observed in sacrificed animals.

After 2 weeks, substantial neoesophageal tubes had been reformed and residual collagen layers of the implants could not be observed macroscopically. The silicone tube had remained in place while epithelial regeneration had started already from both sides of the anastomotic edges beneath the silicone layer. The maximum distance of the epithelialization migration front was 2.5 cm and the minimum 0.5 cm.

At 3 weeks after operation, epithelial continuity was almost restored, although the regenerated epithelial sheet was thin and immature. After 4–5 weeks, the maturation of the newly formed epithelium had progressed and become firm.

In five dogs, immediately after IVH was terminated and oral feeding started, the silicone tube dropped into the stomach spontaneously. These dogs could take the standard dry dog food, but a tendency toward stenosis became apparent. One dog started oral feeding on day 21 but vomited severely and was sacrificed on day 36 because of malnutrition. However, the entire neoesophageal luminal surface was covered with mature mucosa, the transition from the uninjured intact portion to the regenerated area was smooth, and no granulation tissue had formed on the luminal surface.

Two dogs that were not sacrificed underwent a long-term survival study (> 370 and > 70 days) and vomited for several weeks after starting oral feeding (on days 23 and 25, respectively) but improved gradually and later took food easily; no weight reduction was observed. Esophagography was performed in one dog at 6 and 12 months after implantation.

After implantation, an acute inflammatory response was elicited and lasted for 1–2 weeks. The major inflammatory cells were neutrophils and histocytes, and occasionally small lymphocytes. After 2 weeks, the number of neutrophils declined gradually and there was a compensatory increase in mononuclear cells. Within 2 weeks, the replacement site had been massively infiltrated with fibroblasts and endothelial cells. The extracellular matrix, consisting of disorganized fine fibers around the fibroblasts, was strongly stained with Azan, which stains collagen.

After 2 weeks, the epithelial cells advanced from the anastomotic lines of the neoesophageal lumen in all the animals sacrificed on day 14 and the epithelial cell migration front was clearly observed. The fronts of the extending epithelial sheets consisted exclusively of poorly differentiated, spindle-shaped cells. However, the migrating epithelial cells adjacent to the anastomotic line had differentiated, with appearances identical to those of the uninjured intact epithelium. The cells lying on the boundary between the submucosal connective tissue stratified into a cuboidal multi-layered formation and the cell layers at the luminal surface showed terminal differentiation, i.e. flattened and keratinized cells. In the center of the neoesophageal tissue, which was not yet covered with epithelial cells, small residues of implanted collagen sponge were noticed near the luminal surface.

After 4–5 weeks, the neoesophageal lumen was covered completely with a mature epithelial sheet consisting of well-differentiated, multilayered (10–20 layers) stratified cells, which was almost as thick as the normal esophagus.

4.4 Prevention of stenosis

As described in Table 1, the tendency for stenosis was only one complication observed with this artificial esophagus. To confirm whether stenosis occurred because of poor submucosal tissue formation, the stenting time was changed to examine the relationship between the stenting time and stenosis in the late post-operative period [8]. After surgery, the silicone tube was removed in selected animals every week from 2 to 8 weeks, when oral feeding was again initiated.

In 19 cases in which the stent was dislodged within 3 weeks, stenosis developed rapidly and the dog became unable to swallow within 3 days. In two cases in which the stent moved between the 3rd and 4th weeks, stenosis developed gradually after dislodgement, as shown in Table 2. In four cases in which the stents remained in place for more than 4 weeks, stenosis did not occur, and subsequent oral feeding remained possible without weight loss, even at 12 months after surgery. The result is given in Table 3. Endoscopic and macroscopic examination of the post-operative esophagus revealed that no granulation tissue had formed and no infection or anastomotic leakage had occurred at the junction of the prosthesis and

Table 1. Post-operative results of implanted artificial esophagus

Survival (days)	Cause of death	Complications			Silicone tube in replacement site	Epithelial regeneration
		Leak.	Gran.	Sten.		
9	Accid	−	−	−	Present	−
12	Unce.	−	−	−	Present	−
14	Sacri.	−	−	−	Present	−
14	Sacri.	−	−	−	Present	−
14	Sacri.	−	−	−	Present	−
14	Unce.	−	−	−	Present	−
17	Sacri.	−	−	+ −	Present	+
21(15)	Sacri.	−	+	+ −	Not present	−
21	Sacri.	−	−	+ +	Not present	+
21	Sacri.	−	−	−	Present	+
21	Sacri.	−	−	−	Present	−
21	Sacri.	−	−	−	Present	−
21	Unce.	−	−	−	Present	−
23(21)	Unce.	−	−	+	Not present	+
27	Sacri.	−	−	−	Present	+
30	Sacri.	−	−	−	Present	+
36(21)	Sacri.	−	−	+ +	Not present	+
70(23)	Alive	−	−	+	Not present	+
370(22)	Alive	−	−	+	Not present	+

Leak., Leakage; Gran; exbereant granulation tissue development: +, present. −, not present; Sacri., sacrificed; Accid., accident; Unce., uncertain., Sten., degree of stenosis: −, not observed; + −, low grade; +, moderate grade; + + severe. Epithelial regeneration: Continuity of the regenerated epithelial lining was judged by the historical examination: continuous, +; not continuous −; Start of oral feeding (days).

esophagus at time of death. In all dogs, neoesophageal epithelialization was complete in the regenerated esophagus.

Of the cases in which the stent migrated within 4 weeks, radiographic examination showed various degrees of stenosis or shortening, and there was no peristalsis in the neoesophagus (Table 2). Of the cases in which the stent moved after 4 or more weeks, radiographic examination showed no stenosis or shortening, and there was normal peristalsis in the neoesophagus. Endoscopically, the inner surface of the neoesophagus in all cases was completely covered with epithelium from the oral to the anal side.

In cases in which the stent moved 4 or more weeks after surgery and the animals survived for more than 106 days, the neoesophagus was covered with a polylayer of squamous epithelium and had normal esophageal glands and a muscle layer.

Table 2. Early post-operative results of long-term follow-up

| Animal no. | Survival time (days) | Duration of stent (days) | Cause of death | Complications | | | Epithelial regeneration |
				Leakage	Granulation	Stenosis	
1	9	9	Accident	−	−	−	Partial
2	12	12	Hemat.	−	−	−	Partial
3	14	14	Sacrifice	−	−	−	Partial
4	14	14	Sacrifice	−	−	−	Partial
5	14	14	Sacrifice	−	−	−	Partial
6	14	14	Unexplained	−	−	−	Partial
7	17	17	Sacrifice	−	−	−	Complete
8	21	15	Sacrifice	−	+	+ −	Partial
9	21	21	Sacrifice	−	−	−	Partial
10	21	21	Sacrifice	−	−	−	Complete
11	21	14	Sacrifice	−	−	+ +	Complete
12	21	21	Sacrifice	−	−	−	Partial
13	21	21	Unexplained	−	−	−	Partial
14	23	21	Unexplained	−	−	+	Complete
15	27	27	Sacrifice	−	−	−	Complete
16	30	30	Sacrifice	−	−	−	Complete
17	36	21	Sacrifice	−	−	+ +	Complete

+ = present; − = not present; + − = low grade; + + = severe.

4.5 Regeneration of esophageal glands and a muscle layer

As revealed above, stenosis did not occur when the silicone stent remained in place for more than 4 weeks. Tables 4 and 5 show the results of regeneration of neoesophageal submucosal tissue as a function of the stenting time [9]. It is evident that muscle tissue and esophageal glands are able to regrow in the neoesophagus when the portion replaced by the artificial esophagus is stented for at least 4 weeks.

4.6 Replacement of longer portions of the esophagus

The above studies were concerned with the short segment replacement using a 5-cm long bilayered tube. However, it is necessary to confirm that longer segments of esophagus can be replaced safely with this type of artificial esophagus before it is widely used clinically. Therefore, an artificial esophagus was fabricated to replace a 10 cm long gap [10]. This is

Table 3. Late post-operative results of long-term follow-up

Animal no.	Survival time (days)	Duration of stent (days)	Cause of death	Complication: stenosis	Epithelial regeneration
18	730	22	Accident	+	Complete
19	479	24	Alive	−	Complete
23	136	30	Pneumonia	−	Complete
20	239	33	Pneumonia	−	Complete
21	322	36	Alive	−	Complete
22	224	39	Alive	−	Complete
24	441	59	Alive	−	Complete
25	41	30	Alive	−	Complete

Table 4. Results of short-term stenting

Animal no.	Survival time (days)	Duration of stent (days)	Prognosis and death	Regeneration		
				Epithelium	Muscle	Glands
1	9	9	Pneumonia	Partial	−	−
2	12	12	Hematemesis	Partial	−	−
3	14	14	Killed	Partial	−	−
4	14	14	Killed	Partial	−	−
5	14	14	Killed	Partial	−	−
6	14	14	Pneumonia	Partial	−	−
7	17	17	Killed	Complete	−	−
8	21	15	Killed	Partial	−	−
9	21	21	Killed	Partial	−	−
10	21	21	Killed	Complete	−	−
11	21	14	Killed	Complete	−	−
12	21	21	Killed	Partial	−	−
13	21	21	Pneumonia	Partial	−	−
14	23	21	Pneumonia	Complete	−	−
15	36	21	Killed	Complete	−	−

−, not present.

similar in structure and is made of a silicone tube reinforced with a nylon cloth coated on the outside with a 5 mm thick freeze-dried collagen sponge. Seven adult mongrel dogs were used for the implantation. Five of the seven dogs could be fed orally, and their body weight was stable. These

Table 5. Results of long-term stenting

Animal no.	Survival time (days)	Duration of stent (days)	Prognosis and death	Regeneration		
				Epithelium	Muscle	Glands
16	730	22	Pneumonia	Complete	−	−
17	779	24	Alive	Complete		
18	27	27	Killed	Complete	−	−
19	301	28	Alive	Complete		
20	322	28	Alive	Complete		
21	341	28	Alive	Complete		
22	360	28	Alive	Complete		
23	30	30	Killed	Complete	−	−
24	136	30	Pneumonia	Complete	+	+
25	239	33	Pneumonia	Complete	+	+
26	682	36	Pneumonia	Complete	+	+
27	584	39	Alive	Complete		
28	801	59	Alive	Complete		
29	401	30	Alive	Complete		

+, present; −, not present.

five dogs were allowed to live for more than 6 months. The other two dogs died of anesthetic accidents when the silicone stent was removed.

Radiographic examination using radiopaque medium showed no stenosis or shortening of the reconstructed esophagus, and peristalsis was observed in the neoesophagus in all five dogs. Peristaltic waves were transmitted in

Table 6. Post-operative Results

Animal no.	Survival time (days)	Duration of stent (days)	Prognosis and death	Complications		
				Leakage	Infection	Stenosis
1	362	42	Alive	−	−	−
2	42	42	Death	−	−	−
3	231	42	Alive	−	−	−
4	216	42	Alive	−	−	−
5	42	42	Death	−	−	−
6	204	42	Alive	−	−	−
7	182	42	Alive	−	−	−

−, not present.

succession from the oral side of the host esophagus to the anal side through the regenerated segment.

The results of post-operative findings are tabulated in Table 6. Neoesophagus was regenerated promptly in all of the animals and the luminal surface was covered with a polylayer of squamous epithelium. The regenerated esophagus had normal esophageal glands and immature muscle tissue. Takimoto *et al.* therefore concluded that this new artificial esophagus is also applicable to replacement of a 10-cm length of cervical esophagus and that this artificial esophagus may be used clinically in the near future.

References

1. Morfit, H. M. and Karmish, D., Long-term end results in bridging esophageal defects in human beings with teflon prostheses. *American Journal of Surgery*, 1962, **104**, 756–760.
2. Barnes, W. A., Redo, S. F. and Ogata, K., Replacement of portion of canine esophagus with composite prosthesis and greater omentum. *Journal of Thoracic Cardiovascular Surgery*, 1972, **64**, 892–896.
3. Feng-Lin, W., Nieuwenhuis, P., Gogolewski, S., Pennings, A. J. and Wildevuur, C. R. H., *Oesophageal Prosthesis*. Elseveir Science Publishers, New York, 1984, pp. 317–332.
4. Natsume, T., Ike, O., Okada, T., Shimizu, Y., Ikada, Y. and Tamura, K., Experimental studies of a hybrid artificial esophagus combined with autologous mucosal cells. *ASAIO Transactions*, 1990, **36**, M435–M437.
5. Sato, M., Ando, N., Ozawa, S., Nagashima, A. and Kitajima, M., A hybrid artificial esophagus using cultured human esophageal epithelial cells, *ASAIO Journal*, 1993, M554–M557.
6. Sato, M., Ando, N., Ozawa, S., Miki, H. and Kitajima, M., An artificial esophagus consisting of cultured human esophageal epithelial cells, polyglycolic acid mesh, and collagen. *ASAIO Journal*, 1994, **40**, M389–M392.
7. Natsume, T., Ike, O., Okada, T., Takimoto, N., Shimizu, Y. and Ikada, Y., Porous collagen sponge for esophageal replacement. *Journal of Biomedical Materials Research*, 1993, **27**, 867–875.
8. Takimoto, Y., Okumura, N., Nakamura, T., Natsume, T. and Shimizu, Y., Long-term follow-up of the experimental replacement of the esophagus with a collagen-silicone composite tube. *ASAIO Journal*, 1993, **39**, M736–M739.
9. Takimoto, Y., Teramachi, M., Okumura, N., Nakamura, T. and Shimizu, Y., Relationship between stenting time and regeneration of neoesophageal submucosal tissue. *ASAIO Journal*, 1994, **40**, M793–M797.
10. Takimoto, Y., Nakamura, T., Teramachi, M., Kiyotani, T. and Shimizu, Y., Replacement of long segments of the esophagus with a collagen-silicone composite tube. *ASAIO Journal*, 1995, **41**, M605–M608.

CHAPTER IV.4

Engineering Small Intestine

JAMES J. CUNNINGHAM
Department of Chemical Engineering,
The University of Michigan,
Ann Arbor, MI 48109, USA

GREGORY M. ORGAN
Columbia Michael Reese Hospital, and University of Illinois at Chicago,
College of Medicine
Chicago, IL 60680, USA

and

DAVID J. MOONEY
Departments of Chemical Engineering and Biologic and Materials Science,
The University of Michigan,
Ann Arbor, MI 48109, USA

1 Introduction

The small intestine, or small bowel, is one of the most important and complex organs in the human body. It plays a central role in the digestion and absorption of nutrients from ingested foods. The small bowel, as it propels ingested food along its length, functions as a selective barrier. It transports nutrients, electrolytes, water and other compounds, such as drugs, into the bloodstream, while excluding pathogens and non-digestible materials. As such, it is little wonder that defects in, and diseases of, the small bowel can have serious and life-threatening consequences. Engineering intestinal tissue from transplanted cells could potentially meet the clinical need for improved treatment of patients suffering from disorders of the small bowel. At the same time, engineered tissue could provide a novel *in vitro* model system for the study of intestinal development and absorption.

The primary clinical motivation for engineering intestinal tissue is short bowel syndrome (SBS), a condition of malabsorption and malnutrition following resection of an extensive portion of the small bowel [1]. Although SBS can occur in older children and adults, it affects mainly neonates. Impaired absorption of nutrients is particularly harmful for these young patients because of the rapid rate at which their bodies grow and develop. The leading cause of small bowel resection among young children is necrotizing enterocolitis, which occurs most often in premature infants [2]. Another common cause is malrotation of the small bowel during development, which prevents the formation of proper mesenteric attachments. This can then lead to midgut volvulus, resulting in obstruction of the bowel and constriction of the blood supply [3]. Other, less common, causes include congenital defects of the gastrointestinal tract or abdominal cavity, such as gastroschisis, multiple bowel atresias, meconium peritonitis, and Hirschsprung's Disease [3], which is a condition of extensive aganglionosis resulting in a functional blockage. Crohn's disease, irradiation enteritis, and abdominal trauma also contribute to cases of SBS in children and adults [4].

Surgeons have developed several innovative techniques to improve nutrient absorption in the remaining intestine [5–7]. These include increasing intestinal surface area by lengthening, tapering, and growing neomucosa on natural and synthetic membranes grafted into existing intestine, as well as slowing the transit of material through the lumen. Unfortunately, these techniques are usually inadequate to maintain the patient solely on enteral nutrition. Another approach that has aroused considerable interest in recent years is small bowel transplantation [8, 9]. This procedure is, however, still considered experimental. Problems with graft-versus-host disease and rejection have been encountered because of intestinal lymphoid tissue and the abundance of class II MHC cell surface markers. Even if this treatment is eventually successful, it will likely be limited by a shortage of donor tissue, as in other transplant procedures.

Although nutrient absorption can be improved through surgical intervention or natural adaptation [10] of the remaining small bowel, these processes are generally insufficient to meet the nutritional needs of the patient through oral feedings alone. As a result, most patients are placed on Total Parenteral Nutrition (TPN), in which a nutrient solution is infused into the central venous circulation via a catheter [11]. It is primarily advances in TPN which have contributed to the improved survival of SBS patients. Long-term TPN, however, has several major drawbacks. Aside from the considerable expense involved and the impact on the lifestyle of the patient, long-term TPN can lead to serious complications, including catheter-related conditions such as local and systemic infection, venous thrombosis, and loss of vascular access. Progressive liver disease and severe metabolic bone disease can also result [12].

A second, non-clinical, motivation for engineering intestinal tissue is the lack of *in vitro* model systems for the study of intestinal development and function. *In vitro* model systems have many advantages over *in vivo* studies, including better control over experimental conditions, fewer variables, and avoidance of the ethical issues associated with animal testing. A functional, engineered intestinal tissue could be of use in several diverse fields. For example, such a system might be useful for studying intestinal development and pathologies, or for investigations of intestinal drug absorption in order to design drugs with improved absorption characteristics. Furthermore, infant-formula companies would benefit from a better understanding of the effect of certain components of breastmilk on neonatal intestinal development and immunological function. An engineered model tissue could potentially meet these needs.

The motivation for engineering intestinal tissue is thus twofold. The serious shortcomings and complications inherent in current treatments of small bowel disorders such as SBS prompt the need for new alternatives. Furthermore, an *in vitro* intestinal model tissue could serve as a valuable tool for the study of intestinal function and drug absorption. The critical question is how to engineer a complex organ such as the intestine. The final goal is a fully functional tissue that is integrated with the body. There are several approaches used to engineer tissues currently, but the one most likely to have the desired result in this application is transplantation of cultured cells on macroporous matrices [13]. The overall strategy involved in this technique begins with isolating the cell types of interest and expanding them *in vitro*. The cells are then combined with a biomaterial matrix and eventually implanted back into the patient, along with any soluble or insoluble signals that might be necessary. The implanted cells, along with cells induced to grow into the construct from the surrounding tissue, may then develop into a functional and structural equivalent of the natural tissue. Tissue engineering has a number of advantages over the traditional therapies outlined earlier. The problems of tissue availability and rejection associated with whole-organ transplantation can possibly be overcome if healthy autologous cells are isolated and expanded *in vitro*. Patients could be maintained on temporary supports, such as TPN, until the engineered tissue was fully functional. Tissue engineering has the potential to cure patients and allow them to resume their normal lifestyle.

In order to engineer a functional small intestinal tissue, we must first understand the intricacies of intestinal physiology. Then, a number of important design criteria must be considered. The intestine is subject to significant dynamic deformation *in vivo*, and as such, the engineered tissue must be mechanically sound and capable of resisting large strains. Additionally, a central function of the intestine is convective transport of material through the lumen. Thus, all of the elements necessary for peristaltic transport must be present. Another critical role of the intestine is diffusional transport of nutrients across the epithelial cell layer lining the lumen and into the bloodstream. This cell layer must therefore be present and functional in terms of enzymatic digestion and carrier-mediated nutrient transport. Finally, vascularization of the entire tissue will be necessary for the transport of oxygen, nutrients, immune cells, and endocrine factors.

2 Intestinal physiology

2.1 Gross anatomy

The small intestine is a tubular organ that extends a length of about 6 m from the stomach to the large intestine in adult humans (The following discussion is largely adapted from references [14] and [15]). The small intestine can be roughly divided into three functionally and physiologically distinct sections. Proceeding distally, these are the duodenum, the jejunum, and the ileum. Each section plays a role in breaking down and absorbing certain components of the intestinal contents, or chyme, and is specifically adapted to that function. The duodenum is the first and shortest section of the intestine, and is the one into which the bile and pancreatic ducts empty. Carbohydrates and proteins are preferentially digested and absorbed here. The jejunum comprises the first half of the remaining intestine and by its end most of the carbohydrates and amino acids have been absorbed. The distal portion of the intestine, or ileum, is important in the absorption of fats and lipid-soluble vitamins (i.e. A, E, and K) particularly at its distal end. Bile salts and any unabsorbed carbohydrates and amino acids are absorbed here as well. Each section is physiologically adapted to carry out its specific function. These adaptations include differences in the configuration of the interior intestinal surface, as well as differing distributions of cell types in the mucosa.

Just as the small intestine can be subdivided along its length, it can also be subdivided in the radial direction (Fig. 1). The wall of the small intestine, as well as that of the rest of the gastrointestinal tract, is composed of five distinct layers [17]. Proceeding outward, the innermost layer is called the mucosa, which consists of an epithelial cell lining, a connective tissue layer called the lamina propria, and a single layer of smooth muscle cells called the muscularis mucosae. The next layer is the submucosa, which contains part of the enteric nervous system called the submucosal plexus. Next is a layer of smooth muscle aligned circumferentially around the intestine. The fourth layer from the center is the myenteric plexus, another component of the enteric nervous system, and the last layer consists of smooth muscle oriented longitudinally along the intestine. Also, the external surface of the intestine is covered by a mesothelial cell layer called the serosa. The function and importance of all of these layers must be carefully considered when designing an engineered intestinal tissue.

2.2 Convective transport

A primary function of the small intestine is to transport chyme along its length, from one end to the other. At the same time, the contents must be

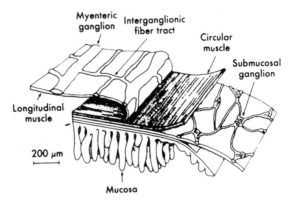

Fig. 1. The intestinal wall is composed of multiple layers of tissue, including two perpendicular smooth muscle layers and a mucosa containing an epithelial cell layer. Note also the presence of distinct projections called villi on the inner surface of the mucosa (Reprinted by permission of Mosby-Year Book, Inc.) [16].

well mixed to maximize their contact with the absorptive inner surface of the SB. There are several components of the small bowel that are absolutely essential for these forms of transport to occur. Each component must function in a coordinated fashion with the others to ensure normal and efficient transport, or motility. The three crucial components involved in the convective transport of chyme are the smooth muscle layers, neural tissue, and the endocrine system.

2.2.1 Smooth muscle layers

The circular and longitudinal smooth muscle layers are the effectors of SB motility, and are controlled by the interplay of neural and endocrine signals [18]. The innermost smooth muscle layer, which is oriented circumferentially, is thick and densely innervated. This layer is involved in mixing and propelling the intestinal contents. When a region of this muscle layer contracts, it constricts that portion of the intestine and results in a segmentation. Segmentation raises the luminal pressure, which in turn distends the intestinal wall and exposes more surface area for absorption. The constricted region also serves as a one-way valve and prevents backflow of the chyme, while the area distal to the contraction expands to accommodate forward flow. The constriction then propagates in the distal direction and propels and mixes the chyme as it does so. These contractions are known as peristalsis and move ingested food throughout the entire gastrointestinal system.

In comparison to the circular layer, the longitudinal smooth muscle layer is much thinner. The contractions of the longitudinal layer coordinate with those of the circular layer to promote motility. This coordination of smooth muscle contraction events in the two muscle layers is extremely important in producing functional peristalsis. The spatial and temporal pattern of individual phasic contractions must be properly adjusted in order for ingested food to travel properly through the intestine [19]. This requirement represents a significant engineering challenge. If any portion of the intestine, natural or engineered, is incapable of these coordinated contractions, a functional blockage may result. This is the case in Hirschsprung's Disease, and that disorder serves as a reminder of the importance of proper neural and hormonal control over smooth muscle contraction.

2.2.2 Innervation

The intestine is a highly innervated organ (The following discussion draws from references [19–21]). A further degree of neural complexity stems from the fact that the neurons involved are of both extrinsic and intrinsic origin. Extrinsic innervation is by the vagi and sympathetics, which are components of the autonomous nervous system. This extrinsic innervation is important in coordinating intestinal function, both internally and with the rest of the body. The autonomic components are particularly important in the gastrointestinal system's response to stress, eating, and behavior [22].

The intrinsic innervation consists of the enteric nervous system, which is a system unto itself and is often referred to as the 'Gut Brain' [19]. The enteric nervous system contains between 10 and 100 million neurons in humans [23], which is roughly the same number as in the spinal cord. The enteric nervous system is made up of several components, including the submucosal (Meissner's) plexus, the myenteric (Aurebach's) plexus, and the interstitial cells of Cajal. The submucosal plexus is the innermost component of the enteric nervous system, and is located between the mucosa and the circular smooth muscle layer. This network of neurons and glial cells exerts secretory control over the mucosa. The myenteric plexus is situated between the two muscle layers and provides motor innervation to both layers as well as secretomotor activity to the mucosa. The two plexuses also play a role in the modulation of blood flow, and water and electrolyte transport, in the small bowel. The interstitial cells of Cajal are found with the circular muscle layer. These fibroblast-like cells are thought to act as pacemakers in the generation of certain types of electrical activity, as well as propagating electrical signals and mediating neurotransmission [24].

Much is known about the electrical activity of the enteric nervous system. There are several distinct types of electrical activity that interact to control intestinal motility. The first of these is slow waves, also known as electrical control activity. These waves are probably generated by the pacemaker activity of the interstitial cells of Cajal and occur at a rate of 3 to 12 per minute in humans [23]. The slow waves depolarize smooth muscle cell membranes nearly to the point of activation, which can then occur through spike bursts, the second type of electrical activity. Spike bursts, or electrical response activity, actually result in muscle contraction, but only in conjunction with a slow wave. Thus, the pattern of slow waves dictates when muscle contraction can occur, and the spike bursts determine whether or not it will occur. A third form of electrical activity, present in fasted animals, is called the Migrating Myoelectric Complex. This process sweeps along the entire length of the intestine and clears the lumen of debris, preventing bacterial overgrowth between meals.

2.2.3 Endocrine control

In addition to these forms of neural control, small intestinal motility is also under endocrine control [23, 25]. Certain endocrine cells in the mucosa secrete peptide hormones in response to varying constituents of the chyme, among other factors. These peptide hormones can function as endocrine factors, entering the blood stream and acting on other organs such as the pancreas and liver, as well as controlling gastrointestinal motility. The hormones can also behave in paracrine, neurocrine, and autocrine manners as well. Several of the hormones important in small bowel motility are cholecystokinin, secretin, neurotensin, and PYY. The action of PYY is known as the 'ileal brake', since it slows small intestinal transit due to high nutrient concentrations at the distal end [25].

2.3 Diffusive transport

Aside from convective transport, or motility, another form of transport takes place in the small intestine. This is the diffusive transport of nutrients through the mucosa and into the bloodstream. The critical components in the diffusional transport of nutrients are the differentiated epithelial cell layer, and the blood vessels and lymphatics, which will be discussed in the following section.

The lining of the intestine consists of an epithelial cell monolayer which rests upon a filamentous basement membrane (The following discussion draws from references [14] and [15]). These epithelial cells (enterocytes) are present along with a small number of neural and lymphoid cells. The enterocytes can be divided into four terminally-differentiated cell types. By far the most abundant of these is the columnar enterocyte, which is specialized for digestion and absorption. Also prevalent, particularly at the distal end of the organ, are goblet cells, which are active in mucus secretion. Paneth cells, the third cell type, produce lysozyme granules and contribute to the barrier function of the intestine. The fourth cell type, the enteroendocrine cell, was encountered above in the discussion of endocrine control. More than 10 types of enteroendocrine cells are present in the epithelium [14]. These cells are situated such that they secrete their peptide hormone products through their basal rather than apical surfaces. Each epithelial cell type is found in varying proportions in the different sections of the intestine, reflecting the specialized functions of these sections. All four differentiated cell types are thought to arise from the same pluripotent mucosal stem cells.

The geometry of the mucosal lining is specifically adapted for absorption. An effective way to increase absorption through a membrane is to increase its surface area. This is manifested on the inner surface of the SB both by folds called plicae circulares, and by many small finger-like projections that extend into the lumen. These projections are called villi, and are present at a density of about 20–40/mm^2 of intestine [15]. This translates into about 5 million lining the entire adult human intestine [15]. Each villus is about 1 mm long, and contains an arteriole, a venule, and a single lacteal, or lymph vessel. The population of epithelial cells lining the surface of the villi is not uniform, but varies from the base to the tip. Epithelial cells arise from the base of each villus and migrate upwards, becoming more differentiated as they progress. The cells are then shed from the tip of the villus. Total cell turnover is accomplished in 3–7 days. The continuous shedding of cells from the villi is fueled by the proliferation and differentiation of stem cells in crevices adjacent to the villi. These crevices are known as the Crypts of Lieberkuhn, and constitute the proliferative unit of the small intestine. Each crypt contains about 250 cells, some 2/3 of which are proliferative [26]. It is estimated that the number of stem cells in an intestinal crypt is 4–16, although 30–40 cells located there might be capable of reverting to a stem cell phenotype following perturbations [27]. A gradual transition from pluripotency to differentiation occurs as the cell moves up and out of the crypts. Two noteworthy exceptions are the Paneth and most of the enteroendocrine cells, which reside permanently in the crypts.

The intestinal surface is also adapted for absorption on the cellular scale. The columnar enterocytes have several characteristics that make them especially well suited to the digestion and absorption of nutrients. Small projections of the plasma membrane, called microvilli, are present on the apical surface of the cells. These are analogous to the villi discussed above, although on a much smaller scale. A typical microvillus is 0.9–1.3 μm tall and up to 0.12 μm wide [15]. Each cell has on the order of 1000 microvilli, which results in a 30-fold increase in the plasma membrane surface area available for reaction and absorption [15]. The apical plasma membrane contains many enzymes, such as hydrolases, alkaline phosphatase, and peptidases, which are important in the breakdown of nutrients. This digestion at the surface is followed by the transport of substances into the cell across the plasma membrane. Depending on the substance, transport may be by simple diffusion or by passive/active carrier-mediated transport. The internalized substances are then transported through the interior of the cell and released on the basal side, where they enter the capillaries and lymph vessels of the lamina propria. The enterocyte cell layer can also modulate its permeability through structures called tight junctions, which firmly connect adjacent cells to one another. The tightness of these junctions can be adjusted, allowing substances, such as water and small ions, to pass between the cells. This mechanism of transport is important for certain drugs and other substances that are not absorbed by the enterocytes [28]. Tight junctions, in addition to mucous and secreted antibodies, are also important in the intestine's role as a barrier to bacterial translocation.

2.4 Vascularization and lymphoid tissue

All the tissues of the intestine discussed thus far are supplied by an extensive vascular network, which transports absorbed nutrients to the rest of the body. The bloodflow to the intestine is considerable, as evidenced by the 1.4 L of blood, on average, that flow through the portal vein each minute [15]. The layout of the intestinal vasculature is specifically designed to accommodate the intestine's absorptive function. Most of the blood entering the small bowel flows from the systemic circulation via the superior mesenteric artery. The blood then leaves the organ through the superior mesenteric vein. This vein joins with the splenic vein to form the portal vein, which carries the blood to the liver for absorption and detoxification of certain components. If nutrients are introduced directly into the systemic circulation without first passing through the liver (e.g. TPN), several complications can result. The mesentery orchestrates the organization and arrangement of the intestinal blood supply, nerves, and lymph vessels.

The intestine is also an important organ of the immune system [8]. Gut associated lymphoid tissue (GALT) is abundant in the intestine and accounts, in part, for the difficulties encountered in intestinal transplantation. Lymphoid tissue is found in several different places within the small bowel, including the mesenteric lymph nodes and Peyer's patches, which are aggregates of lymphoid tissue found under the epithelium throughout the small intestine, but particularly in the distal ileum. Intestinal plasma cells secrete dimeric secretory IgA. This function constitutes an important part of the intestine's role in preventing infection by pathogens that enter the body with ingested food. The omentum, a highly vascularized organ that lies over the intestine in the ventral space of the abdominal cavity, functions in containing a variety of inflammatory conditions that arise in the abdominal cavity.

3 Tissue engineering approach to the small intestine

Despite its complexity and multiple components, the goal in engineering small intestine is to develop a tissue that reproduces as many properties and functions of the natural one as possible. Thus, the natural organ will serve as a guide. The general strategy involves a combination of cells and open-pore matrices which are implanted into the body and become integrated with the surrounding tissues. This process begins with the isolation of the appropriate cell types from a donor, possibly the patient to be treated. These cells are then seeded onto a biomaterial matrix and provided with specific chemical and mechanical cues to guide their growth and development. The cell/material construct is ultimately implanted into the recipient, where it may form a functional, integrated tissue. The biomaterial matrix plays several important roles in this scheme. First, it functions as a vehicle to deliver the cells to the correct anatomic location within the body. Second, it provides mechanical stability to the implanted construct and maintains a potential space for tissue development. The scaffold also acts as a template to specifically guide tissue development, both from the implanted cells as well as fibrovascular ingrowth at the implantation site. The biomaterial matrix may degrade over time, leaving in its place a completely integrated, natural tissue.

It is useful to begin the design process by defining the critical properties, components, and functions that the engineered tissue must have. These can then be incorporated into the design until a functional tissue results. The first requirement is that the engineered tissue be mechanically stable and capable of resisting the strains typically encountered during culture and *in vivo*. This will involve first the properties of the biomaterial matrix, and later those of the engineered smooth muscle layers. Next, the tissue must be capable of the convective transport involved in propelling and mixing chyme, necessitating both smooth muscle and a functioning neural component. The third requirement is that the engineered intestine be capable of nutrient absorption. The critical components in this step are the enterocyte monolayer and the microvasculature. Finally, the engineered tissue should have an appropriately organized vascular supply, as well as the lymphoid tissue found in the natural organ.

Several groups have already begun to tackle certain aspects of small intestinal tissue engineering with some success. Furthermore, much of the work being done with other tissue types will be applicable to the small intestine. For example, efforts to engineer smooth muscle, vascular tissue, and nerves will find application in the design of a number of tissues, including small intestine. The remainder of this chapter will, therefore, focus not only on what has been done with intestine, but also on how work in engineering other tissues can be applied to the intestine. The emphasis will be on what aspects of the physiology need to be considered, and the approaches that might be used.

3.1 Maintenance of mechanical integrity

Among the most important characteristics of an engineered small intestine is that it be mechanically sound. Mechanical properties will be important initially, as the newly implanted tissue will be subject to compressional forces. These forces will tend to collapse the implant if it is not sufficiently resilient. Mechanical properties are also important because of the dynamic nature of intestinal motility. The transport of chyme through the lumen creates regions of raised luminal pressure and increased mechanical strain due to muscle contraction. It is crucial that the engineered intestine remain intact *in vivo* to preserve its role as a barrier. A failure in this regard could lead to serious infection. Thus, throughout all the stages of engineering intestine, the mechanical properties of the tissue need to be considered. The mechanical properties will be dictated at first by the biomaterial matrix and later, if that matrix degrades, by the connective tissue and muscle layers that make up the walls of the organ.

3.1.1 Materials

Many different types of materials have been utilized in tissue engineering. These materials can generally be divided into two groups: naturally-derived materials and synthetic materials. Both types of materials have certain desirable properties that lend themselves to tissue engineering applications. Naturally-derived materials (i.e. collagen) have been used in tissue engineering primarily because of their biological activity [29]. For example, extracellular matrix proteins such as collagen and laminin contain cell-binding sequences in their structures which can specifically interact with cells. One of the best known of these is Arginine–Glycine–Aspartate (RGD), which binds to cell surface receptors of the integrin family. Type I collagen has long been used as a biomaterial, and can be processed into a variety of forms including fibers and hydrogels. Hydrogels of laminin and polysaccharides such as alginate have also been used [30]. A promising new naturally-derived material is an acellular mixture of tissue matrix from the submucosa of the porcine small intestine [31]. This material may have particular relevance for engineering small intestine. Unfortunately, while natural materials may have favorable inter-actions with cells, their mechanical properties are often limiting. This factor may be especially problematic in engineering a dynamic tissue such as intestine. Also, large scale production of natural materials can be difficult to accomplish reproducibly and inexpensively, and immunological responses to animal-derived proteins can provide additional difficulties.

Synthetic materials have therefore been used extensively in tissue engineering applications as well [32]. These materials have several advantages, including relatively simple control of degradation kinetics and physical properties such as strength and elasticity. Furthermore, synthetic materials can be reproducibly synthesized and are easily processed. One class of synthetic polymers that has found common use in tissue engineering is the aliphatic polyesters, polyglycolic acid (PGA) and polylactic acid (PLA) [33]. These polymers have been used for many years as biodegradable sutures and are considered relatively biocompatible. They can be processed into fibers, which can then be formed into non-woven matrices of high porosity [34]. Alternatively, porous films can be produced using a variety of techniques [35]. Furthermore, the polymers degrade by simple hydrolysis to produce non-toxic degradation products. The two monomers can also be co-polymerized to produce poly(lactic-co-glycolic acid) (PLGA) of varying composition. These copolymers can be formulated to have specific mechanical and degradation properties intermediate to those of the pure polymers.

PGA, PLA, and PLGA have been utilized in many of the attempts to engineer intestine to date. Researchers have seeded enterocytes onto tubular non-woven PGA fiber-based scaffolds and implanted them into rats as a first step in engineering intestinal tissue [34, 36]. Porous films of PLA and PLGA have also been seeded with enterocytes and cultured *in vitro* [35] (Fig. 2). It was found that these polymer scaffolds resisted compressional forces *in vivo* and induced the formation of an organized enterocyte layer on their interior surfaces. Furthermore, these constructs became vascularized and remained viable over a period of several months. A variety of other synthetic polymers are available for tissue engineering applications as well. Elastic polymers [33], for example, might find use in engineering muscle and other mechanically dynamic tissues, including intestine. The main limitation of synthetic polymers is their lack of direct interaction with cells. A new approach to this problem is to graft cell-binding domains onto synthetic polymers [37]. This results in a biomaterial that has the desirable physical properties of a synthetic polymer and the biological specificity of a natural material.

3.1.2 Smooth muscle component

Whatever the material used, assuming that it degrades over time, the mechanical properties of the engineered tissue will ultimately be dominated by the tissue itself. In particular, the perpendicular layers of smooth muscle impart strength and allow the native tissue to exert control over its own mechanical state. Thus, it may be critical to incorporate a smooth muscle element into the design of engineered intestine. There are several ways to accomplish this goal. The first approach is to directly seed smooth muscle cells (SMCs) onto three-dimensional matrices, and several research

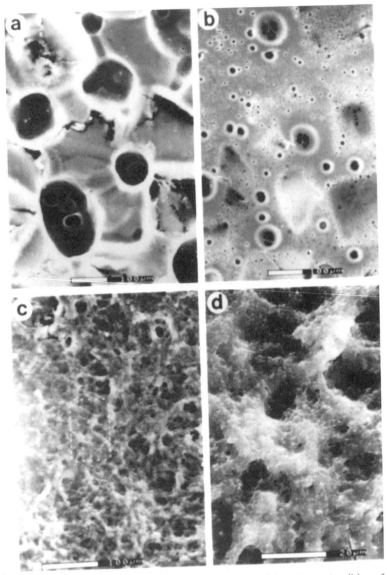

Fig. 2. Scanning electron micrographs of the exterior (a) and interior (b) surfaces of a tubular PLGA porous film matrix. Enterocytes seeded onto this matrix attached *in vitro* and formed a monolayer (c and d). Size bars are shown on photomicrographs (Photo used with permission of the Surgical Forum of the American College of Surgeons) [32].

groups have focused on engineering smooth muscle with this technique [38, 39]. New smooth muscle tissue with a cellularity and matrix composition comparable to that of natural tissue can be obtained [38] (Fig. 3). It is

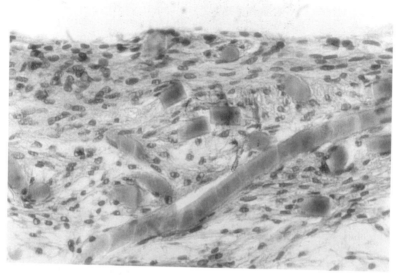

Fig. 3. Hematoxylin and eosin stained cross-section of an engineered smooth muscle tissue *in vitro*. SMCs were seeded onto a PGA fiber-based matrix and cultured for 4 weeks. Note the presence of large fiber fragments interspersed with the tissue (Photo taken by B.S. Kim, used with permission of Nature Publishing Co.) [38].

believed that mechanical stress may be an important factor in determining the orientation and phenotype of SMCs *in vivo*, and several studies have confirmed that statically and cyclically stretching cells *in vitro* causes them to align and exhibit contractile characteristics [39, 40]. However, the perpendicular geometry of the two smooth muscle layers present in small intestine creates an interesting challenge. One way to engineer these two layers may be to utilize two SMC-seeded polymer matrices, aligned perpendicular to one another. This configuration might guide the development of the muscle layers and induce them to adopt the perpendicular orientation seen *in vivo*. An alternative approach might be to wrap pre-existing smooth muscle tissue around the tubular scaffold in perpendicular layers. Both of these approaches overcome the difficulties of developing the muscle layers from cells seeded with enterocytes in the complex environment of a co-culture situation.

3.2 Convective transport

The convective transport of material through the small intestine's lumen is an essential function. This transport allows the intestine to continually process each meals' worth of ingested food, and rids the lumen of debris and desquamated cells between meals. Proper convective transport in an engineered intestine will depend on the productive interaction of the smooth muscle layers and components of the autonomous and enteric nervous systems. Since the engineering of smooth muscle was already discussed, we will now turn briefly to the issue of innervation. As previously mentioned, innervation is by two different types of neurons, extrinsic and intrinsic. Extrinsic innervation of engineered tissue may be accomplished via ingrowth of existing neurons into the tissue. Tubular polymeric guidance channels have been developed to guide the bridging of transected nerves. Release of basic fibroblast growth factor (bFGF) from the polymer material of which these channels are constructed has been shown to further enhance nerve growth [41]. Seeding autologous Schwann cells into the guidance channels has also been shown to induce growth [42]. Furthermore, culturing neurons on electrically dynamic piezoelectric polymers has been shown to promote neurite outgrowth [43]. Any number of these techniques could possibly be combined to guide extrinsic peripheral nerves into the developing tissue. Importantly, limited extrinsic reinnervation has been shown to occur naturally in cases of canine small bowel autotransplantation after 1 year [44], and similar results may be obtained in an engineered tissue. Unfortunately, the development of intrinsic innervation, which is critically important, may present a more difficult challenge due to the complexity and magnitude of the enteric nervous system (ENS). Transplantation or induction of enteric neurons, glial cells, and interstitial cells of Cajal may be required to recreate the ENS. Endocrine control over small bowel motility should also be considered in the design.

3.3 Diffusional transport

Diffusional transport characterizes what is perhaps the small intestine's most basic role: the absorption of nutrients from ingested food. Incorporating diffusional transport into the tissue design will involve both the overall macroscopic geometry of the tissue as well as the microscopic characteristics of the differentiated enterocyte layer. In both cases, the goal is to maximize the surface area available for absorption and to preserve the enzymatic, digestive, and absorptive functions of the columnar enterocytes. Furthermore, it is desirable to engineer a tissue containing the other epithelial cell types and a well-developed microvasculature.

At least two approaches can be taken to develop the optimal macroscopic topography needed for absorption. One possibility is to seed enterocytes onto a scaffold and allow the villi and crypts to develop naturally over time in a process analogous to natural intestinal development. An important issue in this approach is the development of techniques for the isolation of specific enterocyte populations from harvested intestine. Techniques have focused on isolating the rapidly dividing and differentiating cells of the crypts, which are capable of giving rise to the four differentiated cell types. Several protocols have been developed for the isolation of enterocytes enriched for the crypt fraction, including mechanical, enzymatic, and chelation methods [45–47]. Seeding of enterocytes isolated by the chelation method onto polymer constructs has resulted in the formation of a stratified epithelium after implantation [36] (Fig. 4). Mitotically active cells are seen in this layer and the columnar enterocytes remain distinct from invading fibrovascular tissue. Rudimentary villus and crypt formation was also noted in these studies. Defining the proliferative capabilities of the cells *in vitro* and *in vivo* will determine the number of cells that need to be obtained from each original isolation and seeded onto each matrix. The second approach in developing the correct tissue macrostructure may be to fabricate a matrix that contains the folds and projections inherent in natural intestine. Such a matrix may specifically cue seeded cells to form villus and crypt structures.

An engineered intestinal tissue may also need to account for the differing cell populations and distinct functions of the three intestinal sections, rather than being uniform down its length. Certain disorders of the small intestine result in the loss of only a small portion of the organ. For example, some patients may be missing the ileum but may still retain the first two sections of the intestine. While these patients would need at least partial restoration of ileal function, they might not require additional absorptive surface area. It may be possible to treat these patients by stripping the mucosa from a portion of their remaining intestine and repopulating with enterocytes from the ileum.

Once an enterocyte-containing tissue has been engineered, the cells must be induced to adopt the appropriate patterns of gene expression. Differentiation of the enterocyte layer must be considered in two space dimensions, both along the crypt-villus axis and along the lengthwise axis of the whole organ. The mechanisms by which enterocytes differentiate and what cues should be given to reproduce this differentiation in an engineered tissue are unknown. Several clues have, however, been found in the development of a mature enterocyte layer. Evidence suggests that the basement membrane may play a role in the differentiation of enterocytes [48, 49]. Researchers have shown that enterocytes cultured *in vitro* on a layer of EHS sarcoma cell basement membrane adopt a differentiated phenotype as compared to enterocytes cultured on tissue culture plastic, which remain proliferative and undifferentiated [50]. These results suggest that cell–ECM interactions are important in enterocyte differentiation and the formation of a mature epithelial cell layer. The mechanism of differentiation of enterocytes on the lengthwise axis of the small intestine remains unclear, but may result from the influence of varying luminal concentrations of certain substances in the different regions [51].

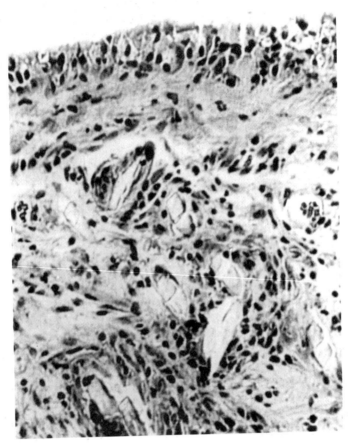

Fig. 4. Cross-section of an engineered intestinal mucosa tissue *in vivo*. ECs were seeded onto a fibrous PGA scaffold and implanted in the omentum of a Lewis rat for 14 days. The section was stained with hematoxylin and eosin. Note the formation of a stratified epithelium with the lumen at the top of the picture. Also note the presence of a blood vessel (V). Original magnification was 16 × (Reprinted by permission of Appleton and Lange, Inc.) [36].

3.4 Vascularization

Another critical aspect of engineering intestinal tissue is vascularization. This will involve both capillary ingrowth at the implantation site and development of the microvasculature, as well as the overall organization of blood flow to the engineered tissue. The initial body response to the implantation of almost any foreign object, including a cell/polymer construct, is an inflammatory response. This is accompanied by capillary invasion if the pores of the material are larger than approximately 10 μm, and the rate of ingrowth can be regulated by the pore size and porosity [52]. Angiogenesis can potentially be enhanced by the selective application of cytokines and growth factors such as basic fibroblast growth factor, epidermal growth factor (EGF) and transforming growth factor-β (TGF-β), among others [53], and it may be possible to use the polymer matrix as a sustained release vehicle [54]. Additionally, matrices could be seeded with endothelial cells to enhance the formation of a vascular bed in the engineered tissue following transplantation [55]. The neovasculature resulting from these steps may continue to adapt and change to meet the needs of the developing tissue over time. However, it is important that the proper macroscopic arterial and especially venous attachments be made initially. It is likely that the mesentery helps to guide the growth and distribution of blood vessels to the intestine and the mesentery may play a central role in the vascular element of an engineered intestine. Ultimately, the issues of lymphatics and lymphoid tissue may need also to be considered.

4 Summary

An engineered small intestine could provide a novel clinical treatment for small bowel disorders, and could also serve as a model system for investigating intestinal function and drug absorption. Considerable progress has been made in engineering intestine and related tissues to date, and a number of important design criteria have evolved from studies of both native and engineered intestinal tissue. The engineered tissue must be mechanically sound and capable of the two forms of transport, convective and diffusive, that are so vital to the intestine's natural role in nutrient digestion and absorption. The design may ultimately involve a variety of cellular and tissue components that contribute to the various aspects of intestinal form and function in the native tissue.

References

1. Vanderhoof, J. A., Short bowel syndrome. *Neonatal Gastroenterology*, 1996, **23**(2), 377–386.
2. Caniano, D. A., Starr, J. and Ginn-Pease, M. E., Extensive short bowel syndrome in neonates: Outcome in the 1980s. *Surgery*, 1989, **105**, 119–124.
3. Kays, D. W., Surgical conditions of the neonatal intestinal tract. *Neonatal Gastroenterology*, 1996, **23**(2), 353–375.

4. Organ, G. M. and Vacanti, J. P., Tissue engineering neointestine. In *Textbook of Tissue Engineering*, ed. R. P. Lanza, R. Langer and W.L. Chick. R.G. Landes, Austin, 1996.

5. Weber, T. R., Vane, D. W. and Grosfeld, J. L., Tapering enteroplasty in infants with bowel atresia and short gut. *Archives of Surgery*, 1982, **117**, 684–688.

6. Porkorny, W. J. and Fowler, C. L., Isoperistaltic intestinal lengthening for short bowel syndrome. *Surgery Gynecology and Obstetrics*, 1991, **172**, 39–43.

7. Diego, M. D., Miguel, E., Lucen, C. M. et al., Short gut syndrome: A new surgical technique and ultrastructural study of the liver and pancreas. *Archives of Surgery*, 1982, **117**, 789–795.

8. Frezza, E. E., Tzakis, A., Fung, J. J. and Van Thiel, D. H., Small bowel transplantation: Current progress and clinical application. *Hepato-Gastoenterology*, 1996, **43**, 363–376.

9. Asfar, S., Atkinson, P., Ghent, C., Duff, J., Wall, W., Williams, S., Seidman, E. and Grant, D., Small bowel transplantation: A life-saving option for selected patients with intestinal failure. *Digestive Diseases and Sciences*, 1996, **41**(5), 875–883.

10. Weber, T. R., Tracy, T. Jr. and Connors R. H., Short-bowel syndrome in children. Quality of life in an era of improved survival. *Archives of Surgery*, 1991, **126**, 841–846.

11. Dorney, S. F., Ament, M. E., Berquist, W. E., Vargas, J. H. and Hassall, E., Improved survival in very short small bowel of infancy with use of long-term parenteral nutrition. *Journal of Pediatrics*, 1985, **107**, 521–525.

12. Foldes, J., Rimon, B., Muggia-Sullam, M., Gimmun, Z., Leichter, I., Steinberg, R., Menczel, J. and Freund, H. R., Progressive bone loss during long-term home total parenteral nutrition. *Journal of Parenteral and Enteral Nutrition*, 1990, **14**(2), 139–142.

13. Langer, R. and Vacanti, J. P., Tissue Engineering. *Science*, 1993, **260**, 920–926.

14. Rubin, W., Small intestine: Anatomy and structural anomalies. In *Textbook of Gastroenterology*, Vol. 2, ed. T. Yamada. J.B. Lippincott Company, Philadelphia, 1991, pp. 1409–1424.

15. Bell, G. H., Emslie-Smith, D., Paterson, C. R., Digestion and absorption in the intestine. *Textbook of Physiology*, 10th edn. Churchill Livingstone, New York, 1980, pp. 66–84.

16. Berne, R. M. and Levy, M. N., eds. *Principles of Physiology*, 2nd edn. Mosby-Year Book, St. Louis, 1996, p 440.

17. Madara, J. L., Epithelia: Biologic principles of organization. In *Textbook of Gastroenterology*, Vol. 1, ed. T. J. B. Yamada. Lippincott Company, Philadelphia, 1991, pp. 102–118.

18. Makhlouf, G. M., Smooth muscle of the gut. In *Textbook of Gastroenterology*, Vol. 1, ed. T. J. B. Yamada. Lippincott Company, Philadelphia, 1991, pp. 61–84.

19. Sarna, S. K. and Otterson, M. F., Small intestinal physiology and pathophysiology. *Gastroenterology Clinics of North America*, 1989, **18**(2), 375–404.

20. Bardakjian, B. L., Gastrointestinal System. In: *Handbook of Biomedical Engineering*, ed. J. Bronzino. CRC Press, 1995, pp 57–69.

21. Wingate, D. L., Neurophysiology of the gastrointestinal tract. In *The Gas-*

Engineering Small Intestine

troenterology *Annual 3*, ed. F. Kern and A. L. Blum. Elsevier Science, 1986, pp. 257–283.

22. Goyal, R. K., and Hirano, I., The enteric nervous system. *The New England Journal of Medicine*, 1996, **334**(17), 1106–1115.

23. Furness, J. B. and Bornstein, J. C., The enteric nervous system and its extrinsic connections. In *Textbook of Gastroenterology*, Vol. 1, ed. T. J. B. Yamada. Lippincott Company, Philadelphia, 1991, pp. 2–24.

24. Sanders, K. M., A case for interstitial cells of Cajal as pacemakers and mediators of neurotransmission in the gastrointestinal tract. *Gastroenterology*, 1996, **111**, 492–515.

25. Taylor, I. L., Mannon, P., Gastrointestinal Hormones. In *Textbook of Gastroenterology*, Vol. 1, ed. T. J. B. Yamada. Lippincott Company, Philadelphia, 1991, pp. 24–49.

26. Gordon, J. I., Intestinal epithelial differentiation: New insights from chimeric and transgenic mice. *The Journal of Cell Biology*, 1989, **108**, 1187–1194.

27. Potten, C. S. and Loeffler, M., Stem cells: Attributes, cycles, spirals, pitfalls, and uncertainties. Lessons for and from the crypt. *Development*, 1990, **110**, 1001–1020.

28. Pappenheimer, J. R. and Reiss, K. Z., Contribution of solvent drag through intercellular junctions to absorption of nutrients by the small intestine of the rat. *Journal of Membrane Biology*, 1987, **100**, 123–136.

29. Cavallaro, J. F., Kemp, P. D. and Kraus, K. H., Collagen fabrics as biomaterials. *Biotechnology and Bioengineering*, 1994, **43**, 781–791.

30. Jen, A. C., Wake, M. C., Mikos, A. G., Review: Hydrogels for cell immobilization. *Biotechnology and Bioengineering*, 1996, **50**, 357–364.

31. Badylak, S. F., Tullius, R., Kokini, K., Shelbourne, K. D., Klootwyk, T., Voytik, S. L., Kraine, M. R., and Simmons, C., The use of xenogenic small intestinal submucosa as a biomaterial for Achilles tendon repair in a dog model. *Journal of Biomedical Materials Research*, 1995, **29**(8), 977–985.

32. Putnam, A. J., Mooney, D. J., Tissue engineering using synthetic extracellular matrices. *Nature Medicine*, 1996, **2**(7), 824–826.

33. Wang, W. H. and Mooney, D. J., Synthesis and properties of biodegradable polymers used in tissue engineering. In *Tissue Engineering*, ed. A. Atala and D. J. Mooney. Birkhauser Press, in press.

34. Mooney, D. J., Mazzoni, C. L., Breuer, C., McNamara, K., Hern, D., Vacanti, J. P. and Langer, R., Stabilized polyglycolic acid fibre-based tubes for tissue engineering. *Biomaterials*, 1996, **17**, 115–124.

35. Mooney, D. J., Breuer, C., McNamara, K., Vacanti, J. P. and Langer, R., Fabricating tubular devices from polymers of lactic and glycolic acid for tissue engineering. *Tissue Engineering*, 1995, **1**(2), 107–118.

36. Organ, G. M., Mooney, D. J., Hansen, L. K., Schloo, B. and Vacanti, J. P., Transplantation of enterocytes utilizing polymer-cell constructs to produce a neointestine. *Transplantation Proceedings*, 1992, **24**, 3009–3011.

37. Hubbell, J. A., Biomaterials in tissue engineering. *Biotechnology*, 1995, **13**, 565–575.

38. Kim, B. S., Putnam, A. J., Kulik, T., Mooney, D. J., Engineering smooth muscle on biodegradable polymer matrices. Submitted paper.

39. Kanda, K., Matsuda, T. and Oka, T., Mechanical stress induced cellular orientation and phenotypic modulation of 3-D cultured smooth muscle cells. *ASAIO Journal*, 1993, M686–M753.

40. Kanda, K. and Matsuda, T., Behavior of arterial wall cells cultured on periodically stretched substrates. *Cell Transplantation*, 1993, **2**, 475–484.

41. Aebischer, P., Salessiotis, A. N. and Winn, S. R., Basic fibroblast growth factor released from synthetic guidance channels facilitates peripheral nerve regeneration across long nerve gaps. *Journal of Neuroscience Research*, 1989, **23**, 282–289.

42. Guenard, V., Kleitman, N., Morrissey, T. K., Bunge, R. P. and Aebischer, P., Syngenic Schwann cells derived from adult nerves seeded in semipermeable guidance channels enhance peripheral nerve regeneration. *The Journal of Neuroscience*, 1992, **12**(9), 3310–3320.

43. Valentini, R. F., Vargo, T. G., Gardella, J. A. Jr. and Aebischer, P., Electrically charged polymeric substrates enhance nerve fibre outgrowth in vitro. *Biomaterials*, 1992, **13**(3), 183–190.

44. Sugitani, A., Reynolds, J. C., Tsuboi, M., Todo, S. and Starzl, T. E., Extrinsic reinnervation of the intestine after small bowel autotransplantation in dogs. *Transplantation Proceedings*, **26**(3), 1640–1641.

45. Harrison, D. D., and Webster, H. L., The preparation of isolated intestinal crypt cells. *Experimental Cell Research*, 1969, **55**, 257–260.

46. Weiser, M. M., Intestinal epithelial cell surface membrane glycoprotein synthesis I. An indicator of cellular differentiation. *Journal of Biological Chemistry*, 1973, **248**(7), 2536–2541.

47. Weiser, M. M., Intestinal epithelial cell surface membrane glycoprotein synthesis II. Glycosyltransferases and endogenous acceptors of the undifferentiated cell surface membrane. *Journal of Biological Chemistry*, 1973, **248**(7), 2542–2548.

48. Tait, I. S., Penny, J. I. and Campbell, C., Does neomucosa induced by small bowel stem cell transplantation have adequate function? *The American Journal of Surgery*, 1995, **169**, 120–125.

49. Simon-Assmann, P., Simo, P., Bouziges, F., Haffen, K. and Kedinger, M., Synthesis of basement proteins in the small intestine. *Digestion*, 1990, **46**(suppl 2), 12–21.

50. Carroll, K. M., Wong, T. T., Drabik, D. L. and Chang, E. B., Differentiation of rat small intestinal epithelial cells by extracellular matrix. *American Journal of Physiology*, 1988, **254**, G355–G360.

51. Roth, K. A., Gordon, J. I., Spatial differentiation of the intestinal epithelium: Analysis of enteroendocrine cells containing immunoreactive serotonin, secretin, and substance P in normal and transgenic mice. *Proceedings of the National Academy of Sciences of the United States of America*, 1990, **87**, 6408–6412.

52. Mooney, D. J. and Langer, R., Engineering biomaterials for tissue engineering: the 10-100 micron scale. In: *Biomedical Engineering Handbook*, ed. J. Bronzino. CRC Press, 1995, pp. 1609–1618.

53. Battegay, E. J., Angiogenesis: Mechanistic insights, neovascular diseases, and therapeutic prospects. *Journal of Molecular Medicine*, 1995, **73**, 333–346.

54. Mooney, D. J., Kaufmann, P. M., Sano, K., McNamara, K., Schwendeman, S., Vacanti, J. P. and Langer, R., Localized delivery of epidermal growth factor improves the survival of transplanted hepatocytes. *Biotechnology and Bioengineering*, 1996, **50**, 422–429.

55. Holder, W. D., Gruber H. E., Roland, W. D., Moore, A. L., Culberson, C. R., Loebsack, A., Burg, K. J. L. and Mooney, D. J., Unique vascular structures occurring in polyglycolide matrices containing aortic endothelial cells implanted in the rat. Submitted paper.

CHAPTER IV.5

Tissue Engineered Kidney For Renal Replacement

H. DAVID HUMES,
DEREK C. BLAKENEY,
WON-KYOUNG LEE,
JANETA NIKOLOVSKI

Departments of Internal Medicine and Biomedical Engineering,
The University of Michigan Medical Center,
Veterans Administration Medical Center and
The University of Michigan at Ann Arbor,
Ann Arbor, MI 48105, USA

1 Physiology

1.1 Fundamentals

The kidneys play critical excretory, regulatory, and endocrinologic roles in maintaining body homeostasis. The filtration unit of the excretory system is the glomerulus. Filtration of toxins, water, and salts present in blood begins at the glomerulus, which is a vascular network of capillaries with high hydraulic conductivity. Blood enters these capillaries from the afferent arteriole and is filtered. Post-filtered blood exits from the glomerulus via the efferent arteriole into a second peritubular capillary system. The filtrate is collected into Bowman's capsule and transits into an epithelial tubule system consisting of the proximal convoluted tubule, loop of Henle, distal convoluted tubule, and the collecting duct. All collecting ducts empty into the renal pelvis, where it then is drained into the ureter. The ureter of each kidney passes the urine into the bladder, where it is stored until release through the urethra. The filtrate is processed along these various tubule segments, so that the tubule unit is the regulator of final urine formulation.

Each tubule segment manifests different solute and fluid transport characteristics to finely regulate urine formation (Fig. 1). The proximal convoluted tubule functions both by reabsorption (returning filtrate components back to blood by active transport through the tubule epithelium and into neighboring capillaries) and secretion. For instance, drugs and toxins are secreted from the surrounding vasculature into the tubular lumen for excretion. Hydrogen ions are secreted into the tubules, thereby maintaining proper blood pH. Water, salts, and nutrients, such as glucose and amino acids, are reabsorbed from the tubule into the interstitial fluid, and then into peritubular capillaries. Dilution or concentration of water continues in the descending loop of Henle, while salt dilution is accomplished in the ascending loop of Henle. Next is the distal convoluted tubule, which secretes potassium, reabsorbs sodium, and helps regulate pH by proton secretion. The final step of urine formation occurs in the collecting duct, which displays a highly refined system for water concentration and dilution depending upon the hydration state of the body. Urine formation can be simply conceived as a two step process: blood ultrafiltration followed by filtrate processing. This two step process is achieved by the glomerulus and tubules of the kidney. The glomerulus and tubule comprise the functional apparatus of the kidney, the nephron. Each human kidney contains approximately one million nephrons, summing to more than 80 km of tubule length. Of the approximately 1000 L of blood that flows through a human kidney each day, 180 L are processed as filtrate, and only 2 L are excreted. The kidney not only serves as an excretory organ, but also has multiple metabolic and endocrine functions. The kidney synthesizes erythropoietin, active forms of vitamin D, renin, angiotensin, prostaglandins, leukotrienes, and kallikrein-kinins.

1.2 Glomerulus: The filtration unit

The process of urine formation begins within the capillary bed of the glomerulus [1]. The glomerular capillary wall has evolved into a structure with the property to separate as much as one-third of the plasma entering the glomerulus into a solution of a nearly ideal ultrafiltrate. This high rate of ultrafiltration across the glomerular capillary is a result of hydraulic pressure generated by the pumping action of the heart and the vascular tone of the preglomerular and post-glomerular vessels (afferent and efferent arterioles) as well as the high hydraulic permeability of the glomerular capillary walls. This hydraulic pressure and hydraulic permeability of the glomerular capillary bed is at least two times and two orders of magnitude higher, respectively, than most other capillary networks within the body [2]. Despite this high rate of water and solute flux across the glomerular capillary wall, this same structure retards the filtration of important circulating macromolecules, especially albumin, so that all but the lower-molecular weight plasma proteins are restricted in their passage across this filtration barrier.

A variety of experimental studies and model systems have been employed to characterize the sieving properties of the glomerulus. Hydrody-

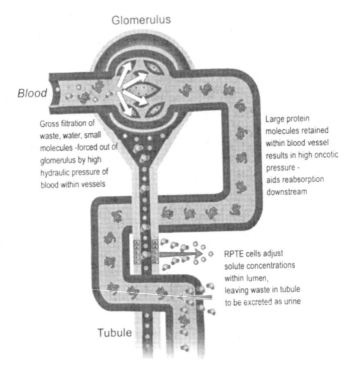

Glomerulus

Blood

Gross filtration of
waste, water, small
molecules -forced out of
glomerulus by high
hydraulic pressure of
blood within vessels

Large protein
molecules retained
within blood vessel
results in high oncotic
pressure -
aids reabsorption
downstream

RPTE cells adjust
solute concentrations
within lumen,
leaving waste in tubule
to be excreted as urine

Tubule

Fig. 1. Schematic of filtration and reabsorption processes in the kidney. RPTE, renal proximal tubule epithelial cells (Figure adapted from [20]).

namic models of solute transport through pores have been successfully used to describe the size selective barrier function of this capillary network to macromolecules [3]. This pore model, in its simplest form, assumes the capillary wall to contain cylindrical pores of identical size and that macromolecules are spherical particles. Based upon the steric hindrances that macromolecules encounter in the passage through small pores, whether by diffusion or by convection, definition of the glomerular capillary barrier can be defined by solving hydrodynamic equations governing the movement of a spherical particle through a fluid-filled cylindrical pore. This modeling characterizes the glomerular capillary barrier as a membrane with uniform pores of 50 Å radius [4]. This pore size predicts that molecules with radii smaller than 14 Å appear in the filtrate in the same concentration as in plasma water. Since there is no restriction to filtration, fractional clearance of this size molecule is equal to one. The filtration of molecules of increasing size decreases progressively, so that the fractional clearance of macromolecules the size of serum albumin (36 Å) is low.

The glomerular barrier, however, does not restrict molecular transfer across the capillary wall only on the basis of size. This realization is based upon the observation that filtration of the circulating protein, albumin, is restricted to a much greater extent than would be predicted from size alone. The realization that albumin is a polyanion at physiologic pH suggests that molecular charge, in addition to molecular size, is another important determination of filtration of macromolecules [5]. The greater restriction to the filtration of circulating polyanions, including albumin is due to the electrostatic hindrance by fixed negatively charged components of the glomerular capillary barrier. These fixed negative charges, as might be expected, simultaneously enhance the filtration of circulating polycations.

Thus, the formation of glomerular ultrafiltrate, the initial step in urine formation, depends upon the pressure and flows within the glomerular capillary bed and the intrinsic permselectivity of the glomerular capillary wall. The permselective barrier excludes circulating macromolecules from filtration based upon size as well as net molecular charge, so that for any given size, negatively charged macromolecules are restricted from filtration to a greater extent than neutral molecules.

1.3 The tubule: filtrate processer

The daily urinary volume of a normal human kidney is roughly 2 litres, however, approximately 100 ml of filtrate is formed every minute, meaning that more than 98% of the glomerular ultrafiltrate must be reabsorbed by the renal tubule. The bulk of the reabsorption, 50–65%, occurs along the proximal tubule. Similar to glomerular filtration, fluid movement across the renal proximal tubule cell is governed by physical forces. Unlike the fluid transfer across the glomerular capillary wall, however, tubular fluid flux is principally driven by osmotic and oncotic pressures rather than hydraulic pressure (Fig. 1). The different cell types residing in the tubule are structurally suited to their function. The proximal tubule cells extensively interdigitate along their basolateral antiluminal side which allows for a high total surface area and is characteristic of salt transporting epithelia. The luminal side of the cells also contains a brush border with extensive microvilli, increasing the surface area for filtrate reabsorption. Also present in these cells are an abundant number of mitochondria providing a high metabolic activity for the purpose of active transport.

Renal proximal tubule fluid reabsorption is based upon active Na^+ transport, requiring the energy-dependent $Na^+K^+ATPase$ located along the basolateral membrane of the renal tubule cell to promote a small degree of luminal hypotonicity [6]. This small degree of osmotic difference (2–3 mOsm/kg H_2O) across the renal tubule is sufficient to drive isotonic fluid reabsorption due to the very high diffusive water permeability of the renal tubule cell membrane. Once across the renal proximal tubule cell, the transported fluid is taken up by the peritubular capillary bed due to the favorable oncotic pressure gradient. A high oncotic pressure within the peritubular capillary is the result of the high rate of protein-free filtrate formed at the proximate glomerular capillary bed [7]. The addition of a renal epithelial monolayer with high rates of active Na^+ transport and high hydraulic permeability assists further in the high rate of salt and water reabsorption along the proximal tubule. Thus, an elegant system has evolved in the nephron to filter and reabsorb large amounts of fluid in bulk to attain high rates of metabolic product excretion while maintaining regulatory salt and water balance.

The kidney not only is important as an excretory organ but provides important reabsorptive, homeostatic, metabolic, and endocrinologic functions. The reabsorptive function of the kidney includes not only sodium and water homeostasis, but also the important reclamation of metabolic substrates, including essential amino acids and glucose. The kidney also serves as a critically important metabolic organ, synthesizing glutathione and free radical scavenging enzymes as well as providing gluconeogenic and ammoniagenic capabilities [8–10]. Catabolism of low molecular weight proteins, including peptide hormones, cytokines and growth factors is also accomplished by the kidney [11]. Finally, the tubule cells of the kidney have an important hormonal function with the production and regulation of erythropoietin, vitamin D, and multiple cytokines, including tumor necrosis factor-α, interleukin-6, and complement factors, critical to inflammation and immunologic regulation [12, 13]. Perhaps the loss of these important metabolic and endocrinologic functions residing in the tubulointerstitial component of the kidney is responsible, in part, for the high rates of hemorrhage, sepsis, and coincident damage to other solid organs frequently observed in patients with acute renal failure. As an example, critically ill patients with acute renal failure are highly catabolic with high rates of urea generation and a propensity to malnourishment. Nutritional support with solutions containing glucose and amino acids appears to enhance renal recovery and, perhaps, patient outcome [14]. The loss of renal tubular function in acute renal failure may aggravate uremic events and catabolic activity, since renal tubule cells excrete 10–15% of nitrogenous products of metabolism via the ammoniagenic pathway [15] and supply up to 50% of circulating blood glucose levels via the gluconeogenic pathway in a starved, malnourished state [9]. These metabolic functions lost in acute renal failure are but singular examples by which lost metabolic renal function may influence the morbidity and mortality rates of this disorder. Accordingly, the development of cell therapy modalities replacing these reabsorptive, metabolic, and endocrinologic functions of the kidney may add significant value to the current suboptimal supportive options available to treat established acute renal failure.

1.4 Endocrine function

The kidney's endocrinologic function of producing both erythropoietin, a hormone that stimulates red blood cell (RBC) formation, and vitamin D, a compound necessary for calcium metabolism, has been well recognized. Prostaglandins, kinins, and renin are manufactured by the kidney too; however, this chapter will focus only on the production and necessity of erythropoietin in contributing to a tissue engineering approach for replacing lost endocrine function.

Erythropoietin enhances erythropoiesis by stimulating formation of proerythroblasts and release of reticulocytes from bone marrow [16]. In adults, greater than 90% of the body's erythropoietin is produced by the interstitial cells of the kidney, while the remainder is produced in the liver. The liver's contribution is not adequate to maintain needed RBC levels, so that patients with chronic renal failure develop persistent anemia.

Due to the large amount of blood that flows through the kidney, erythropoietin-producing cells are ideally situated to sense RBC delivery of oxygen from the bloodstream. As shown in Fig. 2, erythropoietin production is regulated by a feedback loop system. As kidney cells sense hypoxemia, resulting from the concomitant decline in RBC number, the renal interstitial cells increase erythropoietin production. The erythropoietin is carried in the bloodstream to the bone marrow, where it binds to receptors on multipotent stem cells and cues these progenitor cells to differentiate into RBCs. Upon return to normoxia, manufacture of erythropoietin declines to once again establish equilibrium. Of note is that regulation of erythropoietin is controlled by DNA transcription. The mechanism is not yet clear, but appears to depend on a heme protein.

Anemia is an inevitable consequence of renal disease. As the kidney fails, response within the erythropoietin feedback loop greatly diminishes, and the production of erythropoietin declines. This decline is met with an equal decline in hematocrit levels. Thus, renal disease results in the loss of a critical hormone which exacerbates the existing clinical problem of renal failure.

2 Tissue engineering design considerations

The kidney has been the first solid organ whose function was approximated by a machine and a synthetic device. In fact, renal substitution therapy with hemodialysis or chronic ambulatory peritoneal dialysis (CAPD) has been the only successful long-term *ex vivo* organ substitution therapy to date [17]. The kidney was also the first organ to be successfully transplanted from a donor individual to an autologous recipient patient. How-

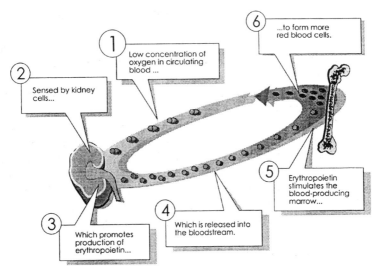

Fig. 2. Endocrinologic feedback loop which regulates erythropoietin production by the kidney (Figure adapted from [20]).

ever, the lack of widespread availability of suitable transplantable organs has kept kidney transplantation from becoming a practical solution for most cases of end-stage renal disease (ESRD).

Although long-term chronic renal replacement therapy with either hemodialysis, hemofiltration, or CAPD has dramatically changed the prognosis of renal failure, it is not a complete replacement therapy, since it only provides filtration function (usually on an intermittent basis) and does not replace the homeostatic, regulatory, metabolic, and endocrine functions of the kidney. Hemodialysis is merely a mass exchange procedure which utilizes a membrane separation device and hemofiltration is accomplished by just rapid ultrafiltration. Because of the non-physiologic manner in which dialysis performs or does not perform the most critical renal functions, patients with ESRD on dialysis continue to have major medical, social, and economic problems [18]. Accordingly, dialysis should be considered as renal substitution rather than renal replacement therapy.

Various forms of artificial kidneys have been utilized over the almost 50 years that artificial kidney systems have been used clinically [19]. Hemodialysis has evolved from the use of large, non-disposable flat-plate dialyzers to disposable parallel-plate dialyzers and coil dialyzers. The widely used dialyzer of today involves a hollow-fiber design. Artificial kidney design has progressed with the aid of versatile biomaterials and the principles of mass transfer and fluid dynamics. The tissue engineering approach to renal replacement can incorporate these engineering principles along with biological components and the latest techniques of cell culture and genetic engineering.

The approach of a tissue engineered construct for renal replacement is to mimic the natural physical forces to duplicate filtration and transport processes in order to attain adequate excretory function lost in renal disorders. In designing an implantable bioartificial kidney for renal replacement function, essential functions of kidney tissue must be retained and utilized to direct the design of the tissue-engineering project [20]. The critical elements of renal function must be replaced, including the excretory, regulatory (transport), and endocrinologic functions. The functioning excretory unit of the kidney, as detailed previously, is composed of the filtering unit, the glomerulus, and the regulatory or reabsorptive unit, the tubule. Therefore, a bioartificial kidney requires two main units, the glomerulus and the tubule, to replace renal excretory function.

2.1 The bioartificial hemofilter

The potential for a bioartificial glomerulus has been achieved with the use of synthetic fibers *ex vivo* with maintenance of ultrafiltration in humans for several weeks with a single device [21, 22]. The availability of hollow fibers with high hydraulic permeability has been an important advancement in biomaterials for replacement function of glomerular ultrafiltration. Conventional hemodialysis for ESRD have used membranes in which solute removal is driven by a concentration gradient of the solute across the membranes and is, therefore, predominantly a diffusive process. Another type of solute transfer also occurs across the dialysis membrane via a process of ultrafiltration of water and solutes across the membrane. This convective transport is independent of the concentration gradient and depends predominantly on the hydraulic pressure gradient across the membrane. Both diffusive and convective processes occur during traditional hemodialysis, but diffusion is the main route of solute movement.

Removal of uremic toxins, predominantly by the convective process, has several distinct advantages, because it imitates the glomerular process of toxin removal with increased clearance of higher-molecular-weight solutes and removal of all solutes (up to a molecular weight cutoff) at the same rate. Development of an implantable device which mimics glomerular filtration will thus depend upon convective transport. This physiologic function has been achieved clinically with the use of polymeric hollow fibers *ex vivo*. Major limitations to the currently available technology for long-term replacement of filtration function include bleeding associated with required anticoagulation, diminution of filtration rate due to protein deposition in the membrane over time or thrombotic occlusion, and large amounts of fluid replacement required to replace the ultrafiltrate formed from the filtering unit. The use of endothelial cell-seeded conduits along filtration surfaces may provide improved long-term hemocompatibility and hemofiltration *in vivo* [23–25].

A potential rate-limiting step in endothelial cell-lined hollow fibers of small caliber is thrombotic occlusion, which limits the functional patency of this filtration unit. In this regard, gene transfer into seeded endothelial cells for constitutive expression of anticoagulant factors can be envisioned to minimize clot formation in these small-caliber hollow fibers. Since gene transfer for *in vivo* protein production has been clearly achieved with endothelial cells [26, 27], gene transfer into endothelial cells for the production of an anticoagulant protein can be accomplished. Therefore, a replication defective, amphotrophic recombinant retrovirus with a selectable marker has been constructed containing the hirudin gene. Hirudin is a protein from the blood sucking leech and is a potent and specific inhibitor of thrombin. Endothelial cells transfected with this construct and grown in cell culture demonstrated an ability to secrete hirudin [28]. Thus, anticoagulant gene transfer into endothelial cells is achievable and can provide added value to maintain a non-thrombogenic surface in an implantable bioengineered hemofilter.

In addition to endothelial cell seeding, modifying the design of a bioartificial hemofilter can possibly reduce one of its potential major limitations, thrombus formation at the header spaces of the cartridge at its entrance and exit of the hollow fiber bundle. In this regard, the transition in the device from a header space of several centimeters to a potted bundle of several hundreds to a thousand hollow fibers with an inner diameter of 0.2–1.0 mm produces significant turbulence and thrombotic potential. When this occurs, swirls in the blood are formed which would activate a coagulation cascade. In other words, even though thrombus formation does not occur at that moment, the activation of the coagulation cascade would eventually end up occluding fibers along the hemofilter. Swirl formation can possibly be reduced by curving the shape of potting material, so that blood turbulence can be minimized as it enters into the hollow fibers.

2.2 The bioartificial renal tubule

The construction of tissue-engineered organs is highly dependent upon the expansion of parenchymal cells of the organ to be replaced, and an ability to grow these cells into a device with differentiated physiologic properties. This cell expansion and differentiation process requires the selection of organ specific stem cells from donor tissue.

Various types of stem, or progenitor, cells exist in different areas of the body as an undifferentiated cell source. Upon receipt of varying cueing signals (such as from growth factors and hormones) these progenitor cells can proliferate and differentiate into varying types of cells as required by the body. If stem cells of a given organ can be found and isolated, there is a chance that these cells could be used as a powerful tissue engineering tool for use in generating natural organ tissues and functions. In particular, the efficiency of kidney tubule transport is dependent on specialized epithelial cells to perform solute transport. Only recently have renal tubule stem cells been identified.

A series of *in vitro* studies was performed by Humes *et al.* [29, 30] which demonstrated a methodology for isolating and growing renal proximal tubule cells from adult mammalian kidneys. The incentive to undertake these studies derived from the fact that tubule epithelium has the ability to regenerate after severe nephrotoxic and ischemic injury [31, 32]. The data demonstrated that transforming growth factor-β (TGF-β), epidermal growth factor (EGF), and retinoic acid promoted tubulogenesis of renal proximal tubule cells in primary tissue culture [29]. Further, these experiments identified non-serum containing growth conditions which select for proximal tubule progenitor cells with a high capacity for replication and terminal differentiation in collagen gels. Clonal expansion of the progenitor cells was seen to emanate from a single genetically tagged stem cell. The results suggest that tubulogeneic cells exist sparsely within the adult kidney in a normally dormant state, yet manifest the ability to rapidly proliferate and differentiate under proper defined conditions. Whether these stem cells are pluripotent, possessing the ability to differentiate into cells of more than one segment of the tubule (the loop of Henle and the distal tubule for example), is still unknown.

With the identification of methods to expand renal tubule cells from kidney tissue, the bioartificial renal tubule is now clearly feasible when conceived as a combination of living cells supported on polymeric substrata. It has been successfully demonstrated that a number of renal tubule cells can be expanded in primary culture from various tissue sources, including human, rabbit, and pig. Active vectorial transport across a confluent renal epithelial monolayer has also been accomplished as well as the production of various metabolic parameters in a renal tubule device containing several hundred to thousands of fibers [33]. With appropriate membranes and biomatricies, immunoprotection of cultured progenitor cells can be achieved concurrent with long-term functional performance as long as conditions support tubule cell viability [33–35]. The technical feasibility of an implantable epithelial cell system derived from cells grown as confluent monolayers along the luminal surface of polymeric hollow fibers has been achieved [34]. These previously constructed devices, however, have used permanent renal cell lines which do not have differentiated transport function. The ability to purify and grow renal proximal tubule progenitor cells with the ability to differentiate morphogenically may provide a capability for replacing renal tubule function.

In designing a bioartificial tubule device, two different schemes can be

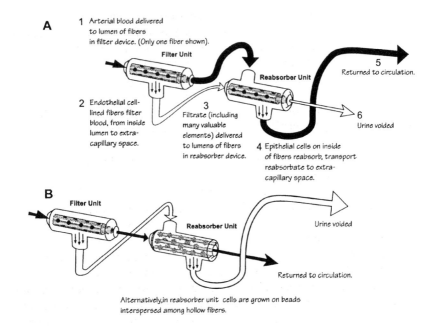

Fig. 3. Conceptual schematization of a tissue-engineered bioartificial kidney with an endothelial-cell-lined hemofilter in series with (A) a proximal tubule cell-lined reabsorber and (B) a reabsorber filled with cells grown within porous microcarrier beads.

envisioned. The first would consist of renal proximal tubule cells grown as a confluent monolayer in the lumen of hollow fibers (Fig. 3). Ultrafiltrate from the upstream hemofilter would be perfused through the hollow fibers, while blood would circulate in the extracapillary space (ECS), leaving the cells immunoprotected as the porous polymer serves as a barrier. Bulk up of such a device can be achieved by increasing the luminal surface area either by lengthening the device or by increasing the number of hollow fibers residing inside. Another design which will afford a higher cell density is one in which the cells are grown within porous microcarrier beads and packed into the ECS of a hollow fiber device. Ultrafiltrate will be circulated through the cells/microspheres and the blood component will be dispersed through the hollow fibers, supplying nutrients such as oxygen while keeping the immunoisolation of the cell compartment.

Construction of such devices poses many challenges in the way of biomaterials, cell sourcing and culture, and nutrient delivery. The design of a bioartificial renal tubule assist device requires modeling and evaluation systems based upon well established chemical and mechanical engineering principles. A fundamental problem which must be solved in designing such a device is to prototype a device which will maintain the viability and performance of the cells within a tissue engineered construct. To achieve this goal, adequate oxygenation to all cells must be maintained. The supply of oxygen will be obtained from the convective flow of blood through a hollow fiber device and the diffusive transport through the permeable fiber walls to the cells. Oxygen is known to be a critical and limiting factor for cell growth and viability and an adequate supply is essential to cell differentiation and function [36]. Knowing the oxygen uptake rate of cultured cells, a mathematical analysis of oxygen depletion and the development of a computational model will serve to assess the effectiveness of the hollow fiber construct, allow for the critical study of specific parameters, and aid in the scale up and overall design of bioartificial tubule reabsorbing unit. Flow visualization techniques can also be used to identify areas of stagnant blood flow in the extracapillary space of a scaled up bioartificial renal tubule cartridge containing bundles of hollow fibers.

A bioartificial proximal tubule satisfies a major requirement of reabsorbing a large volume of filtrate to maintain salt and water balance within the body. The need for additional tubule equivalents to replace other nephronal segment functions, such as the loop of Henle, in order to perform more refined homeostatic elements of the kidney, including urine concentration or dilution, may not be necessary. Patients with moderate renal insufficiency lose the ability to finely regulate salt and water homeostasis because they are unable to concentrate or dilute, yet are able to maintain reasonable fluid and electrolyte homeostasis due to redundant physiologic compensation via other mechanisms. Thus, a bioartificial proximal tubule, which reabsorbs iso-osmotically the majority of the filtrate, may be sufficient to replace required tubular function to sustain fluid electrolyte balance in a patient with end-stage renal disease.

2.3 Endocrine function

The biotechnology industry was founded on the idea of isolating a given protein for use in patient therapy, often to treat the symptoms of a genetic disorder. Genes which code for such proteins have been isolated and transfected into prokaryotes and eukaryotes. Human growth hormone, insulin, and erythropoietin have all been successfully delivered with this strategy. In general however, this approach is limiting due to several reasons. Often, multiple proteins are required to interact in a complex orchestration in order to achieve the desired therapeutic effect. Second, targeted delivery of an adequate dose, while not being too high to cause toxicity, is necessary. Ideally, cells should be involved in the drug delivery system to provide an inherent regulation of protein dosage in response to natural signals.

Cell therapy is an emerging branch of tissue engineering which addresses the need for *in vivo* cueing of protein administration. Cells are transfected with the gene necessary for protein production. Use of biocompatible (or biodegradable depending on the need) porous polymer delivery vehicles can be used to encapsulate the cells, while simultaneously providing protection of the implanted cells from the host's immune defenses (if the cells are not autologous). Conceding certain diffusion distance constraints, oxygen, nutrients, and intrinsic protein production cueing factors from the bloodstream can diffuse in through the pores of the polymer to maintain cell viability. At the same time, cellular metabolic wastes along with the desired protein product could diffuse out into the bloodstream. The blood levels of a circulating protein may be regulated by different mechanisms: at the gene level by transcriptional mechanisms, at the protein level by translational processes, or at the secretory level by cellular processes. Regulation complexity increases from transcriptional to translational to secretory processes. Erythropoietin is regulated by an oxygen sensitive transcriptional mechanism. Thus, the ability to isolate and grow cells with the capability of regulating and secreting erythropoietin may allow for an implantable drug therapy device, and thereby combat anemia associated with renal disease.

In fact, a recent design utilizing hepatocyte (Hep G2) cells which can secrete erythropoietin is currently under study. Porous polysulfone hollow fibers coated with covalently bound heparin will be used as the cell encapsulation device. The fibers will be attached to a Greenfield Filter (originally designed and used to catch pulmonary emboli), which itself is anchored in the bloodstream to the wall of the inferior vena cavae. The pores of the fibers are small enough to avoid an attack of the foreign cells by the host's immune system. Thus, the implanted cells may be able to remain viable and secrete erythropoietin into the bloodstream. This device thereby offers a potential cure to anemia associated with renal failure. It is also conceivable that such a drug delivery/cell therapy device could be utilized to cure the symptoms of other genetic diseases which manifest with the loss or poor function of certain proteins.

3 The bioartificial kidney

The development of a bioartificial filtration device and a bioartificial tubule processing unit would lead to the possibility of an implantable bioartificial kidney, consisting of the filtration device followed in series by the tubule unit (Fig. 3). The filtrate formed by this device will flow directly into the tubule unit. The tubule unit should maintain viability, because metabolic substrates and low-molecular weight growth factors are delivered to the tubule cells from the ultrafiltration unit and the blood in the

ECS. Furthermore, immunoprotection of the cells grown within the hollow fiber is achievable due to the impenetrance of immunologically competent cells through the hollow fiber. Rejection of transplanted cells will, therefore, not occur. This arrangement thereby allows the filtrate to enter the internal compartments of the hollow fiber network, which are lined with confluent monolayers of renal tubule cells for regulated transport function.

This device could be used either extracorporeally or implanted within a patient. In this regard, the specific implant site for a bioartificial kidney will depend upon the final configuration of both the bioartificial filtration and tubule device. As presently conceived, the endothelial-line bioartificial filtration hollow fibers can be placed into a arteriovenous circuit using the common iliac artery and vein, similar to the surgical connection for a renal transplant. The filtrate is connected in series to a bioartificial proximal tubule so that reabsorbate will be transported and reabsorbed into the systemic circulation. The processed filtrate exiting the tubule unit is then connected via tubing to the proximate ureter for drainage and urine excretion via recipient's own urinary collecting system.

4 Conclusion

Three technologies will most likely dominate medical therapeutics in the next century. One is 'cell therapy' — the implantation of living cells to produce a natural substance in short supply from the patient's own cells due to injury and destruction from various clinical disorders. Erythropoietin cell therapy is an example of this approach to replace a critical hormone deficiency in end-stage renal disease. A second therapy is tissue engineering, wherein cells are cultured to replace masses of cells that normally function in a coordinated manner. Growing a functional glomerular filter and tubule reabsorber from a combination of cells, biomaterials, and synthetic polymers to replace renal excretory and regulatory functions is an example of this formulation. Finally, a third technology that will dominate future therapeutics is gene therapy, in which genes are transferred into living cells either to deliver a gene product to a cell in which it is missing or to produce a foreign gene product by a cell to promote a new function. The use of genes which encode for anticoagulant proteins as a means to deliver in a targeted and local fashion an anticoagulant to maintain hemocompatibility of a tissue engineered hemofilter is an example of the application of this third technology.

The kidney was the first organ whose function was substituted by an artificial device. The kidney was also the first organ to be successfully transplanted. The ability to replace renal function with these revolutionary technologies in the past was due to the fact that renal excretory function is based upon natural physical forces which govern solute and fluid movement from the body compartment to the external environment. The need for coordinated mechanical or electrical activities for renal substitution was not required. Accordingly, the kidney may well be the first organ to be available as a tissue-engineered implantable device as a fully functional replacement part for the human body.

References

1. Brenner, B. M. and Humes, H. D., Mechanisms of glomerular ultrafiltration. *New England Journal of Medicine*, 1977, **297**(3), 148–154.

2. Landis, E. M. and Pappenheimer, J. R., Exchange of substances through the capillary walls. In Handbook of Physiology: Circulation, Sec. 2, Vol. 2, ed. W. F. Hamilton and P. Dow. American Physiological Society, Washington DC, 1959, p. 961.

3. Anderson, J. L. and Quinn, J. A., Restricted transport in small pores: A model for steric exclusion and hindered particle motion. *Biophysical Journal*, 1974, **14**(2), 130–150.

4. Chang, R. L. S., Robertson, C. R. and Deen, W. M. et al, Permselectivity of the glomerular capillary wall to macromolecules: 1. theoretical considerations. *Biophysical Journal*, 1975, **15**(9), 861–886.

5. Brenner, B. M., Hostetter, T. H. and Humes, H. D., Molecular basis of proteinuria of glomerular origin. *New England Journal of Medicine*, 1978, **298**(15), 826–833.

6. Andreoli, T. E. and Schafer J. A. Volume absorption in the pars recta: III. Luminal hypotonicity as a driving force for isotonic volume absorption. *American Journal of Physiology,* 1978, **234**(4), F349–F355.

7. Knox, F. G., Mertz J. I and Burnett J. C. et al., Role of hydrostatic and oncotic pressures in renal sodium reabsorption. *Circulation Research*, 1983, **52**(5), 491–500.

8. Deneke, S. M. and Fanburg, B. L., Regulation of cellular glutathione. *American Journal of Physiology*, 1989, **257**(4 Pt 1) (Lung Cell Mol Physiol 1), L163–L173.

9. Kida, K., Nakato, S., Kamiya, F., Tomaya, Y., Nishio, T. and Nakagawa, H., Renal net glucose release in vivo and its contribution to blood glucose in rats. *Journal of Clinical Investigation*, 1978, **62**(4), 721–726.

10. Tannen, R. L. and Sastrasinh, S., Response of ammonia metabolism to acute acidosis. *Kidney International*, 1984, **25**(1), 1–10.

11. Maack, T., Renal handling of proteins and polypeptides. In *Handbook of Physiology*, ed. E. E. Windhager. Oxford University Press, New York, 1992, pp. 2039–2118.

12. Frank, J., Engler-Blum, G., Rodemann, H. P. and Muller, G. A., Human renal tubular cells as a cytokine source: PDGF-b, GM-CSF and IL-6 mRNA expression in vitro. *Experimental Nephrology*, 1993, **1**(1), 26–35.

13. Stadnyk, A. W., Cytokine production by epithelial cells. *FASEB Journal*, 1994, **8**(13), 1041–1047.

14. Kopple, J. D., Dietary considerations in patients with advanced chronic renal failure, acute renal failure, and transplantation. In *Textbook of Nephrology*, ed. S. G. Massry and R. J. Glassock. Williams and Wilkins, Baltimore, 1995, pp. 3387–3436.

15. Pitts, R. F., The renal regulation of acid base balance with special reference to the mechanism for acidifying the urine. I and II. *Science*, 1945, **102**, 49(54), 81–85.

16. Jacobson, L. O., Goldwasser, E. and Fried, W. et al., Role of the kidney in erythropoiesis. *Nature*, 1957, **179**, 633–634.

17. Iglehart, J. K., The American health care system: the end stage renal disease program. *New England Journal of Medicine*, 1993, **328**(5), 366–371.

18. Excerpts from United States renal data system 1991 annual data report. Prevalence and cost of ESRD therapy. *American Journal of Kidney Diseases*, 1991, **18**(5)(suppl)2, 21.

19. Colton, C. K. and Lowrie, E. G., Hemodialysis: Physical principals and technical considerations. In *The Kidney*, Vol. 2, 2nd edn, ed. B. M. Brenner and F. C. Rector. WB Saunders Company, Philadelphia, 1981, pp. 2425–2489.

20. Humes, H. D. Tissue engineering of the kidney. In *The biomedical engineering handbook*, ed. J. D. Bronzino. CRC Press, Boca Raton, 1995, pp. 1807–1824.

21. Golper, T. A., Continuous arteriorvenous hemofiltration in acute renal failure. *American Journal of Kidney Diseases*, 1985, **6**(6), 373–386.

22. Kramer, P., Wigger, W. and Rieger, J. et al., Arteriorvenous haemofiltration: A new and simple method for treatment of over hydrated patients resistant to diuretics. *Klinische Wochenschrift*, 1977, **55**(22), 1121–1122.

23. Kadletz, M., Magometschnigg, H., Minar, E., Konig, G., Grabenwoger, M., Grimm, M. and Wolner, E., Implantation of in vitro endothelialized polytetrafluoroethylene grafts in human beings. *Journal of Thoracic and Cardiovascular Surgery*, 1992, **104**(3), 736–742.

24. Schneider, P. A., Hanson, S. R., Price, T. M. and Harker, L. A., Durability of confluent endothelial cell monolayers of small-caliber vascular prosthesis in vitro. *Surgery*, 1988, **103**(4), 456–462.

25. Shepard, A. D., Eldrup-Jorgensen, J., Keough, E. M., Foxall, T. F., Ramberg, K., Connolly, R. J., Mackey, M. C., Gavris, V., Auger K. R. and Libby, P., Endothelial cell seeding of small-caliber synthetic grafts in the baboon. *Surgery*, 1986, **99**(3), 318–326.

26. Zwiebel, J. A., Freeman, S. M., Kantoff, P. W., Cornetta, K., Ryan, U. S. and Anderson, W. F., High-level recombinant gene expression rabbit endothelial cells transduced by retroviral vectors. *Science*, 1989, **243**(4888), 220–222.

27. Wilson, J. M., Birinyl, L. K., Salomon, R. N., Libby, P., Callow, A. D., and Mulligan, R. C., Implantation of vascular grafts lined with genetically modified endothelial cells. *Science*, 1989, **244**(4910), 1344–1346.

28. Lee, W., Liu, S., Cieslinski, D. A. and Humes, H. D., Development of an implantable bioartificial hemofilter (BHF). In press.

29. Humes, H. D. and Cieslinski, D. A., Interaction between growth factors and retinoic acid in the induction of kidney tubulogenesis in tissue culture. *Experimental Cell Research*, 1992, **201**(1), 8–15.

30. Humes, H. D., Krauss, J. C., Cieslinski, D. A. and Funke, A. J., Tubulogenesis

from isolated single cells of adult mammalian kidney: Clonal analysis with a recombinant retrovirus. *American Journal of Physiology*, 1996, **271**(1 Pt 2), F42–F49.

31. Coimbra, T. M., Cieslinski, D. A. and Humes, H. D., Epidermal growth factor accelerates renal repair in mercuric chloride nephrotoxicity. *American Journal of Physiology*, 1990, **259**(3 Pt 2), F438–F43.

32. Humes, H. D., Cieslinski, D. A., Coimbra, T. M., Messana, J. M. and Galvao, C., Epidermal growth factor enhances renal tubule cell regeneration and repair and accelerates the recovery of renal function in postischemic acute renal failure. *Journal of Clinical Investigation*, 1989, **84**(6), 1757–1761.

33. Nikolovski, J., Poirier, S., Funke, A. J., Cieslinski, D. A. and Humes, H. D., Development of a bioartificial renal tubule for the treatment of acute renal failure (In press).

34. Ip, T. K. and Aebischer, P., Renal epithelial-cell-controlled solute transport across permeable membranes as the foundation for a bioartificial kidney. *Artificial Organs*, 1989, **13**(1), 58–65.

35. Aebischer, P., Whalberg, L., Tresco, P. A. and Winn S. R., Macroencapsulation of dopamine-secreting cells by coextrusion with an organic polymer solution. *Biomaterials*, 1991, **12**(1), 50–56.

36. Colton, C. K., Implantable biohybrid artificial organs. *Cell Transplantation*, 1995, **4**(4), 415–436.

CHAPTER IV.6

Tissue Engineering in Urology

ANTHONY ATALA

Department of Urology, Children's Hospital and
Harvard Medical School,
300 Longwood Ave.,
Boston, MA 02115, USA

1 Introduction

The genitourinary system, composed of the kidneys, ureters, bladder, urethra, and genital organs, is exposed to a variety of possible injury sites from the time the fetus develops. Aside from congenital abnormalities, individuals may also suffer from acquired disorders such as cancer, trauma, infection, iatrogenic injuries or other conditions which may lead to genitourinary organ damage or loss, requiring eventual reconstruction.

A large number of materials, including naturally-derived and synthetic polymers have been utilized to fabricate prostheses for the genitourinary system. Usually, whenever there is a lack of native urologic tissue, reconstruction is performed with native non-urologic tissues, such as gastrointestinal segments, skin or mucosa from multiple body sites [1, 2]. The use of native non-urologic tissues in the genitourinary tract is common, due to a lack of a better alternative, despite the known possible adverse effects. For example, the use of bowel in genitourinary reconstruction is associated with a variety of complications [3]. These include metabolic abnormalities, infection, perforation, urolithiasis, increased mucous production and malignancy. Alternative approaches are constantly being explored in order to overcome the problems associated with the incorporation of non-urologic segments into the urinary tract.

Due to the complications associated with the use of native non-urologic tissues, such as gastrointestinal segments, for genitourinary reconstruction, many investigators have sought to use synthetic materials as an alternative. The most common type of synthetic prostheses for urologic use are made of silicone. Silicone prostheses have been used for the treatment of urinary incontinence [4] (artificial urinary sphincter), vesicoureteral reflux [5, 6] (detachable balloon system, silicone microparticles), and impotence (penile prostheses) [7]. In some disease states, such as urinary incontinence or vesicoureteral reflux, artificial agents (Teflon paste, glass microparticles) have been used as injectable bulking substances, however, these substances are not entirely biocompatible [8, 9]. The literature is replete with a wide array of complications (i.e. device malfunction, infection, etc.) associated with these artificial devices.

Other natural tissues and synthetic materials which have been tried previously in experimental and clinical settings include omentum, peritoneum, seromuscular grafts, de-epithelialized segments of bowel and polyvinyl sponge [3]. These attempts have usually failed. It is evident that urothelial to urothelial anastomoses are preferable functionally. However, the limited amount of autologous urothelial tissue for reconstruction generally precludes this option. There is a critical need for tissues to replace lost and functionally-deficient genitourinary tissues [10]. Engineering tissues using selective cell transplantation [11] may provide a means to create functional new genitourinary tissues.

2 Engineering tissues in urology: Strategy

In tissue engineering, donor tissue is dissociated into individual cells or small tissue fragments and either implanted directly into the autologous host, or attached to a support matrix, expanded in culture, and re-implanted after expansion. Ideally, this approach might allow lost tissue function to be restored or replaced in toto and with limited complications.

The success of using cell transplantation strategies for genitourinary reconstruction depends on the ability to use donor tissue efficiently and to provide the right conditions for long term survival, differentiation and growth. We have achieved an approach to urologic tissue regeneration by patching isolated cells to a support structure which would have suitable surface chemistry for guiding the reorganization and growth of the cells [12, 13]. The supporting matrix is composed of crossing filaments which can allow cell survival by diffusion of nutrients across short distances once the cell-support matrix is implanted. Ideally, the cell-support matrix would become vascularized in concert with expansion of the cell mass following implantation.

3 Cell expansion

One of the initial limitations of applying tissue engineering techniques to the urinary tract, has been the previously encountered inherent difficulty of growing genitourinary associated cells in large quantities. Even as recently as 6 years ago, it was believed that urothelial cells, which line most of the urinary tract, had a natural senescence which was hard to overcome. Normal urothelial cells could be grown in the laboratory setting, but only for a few weeks, and without any demonstrable expansion.

We have recently shown that normal human bladder epithelial cells can be efficiently harvested from surgical material, extensively expanded in culture in serum-free conditions, and their differentiation characteristics, growth requirements and other biological properties studied [14–19]. Using our methods of cell culture we estimate that it would be theoretically possible to expand a urothelial strain from a single specimen which initially covers a surface area of 1 cm^2 to one covering a surface area of 4,202 m^2 (the equivalent area of one football field) within 8 weeks even if it assumed that 50% of the cells would be lost with each passage. This indicates that it should be possible to collect autologous urothelial cells from human patients, expand them in culture, and return them to the human donor in sufficient quantities for reconstructive purposes. Human bladder smooth muscle can be isolated in a similar fashion and studied transiently under serum-free conditions. The ability to greatly expand primary populations of human urothelial-associated cells in serum-free conditions is important because it indicates that frequent or continuous replenishment of *in vitro* stocks of normal cells from fresh surgical material, a requirement that would significantly inhibit the use of these cells for tissue engineering and research purposes, is not needed.

4 Cell delivery vehicles

The support structure chosen for cell delivery is of utmost importance. It is known from previous studies that artificial permanent support structures are lithogenic (Teflon, silicone) [3]. Other investigators have tried permanent homograft or heterograft support structures such as dura, however these contract with time and are problematic in a clinical setting. Natural permanent support structures, such as denuded bowel retain their inherent properties and mucosal regrowth invariably occurs with time. A variety of synthetic polymers, both degradable and non-degradable, have been utilized to fabricate tissue engineering matrices [11, 20–22]. One of the earliest reported studies in the entire tissue engineering field was an attempt to line biodegradable tubular structures with cultured smooth muscle cells obtained from bladder wall biopsies [20]. These constructs were proposed to have wide applicability in genitourinary operations. Another early attempt to engineer urologic structures involved implanting collagen sponge tubes as ureteral replacements [21]. It was hoped that the implanted matrix would induce migration of epithelial cells from the adjacent tissue and the formation of an epithelial cell-lined tubular tissue. However, salt deposits onto the collagen matrix were noted following exposure to urine.

Synthetic polymers can be manufactured reproducibly and can be designed to exhibit the necessary mechanical properties [23]. Among synthetic materials, resorbable polymers are preferable because permanent polymers carry the risk of infection, calcification, and unfavorable connective tissue response. Polymers of lactic and glycolic acid have been extensively utilized to fabricate tissue engineering matrices [11, 24–29]. These polymers have many desirable features; they are biocompatible, processable, and biodegradable. Degradation occurs by hydrolysis and the time sequence can be varied from weeks to over a year by manipulating the ratio of monomers and by varying the processing conditions. These polymers can be readily formed into a variety of structures, including small diameter fibers and porous films.

The porosity, pore size distribution and continuity dictate the interaction of the biomaterials and transplanted cells with the host tissue. Fibrovascular tissue will invade a device if the pores are larger than approximately 10 μm, and the rate of invasion will increase with the pore size and total porosity of a device [30, 31]. This process results in the formation of a capillary network in the developing tissue [31]. Vascularization of the engineered tissue may be required to meet the metabolic requirements of the tissue and to integrate it with the surrounding host. In urologic applications it may also be desirable to have a non-porous luminal surface (e.g. to prevent leakage of urine from the tissue) [32].

5 Formation of urologic tissues

The direction which we have followed to engineer urologic tissue involves the use of biodegradable materials which act as cell delivery vehicles [23]. We have performed a series of *in vivo* urologic associated cell–polymer experiments. Histologic analysis of human urothelial, bladder muscle, and composite urothelial and bladder muscle–polymer scaffolds, implanted in athymic mice and retrieved at different time points, indicated that viable cells were evident in all three experimental groups [25]. Implanted cells oriented themselves spatially along the polymer surfaces. The cell populations appeared to expand from one layer to several layers of thickness with progressive cell organization with extended implantation times (Fig. 1). Polymers alone evoked an angiogenic response by 5 days, which increased with time. Polymer fiber degradation was evident after 20 days. An inflammatory response was also evident at 5 days, and its resolution correlated with the biodegradation sequence. Cell–polymer composite implants of urothelial and muscle cells, retrieved at extended times (50 days), showed extensive formation of multi-layered sheet-like structures and well-defined muscle layers. Polymers seeded with cells and manipulated into a tubular configuration showed layers of muscle cells lining the

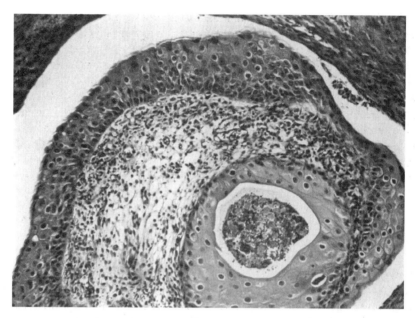

Fig. 1. The polymer scaffold can be manipulated into various shapes. A rolled implant produces tubularized uro-epithelium with cell layering and spatial orientation similar to that of normal urothelium (orig. mag. 100 ×).

multilayered epithelial sheets. Cellular debris appeared reproducibly in the luminal spaces, suggesting that epithelial cells lining the lumina are sloughed into the luminal space (Fig. 2). Cell polymers implanted with human bladder muscle cells alone showed almost complete replacement of the polymer with sheets of smooth muscle at 50 days (Fig. 3). This experiment demonstrated, for the first time, that composite tissue engineered structures could be created *de novo*. Prior to this study, only single cell type tissue engineered structures had been created.

This approach has recently been expanded to engineer new functional urologic structures [33–35]. In one study conducted in dogs, urothelial and smooth muscle cells were harvested, expanded *in vitro* and seeded onto biodegradable polymer scaffolds. These structures where tubularized and used to replace ureteral segments in each animal [33]. The results suggested that the creation of artificial ureters may be achieved *in vivo* using biodegradable polymers as transplanted cell delivery and native cell expansion vehicles. The malleability of the synthetic polymer allowed for the creation of cell–polymer implants manipulated into pre-formed tubular configurations. The combination of both smooth muscle and urothelial cell–polymer scaffolds is able to provide a template wherein a functional ureter may be created *de novo*.

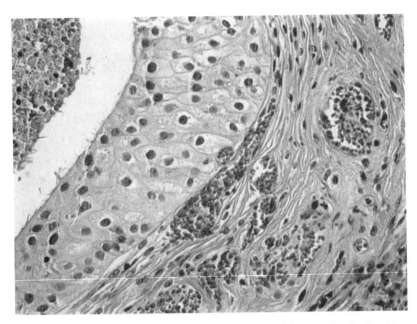

Fig. 2. Formation of multilayered sheet-like structures lining a tube obtained from a cell-polymer implant consisting of human smooth muscle (right) and urothelial cells (center). Sloughed cells (left) appear in the lumen, suggesting that cells may be sloughed from the Luminal surface and collect within it. Abundant vascularity (angiogenesis) is evident (orig. mag. 400 ×).

In other sets of experiments, the same approach was used to augment bladders [34]. Beagle bladder tissue specimens were micro dissected and the mucosal and muscular layers separated. Both urothelial and smooth muscle cells were harvested and expanded separately. The cells were seeded onto 5 × 5 and 10 × 10 cm sized sheets of polyglycolic acid polymers *in vitro*. Cell–polymer scaffolds were created consisting of urothelial and smooth muscle, urothelial, and smooth muscle cell populations. Polymers without cells served as controls. Beagles underwent cruciate cystotomies on the bladder dome and the polymer scaffolds were used to augment the bladder in each animal. Omentum was wrapped over the augmented bladder in order to enhance angiogenesis to the polymer–cell complex. The average increase in bladder capacity was approximately 40%. Furthermore, the average compliance of the augmented bladder increased an average of 30%. Polymer, polymer–urothelial cell, bladder muscle and urothelial/bladder muscle implants were found to contain a normal cellular organization consisting of a urothelial lined lumen surrounded by overlying submucosal tissue and ingrowth of smooth muscle. An angiogenic response and polymer fiber degradation were evident in all animals.

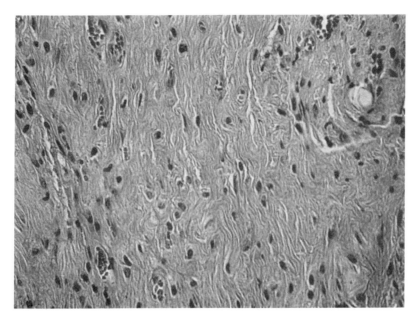

Fig. 3. Cell–polymer implant containing human muscle cells only and retrieved at 50 days. Muscle cells surround the polymer fibers (orig. mag. 40 ×).

These results show that the creation of artificial functional and anatomical bladders may be achieved *in vivo*, however, much work remains to be done in terms of the functional parameters of these implants. The same strategy has been used in trying to achieve urethral reconstruction [35].

Based on the feasibility of tissue engineering techniques in which cells seeded on biodegradable polymer scaffolds form tissue when implanted *in vivo*, the possibility was explored of developing a neo-organ system for *in vivo* gene therapy. The method of gene delivery at present is most effectively achieved by *ex vivo* gene transfer, which includes removal of the target tissue, *in vitro* gene delivery to the target cells, possible selection to enhance the proportion of transfected cells, and reintroduction of the gene-modified cells. Reintroduction of the transformed cells *in vivo* has been a challenging task.

Results indicated that successful gene transfer could be achieved using biodegradable polymer scaffolds as a urothelial cell delivery vehicle. The transfected cell/polymer scaffold formed organ-like structures with functional expression of the transfected genes [36].

End stage renal failure is a devastating disease which involves multiple organs in affected individuals. Although dialysis can prolong survival for many patients with end stage renal disease, only renal transplantation can currently restore normal function. Renal transplantation is severely limited by a critical donor shortage. Augmentation of either isolated or total renal function with kidney cell expansion *in vitro* and subsequent autologous transplantation may be a feasible solution. However, kidney reconstitution using tissue engineering techniques is a challenging task. The kidney is responsible not only for urine excretion but for several other important metabolic functions in which critical kidney by-products, such as renin, erythropoietin, and Vitamin D, play a large role. We explored the possibility of harvesting and expanding renal cells *in vitro* and implanting them *in vivo* in a three dimensional organization in order to achieve a functional artificial renal unit wherein urine production could be achieved [37, 38]. Studies demonstrated that renal cells can be successfully harvested, expanded in culture and transplanted *in vivo* where the single suspended cells form and organize into functional renal structures which are able to excrete high levels of uric acid through a yellow urine-like fluid. These findings suggest that this system may be able to replace transplantation in patients with end stage failure.

Other approaches have also been pursued for renal functional replacement. Polysulphone hollow fibers have been prelined with various extracellular matrix components and seeded with mammalian renal tubular and endothelial cells [39]. Permselective convective fluid transfer and active transport of salt and water were demonstrated. Using this approach, prototypic biohybrid constructs have been developed which are able to replicate the renal excretory functions. In addition, this system is able to facilitate gene and cell therapies by modifying the cells prior to seeding [40].

6 Injectable therapies

Both urinary incontinence and vesicoureteral reflux are common conditions affecting the genitourinary system, wherein injectable bulking agents can be used for treatment. Urinary incontinence may result from a deficient or weak musculature in the bladder neck and urethral area. Primary vesicoureteral reflux results from a congenitally deficient longitudinal submucosal muscle of the distal ureter which leads to an abnormal flow of urine from the bladder to the upper tract [41]. The diagnosis of reflux requires invasive radiographic studies, although less invasive diagnostic modalities are being developed [42, 43]. All of the surgical techniques for the correction of incontinence and vesicoureteral reflux attempt to achieve a normal anatomy.

Although open surgical procedures for the correction of urinary incontinence and reflux have excellent results in the hands of experienced surgeons, it is associated with a well recognized morbidity, including pain and immobilization of a lower abdominal incision, bladder spasms, hematuria, and post-operative voiding frequency. In an effort to avoid open surgical intervention, widespread interest was initiated by Berg's clinical experience with the endoscopic injection of Teflon paste in 1973 [44]. With this technique, a cystoscope is inserted into the bladder, a needle is inserted through the cystoscope and placed under direct vision in the submucosal space, and Teflon paste is injected [45]. The teflon paste, injected endoscopically, treats the incontinence by acting as a bulking material which increases urethral resistance. Vesicoureteral reflux was treated in a similar manner, by injecting Teflon in the subureteral space [46]. However, soon after the introduction of this treatment, a controversy regarding the use of Teflon paste ensued. Teflon particles injected in animals were noted to migrate to distant organs, such as the lungs, liver, spleen and brain, inducing granuloma formation [47]. Polytetrafluoroethylene migration and granuloma formation have also been reported in humans [48]. Teflon's safety for human use was questioned, and the paste was thereafter not approved by the FDA. Due to the problems associated with Teflon, alternate substances have been proposed as bulking agents. Silicone microparticles have been used for the treatment of urinary incontinence and vesicoureteral reflux [6]. However, silicone particles have also been shown to migrate to distant organs [49]. Collagen injections have been used in a similar manner [50]. However, collagen loses its volume over time, leading to treatment failure in the majority of patients [8, 9].

There are definite advantages in treating urinary incontinence and vesicoureteral reflux endoscopically. The method is simple and can be completed in less than 15 minutes, it has a low morbidity and it can be performed in an outpatient basis. The goal of several investigators has been to find alternate implant materials which would be safe for human use.

Laparoscopic approaches for incontinence and reflux have been attempted and are technically feasible [51, 52]. However, at least two surgeons with laparoscopic expertise are needed, the length of the procedure is much longer than with open surgery, the surgery is converted from an extraperitoneal to an intraperitoneal approach, and the cost is higher due to both increased operative time and the expense of the disposable laparoscopic equipment.

Despite the fact that over a decade has transpired since the Teflon controversy, little progress has been made in this area of research. The ideal substance for the endoscopic treatment of reflux should be injectable, non-antigenic, non-migratory, volume stable, and safe for human use. Toward this goal we had previously conducted long-term studies to determine the effect of injectable chondrocytes *in vivo* [53]. We initially determined that alginate, a liquid solution of gluronic and mannuronic acid, embedded with chondrocytes, could serve as a synthetic substrate for the injectable delivery and maintenance of cartilage architecture *in vivo*. Alginate undergoes hydrolytic biodegradation and its degradation time can be varied depending on the concentration of each of the polysaccharides. The use of autologous cartilage for the treatment of vesicoureteral reflux in humans would satisfy all the requirements for an ideal injectable substance. A biopsy of the ear could be easily and quickly performed (analogous to ear piercing) followed by chondrocyte processing and endoscopic injection of the autologous chondrocyte suspension for the treatment reflux.

Chondrocytes can be readily grown and expanded in culture. Neocartilage formation can be achieved *in vitro* and *in vivo* using chondrocytes cultured on synthetic biodegradable polymers [53]. In our experiments, the cartilage matrix replaced the alginate as the polysaccharide polymer underwent biodegradation (Fig. 4). We then adapted the system for the treatment of vesicoureteral reflux in a porcine model (Fig. 5) [54].

Six mini-swine underwent bilateral creation of reflux. All six were found to have bilateral reflux without evidence of obstruction at 3 months following the procedure. Chondrocytes were harvested from the left auricular surface of each mini-swine and expanded with a final concentration of

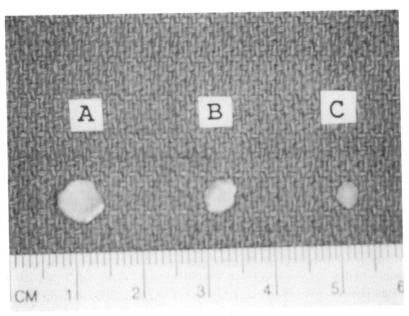

Fig. 4. Gross cartilage structures retrieved 12 weeks after chondrocyte/alginate injection with an initial concentration of 20 (a), 15 (b), and 10 (c) million chondrocytes.

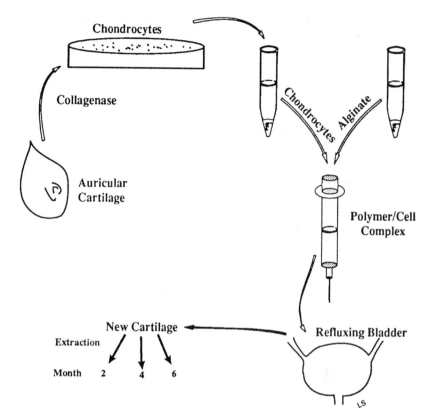

Fig. 5. Outline of overall strategy. Chondrocytes were harvested from the auricular surfaces of mini-pigs. Chondrocytes were mixed with alginate to form a suspension of approximately 20 million cells per cc. Chondrocyte-alginate suspensions were injected endoscopically beneath the refluxing ureter in each animal. Cystoscopic, radiographic and histologic examinations were performed at 2, 4 and 6 months after treatment.

$50-150 \times 10^6$ viable cells per animal. The animals then underwent endoscopic repair of reflux with the injectable autologous chondrocyte solution on the right side only.

Cystoscopic and radiographic examinations were performed at 2, 4, and 6 months after treatment. Cystoscopic examinations showed a smooth bladder wall. Cystograms showed no evidence of reflux on the treated side and persistent reflux in the uncorrected control ureter in all animals. All animals had a successful cure of reflux in the repaired ureter without evidence of hydronephrosis on excretory urography. The harvested ears had evidence of cartilage regrowth within 1 month of chondrocyte retrieval.

At the time of sacrifice, gross examination of the bladder injection site showed a well defined rubbery to hard cartilage structure in the sub-ureteral region. Histologic examination of these specimens using hematoxylin and eosin stains showed evidence of normal cartilage formation The polymer gels were progressively replaced by cartilage with increasing time. Aldehyde fuschin-alcian blue staining suggested the presence of chondroitin sulfate. Microscopic analyses of the tissues surrounding the injection site showed no inflammation. Tissue sections from the bladder, ureters, lymph nodes, kidneys, lungs, liver and spleen showed no evidence of chondrocyte or alginate migration, or granuloma formation.

Our studies showed that chondrocytes can be easily harvested and combined with alginate *in vitro*, the suspension can be easily injected cystoscopically and the elastic cartilage tissue formed is able to correct vesicoureteral reflux without any evidence of obstruction [54]. Using the same line of reasoning as with the chondrocyte technology, our group investigated the possibility of using autologous muscle cells [55].

In vivo experiments were conducted in mini-pigs and reflux was successfully corrected [56]. In addition to its use for the endoscopic treatment of reflux and urinary incontinence, the system of injectable autologous cells may also be applicable for the treatment of other medical conditions, such as rectal incontinence, dysphonia, plastic reconstruction, and wherever an injectable permanent biocompatible material is needed.

Most of the effort expended to engineer genitourinary tissues has occurred within the last 6 years. Recently, the first human application of cell based tissue engineering technology for urologic applications has occurred at our institution with the injection of chondrocytes for the correction of vesicoureteral reflux in children. The clinical trials are currently ongoing. Recent successes suggest that engineered urologic tissues may have continued clinical utility in the near future.

References

1. Atala, A. and Retik, A. B., Hypospadias, In *Reconstructive Urologic Surgery*, eds. J. A. Libertino and L. Zinman. Baltimore: The Williams and Wilkins Co., in press, 1997.

2. Retik, A. B., Bauer, S. B., Mandell, J., Peters, C. A., Colodny, A. and Atala, A., Management of severe hypospadias with a 2-stage repair. *Journal of Urology*, 1994, **152**, 749–751.

3. Atala, A., Retik, A., Pediatric urology — future perspectives. In *Clinical Urology*. eds. R. J. Krane, M. B. Siroky and J. M. Fitzpatrick. J. B. Lippincott, Philadelphia, 1994, pp. 507–524.

4. Levesque, P. E., Bauer, S. B., Atala, A., Zurakowski, D., Colodny, A., Peters, C., Retik, A. B. Ten year experience with the artificial urinary sphincter in children. *Journal of Urology* 1996, **156**, 625.

5. Atala, A., Peters, C. A., Retik, A. B., Mandell, J., Endoscopic treatment of vesicoureteral reflux with a self-detachable balloon system. *Journal of Urology*, 1992, **148**, 724–728.

6. Buckley, J. F., Scott, R., Aitchison, M. et al., Periurethral microparticulate silicone injection for stress incontinence and vesicoureteric reflux. Minimally Invasive Therapy, 1991, 1(suppl 1), 72.

7. Riehmann, M., Gasser, T. C. and Bruskewitz, R. C., The hydroflex penile prosthesis: a test case for the introduction of new urological technology. *Journal of Urology*, 1993, **149**, 1304–1307.

8. Atala, A., ed., Use of non-autologous substances in VUR and Incontinence treatment. *Dialogues in Pediatric Urology*, 1994, **17**, 11.

9. Atala, A., ed., Non-autologous substance in VUR and incontinence therapy. *Dialogues in Pediatric Urology*, 1994, **17**, 12.

10. Atala, A., Tissue engineering in the urinary tract. *Dialogues in Pediatric Urology*, 1995, **18**, 1.

11. Langer, R. and Vacanti, J. P., Tissue engineering. *Science*, 1993, **260**, 920–926.

12. Atala, A., Bauer, S. B., Dyro, F. M., Shefner, J., Shillito, J., Sumeer, S. and Scott, M. R., Bladder functional changes resulting from lipomyelomeningocele repair. *Journal of Urology*, 1992, **148**, 592.

13. Atala, A., Bauer, S. B., Hendren, W. H. and Retik, A. B., The effect of gastric augmentation on bladder function. *Journal of Urology*, 1993, **149**, 1099.

14. Cilento, B. J., Freeman, M. R., Schneck, F. X., Retik, A. B. and Atala, A., Phenotypic and cytogenetic characterization of human bladder urothelia expanded in vitro. *Journal of Urology*, 1994, **152**, 665–670.

15. Tobin, M. S., Freeman, M. R. and Atala, A., Maturational response of normal human urothelial cells in culture is dependent on extracellular matrix and serum additives. *Surgical Forum*, 1994, **45**, 786.

16. Prigent, S. A., The type 1 (EGFR-related) family of growth factor receptors and their ligands. *Progress in Growth Factor Research*, 1992, **4**, 1–24.

17. Freeman, M. R., Schneck, F. X., Soker, S., Raab, G., Tobin, M., Yoo J., Klagsbrun, M. and Atala, A., Human urothelial cells secrete and are regulated by heparin-binding epidermal growth factor-like growth factor (HB-EGF). *Journal of Urology*, 1995, **153**(suppl), 4.

18. Freeman, M. R., Schneck, F. X., Klagsbrun, M. and Atala, A., Growth factor biology of human urothelial cells grown under serum-free conditions. *Journal of Urology*, 1995, **153**(suppl), 4.

19. Atala, A., Yoo, J., Raab, G., Klagsburn, M. and Freeman, M. R., Regulated secretion of an engineered growth factor by human urothelial cells in primary culture. *Journal of Urology*, 1996, **155**(suppl), 5.

20. Thüroff, J. W., Bazeed, M. A., Schmidt, R. A., Luu, D. J. and Tanagho, E. A., Cultured rabbit vesical smooth muscle cells for lining of dissolvable synthetic prosthesis. *Journal of Urology*, 1983, **21**, 155–158.

21. Tachibana, M., Nagamatsu, G. R. and Addonizio, J. C., Ureteral replacement using collagen sponge tube grafts. *Journal of Urology*, 1985, **133**, 866–886.

22. Peppas, N. A. and Langer, R., New challenges in biomaterials. *Science*, 1994, **263**, 1715–1720.

23. Langer, R., New methods of drug delivery. *Science*, 1990, **249**, 1527.

24. Atala, A., Vacanti, J. P., Peters, C. A., Mandell, J., Retik, A. B. and Freeman, M. R., Formation of urothelial structures in vivo from dissociated cells

attached to biodegradable polymer scaffolds in vitro. *Journal of Urology*, 1992, **148**, 658–662.

25. Atala, A., Freeman, M. R., Vacanti, J. P., Shepard, J. and Retik, A. B., Implantation in vivo and retrieval of artificial structures consisting of rabbit and human urothelium and human bladder muscle. *Journal of Urology*, 1993, **150**, 608–612.

26. Mooney, D. J., Organ, G., Vacanti, J. P. and Langer, R., Design and fabrication of biodegradable polymer devices to engineer tubular tissues. *Cell Transplantation*, 1994, **3**, 438–446.

27. Mooney, D. J., Mazzoni, C. L., Breuer, C., McNamara, K., Hern, D., Vacanti, J. P. and Langer, R., Stabilized polyglycolic acid fiber-based devices for tissue engineering. *Biomaterials*, 1996, **17**, 115–124.

28. Mooney, D. J., Breuer, C. and McNamara, K. et al. Fabricating tubular tissues with devices of poly(D,L-lactic-co-glycolic acid). *Tissue Engineering* (in press).

29. Mooney, D., Park, S., Kaufmann, P. M., Sano, S., McNamara, K., Vacanti, J. P. and Langer, R., Biodegradable sponges for hepatocyte transplantation. *Journal of Biomedical Materials Research*, 1995, **29**, 959–965.

30. Wesolowski, S. A., Fries, C. C., Karlson, K. E. et al. Porosity: primary determinant of ultimate fate of synthetic vascular grafts. *Surgury*, 1961, **50**, 91–96.

31. Mikos, A. G., Sarakinos, G., Lyman, M. D. et al. Prevascularization of porous biodegradable polymers. *Biotechnology and Bioengineering*, 1993, **42**, 716–723.

32. Olsen, L., Bowald, S., Busch, C., Carlsten, J. and Eriksson, I., Urethral reconstruction with a new synthetic absorbable device. *Scandanavian Journal of Urology and Nephrology*, 1992, **26**, 323–326.

33. Yoo, J., Satar, N., Retik, A. B. and Atala, A., Ureteral replacement using biodegradable polymer scaffolds seeded with urothelial and smooth muscle cells. *Journal of Urology*, 1995, **153**(suppl), 4.

34. Satar, N., Yoo, J. and Atala, A.: Bladder augmentation using biodegradable polymer scaffolds seeded with urothelial and smooth muscle cells. *Journal of Urology*, 1996, **155**(suppl), 5.

35. Cilento, B. G., Retik, A. B. and Atala, A., Urethral reconstruction using a polymer scaffolds seeded with urothelial and smooth muscle cells. *Journal of Urology*, 1996, **155**(suppl), 5.

36. Yoo, J. and Atala, A., Gene therapy using urothelial tissue engineered neo-organs. *Pediatrics*, 1996, **98S**, 603.

37. Atala, A., Schlussel, R. N. and Retik, A. B., Renal cell growth in vivo after attachment to biodegradable polymer scaffolds. *Journal of Urology*, 1995, **153**(suppl)., 4.

38. Yoo, J., Ashkar, S. and Atala, A., Creation of fuctional kidney structures with excretion of urine-like fluid in vivo. *Pediatrics*, 1996, **98S**, 605.

39. Cieslinski, D. A., Humes, H. D., Tissue engineering of a bioartificial kidney. *Biotechnology and Bioengineering*, 1994, **43**, 678.

40. Humes, H. D., Application of gene and cell therapies in the tissue engineering of a bioartificial kidney. *International Journal of Artificial Organs*, 1996, **19**, 215.

41. Atala, A. and Casale, A. J., Management of primary vesicoureteral reflux. *Information in Urology*, 1990, **2**, 39–42.

42. Atala, A., Wible, J. H., Share, J. C., Carr, M. C., Retik, A. B. and Mandell, J.,

Sonography with sonicated albumin in the detection of vesicoureteral reflux. *Journal of Urology*, 1993, **150**, 756–758.

43. Atala, A., Share, J. C., Paltiel, H. J., Grant, R. and Retik, A., Sonicated albumin in the detection of vesicoureteral reflux in humans. *Society for Pediatric Urology Newsletter*, 1994, **8**, 6–7.

44. Berg, S., Urethroplastie par injection de polytef. *Archives of Surgury*, 1973, **107**, 379.21.

45. Politano, V. A., Periurethral polytetrafluoroethylene injection for urinary incontinence. *Journal of Urology*, 1982, **127**, 439–442.

46. O'Donnell, B., Puri, P., Treatment of vesicoureteric reflux by endoscopic injection of Teflon. *British Medical Journal*, 1984, **289**, 7–9.

47. Malizia, A. A., Reiman, H. M., Myers, R. P., Sande, J. R., Barham, S. S., Benson, R. C., Dewanjee, M. K. and Utz, W. J., Migration and granulomatous reaction after periurethral injection of polytef (Teflon). *Journal of the American Medical Association*, 1984, **251**, 3277–3281.

48. Claes, H., Stroobants, D., Van Meerbeek, J., Verbeken, E., Knockaert, D. and Beart, L., Pulmonary migration following periurethral polytetrafluoroethylene injection for urinary incontinence. *Journal of Urology*, 1989, **142**, 821–822.

49. Henly, D. R., Barrett, D. M., Welland, T. L., O'Connor, M. K., Malizia, A. A. and Wein, A. J., Particulate silicone for use in periurethral injections: local tissue effects and search for migration. *Journal of Urology*, 1995, **153**, 2039–2043.

50. Leonard, M. P., Canning, D. A., Peters, C. A., Gearhart, J. P. and Jeffs, R. D., Endoscopic injection of glutaraldehyde cross-linked bovine dermal collagen for correction of vesicoureteral reflux. *Journal of Urology*, 1991, **145**, 115–119.

51. Atala, A., Kavoussi, L. R., Goldstein, D. S., Retik, A. R. and Peters, C. A., Laparoscopic correction of vesicouretral reflux. *Journal of Urology*, 1993, **150**, 748.

52. Atala, A., Laparoscopic technique for the extravesical correction of vesicoureteral reflux. *Dialogues in Pediatric Urology*, 1993, **16**, 12.

53. Atala, A., Cima, L. G., Kim, W., Paige, K. T., Vacanti, J. P., Retik, A. B. and Vacanti, C. A., Injectable alginate seeded with chondrocytes as a potential treatment for vesicoureteral reflux. *Journal of Urology*, 1993, **150**, 745–747.

54. Atala, A., Kim, W., Paige, K. T., Vacanti, C. A. and Retik, A. B., Endoscopic treatment of vesicoureteral reflux with chondrocyte-alginate suspension. *Journal of Urology*, 1994, **152**, 641–643.

55. Atala, A., Cilento, B. G., Paige, K. T. and Retik, A. B., Injectable alginate seeded with human bladder muscle cells as a potential treatment for vesicoureteral reflux. *Journal of Urology*, 1994, **151**(suppl), 5.

56. Atala, A., Shepard, J. A. and Retik, A. B., Endoscopic treatment of reflux with autologous bladder muscle cells. *Proceedings, Pediatric Urology Section*, American Academy of Pediatrics, Dallas, October 22–34, 1994.

CHAPTER IV.7

Tissue Engineered Skin

J. TEUMER,
J. HARDIN-YOUNG,
N.L. PARENTEAU
Organogenesis Inc.,
150 Dan Road,
Canton, MA 02021, USA

1 Skin structure and function

Tissue engineering applications in skin begin with an understanding of the cells and other components responsible for skin structure and function. The primary function of skin is to serve as a barrier to the outside environment. As a barrier, the skin must be physically tough, yet must also be flexible and elastic to permit free movement. The barrier must be impermeable to protect against entry of toxic substances and to prevent the loss of water out to the dry environment. The skin is more than a passive barrier between the internal and external environment — it performs a crucial immune function by engaging in constant surveillance against the intrusion of foreign substances and microorganisms. In addition, the skin functions to help regulate body temperature by secretion from sweat glands and by regulation of blood flow through superficial capillaries. All these functions are met through specialized cells and structures which are divided between the two main layers of skin: the epidermis and the dermis (see Fig. 1).

1.1 Epidermis

The epidermis is comprised primarily of keratinocytes which form a stratified squamous epithelium. The basal cells are the proliferating cells in the epidermis (see Fig. 1). They anchor the epidermis to the dermis and replenish the terminally differentiated cells lost through normal sloughing. When cells leave the basal layer, they stop proliferating and terminally differentiate as they move up through the suprabasal layers. The most superficial keratinocytes form the stratum corneum, the structure that

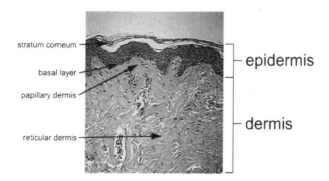

Fig. 1. This figure is a cross section of human skin showing the major structures of the epidermis and dermis.

provides barrier function. In the last stages of differentiation, cells extrude lipids into the intercellular space to form the permeability barrier. The cells also break down their nuclei and other organelles while they form a highly cross-linked protein envelope immediately beneath their cell membranes. These physically and chemically resilient protein envelopes connect to a dense network of intracellular keratin filaments to provide much of the physical strength of the epidermis. In cells of the lower layers, keratin filaments and desmosomes contribute physical strength and help keep the epidermis intact.

Other functions are performed by additional cells and structures in the epidermis. Melanocytes distribute melanin which protects the epidermis and underlying dermis from ultraviolet radiation. Langerhans cells are dendritic cells of the immune system which reside in the epidermis and form a network of dendrites through which they interact with adjacent keratinocytes and nerves [1]. Sweat glands help to regulate body temperature through evaporation of sweat released onto the skin surface. Hair functions to keep the body warm in many mammals, although maintaining body temperature is not an important role for hair in humans. Hair follicles are, however, an important source of growing keratinocytes during re-epithelialization after severe wounds. Sebaceous glands are associated with hair follicles. They secrete sebum, an oily substance which lubricates and moisturizes hair and epidermis.

Keratinocytes produce a large variety of polypeptide growth factors and cytokines which act as signals between cells and help to regulate skin function. For example, cytokine signals regulate keratinocyte migration and proliferation, and they stimulate dermal cells in various ways, such as to promote matrix deposition and neovascularization [2]. Keratinocytes provide the necessary microenvironment in the epidermis by producing cytokines which are thought to regulate Langerhans cell migration and differentiation [3]. Keratinocytes also translate a variety of stimuli into cytokine signals which are transmitted to the other cells of the skin immune system.

1.2 Dermis

The dermis provides physical strength and flexibility to skin as well as scaffolding to support the extensive vasculature, lymphatic system, nerve bundles, and other structures in skin. It is relatively acellular, being comprised predominantly of an extracellular matrix of interwoven collagen fibrils (see Fig. 1). Interspersed among the collagen fibrils are elastic fibers, proteoglycans, and glycoproteins. The dermis is divided into two regions; the papillary dermis, which lies immediately beneath the epidermis, and the deeper reticular dermis. The reticular dermis is more acellular and has a denser meshwork of thicker collagen and elastic fibers than the papillary dermis. It is the reticular dermis which provides skin with most of its strength, flexibility, and elasticity.

Fibroblasts, the major cell type of the dermis, are responsible for producing and maintaining most of the extracellular matrix. Endothelial cells line the blood vessels and play a critical role in the skin immune system by controlling the extravasation of leukocytes. There are cells of hematopoietic origin in the dermis such as macrophages, B cells, and T cells which contribute to the surveillance function. A network of nerve fibers extends throughout the dermis which serves the sensory role in the skin. These nerve fibers also secrete neuropeptides which can influence immune and inflammatory responses in skin through their effects on endothelial cells, leukocytes, and keratinocytes [4]. Like epidermal cells, the dermal cells use cytokines and growth factors as signals to regulate numerous processes critical to skin function.

1.3 Skin immunology

Virtually every cell in the skin plays a role in the surveillance function. The interacting cells in skin comprise a dynamic network capable of sensing a variety of perturbations (including trauma, ultraviolet irradiation, toxic chemicals, and pathogenic microorganisms) in the cutaneous environment and rapidly sending appropriate signals that alert and recruit other branches of the immune system [5, 6]. To restore homeostasis in the skin immune system, the multiple proinflammatory signals generated by skin cells must eventually be counterbalanced by mechanisms capable of promoting resolution of a cutaneous inflammatory response.

The first stage in the induction of a primary immune response is the processing of antigen by dendritic cells, the professional antigen presenting cells in skin. These cells process antigen in the context of MHC-class II molecules [7] and migrate out of the skin to the draining lymph node where they can stimulate proliferation and differentiation of antigen-specific T cells. Under normal conditions, keratinocytes and fibroblasts are unable to stimulate T cells. However, they can be induced by interferon-γ to express MHC-class II molecules and thereby acquire the ability to present antigen. Since the cells are deficient in the necessary co-stimulatory molecules [8, 9], antigen presentation by keratinocytes and fibroblasts does not result in T cell activation. Instead, this antigen presentation can result in T cell non-responsiveness [10, 11] or T cell anergy [12]. The interaction of keratinocytes and fibroblasts with T cells has important implications for the use of allogeneic cells in tissue engineering.

2 Wound healing

Healing of wounds is an important function of skin. It is a complex process which requires the dynamic interaction of many components — cells, soluble factors, and matrix. Upon wounding, the immediate response is to stop bleeding through clot formation. Simultaneously, there is a release of inflammatory cytokines which regulate blood flow to the area, recruit lymphocytes and macrophages to fight infection, and later stimulate angiogenesis and collagen deposition. These latter processes result in the formation of granulation tissue, a highly vascularized and cellular wound connective tissue. Myofibroblasts, rich in actin [13], are recruited through the action of factors such as platelet derived growth factor (PDGF) and transforming growth factor-β (TGF-β) and act to contract the wound, reducing the area to be healed. Keratinocytes are stimulated to proliferate and migrate into the wound bed to restore epidermal coverage. Eventually, the granulation tissue is resorbed leaving dense fibrous scar tissue.

3 Tissue engineering in skin

Tissue engineered skin products have primarily focused on regenerating epidermis and/or repairing dermis. Tissue engineering has not focused on regenerating some structures, like hair follicles and sebaceous glands, whose loss is relatively unimportant compared to epidermal and dermal destruction. For some dermal components, such as blood vessels and immune system cells, there is less need to stimulate regeneration through tissue engineering methods because these components have the ability to quickly repopulate and normalize a wounded area.

Epidermal regeneration to provide wound coverage is an important goal for tissue engineering because, although the epidermis has an enormous capacity to heal, there are situations where large areas of epidermis need to be replaced or where regeneration is deficient. Dermal repair after severe wounds is important to restore skin elasticity, flexibility, and strength. Scar tissue lacks these characteristics and is permanent since the dermis has very little capacity to regenerate. Scar tissue limits movement, causes pain, and is cosmetically undesirable.

3.1 Design considerations

The problem of skin regeneration or, more simply, the establishment of new skin tissue has been approached in several ways. Attempts have been made to use one or more of the growth and other cell factors present in skin to stimulate wound healing. Tissue engineering approaches have focused on providing or mimicking structural and/or biological character- istics of the dermal or the epidermal component. Some techniques have sought to reproduce living, full-thickness tissue for transplantation. Below is a brief synopsis of how and why these approaches have been taken and what we know to date.

3.1.1 The factors of skin
Skin has inherent ability to close and heal itself. The epidermal and dermal response during healing is regulated by inflammatory cytokines and by autocrine and paracrine factors produced by the dermal fibroblasts and epidermal keratinocytes [2, 14]. These factors regulate growth and differ- entiation of keratinocytes, inflammatory reactions, angiogenesis, and extra- cellular matrix formation. Identification of these factors has led to several attempts to speed the healing of wounds by local application of one or more of these factors. TGF-β, epidermal growth factor (EGF), vascular endothelial growth factor (VEGF), and PDGF have been candidates for this purpose [2, 15–18]. However, wound healing involves the interaction of many tissue factors and elements, and research has not yet revealed a single factor or subset of factors which alone are capable of stimulating healing.

The poor healing response in chronic wounds has been attributed to an imbalance of factors rather than an insufficiency of any particular factor [19]. Complex cell extracts have been used in hopes of providing the appropriate mixture of elements. These include the use of platelet extracts to provide primarily PDGF [20], and the use of keratinocyte extracts to provide a complex mixture of elements of rapidly growing keratinocytes [21]. The use of non-viable fibroblasts within a cell-produced matrix of connective tissue proteins has been used to provide both matrix and cell elements [22].

While some of these elements have been shown to promote re-epitheli- alization or to stimulate the formation of granulation tissue, none has had a major impact on regeneration of new skin tissue. This is in part due to the complex nature of the wound healing response [23]. In addition, the use of factors is not a sufficient approach, in and of itself, in situations where there is severe or massive loss of skin tissue. A tissue engineering approach is well-suited to the problem of skin healing and replacement. Living tissue and appropriate scaffolds created through tissue engineering can provide complex temporal control of factor delivery and effect and can be used to provide the needed *combination* of chemical, structural, and cellular elements [24, 25].

3.1.2 Dermal replacement

Approaches to dermal repair and regeneration center around the control of fibroblast repopulation and collagen biosynthesis to limit scar tissue formation. One of the keys to improving dermal repair is control, or redirection, of the wound healing response so that scar tissue does not form. One promising approach derives from the observation that unlike adult tissues, fetal tissues heal without scar formation [26]. TGF-β_1, which is not expressed in the fetus, is a potent stimulator of collagen biosynthesis by fibroblasts in the adult and is thought to be an important inducer of scar formation. In an attempt to inhibit scarring, Ferguson *et al.* have investigated the use of neutralizing TGF-β antibodies to redirect the healing response away from scar tissue formation [27, 28].

3.1.3 Dermal scaffolds

Tissue engineering has looked at redirecting granulation tissue formation through the use of scaffolds and living cells. Yannas *et al.* [29], in one of the earliest tissue engineering approaches to improving dermal healing, designed a collagen–glycosamino glycan sponge to serve as a scaffold or template for dermal extracellular matrix. The goal was to promote fibroblast repopulation in a controlled way which would decrease scarring and wound contraction [30]. A commercial version of this material (Integra™, Integra Life Sciences) is currently approved for use under split-thickness autografts in burns [31, 32].

Several variations on the collagen scaffold have been studied. Efforts have been made to improve fibroblast infiltration and collagen persistence by collagen cross-linking [33–35], inclusion of other matrix proteins [14, 35, 36], hyaluronic acid [34, 37], and by modifying porosity of the scaffold [29, 36]. Decellularized dermal tissue has also been used in an attempt to recapitulate as much of the normal architecture as possible while providing a natural scaffold for re-epithelialization [33, 38–40]. Currently, only the upper papillary layer of dermis is used clinically (Alloderm™, LifeCell). One of the problems with this approach is that deep dermis and the more superficial papillary layer differ in architecture. The deep reticular dermis is needed to prevent wound contraction, and the papillary layer is needed for epidermal adherence and secondary structure. Providing an appropriate scaffold for deep dermal repair remains a challenge for groups investigating native as well as man-made matrices.

While matrix scaffolds have shown some improvement in scar morphology, no acellular matrix has yet been shown to lead to true dermal regeneration. This may be due in part to limits in cell repopulation, the type of fibroblast repopulating the graft (J. Gross, personal communication), and control of the inflammatory and remodeling processes, i.e. the cells' ability to degrade old matrix while synthesizing new matrix. A cellular dermal construct has been fabricated using dermal fibroblasts grown to high density on a resorbable polyglycolic acid mesh (Vicryl^{TM}, Johnson and Johnson). During culture, the fibroblasts span the interstices of the mesh, depositing extracellular matrix components within the spaces. This material may then be implanted [41], and is currently under study for the treatment of diabetic foot ulcers (Dermagraft^{TM}, Advanced Tissue Sciences).

The inflammatory response must be controlled in dermal repair in order to avoid scar tissue formation. Therefore, dermal scaffolds must not be inflammatory and must not stimulate a foreign body reaction. This has been a problem in the past for some glutaraldehyde cross-linked collagen substrates for example [42]. The ability of the matrix to persist long enough to redirect tissue formation must be balanced with effects of the matrix on inflammatory processes.

3.1.4 Epidermal regeneration

Achieving wound closure or re-epithelialization has also been an area of intense interest. Epithelialization of the wound is of paramount concern, particularly in large wounds such as extensive burns. Without epithelial coverage to provide barrier function, no defense exists against infection or fluid loss. The approaches to re-establishing epidermis are numerous, ranging from the use of lyophilized cell extracts (as noted above) to full-thickness skin equivalents possessing a differentiated epidermis (Table 1). Silicone membranes have been used as temporary coverings in conjunction with some of the dermal templates described above [31]. However, living epidermal keratinocytes are necessary to achieve permanent wound closure.

One of the first applications of cultured cells to benefit humans was the use of cultured epidermal cells by Green *et al.* [43, 44]. Epidermal keratinocytes are grown from a small patient biopsy using the co-culture method developed by Rheinwald and Green [45]. The mouse 3T3 fibroblast feeder cell system allows substantial expansion of epidermal keratinocytes. This method can be used to generate enough thin, multi-layered epidermal sheets to resurface the body of a severely burned patient [46]. Once transplanted, the epidermal sheets quickly form epidermis [47] and re-

Table 1. Approaches to skin repair

What has been studied	Purpose	Pertinent references
Lyophilized keratinocyte lysates	Stimulate healing	21
Keratinocyte cell suspensions	Establish epithelium	62
Cultured epithelial autografts (CEA)	Establish epithelium	43, 44, 46–49
Cell-produced matrix and non-viable fibroblasts on a non-resorbable mesh with a silastic membrane barrier	Temporary covering in burns to replace cadaver skin	22
Collagen sponges and modifications	Provide a scaffold for granulation tissue formation and dermal repair	24, 29, 30, 33, 37, 42, 60, 61
Hyaluronate and other matrix materials	Improve dermal healing	34, 37, 60
Collagen–glycosaminoglycan sponge with silastic membrane barrier	Temporary covering in burns; provide a scaffold for dermal repair	31, 32
Allogeneic/xenogeneic dermis	Provide a scaffold for dermal repair and epithelial attachment	14, 35, 38–40, 49
Fibroblasts and cell-produced matrix on resorbable mesh	Stimulate healing	41
Human skin equivalent (HSE); Composite skin cultures; other methods of providing dermis and epidermis	Stimulate healing; serve as a scaffold for regrowth of patient tissue; provide both epidermis and dermis; serve as an alternative to autograft and allograft	14, 25, 36, 40, 52, 54, 55, 59

establish epidermal coverage in these patients. With time, the Cultured Epithelial Autograft (CEA) leads to the formation of new connective tissue or 'neodermis' immediately underlying the epidermis [47]. However, blistering, scarring, and wound contraction are significant problems [48].

Studies have shown that grafting of CEA onto pregrafted cadaver dermis greatly improves graft take [49].

3.1.5 The composite skin grafts

The epidermis and dermis act synergistically [50, 51]. Human autograft, which contains both epidermis and dermis, is the gold standard for resurfacing the body and closing difficult-to-heal wounds. One of the first attempts to mimic a 'full-thickness' skin graft was by Bell *et al.* [52] who described a bi-layered skin construct grown in organotypic culture. The dermal component consisted of a collagen lattice contracted by tractional forces of dermal fibroblasts trapped within the gelled collagen. This contracted lattice was then used as a substrate for epidermal keratinocytes. Over the last 10 years, this technology has advanced to enable the production of large amounts of human skin equivalent (HSE) from a single donor [53]. Using methods of organotypic culture, the resulting HSE develops many of the structural, biochemical, and functional properties of human skin [54–57].

The ability to utilize allogeneic cells rather than autologous cells as in CEA therapy enables the reproducible manufacture of consistent material [53]. The inability of epidermal keratinocytes and dermal fibroblasts to stimulate a T cell response, discussed in a previous section, permits their use in allogeneic applications. The HSE (Apligraf™, Organogenesis Inc.) has now been studied clinically in a number of applications including chronic wounds [25], dermatological excisions, and burns. Its effectiveness in the difficult to treat chronic venous leg ulcer has been attributed to its ability to interact with the wound in multiple ways [25]. Studies in athymic mice also indicate that the use of a differentiated tissue such as the HSE also impacts the ability to engraft successfully [58].

Boyce *et al.* have modified the approach first proposed by Yannas *et al.* to form a bi-layered composite skin made using a modified collagen-glycosaminoglycan substrate seeded with fibroblasts and overlaid with epidermal keratinocytes [55]. An autologous form of this composite skin construct has been used to treat severe burns with some success [36]. An allogeneic form of the construct showed improved healing in a pilot study in chronic wounds [59].

Work with composite grafts suggests a clear benefit in the use of organotypic skin constructs in wound healing, particularly in the hard to heal wound such as the long-standing venous leg ulcer [25]. However, no method has yet approached autograft in its robust performance on the burn wound or in its ability to truly impact dermal repair in humans. Now that these methods have progressed to the clinic, we must direct our focus not only on stimulating wound healing but redirecting it as well for more consistent graft take, better quality of dermal repair, and improved cosmetic outcome.

References

1. Streilein, J. W. and Bergstresser, P. R., Langerhans cells: antigen presenting cells of the epidermis. *Immunobiology*, 1984, **168**, 285–300.
2. McKay, I. A. and Leigh, I. M., Epidermal cytokines and their roles in cutaneous wound healing. *British Journal of Dermatology*, 1991, **124**, 513–518.
3. Lappin, M. B., Kimber, I. and Norval, M., The role of dendritic cells in cutaneous immunity. *Archives of Dermatological Research*, 1996, **288**, 109–121.
4. Williams, I. R. and Kupper, T. S., Immunity at the surface: homeostatic mechanisms of the skin immune system. *Life Sciences*, 1996, **58**, 1485–1507.
5. Streilein, J. W., Skin-associated lymphoid tissues (SALT): origins and functions. *Journal of Investigative Dermatology*, 1983, **80**(Suppl.), 12–16.
6. Bos, J. D. and Kapsenberg, M. L., The skin immune system: progress in cutaneous biology. *Immunology Today*, 1993, **14**, 75–78.
7. Kripke, M. L., Munn, C. G., Jeevan, A., Tang, J. M. and Bucana, C., Evidence that cutaneous antigen-presenting cells migrate to regional lymph nodes during contact sensitization. *Journal of Immunology*, 1990, **145**, 2833–2838.
8. Nickoloff, B. J. and Turka, L. A., Immunological functions of non-professional antigen-presenting cells: new insights from studies of T-cell interactions with keratinocytes. *Immunology Today*, 1994, **15**, 464–469.
9. Phipps, R. P., Roper, R. L. and Stein, S. H., Alternative antigen presentation pathways: accessory cells which down-regulate immune responses. *Regional Immunology*, 1989, **2**, 326–339.
10. Gaspari, A. A. and Katz, S. I., Induction and functional characterization of class II MHC (Ia) antigens on murine keratinocytes. *Journal of Immunology*, 1988, **140**, 2956–2963.
11. Bal, V., McIndoe, A., Denton, G., Hudson, D., Lombardi, G., Lamb, J. and Lechler, R., Antigen presentation by keratinocytes induces tolerance in human T cells. *European Journal of Immunology*, 1990, **20**, 1893–1897.
12. Gaspari, A. A. and Katz, S. I., Induction of in vivo hyporesponsiveness to contact allergens by hapten-modified Ia + keratinocytes. *Journal of Immunology*, 1991, **147**, 4155–4161.
13. Desmouliere, A. and Gabbiani, G., The role of the myofibroblast in wound healing and fibrocontractive diseases. In *The Molecular and Cellular Biology of Wound Repair*, 2nd edn, ed. R. A. F. Clark. Plenum Press, New York, 1996, pp. 391–323.
14. Ansel, J., Perry, P., Brown, J., Damm, D., Phan, T., Hart, C., Lugert, T. and Hefeneider, S., Cytokine modulation of keratinocyte cytokines. *Journal of Investigative Dermatology*, 1990, **94**(Suppl.), 101–107.
15. Martin, P., Hopkinson-Woolley, J. and McCluskey, J., Growth factors and cutaneous wound repair. *Progress in Growth Factor Research*, 1992, **4**, 25–44.
16. Nanney, L. B. and King, L. E., Jr., Epidermal growth factor and transforming growth factor-α. In *The Molecular and Cellular Biology of Wound Repair*, 2nd edn, ed. R. A. F. Clark. Plenum Press, New York, 1996, pp. 171–194.
17. Abraham, J. A. and Klagsbrun, M., Modulation of wound repair by members of the fibroblast growth factor family. In *The Molecular and Cellular Biology of Wound Repair*, 2nd edn, ed. R. A. F. Clark. Plenum Press, New York, 1996, pp. 195–248.

18. Roberts, A. B. and Sporn, M. B., Transforming growth factor-β. In *The Molecular and Cellular Biology of Wound Repair*, 2nd edn, ed. R. A. F. Clark. Plenum Press, New York, 1996, pp. 275–308.

19. Parenteau, N. L., Sabolinski, M. L., Mulder, G. and Rovee, D. T., Wound research. In *Chronic Wound Care: A Clinical Source for Healthcare Professionals*, 2nd edn, eds. D. Krasner and D. Kane. Health Management Publications, Wayne, PA, 1997, pp. 389–395.

20. Knighton, D. R., Fiegel, V. D., Doucette, M. M., Fylling, C. P. and Cerra, F. B., The use of topically applied platelet growth factors in chronic nonhealing wounds: a review. *Wounds: A Compendium of Clinical Research and Practice*, 1989, **1**, 71–78.

21. Duinslaiger, L., Verbeken, G., Reper, P., Delaey, B., Vanhalle, S. and Vanderkelen, A., Lyophilized keratinocyte cell lysates contain multiple mitogenic activities and stimulate closure of meshed skin autograft-covered burn wounds with efficiency similar to that of fresh allogeneic keratinocyte cultures. *Plastic and Reconstructive Surgery*, 1994, **98**, 110–117.

22. Hansbrough, J. F., Norgan, J., Greenleaf, G. and Underwood, J., Development of a temporary living skin replacement composed of human neonatal fibroblasts cultured in Biobrane, a synthetic dressing material. *Surgery*, 1994, **115**, 633–644.

23. Nathan, C. and Sporn, M., Cytokines in context. *Journal of Cell Biology*, 1991, **113**, 981–986.

24. Marks, M. G., Doillon, C. and Silver, F. H., Effects of fibroblasts and basic fibroblast growth factor on facilitation of dermal wound healing by type I collagen matrices. *Journal of Biomedical Materials Research,* 1991, **25**, 683–696.

25. Sabolinski, M. L., Alvarez, O., Auletta, M., Mulder, G. and Parenteau, N. L., Cultured skin as a 'smart material' for healing wounds: experience in venous ulcers. *Biomaterials*, 1996, **17**, 311–320.

26. Mast, B. A., Nelson, J. M., and Krummel, T. M., Tissue repair in the mammalian fetus. In *Wound Healing: Biochemical and Clinical Aspects*, eds. I. K. Cohen, R. F. Diegelmann, and W. J. Lindblad. W.B. Saunders Co., Philadelphia, 1992, pp. 326–341.

27. Shah, M., Foreman, D. M. and Ferguson, M. W., Neutralization of TGF-beta 1 and TGF-beta 2 or exogenous addition of TGF-beta 3 to cutaneous wounds reduces scarring. *Journal of Cell Science*, 1995, **108**, 985–1002.

28. Shah, M., Foreman, D. M. and Ferguson, M. W., Neutralising antibody to TGF-$\beta_{1,2}$ reduces cutaneous scarring in adult rodents. *Journal of Cell Science,* 1994, **107**, 1137–1157.

29. Yannas, I. V., Lee, E., Orgill, D. P., Skrabut, E. M. and Murphy, G. F., Synthesis and characterization of a model extracellular matrix that induces partial regeneration of adult mammalian skin. *Proceedings of the National Academy of Sciences USA*, 1989, **86**, 933–937.

30. Yannas, I. V., Burke, J. F., Orgill, D. P. and Skrabut, E. M., Wound tissue can utilize a polymeric template to synthesize a functional extension of skin. *Science,* 1982, **215**, 174–176.

31. Heimbach, D., Luterman, A., Burke, J. F., Cram, A., Herndon, D., Hunt, J., Jordan, M., McManus, W., Solam, L. and Warden, G., Artificial dermis for

major burns: a multi-center randomized clinical trial. *Annals of Surgery*, 1988, **208**, 313–320.

32. Burke, J. F., Yannas, I. V., Quinby, W. C., Bondoc, C. C. and Jung, W. K., Successful use of a physiologically acceptible artificial skin in the treatment of extensive burn injury. *Annals of Surgery*, 1981, **194**, 413–428.

33. Middelkoop, E., deVries, H. J. C., Ruuls, L., Everts, V., Wildevuur, C. H. R. and Westerhof, W., Adherence, proliferation and collagen turnover by human fibroblasts seeded into different types of collagen sponges. *Cell and Tissue Research*, 1995, **280**, 447–453.

34. Cooper, M. L., Hansbrough, J. F. and Polareck, J. W., The effect of an arginine-glycine-aspartic acid peptide and hyaluronate synthetic matrix on epithelialization of meshed skin graft interstices. *Journal of Burn Care and Rehabilitation*, 1996, **17**, 108–116.

35. vanLuyn, M. J. A., Verheul, J. and vanWachem, P. B., Regeneration of full-thickness wounds using collagen split grafts. *Journal of Biomedical Materials Research*, 1995, **29**, 1425–1436.

36. Hansbrough, J. F., Boyce, S. T., Cooper, M. L. and Foreman, T. J., Burn wound closure with cultured autologous keratinocytes and fibroblasts attached to a collagen-glycosaminoglycan substrate. *Journal of the American Medical Association*, 1989, **262**, 2125–2130.

37. Murashita, T., Nakayama, Y., Hirano, T. and Ohashi, S., Acceleration of granulation tissue ingrowth by hyaluronic acid in artificial skin. *British Journal of Plastic Surgery*, 1996, **49**, 58–63.

38. Langdon, R. C., Cuono, C. B., Birchall, N., Madri, J. A., Kuklinska, E., McGuire J. and Moellmann, G. E., Reconstitution of structure and cell function in human skin grafts derived from cryopreserved allogeneic dermis and autologous cultured keratinocytes. *Journal of Investigative Dermatology*, 1988, **91**, 478–485.

39. Livesey, S. A., Herndon, D. N., Hollyoak, M. A., Atkinson, Y. H. and Nag, A., Transplanted acellular allograft dermal matrix. *Transplantation*, 1995, **60**, 1–9.

40. Cuono, C., Langdon, R. and McGuire, J., Use of cultured epidermal autografts and dermal allografts as skin replacement after burn injury. *Lancet*, 1986, **1**, 1123–1124.

41. Hansbrough, J. F., Dore, C. and Hansbrough, W. B., Clinical trials of a living dermal tissue replacement placed beneath meshed, split-thickness skin grafts on excised burn wounds. *Journal of Burn Care and Rehabilitation*, 1992, **13**, 519–528.

42. deVries, H. J. C., Mekkes, J. R., Middelkoop, E., Hinrichs W. L. J., Wildevuur, C. H. R. and Westerhof, W., Dermal substitutes for full-thickness wounds in a one stage grafting model. *Wound Repair and Regeneration*, 1993, **1**, 244–252.

43. Phillips, T. J., Kehinde, O., Green, H. and Gilchrest, B. A., Treatment of skin ulcers with cultured epidermal allografts. *Journal of the American Academy of Dermatology*, 1989, **21**, 191–199.

44. Green, H., Kehinde, O. and Thomas, J., Growth of cultured human epidermal cells into multiple epithelia suitable for grafting. *Proceedings of the National Academy of Sciences USA*, 1979, **76**, 5665–5668.

45. Rheinwald, J. G. and Green, H., Serial cultivation of strains of human epidermal keratinocytes: the formation of keratinizing colonies from single cells. *Cell*, 1975, **6**, 331–344.

46. Gallico, G. G. III, O'Connor, N. E., Compton, C. C., Kehinde, O. and Green, H., Permanent coverage of large burn wounds with autologous cultured human epithelium. *New England Journal of Medicine*, 1984, **311**, 448–451.

47. Compton, C., Wound healing potential of cultured epithelium, *Wounds: A Compendium of Clinical Research and Practice*, 1993, **5**, 97–111.

48. Sheridan, R. L. and Tompkins, R. G., Cultured autologous epithelium in patients with burns of ninety percent or more of the body surface. *The Journal of Trauma: Injury, Infection and Critical Care*, 1995, **38**, 48–50.

49. Odessey, R., Addendum: multicenter experience with cultured epidermal autograft for the treatment of burns. *Journal of Burn Care and Rehabilitation*, 1992, **13**, 174–180.

50. Parenteau, N. L., Sabolinski, M. L., Mulder, G. and Rovee, D. T., Wound research. In *Chronic Wound Care*, 2nd edn, eds. D. Krasner and D. Kane. Health Management Publications, Wayne, PA, 1997, pp. 389–395.

51. Leary, T., Jones, P. L., Appleby, M., Bugcht, A., Parkinson, K. and Stanley, M., Epidermal keratinocyte selfrenewal is dependent upon dermal integrity. *Journal of Investigative Dermatology*, 1992, **99**, 422–430.

52. Bell, E., Ehrlich, P., Buttle, D. J. and Nakatsuji, T., Living tissue formed in vitro and accepted as skin-equivalent of full-thickness. *Science*, 1981, **221**, 1052–1054.

53. Wilkins, L. M., Watson, S. R., Prosky, S. J., Meunier, S. F. and Parenteau, N. L., Development of a bilayered living skin construct for clinical applications. *Biotechnology and Bioengineering*, 1994, **43**, 747–756.

54. Nolte, C. J. M., Oleson, M. A., Bilbo, P. R. and Parenteau, N. L., Development of a stratum corneum and barrier function in an organotypic skin culture. *Archives of Dermatological Research*, 1993, **285**, 466–474.

55. Boyce, S. T. and Hansbrough, J. F., Biologic attachment, growth, and differentiation of cultured human keratinocytes on a graftable collagen and chondroitin-6-sulfate substrate. *Surgery*, 1988, **103**, 421–431.

56. Bilbo, P. R., Nolte, C. J. M., Oleson, M. A., Mason, V. S. and Parenteau, N. L., Skin in complex culture: the transition from 'culture' phenotype to organotypic phenotype. *Journal of Toxicology — Cutaneous and Ocular Toxicology*, 1993, **12**, 183–196.

57. Parenteau, N. L., Bilbo, P., Nolte, C. J. M., Mason, V. S. and Rosenberg, M., The organotypic culture of human skin keratinocytes and fibroblasts to achieve form and function. *Cytotechnology*, 1992, **9**, 163–171.

58. Parenteau, N., Sabolinski, M., Prosky, S., Nolte, C., Oleson, M., Kriwet, K. and Bilbo, P., Biological and physical factors influencing the successful engraftment of a cultured human skin substitute. *Biotechnology and Bioengineering*, 1996, **52**, 3–14.

59. Boyce, S. T, Glatter, R. and Kitsmiller, J., Treatment of chronic wounds with cultured skin substitutes: a pilot study. *Wounds: A Compendium of Clinical Research and Practice*, 1995, **7**, 24–29.

60. Hanthamrongwit, M., Reid, W. H. and Grant, M. H., Chondroitin-6-sulphate

incorporated into collagen gels for the growth of human keratinocytes: the effect of cross-linking agents and diamines. *Biomaterials*, 1996, **17**, 775–780.

61. Matsui, R., Osaki, K. -I., Konishi, J., Ikegami, K. and Koide, M., Evaluation of an artificial dermis full-thickness skin defect model in the rat. *Biomaterials*, 1996, **17**, 989–994.

62. Stark, G. B., Kaiser, H. W., Cologne Burn Centre experience with glycerol-preserved allogeneic skin: part II: combination with autologous cultured keratinocytes. *Burns*, 1994, **20**(Suppl.), 34–38.

CHAPTER IV.8

Engineering a Bioartificial Liver Support

JULIE R. FRIEND,
WEI-SHOU HU
Department of Chemical Engineering and Materials Science,
University of Minnesota,
421 Washington Ave. SE
Minneapolis, MN 55455-0132, USA

1 Introduction

Liver disease continues to be a serious health problem in the United States with approximately 25,000 people dying of liver failure each year [1]. Depending on the time span required for a given agent to produce clinical symptoms, hepatic failure is grossly classified as either acute or chronic. Chronic liver failure can now be treated with orthotopic liver transplantation [2]. The same approach can be used to treat acute fulminant hepatic failure (FHF), where hepatic regeneration is neither rapid nor sufficient enough to keep the individual alive, although the mortality rates are much higher [3]. However, many patients in hepatic failure do not qualify for transplantation because of metastatic cancer, active alcoholism, or other concurrent medical problems. Even those who qualify for a liver transplant often die while awaiting an allograft due to the scarcity of donor organs [4]. As a result, there is a critical need for improved temporary liver support technology for survival of potential transplant recipients and patients with reversible, acute hepatitis who are excluded from liver transplantation [5].

Development of a bioartificial liver has been a significant challenge because the liver performs many different functions necessary for survival. Over the years, both non-biological and biological approaches to liver support have been attempted. Within the last 15 years, a hybrid approach has evolved, leading to the development of bioartficial liver devices. These devices employ biological components, usually cells, within a synthetic support. While some implantable bioartificial liver systems have been developed and tested in animals [6–8], this chapter will focus on extracorporeal devices. These types of devices are seen as bridge therapies to sustain a patient until a transplantable organ becomes available or until the patient's liver regenerates.

2 Liver physiology

The liver is the largest organ in the body. The average adult liver is 1.5 kg, or roughly 2.5% of the body weight. The liver receives 70% of its blood via the portal vein, which brings nutrient-rich blood from the intestine, and the remainder from the hepatic artery, which brings oxygen-rich blood from the heart. Blood is removed via the hepatic vein. The liver holds approximately 30% of the resting cardiac output at any one time and uses about 20% of the oxygen in the body.

The main structural unit of the liver is the liver lobule. A lobule is hexagonally shaped with portal triads at its corners. Portal triads consist of a portal vein, hepatic artery and bile duct. The central vein, which leads to the hepatic vein, is located at the center of the lobule. Hepatocytes, the most abundant cells in the liver, are arranged in unicelluar plates between the portal triad and the central vein. Blood flows from the portal triad to the central vein through sinusoids between the plates of hepatocytes. Bile canaliculi, which lead to the larger bile ducts, are situated between adjacent hepatocytes.

The liver, unlike some other organs which conduct a single vital function, performs a variety of functions necessary for survival. It synthesizes a wide variety of molecules and regulates their release into the blood stream. It also inactivates hormones and other endogenous molecules and metabolizes xenobiotics for excretion. For example, the liver forms glucose from metabolism of carbohydrates, lactate and amino acids. It stores glucose as glycogen and releases it as necessary. The liver also plays an important role in lipid metabolism. It oxidizes fatty acids and synthesizes storage lipids in the form of triacylglycerols and phospholipids. The liver is an important center of metabolism and/or storage of vitamins A, B, D, and K. Heme is metabolized in the liver to bilirubin, which is secreted in the bile. Ammonia formed through metabolism of proteins and nucleic acids is converted to urea and excreted by the liver. The liver detoxifies drugs, environmental toxins and other foreign chemicals via biotransformation. Biotransformation typically involves oxidation by a cytochrome P450 enzyme (phase I metabolism), followed by conjugation which renders the molecule inactive or excretable. The liver synthesizes 95% of the proteins found in plasma, including albumin, prothrombin and various clotting factors, as well as many of the bile acids.

Hepatocytes, the parenchymal cells of the liver, perform most of the metabolic functions of the liver. They account for approximately 90% of cell mass and 60% of cell number in the liver. Non-parenchymal cells make up the rest. These include sinusoidal endothelial cells which form fenestrated conduits for blood; Kupffer cells which function as macrophages; Ito cells (also known as fat-storing or stellate cells) which secrete extracellular matrix proteins and growth factors such as hepatocyte growth factor; pit cells which exhibit natural killer cell activities; and biliary epithelial cells which line the bile ducts. Hepatocytes are polyhedrally shaped and are typically 20–30 μm in diameter. They are polar cells with three membrane domains: a basal domain in contact with blood, a lateral domain which is in contact with adjacent hepatocytes and an apical domain which opens to bile canaliculi. Although generally considered epithelial cells, hepatocytes are not supported by a traditional basement membrane, although they do contact extracellular matrix proteins such as collagen, laminin and proteoglycans at their basal surfaces. Of the non-parenchymal cells, only the Ito cell, which is found in the space between the sinusoidal cells and hepatocytes, may come into direct contact with the hepatocyte. Sinusoidal cells are separated from hepatocytes by the Space of Disse. Kupffer and pit cells are located within the sinusoidal lumen and do not come into contact with hepatocytes.

The liver is also unique in that it has an amazing capacity to regenerate. In the adult liver, hepatocytes are generally in the quiescent stage of the cell cycle and only divide about once a year. In response to partial hepatectomy or toxic injury, liver cells are quickly stimulated to divide to replace lost cell mass. The process is tightly regulated by growth factors and inhibitors so that only lost mass is restored. In rats, regeneration is complete within 7 days of a 70% partial hepatectomy. Restoration of cell mass in humans may take up to 6 months, depending on the extent of injury [9]. Thus, if adequate support could be provided, it is possible that a patient's liver could recover from acute liver failure.

3 Techniques of liver support

A variety of techniques for artificial liver support have been attempted for treatment of hepatic failure over the past four decades. These include hemodialysis for the removal of ammonia [10], hemoperfusion using a direct cation exchanger (Dowex 50-X8) [11], hemoperfusion through albumin coated Amberlite XAD-7 resin [12], and hemodialysis across polyacrylonitrile (PAN) membranes [13] for the removal of toxic metabolites which build up in the plasma during liver failure. However, none of these therapies proved to be adequate for treating liver failure.

At the same time, other researchers were examining the feasibility of completely biological support. The first generation of such studies included xenogeneic cross-hemodialysis with a healthy dog [14], extracorporeal perfusion of homologous isolated livers [15], extracorporeal perfusion against sliced liver tissue [16], and use of isolated hepatocytes on an extracorporeal circuit [17, 18]. Limited success was reported in human trials using extracorporeal perfusions with livers from pigs [19, 20], dogs [21], baboons [22, 23], and other animal species [24]. Success with these systems was also limited due to problems with maintaining tissue viability and immune compatibility.

Out of this work evolved a hybrid approach to liver support — bioartificial livers (BALs) containing biological components within a synthetic network. These hybrid systems have several advantages over other forms of support. Bioartificial liver systems provide a wider range of specific liver functions than non-biological therapies, cellular components or isolated enzyme systems because they contain intact, metabolically active liver cells [25]. Also, hepatocyte-based systems are less constrained by mass transfer limitations than chunks of liver tissue, and are therefore amenable to scale-up [26]. A major advantage of bioartificial livers is that the system can be constructed using a semipermeable material to protect the hepatocytes from the patient's immune system. This allows for use of hepatocytes from a different species (xenocytes) within the bioartificial liver without need for severe immunomodulation [27, 28]. The main disadvantages lie in the material–membrane interactions, similar to those encountered in blood contacting devices [29], and in the maintenance of hepatocyte viability and function at the high cell densities required for clinical applications. However, some of these problems can be overcome with innovative bioreactor design and use of the latest hepatocyte culture techniques.

4 Cell source for use in a BAL

Both well-differentiated hepatoma cell lines and primary hepatocytes have been explored as the biologically active component of bioartificial liver devices. One advantage to using cell lines is their ability to proliferate to high cell densities. Primary hepatocytes, on the other hand, typically do not divide in culture. Although certain well-differentiated human tumor cells like HepG2 exhibit many of the characteristics of normal hepatocytes [30], cell lines usually lose some of the differentiated functions of primary cells. It is not known which functions are most important to sustain a patient in liver failure. By using primary hepatocytes, all the metabolic functions of the liver are potentially provided. Moreover, primary cells have gained wider acceptance than hepatoma cell lines in applications for human therapy because risk of tumor transfection, though perhaps minimal, cannot be completely ruled out [31]. Because healthy human livers would most likely be used for transplant, hepatocytes used in a BAL will likely come from a liver of a different species. Most groups using primary cells have settled on porcine hepatocytes for use in a BAL because of the large number of cells provided by one liver and the similarity in function to human liver. Techiques to harvest porcine hepatocytes at high viability are well established [32–34].

5 Maintenance of cell viability and function

Primary hepatocytes lose their differentiated function in 3–4 days and viability within 1 week when cultured as a monolayer. Over the past 15 years, many researchers have examined many ways to improve hepatocyte function and viability in culture. Methods to prolong hepatocyte viability and function include addition of growth factors and hormones to the culture medium [35, 36], cultivation of hepatocytes in the presence of attachment factors and extracellular matrix components [37–40], co-culture with liver-derived non-parenchymal and endothelial cells [41, 42], and cultivation within a polymeric membrane or gel matrix [43–45]. Another possibility is to culture the hepatocytes as spheroids. Hepatocyte spheroids exhibit enhanced liver-specific functions and prolonged viability compared to hepatocytes cultured as a monolayer [46–48]. However, good bioreactor design, as well as better cell culture techniques, will be important for maintaining hepatocyte function and viability at the high cell densities that will be necessary for BAL therapy.

6 Examples of bioartificial liver devices

Various BAL devices for potential human therapy have been reported by different research groups in the United States, Europe, and Japan. (For reviews see [49–52]) The basic construction of all systems is similar. Hepatocytes are inoculated into a bioreactor and cultured such that their viability and function can be sustained for an extended duration. Patient blood or plasma enters the bioreactor. An immunobarrier between hepatocytes and host may be incorporated into the design. Nutrients and toxins are metabolized by the hepatocytes as the blood or plasma passes through the reactor, and then the treated blood or plasma is then returned to the patient. Other detoxification methods, such as dialysis or adsorption, may be incorporated within the design [5]. In designing an extracorporeal device, several design criteria must be taken into consideration. These include biocompatibility of the materials in the device, source and supply of functioning cells, immunoprotection for the cells from the host's immune system, maximization of activity in smallest volume, minimization of mass transfer limitations, and maintenance of cell viability and differentiated function. Several different extracorporeal bioartificial liver designs are discussed below.

6.1 The Cedars-Sinai BAL

Demetriou *et al.* designed a device where porcine hepatocytes are attached to microcarriers and injected into the extracapillary space of a hollow fiber

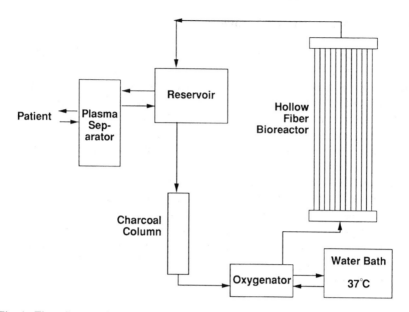

Fig. 1. Flow diagram for the bioartificial liver system designed by Demetriou *et al.*

microfiltration cartridge [53]. A flow diagram of this system is shown in Fig. 1. First, plasma is separated from the patient's whole blood and pumped into a reservoir. This reservoir was added so that the plasma flow rate in the reactor could be increased, thus increasing the transmembrane flow rate. Use of plasma instead of whole blood prevents contact between blood cells and hepatocytes. It also decreases complications with hemolysis, thrombocytopenia, and clot formation and embolization. A column of cellulose-coated charcoal was added before the hollow fiber cartridge to enhance the detoxifying capacity of the BAL and to protect the hepatocytes from some of the toxic substances in the patient's plasma. It was thought that this may help prolong the viability and function of the hepatocytes. Next, the plasma passes through an oxygenator and a water bath to warm the plasma to 37°C. Plasma is then pumped through the lumen of the fibers in the hollow fiber device. The celluose nitrate/diacetate microfiltration membrane of the hollow fibers contains 0.2 μm pores. Because of the large size of the pores, large molecular weight molecules may pass through the membrane. Thus, this membrane does not serve as an immune barrier. Porcine hepatocytes, attached to collagen-coated microcarriers, reside in the extracapillary space. After it passes through the reactor, plasma is mixed with the blood cells and returned to the patient.

Large animal trials were conducted with this system in adult mongrel dogs [54]. A model of ischemic liver failure was used. Two groups of dogs were studied, (1) dogs ($n = 7$) treated with the a BAL containing $5-6 \times 10^9$ porcine hepatocytes and (2) dogs ($n = 6$) treated with activated charcoal alone. Treatment lasted 6 hours. Glucose levels were higher, while lactate and ammonia levels were lower, in dogs treated with the BAL than in dogs treated with charcoal alone. In addition, the dogs treated with the BAL maintained a normal prothrombin time and increased systolic blood pressure. After treatment, hepatocyte viability was greater than 90% as measured by trypan blue exclusion, indicating that hepatocytes were not damaged by exposure to plasma.

Human safety trials have also been completed with this device [55, 56]. Twelve patients suffering from acute liver failure due to various causes were treated one to four times with the BAL. In addition, eight patients with acute on chronic failure were also treated. These patients had underlying chronic liver disease and had developed acute failure. For these experiments, 5×10^9 viable porcine hepatocytes attached to microcarriers were inoculated in the reactor. Patients were treated with the BAL for 7 hours. For patients in acute failure, the most significant finding was that intracranial pressure (ICP) was significantly decreased. In addition, the patients exhibited increased responsiveness to external stimuli. Glucose levels were increased in patients during BAL treatment, while ammonia, bilirubin, aspartate serum transferase, and alanine serum transferase levels were all decreased. These patients were all successfully bridged to transplant. Results were not as good for acute on chronic failure patients. There was no significant decrease in either the ICP or in most of the biochemical parameters measured. However, there was some decrease in bilirubin and ammonia. Two patients recovered from the acute decompensation and had successful transplants. The other patients, who were not candidates for transplant, eventually died. Trials to demonstrate efficacy of this device are underway.

6.2 Kyushu University BAL

A group of researchers at Kyushu University in Japan has taken a novel approach using a packed-bed of spheroids formed in polyurethane foam (PUF) [57]. Freshly harvested hepatocytes are inoculated onto the foam and within 2 days form into spheroids. This has been demonstrated for both rat [58] and dog [59] hepatocytes. The spheroids remain semi-attached and entrapped within the foam; thus, the foam can serve as scaffolding for a packed-bed. A schematic of their reactor is shown in Fig. 2. Blood from the patient is taken from the femoral artery. Plasma is separated from whole blood and pumped to the bioreactor, which is a cylinder of PUF with small capillaries inside where the blood can flow. Plasma then is mixed with the blood and returned to the patient at the

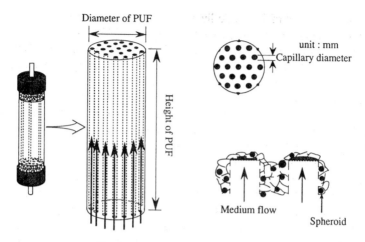

Fig. 2. Hepatocyte entrapment within polyurethane foam bed designed by group at Kyushu University. Reprinted with permission.

femoral vein, or passes through an oxygenator and returned to the reservoir for recirculation.

In *in vitro* experiments, this group has demonstrated that the spheroids maintain albumin synthesis and urea synthesis [57]. Animal experiments were conducted with dogs using an ischemic model of liver failure [59]. Dogs treated with the device ($n = 3$) were compared to dogs treated with a device containing no cells ($n = 3$). Allogeneic dog hepatocytes were used in the bioreactor. The reactor contained $0.7–1.0 \times 10^9$ cells in approximately 300 ml of foam. One-hundred and sixty-eight capillaries, 2.0 mm in diameter, were opened in the foam. Dogs were treated with the device for 6 hours, and blood glucose and ammonia were montiored to determine device performance. Dogs treated with the sham device quickly experienced a rise in blood ammonia and decrease in blood glucose. In dogs treated with a spheroid-containing device, blood glucose was maintained at a higher level, and blood ammonia was maintained at or below 200 μg/dl. While results with this device are encouraging, more experiments should be done to conclusively show the beneficial effects of the device in the animal model. In addition, other functions, particularly biotransformation functions, should be examined.

6.3 University of Minnesota BAL

A team at the University of Minnesota BAL has designed a device where hepatocytes are entrapped within the lumen of the hollow fibers rather than in the extracapillary space [60, 61] (Fig. 3). Freshly harvested hepatocytes are suspended in a cold type I collagen solution, and this hepatocyte-collagen suspension is injected into the hollow fiber lumen. Cell culture medium at 37°C is then perfused in the extracapillary space to raise the temperature of the hepatocyte-collagen suspension, which causes the collagen to gel. This effectively entraps and immobilizes the hepatocytes within the bioreactor. Culture medium is continuously perfused through the extracapillary space for 24 hours, during which time the collagen gel contracts to 60% of its original diameter. Once the gel has contracted, intraluminal medium perfusion is begun to supply hepatocytes with the nutrients and growth factors required to sustain their function. The bioartificial liver is connected to the test animal in liver failure, and the patient's blood is continuously circulated through the extracapillary space. Whole blood is used in this system instead of plasma. Contact between blood cells and hepatocytes is prevented by the membrane of the hollow fibers. This membrane is semi-permeable with a molecular weight cut-off (MWCO) of 100,000 daltons, which should exclude the patient's immunoglobulin and complement proteins from the intraluminal space, while allowing free diffusion of serum albumin and small molecular weight compounds. Thus, toxic components in the patient's blood can diffuse through the hollow fiber membrane into the luminal space where they can be metabolized by the entrapped hepatocytes. The detoxified metabolites can then either diffuse back into the patient's blood stream or exit with the intraluminal stream. The system resides in an incubator which maintains the temperature at 37°C. An oxygenator is also included in the circuit to ensure that the blood contains enough oxygen for the hepatocytes.

In initial *in vitro* studies with rat hepatocytes, oxygen consumption, albumin biosynthesis, lidocaine clearance and bilirubin conjugation were demonstrated for over 7 days of culture in the BAL [62, 63]. *In vivo* function was initially evaluated in an anhepatic rabbit model [64, 65]. As a step toward scale-up for human trials, a large animal liver failure model was developed. Liver failure was induced in canines by administering D-galactosamine at a dosage which caused fatality within 48 hours [66]. Ten kilogram pigs were used as a source of hepatocytes in order to obtain the large number of hepatocytes needed from one animal. To provide enough cells for therapy, two hollow fiber bioreactors were placed in series in the extracapillary loop of the BAL. Each bioreactor was loaded with approximately 30–35 g of porcine liver cells to achieve a total liver cell mass of 60–70 g. Oxygen consumption and ureagenesis were monitored for the first 24 hours to determine metabolic activity of the hepatocytes. For treatment, the right exernal jugular vein was cannulated for BAL access. The experimental animals which received D-galactosamine were divided into two groups: one which received BAL treatment, and one which received no BAL treatment. Control dogs to which no D-galactosamine was injected were included. Dogs were connected to the BAL 24 hours

ENTRAPMENT OF HEPATOCYTES

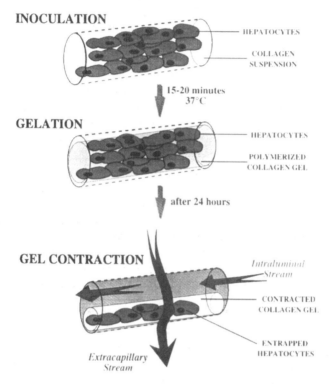

Fig. 3. Entrapment of hepatocytes within collagen gel in the University of Minnesota bioreactor. Reprineted with permission.

after liver failure was induced, and BAL treatment lasted an additional 24 hours. Results indicated increased survival rates for injured canines subsequently treated with the BAL ($n = 5$) compared to injured animals which were not treated ($n = 5$). Hemodynamic stability as well as neurological status were improved in the treated animals. Ammonia and lactate levels were also decreased, although no improvement in glucose homeostasis or prothrombin time were noted. Oxygen consumption was maintained, indicating the hepatocyte viability was still high at the end of the experiment [67]. An IND has been obtained for this device, and human safety trials are pending.

6.4 Other devices

The devices discussed above are only three examples of approaches to bioartificial liver support. Other approaches have been taken as well. For example, Sussman *et al.* have designed a device which is also based on hollow fiber technology. In their design, cells derived from a human hepatoblastoma cell line (C3A) are cultured in the extracapillary space of a hollow fiber dialysis cartridge [68]. Because a cell line is used, cells are able to proliferate once inoculated in the device. The investigators claim that once cells have reached confluence, approximately 200 g of cells are contained within the hollow fiber cartridge. In addition, cartridges may be maintained for 4–8 months, eliminating some of the problems with storage the primary cell-based reactors have. While this reactor has the advantage of being able to hold more cells, it is not clear that this cell line retains all the functions of primary hepatocytes. Whole blood is perfused through the lumen of the fibers. The membrane has a MWCO of 70,000 daltons to protect the cells from antibodies and complement. However, this cutoff may not allow for free diffusion of albumin. Many toxic compounds to be metabolized by hepatocytes are carried by albumin, for example, bilirubin. This device has gone through both large animal [68] and human [69, 70] trials. The device appears to be safe; however, efficacy remains to be demonstrated.

A group in Germany has taken another approach based on coculture and hollow fiber technology. Gerlach *et al.* have designed a bioartificial liver where primary porcine hepatocytes are attached on the outside of a network of woven capillaries coated with Matrigel [71]. There are four types of capillaries: (1) one type for incoming plasma, (2) one type for outgoing plasma, (3) one for oxygenation, (4) and one containing sinusoidal endothelial cells. It is thought that coculture with another liver cell type provides hepatocytes with a more *in vivo* like environment, which may improve or prolong their function. The device can hold 2.5×10^9 hepatocytes, so further scale-up would be necessary for human application. Cytochrome P450 function and galactose metabolism were demonstrated *in vitro* for up to 3 weeks in culture [72]. This device has been tested in a hepatectomized pig model [73], but has yet to be tested on humans.

7 Discussion of devices

While all these devices have had some success in animal, or even human, trials, all have several limitations. First, scale-up for human application will be an issue for all. It is estimated that 100 g of functioning cells will be needed to sustain a patient [54]. The Cedars-Sinai device holds approximately 50 g of cells, the Kyushu device 10 g and the Minnesota device 70 g. Simply increasing the size of the reactors is not really an option because the size of the device is limited by the volume of patient blood that can be out of the body and in the device. For example, the Kyushu group estimates they will have to increase the volume of their reactor to 1.3 L to achieve sufficient cell number in their device. This volume is most likely too large. The Cedars-Sinai group addressed this problem by adding the charcoal columns, which not only protect the hepatocytes from toxic compounds in the plasma but also serve to lighten the metabolic load on the hepatocytes by removing the toxic compounds. The Minnesota group has explored the use of hepatocyte spheroids in their reactor to increase the performance. The spheroid-containing BAL demonstrated at least a two-fold increased metabolic activity on a per cell basis compared to devices entrapped with only dispersed cells [74]. This may provide the necessary increase in activity without an increase in reactor size.

Questions regarding the immune response by the patient also remain with all these devices. These devices incorporate a membrane barrier or plasmapheresis to prevent contact between xenogeneic cells and the blood cells of the patients. This appears to be adequate for protecting the hepatocytes from being damaged by an immune response from the patient. The patient, however, will most likely produce antibodies to porcine proteins released by the hepatocytes, and this may limit the duration or number of treatments. Demetriou et al. have measured an increase in porcine antibodies in patients receiving two or more BAL treatments. However, these levels were lowered by immunosuppression. No increase was seen in patients only receiving one treatment [75]. This issue has yet to be explored in the other devices.

Good mass transfer characteristics are important for good device performance. Both the Cedars-Sinai and Kyushu devices offer little mass transfer resistance to the cells. In the Kyushu device, there is no barrier between the plasma and the cells. In the Cedars-Sinai device, there is only a porous (pore size 0.2 μm) membrane barrier. One criticism of the Minnesota has always been the mass transfer resistance posed by the membrane (MWCO 100,000 daltons) and the collagen gel. However, this resistance has proved to be little detriment to device performance. In fact, the presence of the collagen gel appears to stabilize the activity of the cells for longer culture times [76].

Device storage may also be a problem for these reactors. Trials done with these devices all employed freshly harvested hepatocytes. For practical commercial use, devices will need to be produced and stored for use later. One common method of long-term storage for cells is cryopreservation. However, this often results in a rather significant loss of viability once the cells are thawed. Good cryopreservation methods will be crucial for maintaining adequate function in these bioartificial livers. This also makes maximizing cell number and function in the reactor at the start all the more crucial.

8 Concluding remarks

While most extracorporeal bioartificial liver devices have been developed as a bridge to organ transplant, these devices are also often mentioned as a bridge therapy while a patient's liver regenerates. The time for a human liver to regenerate to the point where it could adequately sustain function for survival may take weeks, or perhaps even months. Most bioartficial liver devices have only been tested in animals for times on the order of hours. Whether they will stand up to the longer duration of therapy needed to allow for regeneration has yet to be seen. However, though there are several obstacles to be overcome, results to this point have been encouraging, and it is likely that these extracorporeal devices will serve as successful 'bridge-to-transplant' therapies. With one device in phase II trails and other devices waiting to start phase I, it is a very exciting time in bioartificial liver therapy.

References

1. National Center for Health Statistics, *Vital Statistics of the United States,* 1993.
2. Starzl, T. E., Demetris, A. J. and van Thiel, D., Liver transplantation (1). *New England Journal of Medicine,* 1989, **321**, 1092–1099.
3. Peleman, R. R., Gavaler, J. S., van Thiel, D. H., Esquivel, C., Gordon, R., Iwatsuki, S. and Starzl, T. E., Orthotopic liver transplantation for acute and subacute hepatic failure in adults. *Hepatology,* 1987, **7**, 484–489.
4. Busuttil, R. W., Colonna, J. O., Hiatt, J. R., Brems, J. J., el Khoury, G., Goldstein, L. I., Quinones-Baldrich, W. J., Abdul-Rasool, I. H. and Ramming, K. P., The first 100 liver transplants at UCLA. *Annals of Surgery,* 1987, **206**, 387–402.
5. Takahashi, T., Malchesky, P. S. and Nose, Y., Artificial liver. State of the art. *Digestive Diseases and Sciences,* 1991, **36**, 1327–1340.
6. Demetriou, A. A., Reisner, A., Sanchez, J., Levenson, S. M., Moscioni, A. D. and Chowdhury, J. R., Transplantation of microcarrier-attached hepatocytes into 90% partially hepatectomized rats. *Hepatology,* 1988, **8**(5), 1006–1009.
7. Saito, S., Sakagami, K., Matsuno, T., Tanakaya, K., Takaishi, Y. and Orita, K., Long-term survival and proliferation of spheroidal aggregate cultured hepatocytes transplanted into the rat spleen. *Transplantation Proceedings,* 1992, **24**(4), 1520–1521.
8. Chang, T. M., Artificial liver support based on artificial cells with emphasis on encapsulated hepatocytes. *Artificial Organs,* 1992, **16**(1), 71–74.
9. Bucher, N. L. and Malt, R. A., *Regeneration of Liver and Kidney,* Little and Brown, Boston, 1971.
10. Kiley, J. E., Welch, H. F., Pender, J. C. and Welch, C. S., Removal of blood ammonia by hemodialysis. *Proceedings of the Society for Experimental Biology and Medicine,* 1956, **91**, 489–490.

11. Schechter, D. C., Nealon, T. F. and Gibbon, J. H., A simple extracorporeal device for reducing elevated blood ammonia levels. *Surgery*, 1958, **44**, 892–897.

12. Ton, H. Y., Hughes, R. D., Silk, D. B. A. and Williams, R., Albumin coated Amberlite XAD-7 resin for hemoperfusion in acute liver failure. Part I: Adsorption studies. *Artificial Organs*, 1979, **3**, 20–22.

13. Opolon, P., Rapin, J. R., Huguet, C., Granger, A., Delorme, M. L., Boschat, M. and Sausse, A., Hepatic failure coma treated by polyacrylonitrile membrane hemodialysis. *Transactions — American Society for Artificial Internal Organs*, 1976, **22**, 701–710.

14. Hori, M., Neto, A. C., Austen, W. G. and McDermott, W. V., Isolated in vivo hepatorenal perfusion in the dog: Circulating and functional response of the kidney to hepatic anoxia. *Journal of Surgical Research*, 1967, **7**, 413–417.

15. Otto, J. J., Pender, J. C., Cleary, J. H., Sensening, D. M. and Welch, D. S., The use of a donor liver in experimental animals with elevated blood ammonia. *Surgery*, 1958, **43**, 301–309.

16. Nose, Y., Mikami, J., Kasai, K., Sasaki, E., Agishi, T. and Danjo, Y., An experimental artificial liver utilizing extracoporeal metabolism with sliced or granulated canine liver. *Transactions — American Society for Artificial Internal Organs*, 1963, **9**, 358–362.

17. Matsumara, K., US Patent No. 3734851, 1973.

18. Eiseman, B., Norton, L. and Kralios, N. C., Hepatocyte perfusion within a centrifuge. *Surgery, Gynecology and Obstetrics*, 1976, **142**, 21–28.

19. Eiseman, B., Liem, D. S. and Raffucci, F., Heterologous liver perfusion in treatment of hepatic failure. *Annals of Surgery*, 1965, **162**, 329–345.

20. Norman, J. C., Saravis, C. A., Brown, M. E. and McDermott, W. V. J., Immunological observations in clinical heterologous (xenogeneic) liver perfusions. *Surgery*, 1966, **60**, 179–190.

21. Kimoto, S., The artificial liver, experiments and clinical application. *Transactions — American Society of Artificial Internal Organs*, 1959, **5**, 102–112.

22. Fischer, M., Botterman, P., Sommoggy, S. P.S. and Ernhardt, W., Functional capacity of extracorporeal baboon liver perfusions. In *Artificial Liver Support*, Brunner, G. and Schmidt, F. W., Eds., Springer-Verlag Publications, Berlin, 1981, pp. 280–285.

23. Lie, T. S., Treatment of acute hepatic failure by extracorporeal hemoperfusion over human and babooon liver. In *Artificial Liver Support*, eds. G. Brunner and F. Schmidt. Springer-Verlag Publications, Berlin, 1981, pp. 274–279.

24. Abouna, G. M., Fischer, L. M., Porter, K. A. and Andres, G., Experience in the treatment of heptic failure by intermittent liver hemoperfusions. *Surgery, Gynecology and Obstetrics*, 1973, **137**, 741–752.

25. Hager, J. C., Carman, R., Porter, L. E., Stoller, R., Leduc, E. H., Galletti, P. M. and Calabresi, P., Neonatal hepatocyte culture on artificial capillaries: A model for drug metabolism and the artificial liver. *ASAIO Journal*, 1983, **6**, 26–35.

26. Matsumura, K. N., Guevara, G. R., Huston, H., Hamilton, W. L., Rikimaru, M., Yamasaki, G. and Matsumura, M. S., Hybrid bioartificial liver in hepatic failure: Preliminary clinical report. *Surgery*, 1987, **101**, 99–103.

27. Neuzil, D., Rozga, J., Moscioni, A., Ro, M., Hakim, R. and Demetriou, A., Use of a xenogeneic liver support system to treat a patient with acute liver failure. *Hepatology*, 1991, **14**, 246A.

28. Nyberg, S. L., Shatford, R. A., Hu, W. -S., Payne, W. D. and Cerra, F. B., Hepatocyte culture systems for artificial liver support: Implications for critical care medicine (bioartificial liver support). *Criticial Care Medicine*, 1992, **20**(8), 1157–1168.

29. Galletti, P. M., Thrombosis in extracorporeal devices. *Annals of the New York Academy of Sciences*, 1987, **516**, 679–682.

30. Kelly, J. H. and Darlington, G. J., Modulation of the liver specific phenotype in the human hepatoblastoma line HepG2. *In Vitro Cellular and Developmental Biology*, 1989, **25**(2), 217–222.

31. Wilson, J. M., Jefferson, D. M., Chowdhury, R., Movikoff, P. M., Johnston, D. E. and Mulligan, R. C., Retrovirus-mediated transduction of adult hepatocytes. *Proceedings of the National Academy of Science USA*, 1988, **85**, 3014–3018.

32. Sielaff, T. D., Hu, M. Y., Rao, S., Groehler, K., Olson, D., Mann, H. J., Remmel, R. P., Shatford, R. A., Amiot, B., Hu, W. S. and Cerra, F. B., A technique for porcine hepatocyte harvest and description of differentiated metabolic functions in static culture. *Transplantation*, 1995, **59**(10), 1459–1463.

33. Morsiani, E., Rozga, J., Scott, H. C., Kong, L. B., Lebow, L. T., McGrath, M. F., Moscioni, A. D. and Demetriou, A. A., Automated large-scale production of porcine hepatocytes for bioartificial liver support. *Transplantation Proceedings*, 1994, **26**(6), 3505–3506.

34. Naik, S., Trenkler, D., Santagnini, H., Pan, J. and Jauregui, H. O., Isolation and culture of porcine hepatocytes for artificial liver support. *Cell Transplantation*, 1996, **5**(1), 107–115.

35. Lanford, R. E., Carey, K. D., Estlack, L. E., Smith, G. C. and Hay, R. V., Analysis of plasma protein and lipoprotein synthesis in long-term primary cultures of baboon hepatocytes maintained in serum-free medium. *In Vitro Cellular and Developmental Biology*, 1989, **25**, 174–182.

36. Dich, J., Vind, C. and Grunnet, N., Long-term culture of hepatocytes: Effect of hormones on enzyme activities and metabolic capacity. *Hepatology*, 1988, **8**(1), 39–45.

37. Akaike, T., Tobe, S., Kobayashi, A., Goto, M. and Kobayashi, K., Design of hepatocyte-specific extracellular matrices for hybrid artificial liver. *Gastroenterologia Japonica*, 1993, **4**, 45–52.

38. Dunn, J., Yarmush, M., Koebe, H. and Tompkins, R., Hepatocyte function and extracellular matrix geometry: Long-term culture in a sandwich configuration. *FASEB Journal*, 1989, **3**, 174–177.

39. Dunn, J., Tompkins, R. and Yarmush, M., Hepatocytes in collagen sandwich: Evidence for transcriptional and translational regulation. *Journal of Cell Biology*, 1992, **116**, 1043–1053.

40. Spray, D. C., Fujita, M., Saez, J., Choi, H., Watanabe, T., Hertzberg, E., Rosenberg, L. and Reid, L., Proteoglycans and glycosaminoglycans induce gap junction synthesis and function in primary liver cultures. *Journal of Cell Biology*, 1987, **105**, 541–551.

41. Yamamoto, N., Imazato, K. and Matsumoto, A., Growth stimulation of adult rat hepatocytes in a primary culture by soluble factor(s) secreted from nonparenchymal liver cells. *Cell Structure and Function*, 1989, **14**, 217–229.

42. Shimaoka, S., Nakamura, T. and Ichihara, A., Stimulation of growth of primary cultured adult rat hepatocytes without growth factors by coculture with nonparenchymal liver cells. *Experimental Cell Research*, 1987, **172**, 228–242.

43. Mikos, A. G., Sarakinos, G., Lyman, M. D., Ingber, D. E., Vacanti, J. P. and Langer, R., Prevascularization of biodegradable polymer scaffolds for hepatocyte transplantation. In *Polymeric Materials Science and Engineering*, Vol. 66, ACS Books and Journals Division Publications, Washington, DC, 1992, pp. 34–35.

44. Cima, L. G., Langer, R. and Vacanti, J. P., Polymers for organ and tissue culture. *Journal of Bioactive and Compatible Polymers*, 1991, **6**, 232–240.

45. Vacanti, J. P., Morse, M. A., Saltzman, W. M., Domb, A. J., Perez-Atayde, A. and Langer, R., Selective cell transplantation using bioabsorbable artificial polymers as matrices. *Journal of Pediatric Surgery*, 1988, **23**, 3–9.

46. Koide, N., Shinji, T., Tanabe, T., Asano, K., Kawaguchi, M., Sakaguchi, K., Koide, Y., Mori, M. and Tsuji, T., Continued high albumin production by multicellular spheroids of adult rat hepatocytes formed in the presence of liver-derived proteoglycans. *Biochemical and Biophysical Research Communications*, 1989, **161**(1), 385–391.

47. Tong, J. Z., De Lagusie, P., Furlan, V., Cresteil. T., Bernard, O. and Alvarez, F., Long-term culture of adult rat hepatocyte spheroids. *Experimental Cell Research*, 1992, **200**, 326–332.

48. Peshwa, M. V., Wu, F. J., Follstad, B. D., Cerra, F. B. and Hu, W. S., Kinetics of hepatocyte spheroid formation. *Biotechnology Progress*, 1994, **10**(5), 460–466.

49. Jauregui, H. O., Chowdhury, N. R. and Chowdhury, J. R., Use of mammalian cells for artificial liver support. *Cell Transplantation*, 1996, **5**(3), 353–367.

50. Yarmush, M. L., Dunn, J. C. and Tompkins, R. G., Assessment of artificial liver support technology. *Cell Transplantation*, 1992, **1**, 323–341.

51. Dixit, V., Development of a bioartificial liver using isolated hepatocytes. *Artificial Organs*, 1994, **18**(5), 371–384.

52. Kasai, S.-i. K., Sawa, M. and Mito, M., Is the biological artifical liver clinically applicable? A historic review of biological artificial liver support systems. *Artificial Organs*, 1994, **18**(5), 348–354.

53. Rozga, J., Williams, F., Ro, M. -S., Neuzil, D. F., Giorgio, T. D., Backfisch, G., Moscioni, A. D., Hakim, R. and Demetriou, A. A., Development of a bioartificial liver: Properties and function of a hollow-fiber module inoculated with liver cells. *Hepatology*, 1993, **17**, 258–265.

54. Rozga, J., Holzman, M. D., Ro, M. -S., Griffin, D. W., Neuzil, D. F., Giorgio, T., Moscioni, A. D. and Demetriou, A. A., Development of a hybrid bioartificial liver. *Annals of Surgery*, 1993, **217**(5), 502–511.

55. Rozga, J., Podesta, L., LePage, E., Morsiani, E., Moscioni, A. D., Hoffman, A., Sher, L., Villamil, F., Woolf, G., McGrath, M., Kong, L., Rosen, H., Lanman, T., Vierling, J., Makowka, L. and Demetriou, A. A., A bioartificial liver to treat severe acute liver failure. *Annals of Surgery*, 1994, **219**(5), 538–546.

56. Chen, S. C., Hewitt, W. R., Watanabe, F. D., Eguchi, S., Kahaku, E., Middleton, Y., Rozga, J. and Demetriou, A. A., Clinical experience with a porcine hepatocyte-based liver support system. *International Journal of Artificial Organs*, 1996, **19**(11), 664–669.

57. Ijima, H., Taniguchi, Y., Matsushita, T. and Funatsu, K., Application of three dimensional culture of adult rat hepatocytes in PUF pores for artificial liver support system. In *Animal Cell Technology: Basic and Applied Aspects*, eds. H. Murakami, S. Shirahata and H. Tachibana. Kluwer Academic Publishers, Dordrecht, 1992, pp. 81–86.

58. Matsushita, T., Ijima, H., Koide, N. and Funatsu, K., High albumin production of multicellular spheroids of adult rat hepatocytes formed in the pores of polyurethane foam. *Applied Microbiology and Biotechnology*, 1991, **36**, 324–326.

59. Ijima, H., Matsushita, T., Nakazawa, K., Koyama, S., Gion, T., Shirabe, K., Shimada, M., Takenaka, K., Sugimachi, K. and Funatsu, K., Spheroid formation of primary dog hepatocytes using polyurethane foam and its application to hybrid artificial liver. In *Animal Cell Technology: From Vaccines to Genetic Medicine*, Kluwer Academic Publishers, Dordrecht, 1997, pp. 577–583.

60. Shatford, R. A., Nyberg, S. L., Meier, S. J., White, J. G., Payne, W. D., Hu, W. S. and Cerra, F. B., Hepatocyte function in a hollow fiber bioreactor: A potential bioartificial liver. *Journal of Surgical Research*, 1992, **53**(6), 549–557.

61. Nyberg, S. L., Shatford, R. A., Peshwa, M. V., White, J. G., Cerra, F. B. and Hu, W. S., Evaluation of a hepatocyte-entrapment hollow fiber bioreactor. A potential bioartificial liver. *Biotechnology and Bioengineering*, 1993, **41**(2), 194–203.

62. Shatford, R. A., Nyberg, S. L., Payne, W. D., Hu, W. S. and Cerra, F. B., A hepatocyte bioreactor as a potential bioartificial liver: Demonstration of prolonged tissue specific functions. *Surgical Forum*, 1991, **42**, 54–56.

63. Nyberg, S. L., Shirabe, K., Peshwa, M. V., Sielaff, T. D., Crotty, P. L., Mann, H. J., Remmel, R. P., Payne, W. D., Hu, W. S. and Cerra, F. B., Extracorporeal application of a gel-entrapment, bioartificial liver: Demonstration of drug metabolism and other biochemical functions. *Cell Transplantation*, 1993, **2**(6), 441–452.

64. Nyberg, S. L., Payne, W. D., Amiot, B., Shirabe, K., Remmel, R. P., Hu, W. S. and Cerra, F. B., Demonstration of biochemical function by extracorporeal xenohepatocytes in an anhepatic model. *Transplantation Proceedings*, 1993, **25**(2), 1994–1945.

65. Nyberg, S. L., Mann, H. J., Hu, M. Y., Payne, W. D., Cerra, F. B., Remmel, R. P. and Hu, W. S., Extrahepatic metabolism of 4-methylumbelliferone and lidocaine in an anhepatic rabbit. *Drug Metabolism and Disposition*, 1996, **24**(6), 643–648.

66. Sielaff, T. D., Hu, M. Y., Rollins, M. D., Bloomer, J. R., Amiot, B., Hu, W. -S. and Cerra, F. B., An anesthetized model of lethal canine galactosamine fulminant hepatic failure. *Hepatology*, 1995, **21**, 796–804.

67. Sielaff, T. D., Hu, M. Y., Amiot, B., Rollins, M. D., Rao, S., McGuire, B., Bloomer, J. R., Hu, W. -S. and Cerra, F. B., Gel-entrapment bioartificial liver therapy in galactosamine hepatitis. *Journal of Surgical Research*, 1995, **59**, 179–184.

68. Sussman, N. L., Chong, M. G., Koussayer, T., He, D. E., Shang, T. A., Whisennand, H. H. and Kelly, J. H., Reversal of fulminant hepatic failure using an extracorporeal liver assist device. *Hepatology*, 1992, **16**(1), 60–65.

69. Sussman, N. L., Gislason, G. T., Conlin, C. A. and Kelly, J. H., The Hepatix extracorporeal liver assist device: Initial clinical experience. *Artificial Organs*, 1994, **18**(5), 390–396.

70. Ellis, A. J., Hughes, R. D., Wendon, J. A., Dunne, J., Langley, P. G., Kelly, J. H., Gislason, G. T., Sussman, N. L. and Williams, R., Pilot-controlled trial of the extracorporeal liver assist device in acute liver failure. *Hepatology*, 1996, **24**(6), 1446–1451.

71. Gerlach, J. C., Encke, J., Hole, O., Muller, C., Courtney, J. M. and Neuhaus, P., Hepatocyte culture between three dimensionally arranged biomatirx-coated independent artificial capillary systems and sinusoidal endothelial cell co-culture compartments. *International Journal of Artificial Organs*, 1994, **17**(5), 301–306.

72. Gerlach, J. C., Encke, J., Hole, O., Muller, C., Ryan, C. J. and Neuhaus, P., Bioreactor for a larger scale hepatocyte in vitro perfusion. *Transplantation*, 1994, **58**(9), 984–988.

73. Gerlach, J., Trost, T., Ryan, C. J., Meissler, M., Hole, O., Muller, C. and Neuhaus, P., Hybrid liver support system in a short term application on hepatectomized pigs. *International Journal of Artificial Organs*, 1994, **17**(10), 549–553.

74. Wu, F. J., Friend, J. R., Mann, H. J., Remmel, R. P., Cerra, F. B. and Hu, W.-S., Hollow fiber bioartificial liver utilizing collagen entrapped porcine hepatocyte spheroids. *Biotechnology and Bioengineering*, 1996, **52**, 34–44.

75. Baquerizo, A., Mhoyan, A., Shirwan, H., Swensson, J., Busuttil, R. W., Demetriou, A. A. and Cramer, D. V., Xenoantibody response of patients with severe acute liver failure exposed to porcine antigens following treatment with a bioartificial liver. *Transplantation Proceedings*, 1997, **29**, 964–965.

76. Lazar, A., Mann, H. J., Remmel, R. P., Shatford, R. A., Cerra, F. B. and Hu, W.-S., Extended liver-specific functions of porcine hepatocyte spheroids entrapped in collagen gel. *In Vitro Cellular and Developmental Biology. Animal*, 1995, **31**(5), 340–346.

Author index

Subject index